The Development of *Dictyostelium discoideum*

The Development of *Dictyostelium discoideum*

William F. Loomis
Editor
Department of Biology
University of California at San Diego
La Jolla, California

1982

ACADEMIC PRESS
A Subsidiary of Harcourt Brace Jovanovich, Publishers
New York London
Paris San Diego San Francisco São Paulo Sydney Tokyo Toronto

ACADEMIC PRESS, INC.
111 Fifth Avenue, New York, New York 10003

United Kingdom Edition published by
ACADEMIC PRESS, INC. (LONDON) LTD.
24/28 Oval Road, London NW1 7DX

Library of Congress Cataloging in Publication Data

Main entry under title:

The Development of Dictyostelium discoideum.

Bibliography: p.
Includes index.
1. Dictyostelium discoideum. 2. Fungi--Development.
I. Loomis, William F.
QK635.D5D49 589.2'9 82-4109
ISBN 0-12-455620-5 AACR2

PRINTED IN THE UNITED STATES OF AMERICA

82 83 84 85 9 8 7 6 5 4 3 2 1

Contents

4. Chemotaxis

Peter N. Devreotes

5. Cell Motility

James A. Spudich and Annamma Spudich

6. Cell Adhesion

Samuel H. Barondes, Wayne R. Springer, and Douglas N. Cooper

7. The Organization and Expression of the *Dictyostelium* Genome

Alan R. Kimmel and Richard A. Firtel

8. Control of Gene Expression

Harvey F. Lodish, Daphne D. Blumberg, Rex Chisholm, Steve Chung, Antonio Coloma, Scott Landfear, Eric Barklis, Paul Lefebvre, Charles Zuker, and Giorgio Mangiarotti

9. Morphogenetic Signaling, Cytodifferentiation, and Gene Expression

Maurice Sussman

10. *Polysphondylium* and Dependent Sequences

David W. Francis

11. Cell Proportioning and Pattern Formation

James H. Morrissey

Contributors

Numbers in parentheses indicate the pages on which the authors' contributions begin.

Eric Barklis (325), Department of Biology, Massachusetts Institute of Technology, Cambridge, Massachusetts 02139

Samuel H. Barondes (195), Department of Psychiatry, University of California at San Diego, La Jolla, California 92093

Daphne D. Blumberg[1] (325), Department of Biology, Massachusetts Institute of Technology, Cambridge, Massachusetts 02139

John Tyler Bonner (1), Department of Biology, Princeton University, Princeton, New Jersey 08544

Robin Cann (451), Department of Biology, University of California at San Diego, La Jolla, California 92093

Rex Chisholm (325), Department of Biology, Massachusetts Institute of Technology, Cambridge, Massachusetts 02139

Steve Chung[2] (325), Department of Biology, Massachusetts Institute of Technology, Cambridge, Massachusetts 02139

Antonio Coloma (325), Department of Biology, Massachusetts Institute of Technology, Cambridge, Massachusetts 02139

Douglas N. Cooper (195), Department of Psychiatry, University of California at San Diego, La Jolla, California 92093

Peter N. Devreotes (117), Department of Physiological Chemistry, Johns Hopkins School of Medicine, Baltimore, Maryland 21205

Richard A. Firtel (233), Department of Biology, University of California at San Diego, La Jolla, California 92093

[1]Present address: National Cancer Institute, Frederick, Maryland 21701.

[2]Present address: Biological Chemistry, University of Illinois Medical Center, Chicago, Illinois 60612

David W. Francis (387), Division of Biological Sciences, University of Delaware, Newark, Delaware 19711

Alan R. Kimmel[3] (233), Department of Biology, University of California at San Diego, La Jolla, California 92093

Scott Landfear (325), Department of Biology, Massachusetts Institute of Technology, Cambridge, Massachusetts 02139

Paul Lefebvre (325), Department of Biology, Massachusetts Institute of Technology, Cambridge, Massachusetts 02139

Harvey F. Lodish (325), Department of Biology, Massachusetts Institute of Technology, Cambridge, Massachusetts 02139

W. F. Loomis (451), Department of Biology, University of California at San Diego, La Jolla, California 92093

Giorgio Mangiarotti (325), Department of Biology, Massachusetts Institute of Technology, Cambridge, Massachusetts 02139

James H. Morrissey[4] (411), Department of Biology, University of California at San Diego, La Jolla, California 92093

Ben A. Murray[5] (71), Department of Biology, University of California at San Diego, La Jolla, California 92093

Peter C. Newell (35), Department of Biochemistry, University of Oxford, Oxford OX1 3QU, England

Wayne R. Springer (195), Department of Psychiatry, University of California at San Diego, La Jolla, California 92093

Annamma Spudich (169), Department of Structural Biology, Stanford University School of Medicine, Stanford, California 94305

James A. Spudich (169), Department of Structural Biology, Stanford University School of Medicine, Stanford, California 94305

Maurice Sussman (353), Department of Biological Sciences, University of Pittsburgh, Pittsburgh, Pennsylvania 15260

Charles Zuker (325), Department of Biology, Massachusetts Institute of Technology, Cambridge, Massachusetts 02139

[3]Present address: Laboratory of Cell and Developmental Biology, National Institutes of Health, Bethesda, Maryland 20205.

[4]Present address: Department of Biochemistry, University of Oxford, Oxford, OX1 3QU, England

[5]Present address: The Rockefeller University, New York, New York 10021.

Preface

Dictyostelium has come of age. Techniques for growing the cells, for constructing genetically marked strains, and for separating and analyzing the different cell types are available. Perhaps more importantly, the concepts around which the rapidly increasing numbers of detailed descriptions can be organized are coming into focus, and the crucial questions are commonly appreciated. As the analyses have become more basic it has become apparent that many of the processes under study are not limited to the biology of *Dictyostelium* but are cases of general problems found in the development of diverse organisms.

If one follows the dictum not to study a process in a system more complicated than necessary, then *Dictyostelium* often becomes the system of choice. This may sound parochial from someone who has studied *Dictyostelium* for many years, but it is increasingly echoed by those who have found this organism useful for basic studies on differential adhesion, coordinate gene expression, cellular movement, pattern formation, and a variety of other problems of current interest in biology. However, the field is far from saturated. The number of ripe unsolved questions continues to expand as the foundations solidify. Now we can come to grips with many of the critical phenomena that have tantalized us for some time. This exciting period of steady improvement in the understanding of molecular and cellular processes underlying complex phenomena should continue for some time.

It has been seven years since the last comprehensive review of work on *Dictyostelium* was completed. Much new information has been accumulated during this period. Since the organism presents favorable material for the study of diverse biological processes, the range of studies has continued to expand. Because it is now difficult for anyone to digest all of the varied aspects, the contributors to this book were encouraged to give their views of the present state of their field of special interest.

This book is meant for those already familiar with the general outlines of *Dictyostelium* biology, which I hope includes all those who have studied biology in the last 10 years. Further details can be found in a predecessor to this volume, *Dictyostelium discoideum,* published by Academic Press in 1975. The present volume is a compendium of thoughts and facts on 11 major fronts at which progress is now being made. Because many authors are represented, there are unavoidable differences of perspective. It is hoped that this will add to, rather than detract from, the effectiveness of the book. From the subjective conflicts may come generally accepted mechanisms that account for developmental processes in this organism. We can then turn our attention to testing whether it is truly a model system for metazoan development.

We can now identify many of the prevalent proteins synthesized at various stages of development of *Dictyostelium,* such as actin, discoidin, heat-shock proteins, and spore-coat proteins. Each has been purified and analyzed. Clones of the genes coding for these proteins have been isolated and are being characterized. In several cases (e.g., actin, discoidin) the degree of resolution extends to the nucleotide level. However, the physiological functions of these proteins have only been guessed at. What we need are well-defined mutations in the respective structural genes so that we can compare the altered phenotypes to those which normally occur. In this analysis the greater the degree of biological and biochemical sophistication, the greater will be the depth of our understanding of the interconnections among these and other components in development.

It seemed appropriate to start the book by considering how *Dictyostelium* fits into evolution and ecology. Bonner describes its place in nature. By comparing the various species, genetically defined processes can be glimpsed. For instance, it is likely that the common precursor to today's slime molds had the ability to encapsulate into spores. Only later did mechanisms resulting in stalk formation evolve. Since prestalk cells must be kept from encapsulating, a signal inhibiting spore formation in the anterior cells is a plausible conjecture.

Since genetic analyses of mutations affecting diverse processes are used in almost all the subjects covered in the book, a chapter by Newell follows, outlining the recent advances in genetic manipulation and mutant isolation techniques. *Dictyostelium* remains one of the few organisms in which rare mutations affecting multicellular processes can be isolated and genetically studied. More than 100 loci have been mapped, affecting a wide variety of processes. Analyses of the phenotypes of single and double mutants in haploids and in heterozygous diploids have given insights that biochemical analyses alone would have taken far longer to show. However, we cannot expect to understand a wild-type function just by describing a mutant strain. There is a causal chain of events from the gene to its product, product to physiology, and physiology to gross phenotype, and each step can be surprisingly complex. These studies have followed two major paths: from the gene outward and from the phenotype inward. That is, in some cases a gene product is implicated in a process and mutations in the structural gene are isolated. This approach can often rule in or rule out the implicated involvement. Quite often it has ruled it out or shown that the adaptive powers of the organism to overcome the lesion physiologically are greater than initially thought. In other cases, mutations are isolated which affect the gross morphology, such as aggregation pattern or proportion of cell types, but there is little hope of soon being able to recognize the mutated gene product. However, techniques are now being developed which allow efficient transformation with cloned sequences: these, when generally available, will allow us to recognize the mutated gene by selecting revertants at the phenotypic level. This will be most helpful in the difficult problem of connecting specific molecules to complex developmental processes.

Although mutations have been recovered at the expected frequencies in the structural genes for a variety of enzymes, none has been found to affect gene-specific regulatory functions. Precedent in bacteria and their viruses led us to believe that either positive or negative regulatory genes should have been recovered. Perhaps many genes are autogenously regulated such that the proteins they specify not only catalyze given reactions but also regulate the genes that make them. We would not expect to recover constitutive mutants easily if this were the case. However, constitutive genes could be constructed *in vitro* by ligating a constitutive or easily controlled promoter onto a cloned sequence of the gene and then reintroducing it into *Dictyostelium* cells. The consequences to out-of-turn expression of a gene could then be followed *in vivo*.

Much of the recent advance in cell biology has been related to a better understanding of the composition and function of the cell membrane. The analyses of *Dictyostelium* plasma membranes are collated by Murray. Although membranes contain only a subset of total cell proteins, there are enough to keep us busy assigning roles for quite some time. Attention has been focused on those that change at the various developmental stages. More than half of the most prevalent developmentally regulated membrane proteins have been named and partly characterized, but a clear understanding of their roles awaits new genetic and biochemical techniques.

Chemotaxis is a dramatic event of early development in *Dictyostelium* and turns out to be beautifully complicated and subtle. Devreotes puts all of the new observations in a coherent framework and points out what we still do not know. Although we know that the chemoattractant is cAMP and that it elicits relay release of more cAMP, we do not know how the signal is perceived, and we know hardly anything about the biochemistry of adenyl cyclase. The black box is still opaque.

The same is true of amoeboid movement; but we do know that there are actin, myosin, and accessory proteins involved. The Spudiches describe how they may interact. This is one case in which biochemistry has forged ahead alone. It is hoped that pertinent mutants will join the story soon. Just as understanding the nature of the excitable response to cAMP has bearing on neurophysiology, these basic studies on a nonmuscle motile system have impact on similar processes throughout the animal kingdom.

Another universal multicellular characteristic is cell–cell association. Barondes and colleagues muse about how the molecular mechanisms can be unequivocally recognized, and then lay out the approaches used with *Dictyostelium.* Molecular species that correlate with altered adhesive properties are good candidates but may yet prove to be red herrings. Immunological assays are only as good as the characterization of the specific immunoactivity. Perhaps only when we can reconstruct functional membranes will we be certain which components are sufficient to account for cell–cell adhesion.

At the descriptive level, definition of the nucleotide sequence of genes and direct assay of their transcripts is hard to beat. Progress in nucleic acid research on *Dictyostelium* is reviewed by Kimmel and Firtel and by Lodish and colleagues. *Dictyostelium* genes look much like other eukaryotic genes complete with TATA boxes and so on. It is the flanking sequences that are abnormal, being uninformative (so far) stretches of A's and T's. Yet there is now no room to doubt that specific genes are controlled at transcription. And the controls involve such complex signals as multicellularity. Much of the hope for the future lies in the reconstruction of accurate representative transcription *in vitro*.

Meanwhile, a body of information has been accumulating on morphogenetic signaling, which is presented here by Sussman. cAMP clearly plays the role of chemoattractant during aggregation, but whether it plays a permissive or an informational role during postaggregative development is still an open question.

Another way to try to gain insights on the regulation of the temporal sequence of biochemical differentiations is to compare a series of mutants to the wild type. Francis reviews such studies in *Polysphondylium* and *Dictyostelium* and presents the concept of a linear or branched sequence of causally related steps. Various other facts of the development of *Polysphondylium* are also presented.

Finally, the phenomenon of cell-type specification and regulation is directly confronted by Morrissey. We all agree that two cell types appear, such that in slugs the anterior will become stalk cells and the posterior will become spores, and that a clearly measurable commitment to this fate is made at this time. The proportion of the cell types is essentially constant in large and small slugs and adjusts itself when altered by the removal of a section of the slug. Thus, it seems clear that there is a field-wide informational system which informs individual cells of the events throughout the whole slug. This seems to be an almost ideal place to look for the mechanism by which patterns are established and regulated in development. It is much simpler than the mechanism for limb buds and is amenable to biochemical and genetic probes. The molecular markers that can be used in cell-type assays have been recognized. Perhaps they are close enough to the original position-determined response to give direct evidence on the nature of the signal.

In several of these analyses we are past the stage of initial excitement at discovering molecules potentially involved and must proceed to the much more difficult task of actually proving that they are or are not involved. The total number of genes we must consider is certainly less than 10,000. Several independent lines of evidence indicate that about 5000 genes are vital to the growth of the cells. These would be the housekeeping enzymes and structural proteins which have been characterized in all cells. The analyses that have bearing on the number of genes needed only during development are more controversial and are discussed in Chapter 8. Some hybridization studies indicate that about 3000 genes, and others that less than 1000 genes, are expressed exclusively during development. Genetic studies show that fewer than 500 genes must function in addition to the 5000 vital genes to give the wild-type behavior and morphogenesis. It is possible that the apparent disagreement results from a large class of genes (several thousand) that are expressed but whose function can be taken over by other gene products if they are missing. The ancillary genes would not be observed in the mutational studies but would be counted by the hybridization experiments. Alternatively, there is a large number of transcripts which are dispensable and result only from lack of rigorous transcriptional control during development. In any case, the initial challenge is to understand the interplay of the few hundred essential developmental genes.

• CHAPTER 1

Comparative Biology of Cellular Slime Molds

John Tyler Bonner

I. Historical

Cellular slime molds were first discovered by the well-known German mycologist O. Brefeld in 1869, but the unique nature of their asexual life cycle was not fully understood until the study of the French mycologist Ph. van Tieghem in 1880. In other words, although these organisms have only been known for a little over a hundred years, today they are common knowledge, described in every freshman biology textbook and presently investigated by hundreds of research workers throughout the world.

The Development of *Dictyostelium discoideum*

ISBN 0-12-455620-5

The work between 1880 and 1940, however, should not be neglected, and even today some of those early papers are rewarding. On the natural history and taxonomic side an important landmark is the 1902 paper of E. W. Olive. This was his doctoral thesis study under Roland Thaxter at Harvard and in it he describes, with Thaxterian clarity and care, the life history and physical characteristics of all the species known at that time. Unfortunately for slime mold biology, Olive shortly afterward returned to his family business where he flourished until a ripe old age. Another early tradition was a developmental study of the life cycle and the first important contribution was that of G. Potts, also in 1902, on the development of *Dictyostelium mucoroides*, a common species and the one first discovered by Brefeld.

Somewhat later R. A. Harper, a professor of botany at Columbia University and a pioneer in lower plant studies in developmental biology, wrote three papers (1926, 1929, 1932) on slime model development. He made some significant contributions slightly marred by remarkably opaque prose. Another important work on the development of *D. mucoroides* was a paper by the German scientist Arthur Arndt, published in 1937. This paper is of great interest because Arndt made the first time-lapse film of slime mold development (which is still extant) and described his observations from the film in this paper. Some years ago, an old friend, originally from Germany, told me he remembered a lecture by Arndt in which he showed the film and from it argued that the organism was so extraordinary that the only possible explanation of its development must require vitalism. Had his view prevailed, many of us would be out of a job.

The modern era began with the discovery of a new species of cellular slime mold, *Dictyostelium discoideum*, by K. B. Raper in 1935. As I shall show in detail presently, and as will be obvious from all the chapters of this volume, this species is especially suited to experimental work. This was shown in a well-known 1940 paper by Raper, which was followed by many additional papers from his laboratory over the years.

What I have called the modern era has a number of trends, all of which are reflected in this volume. Some researchers have been concerned with the discovery of new species, with ecological problems, and with problems involving the relationship of different major groups of slime molds to one another. Others have been concerned with the traditional problems of developmental biology and have experimented on the cellular level. Still others have concentrated on various important aspects of the biochemistry of slime mold development. This molecular approach was given a great boost by the development of axenic strains, thereby eliminating confusion of the chemistry of the bacterial food supply (R. Sussman and M. Sussman, 1967; Ashworth and Watts, 1970; Loomis, 1971). In most studies there has been a combination of these two methods, and they have included areas such as genetics, gene action, chemotaxis, cell adhesion, and differenct aspects of differentiation including pattern formation. There has also been considerable interest in the physiology of phototropism and thermotropism. The reason for such an enormous recent profusion of studies on all these different aspects of cellular slime mold development arises from the conviction that these organisms are ideal for the study of fundamental problems of development. They are multicellular and eukaryotic and have a simple morphogenesis. To some extent we have all been consciously or unconsciously influenced by the success of *E. coli* in molecular biology and hope that the cellular slime molds will illuminate the next level of complexity.

II. The Life History

As has been done so often, I will illustrate the unusual feature of the development by outlining the life history of *Dictyostelium discoideum*. Later I will compare its characteristics with those of other cellular slime molds and consider briefly their classification.

A. Vegetative Stage

D. discoideum has two alternate life cycles: sexual and asexual. The asexual cycle is easy to produce in the laboratory and the one used for almost all experimental studies. The amoebae inhabit soil, humus, and animal dung, and normally feed on bacteria. One of the features of *D. discoideum* is that they feed first as separate amoebae and after a period of starvation enter a social, multicellular stage. In this respect they are quite different from the vast majority of conventional multicellular animals and plants that start as a fertilized egg or spore and that grow at the same time that they differentiate and undergo morphogenesis. In most animals and plants, growth is a major factor in morphogenesis, while in cellular slime molds it is completely separate from it (with the exception of a few late cell divisions; see later). This alone gives the cellular slime molds enormous advantages for the study of developmental processes.

B. Aggregation Stage

The starvation reaction of aggregation involves chemotaxis; the amoebae stream into central collection points (Fig. 1). In *D. discoideum* and a few related species, this process is mediated by cyclic AMP which can orient

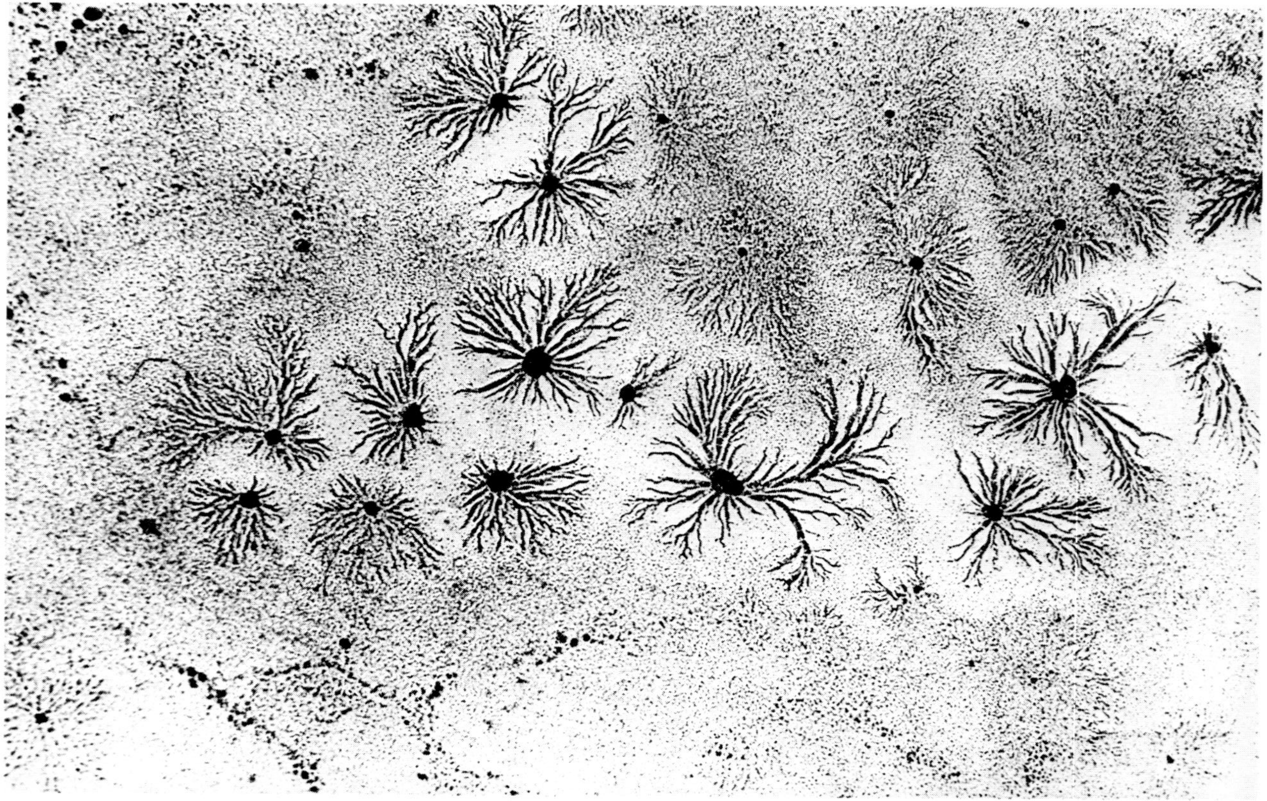

Fig. 1. Aggregation in *Dictyostelium discoideum*. Very low magnification to show many aggregates (photograph by K. B. Raper).

cells in a constant gradient or by pulses which form a relay starting out at some point as a pacemaker and radiating outward. The details of this relay are described in Chapter 4 of this volume; it is an effective way of beginning aggregation when the amoebae are spread evenly over the substratum. Large, overall gradients only come into play when sizeable aggregations of cells have formed (Konijn, 1975).

The amoebae not only increase their production of cyclic AMP at the time of aggregation, but also a number of other substances related to the chemistry of chemotaxis, including the cyclic AMP receptor sites and the phosphodiesterase that converts cyclic AMP to the chemotactically inactive 5′-AMP. There is also an increase in adhesiveness of the cells at that time, a phenomenon (noted by Shaffer, 1957a,b) that has become a subject of great interest and importance (see Chapter 5).

The center of the aggregate is at first rounded but later forms a small tip and rises into the air as an elongate cylinder (sometimes called "first finger"). If the environmental conditions are favorable, this cylinder of cells will soon assume a horizontal orientation along the substratum and migrate (Fig. 2). The migrating slug (or migrating pseudoplasmodium) has distinct anterior and posterior ends, and as it migrates it secretes a slime sheath which is left behind as a collapsed tube, rather like sausage casing. The migrating slug is extraordinarily sensitive to light and heat gradients (Bonner *et al.*, 1950; Francis, 1964; Poff and Skokut, 1977). The pigment responsible for the phototaxis is known (Poff and Butler, 1974), and recently it has been shown that even the individual amoebae are phototactic (Häder and Poff, 1979b,c).

During this period of migration it is soon evident that there are two zones of cells. As first shown by Raper (1940), the anterior quarter of the slug contains cells predestined to become stalk cells. Later, it was discovered that these "prestalk" cells stain darkly with various vital dyes and can be stained differentially with a number of histochemical stains, showing them to be quite distinct from the posterior three quarters of the slug which are "prespore" cells (Bonner, 1952; Bonner *et al.*, 1955). The best methods of showing the prespore region are those of Takeuchi (1963), who uses fluorescent labeled antispore antibodies, or more recently, of Gregg and Karp (1978) with isotope-labeled fucose, which is incorporated only in the prespore cells. The question of how this difference in early differentiation arises has been a matter of considerable research, and it is now thought that probably two processes play a role: (1) the position of a cell can determine its fate—those in the anterior become stalk cells and those in the posterior, spores; (2) at aggregation there is some early differentiation of the cells, which then sort out in the cell mass, so that for these cells the position of a cell is determined by its early differentiation (Bonner, 1959; Takeuchi, 1969; for a review of the most recent work see Tasaka and Takeuchi, 1981).

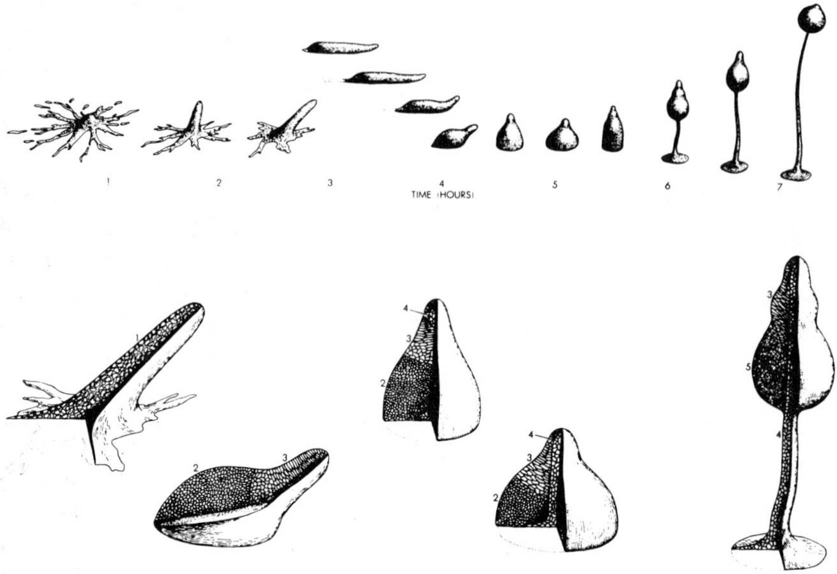

Fig. 2. Development in *Dictyostelium discoideum.* Above: The aggregation, migration, and culmination stages shown in an approximate time scale. Below: Cutaway diagrams to show the cellular structure of different stages. (1) undifferentiated cells at the end of aggregation; (2) prespore cells; (3) prestalk cells; (4) mature stalk cells; (5) mature spores (drawing of L. J. Howard in the *Scientific American*).

There has been considerable speculation and some interesting measurements concerning how the slug moves. The most recent study on this subject is that of Inouye and Takeuchi (1979) who interpret the movement in terms of a motive force that differs in different parts of the migrating slug.

There has been considerable interest in environmental factors that promote migration and in ones that induce culmination or the formation of the final fruiting body. In his early studies, Raper (1940) suggested that desiccation was a factor that induced fruiting, and this was confirmed by Bonner and Shaw (1957). It has also been shown that the higher the solutes in the agar substratum, the shorter the duration of migration, with electrolytes being more effective than nonelectrolytes (Slifkin and Bonner, 1952). More recently, Newell *et al.* (1969) gave evidence that overhead light was a strong stimulus for culmination, but the effect is indirect in that the light being overhead brings the phototactic slug tips away from the surface film of water; and that contact with this film promotes migration, while orientation away from it promotes fruiting (Bonner *et al.*, 1982). Finally, Schindler and Sussman (1977a) have shown that the presence of ammonia favors migration at nontoxic concentrations, while the absence of it favors fruiting.

At the beginning of culmination the slug tip stops forward movement and points upward while the posterior end continues to move so that all the cells are gathered directly under the tip. At this stage, the stalk cells begin to form near the tip at the upper end of the prestalk zone (see Bonner, 1944, and Raper and Fennell, 1952, for further descriptive details of culmination). Stalk cells are formed by a progressive vacuolization of the cells, during which time the cells deposit cellulose cell walls. When the differentiation is complete, the cells die (Wittingham and Raper, 1960). Potential new stalk cells arise from the annulus of prestalk cells around the stalk tip; they move into the open upper end of the stalk by a reverse fountain movement. The cells in this annulus are oriented at right angles to their direction of upward movement and can be seen to be secreting nonstarch polysaccharide at their centripetal ends; they secrete the stalk cylinder itself before they move to the apex to become stalk cells.

At first this reverse fountain movement causes the apical stalk initial to be pushed down through the prespore zone, and as it does, the whole cell mass lowers and even bulges laterally; this is often called the "mexican hat stage." Once the stalk reaches the base, a circle or torus of cells originating from the very posterior end of the previous migrating slug turns into vacuolate, stalk-like cells to form the basal disc (hence *D. discoideum*). There is no interruption of the addition to the tip of the stalk, so the stalk continues to elongate, and as it does the whole prespore mass rises into the air. (This stage is often called a "sorogen" and the mature fruiting body a "sorocarp.") Fairly early on the upward voyage, the prespore mass begins to turn into fully differentiated spores; each prespore amoeba condenses and forms a complex cellulose coat which is capsule shaped. The final fruiting body has a slender, tapering stalk with a terminal, globular or more often lemon-shaped spore mass or sorus. (Fig. 3).

There has been great interest in the factors that induce differentiation. First, there was evidence that cyclic AMP induces stalk cell differentiation in cells that have not undergone normal development (Bonner, 1970; Rossier *et al.*, 1978), but then it was discovered that cyclic AMP also was a factor in spore differentiation (Town *et al.*, 1976; Feit *et al.*, 1978). Since then there is evidence for an additional factor for stalk cell differentiation (Town and Stanford, 1979; review: Gross *et al.*, 1981). Contrary to early work, it is now possible to obtain differentiation without normal morphogenesis in submerged cultures in roller tubes (Sternfeld and Bonner, 1977; Tasaka and Takeuchi, 1979) where good oxygenation appears to be an important factor. This is providing a useful way of studying cell differentiation in this organism.

The rising sorogen is somewhat phototactic, but it is hardly affected by gravity, presumably because it is too small (Rorke and Rosenthal, 1959).

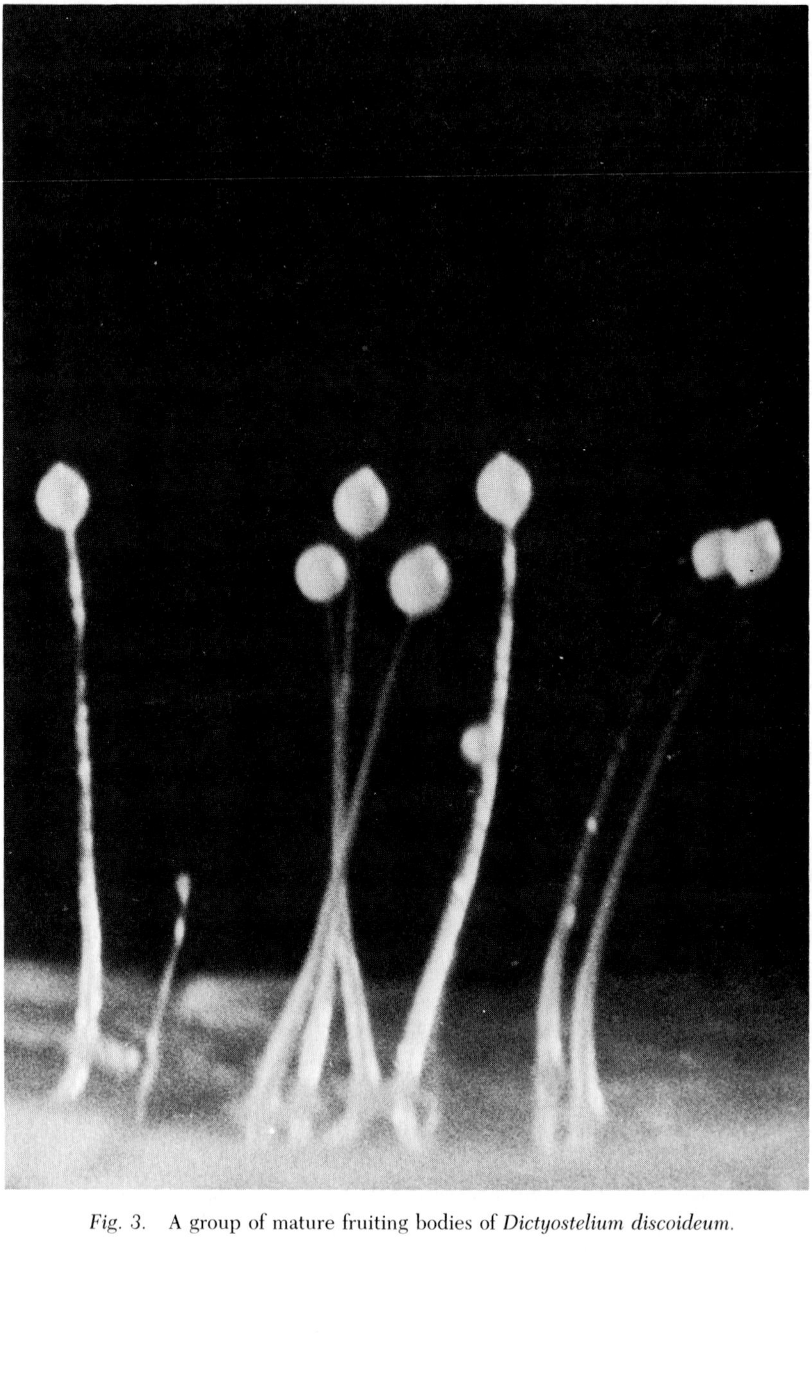

Fig. 3. A group of mature fruiting bodies of *Dictyostelium discoideum.*

Like many microorganisms, the fruiting bodies form at right angles from the substratum, regardless of the orientation of the substratum with respect to gravity. The reason for this turns out to be that the sorogen produces a volatile substance that repels another rising sorogen; it moves away from high concentrations of the gas (Bonner and Dodd, 1962b). Thus, if two sorogens rise side by side they veer away from one another as they rise, and if a sorogen rises near a piece of activated charcoal, the tip will move directly into the charcoal, which apparently actively absorbs the substance. If an isolated fruiting body rises surrounded by a circular concentration of the gas it has produced, then it will rise straight up from the substratum.

To start a new cycle, only one spore is needed. Generally, sori seem to have a germination inhibitor (Russell and Bonner, 1960) whose chemical identity is now known (an adenine derivative—see Taya *et al.*, 1978; review of spore physiology—Cotter, 1975). The result is that in concentrated masses, spores have a very low percent germination, which can be increased by washing the spores. There is one form (*D. mucoroides, var. stoloniferum*) that seems to germinate when the unwashed spores are concentrated. Much work has been done by Cotter (1975, and others) both on the germination reaction and on all the events involved, and his papers should be consulted.

C. Sexual Cycle

Blaskovics and Raper (1957) provided a clear demonstration that there was another resistant stage which they called macrocyst. Both cytological and crossing experiments from Raper's laboratory, along with those of M. F. Filosa and D. W. Francis, suggested that macrocysts are sexual zygotes and that in some species, such as *Polysphondylium pallidum* and *D. discoideum*, there are frequently mating types and macrocysts which are formed only when the cells of opposite mating types are brought together (see Raper, 1982, for a comprehensive review of these discoveries). In contrast to these heterothallic forms, others are homothallic, as in many strains of *D. mucoroides*. It is thought that first the zygotes are formed as giant cells. These become surrounded by many amoebae in a process showing some resemblance to asexual aggregation, and ultimately the giant cell and many attracted cells become surrounded by a spherical wall. The giant cell then phagocytoses the other cells so that eventually it is the only cell within the cyst. This cell subsequently undergoes a series of divisions and later the macrocyst is made up of many cells derived directly from the zygote nucleus. One difficulty with using this cycle for genetic studies has been the extremely poor germination of the macrocysts in the laboratory, a problem that one hopes will be solved soon. (This is the reason for using the parasexual system in genetic studies—see Chapter 3.) It is only through genetic analysis that the details of the sexual stage can be properly understood and determined, for instance, whether in homothallic strains there is still some sexual recombination or whether the giant cell is, in this instance, selfing. There is much to learn concerning both the development of macrocysts and their role in the slime mold life history in nature (for a recent review see O'Day and Lewis, 1981).

III. Classification

Almost all the experimental work done on cellular slime molds has been on species of *Dictyostelium* and *Polysphondylium*, both of which are considered dictyostelids. The difference between the two genera is that, by definition, *Dictyostelium* has simple or irregularly branched fruiting bodies, while *Polysphondylium* has branches coming off in whorls. Those of us who are not born taxonomists are often disturbed by this simple, *ad hoc* morphological distinction and worry that there might be unbranched species, therefore called *Dictyostelium*, that are more closely allied phylogenetically to *Polysphondylium*, as suggested by Traub and Hohl (1976). But there is the danger that we may never be able to give a proper phylogenetic classification until we know all about each species and all variants of those species; we certainly cannot hold off naming species until such a time. Therefore, common sense dictates that we abide by some simple rules of nomenclature, remembering that we may be missing interesting affinities between the forms we name.

The question of where *Dictyostelids* should be placed in relation to other simple ameboid organisms has been examined comprehensively by both L. S. Olive (1975) and K. B. Raper (1973, 1982). They basically agree on all major issues except that Olive prefers to ally slime molds with protozoa, while Raper prefers to consider them fungi. (See Table I for a comparison of their two schemes.) The difficulty is that there is probably a continuum between protozoa and fungi, and slime organisms may be one of the bridges. There is no scientific way of settling this point; again it must be declared by fiat. Considering the fact that Brefeld, van Tieghem, and all the early workers on slime molds were mycologists by training, as are Olive and Raper, tradition is as good a guide as any for a system of classification of primitive forms.

Let me briefly summarize the common points of the combined Olive–Raper system of classification. There are four main groups of slime

TABLE I
Two Systems of Classification

K. B. Raper (1973)
- Kingdom Plantae (plants)
- Kingdom Mycetae (fungi)
  - Division Myxomycota (mycetozoa or slime fungi)
    - Class Acrasiomycetes (cellular slime molds)
      - Subclass Acrasidae (acrasids)
      - Subclass Dictyostelidae (dictyostelids)
    - Class Protosteliomyces (protostelids)
    - Class Myxomycetes (true or plasmodial slime molds)

L. S. Olive (1975)
- Kingdom Protista
  - Phylum Gymnomyxa (mycetozoans and associates)
    - Subphylum Mycetozoa
      - Class Eumycetozoa
        - Subclass Protostelia (protostelids)
        - Subclass Dictyostelia (dictyostelids)
        - Subclass Myxogastria (myxomycetes or true slime molds)
      - Class Acrasea (acrasids)

molds (Myxomycota or Mycetozoa). One contains the true or plasmodial slime molds (Eumycetozoans or Myxomycetes). A second group, the protostelids, was discovered by L. S. Olive and C. Stoianovitch. It consists mostly of single cell forms (or small plasmodia) with minute stalked fruiting bodies. The third group, the acrasids is a mixture of different kinds of aggregative organisms, which differ from dictyostelids in that the pseudopods of the amoebae are lobose rather than filose and aggregation does not occur by streams but by each of the rounded cells moving to the central collection point independently. Some of their multicellular fruiting bodies are branched, but they show weak differentiation between spore and stalk cells. The genera include *Acrasis*, *Copromyxa*, *Guttulina*, *Guttulinopsis*, and *Fonticula*. Little experimental work has been done on acrasids, although they show many interesting possibilities. Dictyostelids, the fourth group, are the concern of this volume. Besides the two common genera previously discussed, there is also *Actyostelium* (Raper and Quinlan, 1958) which has the unusual feature of having an acellular stalk. Because it lacks a division of labor of stalk and spore cells, it is often assumed to be more primitive than *Dictyostelium* and *Polysphondylium*, but its simplicity could, of course, be secondarily derived (Bonner, 1982b).

Besides these four main groups there are two others that are possibly allied: the Plasmodiophorales, which are parasites of higher plants and show affinities to some fungi, and the Labyrinthulales, which have a curious network of gliding cells and are found on eelgrass in coastal waters.

In recent years, largely through the work of Raper, Cavender, and others (see Raper, 1982), there has been an increase in the known number of species of dictyostelids. In 1902, E. W. Olive described 11 species of dictyostelids; now it is estimated that there are about 50 (Raper, 1982).

IV. Ecology

Before examining problems of development in more detail, it may be helpful to ask why, from an ecological or evolutionary point of view, dictyostelids have a life history that seems to set them apart from all other organisms. This view stems from my conviction that it is helpful to understand an organism's behavior, especially if one is interested in revealing the mechanism for that developmental behavior. Unfortunately, if one asks questions about the adaptive nature of any aspect of any organism, one immediately encounters a difficulty: it is very hard to distinguish what is truly adaptive from what is there either by chance or by some developmental constraint. In some cases it may be possible to demonstrate a function for a character, but this is a slightly different matter.

In this section, the life history strategies of cellular slime molds will be examined, as well as the physical effects of the environment, the distribution of dictyostelids in nature, and finally some of the especially interesting life cycle features that may have special relevance to developmental biology.

A. Life History Strategies

One should remember that cellular slime molds live in soil or in dung on the surface of the soil and that they are exposed to raw weather; they must survive and even thrive where temperature varies from freezing to very hot and where rainfall varies from drought to flood. Therefore, the fact that they possess resistant stages—spores, macrocysts and microcysts—to carry them over hard times is not surprising. But the question one might ask is, Why are there three types of resistant bodies; Why is one not enough?

First, let us consider microcysts and spores. In some species (e.g., *Polysphondylium pallidum* and some small species of *Dictyostelium*, but not *D. discoideum* or *D. mucoroides*) besides the spores, each amoeba can, before aggregation, become encapsulated into a microcyst. These cysts are spherical and have quite a different wall structure from the spores (Hohl, 1976). It is presumed that these microcysts are an alternate developmental pathway to meet a sudden onrush of adverse conditions. They can be triggered by environmental cues; Lonski (1976) showed that an increase in ammonia results in an increase in microcysts. Francis (1979) has made a most interesting developmental study of the switch to alternate pathways and has shown that there is a synthesis of different proteins depending on whether development proceeds in the microcyst or the aggregation, spore direction. This safety measure seems so useful that one wonders why all species do not have microcysts. This may be a difference that arose in some ancestor after the separation of two or more branches appeared in the phylogenetic tree of dictyostelids, i.e., the reason might be historical. It is true that different strains of *P. pallidum* isolated from nature show different degrees of microcyst formation, so it is also possible that certain specific ecological conditions, such as periodic catastrophies of a certain kind, favor the existence of microcysts.

Macrocysts, as has already been pointed out, are probably the sexual spores, although it is far from clear if this is universally so and whether there might also be a considerable amount of selfing, especially in homothallic forms. It has been known for some time that submerging the amoebae of a macrocyst-forming, homothallic strain of *D. mucoroides* will induce macrocysts, whereas a relatively dry surface in the culture disk will promote asexual development (Weinkauff and Filosa, 1965). Furthermore, Weinkauff and Filosa also showed (and see Filosa, 1979) that liquid and volatile substances may influence whether or not macrocysts are produced (see also O'Day and Lewis, 1981). The effect of an excess of water in triggering macrocyst formation is interpreted as a possible safety mechanism to form a resistant body in the event of a torrential rain. This possibility is supported by the fact that in a homothallic strain of *D. mucoroides*, Filosa *et al.* (1975) have discovered that quite late in the life cycle one can switch from a migrating slug to macrocyst formation.

The most important question involving life history strategies is, Why does one have fruiting bodies at all; why are not all the cysts solitary as in microcysts or the cysts of many other soil amoebae? The argument I have developed in detail elsewhere (Bonner, 1982b) is that there must be an enormous selection pressure for small fruiting bodies carrying a mass of spores into the air, and this pressure must come from the importance of dispersal, especially when the sources of food will be patchy and ephemeral. My argument is based mainly upon the fact that this kind of fruiting body is found in so many large and diverse groups of organisms, including thousands of fungi, myxobacteria, and, as Olive (1978; see also Olive and Blantin, 1980) has recently described, even ciliate protozoa.

Here let us simply assume the selective advantage of small fruiting bodies, and then examine the fact that in *D. discoideum* roughly 25% of the cells die in forming the stalk of the fruiting body, while in conditions that favor migration, *D. mucoroides* or *D. purpureum* will have as many as 90% of their cells sacrificed to get the few remaining spores into a place favorable for dispersal. This can only mean that the advantage gained in dispersal must be more than enough to offset the cell loss in making a stalk. In this respect cellular slime molds are sociobiological marvels and show a remarkable degree of altruism. (An even more sociobiological aspect is evident in cell masses that contain more than one genotype. See the later discussion of heterocytosis.) In closing this discussion, I remind the reader of *Acytostelium* in which all the cells secrete an acellular stalk, and then all the cells turn into spores. *Acytostelium* gains because it loses no cells, but it also loses in that the spores must give up part of their important resources in the formation of stalk. But these speculations do not resolve the question raised earlier of whether *Acytostelium* is an ancestor or a descendant of the *Dictyostelium* method of development with its division of labor into stalk cells and spores.

There remains the question of why small fruiting bodies on the order of a millimeter or so in height are so effective in dispersal. It is presumed that this is a successful way of adhering spores to passing soil invertebrates such as nematodes, earthworms, larvae of all sorts of insects, and so forth, If a sorus of a dictyostelid is touched, the spores will adhere to the foreign surface by capillarity, and in this way they could be carried some distance. As L. W. Buss has suggested to me, they could also be carried in the gut of the invertebrates, as is known to occur for the spores of many fungi (for a review see Pherson and Beatie, 1979). More widespread dispersal might occur though the aegis of migratory birds; Suthers (in preparation) has demonstrated that bird feces, especially of ground feeding birds, may be rich in dictyostelid propagules.

B. Response to Light, Temperature, and Moisture

Some aspects of the response of dictyostelids to physical environmental factors have already been discussed and will be only briefly summarized here. The migration slugs of many (but not all) species are positively phototactic and the stimulus to form the final, differentiated fruiting body comes not directly from the light, but from the fact that the light is overhead and brings the tip of the slug away from the surface film of water. Let us now put these facts into an appropriate ecological setting. Feeding occurs under at least a film of water, which is required for amoeboid motion and phagocytoses. However, fruiting for optional dispersal must be near the surface of the soil or humus, or at least in some subterranean microchamber. Migration then is a means of traveling from a location suited to feeding to one suited to fruiting. Since this is toward both light and a drier location, phototaxis and the response of the slug to leaving the surface film seem ideal ways of achieving the best dispersal of spores.

There is also an extraordinarily sensitive orientation to temperature gradients (Bonner *et al.*, 1950; Poff and Skokut, 1977). Recently, Whitaker *et al.* (1980) made the interesting extension of these original observations to show that at cool temperatures the migrating slugs of *D. discoideum* are negatively thermotactic; positive thermotaxis only appears at warmer temperatures. One might speculate that this is a mechanism to ensure that the slugs always go toward the surface of the soil. At night the top of the ground will be cooler than below, while in the daytime the temperature gradient will be reversed; a switch from negative to positive thermotaxis in the early morning would mean that slugs always migrated upward to the surface.

From the work of Harper (1932) and Raper (1940) it is known that light induces early aggregation and fruiting. In some strains of *P. pallidum* light is a necessity for the initiation of aggregation; Jones and Francis (1972) have examined the action spectrum of this phenomenon and found that it is different from that of phototaxis. A sudden drop of humidity, as Raper (1940) also showed, has a similar effect. The result in both instances is more numerous but smaller fruiting bodies, a fact explained by Kahn (1964) as due to an increase in the simultaneity of center formation. It is easy to see why the increase in desiccation would be a useful signal to move toward fruiting, but why light should have such an effect is harder to rationalize, unless daytime is the more effective time for aggregation.

C. Biogeography of Dictyostelids

Largely through the work of Cavender and a number of others, it is clear that cellular slime molds are found all over the world, from the colder temperature regions (Alaska) to the tropics. Of special interest is the fact that, as in so many higher plants and animals that have been studied, Cavender (1973, 1978) showed that the number of species, or species diversity, of dictyostelids increases as one moves from northern to tropical latitudes (Fig. 4). More recently, Hagiwara (1976) and Kawabe (1980) have shown this to be true for altitude changes on Japanese mountains; the lower one samples on the mountain, the more species are found to coexist. The arguments for increased diversity of higher animals and plants in the tropics are well

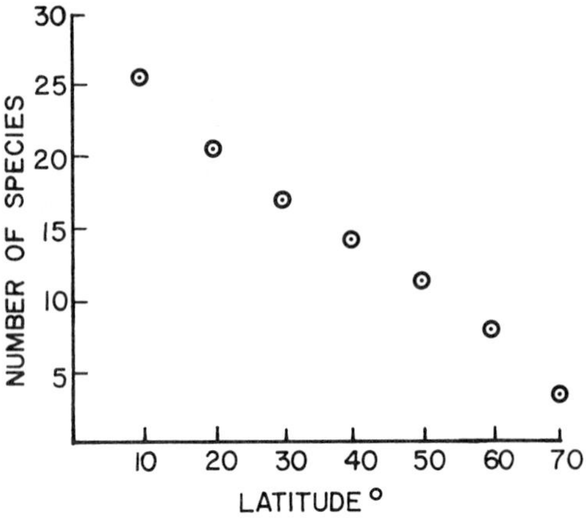

Fig. 4. Species diversity (number of cellular slime mold species) at different latitudes (redrawn from Cavender, 1978).

known, but little has been done on similar problems in microorganisms. There is still much to learn about the microecology of the soil and forest that favor certain species of slime molds (Cavender and Raper, 1956, 1968). In general, dictyostelids are most abundant in forest habitats that are not too moist, such as oak forests in temperate regions. They are rare or absent in dry, desert habitats.

There is an important need to pursue further the question of the microdistribution of cellular slime molds in the soil, especially along the lines initiated by Eisenberg (1976). By microsampling in small patches of soil, as he did, it will be possible eventually to learn more about slime mold microdistribution and how these distributions correlate with soil porosity, invertebrates that might be involved in dispersal, and related properties of the microenvironment.

Another interesting problem is, How can so many species of dictyostelids coexist in a small area of soil; why does not one species out compete all the others according to the exclusion principle of Gauss? The problem was examined by Horn (1971) in a small plot of soil in Princeton; he concluded that each of the four species found there had strong preferences for different species of bacteria thriving in that same plot. So the small area could be subdivided into separate feeding niches, thereby avoiding competition and allowing the coexistence of the four species of slime mold.

D. Incompatibility, Heterocytosis, and Predation

We come finally to a related group of problems which have both ecological and developmental interest. In the case of incompatibility, we have the question of how new species arise in cellular slime molds. If the amoebae always coaggregated, there would never be more than one species; there must be ways of isolating and separating cells. These are shown in an early paper by Raper and Thom (1941). The strongest kind of separation came when they mixed *D. discoideum* with *P. violaceum:* the amoebae did not coaggregate but went to separate central collection points, resulting in the total isolation of the fruiting bodies from one another (Fig. 5). The next level of separation was demonstrated in a mixture of *D. discoideum* and *D. mucoroides*. Here the amoebae of both species coaggregated, but the central cell mass broke up into two tips, one producing a *D. discoideum* slug and fruiting body and the other entirely *D. mucoroides* (Fig. 5).

We now have considerable insight into these different methods of isolation. In the first, the acrasin, that is the aggregation chemoattractant, is different; in *D. discoideum*, it is cyclic AMP, while in *P. violaceum* it is an unusual dipeptide that has just been identified (Shaffer, 1953; Wurster *et al*., 1976; and Shimomura *et al*., 1982). Neither species is sensitive to the ascrasin of the other during aggregation. We now have evidence for at least eight different acrasins among dictyostelids (and there may be more), but we

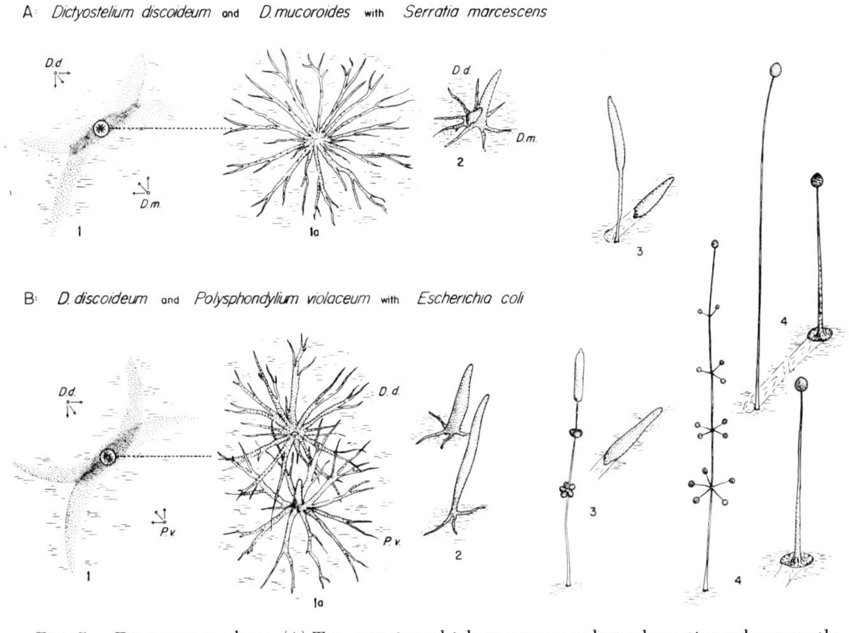

Fig. 5. Diagrams to show: (A) Two species which coaggregate but whose tips subsequently separate at the center of the aggregate. (The pigment from *S. marcesens* is only retained by *D. discoideum* and therefore the cells of the two species can be identified). (B) Two species that have separate acrasins and aggregate with overlapping streams to their individual centers (from Raper and Thom, 1941).

only know the chemical constitution of two (for a review see Bonner, 1982). Obviously, the fact that there are so many acrasins with all the chemical machinery that goes with each one (e.g., an acrasinase, a surface receptor, and no doubt others) raises the interesting question of how they evolved. Also, it is my conviction that if we knew the chemical identity of some of the others, it would be an enormous help to the comparative study of chemotaxis and other developmental processes.

In the second example of Raper and Thom, both species involved used cyclic AMP for their acrasin, which explains their coaggregation. But once the cells of the two species are together in one cell mass, they clearly sort out and form separate tips. For good reasons one might suspect that this sorting out is due to the different adhesive properties of the two species (see Chapter 5), and that the sorting out of the mobile cells can be satisfactorily explained by Steinberg's (1970) differential adhesion hypothesis (Nicol and Garrod, 1978; Sternfeld, 1979). In some earlier experiments we showed that even different strains of the same morphological species showed different surface adhesion properties (Bonner and Adams, 1958).

There are examples of genetically different cell types coexisting in the same fruiting body. This was first conclusively demonstrated by Filosa (1962), who termed the process heterocytosis. Specifically, he showed that one of our laboratory stocks of *D. mucoroides* contained a mutant or variant that when cultured alone would produce a very abnormal fruiting body. However, if as few as 10% of the cells in a mixture were wild type, the phenotype of the fruiting body was normal. If a mixture of spores was cultivated for a series of generations by mass transfer of spores, then invariably both cell types remained in a ratio of 10–15% mutant and the rest of the cells were wild type. At the time it was not known if this was a unique situation or a general phenomenon. This question has been resolved in an especially interesting study of Buss (1982) who set out to see if heterocytosis existed in nature. To do this he plunged a group of small, glass micropipettes into the soil and then separated them and cultivated the slime molds derived from each tube. In more than one instance, he found isolates that produced very abnormal fruiting structures, often hopelessly defective; and these isolates were found in the same micropipette containing cells that produced normal fruiting bodies of the same species. They were so close together in the soil that they necessarily must coaggregate, as did Filosa's laboratory strain; and in 10 passages in Petri dishes in the laboratory they continued to coexist. Buss has a most intriguing interpretation to explain the phenomenon. He suggests that the defective variant is an intraspecific parasite, that cannot compete or survive successfully alone, but by joining wild-type cells it can flourish. Like any good parasite it must not completely take over its wild-type host for it can only exist as long as its host exists. It is Buss' idea that within this story lies a deep seated and important message for the origin of multicellularity and probably for developmental biology as well.

Finally, we come to predation. It has always been assumed that all cellular slime molds are herbivores, which graze on the bacteria in the soil and in decaying leaves, the only exception being an occasional report of cannibalism, including the activity of the giant cell in the macrocyst. However, Waddell (1982b), in our laboratory, has discovered a carnivorous species. He isolated the slime molds from bat dung in a cave and found a form that was morphologically undistinguished: it appeared to be a delicate species similar in form to some of the smaller species of *Dictyostelium*. But when he mixed this with any other species of dictyostelids, after a number of days the new species, *D. caveatum*, was the only one left; the partner had utterly disappeared. This was true even when there were initially very few cells of *D. caveatum* and a large number of the partners. However, the fewer the *D. caveatum* cells, the longer it took for the other species to disappear. It was especially curious that when such a delay was created, the other species was arrested in its development during a period of late aggregation and the conversion apparently took place in the mounds of cells. The key question is, How do the *D. caveatum* cells eliminate the other cells? The *D. caveatum* cells have been examined by Waddell using electron microscopy and there are many large inclusions of a curious sort. Also, one will see the cells of the prey species inside the *D. caveatum*, suggesting that conversion occurs by phagocytosis. If grown alone, the *D. caveatum* can survive on *E. coli* as a food source, but it appears to thrive with greater vigor in the presence of some other species of slime mold. It differs from other known species both in that feeding occurs during and after aggregation in the cell mass and in that it can then fruit shortly afterward. At least from what we know about it so far, it does not appear to require starvation for aggregation when using some other slime mold as its source of energy. Again in this interesting new form we have a combination of possible lessons for slime mold ecology as well as slime mold development.

V. Development

This entire book is about the development of dictyostelids and therefore what I would like to say in this chapter will complement, rather than duplicate, what is said elsewhere. There are so many aspects of development that it will be easy to select ones that are important and that will not repeat later discussion.

A. Cell Cycle and Development

Let me remind the reader of something mentioned earlier. Cellular slime molds are unusual in that they grow first and then undergo all their morphogenetic movements and differentiation at a later stage. There are a number of developmental biologists who think of the cell cycle as the governing factor driving development, and indeed in some systems this is undoubtedly the case. But there is no evidence to support such a contention for dictyostelids, although the possibility cannot be ruled out totally. In an early study we showed that when a slug was first formed, there was an uneven distribution of residual mitoses of low frequency (Bonner and Frascella, 1952). First there was a burst at the tip and later a sparse scattering of mitoses in the posterior prespore region of the migrating slug of *D. discoideum*. This was determined by direct counting of metaphases in stained cells and has been confirmed (Durston and Vork, 1978) by the more elegant method of using tritiated thymidine. They pulse labeled the vegetative cells and then made autoradiographs of the slugs at different times. The original idea was that perhaps the difference between the prespore and prestalk cells is that they are caught at different times in their mitotic cycle; those close to the next mitosis go to the posterior prespore region.

One experiment that seemed to demolish this idea is the work of Katz and Bourguignon (1974) on a temperature sensitive mutant that could stop all the vegetative cells at G2 and then they were released simultaneously by switching to the permissive temperature. One would assume that if the point in the mitotic cycle during the vegetative state of any one cell had any bearing on its subsequent position in the slug, then differentiation would be in severe difficulty if all the cells were the same in this respect. These released cells of Katz and Bourguignon did develop normally, but there are at least two reasons why this experiment is not crucial for the hypothesis: (1) possibly the cells in the slug can compensate for this uniformity by allowing their other method of differentiation to take over, namely cell fate determined by position and (2) possibly the mutant blocking might not equalize old and young cells in all respects, for one can synchronize the DNA synthesis cycle without affecting the growth cycle (for a review see Mitchison, 1971).

There are, however, two experimental facts that make it seem unlikely that differences in the cell cycle during growth have any effect on their subsequent differentiation fate. If slugs are allowed to migrate for 30 hours or more, then all mitotic activity ceases. If they are now cut they will regulate without difficulty, as Raper (1940) showed earlier. However, the transected slugs do not undergo another wave of mitosis (Bonner and Frascella, 1952). So at least one could argue from this that mitosis is not necessary for differentiation, because in the regulation process prestalk cells become prespore cells and vice versa. But this experiment is not airtight either since one might assume that the original anterior–posterior cell distribution has been established by differences in cell-cycle age of the cells and that this polarity once established could somehow manage regulation. Perhaps the best evidence against cell cycle control of differentiation comes from some unpublished experiments we have recently done with *D. mucoroides* var. *stoloniferum*, described by Cavender and Raper (1968) and kindly sent to us by K. B. Raper.

One possibility is that this form might either lack a spore germination inhibitor or is extremely resistant to it; unlike the spores of most species they will show massive germination in dense populations. It is presumed that this is an aid in dispersal, for when a sorus touches the substratum and there is no food, the numerous germinated amoebae will rapidly aggregate and form a new, smaller fruiting body. If one gathers many of these spores and places them on nonnutrient agar, they will produce a second generation of fruiting bodies without growth. (This had been done by us and others years ago for *D. discoideum*.) By repeating the process one can get six such generations with this *stoloniferum* variant. Since no intake of energy is involved there is no growth and the cells become smaller with each generation. One is selecting only spores in each passage, yet the fruiting bodies, which becomes progressively smaller, are always normally proportioned. It seems quite impossible to imagine that the phase of the cell cycle could be playing any role in controlling differentiation six generations or life cycles later.

B. Control of Size

There are two moments in the life cycle where the size of species of *Dictyostelium* is determined. One is at aggregation and the other is in the formation of tips and their accompanying pseudoplasmodia in the central cell mass at the end of aggregation. In *Polysphondylium* there is a third, and that is the breaking up of the posterior end of the rising sorogen into whorls. We will discuss the first only briefly because it is old work and best known. Most of the space will be devoted to the second and third which include interesting recent work.

Within a restricted range of cell densities, and provided the measurements are made from undisturbed growth plates, the aggregation territories remain fixed in size regardless of amoeba density (Bonner and Dodd, 1962a). This means that at low density fewer amoebae will lie inside a territory and, therefore, the resulting fruiting bodies will be correspondingly small. If amoebae, washed free of bacteria, are plated out at different densities, then the amoeba density–aggregation center density relationship becomes more complex (Sussman and Sussman, 1961; Feit, 1969; Hashimoto *et al.*, 1975; Waddell, 1982a). It is clear from direct experiments of Shaffer (1963) on *P. violaceum* that once an initial "founder" cell is formed, it inhibits other cells from becoming founder cells in its vicinity. From observations on *D. discoideum* it is equally clear that once a center is formed it inhibits the formation of centers in the near vicinity; and, as one would expect if there is inhibition, the distribution of centers is nonrandom (Bonner and Dodd, 1962a; Waddell, 1982a). It was mentioned earlier that light affected territory size, and Kahn (1964) had postulated that light caused the centers to appear with greater simultaneity; hence there were more of them because inhibition would have less of a chance to play its role. Finally, Francis (1965) and Konijn (1975) have evidence that a steep acrasin gradient from a large active center would disassemble a weaker peripheral center and draw away its amoebae. This is inhibition by means of a strong cell attractant and quite different from the inhibition described above which suppresses center initiation.

There are, then, three ways in which aggregate size can be controlled, and this control is clearly affected by cell density. But perhaps the most important lesson for us in this discussion is an even simpler one. Despite the fact that there may be a small number of additional mitoses during later development, in broad terms the size of a cellular slime mold is determined by the number of cells that enter into the aggregate. In this respect these organisms are totally unlike the vast majority of all other higher plants or animals where size is almost invariably determined by growth during development. This distinction is important for many reasons. One is the straightforward fact that it means the size range of individual slime molds is unparalleled. So far the smallest individual showing division of labor is one of seven cells (three stalk cells and four spores; Bonner and Dodd, 1962a) whereas the largest will go a few hundred thousand cells. If one includes *Acytostelium*, B. M. Shaffer has told me of seeing a single spore raised on a stalk, very similar to *Protostelium*. Another significant point is that by separating growth from differentiation and morphogenetic movements, it is possible to study those latter two processes without the complication of growth.

The second moment of size control comes at the end of aggregation. This is very easy to see and characteristic of some of the smaller species, since after aggregation the central mass of cells will subdivide into a group of separate pseudoplasmodia, each with its own tip. This problem has been carefully examined by Kopachik (1982) who has been able to provide a clear demonstration that the tip produces a diffusible inhibitor which prevents other tips from appearing in the near vicinity and in this way determines how many cells will follow the tip. His evidence comes from numerous experiments, but the most telling is one in which he puts cells in a minute well, which is then covered with a thin layer of agar, followed by another group of cells on top of the agar above the well. In the control, both groups of cells will come from the posterior, prespore region of a slug of *D. discoideum*, and the upper group will soon form a tip. If, however, the lower group of cells in the well is made up of tip cells, then the group of prespore cells on the upper surface will take a long time to form a tip. This is true even if the upper and lower cells are separated by a dialysis membrane, suggesting that the inhibitor is a small molecule. The earlier experiments of Durston (1976), which provide evidence for an inhibitor of decreasing magnitude along the axis of a migrating slug which prevents the slug from breaking up into smaller units, are undoubtedly a reflection of the same phenomenon.

In addition to evidence for an inhibitor that controls size, Kopachik was also able to show that the responding cells may differ in their susceptibility to the inhibitor. In a variant strain of *D. discoideum* (P4), using reciprocal grafts over wells, he was able to show that the tip of P4 had as much inhibitor as the wild type, but that if a wild tip was put over P4 cells, the latter were resistant to the inhibitor and formed much smaller slugs than normal. He also showed that this response to the tip inhibitor varied at different times in the life cycle, that is, vegetative cells were less susceptible to it than cells in the migrating pseudoplasmodium.

There is some interesting new information concerning the third moment of size control, the breaking up of the cell mass into whorls, which is the characteristic that distinguishes *Polysphondylium*. Byrne and Cox (1982) have shown that if the tip of a rising sorogen of *P. pallidum* is nipped off there is a sudden orgy of tip formation in the remaining cell mass, producing a Medusa-like monster. So, clearly in *Polysphondylium* there is an inhibitor similar to Kopachik's. But we know less about the factors that pinch off the whorls from the posterior end of the rising sorogen. Spiegel and Cox (1980) have demonstrated that there is considerable regularity between whorls (internode distances), but what is especially interesting is their demonstration that the distance is the same for haploid and diploid strains. Despite the fact that the cell size of one is twice that of the other, not only is the distance between whorls the same, but the diameter of the stalks for a similar interval is the same. This means that a haploid internode will have twice as many cells as a diploid one. It is another splendid illustration of the fact that in morphogenesis it is not so much the cell number that is important in producing shape, but the amount of protoplasm, and the cell shape accommodates accordingly (see Frankhauser, 1955).

C. Control of Proportions

This subject borders dangerously on the subject of pattern formation, which will be discussed in detail in Chapter 11, but here I will dwell on a few aspects that also have implications for other aspects of development. The point has already been made that work on the chemical factors that might be involved in the induction of differentiation is in progress in different laboratories. Furthermore, there is copious evidence that prespore cells (characterized by their staining properties and vacuoles; see the review by MacWilliams and Bonner, 1979) can differentiate into prestalk cells and vice versa. The traditional view has been that somehow these two components account for the fact, discovered by Raper (1940), that if a slug of *D. discoideum* is severed, each portion will regulate, although the time required for regulation is much greater for prestalk fragments. This difference is explained by the fact that it takes longer for prestalk cells to convert to prespore cells than the reverse.

The role of sorting out in normal development has always remained somewhat enigmatic even though there was clear evidence of some individual cell movement in the migrating slug (Bonner, 1957; Francis and O'Day, 1971). A new most interesting hypothesis for the mechanism of regulation, which includes this sorting out, has been put forward by Sternfeld and David (1981). They have convincing evidence, confirming the results of Matsukuma and Durston (1979), that chemotaxis to cyclic AMP is occurring in slugs and that prestalk cells are sensitive to its gradients, while cells in the prespore region are relatively insensitive. Using neutral red as a cell marker, they take advantage of the fact that the dye is conspicuous in the prestalk cells and faint in the prespore cells. They note that while all the anterior cells show the stain, only a few (roughly 10%) of the cells in the posterior end stain in this way. They call these cells "anterior-like" cells and normally they lie quiescent among the prespore cells. However, if the anterior prestalk region is removed by amputation, then these anterior-like cells suddenly become active (as thought they were released from an inhibitor), and they sort out by cyclic AMP chemotaxis to the anterior end (see also Takeuchi, 1982).

In this view, the reason regulation is so rapid in a posterior fragment is that there is no dedifferentiation and redifferentiation; the dormant reserve of prestalk cells becomes active and moves up by chemotaxis to form a new prestalk zone. By contrast, the prestalk fragment has to undergo the slow process of changing the direction of differentiation of the posterior cells. A few years ago Sampson (1976) made the discovery that if posterior portions were cut off slugs of different ages, the older the slug, the greater the percentage of spores. This would fit in well with the scheme of Sternfeld and David if one assumes that older slugs have fewer of the dormant, anterior-like cells. The meaning of this new idea is that normal proportions of the prestalk and prespore zones and their regulation must involve cell sorting and chemotaxis as well as redifferentiation. When regulation occurs rapidly it is due to sorting out, when it occurs slowly it is due to dedifferentiation and redifferentation.

There is one other point to make about proportions that will be relevant to our discussion of timing further on. It is well established that both the prestalk and prespore zones and the mature stalk and spore mass show a proportionate relationship over a wide size range (see review by MacWilliams and Bonner, 1979). It would seem logical to assume that the presumptive condition is a necessary half-way step to the final differentiation. Yet the fruiting bodies of *Polysphondylium* show equally precise proportions between stalk and spore, and they have no equivalent to the prestalk and prespore regions (Bonner *et al.*, 1955). Instead, as Hohl *et al.* (1977) have shown, all the sorogen cells of *P. pallidum* are prespore in that they all have the characteristic prespore vacuoles. They disappear only when the cells enter into the tip of the stalk and begin their journey to become a stalk cell. Therefore, we must assume that this half-way mark of differentiation found in some species is not a prerequisite for the control of proportions.

What then could be the significance of the well-regulated presumptive zones? The answer is not known, but there is an interesting correlation which might be significant. Thus far all the species that have visible, proportionate presumptive zones are those that have cyclic AMP as their acrasin (Bonner, 1982b). In all the other species tested, including those of *Polysphondylium*, the acrasin is some other substance. Since, as pointed out earlier, cyclic AMP is a factor in the final differentiation of stalk cells and spores, one might speculate that this is why the early stages of differentiation are different in those species that use the same substance for both chemotaxis and differentiation. The only way in which we will solve this intriguing matter is to learn more about the biochemistry of both differentiation and chemotaxis and make a comparative study between species. (Schaap *et al.* (1981) have shown that *D. minutum* has possible prestalk and prespore cells, which are not separated into zones.)

D. Determination versus Differentiation

There is another feature of cellular slime mold development that also sets them apart from more conventional organisms. In the latter, determination occurs first and later we see the visible manifestations of differentiation. Classic examples are seen, for instance, in insect development. The fate of most parts is already determined in the imaginal disks of the larva in *Drosophila*, and even if they are transplanted to new locations they will give the structure, now improperly placed, for which they were originally determined. In contrast, there are many instances where, as in regulative eggs, regions of the egg or the early embryo appear totipotent and show no sign of determination whatsoever. But when this is so, they show no morphological signs as to their fate, and they remain neutral in their appearance. The prespore–prestalk condition seems to violate these rules. The first sign of differentiation appears first, but there is no evidence of any determination, as a prestalk cell is readily converted into prespore cell and vice versa. In the sense that it is used for insects, there is probably a total absence of determination. The only time the fate of a slime mold cell becomes fixed is when a stalk cell dies; that is indeed an irreversible differentiation. But despite its thick, complex cellulose coat, the spore retains the potentiality of making either stalk cells or spores, for after all, a whole new generation can be propagated from a single spore.

In most of these respects the cellular slime molds show greater affinity with higher plants than animals. In angiosperms, there is a comparable absence of determination; and any cell, if it can escape the prison walls of the cellulose it has deposited, can take on different courses in its development. Xylem cell differentation closely resembles stalk cell differentiation; in both cases there is a heavy deposit of cellulose wall, a large increase in vacuolization, and ultimately death of the cell. Perhaps the moral is that we should shed any narrow conception of determination and remember that the world is not made up of slime molds or insects alone.

There remains an interesting question of the possible presence of some form of determination in the macrocyst cycle of cellular slime molds. In heterothallic forms, when the two mating types are brought together, there is a sensitive period early in development where the cells must be together (Wallace, 1977; Bozzone, 1982). If the cells of opposite mating type fail to touch during this period, they do not form macrocysts when transferred to macrocyst-forming conditions. Normal fruiting bodies will be the only structure present, clearly they are determined (Bozzone, 1982). On the other hand, as mentioned earlier, Filosa *et al.* (1975) showed that in a homothallic form (*D. mucoroides*) it is possible to switch at a late stage from migrating slugs to macrocyst formation, at least with a certain percentage of the slugs. Therefore, the sensitive period and the presence of some sort of determination seems to be a characteristic of at least one set of heterothallic strains, as compared to one homothallic strain.

E. Timing in Development

There are two aspects of the subject of timing in slime mold development. One has been pursued by Soll (1979) who developed some interesting ideas as to how the examination of the variation of timing in specific stages of slime mold development under different conditions might lead to some clues as to the nature and complexity of the chemical machinery responsible. It is not a matter that will be discussed here and I refer the reader to Soll's papers.

The aspect of timing that will be considered here is heterochrony. This ancient subject is concerned with the question of major shifts in the timing of events during development; how they are genetically controlled and how they may result in major evolutionary innovations (de Beer, 1940; Gould, 1977; Bonner, 1982a). One of the best known examples is a shift in the time of appearance of active gonads. In some amphibians, for instance, the larvae will become sexually mature and metamorphosis will be totally bypassed. This so called neoteny is, in contrast to apes, also evident in the development of human beings. We have a relatively prolonged period of development which results in an increase in the duration of brain growth and child care, qualities thought to be fundamental to our evolution. This is assumed to be a relatively easy way, from a genetic point of view, of making a viable developmental change that will produce a major change in the organism; it is thought of as the most obvious mechanism for what is currently called macroevolution.

If we look at the cellular slime molds in this way, we gain some insight into the morphological variation that exists. Let me give three good examples of heterochrony in dictyostelids.

1. The first involves the initiation of stalk formation: it can begin either at the end of aggregation, as in *D. mucoroides* and many other species, or initiation can be delayed until after a period of stalkness migration, as in *D. discoideum*. Looking at it in the light of timing, one could say that the stalkless migration of *D. discoideum* is possible through heterochrony, that is, through the delay in the initiation of stalk formation. This assumes that the *D. mucoroides* form is more primitive; if the reverse is true heterochrony would still come into play.

2. The second period of size control, as discussed in the previous section, is normally at the end of aggregation. However in *D. polycephalum*, stalkless migration of a large slug occurs first, and then before fruiting this slug breaks up into a number of tips which cluster together to form a small cluster of fruiting bodies whose stalks adhere to one another (Raper, 1956). In this case there are two simultaneous examples of heterochrony: one is as before for the initiation of stalk formation and the other is the delay of the second period of size control.

3. The third period of size control, namely the formation of buds to form branches in *Polysphondylium*, has its counterpart in *D. rosarium* (Raper and Cavender, 1968). The difference between these two examples is that in *Polysphondylium* the buds form whorls of small fruiting bodies, whereas in *D. rosarium* all the cells in the bud turn into spores. If one looks for a reason

for this difference, one observes that *D. rosarium* has prestalk and prespore cells (Hamilton, 1981) while a *Polysphondylium* slug is uniformly made up of prespore cells (Hohl *et al*., 1977), and possibly these different early signs of differentiation account for the different developmental behavior. In any event the timing of final differentiation is delayed and remains flexible in *Polysphondylium*, but the buds on the stalk of *D. rosarium* are already committed to the formation of spores only.

Acknowledgments

I would like to thank the following individuals for both rooting out errors and making numerous helpful and stimulating suggestions: D. M. Bozzone, L. W. Buss, K. Inouye, K. B. Raper, H. B. Suthers, and D. Waddell. I am also indebted to and thank K. M. Schenck for help in assembling the references.

References

Arndt, A. (1937). Untersuchungen über *Dictyostelium mucoroides* Brefeld. *Roux' Arch. Entwickl.* **136**, 681–747.

Ashworth, J. M. and Watts, D. J. (1970). Metabolism of the cellular slime mould *Dictyostelium discoideum* grown in axenic culture. *Biochem. J.* **119**, 175–182.

Blaskovics, J. A., and Raper, K. B. (1957). Encystment stages of *Dictyostelium*. *Biol. Bull.* **113**, 58–88.

Bonner, J. T. (1944). A descriptive study of the development of the slime mold *Dictyostelium discoideum*. *Amer. J. Bot.* **31**, 175–182.

Bonner, J. T. (1952). The pattern of differentiation in amoeboid slime molds. *Amer. Natur.* **86**, 79–89.

Bonner, J. T. (1957). A theory of the control of differentiation in the cellular slime molds. *Quart. Rev. Biol.* **32**, 232–246.

Bonner, J. T. (1959). Evidence for the sorting out of cells in the development of the cellular slime molds. *Proc. Natl. Acad. Sci. U.S.A.* **45**, 379–384.

Bonner, J. T. (1970). Induction of stalk cell differentiation by cyclic AMP in the cellular slime mold *Dictyostelium discoideum*. *Proc. Natl. Acad. Sci. U.S.A.* **65**, 110–113.

Bonner, J. T., ed. (1982a). "Evolution and Development." Dahlem Conference, Springer Verlag, New York.

Bonner, J. T. (1982b). Evolutionary strategies and developmental constraints in the cellular slime molds. *Amer. Nat.* **119**, 530–552.

Bonner, J. T., and Adams, M. S. (1958). Cell mixtures of different species and strains of cellular slime moulds. *J. Embryol. Exp. Morphol.* **6**, 346–356.

Bonner, J. T., and Dodd, M. R. (1962a). Aggregation territories in the cellular slime molds. *Biol. Bull.* **122,** 13–24.
Bonner, J. T., and Dodd, M. R. (1962b). Evidence for gas-induced orientation in the cellular slime molds. *Dev. Biol.* **5,** 344–361.
Bonner, J. T., and Frascella, E. B. (1952). Mitotic activity in relation to differentiation in the slime mold *Dictyostelium discoideum. J. Exp. Zool.* **121,** 561–571.
Bonner, J. T., and Shaw, M. J. (1957). The role of humidity in the differentation of the cellular slime molds. *J. Cell. Comp. Physiol.* **50,** 145–154.
Bonner, J. T., Chiquione, A. D., and Kolderie, M. Q. (1955). A histochemical study of differentiation in the cellular slime molds. *J. Exp. Zool.* **130,** 133–158.
Bonner, J. T., Clarke, W. W., Jr., Neely, C. L., Jr., and Slifkin, M. K. (1950). The orientation to light and the extremely sensitive orientation to temperature gradients in the slime mold *Dictyostelium discoideum. J. Cell. Comp. Physiol.* **36,** 149–158.
Bonner, J. T., Davidowski, T. A., Hsu, W.-L., Lapeyrolerie, D. A., and Suthers, H. L. B. (1982). The role of surface water and light on differentiation. *Differentiation* **21,** 123–126.
Bozzone, D. (1982) A comparison of macrocyst and fruiting body development in *Dictyostelium discoideum.* In preparation.
Brefeld, O. (1869). *Dictyostelium mucoroides.* Ein neuer organismus aus der ver-wandschaft der myxomyceten. *Abhandl. Senckenberg, Naturforsch. Ges. Frankfort* **7,** 85–107.
Buss, L. W. (1982). Somatic cell parasites and the evolution of somatic tissue compatibility. In preparation.
Byrne, G., Trujillo, J., and Cox, T. (1982). Tip inhibition and pattern formation in *Polysphondylium pallidum.* In preparation.
Cavender, J. C. (1973). Geographical distribution of Acrasieae. *Mycologia* **65,** 1044–1054.
Cavender, J. C. (1978). Cellular slime molds in tundra and forest soils of Alaska including a new species *Dictyostelium septentrionalis. Can. J. Bot.* **56,** 1326–1332.
Cavender, J. C., and Raper, K. B. (1965b). The Acrasieae in nature. II. Forest soil as a primary habitat. *Amer. J. Bot.* **52,** 297–302.
Cavender, J. C., and Raper, K. B. (1968). The occurrence and distribution of Acrasieae in forests of subtropical and tropical America. *Amer. J. Bot.* **55,** 504–513.
Cotter, D. A. (1975). Spores of the cellular slime mold *Dictyostelium discoideum. In* "Spores VI," (P. Gerhardt, R. N. Costilow, and H. L. Sadoff, eds.), pp. 61–72. American Society of Microbiology.
deBeer, G. R. (1940). "Embryos and Ancestors." Oxford Univ. Press (Clarendon), London and New York.
Durston, A. J. (1976). Tip formation is regulated by an inhibitory gradient in the *Dictyostelium discoideum* slug. *Nature (London)* **263,** 126–129.
Durston, A. J., and Vork, F. (1978). Spatial pattern of DNA synthesis in *Dictyostelium discoideum* slugs. *Exp. Cell Research* **115,** 454–457.
Eisenberg, R. M. (1976). Two-dimensional microdistribution of cellular slime molds in forest soil. *Ecology* **57,** 380–384.
Fankhauser, G. (1955). The role of the nucleus and cytoplasm. *In* "Analysis of Development" (B. H. Willier, P. Weiss, and V. Hamburger, eds.), pp. 126–150. Saunders, Philadelphia.
Feit, I. N. (1969). Evidence for the regulation of aggregate density by the production of ammonia in the cellular slime molds. Ph.D. Thesis, Princeton University, Princeton, New Jersey.
Feit, I. N., Fournier, G. A., Needleman, R. D., and Underwood, M. Z. (1978). Induction of stalk and spore cell differentiation by cyclic AMP in slugs of *Dictyostelium discoideum. Science* **200,** 439–441.

Filosa, M. F. (1962). Heterocytosis in cellular slime molds. *Amer. Nat.* **96,** 79–91.

Filosa, M. F. (1979). Macrocyst formation in the cellular slime mold *Dictyostelium mucoroides:* involvement of light and volatile morphogenetic substance(s). *J. Exp. Zool.* **207,** 491–495.

Filosa, M. F., Kent, S. G., and Gillette, M. M. (1975). The developmental capacity of various stages of a macrocyst forming strain of the cellular slime mold *Dictyostelium mucoroides. Dev. Biol.* **46,** 49–55.

Francis, D. W. (1964). Some studies on phototaxis of *Dictyostelium. J. Cell Comp. Physiol.* **64,** 131–138.

Francis, D. W. (1965). Acrasin and the development of *Polysphondylium pallidum. Dev. Biol.* **12,** 329–346.

Francis, D. W. (1979). True divergent differentiation in a cellular slime mold, *Polysphondylium pallidum. Differentiation* **15,** 187–192.

Francis, D. W., and O'Day, D. H. (1971). Sorting out in pseudoplasmodia of *Dictyostelium discoideum. J. Exp. Zool.* **176,** 265–272.

Gould, S. J. (1977). "Ontogeny and Phylogeny," Harvard University Press, Cambridge.

Gregg, J., and Karp, G. (1978). Patterns of cell differentiation revealed by L[^{3}H]fucose incorporation in *Dictyostelium. Exp. Cell. Res.* **112,** 31–46.

Gross, J. D., Town, C. D., Brookman, J. J., Jermyn, K. A., Peacey, M. J., and Kay, R. R. (1981). Cell patterning in *Dictyostelium. Philos. Trans. R. Soc. London* **B295,** 497–508.

Häder, D.-P., and Poff, K. L. (1979b). Photodispersal from light traps by amoebas of *Dictyostelium discoideum. Exp. Mycol.* **1,** 103–113.

Häder, D.-P., and Poff, K. L. (1979c). Light-induced accumulation of *Dictyostelium discoideum* amoebae. *Photochem. Photobiol.* **29,** 1157–1162.

Hagiwara, H. (1976). Distribution of the *Dictyosteliaceae* (cellular slime molds) in Mt. Ishizuchi, Shikoku. *Trans. Mycol. Soc. Japan* **17,** 226–237.

Hamilton, A. M. (1981). Pattern of differentiation in *Dictyostelium rosarium.* Senior Thesis, Princeton University, Princeton, New Jersey.

Harper, R. A. (1926). Morphogenesis in *Dictyostelium. Bull. Torrey Bot. Club* **53,** 229–268.

Harper, R. A. (1929). Morphogenesis in *Polysphondylium. Bull. Torrey Bot. Club* **56,** 227–258.

Harper, R. A. (1932). Organization and light relations in *Polysphondylium. Bull. Torrey Bot. Club* **59,** 49–84.

Hashimoto, Y., Cohen, M. H., and Robertson, A. (1975). Cell density dependence of the aggregation characteristics of the cellular slime mold *Dictyostelium discoideum. J. Cell. Sci.* **19,** 215–229.

Hohl, H. R. (1976). Myxomycetes. *In* "The Fungal Spore: Form and Function" (D. J. Weber and W. M. Hess, eds.), pp. 464–500. Wiley, New York.

Hohl, H. R., Honegger, R., Traub, F., Markwalder, M. (1977). Influence of cAMP on cell differentiation and morphogenesis in *Polysphondylium. In* "Development and Differentiation in the Cellular Slime Moulds" (P. Cappuccinelli and J. M. Ashworth, eds.), pp. 149–172. Elsevier/North Holland, Amsterdam.

Horn, E. G. (1971). Food competition among the cellular slime molds. *Ecology* **52,** 475–484.

Inouye, K., and Takeuchi, I. (1979). Analytical studies on migrating movement of the pseudoplasmodium of *Dictyostelium discoideum. Protoplasma* **99,** 289–304.

Jones, W. R., and Francis, D. (1972). The action spectum of light induced aggregation in *Polysphondylium pallidum* and a proposed general mechanism for light response in the cellular slime molds. *Biol. Bull.* **142,** 461–469.

Kahn, A. J. (1964). The influence of light on cell aggregation in *Polysphondylium pallidum. Biol. Bull.* **127,** 85–96.

Katz, E. R. and Bourguignon, L. Y. W. (1974). The cell cycle and its relationship to aggregation in the cellular slime mold *Dictyostelium discoideum. Dev. Biol.* **36,** 82–87.

Kawabe, K. (1980). Occurrence and distribution of *Dictyostelid* cellular slime molds in the Southern Alps of Japan. *Jap. J. Ecol.* **30,** 183–188.
Konijn, T. H. (1975). Chemotaxis in the cellular slime molds. *In* "Primitive Sensory and Communication Systems" (M. J. Carlile, ed.), pp. 102–153. Academic Press, New York.
Kopachik, W. (1982). Size regulation in *Dictyostelium*. *J. Embryol. Exp. Morph.* **68,** 23–35.
Lonski, J. (1976). The effect of ammonia on fruiting body size and microcyst formation in the cellular slime molds. *Dev. Biol.* **51,** 158–165.
Loomis, W. F., Jr. (1971). Sensitivity of *Dictyostelium discoideum* to nucleic acid analogues. *Exp. Cell Res.* **64,** 484–486.
MacWilliams, H. K., and Bonner, J. T. (1979). The prestalk-prespore pattern in cellular slime molds. *Differentiation* **14,** 1–22.
Matsukuma, S., and Durston, A. J. (1979). Chemotactic cell sorting in *Dictyostelium discoideum*. *J. Embryol. Exp. Morph.* **50,** 243–251.
Mitchison, J. M. (1971). "The Cell Cycle." Cambridge University Press, Cambridge.
Newell, P. C., Ellingson, J. S., and Sussman, M. (1969). Synchrony of enzyme accumulation in a population of differentiating slime mold cells. *Biochim. Biophys. Acta* **177,** 610–614.
Nicol, A., and Garrod, D. R. (1978). Mutual cohesion and cell sorting-out among four species of cellular slime molds. *J. Cell Sci.* **32,** 377–387.
O'Day, D. H., and Lewis, K. E. (1981). Pheremonal interactions during mating in *Dictyostelium*. *In* "Sexual Interactions in Eukaryotic Microbes" (D. H. O'Day and P. A. Horgen, eds.), pp. 199–221. Academic Press, New York.
Olive, E. W. (1902). Monograph of the Acrasieae. *Proc. Bost. Soc. Natl. Hist.* **30,** 451–513.
Olive, L. S. (1975). "The Mycetozoans." Academic Press, New York.
Olive, L. S. (1978). Sorocarp development by a newly discovered ciliate. *Science* **202,** 530–532.
Olive, L. S. and Blanton, R. L. (1980). Aerial sorocarp development by the aggregative ciliate, *Sorogena stoianovitchae*. *J. Protozool.* **27,** 293–299.
Pherson, D. A., and Beatie, J. (1979). Fungal loads of invertebrates in beech leaf litter. *Rev. Ecol. Biol. Sol.* **16,** 325–335.
Poff, K. L., and Butler, W. L. (1974). Spectral characteristics of the photoreceptor pigment of phototaxis in *Dictyostelium discoideum*. *Photochem. Photobiol.* **20,** 241–244.
Poff, K. L., and Skokut, M. (1977). Thermotaxis by pseudoplasmodia of *Dictyostelium discoideum*. *Proc. Natl. Acad. Sci. U.S.A.* **74,** 2007–2010.
Potts, G. (1902). Zur physiologie des *Dictyostelium mucoroides*. *Flora* **91,** 281–347.
Raper, K. B. (1935). *Dictyostelium discoideum*, a new species of slime mold from decaying forest leaves. *J. Agric. Res.* **50,** 135–147.
Raper, K. B. (1940). Pseudoplasmodium formation and organization in *Dictyostelium discoideum*. *J. Elisha Mitchell Sci. Soc.* **56,** 241–282.
Raper, K. B. (1956). *Dictyostelium polycephalum n. sp.:* a new cellular slime mould with coremiform fructifications. *J. Gen. Microbiol.* **14,** 716–732.
Raper, K. B. (1973). Acrasiomycetes. *In* "The Fungi: An Advanced Treatise" (G. C. Ainsworth, F. K. Sparrow, and A. S. Sussman, eds.), Vol. IVB, pp. 9–36. Academic Press, New York.
Raper, K. B. (1982). "The *Dictyostelids*". In preparation.
Raper, K. B., and Cavender, J. C. (1968). *Dictyostelium rosarium:* a new cellular slime mold with beaded sorocarps. *J. Elisha Mitchell Sci. Soc.* **84,** 31–47.
Raper, K. B., and Fennell, D. I. (1952). Stalk formation in *Dictyostelium*. *Bull. Torrey Bot. Club* **79,** 25–51.
Raper, K. B., and Quinlan, M. S. (1958). *Acytostelium leptosomum:* A unique cellular slime mold with an acellular stalk. *J. Gen. Microbiol.* **18,** 16–32.
Raper, K. B., and Thom, C. (1941). Interspecific mixtures in the Dictyosteliaceae. *Amer. J. Bot.* **28,** 69–78.

Rorke, J., and Rosenthal, G. (1959). Influences on the spatial arrangements of *Dictyostelium discoideum.* Senior Thesis, Princeton University, Princeton, New Jersey.

Rossier, C., Gerisch, G. Malchow, D., and Eckstein, F. (1978). Action of a slowly hydrolysable cyclic AMP analogue on developing cells of *Dictyostelium discoideum. J. Cell Sci.* **35,** 321–338.

Russell, G. K., and Bonner, J. T. (1960). A note on spore germination in the cellular slime mold *Dictyostelium mucoroides. Bull. Torrey Bot. Club* **87,** 187–191.

Sampson, J. (1976). Cell patterning in migrating slugs of *Dictyostelium discoideum. J. Embryol. Exp. Morph.* **36,** 663–668.

Schaap, P., Van Der Molen, L., and Konijn, T. M. (1981). Development of the simple cellular slime mold *Dictyostelium minutum. Dev. Bio.* **85,** 171–179.

Schindler, J., and Sussman, M. (1977). Ammonia determines the choice of morphogenetic pathways in *Dicytostelium discoideum. J. Mol. Biol.* **116,** 161–169.

Shaffer, B. M. (1953). Aggregation in cellular slime moulds: *in vitro* isolation of acrasin. *Nature (London)* **171,** 975.

Shaffer, B. M. (1957a). Aspects of aggregation in cellular slime moulds. 1. Orientation and chemotaxis. *Amer. Nat.* **91,** 19–35.

Shaffer, B. M. (1957b). Properties of slime-mould amoebae of significance for aggregation. *Quart. J. Micro. Sci.* **98,** 377–392.

Shaffer, B. M. (1963). Inhibition by existing aggregations of founder differentiation in the cellular slime mould *Polysphondylium violaceum. Exp. Cell. Res.* **31,** 432–435.

Shimomura, O., Suthers, H. L. B., and Bonner, J. T. (1982). The chemical identity of the acrasin of the cellular slime mold, *Polysphondylium violaceum. Proc. Natl. Acad. Sci. U.S.A.* In press.

Slifkin, M. K., and Bonner, J. T. (1952). The effect of salts and organic solutes on the migration time of the slime mold *Dictyostelium discoideum. Biol. Bull.* **102,** 273–277.

Soll, D. R. (1979). Timers in developing systems. *Science* **203,** 841–849.

Spiegel, F. W., and Cox, E. C. (1980). A one-dimensional pattern in the cellular slime mould *Polysphondylium pallidum. Nature (London)* **286,** 806–807.

Steinberg, M. S. (1970). Does differential adhesion govern self-assembly processes in histogenesis? Equilibrium configurations and the emergence of a hierarchy among populations of embryonic cells. *J. Exp. Zool.* **173,** 395–434.

Sternfeld, J. (1979). Evidence for differential cellular adhesion as the mechanism of sorting-out of various cellular slime mold species. *J. Embryol. Exp. Morph.* **53,** 163–178.

Sternfeld, J., and Bonner, J. T. (1977). Cell differentiation in *Dictyostelium* under submerged conditions. *Proc. Natl. Acad. Sci. U.S.A.* **74,** 268–271.

Sternfeld, J., and David, C. N. (1981). Cell sorting during pattern formation in *Dictyostelium. Differentiation.* **20,** 10–21.

Sussman, M., and Sussman, R. R. (1961). Aggregative performance. *Exp. Cell Res. Suppl.* **8,** 91–106.

Sussman, R. R., and Sussman, M. (1967). Cultivation of *Dictyostelium discoideum* in axenic medium. *Biochem. Biophysic. Res. Commun.* **29,** 53–55.

Suthers, H. B. (1982). Migratory song birds as slime mold distribution vectors. In preparation.

Takeuchi, I. (1963). Immunochemical and immunohistochemical studies on the development of the cellular slime mold *Distyostelium mucoroides. Dev. Biol.* **8,** 1–26.

Takeuchi, I. (1969). Establishment of polar organization during slime mold development. In "Nucleic Acid Metabolism Cell Differentiation and Cancer Growth" (E. V. Cowdry and S. Seno, eds.), pp. 297–304. Pergamon Press, Oxford.

Takeuchi, I. (1982). Pattern formation in the development of *Dictyostelium discoideum. In* "The Proceedings of the IX Congress of the International Society of Developmental Biologists" (M. M. Burger, ed.). Alan R. Liss, New York. In press.

Tasaka, M., and Takeuchi, I. (1979). Sorting out behaviour of disaggregated cells in the absence of morphogenesis in *Dictyostelium discoideum*. *J. Embryol. Exp. Morph.* **49,** 89–102.
Tasaka, M., and Takeuchi, I. (1981). Role of cell sorting in pattern formation in *Dictyostelium discoideum*. *Differentiation* **18,** 191–196.
Taya, Y., Tanalca, Y., and Nishimura, S. (1978). Cell free biosynthesis of discadenine, a spore germination inhibitor of *Dictyostelium discoideum*. *FEBS Lett.* **89,** 326–328.
Town, C., Gross, J. D., and Kay, R. R. (1976). Cell differentiation without morphogenesis in *Dictyostelium discoideum*. *Nature (London)* **262,** 717–719.
Town, C., and Stanford, E. (1979). An oligosaccharide-containing factor that induces cell differentiation in *D. discoideum*. *Proc. Natl. Acad. Sci. U.S.A.* **76,** 308–312.
Traub, F., and Hohl, H. R. (1976). A new concept for the taxonomy of the family Dictyosteliaceae (cellular slime molds). *Amer. J. Bot.* **63,** 664–672.
Van Tieghem, P. (1880). Sur quelques myxomycètes à plasmode agrègè. *Bull. Soc. Bot. de France* **27,** 317–322.
Waddell, D. (1982a). The spatial pattern of aggregation centers in the cellular slime mold. *J. Embr. Exper. Morph.* In press.
Waddell, D. (1982b). A carnivorous slime mold from Arkansas. *Nature (London)*. In press.
Wallace, M. A. (1977). Cultural and genetic studies of the macrocysts of *Dictyostelium discoideum*. Ph.D. Thesis. University of Wisconsin, Madison, Wisconsin.
Weinkauff, A. M., and Filosa, M. F. (1965). Factors involved in the formation of macrocysts by the cellular slime mold, *Dictyostelium mucoroides*. *Can. J. Microbiol.* **11,** 385–387.
Whitaker, B. D., and Poff, K. L. (1980). Thermal adaptation of thermosensing and negative thermotaxis in *Dictyostelium*. *Exp. Cell Res.* **128,** 87–94.
Wittingham, W. F. and Raper, K. B. (1960). Non-viability of stalk cells in *Dictyostelium*. *Proc. Natl. Acad. Sci. U.S.A.* **46,** 642–649.
Wurster, B., Pan, P., Tyan, G. G., and Bonner, J. T. (1976). Preliminary characterization of the acrasin of the cellular slime mold *Polysphondylium violaceum*. *Proc. Natl. Acad. Sci. U.S.A.* **73,** 795–799.

• CHAPTER 2

Genetics

Peter C. Newell

I. Introduction

Mutants of *Dictyostelium* that grow or develop in an aberrant manner have often attracted the attention of investigators seeking to explain the normal working of this organism. In the last decade, the development of a workable system of genetic analysis has allowed an increasingly detailed study of these mutants, and in certain cases their aberations have revealed important insights into the developmental program. Mutants with specific enzyme defects have been found to produce characteristic phenotypes in aggregates of developing amoebae. In addition, certain specific phenotypes have been associated with specific biochemical defects (e.g., loss of phosphodiesterase enzyme activity, loss of discoidin proteins, or altered cyclic GMP metabolism). However, most mutations remain mysterious and pose a silent challenge to anyone inspired to study them.

The Development of *Dictyostelium discoideum*

ISBN 0-12-455620-5

Most genetic analysis has been involved with sorting mutations into complementation groups (presumably affected in the same gene) and locating the mutated loci on one of the seven linkage groups. On several of the linkage groups, the technique of mitotic mapping has allowed the order of markers on the chromosome to be found, and in certain cases the relative distances separating the markers have been estimated. Although the genetic system yields data that must seem crude in comparison to bacterial and viral systems, it is hoped that the study of the distribution of important loci throughout the genome will be the prelude for more detailed molecular work. That work will eventually lead to a better understanding of both the role played by important developmental genes and the system of regulation of these genes. This chapter will be limited to "whole cell" genetic techniques, although such genetics is not fundamentally distinct from the field of molecular genetics considered in Chapter 9. Indeed, the report by Ratner *et al.*, (1981) in which the *bsgA5* mutation was used to select cells transformed by DNA, seems to represent the first bridge established between these two techniques.

Most of the whole cell genetic studies have involved the parasexual cycle, wherein the occasional (and spontaneous) combination of normal haploid strains results in stable or metastable diploids that can be subsequently haploidized. This cycle of events allows the pattern of linkage between markers in the population to be found. Since the diploids that can be isolated are stable for a number of generations, the recessive or dominant characteristics of the mutations, can be examined. The spontaneous mitotic recombination (or "crossing-over") that infrequently occurs in these diploids allows low-resolution mapping. In contrast, the sexual system (the macrocyst cycle) has proved difficult to use analytically because of technical problems. Amoebae of opposite mating types (*matA* and *mata*) will readily form macrocysts in a few days (if kept in the dark in a low phosphate medium containing calcium salts), but their subsequent germination takes weeks or months, and only occurs at low frequency. These problems have yet to be solved for *D. discoideum;* however, with the homothallic *D. mucoroides*, analysis of a few markers has been reported and some evidence for meiosis during formation of the macrocyst has been found. If this system can be adapted to our needs, it should prove useful for more detailed meiotic mapping studies.

II. Chromosomes

Strains of *D. discoideum* possess 7 chromosomes. This has been established from light microscopy (Wilson, 1952, 1953; Wilson and Ross, 1957;

Sussman, 1961; Brody and Williams, 1974; Robson and Williams, 1977; Zada-Hames, 1977a) and from the appearance of 7 pairs of kinetochores in electron micrographs (Moens, 1976). These studies also strongly suggest that all 7 chromosomes have terminal or nearly terminal centromeres. Members of other cellular slime molds genera, however, may have more chromosomes than *D. discoideum*. Studies by Williams (1980) revealed that *Polysphondylium pallidum* had 11 or 12, those isolates with apparently 11 chromosomes possibly having 2 fused.

The DNA content of the cellular slime molds is relatively small compared to animal cells (approximately 1/100 of that of mammals), and the chromosomes are correspondingly small. The chromosome size makes light microscopy difficult, especially when using bacterially grown amoebae which commonly appear to have small metaphase figures compared to amoebae grown axenically. The currently favored staining technique is that of Brody and Williams (1974), using Giemsa staining of air-dried methanol–acetic acid-fixed amoebae, which is technically simpler than the methods using acetoorcein employed previously. Because the period of mitosis is short (approximately 15 min for axenically grown cells and 3–10 min for bacterially grown cells) (Zada-Hames and Ashworth, 1978), the normal mitotic index is around 2%. The use of cell division synchronization techniques can increase this figure, but the number of stained mitotic figures observed in metaphase is still small. Without mitotic arresting agents it can often be difficult to obtain enough well-spread mitotic figures for reliable determination of chromosomal numbers. This is particularly important with bacterially grown cells wherein contaminating bacteria can look sufficiently like chromosomes to be confusing. Mitotic arresting agents found to be effective with *D. discoideum* include colchicine (although only at saturation amounts in axenic medium (Zada-Hames, 1977a) Nocodazole (at 7 μg/ml in axenic medium; (Cappuccinelli *et al.*, 1979), Isopropyl carbamate derivatives (such as IPC and CIPC) at 25 μg/ml (Cappuccinelli and Ashworth, 1976; White *et al.*, 1981, Welker and Williams, unpublished), and the benzimidazole derivatives thiabendazole and cambendazole at 10 and 50 μg/ml, respectively (Welker and Williams, 1980c). The benzimidazoles and CIPC have the advantage that they are active with bacterially grown amoebae as well as the more easily studied axenically grown cells. Amoebae scraped from the growing edge of a clone and incubated for 2.5 h with thiabendazole produce mitotic indices of approximately 30%. Prolonged incubation in the presence of the benzimidazole, however, can produce mitotic figures showing aneuploid numbers of chromosomes (8–13) and diploids (presumably isogenic) with 14 chromosomes (Welker and Williams, 1980c). Karyotypes of *D. discoideum* chromosomes that show each of the chromosomes having a characteristic banding pattern have been reported (Robson and Williams, 1977; Zada-Hames, 1977a) (Fig. 1). Although

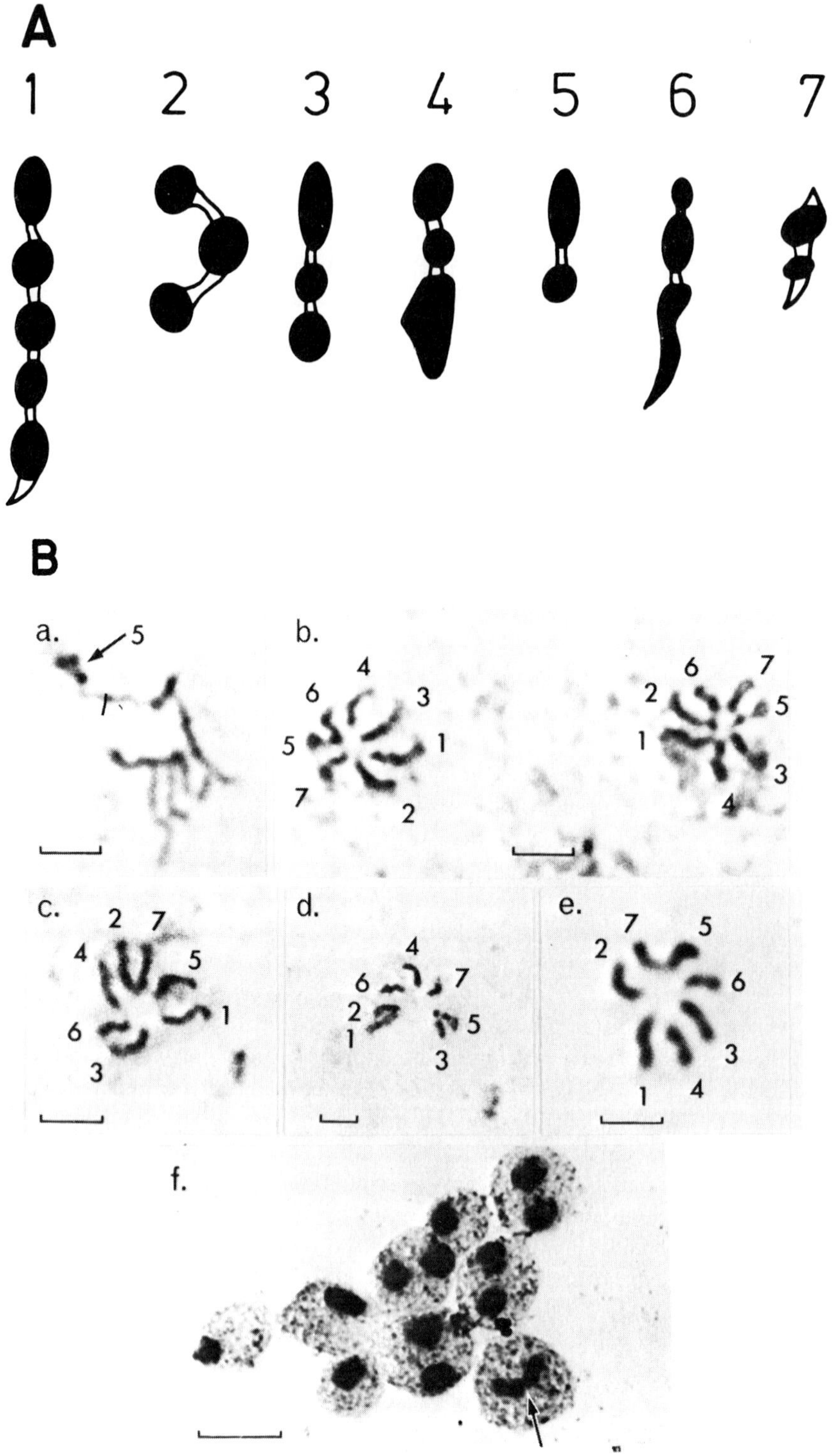
A
1
2
3
4
5
6
7
B
a.
5
b.
6
4
3
5
1
7
2
2
6
7
5
1
3
4
c.
2
7
4
5
1
6
3
d.
4
6
7
2
5
1
3
e.
2
7
5
6
3
1
4
f.

these bands may sometimes be seen directly in some stained preparations, they can be enhanced by trypsin treatment for 60 sec prior to staining and by pretreatment with ribonuclease, which reduces cytoplasmic staining.

Although there have been no convincing reports of stable aneuploid strains having complete extra chromosomes, strains possessing an extra chromosomal fragment have been reported (Williams *et al.*, 1980c). Three independently isolated, but genetically similar, fragment chromosomes were observed in segregants from diploids made between strains of opposite mating types (NC4 and V12). Such diploids are difficult to form because of vegetative incompatibility (see below); once formed the diploids may show imbalance in chromosomal constitution during haploidization. The fragment chromosomes were found (by genetic techniques) to include a region of linkage group II from the V12 parent which encompassed the centromere to the *whiA* locus. Cytological evidence suggested that the fragment chromosomes contained about a half of this linkage group. No deleterious effect on the growth and development of strains bearing the fragment was noticed, nor was their ability to form macrocysts impaired. The mechanism of generation was not conclusively established, but a model involving breakage in the *whiA–acrA* interval associated with mitotic recombination was consistent with the data.

Another chromosomal abnormality reported by Welker and Williams (1981) involves fusion of two chromosomes to form a giant chromosome composed of linkage groups II and VII. This abnormality was observed in an NC4–derived strain bearing a mutation at the *radB* locus (which may be concerned with recombinational DNA repair). Phenotypes of recombinant diploids, obtained after mitotic recombination, established that linkage group II was proximal to linkage group VII in the fused chromsome. There seemed to be no loss of genetic information in this strain; it grew and developed normally, and could only be distinguished by its genetic behavior in recombination experiments and by its possession of only six chromosomes. Although such strains pose problems for geneticists unaware of their presence, they are of considerable interest in considerations of the evolutionary process in eukaryotic organisms.

Fig. 1. (A) Schematic drawing of the karyotype of haploid *D. discoideum* based on observations of Robson and Williams (1977) and Zada-Hames (1977a). In each case the centromeres are probably terminal and located at the top (redrawn from Zada-Hames, 1977b). (B) Giemsa stained, trypsin treated chromosomes of *D. discoideum* showing: (a) The start of chromosome condensation during mitosis with chromosome 5 (arrowed) characteristically condensing earlier than the others; (b–e) Metaphase chromosomes shown numbered. Connections between some of the chromosomes may still be seen; (f) Giemsa-stained, trypsin–treated whole cells (in certain cases binucleate) showing two synchronized telophases in one of the binucleate cells. Bar markers represent 2 μm for (a–e) and 10 μm for (f) (from Robson and Williams, 1977).

III. Genetic Analysis Using the Parasexual Cycle

A. Isolation of Mutants

Cellular slime molds exist predominantly in the haploid state and thus readily express any mutational aberration that they may gain. Various strategies have been used to isolate what has become a rich diversity of mutant phenotypes. The simplest technique is to dilute amoebae, plate them on to nutrient SM plates with a bacterial food source (e.g., *Klebsiella aerogenes*), and allow clones to form. Mutants blocked at various stages in development are immediately obvious by inspection, those that completely fail to aggregate or to form fruiting bodies being the most conspicuous. Nonspecific mutations affecting both growth and development can be avoided if the growth rates of the clones are compared with those of wild-type controls, and only those with a normal growth rate selected. Mutants that are temperature sensitive for growth can be found by transferring amoebae from clones growing at 22°C to bacterially inoculated plates at both 22 and 27°C (using sterile toothpicks) and comparing the growth rates (Loomis, 1969). Mutants unable to grow with *Bacillus subtilis* can be found by similarly transferring amoebae to plates with *K. aerogenes* or *B. subtilis* (Newell *et al.*, 1977a). If the phenotype of interest is only expressed (or can only be measured) in axenic medium, then the equivalent of clonal isolation can be adopted using high dilution of amoebae into multitest well plates. Clones from individual amoebae can grow in axenic medium in such wells, and specific mutant characteristics can be identified by, for example, enzymatic reaction linked to a chromogenic substrate. To overcome the tedium of inoculating and testing thousands of such wells, Brenner *et al.* (1975) introduced the use of a "replicator" consisting of a metal box with 96 tubes that can be held over a multitest tray so as to dispense a drop of medium or reagent into each well. In this manner, mutants were isolated that were deficient in *N*-acetylglucosaminidase, α-1-mannosidase, UDPG pyrophosphorylase, β-glucosidase and alkaline phosphatase (Dimond *et al.*, 1973; Free and Loomis, 1974; Dimond *et al.*, 1976; Dimond and Loomis, 1976; MacLeod and Loomis, 1979).

In some cases, the mutant phenotype is only seen under strict developmental conditions that are not attainable on bacterially inoculated growth plates. In these cases, amoebae from separate clones can be taken and spotted on to nonnutrient plates that allow development to take place both synchronously and under controlled environmental conditions. Mutants that were aberrant in slug phototaxis were isolated in this way (Loomis, 1970), and MacWilliams (personal communication) used this technique for finding slugs showing altered staining reactions to the vital stain neutral red. Highly specific procedures can be devised in this way that could, for example, use immunological reaction to isolate mutants with particular properties (Loomis, personal communication).

Not all spontaneous mutants are sufficiently common, however, for such screening procedures to be feasible. Fortunately, treatment with mutagens may increase the yield of mutants by up to three orders of magnitude. Nitrosoguanidine (NTG) (Yanagisawa *et al.* 1967) and ultraviolet (UV) irradiation (Sussman and Sussman, 1953) have been the most widely used for this purpose, with NTG being generally preferred because of the comparatively little killing required to produce mutants compared to UV irradiation. However, the use of such mutagens is not without its dangers. NTG is known to produce double or multiple mutations and UV for causing chromosomal abnormalities such as translocations, both effects producing problems when genetic analysis is subsequently attempted.

The least laborious strategy for mutant isolation involves mutant selection rather than mutant screening. Mutants resistant to a whole range of drugs have been found simply by plating amoebae on to bacterially inoculated lawns with media supplemented by a particular drug. Spontaneous mutants that are resistant to methanol or cobaltous chloride are redily found at frequencies of approximately 10^{-6} (Williams *et al.*, 1974b; Williams and Newell, 1976). Developmental mutants can sometimes be selected by procedures that use their aberrant abilities (or lack of them). For example, mutants defective in the chemotactic system were found by Henderson (personal communication) by selecting for amoebae unable to pass through nitrocellulose filters in response to cyclic AMP signals. Mutants that develop abnormally rapidly (mutated at the *rdeA* locus) were selected by Kessin (1977) using the ability of such mutants to form spores prematurely, thereby avoiding death when the population was heated after 14 h of development. The type of selection used seems limited only by the investigator's ingenuity; additional strategies should yield other rare and useful mutant phenotypes.

Not all mutant phenotypes that first appear are stable; it is essential to culture potential mutants through two or three clonal passages. The reason for this apparent instability is not well understood. Some mutations are probably mitochondrial and are progressively diluted out by nonmutated mitochondria during growth, mutations producing transient resistance to ethidium bromide or methanol may of this type. Other changes in mutant phenotype with continued growth may result from a change in the organism's pattern of metabolism to compensate for the effects of the introduced mutation. The types of mutant phenotype that have been found to be stable and that have been genetically analyzed (to varying degrees of precision) are summarized in Table I.

TABLE I
Mutant Phenotypes Isolated and Studied in *D. discoideum* with Locus Symbols and Complementation Groups

Phenotype	Locus	Complementation groups	References
1) Growth-related mutations			
Resistance to cycloheximide	*cyc*	*A*	Katz and Sussman (1972)
Resistance to acriflavin	*acr*	*A*[a]*,B,C,D*	Williams *et al.* (1974); Rothman and Alexander (1975); Welker and Williams (1981)
Resistance to cadmium sulfate	*cad*	*A*	Williams and Newell (unpublished)
Resistance to ethidium bromide	*ebr*	*A,B*	Wright *et al.* (1977); Welker and Williams (unpubl.)
Resistance to cobaltous chloride	*cob*	*A*	Williams and Newell (1976); Ratner and Newell (1978)
Resistance to benlate	*ben*	*A*	Williams and Barrand (1978)
Resistance to sodium arsenate	*ars*	*A,B*	Williams (1981)
Resistance to nystatin	*nys*	*A,B,C*	Scandella *et al.* (1980)
Temperature sensitive for growth	*tsg*	*A–T*	Katz and Sussman (1972); Kessin *et al.* (1974); Rothman and Alexander (1975); Free *et al.* (1976); Williams and Newell (1976); Ratner and Newell (1978); Coukell and Cameron (1979); Ross and Newell (1979); Welker and Williams (1980b); Morrissey and Loomis (1981); Welker and Williams (unpublished)
Unable to grow with *Bacillus subtilis*	*bsg*	*A,B,C*	Newell *et al.* (1977a); Morrissey *et al.* (1980); Welker and Williams (1980a)
Axenic growth	*axe*	*A,B,C,*	Williams *et al.* (1974a,b); North and Williams (1978)

TABLE I—*Continued*

Phenotype	Locus	Complementation groups	References
Slow growth on SM agar media	*min*	*A*	Loomis and Ashworth (1968); Williams *et al.* (1980). Note that different loci are termed *minA*
Sensitivity to coumarin	*cou*	*A–F*	Welker and Williams (1980b); Welker and Williams, (1982)
Sensitive to UV and γ-irradiation	*rad*	*A–G*	Welker and Deering (1976; 1978); Coukell and Cameron (1979)
2) Developmental mutations			
Aggregation deficient	*agg*	*A–L*	Williams and Newell (1976); Coukell (1977); Coukell and Roxby (1977); Glazer and Newell (1981)
Large aggregation streams	*stm*	*A–F*	Ross and Newell (1979;1981)
Rapid development	*rde*	*A*	Kessin (1977)
Altered colony morphology	*cly*	*A*	Williams *et al.* (1980)
Sensitivity to ω-amino acids	*oaa*	*A*	North and Williams (1978)
Partial or complete block at slug stage	*slg*	*A–J*	Newell and Ross (1982)
Altered spore shape or viability	*spr*	*A–J*	Katz and Sussman (1972); Mosses *et al.* (1975); Williams and Welker (1980)
White spore mass	*whi*	*A,B,C*	Katz and Sussman (1972); Morrissey *et al.* (1980)
Formation of brown pigment	*bwn*	*A*	Katz and Sussman (1972)
Altered fruiting body morphology	*frt*	*A,B*	Welker and Williams (1980b)
Deficiency in stalk formation	*stl*	*A*	Morrissey and Loomis (1981)
Differentiation into stalk cells only	*stk*	*A*	Morrissey and Loomis (1981)
Altered discoidin (CBP) protein	*cbp*	*A*	Ray *et al.* (1979)
3) Enzymic mutations			
Altered alkaline phosphatase	*alp*	*A*	MacLeod and Loomis (1979)

(*Continued*)

TABLE I—*Continued*

Phenotype	Locus	Complementation groups	References
Deficiency in phosphodiesterase	*pds*	A	Barra *et al.* (1980)
Altered *N*-acetyl-glucosaminidase	*nag*	A	Dimond *et al.* (1973)
Altered α-mannosidase	*man*	A	Free and Loomis (1974)
Altered β-glucosidase	*glu*	A	Loomis (1980)
Glycosidase protein modification	*mod*	A	Free *et al.* (1978)
Altered UDPG-pyrophosphorylase	*upp*	[A]	Dimond *et al.* (1976). (locus not mapped)

[a]Cross-resistant to methanol.

B. Selection of Heterozygous Diploids

When cultures of two strains of the same mating type are mixed they occasionally fuse to form heterozygous diploids. This process involves both fusion of the cell membrane and of the nuclear membrane, and presumably proceeds in a sequential way with a heterokaryon as an intermediate stage. With growing cells, the rate of such fusion is extremely low (less than 5×10^{-7}/24 h) but rises during the developmental phase to about 10^{-5}/24 h (Loomis, 1969). Although original techniques employed development on filters, the effect is just as readily seen in liquid (nonnutrient) shaken suspensions; the use of 24 × 1 ml multiwell plastic trays with development in liquid suspension has made the system easy to use in large scale genetic experiments (Williams *et al.*, 1974a,b).

The most commonly employed technique for selecting the diploids formed in the starving suspensions employs complementation of either two *tsg* mutations (temperature sensitive for growth) (Loomis, 1969) or a *tsg* mutation with a *bsg* mutation (inability to grow on *Bacillus subtilis*) (Newell *et al.*, 1977a). These are recessive mutations that, if present in the original haploids being fused, will prevent growth under the nonpermissive conditions (27°C or the presence of *B. subtilis*, respectively) but allow growth when present in diploids in the heterozygous *tsg*/+ or *bsg*/+ state. Two nonallelic *tsg* mutations may be used in this way, but it is often more convenient to use a *tsg* mutation in one parental haploid with a *bsg* mutation in the other. This not only overcomes the problem of having two similar phenotypes to sort out in later linkage analysis, but also allows the easy introduction of temperature-sensitive developmental mutations into the *bsg* strain. The finding that there are at least three nonallelic *bsg* mutations (Morrissey *et al*. 1980; Welker and Williams, 1980) allows two *bsg* haploids to be fused together at 22°C with *B. subtilis*. This should prove useful, particularly where 27°C has other complications or where a *tsg* mutation has been eliminated by reversion for some particular genetic reason.

Other successful methods of selecting diploids include the use of recessive *rad* mutations conferring sensitivity to γ- and or UV irradiation (the diploid being radiation resistant compared to the sensitive haploids and therefore surviving a dose of UV radiation) (Welker and Deering, 1976), and the use of coumarin sensitivity (*cou*). This is a recessive type of mutation that makes the haploid unable to grow in the presence of coumarin, which can be used in combination with *tsg* or *bsg* mutations (Welker and Williams, 1980b).

Diploid selection methods involving *tsg*, *bsg*, *rad*, or *cou* need the markers already present in the haploid strains before any particular developmental mutants of interest can be selected. If the haploid strains are not so marked, it is difficult to select these markers in the developmental mutants, and mutagenesis used to aid this selection can alter the developmental phenotype. Good parental haploids bearing these selective markers are now freely available, so this problem need not normally arise. However, there are instances where developmental mutations in wild-type strains, found many years ago or found serendipitously in wild-type strains during biochemical experiments, need to be genetically analyzed, but they lack suitable diploid selectors. In some cases, a dominant *cob* mutation can be used to overcome this problem (Williams, 1978), conferring resistance for growth in the presence of cobaltous chloride. A strain that bears both this dominant cobalt marker and a *tsg* mutation can be used to select diploids when mixed with another strain lacking a diploid selector. After incubation under starvation condition, the haploid mixture is plated on plates containing cobaltous chloride at 27°C. Neither strain can grow under these conditions, but heterozygous diploids are selected, since the *tsg* is recessive and the *cob* is dominant. Although this technique has not proved routinely workable in all laboratories, it is clearly of considerable value in certain situations.

C. Complementation Analysis

One of the most useful aspects of the parasexual cycle is the ability to isolate stable heterozygous diploids whose phenotype can be investigated and compared to that of the parental haploids. This allows the determination of the pattern of complementation between sets of mutants possessing similar phenotypes and hence, by inference, the number of genetic loci that may be involved. This type of analysis has been successfully carried out on various classes of mutation, from drug resistances to developmental aberrations (Table I).

It is clearly impracticable to make diploids from every pairwise combination of mutants that has been found, and only mutants with some obvious similarity in their phenotype are generally tested for possible complementation. Several studies have shown that dissimilar phenotypes are generally mutated at different loci. However, similar phenotypes are not necessarily mutated at the same locus, but may often be mutated at several different loci on different linkage groups. Only after more extensive mapping will it become reasonable to look for common loci among groups showing different phenotypes.

If the particular phenotypic class being studied can be "saturated," such that mutations in most of the loci giving the particular phenotype can be isolated, then a clear indication of the number of loci involved can be obtained. For situations in which mutations at less than 10 loci give rise to clearly distinguishable phenotypes that can be identified in clones on agar plates, saturation of a phenotypic class is feasible. For example, 13 mutants giving resistance to the drug nystatin were found by Scandella *et al.*, (1980) to be in three complementation groups (although not all possible fusion combinations were in fact reported).

When saturation is not feasible, a statistical approach can be adopted. For example, when testing for complementation between aggregateless mutants, three groups (Warren *et al.*, 1975, 1976; Williams and Newell, 1976; Coukell, 1975, 1977) found that of approximately 400 diploids formed by fusing 20 mutants with 20 others in pairs, only 16 to 20 diploids showed noncomplementation, and these fell into four or five complementation groups. Statistical analysis of the results suggested that approximately 50 loci would have been expected to have given the observed results. Such statistical treatments are only very approximate, however, and they are strongly influenced by nonrandom mutation (or mutatgenic "hot spots") (Coukell, 1977). Nevertheless, statistical treatments can give a useful indication of the genetic complexity involved.

One of the main technical problems encountered in complementation studies of this type is that, because of the need for specific markers in the mutant haploids (such as *tsg* or *bsg*) to allow fusion, formation of diploids between all pairs of mutants is not straightforward. In the complementation of the aggregateless mutants, this limited the number of pairs that could be fused together. In the study of complementation of the "streamer" phenotype (giving very large aggregation streams) this problem was overcome by exchanging *bsg* for *tsg* mutations after the first round of fusions or (where the linkage of the streamer mutation made this impossible) by reverting the *bsg* mutation after introducing a *tsg* mutation. In this way, all of the 24 mutants found in this study could be fused to form diploids in any pairwise combination, and the pattern of complementation was thereby made clearer.

D. Haploidization and Linkage Analysis

The stability of diploids varies with the genetic background (Sussman and Sussman, 1962; Gingold, 1975). However, most diploids are sufficiently stable to be clonally passaged and can produce haploids spontaneously only at low frequency (about 1×10^{-3} in four days of culture with bacteria on plates). As a consequence, selective methods have been developed that select for the presence of rare haploids in a diploid population.

The most widely used selective technique is based on the presence of a recessive drug resistance marker which allows growth of haploids bearing the marker but inhibits growth of the heterozygous diploids. The first of these resistance markers was *cycA* (resistance to cycloheximide) which maps on linkage group I (Katz and Sussman, 1972). Other useful selectors of this type include *acrB* (resistance to acriflavin) on I (Williams *et al.*, 1974b), *acrA* (resistance to methanol or acriflavin) on II (Williams *et al.*, 1974b), *arcC* (resistance to acriflavin) on III (Rothman and Alexander (1975), *ebrA* (resistance to ethidium bromide) on IV (Wright *et al.*, 1977), and *cobA* (resistance to cobaltous chloride) on VII (Williams and Newell, 1976; Ratner and Newell, 1978).

Plating of heterozygous diploids onto drug-containing plates will not only select for resistant haploid segregants, but for diploids that have become homozygous for the resistance marker by mitotic recombination. These diploids are invaluable for mapping studies (see below) but are a nuisance for linkage analysis. The ratio of haploids to diploids varies (reproducibly) with the diploid and with the selector used, some diploids giving 10 : 90 (haploids : diploids) on cycloheximide plates, while giving 97 : 3 on methanol plates (which select for the *acrA* marker). The clones can be examined for their spore size (if they sporulate) and the diploids discarded, or selective drugs can be used in pairwise combination to give nearly 100% haploids (since a double mitotic recombination event is very rare) (Rothman and Alexander, 1975). Suitable pairs are cycloheximide with methanol or cycloheximide with cobaltous chloride. The combination of methanol with cobaltous chloride has not proven satisfactory and attempts using this combination generally give extremely low plating efficiency.

The axenic genes *axeA* and *axeC* (on II) and *axeB* (on III) may also be used for selecting haploids as, again, these markers are recessive (Williams *et al.*, 1974a, b; Free *et al.*, 1976, North and Williams, 1978). Such selection has not been widely used, probably because of the technical problems with the handling of large numbers of axenic cultures and the protracted nature of the selection.

General agents that either induce haploidization or prevent growth of diploids in preference to haploids have also been successfully used. The first of these was the compound *p*-fluorophenylalanine which when present at 0.1% in the growth medium was found to select for haploids. The effect was best seen if the cultures were maintained at 25.5°C for seven to ten days. During this time the number of haploids in the population would commonly increase by a factor of approximately 100 (Coukell and Roxby, 1977). More recently, the antifungal compound benlate has been used. Benlate is a very effective haploidizing agent; when added to the agar at a concentration of 20 μg/ml in growth plates it causes clones to be almost completely haploid by the time they reach a diameter of about 10 mm (Williams and Barrand, 1978). Plating efficiency is approximately 10% at this concentration of benlate for many diploids, although some show a very much lower efficiency (possibly due to the presence of benlate sensitivity mutations). Use of benlate has proved of great value for haploidizing diploids that do not possess heterozygous drug resistance markers or where drug selection is undesirable. If the plating efficiency is 10% or greater, the benlate method can be used as a general haploidizing method that avoids some of the problems of bias that may be incurred with particular drug selectors. Not all bias is avoided, as methanol-resistant haploids show considerable cross-resistance to benlate, and as a result benlate selection can produce a preponderance of segregants bearing the *acrA* locus if this marker is present.

The pattern of phenotypes that arises in the haploids produced on drug-containing plates can be assessed after transfer on a toothpick, or by pipetting a small drop onto other plates cultured under various selective conditions. Because the chromosomes of *D. discoideum* generally assort randomly with only infrequent genetic exchange between homologues (Mosses *et al.*, 1975), a genetic marker can be assigned to a linkage group from its pattern of segregation. Markers that have been assigned to linkage groups in this way are shown in Table II. Strains used to select developmental mutations of interest are not heavily marked, thus avoiding any side effects of such markers on the phenotype being isolated. Such lightly marked strains (e.g., XP55) can be fused to "tester strains" that bear many markers on a number of linkage groups. Ratner and Newell (1978) have constructed a number of tester strains such as XP144 or XP95 (Fig. 2), marked with easily scored markers on six of the seven linkage groups. More recently, other combinations of markers have been constructed that, for example, use the *nagA* marker in place of *bwnA* on IV (MacLeod and Loomis, 1979) and that can be used with various other lightly marked haploid parentals such as HL100 (Morrissey *et al.*, 1980), which is an NC4 strain bearing a spontaneously derived *bsgB* mutation.

TABLE II
Genetic loci of *D. discoideum* on Linkage Groups I to VII

Linkage group	Gene symbol	Mutant phenotype	Reference
I	*acrB*	Growth in the presence of acriflavin (100 μg/ml)	Williams *et al.* (1974)
	aggB	Aggregation deficient	Williams and Newell (1976)
	benA	Growth in the presence of benlate (600 μg/ml)	Williams and Welker (1980)
	cadA	Growth in the presence of cadmium sulfate (150 μg/ml)	Williams and Newell (unpubl)
	couB	Sensitivity to coumarin (1.3 m*M*)	Welker and Williams (1982)
	cycA	Growth in the presence of cyclohex-imide (500 μg/ml)	Katz and Sussman (1972)
	devA	Developmental mutation	Rothman and Alexander (1975)
	matA	Mating-type locus (also *mata*)	Williams (unpubl)
	modA	Alteration of glycosidase protein modification	Free *et al.* (1978)
	radA	Sensitive to UV and γ-irradiation	Welker and Deering (1976)
	sprA	Round spores	Katz and Sussman (1972)
	tsgE	Temperature sensitive for growth	Katz and Sussman (1972)
	tsgI	Temperature sensitive for growth	Free *et al.* (1976)
	tsgL	Temperature sensitive for growth	Welker and Deering (1978)
	tsgQ	Temperature sensitive for growth	Welker and Williams (1982)
II	*acrA*	Growth with acriflavin (100 μg/ml) or methanol (2%)	Williams *et al.* (1974)
	aggA	Aggregation deficient	Williams and Newell (1976)
	aggF	Aggregation deficient	Coukell (1977)
	aggI	Aggregation deficient	Coukell (1977)
	arsA	Growth in the presence of Na arse-nate (1.5 mg/ml)	Williams (1981)
	arsB	Growth in the presence of Na arse-nate (1.5 mg/ml)	Williams (1981)
	axeA	Axenic growth (provided that *axeB* is present)	Williams *et al.* (1974a,b)
	axeC	Axenic growth (provided *axeA* and *axeB* are present)	North and Williams (1978)
	cbpA	Altered discoidin (CBP) protein	Ray *et al.* (1979)
	clyA	Wide band of unaggregated amoebae in colony	Williams *et al.* (1980)
	couC	Sensitivity to coumarin (1.3 m*M*)	Welker and Williams (1982)
	couE	Sensitivity to coumarin (1.3 m*M*)	Welker and Williams (1982)
	couF	Sensitivity to coumarin (1.3 m*M*)	Welker and Williams (1982)
	devB	Developmental mutation	Rothman and Alexander (1975)
	oaaA	Sensitivity of development to ω-amino acids	North and Williams (1978)

(*Continued*)

TABLE II—*Continued*

Linkage group	Gene symbol	Mutant phenotype	Reference
	radH	Sensitive to UV and γ-irradiation	Welker and Deering (1978)
	slgE	Unable to develop beyond slug stage	Newell and Ross (1982)
	stkA	Differentiation only into stalk cells	Morrissey and Loomis (1981)
	stmB	Formation of large aggregation streams	Ross and Newell (1979)
	stmD	Formation of large aggregation streams	Ross and Newell (1979)
	sprB	Long spores	Mosses *et al.* (1975)
	tsgD	Temperature sensitive for growth	Katz and Sussman (1972)
	tsgF	Temperature sensitive for growth	Rothman and Alexander (1975)
	tsgH	Temperature sensitive for growth	Free *et al.* (1976)
	tsgP	Temperature sensitive for growth	Morrissey and Loomis (1981)
	whiA	White spore mass	Katz and Sussman (1972)
III	*acrC*	Growth in the presence of acriflavin (100 μg/ml)	Rothman and Alexander (1975)
	alpA	Altered alkaline phosphatase activity	MacLeod and Loomis (1979)
	axeB	Axenic growth (provided *axeA* is present)	Williams *et al.* (1974a,b)
	bsgA	Unable to grow on *Bacillus subtilis*	Newell *et al.* (1977a)
	couD	Sensitivity to coumarin (1.3 m*M*)	Welker and Williams (1982)
	radB	Sensitivity to UV and γ-irradiation	Welker and Deering (1976)
	radC	Sensitive to UV irradiation	Welker and Deering (1976)
	radE	Sensitive to UV and γ-irradiation	Coukell and Cameron (1979)
	radG	Sensitive to UV and γ-irradiation	Welker and Deering (1978)
	slgB	Unable to develop beyond the slug stage	Newell and Ross (1982)
	slgD	Unable to develop beyond the slug stage	Newell and Ross (1982)
	slgG	Unable to develop beyond the slug stage	Newell and Ross (1982)
	stmC	Formation of large aggregation streams	Ross and Newell (1979)
	stmE	Formation of large aggregation streams	Ross and Newell (1979)
	tsgA	Temperature sensitive for growth	Kessin *et al.* (1974)
	tsgC	Temperature sensitive for growth	Kessin *et al.* (1974)
	tsgJ	Temperature sensitive for growth	Coukell (unpubl)
	tsgN	Temperature sensitive for growth	Williams and Welker (unpubl)
	tsgR	Temperature sensitive for growth	Welker and Williams (1982)
	whiB	White spore mass	Morrissey *et al.* (1980)
IV	*acrD*	Growth in the presence of acriflavin (100 μg/ml)	Welker and Williams (1981)
	aggJ	Aggregation deficient	Coukell (1977)

TABLE II—*Continued*

Linkage group	Gene symbol	Mutant phenotype	Reference
	aggL	Aggregation deficient	Glazer and Newell (1981)
	bwnA	Formation of brown pigment	Katz and Sussman (1972)
	bsgC	Unable to grow on *Bacillus subtilis*	Welker and Williams (1980a)
	ebrA	Growth in the presence of ethidium bromide (35 μg/ml)	Wright *et al.* (1977)
	ebrB	Growth in the presence of ethidium bromide (35 μg/ml)	Welker and Williams (unpubl)
	frtA	Spore mass tends to slide to base of fruiting body	Welker and Williams (1980b)
	minA	Colonies grow very slowly on SM agar	Williams *et al.* (1980)
	nagA	Altered *N*-acetylglucosaminidase activity	Dimond *et al.* (1973)
	pdsA	Deficient in phosphodiesterase	Barra *et al.* (1980)
	radD	Sensitive to UV and γ-irradiation	Welker and Deering (1978)
	radF	Sensitive to UV and γ-irradiation	Welker and Deering (1978)
	rdeA	Rapid development	Kessin (1977)
	slgC	Long delay at slug stage	Newell and Ross (1982)
	slgI	Unable to develop beyond slug stage	Newell and Ross (1982)
	sprH	Round spores	Williams *et al.* (1980)
	sprJ	Spores fail to mature	Williams and Welker (1980)
	tsgB	Temperature sensitive for growth	Kessin *et al.* (1974)
	whiC	White spore mass	Morrissey *et al.* (1980)
V			
VI	*gluA*	Altered β-glucosidase activity	Dimond and Loomis (1976)
	manA	α-mannosidase-1 deficient	Free and Loomis (1974)
	stlA	Deficient in stalk formation	Morrissey and Loomis (1981)
VII	*bsgB*	Unable to grow on *Bacillus subtilis*	Morrissey *et al.* (1980)
	cobA	Growth with cobaltous chloride (300–360 μg/ml)	Ratner and Newell (1978); Welker and Williams (1980b)
	couA	Sensitivity to coumarin (1 m*M*)	Welker and Williams (1980b)
	frtB	Formation of aggregates in rings in clones on agar	Welker and Williams (1980b)
	slgJ	Unable to develop beyond slug stage	Newell and Ross (1982)
	stmA	Formation of large aggregation streams	Ross and Newell (1979)
	tsgG	Temperature sensitive for growth	Rothman and Alexander (1975); Welker and Williams (1980b)
	tsgK	Temperature sensitive for growth	Ross and Newell (1979)
	tsgM	Temperature sensitive for growth	Welker and Williams (1980b)

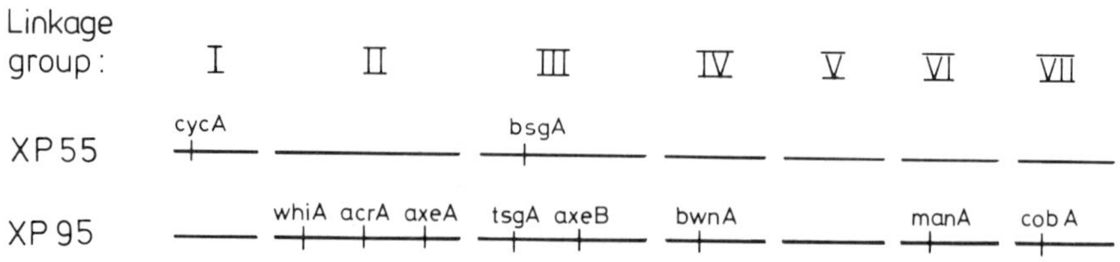

Fig. 2. Markers commonly used to define linkage groups in *D. discoideum* using parental strain XP55 and the appropriate tester strain XP95. Reproduced, with permission, from the Annual Review of Genetics, Volume 12. © 1978 by Annual Reviews, Inc.

Although the use of tester strains can avoid the need for multiple fusions and haploidizations during linkage assignments, difficulties may still be caused by secondary mutations or by the primary markers used in the tester strains themselves if these affect developmental phenotypes. Developmental mutations are particularly subject to the genetic background in which they are placed (compared to mutations affecting growth properties or drug resistance), and until markers can be found for use in tester strains that have minimal effects on such mutations, the interpretation of linkage data may be more complex than is immediately apparent.

Another cause for concern with linkage analysis is the appearance of nonrandom segregation patterns that can defy simple interpretation. One of the major causes of nonrandomness is the possession, by one or both of the haploid strains, of markers that cause slightly lowered growth rate. This may bias the segregation by allowing the faster haploids to predominate. The stage at which the haploidization event occurs may also influence the segregation pattern. Very early haploidization, in a clone being grown prior to plating under selective conditions, can dominate the population with a single type of haploid phenotype, even in the absence of any differential growth rate. Such a "clonal growth" effect can be obviated by using a technique in which a number of the heterozygous diploid clones are completely cut out of the agar while they are still tiny (2 mm diameter) and each transferred as an inoculum on to individual drug selective plates. This procedure tends to reduce the incidence of bias in the pattern of phenotypes of the resultant haploids, and any drug plates that produce atypical numbers of haploid clones are easily observed and discarded (Wallace and Newell, 1982). The use of benlate may also help in this regard because, with the high plating efficiency compared to other drug plates, unless the inoculum contains very high numbers of haploids, the majority of the haploidization events must occur on the benlate plates and yield haploid clones containing descendents of these isolated events.

Some markers seem to be incompatible in the same haploid cell, and this can produce strongly biased results such that whole classes of segregant are missing. The basis of such incompatibility is often unknown, and may not be related to the markers known to be present on the affected linkage groups. This effect is seen with the two markers *acrA* and *cobA* in certain diploids. Haploids bearing both these resistances are only rarely segregated, and where they are found they are very slow growers. The effect of such incompatibility is to give three rather than four classes of segregant when examining markers on both of the affected linkage groups (Newell *et al.*, 1977b).

Translocations can also disturb the phenotypic pattern by disallowing the segregation of one linkage group without another linkage group from the same parental. As more mutagenic treatments are given to the organism, the risk of such translocation increases. In some diploids, linkage groups III and VI show this type of association, although whether it is due to a true translocation remains to be ascertained. In some diploids the effect is not absolute and 2–3% of these linkage groups segregate (Ross and Newell, 1979; Welker and Williams, 1980b). In other diploids, however, linkage groups III and VI did not segregate at all in several hundred segregants tested.

An occasionally incorrect assumption made with linkage analysis is that the segregation pattern is due to segregation of a single marker. Sometimes it is found that the pattern of phenotypes can best be explained if two markers are needed to give the characteristic mutant phenotype. The pattern given by such a double marker is usually that of no apparent segregation with any of the marked linkage groups but three classes of segregant on two of the linkage groups. This arises because in the absence of one of the markers the phenotype is wild-type (or nearly wild-type). A clear example of this effect was seen when mapping the slugger mutants (Newell and Ross, 1982), wherein linkage group VII and linkage group III were needed for the particular slugger phenotype. If the effect is such that one of the two markers alone gives most of the characteristics of the mutant and the other marker alone seems almost wild-type, but together the mutant phenotype is stronger, then the mutation can be considered to lie with the dominant marker and the other marker to be a secondary or "enhancer" gene. The most difficult situation arises, of course, when the effect is not clear cut and the possession of one of the two markers gives an unsatisfactory intermediate phenotype that is hard to categorize.

E. Mitotic Recombination and Mapping

In addition to selecting segregant haploids from heterozygous diploids, drug-containing plates can select diploids that have become homozygous for the drug selection marker being used. These arise by mitotic recombination between the centromere and the drug marker. Compared to the recombination seen with meiosis, the frequency of recombination occurring during mitosis is very low; however, with the aid of drug selection, enough recombinant diploids can be isolated to carry out low-resolution mapping of the markers on the chromosomes. Since the frequency of mitotic recombination depends upon the distance between the selective marker and the centromere, the relative distances between markers can be estimated from the relative incidence of recombination to give homozygosity.

Drug markers that can be used for mapping include *cycA* and *acrB* on Linkage group I, *acrA* on II, *acrC* on III, *ebrA* on IV, and *cobA* on VII. Linkage groups V and VI currently lack suitable markers for selecting mitotic recombinants. Linkage groups that have been investigated by mitotic recombination include linkage group I (Williams *et al*., 1974; Williams and Newell, 1976) linkage group II (Williams *et al*., 1974b; Katz and Kao, 1974; Gingold and Ashworth, 1974; Mosses *et al*., 1975) and linkage group VII (Wallace and Newell, 1982a) (Fig. 3). In the studies of Mosses *et al*., along with those of Wallace and Newell, not only were the order of the markers found for linkage groups II and VII, but the approximate relative mitotic map intervals between some of the markers was also determined.

Because the drug marker will only select recombination that has occurred between it and the centromere, only markers that are proximal to the selective drug marker can be mapped. This was a serious limitation with linkage group I using the *cycA* marker because it was proximal to most of the markers investigated. Fortunately, *acrB* (which is also on I) proved to be distal to these markers and allowed their order to be determined. In the case of linkage group II the selective marker *acrA* was found to lie distal to the *whiA* marker and to two aggregateless loci, but proximal to the markers *tsgD* and *sprB*. Fortunately, the five markers studied in linkage group VII were proximal to the selector *cobA*.

Selection of new mutations at the selective drug resistance locus is clearly a factor that can limit the resolution of mapping by mitotic recombination. Mosses *et al*. (1975) found with linkage group II that although the frequencies of mitotic recombination between the centromere and *whiA* and between *whiA* and *acrA* were large enough to allow their relative positions to be determined, the apparent recombination frequency in the postulated interval between *acrA* and *tsgD* was low (if *tsgD* were proximal to *acrA*) and approximated the frequency of isolation of new mutations at the selector *acrA*. This made the position of *tsgD* ambiguous with respect to *acrA*; the position shown in Fig. 3 was derived from other experiments.

This problem may, in principle, be encountered whenever mutations that are distal to the selector are being mapped, as any new mutations selected at low frequency at the selector locus will mimic recombinants with cross-overs close to and proximal to the selector. Fortunately, in the case of linkage group VII, the markers studied were all proximal to the *cobA* selector, and the

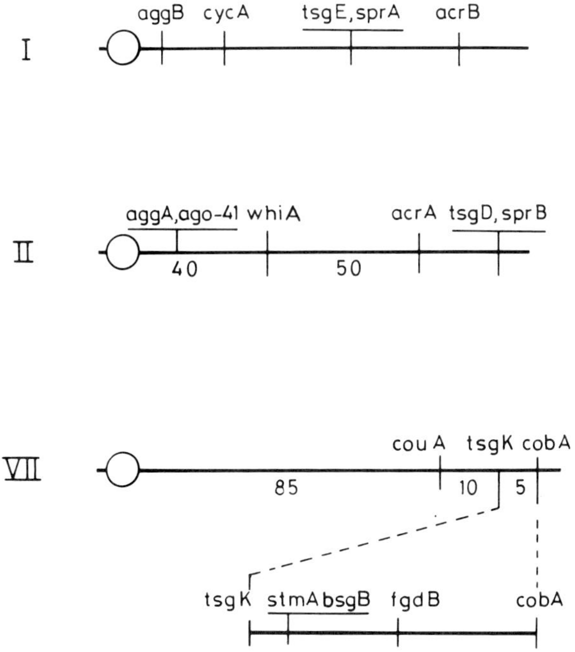

Fig. 3. Map of genetic markers of *D. discoideum* linkage groups I, II, and VII. The figure shows the relative order of markers and in the case of *whiA* and *acrA* (on II) and *couA*, *tsgK*, and *cobA* (on VII), their approximate relative position found from mitotic recombination studies. Data for linkage groups I and II are derived from Williams *et al.* (1974); Gingold and Ashworth (1974); Katz and Kao (1974), Mosses *et al.* (1975), Williams and Newell (1976) and for linkage group VII from Wallace and Newell (1982a).

frequency of isolation of new mutations at *cobA* was five- to tenfold lower than the frequency of recombination at the smallest of the intervals investigated.

Although the resolution of mitotic mapping is strictly limited compared to that of mapping using meiosis, it has been possible to increase the mitotic recombination frequency by 60–100-fold in certain circumstances, by a technique employing UV irradiation under conditions that do not greatly affect the haploidization or mutation frequencies (Wallace and Newell, 1982b). This may prove of value where increased resolution is required.

F. Usefulness of Parasexual Genetics for Developmental Studies

The parasexual cycle of *D. discoideum* described earlier provides a means of performing a number of intricate genetic maneuvers that can provide a very valuable adjunct to developmental studies. Some of the ways in which this system can be employed are given next.

1. *Examination of Loci*

Before embarking on a time-consuming series of biochemical experiments aimed at comparing a collection of newly acquired mutants, it is useful to ascertain that the mutants bear single-site mutations rather than double or multiple hits from the mutagen employed. Fusion to form diploids followed by haploidization can be used to show that the mutant phenotypes segregate in a simple manner and do not require several linkage groups for their characteristic lesion. Further study of an interesting lithium resistant mutant was abandoned when it was found that the resistance was due to several mutations scattered over at least three linkage groups (Newell, unpublished). One of the slugger mutants studied by Newell and Ross (1982) was also found to require two linkage groups for its slugger phenotype. Although such a finding does not necessarily rule out the possibility of biochemical analysis, it does warn of a greater complexity than originally envisaged. Sometimes segregation reveals a mutation that was present but was masked, an example of this effect is the presence of a developmental mutation along with an aggregation (*agg*) mutation, the developmental mutation being epistatic to the *agg* mutation (Wallace and Newell, 1982a).

2. *Determination of the Effects of Genetic Background*

The mutation of interest may be affected by the genetic "background" such that minor (often unknown) mutations present in a particular strain may modify the mutant phenotype. One way of reducing this effect when comparing mutations is to use an identical (isogenic) parental background for isolating the mutants or, if the mutations are already in different parental strains, to back-cross them to some suitably marked strain from which a particular chromosomal set can be selected.

3. *Assessment of Genetic Complexity*

From a biochemical viewpoint it may often be useful (or at least reassuring) to have some idea of the number of genes that may be mutated to give rise to a particular phenotype. Complementation analysis can give a rough idea of the "genetic complexity" involved. This may vary from unity, as for the stalky phenotype StkA (stalk formation without spores, Morrissey and Loomis, 1982) to between 50 and 150 for the Agg phenotype (Warren,

Warren and Cox, 1975, 1976; Williams and Newell, 1976; Coukell, 1975, 1977).

4. *Assignment of Mutants to Groups for Biochemical Analysis*

An important use of the ability to assign mutations of a particular phenotype to complementation groups is that it makes it much easier to explore the biochemical malfunctions associated with the genetic lesions. In the study of the Stm phenotype (Ross and Newell, 1979, 1981), for example, the 24 mutants isolated were finally placed in six complementation groups and (in most cases) only one representative of each group (in the same genetic background) was examined in detail biochemically.

5. *Assessment of Gene Clustering and Arrangement of Loci*

Linkage analysis and mitotic mapping may at times appear to give very little information for a great deal of effort. However, a knowledge of the physical relationship between genes is of great potential importance. It is, for example, instructive to know if genes for a particular function are clustered or widely scattered on several linkage groups (all of those so far examined showing no sign of clustering). The transfer of loci between strains also requires a knowledge of locus positions, and for certain complex strain constructions the rough map position on a particular linkage group is essential information. Fine structure mapping, which will undoubtedly be necessary for unravelling genetic regulatory mechanisms, will not, however, be directly possible with the presently available resolution of mitotic recombination. A combination of this technique with that of gene cloning of selected regions followed by DNA sequencing seems a powerful method to adopt.

6. *Study of Gene Interactions with Combined Mutants and Suppressor Mutations*

The complex interplay between genes is one of the problems that must be solved before the regulation of development can be understood. Although study of isolated cloned DNA fragments may be ideal for examining the interaction of genes that are contiguous or close together, the interaction of distantly positioned genes may require primary studies at the whole cell level. With parasexual genetics double mutants may be simply constructed if the genes are on different linkage groups (or with more difficulty if they are linked) and their interaction examined. Double mutant haploids bearing radiation-sensitive mutations were constructed by Welker and Deering (1979) in order to probe the radiation pathways involved. Morrissey and Loomis (1981) reported the construction of a double mutant bearing mutations at both the stalkless locus *stlA* (producing very little stalk) and the stalky locus *stkA* (producing only stalk). Further analysis of such mutants should clarify the hierarchy of regulative commands that operate for particular parts of the developmental program.

Another technique that can be used for studying gene interaction is the isolation of suppressors (second site revertants) of developmental mutants. Such suppressors must somehow interact (directly or indirectly) with the primary mutant locus, and if the suppressors are present on a different linkage group they may be isolated and studied independently. In other systems, such suppressors have been invaluable, although with *Dictyostelium* only suppressors of growth-related mutations such as *tsgD* (Ratner and Newell, 1978) or brown pigment formation *bwnA* (Williams and Newell, 1976) have been reported. Not all developmental loci will be simply suppressed in this way [indeed Morrissey and Loomis (1981) failed to find such mutations for the *stkA* locus], but where suppressors are obtainable they could lead to valuable insights into the normal genetic interaction that may be operating.

IV. The Macrocyst Cycle

A. Macrocyst Formation

Macrocysts are large, thick walled, resting structures formed after the fusion of starving amoebae that (in heterothallic strains) are of opposite mating type (Fig. 4). Some strains (e.g., *D. mucoroides*) are homothallic and do not require the presence of opposite mating types to form macrocysts. Because of the ease of inducing populations of *D. mucoroides* to initiate the macrocyst cycle, this strain has been studied in considerable detail. The currently accepted view of macrocyst formation, based mainly on the work of K. B. Raper (1951), Blaskovics and Raper (1957), Dengler *et al.*, (1970), Filosa and Dengler (1972), and Erdos *et al.* (1972) is summarized in Fig. 5.

Macrocysts are formed by many members of the *Dictyostelia* including *D. mucoroides* and *D. minutum* (Blaskovics and Raper, 1957), *P. violaceum* (Erdos *et al.*, 1972), *D. discoideum* (Nickerson and Raper, 1973; Wallace and Raper, 1979; Robson and Williams, 1980), *D. purpureum* (Nickerson and

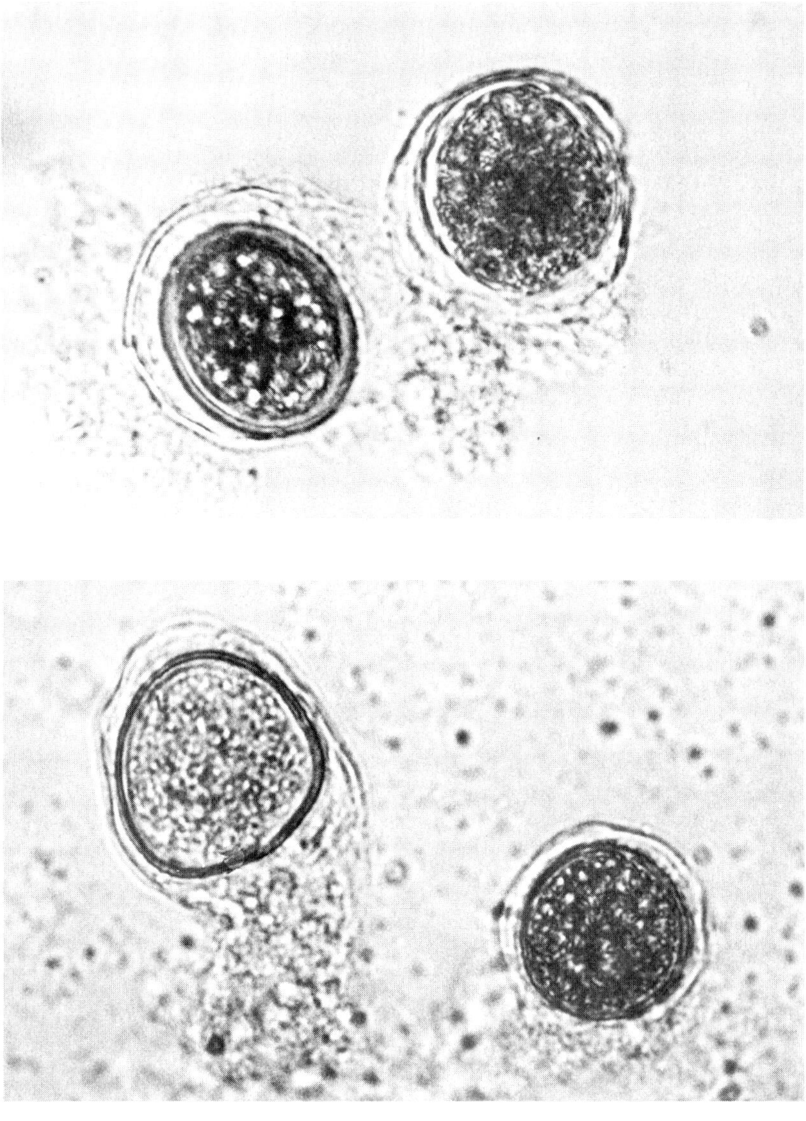

Fig. 4. Macrocysts formed from strain NP486 (mating type *mata*2) and strain XP251 (mating type *mat*A1). Each macrocyst shows a multilayered wall structure enclosing an engorged cytophagic cell containing partially digested endocytes.

Raper, 1973) and *P. pallidum* (Francis, 1975, 1980). The environmental conditions that favor formation of macrocysts differ from those favoring production of sorocarps, and include darkness, high humidity, low phosphate

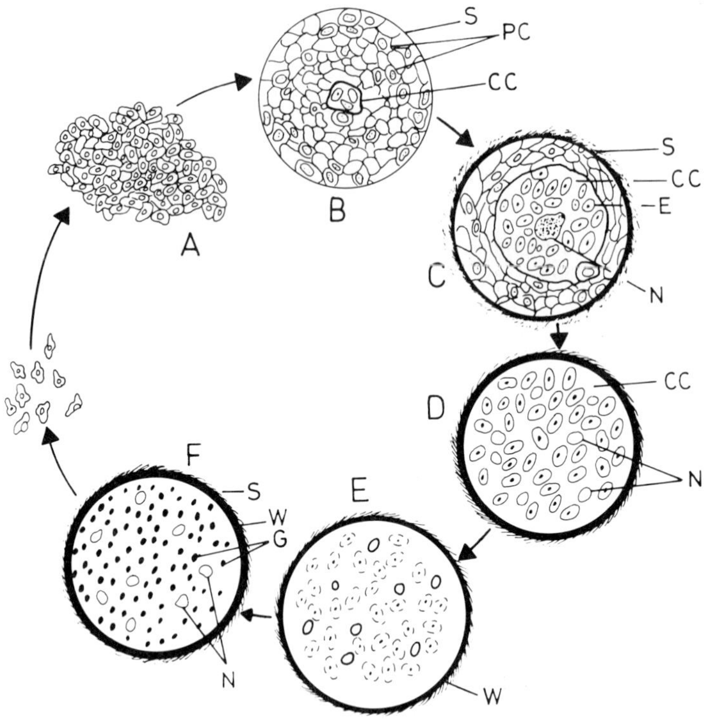

Fig. 5. Macrocyst formation in *D. mucoroides.* (A) Cells clump or aggregate. (B) A cytophagic cell [CC] appears in the center and begins to engulf the peripheral cells [PC]. A sheath is formed around the macrocyst. (C) The engulfed cells are transformed into endocytes [E], and the nucleus [N] of the cytophagic cell enlarges. (D) All the peripheral cells are engulfed, and the outer membrane of the cytophagic cell is now next to the surrounding sheath, which has increased in thickness. (E) A wall [W] is formed between the outer membrane of the cytophagic cell and the sheath. The endocytes are gradually digested and the cytophagic cell nucleus divides to form several nuclei. (F) The endocytes now appear as granules [G] as the macrocyst finally matures [based on a drawing of Filosa & Dengler (1972)].

concentration, temperatures above 20°C, high population density, non-porous sealed culture vessels and high calcium ion concentration (Blaskovics and Raper, 1957; Nickerson and Raper, 1973; Weinkauff and Filosa; 1965; Weinkauff and Raper, 1972; O'Day, 1979; Robson and Williams, 1979).

The initial sign of macrocyst formation by starving cells is the formation of loose aggregates that become surrounded by a covering resembling a slime sheath. Observations of O'Day (1979) suggest that the aggregates are formed by a chemotactic system that is similar to that used in aggregation for sorocarp formation. Pulsatile streaming may be seen and the aggregates can be shown to act as initiation centers if transferred to starving lawns of amoebae about to form sorocarps. Moreover, amoebae that are forming macrocysts respond chemotactically to an exogenous source of cyclic AMP. Under macrocyst-forming conditions the aggregates of amoebae do not form the slug stage characteristic of the sorocarp-forming pathway, but instead secrete a thick secondary cellulose-like wall around themselves inside the slime-like covering. These bodies, which may be from 25 to 50 μm in diameter, then lay down a tertiary membrane inside the cellulosic wall. As the macrocyst develops, the zygote in the center of the cell aggregate engulfs and slowly digests the amoebae around it. As it does so, this cell enlarges greatly in size, and endocytes, which are thought to be partly digested amoebae, can be seen inside it. This cell, called the "giant" or "cytophagic" cell, eventually occupies the entire space inside the walls of the macrocyst; meanwhile the endocytes are gradually digested. The peculiar nature of the cytophagic cell can be observed if it is separated from the aggregate at an early stage. Fukui (1976) found that the cytophagic cell had an unusually high buoyant density (possibly due to the endocytes inside it), and this enabled him to isolate this cell from enzymatically ruptured macrocysts using dextrin gradient separation. He found that the cytophagic cells were not only highly active in ingesting amoebae on agar surfaces, but in shaken liquid cultures they readily reconstituted macrocysts.

Evidence of O'Day *et al.* (1981) suggests that the formation of these cytophagic cells is regulated by an inhibitor (possibly produced by the cytophagic cells) that blocks further cytophagic cell formation. This inhibitor can be extracted from culture supernatants of *Dictyostelium discoideum* after cytophagic cell formation. It appears to have a molecular weight of under 10,000 and to be highly active; macrocyst formation is prevented even in 500-fold dilutions of macrocyst culture supernatant.

The early events involved in the formation of the zygote nucleus in the cytophagic cells of *D. discoideum* have been examined by Szabo *et al.* (1982), using the fluorescent nuclear stain Hoechst 33258. It was found that about 11 h after sexually compatible strains were mixed, cell fusion to form binucleate cells started to occur and continued for approximately 15 h. At this time, about 37% of the cells appeared to have fused. After approximately 20 h (in the presence of Ca^{2+} ions) the binucleate cells visibly enlarged to form the giant cytophagic cells whereafter near 22 h the two nuclei of these cells also swelled and then fused to form a large, irregularly shaped zygote nucleus. Another interesting observation was that a small fraction (up to 1.6%) of the population was seen to be multinucleate, although only two of these nuclei fused in any cell to form the irregular zygote nucleus. The presence of nuclei other than the zygote nucleus is a possible explanation of some of the complex segregation patterns from macrocysts (Francis, 1980; Wallace and Raper, 1979) discussed later.

At some point in the life of the cytophagic cell, the nucleus divides to form several smaller nuclei. As the macrocysts mature over a period of days, the acellular contents gradually shrink to form a compacted brownish mass that occupies the central position within the macrocyst wall. The macrocyst is then heat resistant (tolerating 42°C for 10 min) and capable of remaining viable and dormant until at some time it germinates.

B. Macrocyst Germination

Studies by Nickerson and Raper (1973) on five strains of cellular slime mold demonstrated that germination occurs in three distinct stages. In the first stage, the contracted and darkly pigmented contents of the macrocyst underwent swelling. In the second stage, the pigment was lost and cells were once more visible in what had previously looked like a homogeneous cytoplasm. In the third stage, the heavy cellulosic wall split and liberated active amoebae. Using electron microscopy, Erdos *et al.* (1973a) investigated germination of *D. mucoroides* at the fine structure level. They found that the inner part of the tertiary wall of the macrocyst separated into two zones during germination, and that it was the outer of these that split first, together with the secondary wall, leaving the inner part of the tertiary wall intact. The newly formed amoebae were inside this inner wall and were liberated when it ruptured.

The conditions that bring about germination are not thoroughly understood, and reliably inducing germination at high frequency must be accomplished before the macrocyst cycle can be useful for genetic studies. What is currently known is that germination occurs more efficiently if the macrocysts are placed at a particular (strain-specific) temperature, under illumination, after a period of aging lasting weeks or months. This requirement for aging may involve changes in the thick cellulosic wall of the macrocyst. Francis (1980) found that contamination of the macrocysts of *P. pallidum* with the fungus *Trichoderma* and incubation for 5–6 weeks at 20–30°C produced reliable germination at a frequency of about 10% compared to less than 1% in the uncontaminated controls. If this technique can be extended to other slime mold species it should be of great benefit in genetic studies.

C. Mating Types and Incompatibility

The most studied of the macrocyst-forming slime molds, *D. mucoroides*, is homothallic but other strains investigated have mating types that limit the possibilities for macrocyst formation. In a survey of the naturally occurring genera in the United States and Mexico, Clark *et al.* (1973) and Erdos *et al.*

1973b) found evidence for at least two mating types in *D. discoideum*, *matA* typified by strain NC4 and *mata* typified by strain V12. Only one of the isolates studied was homothallic (strain AC4) but two others were bisexual and five of the isolates failed to form macrocysts with either mating type, suggesting the possibility that other mating types exist. Robson and Williams (1980) found evidence for two of the strains isolated by Erdos *et al*. being homothallic (AC4 and ZA3A) and confirmed the bisexual nature of two of the strains. Although the existence of a third mating type could not be ruled out, these authors concluded that their results were compatible with a single locus, two allele mating-type system. Studies by Erdos *et al*. (1975), Clark and Speight (1973), and Clark (1974) indicate that the situation may be more complex with other species and genera. For example, both of the species *P. pallidum* and *P. violaceum* possess two syngens (or mating-type groups) each of which contains two mating types. *D. giganteum* has been found to have one syngen containing four mating types; mating and macrocyst formation in this species is probably controlled by a single locus, multiple allele incompatibility system.

The relationship between sexual mating types and vegative incompatibility (the latter influencing the degree of parasexual fusion) has been investigated in detail by Robson and Williams (1979). Under favorable conditions, the frequency of fusion between members of the same mating type in the parasexual cycle was found to be on the order of 2×10^{-5} in 24 h. However, between members of opposite mating types this frequency was two or three orders of magnitude lower. The only previously reported diploid (DP72) formed between opposite mating types (Mosses *et al*., 1975) was found by Robson and Williams to be homozygous (*mata2*/*mata2*) for the mating type locus. It had presumably "escaped" from vegetative incompatibility by mitotic recombination at this locus. Two other newly formed diploids between *matA1* and *mata2* strains, DU260 and DU454, were also found to have become homozygous at the mating-type locus and were shown to be *matA1*/*matA1*. Because the diploid DU454 was still heterozygous for all other markers examined on six linkage groups, it appears likely that the mating type locus (or a locus linked to it) is the only site causing vegetative incompatibility and the inability of opposite mating types to fuse.

This situation differs from the rather more complex incompatibility systems found with other fungi (reviewed by Esser and Blaich, 1973). For example, in *Neurospora crassa* incompatibility is controlled by alleles of at least ten loci (Mylyk, 1975); such complexity may be related to the special problems of defense of indentity and integrity in hyphal systems. The *Dictyostelium* system may more closely resemble the system for the Myxomycete *Physarum polycephalum* wherein preliminary evidence suggests that cell fusion by haploid amoebae prior to plasmodium formation is governed by alleles at the single *matB* locus. (Youngman *et al*., 1979).

An important conclusion from the study of Robson and Williams is that it will be difficult in *Dictyostelium* to use both the parasexual cycle and the macrocyst cycle sequentially in genetic experiments, unless the mating type locus can be mutated in some way, so as to inactivate the vegetative incompatibility function without impairing the mating function, as has indeed been achieved with *Neurospora crassa* (Griffiths and Delange, 1978).

D. Evidence for Pheromones

Pheromones are compounds produced by strains undergoing sexual union which promote the formation of the zygote. O'Day and Lewis (1975) provided evidence that strain V12 of *D. discoideum* could be induced to form macrocysts in the presence of the culture *supernatant* of the opposite mating type strain NC4. Consequently, contact between cells of opposite mating type was not essential. The relationship seemed not to be reciprocal, as culture supernatants of V12 could not induce macrocyst formation in NC4 strains. Existence of such a pheromone was confirmed by Machac and Bonner (1975) who found that it was able to diffuse through a dialysis membrane separating NC4 and V12 cultures. The nature of the pheromone is unknown but Lewis and O'Day (1977) suggested that it was volatile at room temperature. Some of the data on the apparent properties of the pheromone may need reevaluation following the recent finding of T. Yamada (personal communication) that V12 exhibits homothallism in aging cultures.

In other species such as *D. purpureum*, a similar nonreciprocal interaction has been found, wherein a pheromone is released by strains of Dp6 and Dp7 that is able to induce macrocyst formation in Dp2, but Dp2 supernatants have no effect on Dp6 or Dp7. Such interactions resemble those by pheromones in the water mold *Achlya* (Raper, 1951), although with *Achlya* a more complex sequential induction was observed, for which there is little evidence in the cellular slime molds.

E. Evidence for Meiosis

Although the evidence for the involvement of meiosis in the macrocyst cycle is not definitive, several observations strongly suggest that the macrocyst is a true sexual system with zygote formation and transient diploidy. The most convincing evidence for meiosis comes from the work of MacInnes and Francis (1974), who studied the segregation pattern of three developmental mutations in the homothallic strain DM7 of *D. mucoroides*. Macrocysts were allowed to form between pairs of parentals bearing the mutations *tag*, *cactus*, or *plaque*, whereafter they germinated and the progeny of individual macrocysts was examined. Both of the parental phenotypes were observed in progeny of single macrocysts (parental ditypes) as well as progeny of re-segregant wild-type or doubly mutant types (nonparental ditypes). The simplest explanation of such findings is that a heterozygous diploid was transiently formed in the macrocyst. Evidence that meiosis can occur in such diploids comes from the finding that both parental and nonparental types sometimes occurred together at similar frequencies in the same macrocyst (tetratypes), but that mixed ditypes consisting of one parental and one re-segregant (wild-type or double mutant) were not seen. Such findings are readily explained by meiotic reduction divisions, but would not be expected by parasexual haploidization involving progressive aneuploidy.

Using the species *D. giganteum*, Erdos *et al.* also concluded that the progeny of macrocysts formed by opposite mating types could best be explained by a meiotic pattern of segregation. In a study of macrocyst formation by various strains of *D. discoideum* bearing cycloheximide and methanol resistance markers, Wallace and Raper (1979) found more complex results. By analyzing the progeny of many macrocysts, they concluded that their pattern of results could have come from single meioses occurring in each macrocyst followed by random selection and proliferation of some or all of the four meiotic nuclei. Using *Polysphondylium pallidum*, Francis (1980) concluded that the pattern of segregation of several unlinked drug resistance markers could best be explained if the macrocyst contained a single diploid zygote that underwent meiosis before germination, with three of the four haploid nuclei disintegrating. In both the studies of Wallace and Raper on *D. discoideum* and of Francis on *P. pallidum;* however, the results were not thought to be conclusive evidence for meiosis because it was conceivable that similar results could have arisen from some unknown parasexual mechanism involving high frequency formation of diploids followed by their haploidization within one or two generations. Although such a mechanism seems unlikely, it cannot be easily disproved and final judgement on the involvement of meiosis may have to await the demonstration of high-frequency recombination between markers previously shown to be on the same linkage group.

References

Barra, J., Barrand, P., Blondelet, M. H., and Brachet, P. (1980). *pdsA*, a gene involved in the production of active phosphodiesterase during starvation of *Dictyostelium discoideum* amoebae. *Mol. Gen. Genet.* **177**, 607–613.

Blaskovics, J. A., and Raper, K. B. (1957). Encystment stages of *Dictyostelium. Biol. Bull.* **113,** 58–88.

Brenner, M., Tisdale, D., and Loomis, W. F. (1975). Techniques for rapid biochemical screening of large numbers of cell clones. *Exp. Cell Res.* **90,** 249–252.

Brody, T., and Williams, K. L. (1974). Cytological analysis of the parasexual cycle in *Dictyostelium discoideum. J. Gen. Microbiol.* **82,** 371–383.

Cappuccinelli, P., and Ashworth, J. M. (1976). Effect of inhibitors of microtubule and microfilament function on the cellular slime mold *Dictyostelium discoideum. Exp. Cell Res.* **103,** 387–393.

Cappuccinelli, P., Fighetti, M., and Rubino, S. (1979). A mitotic inhibitor for chromosomal studies in slime moulds. *FEMS Lett.* **5,** 25–27.

Clark, M. A. (1974). Syngenic division of cellular slime mold *Polysphondylium violaceum. J. Protozool.* **21,** 755–757.

Clark, M. A., Francis, D., and Eisenberg, R. (1973). Mating types in the cellular slime molds. *Biochem. Biophys. Res. Commun.* **52,** 672–678.

Clark, M. A., and Speight, S. E. (1973). Macrocyst-forming ability of morphogenetic mutants of *Polysphondylium violaceum* (abstract) *J. Protozool.* **20,** 507.

Coukell, M. B. (1975). Parasexual genetic analysis of aggregation-deficient mutants of *Dictyostelium discoideum. Mol. Gen. Genet.* **142,** 119–135.

Coukell, M. B. (1977). Evidence against mutational "hot-spots" at aggregation loci in *Dictyostelium discoideum. Molec. Gen. Genet.* **151,** 269–273.

Coukell, M. B., and Cameron, A. M. (1979). *radE*, a new radiation-sensitive locus in *Dictyostelium discoideum.J. Gen. Microbiol.* **114,** 247–256.

Coukell, M. B., and Roxby, N. M. (1977). Linkage analysis of developmental mutations in aggregation-deficient mutants of *Dictyostelium discoideum. Mol. Gen. Genet.* **151,** 275–288.

Dengler, R. E., Filosa, M. F., and Shao, Y. Y. (1970). Ultrastructural aspects of macrocyst development in *Dictyostelium mucoroides* (abstract). *Amer. J. Bot.* **57,** 737.

Dimond, R. L., Brenner, M., and Loomis, W. F. (1973). Mutations affecting N-acetylglucosaminidase in *Dictyostelium discoideum. Proc. Natl. Acad. Sci. U.S.A.* **70,** 3356–3360.

Dimond, R. L., Farnsworth, R. A., and Loomis, W. F. (1976). Isolation and characterization of mutations affecting UDP-glucose pyrophosphorylase activity in *Dictyostelium discoideum. Dev. Biol.* **50,** 169–181.

Dimond, R. L., and Loomis, W. F. (1976). Structure and function of β-glucosidases in *Dictyostelium discoideum. J. Biol. Chem.* **251,** 2680–2687.

Erdos, G. W., Nickerson, A. W., and Raper, K. B. (1972). Fine structure of macrocysts in *Polysphondylium violaceum. Cytobiologie* **6,** 351–366.

Erdos, G. W., Nickerson, A. W., and Raper, K. B. (1973a). The fine structure of macrocyst germination in *Dictyostelium mucoroides. Dev. Biol.* **32,** 321–330.

Erdos, G. W., Raper, K. B., and Vogen, L. K. (1973b). Mating types and macrocyst formation in *Dictyostelium discoideum. Proc. Natl. Acad. Sci. U.S.A.* **70,** 1828–1830.

Erdos, G. W., Raper, K. B., and Vogen, L. K. (1975). Sexuality in the cellular slime mold *Dictyostelium giganteum. Proc. Natl. Acad. Sci. U.S.A.* **72,** 970–973.

Esser, K., and Blaich, R. (1973). Heterogenic incompatibility in plants and animals. *Advan. Genet.* **17,** 107–152.

Filosa, M. F., and Dengler, R. E. (1972). Ultrastructure of macrocyst formation in cellular slime mold *Dictyostelium mucoroides.* Extensive phagocytosis of amoebae by a specialized cell. *Dev. Biol.* **29,** 1–16.

Francis, D. (1975). Cyclic-AMP induced changes in protein synthesis in a cellular slime mold *Polysphondylium pallidum. Nature (London)* **258,** 763–765.

Francis, D. (1980). Techniques and marker genes for use in macrocyst genetics with *Polysphondylium pallidum. Genetics*, **96,** 125–136.

Free, S. J., and Loomis, W. F. (1974). Isolation of mutations in *Dictyostelium discoideum* affecting α-mannosidase. *Biochimie*, **56,** 1525–1528.

Free, S. J., Schimke, R. T., and Loomis, W. F. (1976). Structural gene for α-mannosidase-1 in *Dictyostelium discoideum. Genetics*, **84,** 159–174.

Free, S. J., Schimke, R. T., Freeze, H., and Loomis, W. F. (1978). Characterization and genetic mapping of *modA* mutation in post-translational modification of glycosidases of *Dictyostelium discoideum. J. Biol. Chem.* **253,** 4102–4106.

Fukui, Y. (1976). Enzymatic dissociation of nascent macrocysts and partition of the liberated cytophagic cells in *Dictyostelium mucoroides. Dev. Growth Differ.* **18,** 145–155.

Gingold, E. B. (1975). Stability of diploid clones of cellular slime mold *Dictyostelium discoideum. Heredity*, **33,** 419–423.

Gingold, E. B. and Ashworth, J. M. (1974). Evidence for mitotic crossing-over during the parasexual cycle of the cellular slime mould. *J. Gen. Microbiol.* **84,** 70–78.

Glazer, P. M., and Newell, P. C. (1981). Initiation of aggregation by *Dictyostelium discoideum* in mutant populations lacking pulsatile signalling. *J. Gen. Microbiol.* **125,** 221–232.

Griffiths, A. J. F., and Delange, A. M. (1978). Mutations of the α-mating type gene in *Neurospora crassa. Genetics*, **88,** 239–254.

Katz, E. R., and Kao, V. (1974). Evidence for mitotic recombination in cellular slime mold *Dictyostelium discoideum. Proc. Natl. Acad. Sci. U.S.A.* **71,** 4025–4026.

Katz, E. R., and Sussman, M. (1972). Parasexual recombination in *Dictyostelium discoideum:* Selection of stable diploid heterozygotes and stable haploid segregants. *Proc. Natl. Acad. Sci. U.S.A.* **69,** 495–498.

Kessin, R. H. (1977). Mutations causing rapid development of *Dictyostelium discoideum. Cell* **10,** 703–708.

Kessin, R. H., Williams, K. L., and Newell, P. C. (1974). Linkage analysis in *Dictyostelium discoideum* using temperature-sensitive growth mutants selected with bromodeoxyuridine. *J. Bacteriol.* **119,** 776–783.

Lewis, K. E., and O'Day, D. H. (1977). Sex hormone of *Dictyostelium discoideum* is volatile. *Nature (London)* **268,** 730–731.

Loomis, W. F. (1969). Temperature-sensitive mutants of *Dictyostelium discoideum. J. Bacteriol.* **99,** 65–69.

Loomis, W. F. (1970). Mutants in phototaxis of *Dictyostelium discoideum. Nature (London)* **227,** 745–746.

Loomis, W. F. (1980). A β-glucosidase gene of *Dictyostelium discoideum. Dev. Genet.* **1,** 241–246.

Loomis, W. F., and Ashworth, J. M. (1968). Plaque-sized mutants of the cellular slime mold *Dictyostelium discoideum. J. Gen. Microbiol.* **53,** 181–196.

Machac, M. A., and Bonner, J. T. (1975). Evidence for a sex hormone in *Dictyostelium discoideum. J. Bacteriol.* **124,** 1624–1625.

MacInnes, M. A., and Francis, D. (1974). Meiosis in *Dictyostelium mucoroides. Nature (London)* **251,** 321–324.

Macleod, C. L., and Loomis, W. F. (1979). Biochemical and genetic analysis of a mutant with altered alkaline phosphatase activity in *Dictyostelium discoideum. Dev. Genet.* **1,** 109–121.

Moens, P. B. (1976). Spindle and kinetochore morphology of *Dictyostelium discoideum. J. Cell Biol.* **68,** 113–122.

Morrissey, J. H., and Loomis, W. F. (1981). Parasexual analysis of cell proportioning mutants of *Dictyostelium discoideum. Genetics.* In press.

Morrissey, J. H., Wheeler, S., and Loomis, W. F. (1980). New loci in *Dictyostelium discoideum* determining pigment formation and growth on *Bacillus subtilis. Genetics* **96,** 115–123.

Mosses, D., Williams, K. L., and Newell, P. C. (1975). The use of mitotic crossing-over for genetic analysis in *Dictyostelium discoideum:* mapping of linkage group II. *J. Gen. Microbiol.* **90,** 247–259.

Mylyk, O. M. (1975). Heterokaryon incompatibility genes in *Neurospora crassa* detected using duplication-producing chromosome rearrangements. *Genetics* **80,** 107–124.

Newell, P. C. (1978). Genetics of the cellular slime molds. *Ann. Rev. Genet.* **12,** 69–93.

Newell, P. C., and Ross, F. M. (1982). Genetic analysis of the slug stage of *Dictyostelium discoideum J. Gen. Microbiol.* **128,** 1639–1652.

Newell, P. C., Henderson, R. F., Mosses, D. G., and Ratner, D. I. (1977a). Sensitivity to *Bacillus subtilis:* A novel system for selection of heterozygous diploids of *Dictyostelium discoideum. J. Gen. Microbiol.* **100,** 207–211.

Newell, P. C., Ratner, D. I., and Wright, D. (1977b). New techniques for cell fusion and linkage analysis of *Dictyostelium discoideum. In* "Development and Differentiation in the Cellular Slime Moulds" (P. Cappuccinelli and J. M. Ashworth, eds.), pp. 51–61. Elsevier, Amsterdam.

Nickerson, A. W., and Raper, K. B. (1973). Macrocysts in the life cycle of the *Dictyosteliaceae:* Formation of the macrocysts. *Amer. J. Bot.* **60,** 190–197.

North, M. J., and Williams, K. L. (1978). Relationship between the axenic phenotype and sensitivity to *w*-aminocarboxylic acids. *J. Gen. Microbiol.* **107,** 223–230.

O'Day, D. H., (1979). Aggregation during sexual development in *Dictyostelium discoideum. Can. J. Microbiol.* **25,** 1416–1426.

O'Day, D. H., and Lewis, K. E. (1975). Diffusible mating-type factors induce macrocyst development in *Dictyostelium discoideum. Nature (London)* **254,** 431–432.

O'Day, D. H., Szabo, S. P., and Chagla, A. H. (1981). An autoinhibitor of zygote giant cell formation in *Dictyostelium discoideum. Exp. Cell Res.* **131,** 456–458.

Raper, J. R. (1951). Sexual hormones in *Achlya. Amer. Sci.* **39,** 110–120.

Raper, K. B. (1951). Isolation, cultivation and conservation of simple slime molds. *Quart. Rev. Biol.* **26,** 169–190.

Ratner, D. I., and Newell, P. C. (1978). Linkage analysis in *Dictyostelium discoideum* using multiply marked tester strains: establishment of linkage group VII and reassessment of earlier linkage data. *J. Gen. Microbiol.* **109,** 225–236.

Ratner, D. I., Ward, T. E., and Jacobson, A. (1981). Evidence for the transformation of *Dictyostelium discoideum* with homologous DNA. *Dev. Biol.* In press.

Ray, J., Shinnick, T., and Lerner, R. (1979). Mutation altering the function of a carbohydrate binding protein blocks cell-cell cohesion in developing *Dictyostelium discoideum. Nature (London)* **279,** 215–221.

Robson, G. E., and Williams, K. L. (1977). The mitotic chromosome of the cellular slime mould *Dictyostelium discoideum:* A karyotype based on Giemsa banding. *J. Gen. Microbiol.* **99,** 191–200.

Robson, G. E., and Williams, K. L. (1979). Vegetative incompatibility and the mating-type locus in the cellular slime mould *Dictyostelium discoideum. Genetics* **93,** 861–875.

Robson, G. E., and Williams, K. L. (1980). The mating system of the cellular slime mold *Dictyostelium discoideum. Current Genetics* **1,** 229–232.

Ross, F. M., and Newell, P. C. (1979). Genetics of aggregation pattern mutations in the cellular slime mould *Dictyostelium discoideum. J. Gen. Microbiol.* **115,** 289–300.

Ross, F. M., and Newell, P. C. (1981). Streamers : chemotactic mutants of *Dictyostelium* with altered cyclic GMP metabolism. *J. Gen. Microbiol.* **127,** 339–350.

Rothman, F. G., and Alexander, E. T. (1975). Parasexual genetic analysis of the cellular slime mold *Dictyostelium discoideum* A3. *Genetics* **80,** 715–731.

Scandella, D., Rooney, R., Katz, E. R. (1980). Genetic, biochemical and developmental studies of nystatin resistant mutants in *Dictyostelium discoideum. Mol. Gen. Genet.* **180,** 67–75.

Sussman, M., and Sussman, R. R. (1962). Ploidal inheritance in *Dictyostelium discoideum*: Stable haploid, stable diploid and metastable strains. *J. Gen. Microbiol.* **28,** 417–429.

Sussman, R. R. (1961). A method for staining the chromosomes of *Dictyostelium disçoideum* myxamoebae in the vegetative stage. *Exp. Cell. Res.* **24,** 154–155.

Sussman, R. R., and Sussman, M. (1953). Cellular differentiation in *Dictyostelium:* heritable modifications of the developmental pattern. *Ann. N.Y. Acad. Sci.* **56,** 949–960.

Szabo, S. P., O'Day, D. H., and Chagla, A. H. (1982). Cell fusion, nuclear fusion, and zygote differentiation during sexual development of *Dictyostelium discoideum.* Dev. Biol. In press.

Wallace, J. S., and Newell, P. C. (1982a). Genetic analysis by mitotic recombination in *Dictyostelium* of growth and developmental markers on linkage groups VII. *J. Gen. Microbiol.* **128,** 953–964.

Wallace, J. S., and Newell, P. C. (1982b). Stimulation of mitotic recombination to *Dictyostelium discoideum* by ultraviolet irradiation. *FEMS Microbiol. Lett.* **14,** 37–44.

Wallace, M. A., and Raper, K. B. (1979). Genetic exchanges in the macrocysts of *Dictyostelium discoideum. J. Gen. Microbiol.* **113,** 327–337.

Warren, J. A., Warren, W. D., and Cox, E. C. (1975). Genetic complexity of aggregation in cellular slime mold *Polysphondylium violaceum. Proc. Natl. Acad. Sci. U.S.A.* **72,** 1041–1042.

Warren, J. A., Warren, W. D., and Cox, E. C. (1976). Genetic and morphological study of aggregation in cellular slime mold *Polysphondylium violaceum. Genetics* **83,** 25–47.

Weinkauff, A. M., and Filosa, M. F. (1965). Factors involved in the formation of macrocysts by the cellular slime mold *Dictyostelium mucoroides. Can. J. Microbiol.* **11,** 385–387.

Weinkauff, A. M., and Raper, K. B. (1972). Formation and germination of macrocysts in cellular slime molds (abstract). *Amer. J. Bot.* **59,** 668.

Welker, D. L., and Deering, R. A. (1976). Genetic analysis of radiation-sensitive mutations in the slime mould *Dictyostelium discoideum. J. Gen. Microbiol.* **97,** 1–10.

Welker, D. L., and Deering, R. A. (1978). Genetics of radiation sensitivity in the slime mould *Dictyostelium discoideum. J. Gen. Microbiol.* **109,** 11–23.

Welker, D. L., and Deering, R. A. (1979). Interaction between radiation-sensitive mutations in double-mutant haploids of *Dictyostelium discoideum. Mol. Gen. Genet.* **167,** 265–270.

Welker, D. L., and Williams, K. L. (1980a). *Bacillus subtilis* sensitivity loci in *Dictyostelium discoideum. FEMS Microbiol. Lett.* **9,** 179–183.

Welker, D. L., and Williams, K. L. (1980b). The assignment of four new loci, including the coumarin sensitivity locus *couA*, to linkage group VII of *Dictyostelium discoideum. J. Gen. Microbiol.* **120,** 149–159.

Welker, D. L., and Williams, K. L. (1980c). Mitotic arrest and chromosome doubling using thiabendazole, cambendazole, nocodazole and benlate in the slime mould *Dictyostelium discoideum. J. Gen. Microbiol.* **116,** 397–407.

Welker, D. L., and Williams, K. L. (1981). Genetic and cytological characterisation of fusion chromosomes of *Dictyostelium discoideum. Chromosoma* **82,** 321–332.

Welker, D. L., and Williams, K. L. (1982). Coumarin-sensitivity mutations in *Dictyostelium discoideum* affect the cytoskeleton. *J. Gen. Microbiol.* **128,** 1329–1343.

White, E., Scandella, D., and Katz, E. R. (1981). Inhibition by CIPC of mitosis and development in *Dictyostelium discoideum* and the isolation of CIPC-resistant mutants. *Dev. Genet.* **2,** 99–111.

Williams, K. L. (1978). Characterisation of dominant resistance to cobalt chloride in *Dictyostelium discoideum* and its use in parasexual genetic analysis. *Genetics,* **90,** 37–47.

Williams, K. L. (1980). Examination of the chromosomes of *Polysphondylium pallidum* following metaphase arrest by benzimidazole derivatives and colchicine. *J. Gen. Microbiol.* **116,** 409–415.

Williams, K. L. (1981). Two arsenate resistant loci in the cellular slime mould *Dictyostelium discoideum. FEMS Microbiol.* Lett. **11,** 317–320.

Williams, K. L., and Barrand, P. (1978). Parasexual genetics in the cellular slime mould *Dictyostelium discoideum:* Haploidization of diploid strains using benlate. *FEMS Microbiol.* Lett. **4,** 155–159.

Williams, K. L., and Newell, P. C. (1976). A genetic study of aggregation in the cellular slime mould *Dictyostelium discoideum* using complementation analysis. *Genetics* **82,** 287–307.

Williams, K. L., and Welker, D. L. (1980). Mutations specific to spore maturation in the asexual fruiting body of *Dictyostelium discoideum. Dev. Genet.* **1,** 355–362.

Williams, K. L., Kessin, R. H., and Newell, P. C. (1974a). Genetics of growth in axenic medium of the cellular slime mould *Dictyostelium discoideum.* Nature *(London)* **247,** 142–143.

Williams, K. L., Kessin, R. H., and Newell, P. C. (1974b). Parasexual genetics in *Dictyostelium discoideum:* Mitotic analysis of acriflavine resistance and growth in axenic medium. *J. Gen. Microbiol.* **84,** 59–69.

Williams, K. L., Robson, G. E., and Welker, D. L. (1980). Chromosome fragments in *Dictyostelium discoideum* obtained from parasexual crosses between strains of different genetic background. *Genetics* **95,** 289–304.

Wilson, C. M. (1952). Sexuality in the Acrasiales. *Proc. Natl. Acad. Sci. U.S.A.* **38,** 659–662.

Wilson, C. M. (1953). Cytological study of the life cycle of *Dictyostelium. Amer. J. Bot.* **40,** 714–718.

Wilson, C. M., and Ross, I. K. (1957). Further cytological studies in the Acrasiales. *Amer. J. Bot.* **44,** 345–350.

Wright, M. D., Williams, K. L., and Newell, P. C. (1977). Ethidium bromide resistance: A selective marker located on linkage group IV of *Dictyostelium discoideum. J. Gen. Microbiol.* **102,** 423–426.

Yanagisawa, K., Loomis, W. F., and Sussman, M. (1967). Developmental regulation of the enzyme UDP-galactose polysaccharide transferase. *Exp. Cell Res.* **46,** 328–334.

Youngman, P. J., Pallotta, D. J., Hosler, B., Struhl, G., and Holt, C. E. (1979). A new mating compatibility locus in *Physarum polycephalum. Genetics* **91,** 683–693.

Zada-Hames, I. M. (1977a). Analysis of karyotype and ploidy of *Dictyostelium discoideum* using colchicine-induced metaphase arrest. *J. Gen. Microbiol.* **99,** 201–208.

Zada-Hames, I. M. (1977b). Study of karyotype and the cell cycle in *Dictyostelium discoideum.* Ph.D. Thesis, University of Essex, Colchester, United Kingdom.

Zada-Hames, I. M., and Ashworth, J. M. (1978). Cell-cycle during vegetative stage of *Dictyostelium discoideum* and its response to temperature change. *J. Cell Sci.* **32,** 1–20.

• CHAPTER 3

Membranes

Ben A. Murray

I. Introduction

The interaction of a cell with its environment plays a major role both in its routine activities and in the changes that it may undergo during processes of development and differentiation. The plasma membrane marks the interface between the cell and its environment, hence, the properties of the membrane have great influence on this interaction. The properties of specific macromolecular materials that are external to the cell membrane (the "extracellular matrix"), some of which may be products of the cell itself and some

The Development of *Dictyostelium discoideum*

ISBN 0-12-455620-5

of which may be supplied from the outside, are also frequently important in shaping and modulating the cell's interaction with its environment. Considerable attention, therefore, has been focused on the cell membrane of *Dictyostelium discoideum* and of related cellular slime mold species, in the hope that such investigations will help us understand the problems of cell–cell communication and histiotypic development, for which these organisms provide such good models. In this chapter, I will discuss the function, structure, and composition of the *Dictyostelium* cell membrane and its changes during development. Some of these topics have also been discussed by Bakke and Lerner (1981).

II. Membrane Function

A. Phagocytosis and Pinocytosis

1. Vegetative Cells

The plasma and associated membrane systems of *Dictyostelium* must function to mediate the transport of material between the outside and the inside of the cell. *Dictyostelium*, like other amoebae, actively obtains nutrients by bulk engulfment of extracellular fluid (pinocytosis) and of particulate nutrients (phagocytosis); and it is generally assumed that *Dictyostelium* obtains its nutrients exclusively through these routes. Phagocytosis and pinocytosis (collectively termed endocytosis) in a variety of cell types have been the subjects of several reviews (Silverstein *et al*., 1977; Stossel, 1977; Chapman-Andresen, 1977; Edelson and Cohn, 1978). Wild isolates of *Dictyostelium* must phagocytose particulate nutrients (e.g., living or killed bacteria or yeast or particulate fractions prepared from such organisms) in order to grow (Raper, 1935; Hohl and Raper, 1963). Mutants have been described that are capable of growing in axenic or defined media that lack large particulate material (Sussman and Sussman, 1967; Watts and Ashworth, 1970; Loomis, 1971; Franke and Kessin, 1977); these mutants can presumably obtain all necessary nutrients by pinocytosis. Cells must be simultaneously mutant in at least two loci (*axeA* and *axeB*) in order to grow well axenically (i.e., aphagocytically) (Williams *et al*., 1974; North and Williams, 1978); the mechanism of action of these mutations is unknown.

In those few cases that have been examined, there is no evidence for active transport of small molecules such as amino acids (Lee, 1972a; North and Williams, 1978). Thus, *Dictyostelium* takes up amino acids to a much lesser extent than does a bacterium (e.g., *Escherichia coli* or *Klebsiella aerogenes*) which possesses amino acid transport systems. Hence, care is required when bacterially grown cells are used for metabolic labeling studies; unless the cells are washed extensively and appropriate antibiotics (commonly 500 μg/ml streptomycin) are added, remaining bacteria, numbering only a few percent of the number of *Dictyostelium* amoebae present, can account for the majority of the added label incorporated!

Most quantitative studies of phagocytosis and pinocytosis have employed axenically grown cells. Since axenically grown and bacterially grown cells are known to differ in a number of physiological properties (reviewed by Loomis, 1975; see also Burns *et al.*, 1981), results using bacterially grown cells may vary somewhat. In particular, comparisons of the rates of pinocytosis of axenic and nonaxenic strains have not been reported.

Pinocytosis in strain AX2 or AX3 cells growing axenically has been measured to be 1.4×10^{-14} liter/min/cell, using FITC–dextran as a fluid phase marker (Thilo and Vogel, 1980; Rossomando *et al.*, 1981), or about 3.3% of the cell volume per minute for a cell volume of 4.2 pl/cell (Thilo and Vogel, 1980). Rossomando *et al.* (1981) suggest that this is the sum of a nutrient-dependent and a nutrient-independent component, each accounting for about half of the total uptake rate. Under the same conditions, the cells can ingest (by phagocytosis) about four *E. coli* B/r cells, eight latex beads (1.08 μm diameter), or 0.2 sheep erythrocytes per min per cell (Vogel *et al.*, 1980; Glynn 1981). De Chastellier and Ryter (1977) reported that about 0.2 killed yeast cells (*Saccharomyces cerevisiae*) (which are considerably larger than bacteria) are taken up per cell per minute under these conditions. Axenically grown cells of the related species *Polysphondylium pallidum* take up about three polystyrene spheres (1.1 μm diameter) per min per cell (Githens and Karnovsky, 1973).

The cell membrane (or specialized areas of the membrane) must presumably recognize a suitable prey object for phagocytosis. Evidence has been presented showing that there are at least two distinct surface recognition systems for phagocytosis (Vogel *et al.*, 1980; Vogel, 1981). Vogel *et al.* have isolated several mutants wherein one of these systems has been altered. They have interpreted their results to indicate that one recognition system involves a glucose-binding component (presumably a carbohydrate-binding protein or lectin) whereas the other component is less specific. In the mutants, the nonspecific system has been altered such that it will recognize only relatively hydrophobic surfaces such as those of latex beads.

Ryter and Hellio have presented histochemical evidence which also supports the proposal of different cell surface receptors for different phagocytosed particles. Membranes of phagocytic cups phagocytosing yeast cells do not bind wheat germ agglutinin (WGA) whereas such binding is easily demonstrated on cups phagocytosing bacteria or latex beads (Ryter and Hellio, 1980). Furthermore, removal of cell surface WGA receptors by pretreatment with WGA impairs the ability of the cells to phagocytose yeast cells but not the ability to phagocytose bacteria (*Klebsiella aerogenes*) or latex beads (Hellio and Ryter, 1980).

Engulfment of the external medium during phagocytosis or pinocytosis implies that significant amounts of the cell's own membrane are also internalized (Silverstein *et al.*, 1977; Edelson and Cohn, 1978), and indeed this has been shown to be the case in *Dictyostelium* by Thilo and Vogel (1980). By surface labeling *Dictyostelium* cells with galactosyl transferase and tritiated UDP–galactose, and examining the location of the label by its sensitivity to release by β-galactosidase, they were able to show that an area approximately equivalent to that of the plasma membrane is internalized every 45 min. Similar rates of internalization have been measured in other systems (Silverstein *et al.*, 1977; Edelson and Cohn, 1978). Most of the internalized membrane is later returned to the cell surface. These findings are consistent with earlier ultrastructural results, indicating that plasma membrane is internalized within pinocytic and phagocytic vesicles and later returned to the cell surface by exocytosis of vesicles containing indigestible remnants of the prey (Hohl, 1965; Ryter and de Chastellier, 1977). Coated vesicles, which may be involved in membrane transport (Silverstein *et al.*, 1977), have recently been identified in *Dictyostelium* (Swanson *et al.*, 1981). The ultrastructural and biochemical data suggest that the total membrane area at any one time in the internal digestive membrane system amounts to 50–100% of the area of the plasma membrane (Ryter and de Chastellier, 1977; Thilo and Vogel, 1980; Glynn, 1981).

Considerable complexity in the control of the cellular digestive system is indicated by the observation that both the accumulation and the secretion of many lysosomal hydrolases, which share common antigenic determinants (Knecht and Dimond, 1981), vary considerably on different growth media and at different cell densities (Burns *et al.*, 1981). Accumulation and secretion appear to be regulated separately. In early development, where the most detailed studies have been made, different classes of lysosomal hydrolases are secreted with different kinetics, all of which differ from the kinetics of egestion of phagocytosed beads (Dimond *et al.*, 1981). These observations suggest that there may be different classes of secretory vesicles for different groups of secreted enzymes, and that secretion may not simply reflect the release of indigestible material to the outside. The function of enzyme secretion thus remains unknown.

2. *Changes during Development*

Both ultrastructural evidence (de Chastellier and Ryter, 1977; Ryter and Brachet, 1978) and direct measurements of phagocytosis (de Chastellier and

Ryter, 1977) suggest that the phagocytic capacity of the cell (measured by adding yeast cells to *Dictyostelium* amoebae after different periods of starvation) is significantly enhanced during the first few hours of starvation. According to de Chastellier and Ryter (1977), phagocytosis of yeast increases from about 0.1 to 0.4 yeast/min/cell by 1.5 h of starvation. At the same time, enhanced secretion of lysosomal enzymes is seen (Dimond *et al.*, 1981; see also Section I,A,1).

After a few more hours of starvation, the phagocytic capacity declines (de Chastellier and Ryter, 1977), secretion stops (Dimond *et al.*, 1981), and the digestive system becomes increasingly autophagic (de Chastellier and Ryter, 1977; George *et al.*, 1972; Yamamoto *et al.*, 1981). Areas of amoebal cytoplasm, sometimes including mitochondria, are pinched off into digestive vacuoles. This may represent at least part of the mechanism by which the cells maintain themselves during development and by which they reduce their mass by about half during development (White and Sussman, 1961; Lee, 1972a; Loomis, 1975).

Autophagic vacuoles persist throughout much of development and can be found in culminants (George *et al.*, 1972). They are much more prominent in anterior (prestalk) cells of slugs than in posterior (prespore) cells (Maeda and Takeuchi, 1969; Yamamoto *et al.*, 1981); this apparently accounts for the differences that these cell types display in staining with vital dyes such as neutral red or Nile blue (MacWilliams and Bonner, 1979; Sternfeld and David, 1981). The physiological significance of this difference is not understood.

B. Osmoregulatory Apparatus

In addition to regulating traffic in nutrients, *Dictyostelium*, like other freshwater microorganisms that lack a rigid cell wall, must continually expel water to counter the osmotic influx from the relatively hypotonic environment. Like other microorganisms (Patterson, 1980), *Dictyostelium* possesses a contractile vacuole for this purpose (Loomis, 1975; de Chastellier *et al.*, 1978; Quiviger *et al.*, 1978). Liquid is apparently collected in small vacuoles which then merge to form a main vacuole; the vacuolar membrane fuses with the cell membrane and expels its contents, whereafter the cycle repeats itself. The contractile vacuole pulses once every 3 to 5 min during axenic growth; its rate of pulsation increases to about once per minute during the first hour of starvation (Quiviger *et al.*, 1980). The contractile vacuole becomes less prominent as development proceeds and is rarely found in aggregates or later stages (Quiviger *et al.*, 1980; George *et al.*, 1972).

The membrane of the contractile vacuole is structurally (and probably also functionally) distinct from the remainder of the plasma membrane. The contractile vacuole membrane does not stain with silver proteinate or with phosphotungstate (which react with some carbohydrates), whereas the plasma membrane and digesting apparatus do react (Ryter and de Chastellier, 1977; Quiviger *et al.*, 1978, 1980). Furthermore, the alkaline phosphatase activity that is resistant to glutaraldehyde fixation (15% of the total cellular alkaline phosphatase activity; Quiviger *et al.*, 1980) is found in the membrane of the main contractile vacuole and in no other membrane system of vegetative or early development cells, including the small collecting vacuoles which feed the main vacuole (Quiviger *et al.*, 1978, 1980) (see also Section III,A,1). It has been suggested that alkaline phosphatase is involved in ion transport across the contractile vacuole membrane (Bowers and Korn, 1973; Quiviger *et al.*, 1978).

C. Chemical and Mechanical Interactions

The plasma membrane is involved in communications between the cell and its exterior that are mediated both by soluble extracellular molecules and by direct cell contact. The best-studied examples involving soluble molecules are the chemotactic systems, by which the cell moves toward food during vegetative growth and toward other cells during development. Membrane-associated enzymatic activities involved in these processes include folate deaminase, adenylate cyclase, and a certain portion of the cyclic AMP phosphodiesterase. These systems are discussed in detail by Devreotes in Chapter 4.

Vegetatively growing cells adhere weakly to other cells and to extracellular substrata (Gerisch, 1961; Beug *et al.*, 1970, 1973a,b). This interaction, mediated by "contact sites B," is sensitive to disruption by EDTA (Gerisch, 1961). The phagocytosis-deficient mutants described by Vogel *et al.* (1980) (see Section I,A,1) also seem to be deficient in "contact sites B"-mediated adhesion to other cells in the presence or absence of axenic medium. This suggests that the "nonspecific" component of the phagocytic system, which is affected in these mutants, is also required for contact sites B-mediated adhesion. The mutants are also defective in adhesion to foreign substrata such as plastic tissue–culture dishes, but only in the presence of axenic medium (containing glucose), suggesting that, unlike contact sites B, adhesion to these substrata can employ either of the two binding mechanisms (glucose binding and nonspecific binding) used for phagocytosis (Vogel *et al.*, 1980; see also Section I,A,1).

As the cells develop, they gain the ability to form strong EDTA-resistant adhesions (mediated by contact sites A, which are serologically distinguishable from contact sites B) (Beug *et al.*, 1970, 1973a,b). These adhesive properties reflect properties of the plasma membrane, since membrane ghosts possess adhesive properties corresponding to the properties of the cells from which they were prepared (Sussman and Boschwitz, 1975a,b). Several groups have suggested that, after aggregation, the contact sites A-mediated adhesion system is replaced by a serologically distinct system (Steinemann and Parish, 1980; Wilcox and Sussman, 1981a,b). Studies implicating particular membrane glycoproteins in some of these adhesive systems are discussed in Section IV,A,3. The adhesion system of *Dictyostelium* is discussed more fully in Chapter 6 by Barondes and his colleagues.

Cells must normally form multicellular aggregates before proceeding further in development (Alton and Lodish, 1977; Landfear and Lodish, 1980; Wilcox and Sussman, 1978, 1981a; Chung *et al.*, 1981; earlier work reviewed by Loomis, 1975), although under certain conditions isolated cells of appropriate sporogenous mutant strains can form stalk and spore cells (Gross *et al.*, 1981). Although cell adhesion would clearly help in the establishment and maintenance of aggregates, it has been shown that cell adhesion is not sufficient for further development (Schmidt *et al.*, 1978; Kilpatrick *et al.*, 1980). Chymotrypsin digestion of preaggregative cells blocks their development at the aggregate stage without affecting EDTA-resistant adhesion. It may be possible to study cell–cell interactions by making use of the observations that isolated membrane fractions from cells at appropriate stages of development are able to block aggregation and to perturb the expression of certain developmentally regulated enzymes (McMahon *et al.*, 1975; Smart and Tuchman, 1976; Tuchman *et al.*, 1976; Jaffe and Garrod, 1979). Similar approaches have been fruitful in other systems (Whittenberger *et al.*, 1978; Gottlieb and Glaser, 1980).

Related to adhesion is the necessity for generation of traction (against other cells, the substratum, or extracellular materials such as the slime sheath) involved in the motility of *Dictyostelium*. Motility has been reviewed by Poff and Whitaker (1979) and is discussed in detail by Spudich and Spudich in Chapter 5. Traction probably involves the association of internal cytoskeletal elements, which provide structural strength and motive force, with integral membrane proteins (Clarke *et al.*, 1975; Jacobson, 1980b; Luna *et al.*, 1981), which in turn are coupled (perhaps by adhesive interactions) to the outside world.

D. Electrical Properties

Like other cells, *Dictyostelium* maintains a voltage gradient across its membrane, with the inside approximately 10–20 mV negative with respect to the outside (Hülser and Webb, 1973, Weijer and Durston, personal communication). No change in membrane potential was seen between vegetative and 6 h developed cells (J. Miller, A. Willard and W. F. Loomis, personal communication). Cells later in development were too small for successful impalement with a microelectrode.

Cells from the aggregation stage and cells in slugs have been examined for low-resistance electrical coupling and for coupling detected by passage of low molecular weight fluorescent markers from cell to cell (Weijer and Durston, personal communication). No coupling was detected by either method. The surface of vegetative cells contains a surplus of anionic groups, as assayed by cell electrophoresis (Lee, 1972b,c; Garrod and Gingell, 1970; Yabuno, 1970) and by electron microscopic histochemistry using anionic and cationic ferritin as markers (Maeda, 1980). Lee (1972b) concluded from chemical modification studies that the majority of these anionic groups represent carboxyl groups on proteins. During development, the net density of anionic groups on the membrane declines (Garrod and Gingell, 1970; Yabuno, 1970; Lee, 1972b,c; Maeda, 1980) and shifts from a uniform to a localized distribution on the membrane, with the advancing surface of aggregation competent cells devoid of anionic sites detectable by staining with cationic ferritin (Maeda, 1980). The physiological relevance of these findings is presently unclear. The reduction in charge density may play a permissive role in allowing close cell approach for adhesion, but many other factors are also clearly involved (Gingell and Garrod, 1969; Garrod and Gingell, 1970; Lee, 1972c; see Chapter 6). The surface charge is probably the result of a heterogeneous collection of charged surface molecules, different species of which may be varying in different fashions, and for different functional reasons during development.

E. Membrane Fusion

Dictyostelium cells undergo fusion on rare occasions to form diploids (Loomis and Ashworth, 1968; Sinha and Ashworth, 1969; Ono *et al.*, 1972; Brody and Williams, 1974; Newell, 1978; Loomis, 1980); this phenomenon is the basis for the parasexual genetic system in this organism (Newell, 1978; Loomis, 1980; see Chapter 2). Since the phenomenon is so rare it has not been possible to analyze it in detail. Cell fusion also occurs during macrocyst development, which forms the sexual cycle of *Dictyostelium* (O'Day and Lewis, 1981; see Chapter 2). Because this type of fusion can be made to occur with reasonable frequency, some work on its mechanism has begun (O'Day and Lewis, 1981). As in many other membrane fusion systems, Ca^{2+} is required for fusion (Chagla *et al.*, 1980), but little else is known of the details of the process.

Several laboratories have attempted to increase the natural fusion rate of

Dictyostelium amoebae, usually in attempts to improve the parasexual system for genetic analysis. Brief reports announcing that fusion can be achieved using poly(ethylene glycol) (Kuhn and Parish, 1981) or high electric fields (Neumann *et al.*, 1980) have appeared, but it is still too early to assess the usefulness of these methods.

III. Membrane Ultrastructure

The transmission electron microscopic ultrastructure of the various membrane systems of *Dictyostelium* has been described (reviewed by Loomis, 1975; see also McMillan and Luftig, 1975; Maeda and Eguchi, 1977) and is typical of eukaryotic cells. The plasma membrane shows the usual trilaminar form but the outer layer is considerably more electron-dense than the inner layer. The endoplasmic reticulum frequently appears as long straight profiles, perhaps representing flat sheets, which are uncommon in other organisms. The mitochondria display tubular cristae.

Several freeze-fracture studies have demonstrated intramembranous particles in the plasma membrane (Aldrich and Gregg, 1973; Gregg and Nesom, 1973; Gregg and Yu, 1975; Yu and Gregg, 1975; Hohl *et al.*, 1978; Favard-Sereno and Livrozet, 1979; Erdos and Hohl, 1980). In developing cells the particles are larger (Aldrich and Gregg, 1973). Both decreases (Aldrich and Gregg, 1973) and increases (Favard-Sereno and Livrozet, 1979) of particle density in developing cells have been reported. Favard-Sereno *et al.* (1981) found an increase in density and a shift in the size distribution of particles in phagosome membranes after phagosome internalization. As yet the chemical bases for these changes are unknown.

When examined by scanning electron microscopy, the outer surface of the cell shows a series of dramatic changes during early development (Rossomando *et al.*, 1974; de Chastellier and Ryter, 1977, 1980; Ryter and Brachet, 1978; Ryter *et al.*, 1979). Early in development cells display many large phagocytic cups. Later, aggregation-competent cells show fewer cups but become covered with numerous microvilli, which have been suggested to be involved in cell–cell adhesion (Jones *et al.*, 1977).

IV. Isolation of Membranes

A. Membrane Markers

Three types of membrane markers have been employed in this system (Table I): (1) assay of enzymatic activities known or assumed to be localized

TABLE I
Membrane Markers

Marker	References
Enzymatic markers	
Alkaline phosphatase	Spudich, 1974; Green and Newell, 1974; Lee *et al.* 1975; Rossomando and Cutler, 1975; Cutler and Rossomando, 1975; Parish and Müller, 1976; Siu *et al.* 1977; Gilkes and Weeks, 1977a,b; Ono *et al.* 1978; Jacobson, 1980a
5′-Nucleotidase	Green and Newell, 1974; Lee *et al.* 1975; Rossomando and Cutler, 1975; Gilkes and Weeks, 1977a,b; McMahon *et al.*, 1977
External labeling	
Lactoperoxidase-mediated radioiodination	Green and Newell, 1974; Lee *et al.*, 1975; Siu *et al.*, 1977
^{125}I-labeled concanavalin A prebound to cells	Jacobson, 1980a
[^{14}C]dinitrofluorobenzene labeling	Sievers *et al.*, 1978
Enzymatic labeling with [^{3}H]galactose	Thilo and Vogel, 1980
Metaperiodate oxidation and $NaB^{3}H_4$ reduction	Toda *et al.*, 1980, 1981
Transmission electron microscopy	Green and Newell, 1974; Rossomando and Cutler, 1975; Cutler and Rossomando, 1975; Parish and Müller, 1976; Siu *et al.*, 1976, 1977; Gilkes and Weeks, 1977a; McMahon *et al.*, 1977; Sievers *et al.*, 1978; Condeelis, 1979; Luna *et al.*, 1981
Scanning electron microscopy	Jacobson, 1980a

primarily or exclusively in particular membranes or organelles, (2) labeling of the external membrane surface using reagents known or assumed not to penetrate into the cell's interior, and (3) morphological examination using the electron microscope. Each type of marker has its advantages and disadvantages.

1. *Enzymatic Activity*

Specific activities of appropriate enzymatic markers provide a convenient and easily quantifiable measure of enzyme purity. Many different markers are available with differing patterns of distribution between membranes and organelles. However, particular markers are often not restricted to a particular structure and the degree of localization is frequently difficult to assess accurately (Newell, 1977). Also, the enzymatic activity may be inhibited or activated by the procedures used for membrane isolation (see Gilkes and Weeks, 1977a, 1977b; Mohan Das and Weeks, 1980, 1981).

The most frequently employed plasma membrane enzymatic marker is alkaline phosphatase (Table I). Even the status of this enzyme is not completely clear (Newell, 1977). The enzyme is enriched in plasma membrane preparations in parallel with the enrichment of externally applied ^{125}I label (Green and Newell, 1974; Lee *et al.*, 1975; Siu *et al.*, 1976, 1977), suggesting that it is indeed a plasma membrane protein. Conversely, cytochemical studies of Quiviger *et al.* (1978, 1980) showed that the enzymatic activity could be detected on the contractile vacuole membrane but not on other membrane systems of the cell, suggesting that it may be a marker enzyme for the contractile vacuole and not for the plasma membrane (see also Section I,B). However, these histochemical studies were performed on glutaraldehyde-fixed material, and the same workers (Quiviger *et al.*, 1980) have shown that only 15% of the total cellular activity is stable to the fixation conditions used. Thus, the majority of the alkaline phosphatase activity may be associated with the plasma membrane but may be sensitive to glutaraldehyde in this structure.

Mohan Das and Weeks have shown that the amount of detectable activity in vegetative cell membranes can vary considerably depending on the storage conditions of the membranes. The activity (assayed at 30°C) is reversibly inhibited by storage at 4°C and is reversibly activated by incubation at room temperature or (even more dramatically) at elevated temperatures (50°C) (Mohan Das and Weeks, 1980). Prolonged dialysis also activates the enzyme, perhaps by removal of a low molecular weight inhibitor (Mohan Das and Weeks, 1981). It has not been determined whether or not this inhibition can account for the failure to demonstrate alkaline phosphatase histochemically on plasma membranes of vegetative cells (Quiviger *et al.*, 1978, 1980).

The specific activity of alkaline phosphatase increases dramatically late in development, during culmination (Krivanek, 1956; Gezelius and Wright, 1965; Loomis, 1969). Both RNA and protein synthesis are required for this increase (Loomis, 1969). However, Mohan Des and Weeks (1980, 1981) have suggested that the increase reflects differing degrees of inhibition of a constant amount of enzyme protein, since the enzyme from culminants cannot be further activated by heat or dialysis and since the total amount of activity recovered after heating or dialysis is quite similar at all stages of development. MacLeod and Loomis (1979) have presented genetic evidence that the same gene product is responsible for both the vegetative and developmentally regulated activities.

Unlike the situation for vegetative cells, plasma-membrane-associated alkaline phosphatase activity is easy to demonstrate in culminants (Quiviger *et al.*, 1980). The activity is associated only with prestalk cells, with the most intense activity found on the cells adjacent to the prestalk–prespore boundary (Krivanek, 1956; Quiviger *et al.*, 1980). Quiviger *et al.* suggest that the enzyme is involved in ion and water movements required for the final differentiation of spores (which must pump water out) and stalk cells (which must pump water in).

5′-Nucleotidase activity, which has also been used as a plasma membrane marker (Table I), shows a pattern of developmental regulation and of localization at culmination similar to that of alkaline phosphatase (Krivanek and Krivanek, 1958; Gezelius and Wright, 1965; Lee *et al.*, 1975; Armant and Rutherford, 1979; Armant *et al.*, 1980). The two activities copurify and may reflect different properties of the same molecular species (Armant and Rutherford, 1981; G. Weeks, personal communication).

Unlike many other alkaline phosphatases, the specificty of the enzyme from *D. discoideum* is quite high. Out of 18 potential substrates tested, the purified enzyme can hydrolyze only *p*-nitrophenylphosphate (the usual substrate in the alkaline phosphatase assay), 5′-AMP (the usual substrate in the 5′-nucleotidase assay), 2′-deoxy-5′-AMP, and (slightly) ADP (Armant and Rutherford, 1981).

These enzyme activities thus seem to be associated with the plasma membrane, but they may not be present on plasma membranes of all cell types. Assay conditions are critical; preparations to be used for evaluating membrane purity should probably be activated at 50°C immediately before analysis (Mohan Das and Weeks, 1980).

No study of purified membranes has examined the possible extent of contamination by contractile vacuole membranes, although such an examination should be possible by employing the difference in staining of these membranes using histochemical stains (silver proteinate and phosphotungstic acid) for carbohydrate (de Chastellier and Ryter, 1977). The plasma membrane stains strongly whereas the contractile vacuole does not.

2. *External Labeling*

External labeling probably provides the most definitive marker for the external membrane. The most common labeling technique has been lactoperoxidase-catalyzed radioiodination of cell surface proteins, but other markers have been employed (Table I). Unless the reaction conditions are carefully controlled, this method may suffer from penetration of label into the cells, resulting in nonspecific labeling of internal components.

3. *Electron Microscopy*

Electron microscopy provides a sensitive (but only semiquantitative) method for evaluating the presence of certain types of contaminants (e.g., mitochondria, ribosomes, nonmembranous but ultrastructurally visible material), as well as providing an evaluation of the physical state of the isolated membranes themselves (large or small vesicles, large sheets, etc.). As such, it is an extremely useful and necessary tool. However, it cannot distinguish all possible contaminating membrane types (e.g., smooth endoplasmic reticulum).

B. Cell Breakage and Membrane Isolation

Gilkes and Weeks (1977a,b) have pointed out that the conditions of lysis of the cells can be critical in determining the partitioning of marker enzymes during subsequent steps of purification (and thus in determining the success of purification itself). Methods of cell breakage which have been employed in *Dictyostelium* are listed in Table II. Each has its advantages and disadvantages, and different methods may be appropriate for different experimental questions. It has been noted that *Dictyostelium* cells are considerably more difficult to mechanically disrupt than are mammalian cells (Green and Newell, 1974). Cell rupture using a Dounce homogenizer is gentle to the cells but tedious for the scientist, as the cells can require from forty to several hundred strokes for breakage, and even then breakage may not be complete (McMahon *et al.*, 1977; Ono *et al.*, 1978). Other methods of mechanical disruption of the cells also suffer from a lack of complete lysis. Freezing and thawing is more effective but harsher, and generally makes it more difficult to purify the plasma membrane away from other cellular components; cell organelles such as lysosomes will be ruptured, releasing degradative enzymes which may compromise the structural and functional integrity of the isolated membranes. Cell rupture by treatment with digitonin (Riedel and Gerisch, 1968) or amphotericin B (Rossomando and Cutler, 1975) is rapid and the membranes are released as large intact "ghosts," but the complexing action of these agents with steroids means that the lipid component of the membrane, at least, will definitely be perturbed. The same is true of the concanavalin A–Triton purification procedure, wherein the lipid component would be expected to be largely extracted (Luna *et al.*, 1981). In addition, concanavalin A is known to perturb (by "patching" and "capping") the distribution of the cell surface glycoconjugates with which it reacts, and it probably also induces changes in the linkages of these and other components with underlying cytoskeletal elements (Condeelis, 1979; Luna *et al.*, 1981).

The methods used to isolate plasma membranes once the cells have been broken are listed in Table III. Purifications of 12- to 15-fold, as assayed by increases in the specific activities of marker enzymes or external labels (Table I), are routinely obtained using most of these methods. In several cases (Gilkes and Weeks, 1977a,b; Jacobson, 1980a), 36-fold purifications

TABLE II
Methods for Breaking Cells

Method	References
Mechanical disruption	
Mechanical homogenization (Potter-Elvejhem, Dounce, Ten Broeck)	Green and Newell, 1974; Siu *et al.*, 1977; McMahon *et al.*, 1977; Sievers *et al.*, 1978; Ono *et al.*, 1978; Luna *et al.*, 1981
Freeze–thaw	Green and Newell, 1974; Spudich, 1974; Sussman and Boschwitz, 1975a,b; Gilkes and Weeks, 1977a; Condeelis, 1979; Luna *et al.*, 1981
Agitation with glass beads	Gilkes and Weeks, 1977a,b; Luna *et al.*, 1981
Forcing through fine filter	Gilkes and Weeks, 1977a
Vortexing and sonication of bead–bound cells	Jacobson, 1980a
Explosive decompression	Luna *et al.*, 1981
Chemical disruption	
Digitonin treatment (500 μg/ml, 15–20 min)	Riedel and Gerisch, 1968; Müller *et al.*, 1979
Amphotericin B treatment (360 μg/ml, 60 min)	Rossomando and Cutler, 1975; Cutler and Rossomando, 1975
Detergent disruption	
Triton treatment of conA-stabilized membranes	Parish and Müller, 1976; Condeelis, 1979; Luna *et al.*, 1981

TABLE III
Methods for Isolating Membranes

Method	References
Differential centrifugation	Green and Newell, 1974; Rossomando and Cutler, 1975; Cutler and Rossomando, 1975; Gilkes and Weeks, 1977b
Sucrose gradient flotation	Green and Newell, 1974; Gilkes and Weeks, 1977a
Sucrose gradient sedimentation	Green and Newell, 1974; Spudich, 1974; Rossomando and Cutler, 1975; Cutler and Rossomando, 1975; McMahon *et al.*, 1977; Sievers *et al.*, 1978; Gilkes and Weeks, 1977a,b; Luna *et al.*, 1981
Renografin gradient flotation	McMahon *et al.*, 1977
Two-phase polymer partitioning (Brunette and Till, 1971)	Siu *et al.*, 1977; Ono *et al.*, 1978; Müller *et al.*, 1979
Binding intact cells to polycationic beads	Jacobson, 1980a
Detergent lysis of conA-treated cells	Parish and Müller, 1976; Condeelis, 1979; Luna *et al.*, 1981

have been reported. The degree of purification achieved may reflect the removal of loosely bound (but perhaps functionally significant) membrane components as well as of contaminating non-membrane-associated material.

The most appropriate method to use depends on the particular problem being investigated. The bead isolation method of Jacobson (1980), wherein cells are first bound to polycation-coated beads and then broken, leaving only the attached surface material on the beads, is rapid and simple to use, and it provides one of the highest membrane purifications achieved for *Dictyostelium* (36-fold, based on specific activity of prebound ^{125}I-labeled concanavalin A). However, the critical dependence of this method on isolation conditions (e.g., pH of binding and neutralization, nature and concentration of polyanion used) and the need for harsh reagents (e.g., sodium dodecyl sulfate) for removal of the membrane components from the beads may limit its application to certain biochemical and structural studies and render it of little use for preparation of significant quantities of undenatured membrane proteins. The method should be very useful for studying the interaction of cytoplasmic components such as actin with the cytoplasmic face of the plasma membrane (Jacobson, 1980b). The two-phase polymer system (Brunette and Till, 1971; Siu *et al.*, 1977) is also simple to use and gives high yields of membranes with about a 12-fold purification. More extensive purification schemes involving several gradient steps can give greater purification (to 36-fold), but yields may suffer (Green and Newell, 1974; Gilkes and Weeks, 1977a,b). The simplest method is a freeze–thaw lysis followed by centrifugal washing (Sussman and Boschwitz, 1975a), but such "ghosts" retain significant amounts of contaminating intracellular organellar material.

Standardization and evaluation of membrane isolation techniques is difficult. This situation is by no means limited to *Dictyostelium*. Seemingly minor differences in cell lysis or membrane isolation procedures—or even differences between laboratories employing the same techniques (Green and Newell, 1974; Gilkes and Weeks, 1977a)—can and do give rise to major differences in results. In many cases it is difficult or impossible to assess which procedure most accurately reflects the situation in the unperturbed membrane.

V. Composition of Membranes

As in other cells, *Dictyostelium* plasma membranes contain protein, lipid, and carbohydrate, but little or no nucleic acid. The detailed analysis of these components and of their changes during development will now be consid-

TABLE IV
Plasma Membrane Composition[a]

	mg component per mg protein	mg component per 10^8 cells[b]	Percentage of membrane mass	Percentage of total–cell component in membrane[c]
Protein	(1.00)	0.308	63.3	2.8
Phospholipid	0.435±0.046	0.134	27.5	17.1
Sterol[d]	0.099±0.011	0.030	6.3	20.6
Carbohydrate	0.046±0.005	0.014	2.9	3.7–6.1

[a]Data (mean ± standard deviation) for membranes prepared from vegetative AX2 cells grown axenically in HL5 medium (Gilkes and Weeks, 1977a; Gilkes *et al.*, 1979 with permission).

[b]Calculated assuming plasma membrane contains 2.8% of total cell protein (Green and Newell, 1974; Gilkes and Weeks, 1977b) and using a value of 11.0 mg protein per cell (Ashworth and Watts, 1970; Loomis, 1975).

[c]Calculated from specific activities in whole cell homogenate and plasma membrane fraction (Gilkes and Weeks, 1977a) using the formula:

$$\frac{\text{mg component/mg protein (membrane)}}{\text{mg component/mg protein (homogenate)}} \times 2.8\% \text{ (percent of cell protein in membranes).}$$

For carbohydrate, percentages were calculated for total neutral and amino sugar, excluding ribose and glucose, using membrane carbohydrate compositions given in Gilkes and Weeks (1977a) and Gilkes *et al.* (1979), respectively.

[d]88% of membrane sterol (.087 mg/mg protein or .027 mg/10^8 cells) is Δ^{22}-stigmasten-3β-ol (Gilkes and Weeks, 1977a).

ered. The gross composition of the plasma membrane of vegetative strain AX2 cells grown axenically in HL5 medium is given in Table IV. Detailed comparative data are not available, but the composition of membranes of other strains is probably similar. Some differences might be expected for bacterially grown cells in light of their known differences from axenically grown cells in volume, mass, and metabolism (Ashworth and Watts, 1970; reviewed by Loomis, 1975).

A. Proteins

1. *Major Membrane Proteins of Vegetative and Developing Cells*

Plasma membrane proteins at various stages of development have been the objects of considerable study. The major membrane proteins of vegetative cells (grown either on bacteria or axenically) have been analyzed by many groups using various methods of cell lysis and membrane purification (Siu *et al.*, 1977; Hoffman and McMahon, 1977; Müller and Gerisch, 1978; Gilkes *et al.*, 1979; Müller *et al.*, 1979; Condeelis, 1979; Luna *et al.*, 1981). The major polypeptides detected, using sodium dodecyl sulfate–polyacrylamide gel electrophoresis and protein stains such as Coomassie brilliant blue, are quite similar in the different studies, although the relative amounts of the different components vary somewhat. The exact molecular weight values obtained also vary among different reports, probably because different laboratories use different gel systems and different standard proteins for molecular weight calibration.

A representative analysis is presented in Fig. 1 (Siu *et al.*, 1977). Between 10 and 15 major bands are detected, a decided simplification of the pattern seen from whole cell lysates. In addition, numerous minor bands are usually seen. Comparisons of minor bands between reports is difficult. Some represent contaminants in the membrane preparations, but many probably represent membrane components that, although less prominent than others on a mass basis, may nevertheless be crucial in membrane function and its changes during development. Higher resolution techniques such as two-dimensional polyacrylamide gel electrophoresis will be essential for studies of these components. Actin and myosin are the most notable major proteins to vary in relative amounts between different preparations. These proteins are particularly prominent in membranes from concanavalin A-stabilized preparations, since treatment with concanavalin A causes cytoplasmic cytoskeletal elements to bind to the underside of the membrane (Condeelis, 1979; Luna *et al.*, 1981).

The major membrane proteins change remarkably little during development (Fig. 2). Similarly, only a few of the major surface proteins detectable by lactoperoxidase-mediated radioiodination are developmentally regulated (Smart and Hynes, 1974; Geltosky *et al.*, 1976; Siu *et al.*, 1977). The most prominent changes in Fig. 2 involve a group of proteins with apparent molecular weights of 38,000, 36,500, 18,000, and a small group with molecular weights of less than 12,000. These proteins coordinately increase greatly in relative abundance early in development (Siu *et al.*, 1977). By 6 h of development (in bacterially grown NC4 cells) they form approximately 35% of the plasma membrane protein mass. By 12 h, the proteins have disappeared from the membranes. These proteins have been shown by tryptic peptide analysis to be species resulting from differing numbers of proteolytic cleavages of the 38,000-dalton protein (Elder *et al.*, 1977; Bordier *et al.*, 1978). The protein is probably an integral membrane protein (as assayed by charge–shift electrophoretic analysis), and it is found in the membrane as a trimer. Four different species of trimer can be found in the membrane, differing in the number of component chains which have undergone pro-

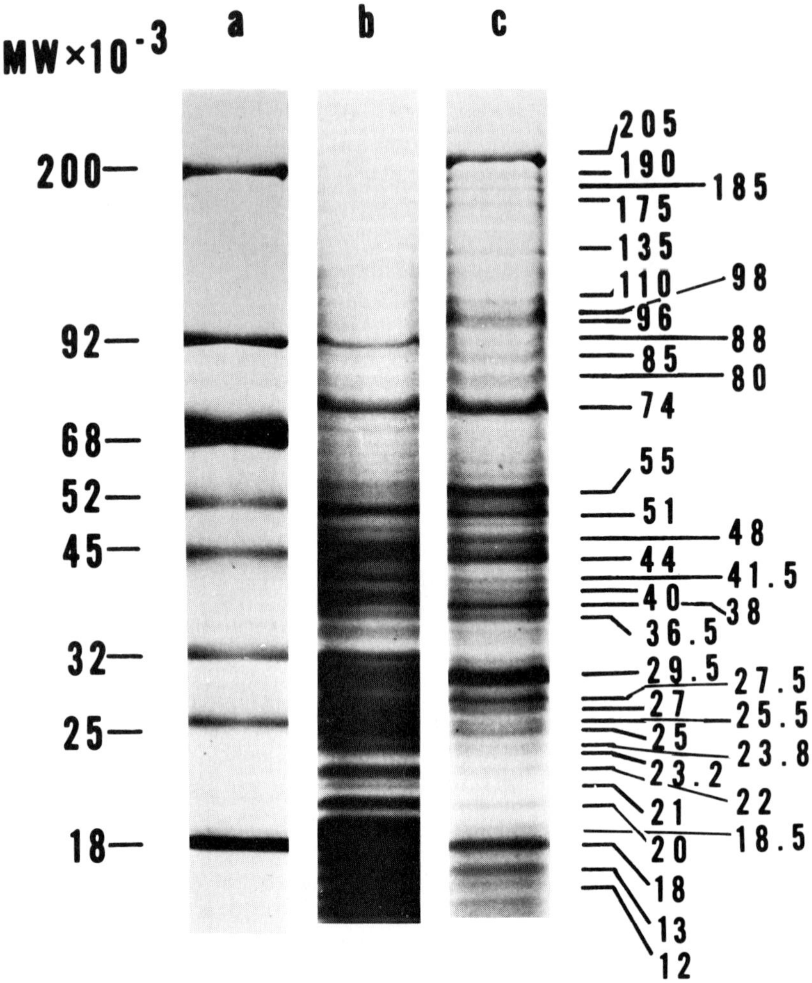

Fig. 1. Electrophoretograms of vegetative NC4 plasma membrane proteins and whole cell proteins. Proteins were run on 5–15% gradient sodium dodecyl sulfate (SDS)–polyacrylamide gels. (a) Protein molecular weight markers stained with Coomassie brilliant blue; (b) Coomassie stained pattern of whole cell proteins from zero–hour cells; and (c) Coomassie stained pattern of plasma membrane proteins of zero–hour cells. Marker and plasma membrane protein bands are identified by numbers designating their apparent molecular weights × 10^{-3}. From Siu *et al.* (1977), with permission.

teolytic processing (Bordier *et al.*, 1978). Surprisingly, in isolated membranes, the complex is more resistant than most other membrane proteins to proteolysis by endogenous or exogenously added proteases, and no *in vitro* proteolytic processing of the subunits can be demonstrated. (C. Bordier and W. F. Loomis, unpublished). Thus, the predominance of this protein in membrane fractions will depend to some degree on the care with which proteolysis has been avoided. The function of this protein complex is unknown. In light of the results discussed in Section III,A,2, it might be interesting to investigate a possible function of this complex in the increased phagocytic effort of the cells following starvation.

The other major change apparent in Fig. 2 is the appearance of a 60,000-dalton protein at 12 h of development that accumulates during culmination. Orlowski and Loomis (1979) have suggested that this is one of the proteins that will eventually be found in the spore coat.

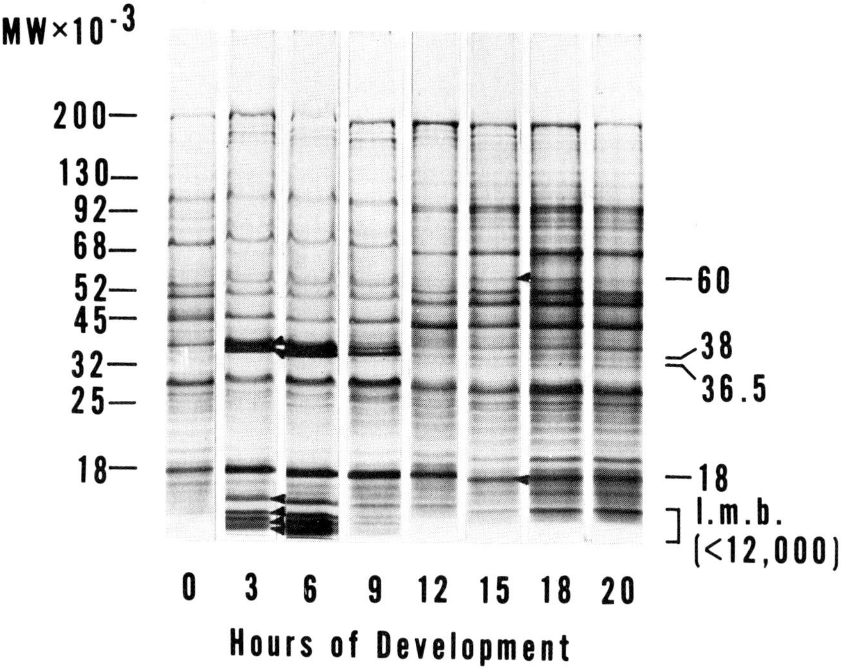

Fig. 2. Electrophoretograms of NC4 plasma membrane proteins at different times of development. Plasma membranes were isolated from NC4 cells every 3 h after the initiation of development. Proteins were separated on a 5–15% gradient SDS polyacrylamide gel followed by staining with Coomassie brilliant blue. Molecular weight values of protein standards are included on the left. Protein bands that changed during development are marked with an arrowhead and their corresponding molecular weight values are indicated on the right (l.m.b., low molecular weight protein bands). From Siu *et al.* (1977), with permission.

2. *Changes in Membrane Protein Synthesis during Development*

When synthesis of new proteins is examined by following incorporation of radioactively labeled amino acids, a larger proportion of bands (but still a minority) exhibit major changes in labeling during development (Fig. 3; Siu *et al.*, 1977; Parish and Schmidlin, 1979a). The proteins of the early complex

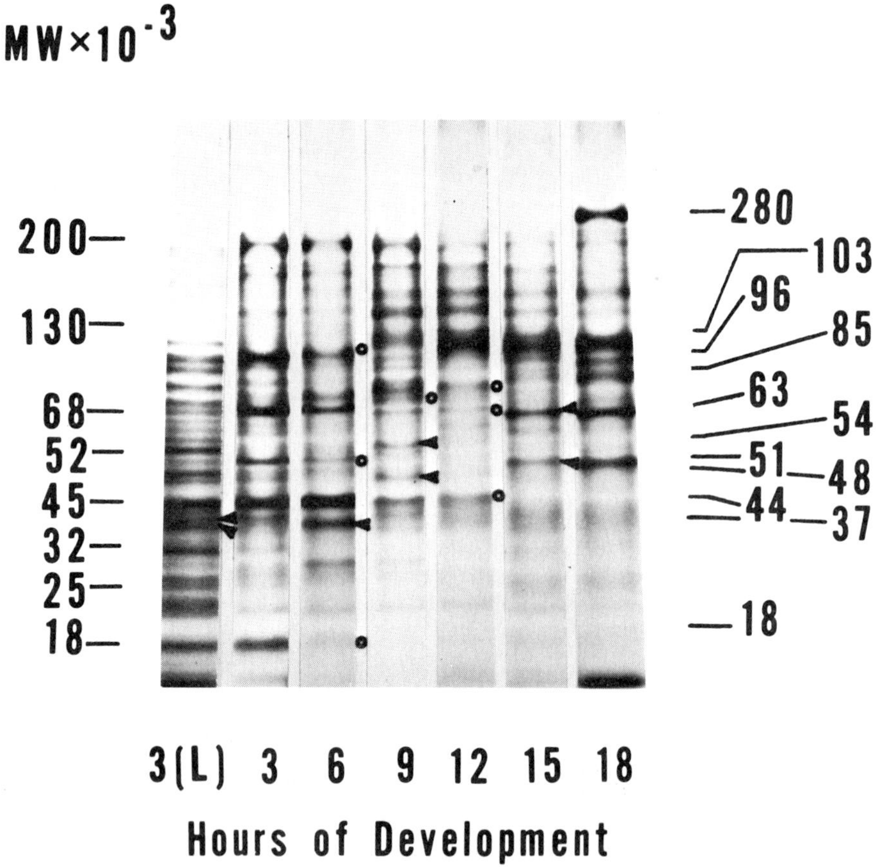

Fig. 3. Fluorograms of ^{35}S-labeled NC4 plasma membrane proteins. NC4 cells were labeled with [^{35}S]methionine for 2 h prior to the time of collection for the isolation of plasma membranes. Membrane proteins were separated on a gradient SDS–polyacrylamide slab gel and the gel was processed for fluorography. The appearance of a band is indicated by an arrowhead and the disappearance of a band is indicated by an open circle. Some of the predominantly labeled bands are indicated on the right by a number designating their approximate molecular weights × 10^{-3}. A [^{35}S]methionine-labeled whole cell lysate from 3 h cells is included in the gel for comparison [3(L)]. Molecular weights of standards are indicated on the left. From Siu *et al.* (1977), with permission.

are synthesized only during the first six hours of development, as would be expected if control of their synthesis is a major factor in the control of their abundance in the membrane (Siu *et al.*, 1977). Several other changes (both initiation and cessation) in synthesis of specific bands can be detected at other times during development. Late in development (after 15 h) a new class of major synthesized membrane proteins appears, with molecular weights of 280,000, 103,000, 96,000, 85,000, 63,000, and 51,000. The 103,000, 85,000, 63,000, and 51,000-dalton species are also the major synthesized (that is, labeled) species in whole cell lysates at these times. Some or all of these proteins are probably spore coat proteins (Orlowski and Loomis, 1979) in the process of synthesis.

One of the developmentally regulated membrane-associated proteins seems to be actin (Siu *et al.*, 1977). This protein is probably associated with the inner surface of the membrane as a peripheral protein (Condeelis, 1979; Jacobson, 1980b; Luna *et al.*, 1981; see Chapter 5).

3. *Membrane Glycoproteins*

Several laboratories (Hoffman and McMahon, 1977, 1978a,b; West and McMahon, 1977, 1979; Gilkes *et al.*, 1979; Orlowski and Loomis, 1979; Parish and Schmidlin, 1979a; Toda *et al.*, 1980, 1981; Ono *et al.*, 1981; Lam and Siu, 1981) have examined membrane glycoproteins using a variety of labeling and staining techniques. A significant fraction of these glycoproteins exhibit changes in synthesis or accumulation during development. The importance of these changes will become more apparent as particular membrane glycoproteins are studied and their functions are elucidated.

Glycoprotein expression in the membrane has also been examined by staining gel-separated proteins using appropriately labeled lectins (Burridge and Jordan, 1979; West and McMahon, 1977, 1979; West *et al.*, 1978). Lectins with different sugar specificities reveal different patterns of change during development. For example, concanavalin A reacts with many bands, most of which do not change during development (Burridge and Jordan, 1979; West and McMahon, 1977), in agreement with the results of Geltosky *et al.* (1976). Some glycoproteins that bind the lectin wheat germ agglutinin are nonuniformly distributed along the axis of the slug stage of *D. discoideum* (West and McMahon, 1979), raising the possibility that they may be involved in, or as least serve as markers of, the histiotypic differentiation of prestalk (anterior) and prespore (posterior) cells in the slug (see Chapter 11).

Several studies have examined the binding of lectins such as concanavalin A to the cell surface at different stages of development (Weeks, 1973, 1975; Rossomando *et al.*, 1974; Gillette *et al.*, 1974; Weeks and Weeks, 1975; Molday *et al.*, 1976; Darmon and Klein, 1976; Grabel and Farnsworth, 1977;

Filosa, 1978). Such studies have indicated that most binding sites for concanavalin A on developed cells are of lower affinity than those on vegetative cells (Weeks, 1975), indicating that there are changes in the cell surface carbohydrates during development. Binding of concanavalin A (Gillette and Filosa, 1973; Yu and Gregg, 1975; Weeks and Weeks, 1975; Darmon and Klein, 1976; Saito and Yanagisawa, 1978) or wheat germ agglutinin (West and McMahon, 1981) can interfere with development. Since lectins bind to many cell surface proteins, some of which are developmentally regulated and some of which are not (Geltosky *et al.*, 1976; West and McMahon, 1977, 1979; Burridge and Jordan, 1979), it is difficult to interpret these results in greater detail. Lectins may also bind to nonprotein glycoconjugates on the cell surface.

Several membrane proteins and glycoproteins believed to be involved in the adhesion system of *Dictyostelium* have been studied in greater detail. Müller *et al.* (Müller and Gerisch, 1978; Müller *et al.*, 1979) have used an immunological assay involving the blocking of adhesion by Fab antibody fragments directed against aggregation-competent cell membranes to purify a glycoprotein with an apparent molecular weight of 80,000 from aggregation-competent cell membranes. This glycoprotein (gp80) seems to provide the major target (contact sites A) in these membranes for the adhesion-blocking activity of the Fab fragments. Fab fragments of antibodies directed against the purified glycoprotein are also capable of blocking contact sites A-mediated adhesion (Murray *et al.*, 1981). Gp80 is a minor component of the membrane, containing about 1% of the membrane protein of aggregation-competent cells (Müller and Gerisch, 1978). Gp80 contains about 25% carbohydrate (Müller *et al.*, 1979), has an acidic isoelectric point (pI 3.5–4.0) (Murray *et al.*, 1981), and is phosphorylated (Coffman *et al.*, 1981; Schmidt and Loomis, 1982). A number of other membrane proteins of *Dictyostelium* are also phosphorylated (Parish *et al.*, 1977), including some whose appearance and disappearance is developmentaly regulated (Coffman *et al.*, 1981). A glycoprotein with an apparent molecular weight of 70,000 is thought to be involved, in a fashion similar to gp80, in the adhesion of *Polysphondylium pallidum* (Hintermann and Parish, 1979; Steinemann *et al.*, 1979). Parish and co-workers (Parish *et al.*, 1978a,b; Parish and Schmidlin, 1979a) have used antibody staining of polyacrylamide gels and metabolic labeling to examine membrane proteins during development, and have argued that a prominent 80,000-dalton band which is synthesized during aggregation of *D. discoideum* is in fact the "contact sites A" glycoprotein (gp80). Parish *et al.* claim that the protein is removed from the membrane later in development, but Murray *et al.* (1981) have shown, using antibodies raised specifically against the purified glycoprotein, that gp80 persists throughout development and can even be detected in mature spores. The *synthesis* of gp80 is limited to a defined period during cell aggregation (Murray *et al.*, 1981). There is some evidence that antibodies directed against gp80 may no longer be effective in blocking adhesion after the aggregation stage (Steinemann and Parish, 1980; Wilcox and Sussman, 1981a,b), but that a serologically distinct adhesion system then comes into play, perhaps involving a glycoprotein with an apparent molecular weight of 95,000 (Steinemann and Parish, 1980).

Concanavalin A-binding glycoproteins have been studied by Geltosky *et al.* (1976, 1979, 1980). As confirmed by other workers (Burridge and Jordan, 1979; West and McMahon, 1977; Crean and Rossomando, 1977b), only a few of these proteins change with development. One band with an apparent molecular weight of 150,000 dramatically increases in intensity late in development. Fab fragments of antibodies prepared against this glycoprotein are also capable of blocking the adhesion of developed *D. discoideum* cells (Geltosky *et al.*, 1979). Blocking of adhesion is more effective for prespore cells than for prestalk cells (Lam *et al.*, 1981). The relationship, if any, between this glycoprotein and the 80,000-dalton glycoprotein is unclear.

The endogenous lectins discoidin (Reitherman *et al.*, 1975; Bartles and Frazier, 1980) and pallidin (Chang *et al.*, 1977; Rosen and Kaur, 1979), which are not glycoproteins, can bind to cells of their respective species of slime mold (*D. discoideum* and *Polysphondylium pallidum*). Cell-surface-associated lectins (both exogenously added and endogenously synthesized) can be eluted from the cell surface by incubation with appropriate competitive sugars, suggesting that the lectin is attached to the cell by its carbohydrate binding sites (Springer *et al.*, 1980; see Chapter 6). Receptors for discoidin (presumably containing carbohydrate) are developmentally regulated and accumulate at approximately the same time as does discoidin itself (Reitherman *et al.*, 1975). It has been difficult to isolate the receptor species and characterize them chemically; different laboratories have reported significantly different results (Burridge and Jordan, 1979; Bartles *et al.*, 1981; Breuer and Siu, 1981; Ray and Lerner, 1982).

4. *Control of Membrane Protein Expression*

The developmentally regulated changes in membrane protein synthesis and accumulation appear, in many cases, to be under the control of the dependent sequence of development (Loomis *et al.*, 1978; see Chapter 10). For example, the major membrane complex of cells early in development is synthesized in large amounts but is *not* removed from the membrane in developing cells of the early mutant WL3 (Siu *et al.*, 1977); this fact was employed to obtain large amounts of the complex for biochemical study (Bordier *et al.*, 1978). Pulsing of early developing cells with cyclic AMP causes the premature appearance of several regulated membrane proteins (Parish *et al.*, 1978b; Parish, 1979), as well as of many other developmental functions (see Chapter 4). Disaggregated cells which have been allowed to reaggregate will resynthesize many of the membrane proteins which they had originally synthesized during aggregation (Parish and Schmidlin, 1979b), similar to the recapitulation of other developmental functions which has been known for many years (reviewed in Loomis, 1975). Synthesis and accumulation of membrane proteins thus appears to be subject to many of the same modes of control which are already known for other markers of development.

Thus, as might be expected, there is considerable change in the synthesis and accumulation of specific proteins of the plasma membrane during development. Most but not all of this change involves quantitatively minor components of the membrane. Further understanding of their roles will require the detailed investigation of the structure, function, and control of particular membrane species.

B. Lipids

The lipid composition of whole cells and its change with development have been the subject of reports from several laboratories (Heftmann *et al.*, 1959, 1960; Johnson *et al.*, 1962; Davidoff and Korn, 1962, 1963a,b; Gerisch *et al.*, 1969; Lenfant *et al.*, 1969; Ellouz and Lenfant, 1971; Wilhelms *et al.*, 1974; Long and Coe, 1974; Ellingson, 1974, 1980; Weeks, 1976; Gilkes and Weeks, 1977a; Herring and Weeks, 1979; Weeks and Herring, 1980; Herring *et al.*, 1980; Mohan Das *et al.*, 1980). Lipid composition of isolated plasma membranes has been examined (Gilkes and Weeks, 1977a; Herring and Weeks, 1979; Weeks and Herring, 1980; Herring *et al.*, 1980, Mohan Das *et al.*, 1980; Table IV). The isolated membrane is about two-thirds protein and one-third lipid, with small amounts (about 3%) of carbohydrate. Lipid compositions of vegetative, aggregation-competent, and pseudoplasmodial cell membranes are not significantly different, with the possible exception of a slight (20%) decline in the total sterol content in pseudoplasmodial cell membranes (Weeks and Herring, 1980).

1. *Sterols*

The molar ratio of sterol to phospholipid in the membrane is 0.410 (Gilkes and Weeks, 1977a). Free sterol and sterol esters were not separately quantitated in studies of isolated membranes, but Long and Coe (1974) report that at least 90% of the total cell sterol in vegetative (bacterially grown) NC4 cells is free sterol. The major species in *D. discoideum* is Δ^{22}-stigmasten-3β-ol (Heftmann *et al.*, 1959, 1960; Johnson *et al.*, 1962; Long and Coe, 1974; Scandella *et al.*, 1980), which accounts for 88% of the membrane sterol (Gilkes and Weeks, 1977a).

Hase (1981) has reported the identification of steryl glycosides in lipid extracts from NC4 cells. Δ^{22}-stigmastenyl-D-glucoside accounts for 90% of the steryl glycoside at all stages of development. Steryl glycoside is a minor component of the total cell lipids, accounting for about 0.05% by weight of all cell lipids during most of development and increasing to 0.15% of cell lipids in the mature sorocarp (Hase, 1981).

Recent results of Favard-Sereno *et al.* (1981) suggest that not all membranes of *Dictyostelium* contain large quantities of sterol. Complexes of sterol with digitonin or filipin can be detected (by thin-section and freeze-fracture electron microscopy) in plasma membranes and in membranes of phagosomes, digestive vacuoles and autophagic vacuoles. As discussed in section I,A,1, these membrane compartments (except perhaps that of autophagic vacuoles) may be functionally and structurally connected by membrane recycling. In contrast, complexes indicative of the presence of sterols are not detected in endoplasmic reticulum, nuclear membranes, or mitochondrial membranes.

Scandella *et al.* (1980) have isolated spontaneous nystatin-resistant mutants of *D. discoideum* which fall into at least three complementation groups (*nysA*, *nysB*, and *nysC*). Nystatin is a polyene antibiotic which acts by complexing with sterols in cell membranes. Mutants in *nysB* and *nysC* accumulate different (as yet unidentified) sterols from the stigmastenol found in the wild type, whereas the sterol composition of *nysA* mutants appears unchanged from wild type. The mutants have no effect on vegetative growth rates, and development is completely normal in *nysA* and *nysB* mutants. *NysC* mutants aggregate somewhat more slowly than wild-type; the final fruits have short stalks but are otherwise normal. Further analysis of these mutants may provide more information on the function of sterols in *Dictystelium* development.

2. *Phospholipids*

The phospholipid content of membranes from vegetative axenically grown AX2 cells is given in Table V (Weeks and Herring, 1980). The phospholipid content of whole cells is not significantly different from that of membranes except that the level of the plasmalogen phosphatidalethanolamine (in which the linkage of fatty acid to the 1 position of glycerol is through a vinyl ether rather than through an ester linkage) is lower in the whole cell homogenate (14.3 versus 27.3%). Similar data for whole cell phospholipid (with the ex-

TABLE V
Phospholipid Composition of Plasma Membranes[a]

Phospholipid	Percentage total phospholipid
Lysophosphatidylcholine	1.5±0.3
Lysophosphatidylethanolamine	1.4±0.6
Phoshatidylinositol	7.8±2.0
Phosphatidylserine	1.5±0.8
Phosphatidylglycerol	1.2±0.1
Phosphatidic acid	0.3±0.3
Cardiolipin	2.0±0.8
Phosphatidylcholine	28.8±2.2
Phosphatidylethanolamine	27.8±4.2
Phosphatidalethanolamine (plasmalogen form)	27.3±3.8
Others[b]	1.2±0.8

[a]Data (mean ± standard deviation) from Weeks and Herring (1980), with permission, for membranes prepared from vegetative AX2 cells grown axenically in HL5 medium.

[b]Several minor unidentified constituents.

ception of lower levels of phosphatidylcholine) have been reported by Ellingson (1974). Lysolipid levels are low and are not increased by prolonged incubation of the purified plasma membrane fraction at 4°C (Weeks and Herring, 1980). This suggests that phospholipases, which are present at high levels in *D. discoideum* (presumably for digestion of ingested bacterial lipids) (Ferber *et al.*, 1970; Irvine *et al.*, 1980), have been successfully removed. Crude membrane preparations *are* degraded under these conditions (Weeks and Herring, 1980).

The fatty acid content of vegetative axenically grown AX2 cell membranes is given in Table VI (Herring and Weeks, 1979; Weeks and Herring, 1980; Herring *et al.*, 1980). These results agree reasonably well with previously reported data for whole cell fatty acid content (Davidoff and Korn, 1963a). Both groups agree that the amount of unsaturated fatty acids is extremely high, making up 75–90% of the fatty acids of the organism and of the membrane. This may provide the membrane with sufficient fluidity for the organism to grow at 22°C (Davidoff and Korn, 1963a; Herring and Weeks, 1979; Weeks and Herring, 1980; Herring *et al.*, 1980).

There is no significant difference in the phospholipid or fatty acid composition of membranes prepared from vegetative, aggregation-competent, and pseudoplasmodial cells of strain AX2 (Weeks and Herring, 1980). Several groups have used electron spin resonance (Von Dreele and Williams, 1977; Kawai and Tanaka, 1978; Herring and Weeks, 1979; Weeks and Herring, 1980) and fluorescence depolarization (Herring *et al.*, 1980) to examine the fluidity of membranes in intact cells or of isolated membranes at various stages of development. Again, there is no significant change in fluidity at any of the stages examined. The temperature at which the cells are grown also does not affect the fatty acid composition or fluidity of the membranes (Mohan Das *et al.*, 1980). Thus, the gross environment of the lipid bilayer does not seem to be significantly altered during development. One caution that must be kept in mind is that it is not certain that the reporter molecules used will adequately sample all areas of the membranes (Von Dreele and Williams, 1977), and indeed Conrad and Singer (1979, 1981; M. Conrad, personal communication) have reported data suggesting that the presence of internal membrane proteins and/or sterols in a bilayer causes the exclusion of many common reporter molecules from the bilayer.

The level of unsaturation of lipid fatty acids can be increased still further by growing the cells in HL5 medium supplemented with high levels of

TABLE VI
Fatty Acid Composition of Plasma Membranes[a]

Fatty acid	Percentage of total fatty acids
14:0[b]	1.2±.2
Palmitaldehyde[c]	3.5±.9
16:0	7.5±.5
16:1 (Δ^9)	2.4±.2
16:2 ($\Delta^{5,9}$) and 17:0[d]	1.3±.4
18:0	2.7±1.0
18:1 (Δ^9 and Δ^{11})[c]	34.6±1.7
18:2 ($\Delta^{5,9}$ and $\Delta^{5,11}$)[d]	41.4±0.3
Others[e]	4.7±1.8
Saturated fatty acids	14.9–20.9[f]
Unsaturated fatty acids	79.1–85.1[f]

[a]Data (mean ± standard error of five determinations) from Weeks and Herring (1980), with permission, for membranes prepared from vegetative AX2 cells grown axenically in HL5 medium.

[b]In all fatty acid abbreviations, the number preceding the colon is the chain length, the number following the colon is the number of double bonds, and the numbers following Δ denote the positions of the double bonds. Thus, 5,9-octadecadienoic acid is abbreviated to 18:2$\Delta^{5,9}$ etc.

[c]Derived from the vinyl ether linkage in the plasmalogen form of phosphatidylethanolamine (Davidoff and Korn, 1963a).

[d]Not resolved by the techniques used by Weeks and Herring (1980). Identified in whole cells by Davidoff and Korn (1963a).

[e]Several unidentified minor components.

[f]Range of possible values calculated from upper part of table.

polyenoic fatty acids along with bovine serum albumin to prevent fatty acid toxicity (Weeks, 1976). These cells grow normally but exhibit aberrant development. Aggregate formation is normal, but slug formation and subsequent development is grossly deranged (Weeks, 1976; Mohan Das and Weeks, 1979). Electron spin resonance and fluorescence depolarization analysis indicate that, although the membranes from such supplemented cells incorporate large amounts of the polyenoic acids, the membrane fluidity is not significantly different from that of unsupplemented cells (Herring and Weeks, 1979; Herring *et al.*, 1980). The same result was obtained with aqueous dispersions of phosphatidylcholine purified from supplemented and nonsupplemented cells, suggesting that further increases in the already large degree of unsaturation of these phospholipids has little effect on fluidity. Thus, the mechanism of the derangement of development is presently unknown; possibilities include disruption of specific membrane lipid–protein interactions (Herring *et al.*, 1980).

De Silva and Siu (1980, 1981) have reported that phospholipid synthesis and incorporation into the plasma membrane increases dramatically for a limited period of time during cell aggregation. Phosphatidylethanolamine and phosphatidylserine are synthesized in a ratio comparable to that found in the plasma membrane and are apparently transported to the membrane in small, low-density (high-phospholipid) vesicles (De Silva and Siu, 1981). The authors suggest that this may reflect a need for increased membrane biosynthesis at this time, perhaps to allow for the formation of filopodia which may be required for cell adhesion (Rossomando *et al.*, 1974; Jones *et al.*, 1977; Ryter and Brachet, 1978).

A possible functional role for phospholipids in chemotaxis (see Chapter 4) may emerge from results from Mato's laboratory. Suspended cyclic AMP sensitive cells respond to a cyclic AMP pulse with a transient methylation of a protein of molecular weight 120,000 and demethylation of phospholipid (Mato and Marin-Cao, 1979). Methyl groups can be incorporated from S-adenosylmethionine into mono- and dimethylated phosphatidylethanolamine and phosphatidylcholine in *D. discoideum* lysates (Garcia-Gil *et al.*, 1980); the *in vitro* reaction is enhanced by added calmodulin (Garcia-Gil *et al.*, 1980) and physiological concentrations of cyclic GMP (Alemany *et al.*, 1980). Methylation and demethylation of phospholipids and/or proteins thus may be involved in the machinery of response to and relay of the chemotactic signal, as has been demonstrated in other bacterial and eukaryotic systems (Hirata and Axelrod, 1980; Armitage, 1981).

3. *Glycolipids*

Several other lipid components have been found in whole cells of *D. discoideum* and presumably are found in the plasma membrane. The major glycolipid of *D. discoideum* is a glycosphingolipid containing C_{18}-phytosphingosine (CH_3—$(CH_2)_{13}$—CHOH—CHOH—$CHNH_2$—CH_2OH) and behenic acid (docosanoic acid, $CH_3(CH_2)_{20}COOH$) (Wilhelms *et al.*, 1974), presumably linked to C_{18}-phytosphingosine by an amide linkage with the sphingosine nitrogen. *N*-acetylglucosamine, fucose, and mannose are found in a molar ratio of 12 : 5 : 2. Phosphate and ethanolamine are also present (Wilhelms *et al.*, 1974). The exact structure(s) and extent of heterogeneity of these glycolipid fractions has not been determined. These glycolipids carry a major antigenic determinant or determinants ("antigen I") recognized by antisera raised against whole cells (Gerisch *et al.*, 1969; Wilhelms *et al.*, 1974); however, Fab fragments derived from these antibodies, although they bind well to vegetative and aggregation competent cells, will not block the developmentally regulated EDTA-resistant cell adhesion (contact sites A) (Beug *et al.*, 1970). Another glycolipid, tentatively identified as the sphingolipid glucosylceramide, has been observed (Rössler *et al.*, 1978; Hoffman and McMahon, 1978a; Crean and Rossomando 1979). Hase (1981) has presented evidence that this component is not glucosylceramide but is an approximately equimolar mixture of a (nonglycosylated) ceramide containing hydroxy fatty acids and of a steryl glycoside (see Sec. IV,B,1).

Crean and Rossomando (1977a) and Rössler *et al.* (1978) have identified several mannolipids which appear to be polyprenylphosphate derivatives involved in the membrane-associated synthesis of oligosaccharide side chains for glycoproteins. The rates of synthesis of two of these mannolipids are developmentally regulated (Rössler *et al.*, 1978). The same group has also reported the existence of similar fucosyllipids and *N*-acetylglucosaminyllipids which appear to be involved in oligosaccharide synthesis (Rössler *et al.*, 1981).

C. Carbohydrates

Carbohydrates make up about 3% by weight of isolated plasma membranes of *D. discoideum* (Table IV). The carbohydrate consists about equally of neutral and amino sugars. Individual sugars have been determined by two groups (Table VII). The values obtained are in reasonable agreement except for galactose. Both groups agree that no sialic acid can be detected in membranes or in whole cell lysates. Ribose and glucose may arise from membrane contamination by RNA and by residual glucose from sucrose gradients used to isolate the membranes, repectively. Kilpatrick and Stirling (1978), however, used no sucrose or glucose in preparing membrane ghosts for analysis, and they suggest that glucose may be a true and major component of the plasma membrane polysaccharides. Glucose has also been detected in the purified contact sites A glycoprotein (Müller *et al.*, 1979).

TABLE VII
Sugar Composition of Plasma Membranes

Sugar	nmol/mg protein[a]	mg/mg protein[b]	nmol/10^8 cells[c]	nmol/10^8 cells[d]	nmol/10^8 cells[e]
Neutral sugars					
Fucose	38.4±4.8	0.006	10.8	11.3	11
Ribose	21.0±19.9	0.003	6.5	2.9	7
Mannose	61.8±11.0	0.011	19.0	16.7	19
Galactose	3.5±1.8	0.001	1.1	<0.3	41
Glucose				14.6	21
Rhamnose					3
Unknown					18
Amino sugars					
Glucosamine	135.1±14.4	0.023	41.6	22.9	26
Galactosamine	9.3±3.8	0.002	2.9	N.D.[f]	Trace
Total neutral sugars	124.7	0.021	37.4	45.5	120
Total amino sugars	144.4	0.025	44.5	22.9	26

[a]Data (mean ± standard deviation) from Gilkes *et al.* (1979), with permission, for membranes prepared from vegetative AX2 cells grown axenically in HL5 medium.
[b]Calculated from data in previous column.
[c]Calculated from data in first column as described in Table IV, footnote *b*.
[d]Data calculated from Gilkes *et al.* (1977a) as described in Table IV, footnote *b*.
[e]Data from Kilpatrick and Stirling (1978), with permission, for ghosts (Sussman and Boschwitz, 1975a) prepared from vegetative NC4 cells grown on *Escherichia coli* B/r.
[f]N.D., not detected. Blank means no information given.

Glycoproteins and glycolipids have been discussed in Sections IV,A,3 and IV,B,3 respectively, and it is likely that these species account for most or all of the membrane carbohydrate. Glycosaminoglycans (acid mucopolysaccharides) are found late in development of *Dictyostelium discoideum* (White and Sussman, 1963). They have not been extensively characterized.

Phosphorylated and sulfated sugars have been found in whole *Dictyostelium* extracts and, in particular, on lysosomal hydrolases (Freeze *et al.*, 1980; Freeze and Miller, 1980; Gustafson and Milner, 1980a,b). Ivatt *et al.* (1981) reported that the level of sulfation of whole cell glycopeptides declines dramatically after aggregation, in conjunction with a shift in oligosaccharide processing to more extensive trimming of the precursor oligosaccharide and to fucosylation of peripheral rather than core sugars. Chiarugi *et al.* (1978) reported that sulfate-containing polysaccharides can be released from the surface of *Dictyostelium* cells by trypsinization, although the polysaccharides were not further characterized. The possible occurrence and functional role of modified sugars in membrane carbohydrates clearly requires more study. Mutations which affect the phosphorylation and sulfation of these molecules

(Free *et al.*, 1978; Free and Schimke, 1978; Freeze and Miller, 1980) should be useful for such studies.

Although carbohydrate forms a quantitatively minor fraction of the membrane, its functional importance may be great. For example, interactions between the discoidin(s) and the developmentally regulated cell surface receptors (presumably carbohydrate) have been suggested to be important in cell–cell adhesion (see Chapter 6) or in cell–cell interactions which may be a prerequisite for development (Rahmsdorf *et al.*, 1976; Darmon and Klein, 1978; Marin *et al.*, 1980). There is evidence that glycosyltransferase activities that are involved in synthesis of membrane oligosaccharides are developmentally regulated (Sievers *et al.*, 1978; Rossler *et al.*, 1978; Crean and Rossomando, 1979). The "contact sites A" glycoprotein, which is a target for adhesion-blocking Fab antibody fragments, contains about 25% carbohydrate (Müller *et al.*, 1979), and the target sites *in vivo* for the adhesion-blocking Fab fragments are sensitive to periodate digestion (Beug *et al.*, 1970). The possible role of carbohydrates in cell–cell interaction is currently a very active field of research.

VI. Extracellular Material

A. Extracellular Components of *Dictyostelium* Development

The generation of form during development involves not only autonomous development of the cells and interaction of cells with one another, but also the interaction of cells with the external inanimate environment. An important part of that environment is often provided by materials secreted by the cells themselves. In *Dictyostelium*, the vegetative amoebae seem to have no detectable extracellular material per se; although Dykstra and Aldrich (1978) have demonstrated the staining of a thin cell coat on vegetative cells using Alcian blue and ruthenium red, this may reflect staining of charged carbohydrate groups on integral membrane glycoproteins and glycolipids. In contrast, extracellular materials are of great importance for development. Four types of extracellular structures can be identified. The slime sheath appears immediately following aggregation and covers the migrating pseudoplasmodium (Raper, 1935, 1940). The sheath remains surrounding the stalk (but not the spores) after culmination (Raper and Fennel, 1952; Murata and Ohnishi, 1980). The structure and composition of the sheath, which has received more attention recently than the other extracellular materials, is discussed later. Stalk cells become encased in thick, cellulosic walls. The spore walls are also thick, contain cellulose, and serve to protect the encapsulated spore from harsh environmental conditions (Cotter, 1975; Hohl, 1976). Finally, the spores are encased in an aqueous matrix which holds them in a globular mass, perhaps by surface tension (Raper and Fennel, 1952). The ultrastructure and biology of these structures have been reviewed by Loomis (1975); spore structure has also been reviewed by Cotter (1975) and Hohl (1976).

B. Structure and Composition of the Slug Sheath

Ultrastructurally, the surface sheath of the slug consists of at least two distinct components (Farnsworth and Loomis, 1975). The first is a homogeneous electron dense layer with a trilaminar appearance in favorable sections. It is approximately 75 Å thick at the tip of the pseudoplasmodium and increases in thickness as a linear function of distance from the tip. The second component consists of many electron-dense circular profiles, probably crosssections of fibrillar elements, with diameters of 125–150 Å and space about 10 μm apart in a single layer. This latter component probably represents the urea and sodium dodecyl sulfate (US) insoluble material isolated by Freeze and Loomis (1977a), since aggregates of the mutant strain U1 (which lacks the enzyme UDP–glucose pyrophosphorylase, a necessary enzyme for cellulose biosynthesis) (Dimond *et al.*, 1976) lack cellulose, have no US-insoluble material, and possess the continuous but not the fibrillar component of the sheath (Freeze and Loomis, 1977b). Cellulose accounts for about 95% of the total carbohydrate and about 58% of the dry weight of the US-insoluble component (Freeze and Loomis, 1977a). About 3% of the dry weight of the US-insoluble portion of the sheath is accounted for by a carbohydrate heteropolymer containing *N*-acetylglucosamine, fucose, xylose, mannose, and galactose. Protein accounts for about 15%, sulfate for 1–1.5%, and lipid (fatty–acid containing material) for another 3–5% of the dry weight. The remainder of the dry weight is thought to be due to contaminating insoluble material of nonbiological origin (Freeze and Loomis, 1978). The composition of US-insoluble material from slime trails left behind migrating slugs (Freeze and Loomis, 1977a) and from mature stalks (Freeze and Loomis, 1978) is very similar. The crystallinity of the cellulose in the US-insoluble component is decreased in mutant strains deficient in *N*-acetylglucosaminidase, but not in other strains deficient in α-mannosidase or β-glucosidase. *N*-Acetylglucosaminidase may be involved in the modification (maturation?) of the sheath after its deposition (Watts and Treffry, 1975). *N*-Acetylglucosaminidase-deficient strains form small slugs which migrate poorly (Dimond *et al.*, 1973). Thus, *N*-acetylglucosaminidase, an enzyme which accumulates very early in development and whose accumulation is regulated by the concentration of a soluble factor released by the cells (Grabel and Loomis, 1977, 1978), may be required only much later during development, at the stage of slug migration.

Little is known of the composition and structure of the ultrastructurally-continuous component of the sheath, although preliminary evidence suggests that it may be highly proteinaceous (Smith and Williams, 1979). It is completely solubilized by US (Freeze and Loomis, 1977b). It increases in thickness as a linear function of the length of the slug, suggesting that it may be continuously synthesized by all peripheral cells in the slug. Thus, as the slug moves forward with respect to the sheath (Shaffer, 1965), the sheath at the rear will be the oldest and hence the thickest. This could create a gradient in chemical or physical properties of the sheath which might be important in pattern formation (Ashworth, 1971; Loomis, 1972; Farnsworth and Loomis, 1974, 1975).

VII. Concluding Remarks

There are two main conclusions which stand out from the work discussed earlier. One is that the plasma membrane plays a critical role in many aspects of cell development and interaction in *Dictyostelium*. The other is that the mechanisms by which these functions are carried out are not likely to be found by examining changes in the gross physical and chemical properties of the membrane. Most such properties do not change significantly during development. It seems likely that future fruitful investigations will concern specific molecular species in the membrane, their functions, and their changes during development. Our knowledge of the molecular composition of the plasma membrane at various stages of development is now quite extensive and should provide a solid foundation for this research. Immunological, biochemical, and genetic techniques all promise to be very useful in such studies in the future, as they have been in the past. Indeed, by such means a considerable amount of information has been gained in several specific areas of *Dictyostelium* development related to membranes. These areas will be discussed in the next three chapters.

Acknowledgments

I would like to thank all my colleagues, cited in the text, who have provided me with unpublished work. I would also like to thank Bill Loomis, Jim Morrissey, Jerzy Schmidt, and

Kevin Devine for critical comments on the manuscript. I am supported by NIH Posdoctoral Fellowship GM07241.

References

Aldrich, H. C., and Gregg, J. H. (1973). Unit membrane structural changes following cell association in *Dictyostelium*. *Exp. Cell Res.* **81,** 407–412.

Alemany, S., Garcia-Gil, M., and Mato, J. M. (1980). Regulation by guanosine 3′,5′-cyclic monophosphate of phospholipid methylation during chemotaxis in *Dictyostelium discoideum*. *Proc. Natl. Acad. Sci. U.S.A.* **77,** 6996–6999.

Alton, T. H., and Lodish, H. F. (1977). Synthesis of developmentally regulated proteins in *Dictyostelium discoideum* which are dependent on continued cell–cell interaction. *Dev. Biol.* **60,** 207–216.

Armant, D. R., and Rutherford, C. L. (1979). 5′-AMP nucleotidase is localized in the area of cell–cell contact of prespore and prestalk regions during culmination of *Dictyostelium discoideum*. *Mech. Aging Dev.* **10,** 199–217.

Armant, D. R., and Rutherford, C. L. (1981). Copurification of alkaline phosphatase and 5′-AMP specific nucleotidase in *Dictyostelium discoideum*. *J. Biol. Chem.* **256,** 12710–12718.

Armant, D. R., Stetler, D. A., and Rutherford, C. L. (1980). Cell surface localization of 5′AMP nucleotidase in prestalk cells of *Dictyostelium discoideum*. *J. Cell Sci.* **45,** 119–129.

Armitage, J. (1981). Multiple methylation and bacterial adaptation. *Nature (London)* **289,** 121–122.

Ashworth, J. M. (1971). Cell development in the cellular slime mould *Dictyostelium discoideum*. *Symp. Soc. Exp. Biol.* **25,** 27–49.

Ashworth, J. M., and Watts, D. J. (1970). Metabolism of the cellular slime mould *Dictyostelium discoideum* grown in axenic culture. *Biochem. J.* **119,** 175–182.

Bakke, A. C., and Lerner, R. A. (1981). The cascade of membrane events during development of *Dictyostelium discoideum*. *Subcellular Biochem.* **8,** 75–122.

Bartles, J. R., and Frazier, W. A. (1980). Preparation of ^{125}I-discoidin I and the properties of its binding to *Dictyostelium discoideum* cells. *J. Biol. Chem.* **255,** 30–38.

Bartles, J. R., Santoro, B. C., and Frazier, W. A. (1981). Purification of a high-affinity discoidin I-binding proteoglycan from axenic *Dictyostelium discoideum* growth medium. *Biochim. Biophys. Acta* **674,** 372–382.

Beug, H., Gerisch, G., Kempff, S., Riedel, V., and Cremer, G. (1970). Specific inhibition of cell contact formation in *Dictyostelium* by univalent antibodies. *Exp. Cell Res.* **63,** 147–158.

Beug, H., Katz, F. E., and Gerisch, G. (1973a). Dynamics of antigenic membrane sites relating to cell aggregation in *Dictyostelium discoideum*. *J. Cell Biol.* **56,** 647–658.

Beug, H., Katz, F. E., Stein, A., and Gerisch, G. (1973b). Quantitation of membrane sites in aggregating *Dictyostelium* cells by use of tritiated univalent antibody. *Proc. Natl. Acad. Sci. U.S.A.* **70,** 3150–3154.

Bordier, C., Loomis, W. F., Elder, J., and Lerner, R. (1978). The major developmentally regulated protein complex in membranes of *Dictyostelium*. *J. Biol. Chem.* **253,** 5133–5139.

Bowers, B., and Korn, E. D. (1973). Cytochemical identification of phosphatase activity in the contractile vacuole of *Acanthamoeba castellani*. *J. Cell Biol.* **59,** 784–791.

Breuer, W., and Siu, C.-H. (1981). Identification of endogenous binding proteins for the lectin discoidin-I in *Dictyostelium discoideum*. *Proc. Natl. Acad. Sci. U.S.A.* **78**, 2115–2119.

Brody, T., and Williams, K. L. (1974). Cytological analysis of the parasexual cycle in *Dictyostelium discoideum*. *J. Gen. Microbiol.* **82**, 371–383.

Brunette, D. M., and Till, J. E. (1971). A rapid method for the isolation of L-cell surface membranes using an aqueous two-phase polymer system. *J. Membr. Biol.* **5**, 215–224.

Burns, R. A., Livi, G. P., and Dimond, R. L. (1981). Regulation and secretion of early developmentally controlled enzymes during axenic growth in *Dictyostelium discoideum*. *Dev. Biol.* **83**, 407–416.

Burridge, K., and Jordan, L. (1979). The glycoproteins of *Dictyostelium discoideum*. Changes during development. *Exp. Cell Res.* **124**, 31–38.

Chagla, A. N., Lewis, K. E. and O'Day, D. H. (1980). Ca^{2+} and cell fusion during sexual development in liquid cultures of *Dictyostelium discoideum*. *Exp. Cell Res.* **126**, 501–505.

Chang, C.-M., Rosen, S. D., and Barondes, S. H. (1977). Cell surface location of an endogenous lectin and its receptor in *Polysphondylium pallidum*. *Exp. Cell Res.* **104**, 101–109.

Chapman-Andresen, C. (1977). Endocytosis in freshwater amebas. *Physiol. Rev.* **57**, 371–385.

Chiarugi, V. P., Del Rosso, M., Cappelletti, R., Vannucchi, S., Cella, C., Fibbi, G., and Urbano, P. (1978). Sulphated polysaccharides and the differentiation of the cellular slime mould *Dictyostelium discoideum*. *Caryologia* **31**, 183–190.

Chung, S., Landfear, S. M., Blumberg, D. D., Cohen, N. S., and Lodish, H. F. (1981). Synthesis and stability of developmentally regulated *Dictyostelium* mRNAs are affected by cell–cell contact and cAMP. *Cell* **24**, 785–797.

Clarke, M., Schatten, G., Mazia, D., and Spudich, J. A. (1975). Visualization of actin filaments associated with the cell membrane in amoebae of *Dictyostelium discoideum*. *Proc. Natl. Acad. Sci. U.S.A.* **72**, 1758–1762.

Coffman, D. S., Leichtling, B. H., and Rickenberg, H. V. (1981). Phosphoproteins in *Dictyostelium discoideum*. *J. Supramol. Struct. Cell. Biochem.*, **15**, 369–385.

Condeelis, J. (1979). Isolation of concanavalin A caps during various stages of formation and their association with actin and myosin. *J. Cell Biol.* **80**, 751–758.

Conrad, M. J., and Singer, S. J. (1979). Evidence for a large internal pressure in biological membranes. *Proc. Natl. Acad. Sci. U.S.A.* **76**, 5202–5206.

Conrad, M. J., and Singer, S. J. (1981). The solubility of amphipathic molecules in biological membranes and lipid bilayers and its implications for membrane structure. *Biochemistry* **20**, 808–818.

Cotter, D. A. (1975). Spores of the cellular slime mold *Dictyostelium discoideum*. *In* "Spores VI" (P. Gerhardt, R. N. Costilow, and H. L. Sadoff, eds.), pp. 61–72. American Society for Microbiology, Washington, D.C.

Crean, E. V., and Rossomando, E. F. (1977a). Synthesis of a mannosyl phosphoryl polyprenol by the cellular slime mold *Dictyostelium discoideum*. *Biochim. Biophys. Acta* **498**, 439–441.

Crean, E. V., and Rossomando, E. F. (1977b). Developmental changes in membrane-bound enzymes of *Dictyostelium discoideum* detected by concanavalin A-Sepharose affinity chromatography. *Biochem. Biophys. Res. Commun.* **75**, 488–495.

Crean, E. V., and Rossomando, E. F. (1979). Glucosphingolipid synthesis in the cellular slime mold *Dictyostelium discoideum*. *Arch. Biochem. Biophys.* **196**, 186–191.

Cutler, L. S., and Rossomando, E. F. (1975). Localization of adenylate cyclase in *Dictyostelium discoideum*. II. Cytochemical studies on whole cells and isolated plasma membrane vesicles. *Exp. Cell Res.* **95**, 79–87.

Darmon, M., and Klein, C. (1976). Binding of concanavalin A and its effect on the differentiation of *Dictyostelium discoideum*. *Biochem. J.* **154**, 743–750.

Darmon, M., and Klein, C. (1978). Effects of amino acids and glucose on adenylate cyclase and cell differentiation of *Dictyostelium discoideum*. *Dev. Biol.* **63,** 377–389.

Davidoff, F., and Korn, E. D. (1962). Lipids of *Dictyostelium discoideum:* phospholipid composition and the presence of two new fatty acids: *cis,cis*-5,11-octadecadienoic and *cis,cis*-5,9-hexadecadienoic acids. *Biochem. Biophys. Res. Commun.* **9,** 54–58.

Davidoff, F., and Korn, E. (1963a). Fatty acid and phospholipid composition of the cellular slime mold, *Dictyostelium discoideum. J. Biol. Chem.* **238,** 3199–3209.

Davidoff, F., and Korn, E. (1963b). The biosynthesis of fatty acids in the cellular slime mold, *Dictyostelium discoideum. J. Biol. Chem.* **238,** 3210–3215.

De Chastellier, C., and Ryter, A. (1977). Changes of the cell surface and of the digestive apparatus of *Dictyostelium discoideum* during the starvation period triggering aggregation. *J. Cell Biol.* **75,** 218–236.

De Chastellier, C., and Ryter, A. (1980). Characteristic ultrastructural transformation upon starvation of *Dictyostelium discoideum* and their relations with aggregation. Study of wild type amoebae and aggregation mutants. *Biol. Cellulaire* **38,** 121–128.

De Chastellier, C., Quiviger, B., and Ryter, A. (1978). Observation on the functioning of the contractile vacuole of *Dictyostelium discoideum* with the electron miscroscope. *J. Ultrastruct. Res.* **62,** 220–227.

De Silva, N. S., and Siu, C.-H. (1980). Preferential incorporation of phospholipids into plasma membranes during cell aggregation of *Dictyostelium discoideum. J. Biol. Chem.* **255,** 8489–8496.

De Silva, N. S., and Siu, C.-H. (1981). Vesicle-mediated transfer of phospholipids to plasma membrane during cell aggregation of *Dictyostelium discoideum. J. Biol. Chem.* **256,** 5845–5850.

Dimond, R. L., Brenner, M., and Loomis, W. F. (1973). Mutations affecting *N*-acetylglucosaminidase in *Dictyostelium discoideum. Proc. Natl. Acad. Sci. U.S.A.* **70,** 3356–3360.

Dimond, R. L., Farnsworth, P. A., and Loomis, W. F. (1976). Isolation and characterization of mutations affecting UDPG pyrophosphorylase activity in *Dictyostelium discoideum*. *Dev. Biol.* **50,** 169–181.

Dimond, R. L., Burns, R. A., and Jordan, K. B. (1981). Secretion of lysosomal enzymes in the cellular slime mold, *Dictyostelium discoideum. J. Biol. Chem.* **256,** 6565–6572.

Dykstra, M. J., and Aldrich, H. C. (1978). Successful demonstration of an elusive cell coat in amebae. *J. Protozool.* **25,** 38–41.

Edelson, P. J., and Cohn, Z. A. (1978). Endocytosis: regulation of membrane interactions. *Cell Surf. Rev.* **5,** 387–405.

Elder, J. H., Pickett, R. A., Hampton, J., and Lerner, R. A. (1977). Radioiodination of proteins in single polyacrylamide gel slices. Tryptic peptide analysis of all the major members of complex multicomponent systems using microgram quantities of total protein. *J. Biol. Chem.* **252,** 6510–6515.

Ellingson, J. S. (1974). Changes in the phospholipid composition in the differentiating cellular slime mold, *Dictyostelium discoideum. Biochim. Biophys. Acta* **337,** 60–67.

Ellingson, J. S. (1980). Identification of *N*-acylethanolamine phosphoglycerides and acylphosphatidylglycerol as the phospholipids which disappear as *Dictyostelium discoideum* cells aggregate. *Biochemistry* **19,** 6176–6182.

Ellouz, R., and Lenfant, M. (1971). Biosynthese de la chaine laterale ethyle du stigmastanol et du stigmasten-22-ol-3β du myxomycete *Dictyostelium discoideum. Eur. J. Biochem.* **23,** 544–550.

Erdos, G. W., and Hohl, H. R. (1980). Freeze-fracture examination of the plasma membrane of the cellular slime mould *Polysphondylium pallidum* during microcyst formation and germination. *Cytobios* **29,** 7–16.

Farnsworth, P. A., and Loomis, W. F. (1974). A barrier to diffusion in pseudoplasmodia of *Dictyostelium discoideum*. *Dev. Biol.* **41,** 77–83.

Farnsworth, P. A., and Loomis, W. F. (1975). A gradient in the thickness of the sheath in pseudoplasmodia of *Dictyostelium discoideum*. *Dev. Biol.* **46,** 349–357.

Favard-Sereno, C., and Livrozet, M. (1979). Plasma membrane structural changes correlated with the acquisition of aggregation competence in *Dictyostelium discoideum*. *Biol. Cellulaire* **35,** 45–54.

Favard-Sereno, C., Ludosky, M.-A., and Ryter, A. (1981). Freeze-fracture study of phagocytosis in *Dictyostelium discoideum*. *J. Cell Sci.* **51,** 63–84.

Ferber, E., Munder, P. G., Fischer, H., and Gerisch, G. (1970). High phospholipase activities in amoebae of *Dictyostelium discoideum*. *Eur. J. Biochem.* **14,** 253–257.

Filosa, M. F. (1978). Concanavalin–mediated attachment to membranes of a cellular slime mould cyclic AMP phosphodiesterase. *Differentiation* **10,** 177–180.

Franke, J., and Kessin, R. (1977). A defined minimal medium for axenic strains of *Dictyostelium discoideum*. *Proc. Natl. Acad. Sci. U.S.A.* **74,** 2157–2161.

Free, S. J., and Schimke, R. T. (1978). Effects of a post-translational modification mutation on different developmentally regulated glycosidases in *Dictyostelium discoideum*. *J. Biol. Chem.* **253,** 4107–4111.

Free, S. J., Schimke, R. T., Freeze, H., and Loomis, W. F. (1978). Characterization and genetic mapping of *modA*. A mutation in the post–translational modification of the glycosidases of *Dictyostelium discoideum*. *J. Biol. Chem.* **253,** 4102–4106.

Freeze, H., and Loomis, W. F. (1977a). Isolation and characterization of a component of the surface sheath of *Dictyostelium discoideum*. *J. Biol. Chem.* **252,** 820–824.

Freeze, H., and Loomis, W. F. (1977b). The role of the fibrillar component of the surface sheath in the morphogenesis of *Dictyostelium discoideum*. *Develop. Biol.* **56,** 184–194.

Freeze, H., and Loomis, W. F. (1978). Chemical analysis of stalk components of *Dictyostelium discoideum*. *Biochim. Biophys. Acta* **539,** 529–537.

Freeze, H. H., and Miller, A. L. (1980). ModA: a post-translational mutation affecting phosphorylated and sulfated glycopeptides in *Dictyostelium discoideum*. *Mol. Cell. Biochem.* **35,** 17–27.

Freeze, H. H., Miller, A. L., and Kaplan, A. (1980). Acid hydrolases from *Dictyostelium discoideum* contain phosphomannosyl recognition markers. *J. Biol. Chem.* **255,** 11081–11084.

Garcia-Gil, M., Alemany, S., Marin Cao, D., Castano, J. G., and Mato, J. M. (1980). Calmodulin modulates phospholipid methylation in *Dictyostelium discoideum*. *Biochem. Biophys. Res. Commun.* **94,** 1325–1330.

Garrod, D. R., and Gingell, D. (1970). A progressive change in the electrophoretic mobility of preaggregation cells of the slime mould, *Dictyostelium discoideum*. *J. Cell Sci.* **6,** 277–284.

Geltosky, J. E., Siu, C.-H., and Lerner, R. A. (1976). Glycoproteins of the plasma membrane of *Dictyostelium discoideum* during development. *Cell* **8,** 391–396.

Geltosky, J. E., Weseman, J., Bakke, A., and Lerner, R. A. (1979). Identification of a cell surface glycoprotein involved in cell aggregation in *D. discoideum*. *Cell* **18,** 391–398.

Geltosky, J. E., Birdwell, C. R., Weseman, J., and Lerner, R. A. (1980). A glycoprotein involved in aggregation of *D. discoideum* is distributed on the cell surface in a nonrandom fashion favoring cell junctions. *Cell* **21,** 339–345.

George, R. P., Hohl, H. R., and Raper, K. B. (1972). Ultrastructural development of stalk-producing cells in *Dictyostelium discoideum*, a cellular slime mould. *J. Gen. Microbiol.* **70,** 477–489.

Gerisch, G. (1961). Zellfunktionen und zellfunctionswechsel in der entwicklung von *Dictyostelium discoideum*. *Exp. Cell Res.* **25,** 535–554.

Gerisch, G., Malchow, D., Wilhelms, H., and Lüderitz, O. (1969). Artspezifität polysaccharid–haltiger zellmembran-antigen von *Dictyostelium discoideum*. *Eur. J. Biochem.* **9**, 229–236.

Gezelius, K., and Wright, B. E. (1965). Alkaline phosphatase in *Dictyostelium discoideum*. *J. Gen. Microbiol.* **38**, 309–327.

Gilkes, N. R., and Weeks, G. (1977a). The purification and characterization of *Dictyostelium discoideum* plasma membrane. *Biochim. Biophys. Acta* **464**, 142–156.

Gilkes, N. R., and Weeks, G. (1977b). An improved procedure for the purification of plasma membranes from *Dictyostelium discoideum*. *Can. J. Biochem.* **55**, 1233–1236.

Gilkes, N. R., Laroy, K., and Weeks, G. (1979). An analysis of the protein, glycoprotein and monosaccharide composition of *Dictyostelium discoideum* plasma membranes during development. *Biochim. Biophys. Acta* **551**, 349–362.

Gillette, M., and Filosa, M. (1973). Effect of concanavalin A on cellular slime mold development: premature appearance of membrane-bound cyclic AMP phosphodiesterase. *Biochem. Biophys. Res. Commun.* **53**, 1159–1166.

Gillette, M. U., Dengler, R. E., and Filosa, M. F. (1974). The localization and fate of concanavalin A in amoebae of the cellular slime mold, *Dictyostelium discoideum*. *J. Exp. Zool.* **190**, 243–248.

Gingell, D., and Garrod, D. R. (1969). Effect of EDTA on electrophoretic mobility of slime mould cells and its relationship to current theories of cell adhesion. *Nature (London)* **221**, 192–193.

Githens, S., and Karnovsky, M. L. (1973). Phagocytosis by the cellular slime mold *Polysphondylium pallidum* during growth and development. *J. Cell Biol.* **58**, 536–548.

Glynn, P. J. (1981). A quantitative study of the phagocytosis of *Escherichia coli* by myxamoebae of the slime mould *Dictyostelium discoideum*. *Cytobios* **30**, 153–166.

Gottlieb, D. I., and Glaser, L. (1980). Cellular recognition during neural development. *Annu. Rev. Neurosci.* **3**, 303–318.

Grabel, L. B., and Farnsworth, P. A. (1977). The endocytosis of concanavalin A by *Dictyostelium discoideum* cells. *Exp. Cell Res.* **105**, 285–289.

Grabel, L., and Loomis, W. F. (1977). Cellular interaction regulating early biochemical differentiation in *Dictyostelium*. *In* "Development and Differentiation in the Cellular Slime Moulds" (P. Cappuccinelli and J. Ashworth, eds.), pp. 189–199. Elsevier, North-Holland, New York.

Grabel, L., and Loomis, W. F. (1978). Effector controlling accumulation of N-acetylglucosaminidase during development of *Dictyostelium discoideum*. *Dev. Biol.* **64**, 202–209.

Green, A. A., and Newell, P. C. (1974). The isolation and subfractionation of plasma membrane from the cellular slime mould *Dictyostelium discoideum*. *Biochem. J.* **140**, 313–332.

Gregg, J. H. (1971). Developmental potential of isolated *Dictyostelium* myxamoebae. *Dev. Biol.* **26**, 478–485.

Gregg, J. H., and Nesom, M. G. (1973). Response of *Dictyostelium* plasma membrane to adenosine 3′,5′-cyclic monophosphate. *Proc. Natl. Acad. Sci. U.S.A.* **70**, 1630–1633.

Gregg, J. H., and Yu, N. Y. (1975). *Dictyostelium* aggregate-less mutant plasma membranes. *Exp. Cell Res.* **96**, 283–286.

Gross, J. D., Town, C. D., Brookman, J. J., Jermyn, K. A., Peacey, M. J., and Kay, R. R. (1981). Cell patterning in *Dictyostelium*. *Phil. Trans. Roy. Soc. Lond.* **B295**, 497–508.

Gustafson, G. L., and Milner, L. A. (1980a). Immunological relationship between β-N-acetylglucosaminidase and proteinase I from *Dictyostelium discoideum*. *Biochem. Biophys. Res. Commun.* **94**, 1439–1444.

Gustafson, G. L., and Milner, L. A. (1980b). Occurrence of N-acetylglucosamine-1-phosphate in proteinase I from *Dictyostelium discoideum*. *J. Biol. Chem.* **255**, 7208–7210.

Hase, A. (1981). Isolation and characterization of glycolipid from *Dictyostelium discoideum*. *Arch. Biochem. Biophys.* **210,** 280–288.

Heftmann, E., Wright, B. E., and Liddel, G. U. (1959). Identification of a sterol with acrasin activity in a slime mold. *J. Amer. Chem. Soc.* **81,** 6525–6526.

Heftmann, E., Wright, B. E., and Liddel, G. U. (1960). The isolation of Δ^{22}-stigmasten-3β-ol from *Dictyostelium discoideum*. *Arch. Biochem. Biophys.* **91,** 266–270.

Hellio, R., and Ryter, A. (1980). Relationships between anionic sites and lectin receptors in the plasma membrane of *Dictyostelium discoideum* and their role in phagocytosis. *J. Cell Sci.* **41,** 89–104.

Herring, F. G., and Weeks, G. (1979). Analysis of *Dictyostelium discoideum* plasma membrane fluidity by electron spin resonance. *Biochim. Biophys. Acta* **552,** 66–77.

Herring, F. G., Tatischeff, I., and Weeks, G. (1980). The fluidity of plasma membranes of *Dictyostelium discoideum*. The effects of polyunsaturated fatty acid incorporation assessed by fluorescence depolarization and electron paramagnetic resonance. *Biochim. Biophys. Acta* **602,** 1–9.

Hintermann, R., and Parish, R. W. (1979). Synthesis of plasma membrane proteins and antigens during development of the cellular slime mold *Polysphondylium pallidum*. *FEBS Lett.* **108,** 219–225.

Hirata, F., and Axelrod, J. (1980). Phospholipid methylation and biological signal transmission. *Science* **209,** 1082–1090.

Hoffman, S., and McMahon, D. (1977). The role of the plasma membrane in the development of *Dictyostelium discoideum*. II. Developmental and topographic analysis of polypeptide and glycoprotein composition. *Biochim. Biophys. Acta* **465,** 242–259.

Hoffman, S., and McMahon, D. (1978a). Defective glycoproteins in the plasma membrane of an aggregation minus mutant of *Dictyostelium discoideum* with abnormal cellular interactions. *J. Biol. Chem.* **253,** 278–287.

Hoffman, S., and McMahon, D. (1978b). The effects of inhibition of development in *Dictyostelium discoideum* on changes in plasma membrane composition and topography. *Arch. Biochem. Biophys.* **187,** 12–24.

Hohl, H. R. (1965). Nature and development of membrane systems in food vacuoles of cellular slime molds predatory upon bacteria. *J. Bacteriol.* **90,** 755–765.

Hohl, H. R. (1976). Myxomycetes. *In* "The Fungal Spore: Form and Function" (D. J. Weber and W. M. Hess, eds.), pp. 463–500. Wiley, New York.

Hohl, H. R., and Raper, K. B. (1963). Nutrition of cellular slime molds. I. Growth on living and dead bacteria. *J. Bacteriol.* **85,** 191–198.

Hohl, H. R., Bühlmann, M., and Wehrli, E. (1978). Plasma membrane alterations as a result of heat activation in *Dictyostelium* spores. *Arch. Microbiol.* **116,** 239–244.

Hülser, D. F., and Webb, D. J. (1973). The use of the tip potential of glass microelectrodes in the determination of low cell membrane potentials. *Biophysik* **10,** 273–280.

Irvine, R. F., Letcher, A. J., Brophy, P. J., and North, M. J. (1980). Phosphatidylinositol-degrading enzymes in the cellular slime mould *Dictyostelium discoideum*. *J. Gen. Microbiol.* **121,** 495–497.

Ivatt, R., Das, P., Henderson, E., and Robbins, P. (1981). Developmental regulation of glycoprotein biosynthesis in *Dictyostelium*. *J. Supramol. Struct. Cell. Biochem.* **17,** 359–368.

Jacobson, B. S. (1980a). Improved method for isolation of plasma membrane on cationic beads. Membranes from *Dictyostelium discoideum*. *Biochim. Biophys. Acta* **600,** 769–780.

Jacobson, B. S. (1980b). Actin binding to the cytoplasmic surface of the plasma membrane isolated from *Dictyostelium discoideum*. *Biochem. Biophys. Res. Commun.* **97,** 1493–1498.

Jaffe, A. R., and Garrod, D. R. (1979). Effect of isolated plasma membranes on cell cohesion in the cellular slime mould. *J. Cell Sci.* **40**, 245–256.

Johnson, D. F., Wright, B. E., and Heftmann, E. (1962). Biogenesis of Δ^{22}-stigmasten-3β-ol in *Dictyostelium discoideum*. *Arch. Biochem. Biophys.* **97**, 232–235.

Jones, G. E., Pacy, J., Jermyn, K., and Stirling, J. (1977). A requirement for filopodia in the adhesion of pre-aggregative cells of *Dictyostelium discoideum*. *Exp. Cell Res.* **107**, 451–455.

Kawai, S., and Tanaka, K. (1978). Spin–labeling studies on the membranes of differentiating cells of *Dictyostelium discoideum*. *Cell Struct. Funct.* **3**, 31–37.

Kay, R. R., Town, C. D., and Gross, J. D. (1979). Cell differentiation in *Dictyostelium discoideum*. *Differentiation* **13**, 7–14.

Kilpatrick, D. C., and Stirling, J. L. (1978). Carbohydrate composition of cells and plasma membranes of *Dictyostelium discoideum* at selected stages of development. *Biochim. Biophys. Acta* **543**, 357–363.

Kilpatrick, D. C., Schmidt, J. A., Stirling, J. L., Pacy, J., and Jones, G. E. (1980). The effect of chymotrypsin on the development of *Dictyostelium discoideum*. *J. Embryol. Exp. Morphol.* **57**, 189–201.

Knecht, D. A., and Dimond, R. L. (1981). Lysosomal enzymes possess a common antigenic determinant in the cellular slime mold. *Dictyostelium discoideum*. *J. Biol. Chem.* **256**, 3564–3575.

Krivanek, J. O. (1956). Alkaline phosphatase activity in the developing slime mold, *Dictyostelium discoideum* Raper. *J. Exp. Zool.* **133**, 459–480.

Krivanek, J. O., and Krivanek, R. C. (1958). The histochemical localization of certain biochemical intermediates and enzymes in the developing slime mold, *Dictyostelium discoideum* Raper. *J. Exp. Zool.* **137**, 89–115.

Kuhn, H., and Parish, R. W. (1981). Fusion of *Dictyostelium* cells with one another and with erythrocyte ghosts. *Exp. Cell Res.* **131**, 89–96.

Lam, T. Y., and Siu, C-H. (1981). Synthesis of stage–specific glycoproteins in *Dictyostelium discoideum* during development. *Dev. Biol.* **83**, 127–137.

Lam, T. Y., Pickering, G., Geltosky, J., and Siu, C.-H. (1981). Differential cell cohesiveness expressed by prespore and prestalk cells of *Dictyostelium discoideum*. *Differentiation* **20**, 22–28.

Landfear, S. M., and Lodish, H. F. (1980). A role for cyclic AMP in expression of developmentally regulated genes in *Dictyostelium discoideum*. *Proc. Natl. Acad. Sci. U.S.A.* **77**, 1044–1048.

Lee, A., Chance, K., Weeks, C., and Weeks, G. (1975). Studies on the alkaline phosphatase and 5′-nucleotidase of *Dictyostelium discoideum*. *Arch. Biochem. Biophys.* **171**, 407–417.

Lee, K-C. (1972a). Permeability of *Dictyostelium discoideum* towards amino acids and inulin: a possible relationship between initiation of differentiation and loss of "pool" metabolites. *J. Gen. Microbiol.* **72**, 457–471.

Lee, K-C. (1972b). Cell electrophoresis of the cellular slime mould *Dictyostelium discoideum*. I. Characterization of some of the cell surface ionogenic groups. *J. Cell Sci.* **10**, 229–248.

Lee, K-C. (1972c). Cell electrophoresis of the cellular slime mould *Dictyostelium discoideum*. II. Relevance of the changes in cell surface charge density to cell aggregation and morphogenesis. *J. Cell Sci.* **10**, 249–265.

Lenfant, M., Ellouz, R., Das, B. C., Zissman, E., and Lederer, E. (1969). Sur la biosynthese de la chaine laterale ethyle des sterols du myxomycete *Dictyostelium discoideum*. *Eur. J. Biochem.* **7**, 159–164.

Long, B. H., and Coe, E. L. (1974). Changes in neutral lipid constituents during differentiation of the cellular slime mold, *Dictyostelium discoideum*. *J. Biol. Chem.* **249**, 521–529.

Loomis, W. F. (1969). Developmental regulation of alkaline phosphatase in *Dictyostelium discoideum*. *J. Bacteriol.* **100,** 417–422.

Loomis, W. F. (1971). Sensitivity of *Dictyostelium discoideum* to nucleic acid analogues. *Exp. Cell Res.* **64,** 484–486.

Loomis, W. F. (1972). Role of the surface sheath in the control of morphogenesis in *Dictyostelium discoideum*. *Nature (London) New Biol.* **240,** 6–9.

Loomis, W. F. (1975). "*Dictyostelium discoideum*, A Developmental System." Academic Press, New York.

Loomis, W. F. (1980). Genetic analysis of development in *Dictyostelium*. *In* "The Molecular Genetics of Development" (T. Leighton and W. F. Loomis, eds.), pp. 179–212. Academic Press, New York.

Loomis, W. F., and Ashworth, J. M. (1968). Plaque–size mutants of the cellular slime mould *Dictyostelium discoideum*. *J. Gen. Microbiol.* **53,** 181–186.

Loomis, W. F., Morrissey, J., and Lee, M. (1978). Biochemical analysis of pleiotropy in *Dictyostelium*. *Dev. Biol.* **63,** 243–246.

Luna, E. J., Fowler, V. M., Swanson, J., Branton, D., and Taylor, D. L. (1981). A membrane cytoskeleton from *Dictyostelium discoideum*. I. Identification and partial characterization of an actin–binding activity. *J. Cell Biol.* **88,** 396–409.

MacLeod, C. L., and Loomis, W. F. (1979). Biochemical and genetic analysis of a mutant with altered alkaline phosphatase activity in *Dictyostelium discoideum*. *Dev. Genet.* **1,** 109–121.

MacWilliams, H. K., and Bonner, J. T. (1979). The prestalk–prespore pattern in cellular slime molds. *Differentiation* **14,** 1–22.

Maeda, Y. (1980). Changes in charged groups on the cell surface during development of the cellular slime mold *Dictyostelium discoideum:* an electron microscopic study. *Dev. Growth Diff.* **22,** 679–685.

Maeda, Y., and Eguchi, G. (1977). Polarized structures of cells in the aggregating cellular slime mold *D. discoideum:* an electron microscope study. *Cell Struct. Funct.* **2,** 159–169.

Maeda, Y., and Takeuchi, I. (1969). Cell differentiation and fine structures in the development of the cellular slime molds. *Dev. Growth Diff.* **11,** 232–245.

Marin, F. T., Goyette-Boulay, M., and Rothman, F. G. (1980). Regulation of development in *Dictyostelium discoideum*. III. Carbohydrate–specific intercellular interactions in early development. *Dev. Biol.* **80,** 301–312.

Mato, J. M., and Marin-Cao, D. (1979). Protein and phospholipid methylation during chemotaxis in *Dictyostelium discoideum* and its relationship to calcium movements. *Proc. Natl. Acad. Sci. U.S.A.* **76,** 6106–6109.

McMahon, D., Hoffman, S., Fry, W., and West, C. (1975). The involvement of the plasma membrane in the development of *Dictyostelium discoideum*. *In:* "Development Biology", (W. A. Benjamin, ed.), pp. 60–75. ICN/UCLA Symp. Mol. Cell. Biol. 2, Menlo Park, CA.

McMahon, D., Miller, M., and Long, S. (1977). The involvement of the plasma membrane in the development of *Dictyostelium discoideum*. I. Purification of the plasma membrane. *Biochim. Biophys. Acta* **465,** 224–241.

McMillan, P. N., and Luftig, R. B. (1975). Preservation of membrane ultrastructure with aldehyde or imidate fixatives. *J. Ultrastruct. Res.* **52,** 243–260.

Mohan Das, D. V., and Weeks, G. (1979). Effects of polyunsaturated fatty acids on the growth and differentiation of the cellular slime mould, *Dictyostelium discoideum*. *Exp. Cell Res.* **118,** 237–243.

Mohan Das, D. V., and Weeks, G. (1980). Reversible heat activation of alkaline phosphatase of *Dictyostelium discoideum* and its developmental implication. *Nature (London)* **288,** 166–167.

Mohan Das, D. V., and Weeks, G. (1981). The inhibition of *Dictyostelium discoideum* alkaline phosphatase by a low molecular weight factor and its implication for the developmental regulation of the enzyme. *FEBS Lett.* **130,** 249–252.

Mohan Das, D. V., Herring, F. G., and Weeks, G. (1980). The effect of growth temperature on the lipid composition and differentiation of *Dictyostelium discoideum. Can. J. Microbiol.* **26,** 796–799.

Molday, R., Jaffe, R., and McMahon, D. (1976). Concanavalin A and wheat germ agglutinin receptors on *Dictyostelium discoideum. J. Cell Biol.* **71,** 314–322.

Müller, K., and Gerisch, G. (1978). A specific glycoprotein as the target site of adhesion blocking Fab in aggregating *Dictyostelium* cells. *Nature (London)* **274,** 445–449.

Müller, K., Gerisch, G., Fromme, I., Mayer, H., and Tsugita, A. (1979). A membrane glycoprotein of aggregating *Dictyostelium* cells with the properties of contact sites. *Eur. J. Biochem.* **99,** 419–426.

Murata, Y., and Ohnishi, T. (1980). *Dictyostelium discoideum* fruiting bodies observed by scanning electron microscopy. *J. Bacteriol.* **141,** 956–958.

Murray, B. A., Yee, L. D., and Loomis, W. F. (1981). Immunological analysis of a glycoprotein (contact sites A) involved in intercellular adhesion of *Dictyostelium discoideum. J. Supramol. Struct. Cell. Biochem.* **17,** 197–211.

Neumann, E., Gerisch, G., and Opatz, K. (1980). Cell fusion induced by high electric impulses applied to *Dictyostelium. Naturwissenschaften* **67,** 414–415.

Newell, P. C. (1977). Aggregation and cell surface receptors in cellular slime molds. *In* "Microbial Interactions. Receptors and Recognition Series B, Vol. 3", pp. 1–57. Chapman and Hall, London.

Newell, P. C. (1978). Genetics of the cellular slime molds. *Annu. Rev. Genet.* **12,** 69–93.

North, M. J., and Williams, K. L. (1978). Relationship between the axenic phenotype and sensitivity to ω-aminocarboxylic acids in *Dictyostelium discoideum. J. Gen. Microbiol.* **107,** 223–230.

O'Day, D. H., and Lewis, K. E. (1981). Pheromonal interactions during mating in *Dictyostelium. In* "Sexual Interactions in Eukaryotic Microbes", pp. 199–221. Academic Press, New York.

Ono, H., Kobayashi, K., and Yanagisawa, K. (1972). Cell fusion in the cellular slime mold, *Dictyostelium discoideum. J. Cell Biol.* **54,** 665–666.

Ono, K.-I., Ochiai, H., and Toda, K. (1978). "Ghosts" formation and their isolation from the cellular slime mold *Dictyostelium discoideum. Exp. Cell Res.* **112,** 175–185.

Ono, K.-I., Toda, K., and Ochiai, H. (1981). Drastic changes in accumulation and synthesis of plasma-membrane proteins during aggregation of *Dictyostelium discoideum. Eur. J. Biochem.* **119,** 133–143.

Orlowski, M., and Loomis, W. F. (1979). Plasma membrane proteins of *Dictyostelium:* the spore coat proteins. *Dev. Biol.* **71,** 297–307.

Parish, R. W. (1979). Cyclic AMP induces the synthesis of developmentally regulated plasma membrane proteins in *Dictyostelium. Biochim. Biophys. Acta* **553,** 179–182.

Parish, R. W., and Müller, U. (1976). The isolation of plasma membranes from the cellular slime mold *Dictyostelium discoideum* using concanavalin A and Triton X-100. *FEBS Lett.* **63,** 40–44.

Parish, R. W., and Schmidlin, S. (1979a). Synthesis of plasma membrane proteins during development of *Dictyostelium discoideum. FEBS Lett.* **98,** 251–256.

Parish, R. W., and Schmidlin, S. (1979b). Resynthesis of developmentally regulated plasma membrane proteins following disaggregation of *Dictyostelium* pseudoplasmodia. *FEBS Lett.* **99,** 270–274.

Parish, R. W., Müller, U., and Schmidlin, S. (1977). Phosphorylation of plasma membrane proteins in *Dictyostelium discoideum. FEBS Lett.* **79,** 393–395.

Parish, R. W., Schmidlin, S., and Parish, C. R. (1978a). Detection of developmentally controlled plasma membrane antigens of *Dictyostelium discoideum* cells in SDS-polyacrylamide gels. *FEBS Lett.* **95,** 366–370.

Parish, R. W., Schmidlin, S., and Weibel, M. (1978b). Effect of cyclic AMP pulses on the synthesis of plasma membranes proteins in aggregateless mutants of *Dictyostelium discoideum. FEBS Lett.* **96,** 283–286.

Patterson, D. J. (1980). Contractile vacuoles and associated structures: their organization and function. *Biol. Rev.* **55,** 1–46.

Poff, K. L., and Whitaker, B. D. (1979). Movement of slime molds. *In* "Physiology of Movements." Encyclopedia of Plant Physiology, New Series, Vol. 7 (W. Haupt and M. E. Feinleib, eds.), pp. 355–382. Springer–Verlag, New York.

Quiviger, B., de Chastellier, C., and Ryter, A. (1978). Cytochemical demonstration of alkaline phosphatase in the contractile vacuole of *Dictyostelium discoideum. J. Ultrastruct. Res.* **62,** 228–236.

Quiviger, B., Benichou, J.-C., and Ryter, A. (1980). Comparative cytochemical localization of alkaline and acid phosphatases during starvation and differentiation of *Dictyostelium discoideum. Biol. Cellulaire* **37,** 241–250.

Rahmsdorf, H. J., Cailla, H. L., Spitz, E., Moran, M. J., and Rickenberg, H. V. (1976). Effect of sugars on early biochemical events in development of *Dictyostelium discoideum. Proc. Natl. Acad. Sci. U.S.A.* **73,** 3183–3187.

Raper, K. B. (1935). *Dictyostelium discoideum,* a new species of slime mold from decaying forest leaves. *J. Agric. Res.* **50,** 135–147.

Raper, K. B. (1940). Pseudoplasmodium formation and organization in *Dictyostelium discoideum. J. E. Mitchell Sci. Soc.* **56,** 241–282.

Raper, K. B., and Fennell, D. I. (1952). Stalk formation in *Dictyostelium. Bull. Torrey Bot. Club* **79,** 25–51.

Ray, J., and Lerner, R. A. (1982). A biologically active receptor for the carbohydrate-binding protein(s) of *Dictyostelium discoideum. Cell* **28,** 91–98.

Reitherman, R. W., Rosen, S. D., Frazier, W. A., and Barondes, S. H. (1975). Cell surface species–specific high affinity receptors for discoidin: developmental regulation in *Dictyostelium discoideum. Proc. Natl. Acad. Sci. U.S.A.* **72,** 3541–3545.

Riedel, V., and Gerisch, G. (1968). Isolierung der Zellmembranen von kollektiven Amöben (*Acrasina*) mit Hilfe von Digitonin und Filipin. *Naturwissenschaften* **55,** 656.

Rosen, S. D., and Kaur, J. (1979). Intercellular adhesion in the cellular slime mold *Polysphondylium pallidum. Amer. Zool.* **19,** 809–820.

Rössler, H., Peuckert, W., Risse, H.-J., and Eibl, H. J. (1978). The biosynthesis of glycolipids during the differentiation of the slime mold *Dictyostelium discoideum. Mol. Cell. Biochem.* **20,** 3–15.

Rössler, H. H., Schneider-Seelbach, E., Malati, T., and Risse, H.-J. (1981). The dependence of glycosyltransferases in *Dictyostelium discoideum* on the structure of polyisoprenols. *Mol. Cell. Biochem.* **34,** 65–72.

Rossomando, E. F., and Cutler, L. S. (1975). Localization of adenylate cyclase in *Dictyostelium discoideum.* I. Preparation and biochemical characterizations of cell fractions and isolated plasma membrane vesicles. *Exp. Cell Res.* **95,** 67–78.

Rossomando, E. F, Steffek, A. J., Mujwid, D. K., and Alexander, S. (1974). Scanning electron microscopic observations on cell surface changes during aggregation of *Dictyostelium discoideum. Exp. Cell Res.* **85,** 73–78.

Rossomando, E. F., Jahngen, E. G., Varnum, B., and Soll, D. R. (1981). Inhibition of a nutrient-dependent pinocytosis in *Dictyostelium discoideum* by the amino acid analogue hadacidin. *J. Cell Biol.* **91,** 227–231.

Ryter, A., and Brachet, P. (1978). Cell surface changes during early development stages of *Dictyostelium discoideum:* a scanning electron microscopic study. *Biol. Cellulaire* **31,** 265–270.

Ryter, A., and de Chastellier, C. (1977). Morphometric and cytochemical studies of *Dictyostelium discoideum* in vegetative phase. Digestive system and membrane turnover. *J. Cell Biol.* **75,** 200–217.

Ryter, A., and Hellio, R. (1980). Electron–microscope study of *Dictyostelium discoideum* plasma membrane and its modifications during and after phagocytosis. *J. Cell Sci.* **41,** 75–88.

Ryter, A., Klein, C., and Brachet, P. (1979). *Dictyostelium discoideum* surface changes elicited by high concentrations of cAMP. *Exp. Cell Res.* **119,** 373–380.

Saito, M., and Yanagisawa, K. (1978). Participation of the cell surfaces in determining the developmental courses in the cellular slime mold *Dictyostelium discoideum. J. Embryol. Exp. Morph.* **48,** 153–160.

Scandella, D., Rooney, R., and Katz, E. R. (1980). Genetic, biochemical, and developmental studies of nystatin resistant mutants in *Dictyostelium discoideum. Mol. Gen. Genet.* **180,** 67–75.

Schmidt, J. A., and Loomis, W. F. (1982). Phosphorylation of the contact site A glycoprotein (gp80) of *Dictyostelium doscoideum. Dev. Biol.* **91,** 296–304.

Schmidt, J. A., Stirling, J. L., Jones, G. E., and Pacy, J. (1978). A chymotrypsin-sensitive step in the development of *Dictyostelium discoideum. Nature (London)* **274,** 400–401.

Shaffer, B. M. (1965). Cell movement within aggregates of the slime mould *Dictyostelium discoideum* revealed by surface markers. *J. Embroyl. Exp. Morph* **13,** 97–117.

Sievers, S., Risse, H.-J., and Sekeri-Pataryas, K. H. (1978). Localization of glycosyl transferases in plasma membranes from *Dictyostelium discoideum. Mol. Cell. Biochem.* **20,** 103–110.

Silverstein, S. C., Steinman, R. M., and Cohn, Z. A. (1977). Endocytosis. *Annu. Rev. Biochem.* **46,** 669–722.

Sinha, U., and Ashworth, J. M. (1969). Evidence for the existence of elements of a para-sexual cycle in the cellular slime mould, *Dictyostelium discoideum. Proc. Roy. Soc. B* **173,** 531–540.

Siu, C.-H., Lerner, R. A., Ma, G., Firtel, R. A., and Loomis, W. F. (1976). Developmentally regulated proteins of the plasma membrane of *Dictyostelium discoideum.* The carbohydrate–binding protein. *J. Mol. Biol.* **100,** 157–178.

Siu, C.-H., Lerner, R. A., and Loomis, W. F. (1977). Rapid accumulation and disappearance of plasma membrane proteins during development of wild-type and mutant strains of *Dictyostelium discoideum. J. Mol. Biol.* **116,** 469–488.

Smart, J. E., and Hynes, R. O. (1974). Developmentally regulated cell surface alterations in *Dictyostelium discoideum. Nature (London)* **251,** 319–321.

Smart, J. E., and Tuchman, J. (1976). Inhibition of the development of *Dictyostelium discoideum* by isolated plasma membranes. *Dev. Biol.* **51,** 63–76.

Smith, E., and Williams, K. L. (1979). Preparation of slime sheath from *Dictyostelium discoideum. FEMS Microbiol. Lett.* **6,** 119–122.

Springer, W. R., Haywood, P. L., and Barondes, S. H. (1980). Endogenous cell surface lectin in *Dictyostelium:* quantitation, elution by sugar, and elicitation by divalent immunoglobulin. *J. Cell Biol.* **87,** 682–690.

Spudich, J. A. (1974). Biochemical and structural studies of actomyosin-like proteins from non-muscle cells. II. Purification, properties, and membrane association of actin from amoebae of *Dictyostelium discoideum. J. Biol. Chem.* **249,** 6013–6020.

Steinemann, C., and Parish, R. W. (1980). Evidence that a developmentally regulated glycoprotein is target of adhesion–blocking Fab in reaggregating *Dictyostelium. Nature (London)* **286,** 621–623.

Steinemann, C., Hintermann, R., and Parish, R. W. (1979). Identification of a developmentally regulated plasma membrane glycoprotein involved in adhesion of *Polysphondylium pallidum* cells. *FEBS Lett.* **108,** 379–384.

Sternfeld, J., and David, C. N. (1981). Cell sorting during pattern formation in *Dictyostelium*. *Differentiation* **20,** 10–21.

Stossel, T. P. (1977). Endocytosis. *In* "Receptors and Recognition Series A" (P. Cuatrecasas and M. F. Greaves, eds.), Vol. 4, pp. 103–141. Chapman and Hall, London.

Sussman, M., and Boschwitz, C. (1975a). Adhesive properties of cell ghosts derived from *Dictyostelium discoideum*. *Dev. Biol.* **44,** 362–368.

Sussman, M., and Boschwitz, C. (1975b). An increase of calcium/magnesium binding sites in cell ghosts associated with the acquisition of aggregative competence in *Dictyostelium discoideum*. *Exp. Cell Res.* **95,** 63–66.

Sussman, R., and Sussman, M. (1967). Cultivation of *Dictyostelium discoideum* in axenic medium. *Biochem. Biophys. Res. Commun.* **29,** 53–55.

Swanson, J. A., Taylor, D. L., and Bonner, J. T. (1981). Coated vesicles in *Dictyostelium discoideum*. *J. Ultrastruct. Res.* **75,** 243–249.

Thilo, L., and Vogel, G. (1980). Kinetics of membrane internalization and recycling during pinocytosis in *Dictyostelium discoideum*. *Proc. Natl. Acad. Sci. U.S.A.* **77,** 1015–1019.

Toda, K., Ono, K.-I., and Ochiai, H. (1980). Surface labeling of membrane glycoproteins and their drastic changes during development of *Dictyostelium discoideum*. *Eur. J. Biochem.* **111,** 377–388.

Tuchman, J., Smart, J. E., and Lodish, H. F. (1976). Effects of differentiated membranes on the developmental program of the cellular slime mold. *Dev. Biol.* **51,** 77–85.

Vogel, G. (1981). Recognition mechanisms in phagocytosis in *Dictyostelium discoideum*. *Monogr. Allergy* **17,** 1–11.

Vogel, G., Thilo, L., Schwarz, H., and Steinhart, R. (1980). Mechanism of phagocytosis in *Dictyostelium discoideum:* phagocytosis is mediated by different recognition sites as disclosed by mutants with altered phagocytic properties. *J. Cell Biol.* **86,** 456–465.

Von Dreele, P. H., and Williams, K. L. (1977). Electron spin resonanace studies of the membranes of the cellular slime mold *Dictyostelium discoideum*. *Biochim. Biophys. Acta* **464,** 378–388.

Watts, D. J., and Ashworth, J. M. (1970). Growth of myxamoebae of the cellular slime mould *Dictyostelium discoideum* in axenic culture. *Biochem. J.* **119,** 171–174.

Watts, D. J., and Treffry, T. E. (1975). Incorporation of N-acetylglucosamine into the slime sheath of the cellular slime mould *Dictyostelium discoideum*. *FEBS Lett.* **52,** 262–264.

Weeks, C., and Weeks, G. (1975). Cell surface changes during the differentiation of *Dictyostelium discoideum*. *Exp. Cell Res.* **92,** 372–382.

Weeks, G. (1973). Agglutination of growing and differentiating cells of *Dictyostelium discoideum* by concanavalin A. *Exp. Cell Res.* **76,** 467–470.

Weeks, G. (1975). Studies of the cell surface of *Dictyostelium discoideum* during differentiation. The binding of ^{125}I-concanavalin A to the cell surface. *J. Biol. Chem.* **250,** 6706–6710.

Weeks. G. (1976). The manipulation of the fatty acid composition of *Dictyostelium discoideum* and its effect on cell differentiation. *Biochim. Biophys. Acta* **450,** 21–32.

Weeks, G., and Herring, F. G. (1980). The lipid composition and membrane fluidity of *Dictyostelium discoideum* plasma membranes at various stages during differentiation. *J. Lipid Res.* **21,** 681–686.

West, C. M., and McMahon, D. (1977). Identification of concanavalin A receptors and galactose–binding proteins in purified plasma membranes of *Dictyostelium discoideum*. *J. Cell Biol.* **74,** 264–273.

West, C. M., and McMahon, D. (1979). The axial distribution of plasma membrane molecules in pseudoplasmodia of the cellular slime mold *Dictyostelium discoideum*. *Exp. Cell Res.* **124,** 393–401.

West. C. M., and McMahon, D. (1981). The involvement of a class of cell surface glycoconjugates in pseudoplasmodial morphogenesis of *Dictyostelium discoideum*. *Differentiation* **20,** 61–64.

West, C. M., McMahon, D., and Molday, R. S. (1978). Identification of glycoproteins, using lectins as probes, in plasma membranes from *Dictyostelium discoideum* and human erythrocytes. *J. Biol. Chem.* **253,** 1716–1724.

White, G. J., and Sussman, M. (1961). Metabolism of major cell constituents during slime mold morphogenesis. *Biochim. Biophys. Acta* **53,** 285–293.

White, G. J., and Sussman, M. (1963). Polysaccharides involved in slime–mold development. II. Water–soluble acid mucopolysaccharide(s). *Biochim. Biophys. Acta* **74,** 179–187.

Whittenberger, B., Raben, D., Lieberman, M. A., and Glaser, L. (1978). Inhibition of growth of 3T3 cells by extract of surface membranes. *Proc. Natl. Acad. Sci. U.S.A.* **75,** 5457–5461.

Wilcox, D, K., and Sussman, M. (1978). Spore differentiation by isolated *Dictyostelium discoideum* cells, triggered by prior cell contact. *Differentiation* **11,** 125–131.

Wilcox, D. K., and Sussman, M. (1981a). Defective cell cohesivity expressed late in the development of a *Dictyostelium discoideum* mutant. *Dev. Biol.* **82,** 102–112.

Wilcox, D. K., and Sussman, M. (1981b). Serologically distinguishable alterations in the molecular specificity of cell cohesion during morphogenesis in *Dictyostelium discoideum*. *Proc. Natl. Acad. Sci. U.S.A.* **78,** 358–362.

Wilhelms, O.-H., Lüderitz, O., Westphal, O., and Gerisch, G. (1974). Glycosphingolipids and glycoproteins in the wild–type and in a non-aggregating mutant of *Dictyostelium discoideum*. *Eur. J. Biochem.* **48,** 89–101.

Williams, K. L., Kessin, R. H., and Newell, P. C. (1974). Parasexual genetics in *Dictyostelium discoideum:* mitotic analysis of acriflavin resistance and growth in axenic medium. *J. Gen. Microbiol.* **84,** 59–69.

Yabuno, K. (1970). Changes in electronegativity of the cell surface during the development of the cellular slime mold, *Dictyostelium discoideum*. *Dev. Growth Diff.* **12,** 229–239.

Yamamoto, A., Maeda, Y., and Takeuchi, I. (1981). Development of an autophagic system in differentiating cells of the cellular slime mold *Dictyostelium discoideum*. *Protoplasma* **108,** 55–69.

Yu, N. Y., and Gregg, J. H. (1975). Cell contact mediated differentiation in *Dictyostelium*. *Dev. Biol.* **47,** 310–318.

• CHAPTER 4

Chemotaxis

Peter N. Devreotes

I. Introduction

Exciting areas for future research include the biochemical basis of chemotaxis, cell–cell communication, signal transduction across membranes, biological oscillations, morphogenesis, and differentiation. At distinct phases in a well-defined developmental program, *D. discoideum* provides an experimentally accessible model for investigation of each of these phenomena. The organism grows rapidly (generation times from 3 to 10 h) on simple media,

The Development of *Dictyostelium discoideum*

ISBN 0-12-455620-5

facilitating the isolation of mutants. Since development can be synchronously activated by removal of exogenous nutrients, large numbers of cells at the same stage can be obtained. This chapter focuses on early aggregation—the role of chemoattractants in organizing cell movements, the intracellular events triggered by occupancy of chemoreceptors, and the current understanding of the macromolecular components of the chemosensory system. The effects of chemoattractants on the developmental program, is briefly outlined. The reader is also referred to the reviews by Loomis (1979), Newell (1977a, b), Gerisch and Malchow (1976), and Mato and Konijn (1979).

A variety of chemoattractants and chemorepellents are active in *D. discoideum* and related cellular slime molds. In *D. discoideum, D. rosarium, D. mucoroides*, and *D. purpureum*, the attractant is cyclic adenosine 3′5′-monophosphate (cAMP). Other species and strains, containing cAMP but not attracted to it, utilize independent species-specific attractants (Hanna *et al.*, 1979; Francis, 1975; Bonner *et al.*, 1972). One of these, purified from *Polysphondylium violaceum*, appears to be a small peptide, blocked at each end or cyclized (Wurster *et al.*, 1976; Pan, 1977). *Dictyostelium laceteum* and *Dictyostelium minutum*, both insensitive to cAMP, are specifically attracted to two different compounds that contain an aromatic ring (Mato *et al.*, 1977c; Kakebeeke *et al.*, 1978).

Most of these species-specific attractants are only effective during the aggregative phase of development. Conversely folic acid and related pteridine derivatives are attractants during the feeding or vegetative stage of all species and strains tested (Pan *et al.*, 1972, 1975). Species-specific chemorepellents are released by vegetative cells and may function to disperse the cells, thus aiding in foraging (Samuel, 1961; Keating and Bonner, 1977; Bonner, 1977; Kakabeeke *et al.*, 1979). This spectrum of attractants and repellents provides useful tools for studying the underlying biochemical mechanisms of chemotaxis. The chemotactic movement of slime mold amoebae is similar to that of leukocytes or macrophages. Unlike bacteria, these eukaryotes move in straight paths toward the source of attractant.

Chemoattractants are not solely distributed in stable gradients that continuously attract cells to the center of aggregation. Instead, these molecules are deployed in a highly organized pattern (Tomchik and Devreotes, 1981). In *D. discoideum*, secretory signals are *relayed* through the population. Cells stimulated by chemoattractant not only move toward its source but, in a precisely timed fashion, secrete additional molecules to attract more distal cells (Shaffer, 1962; Shaffer, 1975; Roos *et al.*, 1975). Communication among cells by signal relay, often viewed as a curiosity since it is unprecedented in biology, is actually a subtle and effective mechanism for a system of *identical* cells to communicate via a secretory mechanism. Similar mechanisms of signal relay may function in other tissues where form arises spontaneously in a population of identical cells.

The spontaneous periodic propagation of chemical waves during slime mold aggregation is a prototypcal biological oscillator that can be easily manipulated. The phase of the oscillations can be shifted by perturbations of the extracellular medium, and variants with different periods have been isolated (Gerisch and Hess, 1974; MacInnes and Francis, 1977; Gerisch *et al.*, 1977a; see Berridge and Rapp, 1978).

Research on cellular slime molds is especially exciting in view of recent advances in signal transduction across membranes. In *D. discoideum*, binding of chemoattractants not only elicits chemotaxis but triggers at least six other cellular reactions. Similar sets of reactions appear to be associated with chemotactic and/or secretory responses in higher organisms (see below). It is possible that the mechanisms employed by this primitive organism may have been conserved in evolution.

II. Analysis of Early Aggregation

A. Movements of Cells

During this phase of the life cycle, separated amoebae begin to communicate and aggregate to form a multicellular structure that later undergoes a series of shape changes. Early in aggregation, monolayers of cells on an agar surface segregate into territories encompassing about 10^5 to 10^6 cells; cells within each territory are attracted to its center. The movement of the cells is periodic and highly organized (Arndt, 1937; Bonner, 1947; Shaffer, 1957, 1962, Gerisch, 1968). Time-lapse films of individual cells reveal that they move inwardly in steps, advancing for about 2 min and then stopping for about 5 to 7 min before moving again (Arndt, 1937; Cohen and Robertson, 1971; MacKay, 1979). All cells at a given radial distance from the center simultaneously begin a movement step. The coordinated movements can be visualized as bands that form spiral or concentric circular patterns about a center (Gerisch, 1971; Alcantara and Monk, 1974; see Fig. 1A). In time-lapse films, each band can be seen spreading outward from a center as an enlarging ring. The period for initiation of successive rings is about 7–10 min, and each ring takes about 30 min to reach the edge of the territory (Alcantara and Monk, 1974; Gross *et al.*, 1976). Since the territories are radially symmetric, to understand the dynamics of aggregation, it is necessary to consider only a

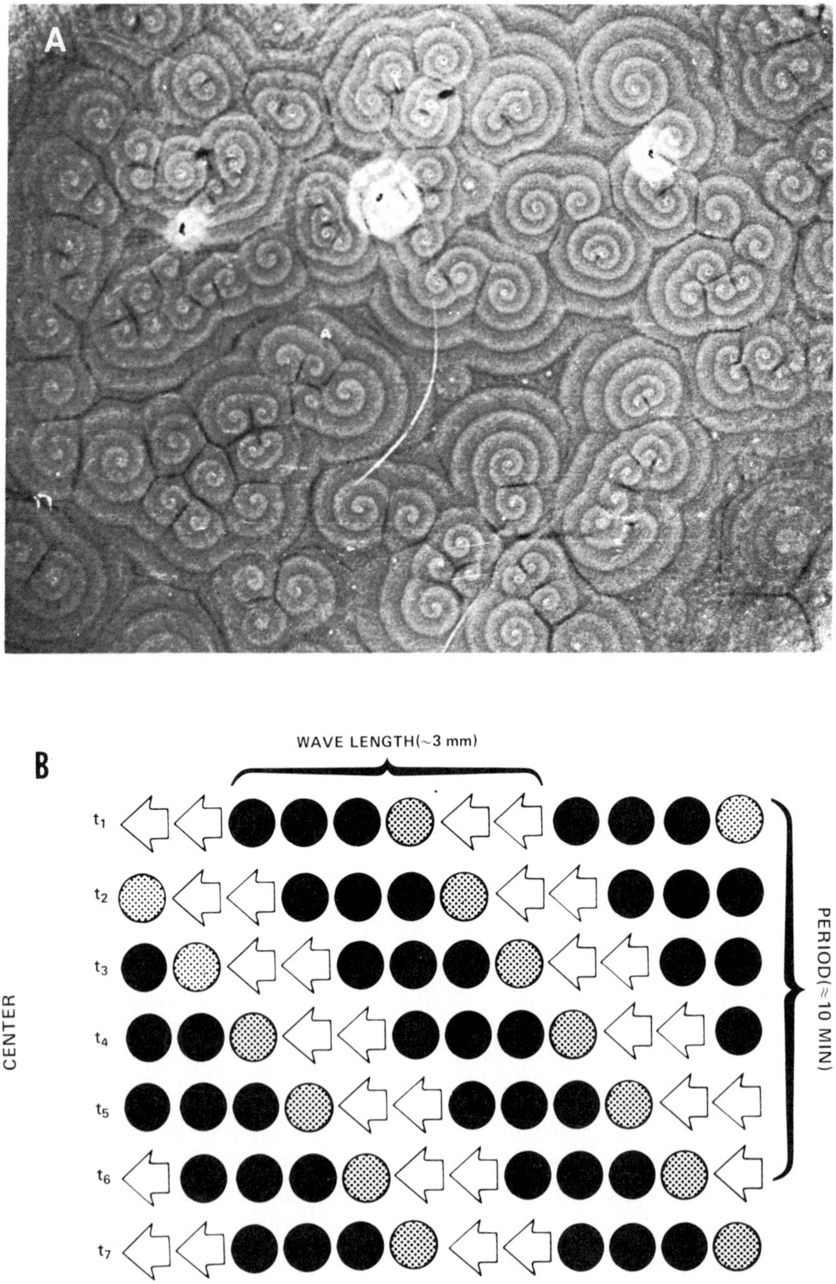
A
B
WAVE LENGTH(~3 mm)
CENTER
t_1
t_2
t_3
t_4
t_5
t_6
t_7
PERIOD(≈10 MIN)

single line of cells along a radius. A schematic diagram of how this line of cells appears at successive times is shown in Fig. 1B.

Although the bands themselves move outward (note this outward progression in the diagram) cells within the movement bands (white bands in Fig. 1A and 1B) move inward. The band maintains a nearly constant width of about 100 cells, since as a cell at the distal edge begins its inward step and elongates, one at the proximal edge ceases and becomes more rounded. The outward velocity of the bands (~300 μm/min) is 10–30 times that of the inward cellular motion (~12 μm/min) (Alcantara and Monk, 1974; Gross *et al.*, 1976; Cohen and Robertson, 1971; MacKay, 1979). Each movement step covers a distance of about 20 μm. Since the cells can aggregate over a distance as great as 1 cm, many steps would be required to bring the cells to the center.

After the passage of perhaps 20 waves, the cells make end-to-end contacts forming strings or "streams" (Shaffer, 1957, 1962; Beug *et al.*, 1973). Aggregation then proceeds by a shortening toward and, especially in spiral aggregates, winding of the streams about the aggregates (see Clark and Steck, 1979). In time-lapse films, waves are observed propagating along the streams, but the period drops to about 2 min (Gross *et al.*, 1977). The velocity of stream retraction is greater than that of individual cell motion, suggesting that the adhering cells may be mechanically drawn toward the aggregate.

B. Chemotaxis Is Guided by cAMP Waves

There is little doubt that in *D. discoideum* these organized cell movements are directed by extracellular cAMP. Cells are chemotactically at-

Fig. 1. Dynamics of cell aggregation. (A) Dark-field photograph of cell monolayer during early aggregation. Amoebae were freed of bacteria by centrifugation and spread on thin agar supports at a density of 10^6 cells/cm^2. After 18 h of incubation at 7°C, the cultures were warmed to 22°C and photographed within 30 min. The characteristic spiral patterns of alternate light and dark bands are visible within each aggregation territory. Illumination was provided with a dark-field condensing lens system similar to that described in Gross *et al.* (1977). Aggregation territories are about 1 cm. (B) Schematic diagram of a single radial line of cells during aggregation. Represented is the same radial line of amoebae at subsequent times, t_1, t_2, t_3 through t_7. The inward movement of the cells, which is 20–40 times slower than the outward progression of the movement band, is not detectable on the scale of the diagram. It is represented by the transient, center-directed orientation of the moving cells. ◊ chemotactically moving cells; ● randomly oriented cells; ⊛ cells in transition between chemotaxis and random orientation. Symbols, each representing about 50 cells, are colored white or black to represent the light and dark bands in Fig. 1A.

tracted to cAMP, isolated cells secrete cAMP in response to cAMP stimuli (see below), and an uniform background level of the nonhydrolyzable cAMP analog, adensine 3′,5′-cyclic phosphorothioate (cAMP-S), blocks aggregation (Konijn *et al.*, 1968; Konijn, 1970; Robertson *et al.*, 1972; Shaffer, 1975; Roos *et al.*, 1975; Rossier *et al.*, 1979). The distribution of cAMP at a given moment can be determined by an isotope dilution–fluorographic technique (Tomchik and Devreotes, 1981). As shown in Fig. 2A, cellular cAMP is visualized by its competition with exogenous [^{3}H]cAMP for high-affinity binding sites on protein kinase immobilized on a Millipore filter used to blot the monolayer. The cAMP pattern revealed by fluorography resembles the patterns of cell morphology seen in dark-field photographs (i.e., Fig. 1A).

Figure 2B shows that the dark bands (high concentrations of cAMP) form spiral or concentric circular patterns on a lighter background. Bands intersect but do not cross; overlapping arcs are never observed. Rather, V-shaped structures face each other at the two points of intersection of the arcs. The highest concentrations of cAMP are estimated to be about 10^{-6} *M*; the concentration in the interpeak regions is below detectable levels (10^{-8} *M*). Total cAMP is being measured; extracellular cAMP levels may be considerably lower (see Tomchik and Devreotes, 1981).

A scan of the optical density along the direction normal to the ridges of several waves is shown in Fig. 2C. The wave crosssection has a symmetrical "sawtooth" profile, and the changes in optical density are linear. The valley-to-valley distance is about 1.5 mm; the width at half maximum is about 0.7 mm. Although the majority of the waves detected have this profile, a few waves at the centers of small spirals are considerably weaker and narrower (Devreotes *et al.*, in preparation). As an isotope dilution method was employed to measure cAMP, linear changes in optical density suggest that cAMP levels change geometrically or exponentially.

A rough estimate of the change in cAMP concentration as a function of time can be made from these profiles. The cells' motion is negligible compared to that of the wave, the wave travels at uniform velocity, and its amplitude must be roughly constant over several minutes during steady-state propagation. The profiles of concentration versus time are similar to those of concentration versus distance. The dimensions on the x-axis can be replaced by: time (min) = distance (mm)/velocity (mm/min). For example, given a wave velocity of 300 μ/min, 1 mm corresponds to a time of 3⅓ min. A geometric or exponential rise in external cAMP concentration corresponds to an approximately linear rise in receptor occupancy versus time (Green and Newell, 1975; Henderson, 1975). The cAMP secretion rate is controlled by an adaptation process; therefore, continued secretion depends on a continuous rise in receptor occupancy (see following sections).

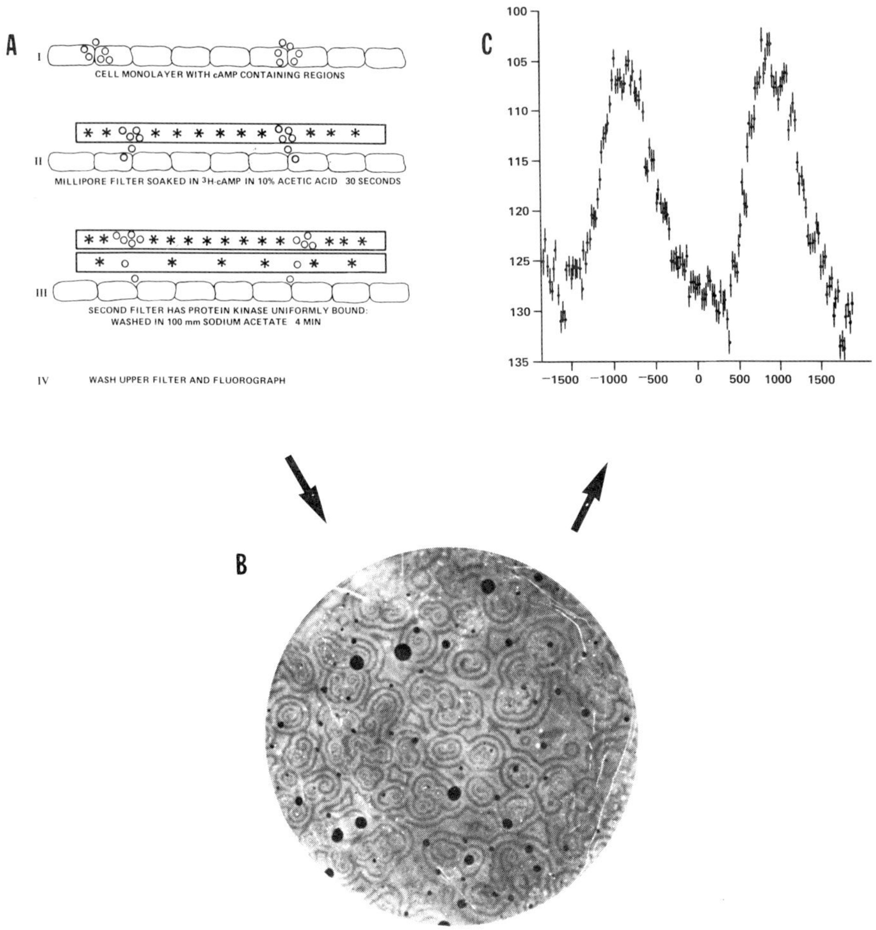

Fig. 2. Fluorographic image of cyclic AMP waves. Amoebae were developed to an early aggregation stage as described in Fig. 1(A). A millipore filter was soaked in acetic acid containing [^{3}H]cAMP, blotted and layered onto the cell monolayer. After 1 min a second filter, which had previously been soaked in a solution of protein kinase, was layered onto the first. After 4 min, the upper filter was removed, washed, blotted, and left to dry. (B) The filter was pressed against X-ray film and the resulting fluorograph was used as a negative to make this print. Calibration bar, 1 cm. (Completely black circles are artifacts caused by air bubbles trapped between the two filters.) (C) The fluorographic image was scanned within an Optronics X-ray film scanner. The graph shows the optical density (in relative units) as a function of distance (in micrometers) along a line perpendicular to several successive waves.

Dark-field photographs of the cells (e.g., Fig. 1A) were aligned with corresponding fluorographic images to determine the relative positions of the cAMP wave and the chemotactic movement steps. A cell begins a movement step as the foot of the leading edge of the wave meets the cell. The movement step may end on the leading edge if surface cAMP receptors become saturated or later, as the peak of the wave reaches the position of the cell, thereby reversing the gradient. In the latter case, the duration of a cell's movement step can be predicted to be equal to the half-width of the wave divided by the wave velocity. For the waves shown in Fig. 2C the corresponding movement steps are about 120 sec, within the reported range (60–180 sec) of movement step durations (Alcantara and Monk, 1974; MacKay, 1979).

The fact that the movement step duration depends on the wave profile and velocity and not on internal parameters is consistent with the observations of Futrelle *et al.* (1980, 1981). Four seconds after a cAMP-filled pipette is touched to the agar surface adjacent to a chemotactically sensitive cell, the cell rounds up. Within 15 sec, it begins to move toward the pipette. This movement ceases if the pipette is lifted from the surface. The duration of movement depends closely on the duration of pipette application. In other experiments, cells move continuously for up to an hour, as long as an external gradient is maintained (Alcantara and Monk, 1974; Cornejo, H., 1980). Clearly the cells do not have a preprogrammed chemotactic response of fixed duration; they depend continuously on an external gradient.

Eukaryotic cells might sense a chemical gradient by a temporal or a spatial mechanism. In a temporal mechanism, the detector compares measurements made at different times at different points in the gradient. In a spatial mechanism, the detector compares measurements made at different points at the same time. A scheme has been proposed whereby a cell might measure a gradient by utilization of pilot pseudopods as temporal detectors (Gerisch *et al.*, 1975c; Rossier *et al.*, 1980). Pseudopods extended up the gradient would sense a positive change in concentration with respect to time ($+\Delta C/\Delta t$) and be reinforced. Those extended down the gradient would detect a negative change ($-\Delta C/\Delta t$) and be extinguished. This scheme is reasonable for stable gradients. However, the gradients illustrated in Fig. 2C are sweeping rapidly (5 μm/sec) past the responding cell. Pseudopods extended in any direction (at a velocity less than 5 μm/sec) would sense a positive increase with respect to time ($+\Delta C/\Delta t$). Moreover, since pseudopods normally move much more slowly than 5 μm/sec, the temporal signal detected by extending a pseudopod either up or down the gradient would be nearly identical; most of it would derive from the rising gradient. The pilot–pseudopod model would need to be modified to apply to this case. These considerations suggest that the cells use spatial detectors; that is, simultaneously compare concentrations at two ends of the cell. Mato *et al.* .(1975) arrived at a similar conclusion from the way the range for threshold attraction depended on cAMP concentration.

The necessary sensitivity of the detectors can be estimated. The wave profiles indicate that the total cAMP concentration changes nearly 100-fold over a distance of about 1 mm (Tomchik and Devreotes, 1981). If the changes are geometric or exponential, the concentration would approximately double across a cell. This corresponds to at most a 7% difference in receptor occupancy (the largest change in receptor occupancy occurs around the K_d). Estimates of threshold chemotactic sensitivity from the orientation of cells in stable gradients suggest that about a 1% difference in occupancy can be detected (Mato *et al.*, 1975).

This, as any detection system, must face two apparent problems presented by the situation. First, the time profile of the cAMP concentration at the front and back of a cell are similar but shifted by about 2 sec—the time required for the wave to advance one cell diameter. That is, the concentration at the back of a cell rises to the previous value at the front within 2 sec. Can the detectors operate this rapidly (see Berg and Purcell, 1977; Nanjundiah, 1978)? Second, during the rising phase of the wave, the concentration of cAMP at the front of the cell is higher than at the back; on the falling phase the situation is symmetrically reversed. Why don't the cells reorient in the reversed gradient presented by the descending limb of the wave? Reorientation may not occur because the cAMP concentrations are decreasing as a function of time (Futrelle *et al.*, 1981). Some evidence, however, indicates that cells can orient in such falling gradients (Futrelle, personal communication). Alternatively, the high cAMP concentrations encountered at the peak of the wave may temporarily inhibit cell motility; there is some evidence that high concentrations of cAMP immobilize the cells (Ryter *et al.*, 1979; Cornejo, 1980).

Inherent cell polarity may play an important role here and in eukaryotic chemotaxis in general. Leukocytes, when oriented in a gradient of the chemotactic peptide, f-Met-Leu-Phe, are insensitive to peptide-containing pipettes applied to the tail region. When the gradient is reversed or a reversed gradient of another attractant applied, the cells move around in a circle rather than switching head and tail regions (Zigmond, 1981). Such polarity could reinforce an orientation initially imposed by the external gradient. Once the cells were oriented the rapid temporal increase experienced by *D. discoideum* could primarily provide a chemokinetic signal (cells do move more rapidly than normal during a movement step) (MacKay, 1979; Cornejo, 1980). By employing such a mechanism, a cell need not reassess the direction of the gradient as often. A strong polarization might also prevent the cells from reorienting in the reversed gradient. *D. discoideum* cells attracted to a pipette-imposed gradient can reverse orientation when the pipette is repositioned at the back (Futrelle *et al.*, 1981). However, when the amoebae are strongly oriented, pseudopods are more easily elicited from the anterior than from the posterior regions of the cell (Swanson, personal communication).

C. Cell–Cell Relay of cAMP Signals

The cAMP waves do not dissipate, like ripples on the surface of water, as they travel through the monolayer. In fact, in some instances they appear to build up as they move out (Tomchik and Devreotes, 1981). The stable wave results from the cells' capacity to *relay* (amplify, regenerate) cAMP signals. When stimulated by extracellular cAMP, they synthesize and secrete additional cAMP that serves as a signal for more distal cells (Shaffer, 1975; Roos *et al.*, 1975). Thus, aggregation is brought about by two coordinated responses to extracellular cAMP: chemotaxis and signal relay.

Both of these processes are reflected in the behavior of synchronized populations of cells. The classic experiments of Gerisch and coworkers demonstrated that spontaneous oscillations in cAMP levels in oxygenated cell suspensions were accompanied by decreases in optical density (Gerisch and Hess, 1974). The light scattering changes are probably due to cell shape changes or to the state of cell–cell aggregation and are probably related to the chemotactic response of the cells (see below). Responses can also be elicited by addition of cAMP to a suspension of synchronized cells (Shaffer, 1975; Roos *et al.*, 1975; Geller and Brenner, 1978a; Dinauer *et al.*, 1980a), resulting in a dramatic increase in cAMP synthesis, intracellular accumulation and secretion. This response will be referred to as a cAMP signaling response. cAMP levels have been measured by isotope dilution and by prelabeling cells with [^{3}H]adenine, [^{3}H]adenosine, or ^{32}P. Isotope dilution allows a clear estimate of the amplification of small cAMP stimuli, but can lead to artifacts when larger stimuli are applied since the input stimulus must be substracted from the total cAMP recovered (Grutsch and Robertson, 1978).

Most investigators agree that intracellular cAMP levels rise to a value of about 10^{-6} *M*, about 10^7 molecules per cell per response are secreted, and the duration of the response is several minutes (Roos *et al.*, 1975; Gerisch and Wick, 1975). A time course of several minutes for cAMP secretion is consistent with the dimensions of the cAMP waves detected in the monolayer. Mutants defective in the generation of cAMP signaling responses do not aggregate, and specific inhibitors of the response also block aggregation

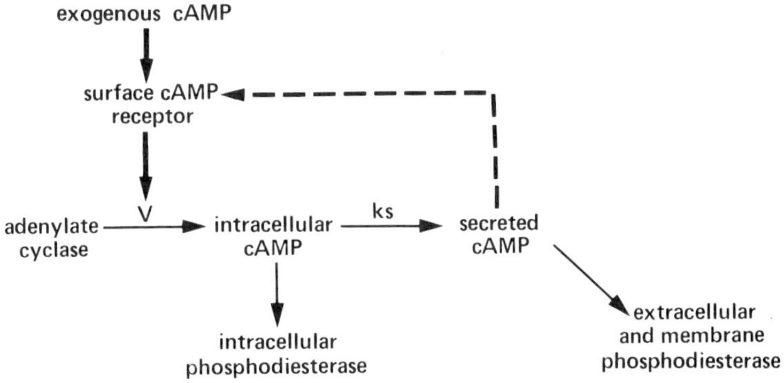

Fig. 3. Interaction of components essential in generation of the cAMP signaling response. Dotted line indicates positive feedback loop created when secreted cAMP serves as a stimulus by binding to surface cAMP receptors.

(Konno, 1980; Frantz, 1980; unpublished observations). There is little question that the responses of synchronized cell populations to exogenous cAMP correspond to the signals guiding morphogenesis *in situ*.

Figure 3 illustrates the interaction of the components which generate the cAMP signaling response. A change in occupancy of surface cAMP receptors leads to activation of adenylate cyclase and an ensuing accumulation of intracellular cAMP (Roos *et al.*, 1975; Dinauer *et al.*, 1980a; Gerisch and Wick, 1975). Not all of the newly synthesized cAMP can be accounted for in the extracellular medium, suggesting that a fraction is degraded by intracellular phosphodiesterase (Dinauer *et al.*, 1980a). The cAMP secretion rate is directly proportional to the intracellular cAMP level; although there may be a specific mechanism for its transport, there is no long term storage of newly synthesized cAMP (Gerisch and Wick, 1975; Dinauer *et al.*, 1980a). After several minutes, the rate of cAMP synthesis decreases and the remaining cAMP is degraded by extracellular and membrane bound phosphodiesterase, the latter probably more significant *in situ* (Nanjundiah and Malchow, 1978; Gerisch, 1976). The cAMP secretion rate, intracellular and extracellular phosphodiesterase activity, and receptor affinity or number do not seem to be regulated during the transient response. Thus, the single control point determining the magnitude and duration of the cAMP signaling response is at the level of cAMP synthesis (v, in Fig. 3).

Interaction of extracellular cAMP with surface cAMP receptors can create a positive feedback loop since the secreted cAMP adds to the stimulus. The action of the extracellular phosphodiesterase is to counteract this feedback loop by decreasing extracellular cAMP; removal (or decrease) of the cAMP stimulus during a cAMP signaling response leads to the abrupt termination of the response (Devreotes and Steck, 1979). The opposing effects of the positive feedback loop and phosphodiesterase suggest the system can exhibit excitable behavior. Evidence for this excitability is found when the cells are studied in suspension (and presumably occurs *in situ* on an agar surface). For instance, at critical times in development cAMP levels oscillate spontaneously (Gerisch and Hess, 1974; Gerisch and Wick, 1975). In suspension, there is a discrete threshold for initiation of the response and nearly maximal responses (i.e., "all-or-nothing") are obtained from small superthreshold stimuli (M. Brenner, personal communicaton). Furthermore, there is a discrete "refractory period" before a second response can be initiated.

D. Response–Adaptation Control of Adenylate Cyclase

To eliminate the effect of the extracellular phosphodiesterase on the stimulus, cells in suspension can be stimulated with high concentrations of cGMP or cAMP-S, a nonhydrolysable cAMP analog, or rapidly perfused with cAMP. It was found that, in spite of the continued presence of a saturating stimulus and full occupancy of surface receptors, the response terminated after several min (Gerisch *et al.*, 1977a; Rossier *et al.*, 1980; Devreotes and Steck, 1979). The effects of both the positive feedback loop and extracellular phosphodiesterase can be minimized if the cells are stimulated under conditions of rapid perfusion. Secreted cAMP is rapidly removed; degraded cAMP is rapidly renewed. In this way, extracellular cAMP is "clamped" at the concentration of the applied stimulus (Devreotes *et al.*, 1979).

Under these conditions, any increment in stimulus concentration from C_1 to C_2 leads to a response which subsides after several minutes. Persistent stimulation with concentration C_2 fails to elicit further response. However, an increment from C_2 to C_3 will elicit another response. An important property of the system is that the extracellular cAMP concentration continues to be monitored after the response has subsided. Both the initial and final concentrations of a cAMP stimulus determine the magnitude of the response. For example, the sum of the magnitudes of the two responses elicited by the increments, C_1 to C_2 and C_2 to C_3, equals the magnitude of the single response elicited by the larger increment, C_1 to C_3. The magnitude of the elicited response saturates at a stimulus concentration of 10^{-5} *M* cAMP, and increments originating at 10^{-5} *M* cAMP elicit no response. These properties suggest that cells respond to changes in the fractional occupancy of surface cAMP-binding sites. The extinction of each response by the adjustment of cellular sensitivity to the level of the current stimulus is referred to as "adaptation" (Devreotes and Steck, 1979).

There is a rapid recovery of responsiveness (deadaptation) after the removal of cAMP stimuli (Devreotes and Steck, 1979; Dinauer *et al.*, 1980b). A level of adaptation was established by an initial stimulus of defined magnitude and duration. After a variable recovery period, the magnitude of the response to a second stimulus was determined and its attenuation taken as a measure of residual adaptation to the first stimulus. The magnitude of the response to the second stimulus increased with the recovery time in a first-order fashion, with a $t_{1/2}$ = 3–4 min for stimuli of 10^{-8} *M* to 10^{-5} *M* cAMP; recovery was complete in 12 to 15 min. The time course of the rise in adaptation during stimulation with 10^{-6} *M* cAMP, a saturating dose, was also assessed (Dinauer *et al.*, 1980c). Amoebae were pretreated with 10^{-6} *M* cAMP for periods of 0.33–12 min, and then immediately given test stimuli of 10^{-8} *M* to 2.5×10^{-7} *M* cAMP. The response to a given test stimulus was progressively attenuated and finally extinguished as the duration of the pretreatment stimulus increased; the rates of attenuation could be ranked according to the concentration of the test stimulus. The responses to test stimuli of 10^{-8}, 5×10^{-8}, 10^{-7}, or 2.5×10^{-7} *M* cAMP were extinguished after 1, 2.25, 2.5, and 10 min, respectively. These data suggested that adaptation begins within 20 sec of stimulation, rises rapidly for 2.5 min, and reaches a plateau by 10 min.

These observations of the cAMP signaling response can be summarized by the scheme shown in Fig. 4 (Dinauer *et al.*, 1980b; see Gerisch, 1978; MacNab and Koshland, 1972). The occupancy of surface cAMP receptors controls the extent or level of both an excitation process and an adaptation process that counteracts excitation. The relative excess of excitation compared to adaptation at any instant is reflected in the value of a parameter *X* that determines the activity of adenylate cyclase. An increase in the extracellular cAMP concentration increases the extent or level of both excitation and adaptation. Both processes rise from prestimulus levels at a rate proportional to the increment in receptor occupancy, but the excitation level changes more rapidly than the level of adaptation. A transient response ensues, which terminates when the adaptation level matches the excitation level at the new value specified by receptor occupancy. Serial increases in cAMP receptor occupancy merely repeat this sequence of events; conversely, both excitation and adaptation decay when receptors are vacated. The magnitude of a signaling response is determined by the change in the levels of excitation and adaptation from their values at the onset of a stimulus to their final occupancy-specified value. The size of the response elicited by a single increase to a given concentration of cAMP equals that elicited by a

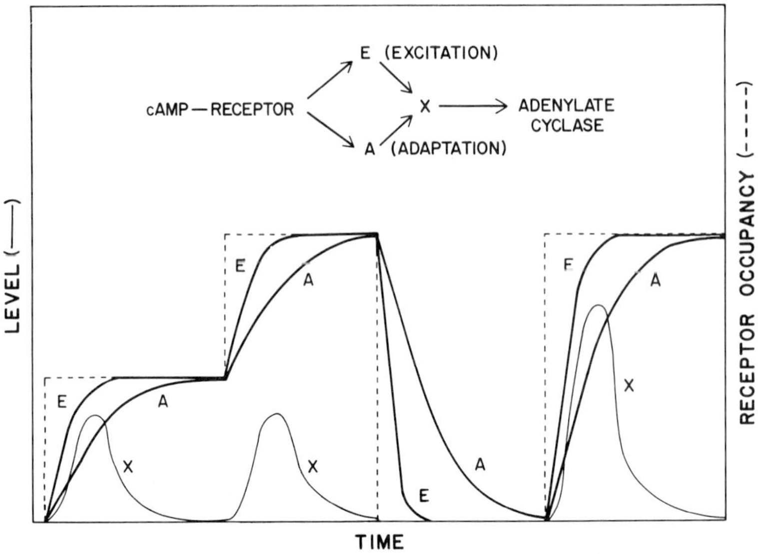

Fig. 4. Excitation–adaptation scheme for control of the cAMP signaling response. Upper part of the figure indicates that cAMP receptor occupancy specifies changes in two hypothetical processes, E (excitation) and A (adaptation). X represents a process that continuously compares the levels of E and A and directly controls the activity of the adenylate cyclase.

Lower portion of the figure shows a specific example. Dashed bars represent the cAMP stimulus. A small increment in the cAMP stimulus concentration leads to a rapid increase in E and a slower increase in A. The value of X is determined by the relative excess of E over A. It rises to a peak value and then subsides as the value of A approaches that of E. A subsequent increment in the stimulus concentration causes further increases in E and A and another transient change in X. When the stimulus is lowered, E and A fall to the initial value and sensitivity is regained. The final large increment elicits greater increases in both E and A and a greater transient change in X. The sum of the amount of cAMP production (controlled by X) elicited by the two smaller increments equals that elicited by the single larger increment.

stepwise increase to this same concentration. In the latter case, the response to a stimulus increment is appropriately scaled because amoebae keep track of the preceding cAMP concentration, even when the response had ceased, through the altered levels of excitation and adaptation.

Reemphasis of two key features of the adaptation process illustrated by this scheme is in order. (These do not depend on the scheme but are derived from experimental observations). First, although the cAMP signaling response is transient, adaptation persists as long as the stimulus is applied. Adaptation therefore cannot depend on any transient component of the response such as the rise in cAMP or cGMP levels, or a transient decrease in some other metabolite (see Chung and Coe, 1977, 1978; Geller and Brenner, 1978a). To be a candidate for adaptation, a given biochemical process must not be transient; it must continue after the response has ceased. Second, adaptation is graded—small stimuli elicit low levels of adaptation. This would argue against certain potential mechanisms of response attenuation such as inactivation of functional cAMP surface binding sites or adenylate cyclase or depletion of a substrate or cofactor (Goldbeter and Segal, 1977; Goldbeter and Martiel, 1980; Martiel and Goldbeter, 1981). Such mechanisms cannot lead to complete extinction of the initial response and allow responses to subsequent increments in the stimulus. The illustrated scheme (Fig. 4) is the simplest one that incorporates the unique features of adaptation.

The scheme can be put into quantitative form by using first order differential equations stating that the rate of change in the excitation (E) and adaptation (A) processes is proportional to the change in receptor occupancy. (More precisely, the new level of receptor occupancy minus the previous level of excitation or adaptation.)

$$\frac{dE}{dt} = \alpha_1\,(O-E) \qquad \frac{dA}{dt} = \alpha_2\,(O-A)$$

The parameters E and A are the instantaneous levels of the excitation and adaptation processes and have the same dimensions (a fraction between 0 and 1) as receptor occupancy O. The rate constant, $\alpha_2 = 0.25$/min ($t_{1/2} \cong 3$ min), is derived from experimental measurements of the half-time for the rise and fall in adaptation. A value of 1/min for the rate constant α_1 forces the response to have a time course similar to that experimentally observed. The rate of cAMP synthesis is taken to be proportional to the difference in excitation and adaptation levels [i.e., $v = K(E - A)$]. The parameter K has the value 10^7 molecules/min cell so that the integrated response to a maximal stimulus will be about 5×10^7 molecules/cell.

By applying this simple expression for v to the pathway shown in Fig. 3, it is possible to simulate by computer analysis most of the experimental observations. The simplest case occurs when the effects of both the positive feedback loop and extracellular phosphodiesterase are ignored (corresponding to data from rapid perfusion experiments). Figure 5A shows an example in which calculated kinetics of cAMP secretion in response to a complex stimulus are compared to corresponding experimental data (MacKay, S. *et al.*, in preparation).

A more powerful test of the scheme is to use it to analyze the system representing cell suspensions where the effects of the positive feedback loop and extracellular phosphodiesterase must be taken into account (see Fig. 3). In this case, the solutions have properties which are consistent with those from experiments carried out in suspension. For instance, there is a sharp

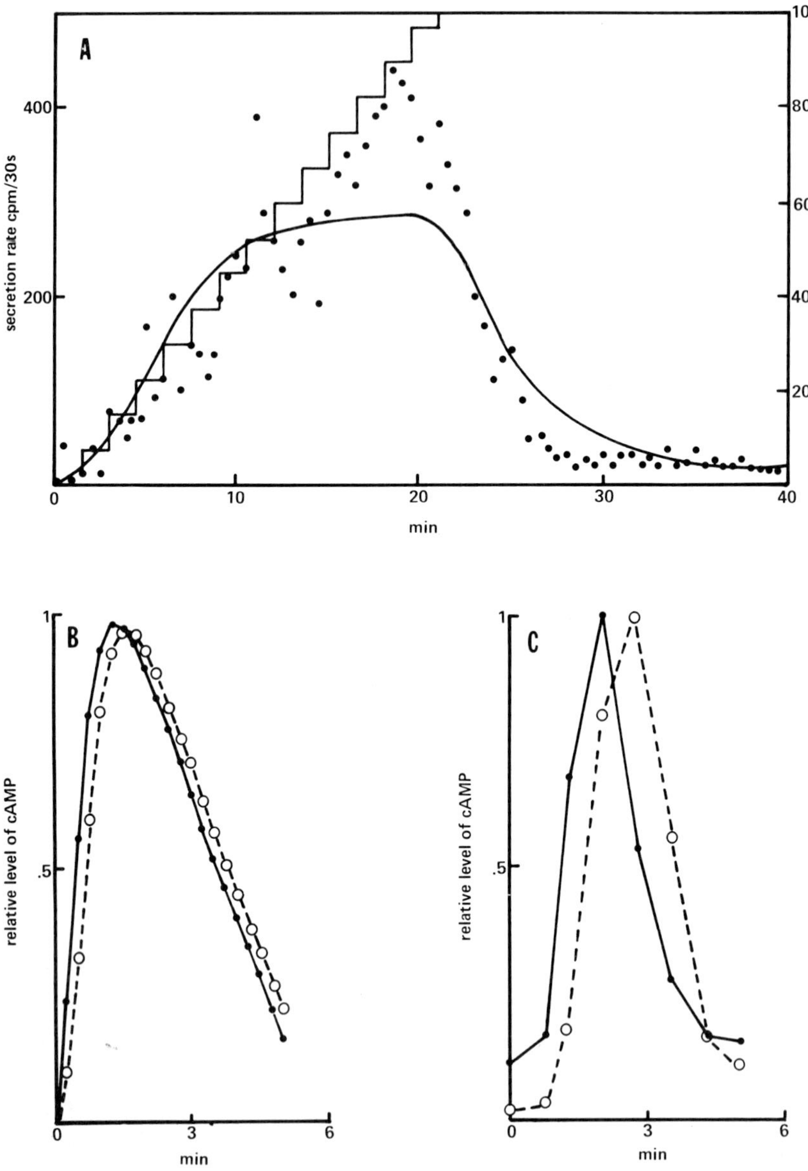

Fig. 5. Comparison of computer analysis with experiment. (A) Rate of cAMP secretion in response to "clamped" stimulus doubled every 90 sec; the stepped line shows approximate changes in receptor occupancy. Points are experimentally observed data (Devreotes and Steck, 1979); solid line was calculated from equations given in text. (B) Calculated changes in intracellular (—●—) and extracellular (—○—) cAMP levels in cell suspension. (C) Experimentally observed changes in intracellular (—●—) and extracellular (—○—) cAMP levels in cell suspension (Gerisch and Wick, 1975).

threshold above which an all-or-none response is triggered; and a distinct refractory period appears. Note that these are not intrinsic properties of a cell but arise from the opposed effects of the positive feedback loop and phosphodiesterase. This treatment also resolves an apparent discrepancy in the literature. Gerisch and Wick (1975) observe that intracellular cAMP levels peak about 30 sec prior to extracellular levels while Dinauer *et al.* (1980a) show that the secretion rate is directly proportional to intracellular levels. As illustrated in Fig. 5B and 5C, these observations are completely consistent. When the secretion *rate* is directly proportional to intracellular levels, the extracellular cAMP *levels* peak about 15 to 30 sec after the intracellular level (Knox and Devreotes, manuscript in preparation).

E. Oscillations and the Origin of Aggregation Centers

The spontaneous oscillations in cAMP levels in aerated or shaken cell suspensions probably reflect the process underlying the periodic initiation of cAMP waves from aggregation centers. The period of these oscillations is about 8 min in strain AX2 while in an AX2 derived mutant, *Ap66*, it is about 17 min (Gerisch and Hess, 1974; Gerisch *et al.*, 1977a). Independent isolates of strain AX3 have periods ranging from 6 to 11 min (unpublished observation). NC4 cells do not usually oscillate in suspension but they have occasionally been observed to do so; a fast developing mutant of NC4, *HH201*, has a period of about 4 min (Coukell and Chan, 1980). Although conditions other than temperature that affect the period have not been defined, the phase can be shifted by addition of small doses of cAMP or folic acid. Applied in the initial half of the oscillatory cycle, these cause a phase delay; application during the latter half leads to a phase advance. Continuous application of 5 n*M* cAMP extinguishes the oscillations (Gerisch and Hess, 1974; Gerisch *et al.*, 1977a; Malchow *et al.*, 1978a; Wurster, 1976).

Several different schemes have been proposed to explain the spontaneous oscillations and center formation on two-dimensional surfaces. There could be a distinct set of cells that oscillates autonomously and drives the other cells via extracellular cAMP (or possibly some other compound—see Klein, 1980). Some experiments suggest that there are no such rare "autonomous" pacemaker cells in *D. discoideum* (Newell, 1981; personal communication). Another class of models arises from a set of interdependent differential equations (Goldbeter, 1975; Goldbeter and Segal, 1977; Goldbeter *et al.*, 1978; Cohen, 1978; Goldbeter and Segal, 1980). In different versions of this model the parameter chosen to attenuate the signaling response has been either reduction of the substrate, intracellular ATP, desensitization (reduction) of surface receptors, or covalent modification of the adenylate cyclase

(Goldbeter and Segal, 1977; Goldbeter and Martiel, 1980; Martiel and Goldbeter, 1981). Only for a given set of parameters does the system of equations have an oscillatory solution. Outside this range, however, the system is still excitable and will respond to exogenous stimuli. This is an attractive feature consistent with experimental observations.

Many inhibitors of cAMP synthesis at low concentrations can block oscillations, but only slightly alter responses to exogenously applied cAMP stimuli (Geller and Brenner, 1978b; M. Brenner, personal communication). Furthermore, early in development, cells can amplify exogenous cAMP stimuli; but spontaneous oscillations only occur at critical times, presumably when the factors that amplify and attenuate cAMP production are appropriately balanced (Goldbeter and Segal, 1980; Goldbeter, 1981; Cohen, 1977; Cohen, 1978). These models give an excellent representation of early observations of cAMP signaling responses; however, it is not clear they can fully account for observations of changes in cAMP secretion rates in response to fixed eternal cAMP stimuli (see above). An alternative view is offered by the excitation–adaptation scheme described above. Derived from observations of responses to fixed stimuli, it also simulates most of the behavior of cell suspension. For instance, it successfully explains the phase-shift data. Stimuli applied in the first half of the cycle (just after the cells have responded spontaneously) add to the cAMP level, delay the onset of deadaptation, and cause a phase delay. Once the level of adaptation has decayed, a sufficiently large stimulus can trigger an early response advancing the phase. It is not clear whether this scheme can account for oscillations since no oscillatory solution has been found. However, within 10 min after a response subsides, the cells become so exquisitely sensitive to subsequent stimuli that oscillations can arise from very small fluctuations in the background levels of excitation and adaptation (Knox and Devreotes, manuscript in preparation).

III. Chemosensory Transduction

A. Reactions Triggered by Chemoattractants

Two cellular responses to cAMP, chemotaxis and cAMP secretion, are of obvious biological significance in *D. discoideum.* How is occupancy of surface receptors coupled to cell motility and activation of adenylate cyclase in *D. discoideum* or to synthesis of other chemoattractants in cAMP insensitive strains? Some hints may lie in analysis of other reactions triggered during a response: a decrease in the optical density of a cell suspension is observed;

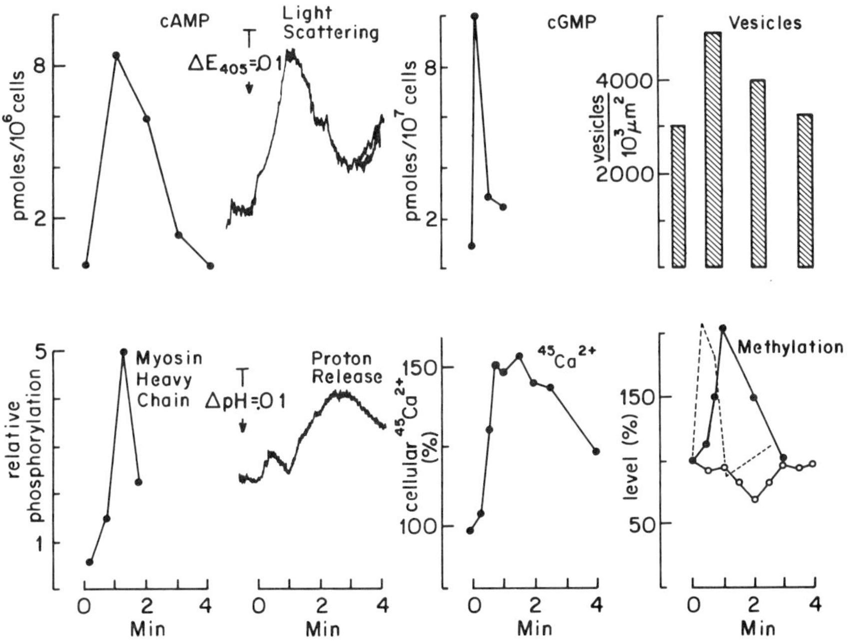

Fig. 6. Summary of reactions triggered by cAMP. Data from several sources replotted on the same time scale. In each case cAMP stimulus was added to $t = 0$. For methylation experiment, dotted line is for level of methylation of protein, closed circles are for phosphatidyl L-[*methyl*-^{3}H]choline, open circles are for lysophosphatidyl L-[*methyl*-^{3}H]choline. Light scattering response from Gerisch and Wick, 1975; cAMP accumulation from unpublished observations; cGMP accumulation from Mato *et al.*, 1977a; vesicle formation from Maeda and Gerisch, 1977; proton release from Malchow *et al.*, 1978; methylation experiments from Mato and Marin-Cao, 1979; $^{45}Ca^{2+}$ influx from Wick *et al.*, 1978; myosin phosphorylation from Malchow *et al.*, 1981.

guanylate cyclase is activated, leading to a rapid accumulation of intracellular cyclic guanosine 3′,5′-monophosphate (cGMP); the methylation states of proteins and several phospholipids change; the rate of $^{45}Ca^{2+}$ influx increases; dephosphorylation of the heavy chain of myosin occurs; the medium becomes more acidic, presumably due to the release of a weak acid; and there is a small increase in the number of intracellular vesicles as seen by electron microscopy (see Fig. 6).

Most of these phenomena were originally discovered in *D. discoideum* using cAMP as a stimulus. However, folic acid and related pteridine derivatives, as well as the species-specific attractants, elicit light scattering changes and cGMP accumulation, and probably also trigger a number of the other events listed above. Are any of these reactions sequential events in transduction pathways leading to chemotaxis, activation and/or inactivation of adenylate cyclase (in cAMP sensitive strains), or synthesis and secretion of other species-specific attractants (in cAMP insensitive strains)? Or do they all merely occur in parallel? The products of some of these reactions do influence some of the other reactions *in vitro,* but a causal interaction of any of these events has not been clearly demonstrated. What is the significance of the effects of pteridine compounds? Although there is, at present, insufficient experimental evidence to allow speculation on a specific transduction pathway, the relevant findings are listed and generalities about the data are pointed out.

1. *The Light Scattering Response*

The transient decrease in the optical density of suspensions of activated cells was first reported in 1974 (Gerisch and Hess, 1974) (see Fig. 6). It has not been definitely demonstrated whether this decrease is caused by changes in cell shape or by the state of aggregation of cells in suspension. There is a biphasic change in cell shape when cells are suddenly exposed to a cAMP gradient. Cells round up or "cringe" and subsequently elongate for chemotaxis; this behavior may account for the biphasic light scattering response in cell suspensions (Futrelle *et al.*, 1981). Addition of EDTA to suspensions of *Polysphondylium violaceum* cells prevents formation of stable cell–cell contacts; but light scattering responses are unaffected, suggesting that a shape change is responsible for the decrease in optical density (Wurster *et al.*, 1978). However, cells were not examined during the course of the response and there may have been a transient clumping. *D. discoideum* amoebae do display an increased adhesiveness to the agar substrate during the chemotactic movement step *in situ* and conceivably are also transiently more adhesive to each other (Dinauer and Devreotes, unpublished observation). In any case, the transient decreases in optical density are extremely useful since they provide a simple, immediate measure of cell responsiveness.

2. *Increases in cGMP Levels*

cGMP increases about tenfold from a basal level of approximately 1 pmole per 10^7 cells (see Fig. 6). In contrast to cAMP, cGMP is not secreted, since it is intracellularly degraded within a few minutes following activation (Mato *et al.*, 1977a,b). The increase in cGMP is probably due to transient activation of guanylate cyclase (rather than inhibition of phosphodiesterase) since sonicates of activated cells synthesize [^{32}P]cGMP four times faster than those from control cells (Mato and Malchow, 1978). A reported dose–response curve for cAMP-elicited cGMP accumulation covers the range from 5 $\times$

10^{-10} *M* to 5×10^{-7} *M* (Mato *et al.*, 1977a). At a stimulus concentration of 5×10^{-8} *M* cAMP, cGMP accumulation is 85% of maximal while cAMP accumulation is only 50% saturated. The dose–response data for cGMP should be viewed with caution since the effects of positive feedback were not controlled for (cells were stimulated in suspension) and the highest concentration tested was 5×10^{-7} *M* cAMP. Since such a discrepancy could be physiologically meaningful, the dose dependence of cAMP-elicited cGMP accumulation should be reinvestigated.

The increase in cGMP can be enhanced about twofold by a 10 min pretreatment with concanavalin A and 0.3 m*M* to 3 m*M* caffeine and by about 50% by 0.1 m*M* ATP pretreatment (Mao *et al.*, 1978c; Mato, 1978; M. Brenner, personal communication). Histochemical staining of the cGMP remaining in sections after fixation is diffuse in early cells but distinctly nuclear in aggregation competent cells (Mato and Steiner, 1980; Pan and Wedner, 1979).

There is a close correlation between chemotaxis, the light scattering response in cell suspensions, and cGMP accumulation. It has been suggested that cGMP may be an important regulatory molecule for chemotaxis. Each of these responses is elicited by similar doses of folic acid in vegetative cells of all species, by cAMP and a series of cAMP analogs in differentiated *D. discoideum* cells, and by species-specific chemoattractants in differentiated cells of *Polysphondylium violaceum* and *D. lacteum* (Wurster *et al.*, 1978; Mato and Konijn, 1977a,b; Mato *et al.*, 1977a,b,c,d).

3. *Fluctuations in Methylation States of Proteins and Phospholipids*

cAMP binding has been reported to elicit changes in the methylation states of proteins and phospholipids. The degree of methylation of proteins near 120 kdaltons increases twofold within 30 sec of stimulation (Mato and Marín-Cao, 1979). When cells preincubated with L-[*methyl*-^{3}H]methionine, are activated by cAMP, the level of phosphatidyl [*methyl*-^{3}H]choline transiently increases. No fluctuations are observed in cells preincubated with [*methyl*-^{14}C]choline, so that the changes represent lipid methylation, not turnover. A similar transient increase in phosphatidyl L-[*methyl*-^{3}H]choline can be brought about by high concentrations of 8-Br-cGMP, a poor chemoattractant which presumably acts intracellularly. This is consistent with the observation that physiological concentrations (1–10 μ*M*) of cGMP enhance phospholipid methyltransferase *in vitro*. 8-Br-cGMP has no effect on the level of lysophosphatidyl [*methyl*-^{3}H]choline, whereas activation by cAMP induces a transient decrease in the amount of radioactivity in this molecule. Interestingly, folic acid does not elicit this decrease (Alemany *et al.*, 1980). It will be important to evaluate the relevance of these reactions to chemotaxis and secretion. The presence of protein carboxymethylase and esterases and phospholipid methylases appears to be correlated with the presence of secretory and chemotactic functions in higher organisms (Gagnon and Heisler, 1981; and see below).

4. *Dephosphorylation of Myosin*

When activated cells are lysed in detergent and incubated with [γ-^{32}P]ATP the heavy chain of myosin is phosphorylated (Rahmsdorf *et al.*, 1978). Apparently, during the activated phase *in vivo* there is a transient inhibition of myosin heavy chain kinase, and heavy chains are dephosphorylated, liberating more substrate for the subsequent reaction *in vitro* (Malchow *et al.*, 1981). The phosphorylation of purified *D. discoideum* myosin heavy chains inhibits both actin-activated ATPase activity and self assembly of myosin into thick filaments *in vitro* (Kuczmarski and Spudich, 1980). The dephosphorylation of myosin during chemotactic stimulation can thus be correlated with increased *in vivo* activity of myosin.

5. *Proton Efflux*

There is an acidification of the medium associated with the response. With cells in suspension the change is about 0.01 to 0.02 pH units and is readily observable by continuous recording with a pH electrode in unbuffered cell suspensions. This response is elicited by very low cAMP concentrations (10^{-10} *M*) and saturates at a level of about 10^{-8} *M*. The number of protons released is three orders of magnitude greater than the number of cAMP molecules and cannot merely result from cAMP hydrolysis (Malchow *et al.*, 1978a,b). There is also increased oxygen consumption and accelerated CO_2 production during the activated phase (Coe and Chung, 1978; Gerisch *et al.*, 1977a).

6. *Possible Role of Cytoplasmic Calcium*

Cytoplasmic calcium has been cited as a second messenger linking occupancy of surface receptors to activation of intracellular enzymes. Although it has not been determined in *D. discoideum*, in many cell types the resting level of cytoplasmic calcium is about 10^{-7} *M*. During activation it is thought to rise to close to 10^{-5} *M*, a value above the K_d of intracellular calcium binding proteins (Kretsinger, 1979). One hypothesis is that the transduction event(s) in *D. discoideum* involve(s) such a transient increase in cytoplasmic calcium. Although a rapid increase in the amount of cell associated $^{45}Ca^{2+}$

occurs upon chemotactic stimulation (see Fig. 6), extracellular calcium levels do not have rapid, dramatic effects on cellular responsiveness. Chemotaxis and light scattering oscillations take place in 10 m*M* EDTA or 1 m*M* EGTA, and ionophore A23187 in the presence or absence of calcium has little effect on resting levels of cAMP or cGMP (Mato, 1977a,b). Lowering extracellular calcium with EGTA–Ca^{2+} buffer does not affect the onset of aggregation (Saito, 1979; Mato *et al.*, 1977a) although there may be an EGTA-specific inhibition above 10 m*M* EGTA (Mason *et al.*, 1971). cAMP stimulated cAMP secretion is only slightly enhanced by 1 m*M* EGTA and slightly inhibited by 1 m*M* $CaCl_2$ or 1 m*M* $CaCl_2$ plus 10^{-6} *M* ionophore A23187 (unpublished observation). The absence of an effect of extracellular calcium does not preclude a role for cytoplasmic calcium; cells may have mechanisms of controlling cytoplasmic calcium in diverse environments.

Divalent cations have been reported to have long term effects. In *Polysphondylium violaceum*, ionophore A23187-coated glass fibers become foci for center formation (Cone and Bonner, 1979). Under certain conditions, 1 m*M* EDTA plus 5 m*M* $MgCl_2$, ionophore A23187, and $MnCl_2$ can accelerate aggregation in wild-type cells and "rescue" certain aggregationless mutants (Klein and Brachet, 1975; Loomis *et al.*, 1978; Brachet and Klein, 1977; Brachet 1976). Calcium above 10 μ*M* is required for an optimal rate of differentiation, as measured by appearance of contact sites A (Marin and Rothman, 1980). Gerisch *et al.* (1979) reported two experiments in which long pretreatments with $CaCl_2$ or EGTA inhibited changes in cyclic nucleotide levels, but the reproducibility of this effect was not discussed.

A calmodulin-like protein has been purified from vegetative cells by its ability to activate bovine-brain phosphodiesterase (Clarke *et al.*, 1980; Bazari and Clarke, 1981). Antiserum against bovine-brain calmodulin stains cells uniformly throughout the cytoplasm (Mato and Steiner, 1980). Although calmodulin has been demonstrated to activate a large number of intracellular enzymes including phosphodiesterase, adenylate cyclase, and guanylate cyclase (Kretsinger, 1979), it does not activate extracellular phosphodiesterase or adenylate cyclase in *D. discoideum* (Clarke *et al.*, 1980 and unpublished observation). Calmodulin has been shown to inhibit myosin heavy chain kinase *in vitro* (Malchow *et al.*, 1981). A partially purified calcium dependent factor was previously demonstrated to enhance actin-dependent myosin ATPase activity (Mockrin and Spudich, 1976; Clarke and Spudich, 1977). Transfer of [^{3}H]methyl from *S*-adenosyl L-*methyl*-[^{3}H]methionine to phosphatidylcholine and to phosphatidylethanolamine is inhibited by EGTA, chlorpromazine, and antiserum directed against bovine brain calmodulin. This reaction is enhanced by addition of exogenous calmodulin which also reverses the inhibition by chlorpromazine. Some methylation reaction may, in turn, affect calcium movements. ATP dependent

$^{45}Ca^{2+}$ uptake into unfractionated cell homogenates is inhibited by addition of S-adenosylmethionine (Gil *et al.*, 1980).

7. *Vesicle Formation*

Maeda and Gerisch (1977) have studied cells fixed during the transient activation phase by transmission electron microscopy. They report a transient increase in the number of small smooth vesicles observed in sections and an increase in the number undergoing fusion with the plasma membrane, suggesting there is an exocytotic or endocytotic event associated with the response (see Fig. 6). It was suggested that the vesicles may contain cAMP. High concentrations (2 mg/ml) of colchicine have no effect on cAMP secretion and aggregation proceeds in the presence of the drug nocadazole, an effective inhibitor of mitosis in vegetative cells (Cornejo, 1980; unpublished observation). It would be interesting to test the effects of these drugs on fusion of the presumptive cAMP-containing vesicles.

B. Comparison of Kinetics of These Responses

Close examination of the kinetics of cAMP secretion or production during a cAMP signaling response reveals rather complex changes in secretion rate. Often there is a shoulder or inflection point as the secretion rate increases (Gerisch *et al.*, 1977a; Dinauer *et al.*, 1980b). By working at 11–12°C, Gerisch *et al.* were able to clearly resolve a biphasic response of intracellular cAMP. The first peak occurs at about 15 to 30 sec and the second (the major peak) at about 2.5 min. In certain instances the second peak was selectively inhibited (Gerisch *et al.*, 1979). The early peak is not always observed; the exact conditions eliciting it are not clear. The critical factor may be recovery time; that is, the interval since removal of the last stimulus. Dinauer *et al.* (1980b) showed that the rising phase of the secretion rate was accelerated if the interval since the last stimulus was less than 15 min, although two distinct peaks were not always resolved. The falling phase of the major change in the cAMP secretion rate has two distinct kinetic components and occasionally there are two separate peaks (Devreotes and Steck, 1979). These two components are clearly observable when the stimulus is maintained for 10 min. When cells are stimulated in suspension, removal of extracellular cAMP by phosphodiesterase accelerates the falling phase and little of this fine structure is apparent.

Actually, the kinetics of many of the responses to cAMP outlined above are biphasic and often multiphasic. In contrast to cAMP, the initial cGMP peak is always predominant. The second kinetic component often appears as a biphasic decline in the intracellular cGMP level. In several instances two clearly defined peaks have been observed; the first occurring at 15 to 30 sec, the second at about 2.5 min, close to the major cAMP peak (Gerisch *et al.*, 1977a, 1979; Wurster *et al.*, 1977, 1979). The form of the light scattering response depends on developmental age. Early in development, it consists of a single rapid peak at about 30 sec (peak I). As development proceeds, a peak at 3.5 min (peak II) is seen whereafter an intermediate peak at 2.5 min (peak III) appears. The slowest peak then subsides and peaks I and III persist as a biphasic response (Gerisch *et al.*, 1975b; Lax, 1979). The increase in phosphatidyl [*methyl*-^{3}H]choline peaks rapidly at about 60 sec and rises again by 5 min (Mato and Marín-Cao, 1979). The initial decrease in the pH of the medium peaks at about 20–30 sec and a second decrease occurs at about 3 to 4 min, delayed somewhat compared to the second peaks in cAMP and light scattering (Malchow *et al.*, 1978).

It is difficult to closely compare the kinetics of these reactions. The experiments were carried out under variable conditions using a number of strains or subclones of *D. discoideum* and cells of different developmental ages. The similar biphasic kinetics of all the responses may indicate that they are all driven by changes in the level of yet another metabolite which fluctuates biphasically. Alternatively, the biphasic responses could be an artifact of the stimulus delivery; initially there is a sudden change in receptor occupancy and this is followed by a secondary increase due to cAMP secretion. It would be interesting to study the kinetics in response to the type of stimulus that is encountered *in situ* during aggregation, that is, a gradually starting, exponentially rising cAMP level.

As previously explained, cAMP-stimulated cAMP production stops within several minutes because of adaptation and in cell suspensions, extracellular cAMP is normally cleared away by phosphodiesterase. Under these conditions any response that depends on continuous receptor occupancy would necessarily appear to be transient. A distinction can be made by choosing conditions where the stimulus is continuously applied, for instance by employing high cAMP, cGMP, or cAMP-S as a stimulus. Under these conditions, the light scattering response and cGMP accumulation are slightly prolonged (Gerisch *et al.*, 1975c; Mato and Konijn, 1977a). These appear to be adapting responses, although this point could be tested more rigorously. The time course of proton release is greatly extended by indefinitely maintaining receptor occupancy (Malchow *et al.*, 1978b). (There is not sufficient evidence to decide whether the increase in phosphatidyl [*methyl*-^{3}H]choline, the decrease in lysophosphatidyl [*methyl*-^{3}H]choline, myosin dephosphorylation, $^{45}Ca^{2+}$ influx, or vesicle formation reactions are transient.) As argued above, none of the transient reactions are kinetically consistent with the adaptation process for cAMP production. On the other hand, the kinetics for acidification of the medium are prolonged when continuous cAMP stimuli are applied and this reaction could be related to the adaptation process.

C. Sensitivity to Folic Acid and Pteridine Derivatives

Investigation of the effects of the pan-species chemoattractant folic acid (pteroylglutamic acid) may provide clues to transduction mechanisms. In addition to acting as a chemoattractant, folic acid elicits the light scattering response, cGMP accumulation, and in *D. discoideum*, cAMP production and secretion. Vegetative cells are most sensitive; there is a gradual decrease in sensitivity as cells differentiate. Responses are first detected at concentrations of about 10^{-8} *M*; but 10^{-4} *M* is also effective (Pan *et al.*, 1972; Mato *et al.*, 1977d; Wurster *et al.*, 1979).

The pteridine derivatives, pterin (2-amino-4-hydroxy-6-methylpterin) and xanthopterin (2-amino-4,6-dihydroxypteridine) are at least as effective as folic acid in chemotaxis assays while *p*-aminobenzoic acid is inactive (Pan *et al.*, 1975). There are extracellular and membrane-bound folate deaminases which presumably serve to inactivate folic acid since the product, D-2-deamino-2-hydroxyfolic acid (DAFA), is not a chemoattractant (Pan and Wurster, 1978; Bernstein and Van Driel, 1980a,b; Kakabeeke *et al.*, 1980a). Lumazine (2,4,6-hydroxylpterdine) is also inactive. From these results it would appear that the 2-position NH_2 of the pterin moiety of folic is essential for function. An exception to this general rule is found in the strain *D. minutum* where DAFA (which lacks the 2-position NH_2) attracts these cells at the same threshold concentration as folic acid. An alternate extracellular enzymatic activity has been described in *D. minutum* which apparently inactivates folic acid by splitting the C-9—N-10 bond to yield a 6-hydroxymethylpterin (which is apparently inactive) and *p*-aminobenzoylglutamic acid (Kakebeeke *et al.*, 1980b).

Aggregation competent cells of strain NC4 and AX3 secrete cAMP when stimulated with folate, pterin, xanthopterin, amethopterin, and aminopterin (see Fig. 7). The maximum responses to amethopterin and aminopterin are weaker than those to folate or xanthopterin. In NC4 the largest responses to folate are only about 5% of the maximum response to cAMP. In AX3 maximal cAMP secretion responses induced by folate or xanthopterin are often greater than those elicited by cAMP. However, since the positive feedback loop described in Section II,C appears to be much stronger in AX3 than NC4, there may not be a significant difference in the intrinsic properties of the cells. The dose–response curve for all the folate analogs tested is similar

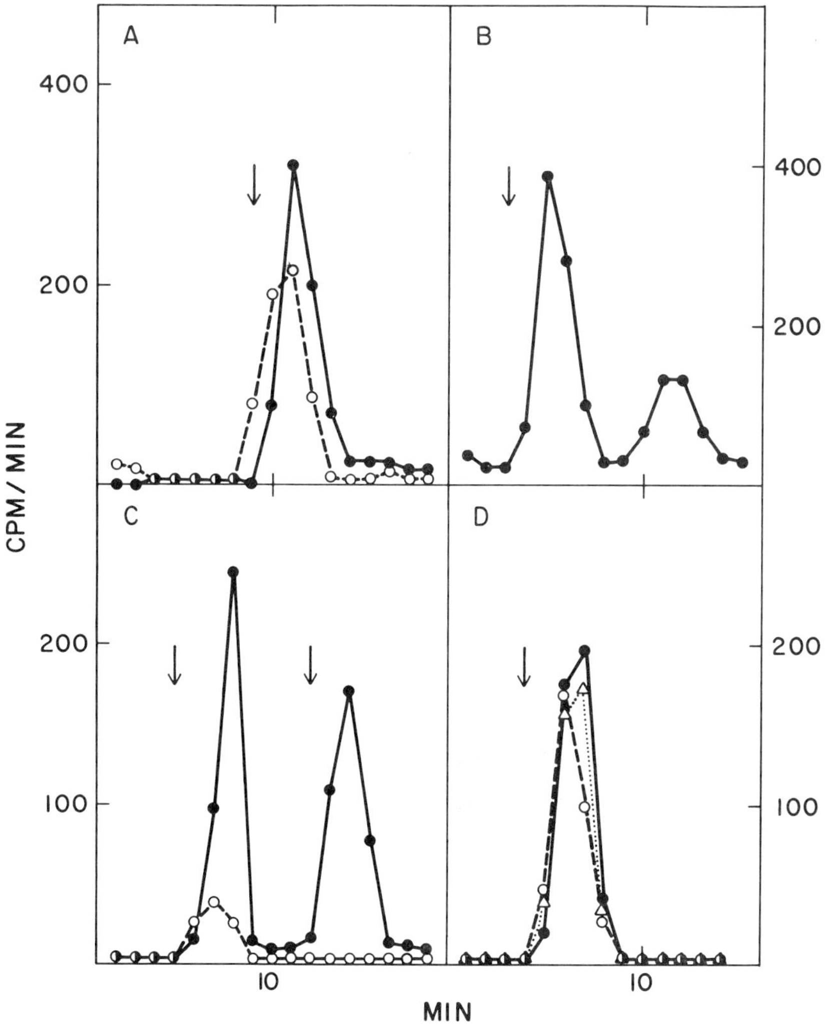

Fig. 7. cAMP signaling responses elicited by folic acid. AX3 amoebae were labeled with [^{3}H]adenosine, developed to the early aggregation stage, and placed in a perfusion apparatus (Devreotes *et al.*, 1979). The experiments illustrated in the different panels in this figure were carried out on different days. (A) Responses elicited from identical filters of amoebae stimulated in parallel with 10^{-5} *M* cAMP (—○—) or 10^{-4} *M* folic acid (—●—); each stimulus was initiated at the time indicated by the arrow and sustained. (B) Another response elicited by a sustained stimulus of 10^{-4} *M* folic acid. (C) One filter was stimulated with 10^{-4} *M* folic acid from the time indicated by the first arrow; at the time indicated by the second arrow, the stimulus was switched to 10^{-5} *M* cAMP (—●—). The other filter was stimulated with 10^{-8} *M* cAMP from the time indicated by the first arrow; at the time indicated by the second arrow, the stimulus was switched to 10^{-4} *M* folic acid (—○—). (D) At the time indicated by the arrow, three identical filters of amoebae were stimulated with a sustained stimulus of either 10^{-4} *M* folic acid (—●—), 10^{-4} *M* pterin (—○—), or 10^{-4} *M* xanthopterin (—Δ—).

and covers the range from 10^{-7} *M* to 10^{-4} *M*, similar to that for folate-elicited cGMP accumulation. Neither curve saturates before the solubility limits are reached.

Since it elicits significant cAMP release, it is not surprising that folic acid mimics some of the effects of applied cAMP pulses. For instance, pulses of folic acid delivered at 7 min, but not 2 min intervals, induce a precocious initiation of spontaneous oscillations (Wurster and Schubiger, 1977; see Section V) and, added just after the active phase, cause a phase delay (Gerisch *et al.*, 1977a; see Section II,E).

There is a significant asymmetric interaction between the cAMP secretion responses stimulated by folate and by cAMP. Once a response to folate has terminated, the cells will still respond normally to cAMP. However, following a response to a low concentration of cAMP (10^{-8} *M*), even the highest concentrations of folate (10^{-4} *M*) elicit no response (see Fig. 7). These observations show that stimulation by folate leads to cAMP production; but, unlike cAMP stimuli, does not elicit the adaptation reaction. Although folate does not inhibit extracellular phosphodiesterase (unpublished observation), it is technically difficult to demonstrate that, alone, it activates adenylate cyclase. Since it induces cAMP release, activation of the enzyme should necessarily occur when secreted cAMP binds to surface cAMP receptors. Folic acid may elicit cAMP production by increasing cellular sensitivity to low endogenous levels of cAMP. Kawai (1980) has reported that folic acid increases [^{3}H]cAMP binding in *D. discoideum*.

D. Phamacological Studies of Intact Cells

The effects of a variety of treatments on intact cells are summarized in this section. Although it is difficult to interpret the effects of drugs applied to intact cells, these observations rule out certain possible mechanisms of transduction processes and provide some clues for further investigation. Puromycin and cycloheximide have no immediate effect on either cGMP or cAMP accumulation but inhibit the latter by about 45% after 45 min of pretreatment (unpublished observation). This suggests that protein synthesis is not directly involved in these events but at least one of the components involved (perhaps cAMP receptors or adenylate cyclase) turns over rapidly. Chloroquine, which inhibits protein degradation by raising intralysosomal pH, also has no immediate effect on cAMP accumulation but inhibits by 50% after 1 h. The protease inhibitor *p*-tosyl-L-arginine methyl ester (TAME) inhibits the cAMP signaling responses. The trypsin inhibitor, N$^{\alpha}$-*p*-tosyl-L-lysine chloromethyl ketone (TLCK), has little effect on responsiveness but does lower the background cAMP secretion rate (unpublished observations). Both TLCK and chloroquine inhibit cell aggregation in the early stages (Fong and Bonner, 1979).

Chlorpromazine and other phenothiazines that bind specifically to calmodulin block cAMP and cGMP accumulation, and the light scattering response (unpublished observations; M. Brenner, personal communication). The order of potency for the phenothiazines in inhibiting cAMP secretion is consistant with that for binding to calmodulin. However, these drugs also have other membrane perturbing activities and a direct effect on calmodulin is not necessarily indicated. Sodium azide, potassium cyanide, and carbonyl cyanide *m*-chlorophenylhydrazone (CCCP), inhibitors of oxidative phosphorylation, rapidly block cAMP accumulation (Dinauer *et al.*, 1980a). In the case of azide and cyanide, it has been shown that these drugs actually block activation of adenylate cyclase and inhibition does not result merely from depletion of the substrate, ATP (Dinauer *et al.*, 1980a). Dinitrophenol (DNP) and rotenone have only slight effects on cAMP, although these drugs do block spontaneous oscillations in cell suspensions (Geller and Brenner, 1978b). Ammonia blocks cAMP accumulation, which may have significant effects on morphogenesis since cells produce ammonia throughout development (Schindler and Sussman, 1977; Kay, 1979; Thadani *et al.*, 1977). Raising the osmotic strength to about 150 mosmol, a nonphysiologically high concentration for these organisms, also completely inhibits cAMP-elicited cAMP production (unpublished observation; Darmon and Klein, 1978). Concanavalin A, which when bound rapidly redistributes on the cell surface, increases cGMP accumulation, [^{3}H]cAMP binding, and phosphodiesterase activity, but inhibits cAMP secretion responses (Gilette *et al.*, 1974; Mato *et al.*, 1978c; Juliani and Klein, 1977; Filosa, 1978; unpublished observations).

M. Brenner has found that caffeine and adenosine are extremely interesting inhibitors (personal communication). Each inhibits cAMP accumulation and, in the same dose range, enhances cGMP accumulation and light scattering responses. The basal activities of guanylate or adenylate cyclase are not affected by caffeine. Chemotaxis is not inhibited by caffeine. This demonstrates definitively that the increase in intracellular cAMP is not necessary for chemotaxis. The parallel enhancement of both cGMP and light scattering changes is consistent with the hypothesized role of cGMP in chemotaxis. These drugs provide a method for studying the intertwined roles of chemotaxis and cAMP signal relay. For instance, it has recently been demonstrated that appropriate doses of adenosine or caffeine block center formation with little effect on cell–cell relay of cAMP signals or chemotaxis, resulting in abnormally large aggregation territories (Newell, 1981; unpublished observation).

E. Chemosensory Transduction in Higher Organisms

There is a striking similarity of the set of reactions described in this section to those triggered by chemoattractants and agents which elicit secretion responses in leukocytes, mast cells, platelets, and macrophages. For example, in mast cells, cross linking of IgE receptor leads to increases in intracellular cGMP and cAMP, increases in calcium influx, increases in methylation of phospholipids, and exocytosis of histamine-containing vesicles (Ishizaka, 1981). Similar sets of reactions are triggered in leukocytes and macrophages by f-Met-Leu-Phe and C5a, and in activated platelets. Most of the reactions are rapid, lasting no more than a few minutes. The cellular slime molds may provide one of the most easily manipulated systems for study of the interrelationship of this common set of reactions.

IV. Molecular Components of the Chemosensory System

A. Surface cAMP Receptors

As outlined in the preceding sections, each of the reactions that are activated by low concentrations of extracellular cAMP could theoretically be connected to different cell surface cAMP receptors. Alternatively, a single receptor class could be coupled to all these intracellular events. cGMP accumulation, chemotaxis, and light scattering seem to be mediated by the same receptor, but there is little evidence to indicate whether or not this receptor is linked to the other reactions.

There is little characterization of surface cAMP receptors at the molecular level. The most significant obstacles are the rapid dissociation rate ($t_{1/2}$ = 1–5 sec) of the cAMP–receptor complex and the absence of a slowly dissociating ligand (Mullens and Newell, 1978; King and Frazier, 1977). Accordingly, most studies have been designed to measure equilibrium binding. [^{3}H]cAMP, along with cGMP or dithiothreitol (DTT), inhibitors of *D. discoideum* phosphodiesterase, are mixed with intact cells in the presence and absence of a saturating concentration (10^{-4} *M*) of unlabeled cAMP. Cells are collected on millipore or nucleopore filters, centrifuged through silicone, or collected in pellets (i.e., there is no washing step). Saturable binding is taken as radioactivity bound in the absence minus that bound in the presence of unlabeled cAMP (Malchow and Gerisch, 1974; Green and Newell, 1975; Henderson, 1975).

Scatchard plots of saturable binding are curvilinear upwards indicating either heterogeneity in binding affinities or negatively cooperative interactions between binding sites—10,000 sites with a dissociation constant of about 5 n*M* and about 200,000 sites with a dissociation constant of about 100 n*M*. This range of affinities is consistent with that over which chemotaxis and cAMP elicited cAMP secretion is observed. In the latter case, responses can be elicited with doses as low as 10^{-10} *M* cAMP (Devreotes and Steck, 1979; unpublished observations). Cells are attracted chemotactically to a similar range of cAMP concentrations; the threshold of sensitivity is between 10^{-9} *M* and 10^{-8} *M* cAMP and cells are disoriented above 10^{-6} *M* cAMP (presumably because receptors are saturated and there is no gradient of receptor occupancy across the cell) (Konijn, 1970). The rapid dissociation rate is also consistent with physiological experiments—chemotaxis abruptly terminates when the cAMP gradient is removed; and when a cAMP stimulus is removed during a cAMP signaling response, intracellular cAMP levels begin to fall within 5 sec (Futrelle *et al*., 1980, 1981; Dinauer *et al*., 1980a).

Such curvilinear Scatchard plots have been observed for *D. discoideum* NC4 (K_d = 10–100 n*M*), *D. purpureum* (K_d = 10–580 n*M*), and *D. mucoroides* ($K_d \cong$ 10–95 n*M*) (Mullens and Newell, 1978; Newell and Mullens, 1978). A number of mutants characterized by aberrant aggregation patterns have been described for which the Scatchard plots are linear—the high or low affinity sites are absent (Ross and Newell, 1980; Barclay and Henderson, 1977, personal communication; Juliani and Klein, 1978). Such observations do not discriminate between site heterogeneity and site–site interaction hypotheses; the data could be accounted for by either scheme. The rate of dissociation of [^{3}H]cAMP is accelerated in the presence of unlabeled cAMP, theoretically favoring a site–site interaction model. However, alternate interpretations of similar observations of insulin binding have been offered (Pollet *et al*., 1977). Ca^{2+} ion, brief concanavalin A treatment, and folic acid dramatically increase the number of cAMP binding sites without altering the ratio of high to low affinity sites (Juliani and Klein, 1977; Kawai, 1980).

A series of cAMP analogs have been tested in a standardized chemotaxis assay. When the threshold for chemotaxis is raised 1000-fold by substitution of a given atom or group, that part of the molecule is considered essential for activity. This series of analogs, in a similar order of potency, elicit cGMP accumulation and light scattering responses. Substitutions in the purine ring, including those of the 8 position which tend to stabilize the syn conformation of the nucleotide, lower activity 10- to 100-fold. Minor substitution of the cyclized 3′-oxygen drastically reduces activity, while bulky substitutions at the 2′ position of the ribose moiety have only a slight effect. A model for cAMP–surface receptor interaction has been proposed in which cAMP is held in the anti conformation. Hydrogen bonding at the 6-amino, N-7, and 3′-oxygen, as well as a hydrophobic interaction of the purine ring are postulated to be important for activity (Jastorff, 1978; Jastorff *et al.*, 1978; Mato and Konijn, 1977a; Mato *et al.*, 1978a).

King and Frazier (1977, 1979; and see King *et al.*, 1978) have employed a technique to measure [^{3}H]cAMP binding in which a cell suspension is mixed with [^{3}H]cAMP and DTT, aliquots are removed, placed on nucleopore filters and rapidly washed. Their unprecedented finding is that, under these conditions, the number of binding sites or the binding affinity oscillates with a period of exactly two min. Oscillations are observed at 0°C and in membrane preparations at 0°C. These authors suggested that the transient decreases in binding could account for the adaptive behavior of many of the cAMP-elicited responses described. In addition, receptor desensitization has been employed as a means of response attenuation in a computer simulation of cAMP relay responses (Goldbeter and Martiel, 1980). There are several arguments against this possibility. First, under the conditions used by King *et al.*, cellular levels of cAMP do not oscillate; DTT blocks oscillations since it inhibits extracellular phosphodiesterase; cells at 0°C do not synthesize and secrete cAMP; and membrane preparations are completely devoid of cAMP-elicited cAMP production. The oscillatory [^{3}H]cAMP binding appears to be unrelated to other physiological behavior. Second, while several other investigators have measured transient decreases in binding of [^{3}H]cAMP during the active secretory phase, the decreases could be attributed to isotope dilution effects by secreted cAMP (Klein *et al.*, 1977; Gerisch *et al.*, 1979; unpublished observation).

When cells are preincubated with unlabeled cAMP, washed, and rechallenged with [^{3}H]cAMP there is a gradual decrease in the level of [^{3}H]cAMP binding. The amount of bound [^{3}H]cAMP depends on the concentration and duration of the pretreatment. The remaining sites have the same Scatchard plot as the original sites, but there is a slight shift in the electrophoretic mobility of a protein which can be covalently modified by 8-azido[^{32}P]cAMP (see below). Evidence suggests that this "desensitization" results from conversion of binding sites to a slowly dissociable form so that fewer sites are available for binding of [^{3}H]cAMP (Klein and Juliani, 1977; Klein, 1979). The desensitization phenomenon does not correlate kinetically or in a dose dependent manner with the adaptation process(es) which attenuate(s) cAMP or cGMP production. Pretreatment with 10^{-9} *M* cAMP barely reduces the amount of [^{3}H]cAMP subsequently bound. When this concentration is persistently applied to sensitive cells, cAMP-elicited cAMP production and cGMP production are completely extinguished within a few min (Devreotes and Steck, 1979; Mato *et al.*, 1977a). However, the slow loss of binding sites does correlate with a gradual, irreversible inhibition of subsequent cAMP signalling responses during pretreatment with high cAMP concentrations. For instance, cells treated with 10^{-5} *M* cAMP for 30 min recover about 55% of their original sensitivity within 10 min but there is no further recovery by 45 min (Dinauer *et al.*, 1980b).

Several investigators have reported photoaffinity labeling of surface receptors by 8-azido[^{32}P]cAMP (Hahn *et al.*, 1977; Juliani and Klein, 1981; Wallace and Frazier, 1979). 8-azido[^{32}P]cAMP is covalently attached to material which migrates as 10 to 15 bands on SDS gels, most of which are not affected by addition of unlabeled cAMP to the reaction. The large degree of nonspecific labeling is not surprising. Modification of the 8 position of the adenine ring reduces chemotactic sensitivity about 100-fold; and high concentrations of the photolabel must be applied to cells. One band is protected by including unlabeled cAMP in the reaction but there is a discrepancy concerning which one. Wallace and Frazier (1979) report on a 40,000 molecular weight band while Juliani and Klein (1981), working at lower cell densities, observe that a 45,000 molecular weight band is protected. Photoaffinity labeling of any of these bands has not been demonstrated to alter chemotactic sensitivity or binding of [^{3}H]cAMP. In addition, perfusion of cells with active 8-azido-cAMP under intense UV irradiation has no effect on subsequent assays of cAMP-elicited cAMP secretion (unpublished observation). These latter observations do not rule out that functional receptors are labeled but suggest that, if so, only a small fraction (<5%) is modified.

B. Adenylate Cyclase

None of the reactions that are activated by cAMP in intact cells are stimulated when a physiological concentration of cAMP is added to broken cells (but see Klein and Darmon, 1979). This uncoupling appears to be instantaneous, but several of the key enzymes from stimulated cells remain active *in vitro*. When cells taken at the peak of a cAMP signaling response are quickly lysed and added to the reaction mixture for adenylate cyclase, there is an elevated rate of [^{32}P]cAMP synthesis which gradually slows over a period of one min to a stable basal rate (Roos and Gerisch, 1976; Roos *et al.*, 1977b; see Fig. 8A). This decay can be retarded for about 15 min by holding the activated homogenate on melting ice and preserved indefinitely by freezing (Roos *et al.*, 1977a; Klein *et al.*, 1977).

The residual activation measurable after lysis can be exploited to follow the time course of enzyme activation of intact cells. The activity of samples taken during the course of response are assayed for one min *in vitro*. The

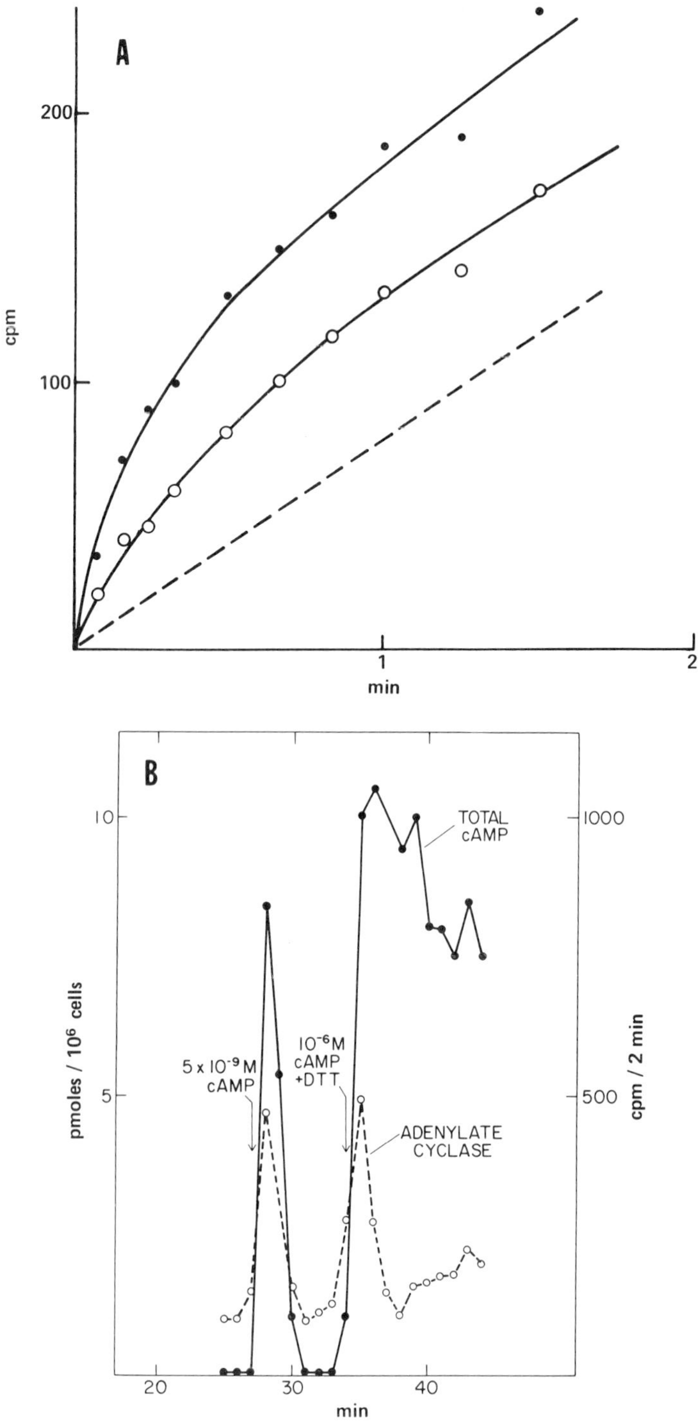

A
200
100
cpm
1
2
min
B
10
5
1000
500
pmoles / 10^6 cells
cpm / 2 min
TOTAL cAMP
ADENYLATE CYCLASE
5 x 10^{-9}M cAMP
10^{-6}M cAMP +DTT
20
30
40
min

activity of the enzyme *builds up* for about 1 min, reaches a peak value and returns to basal levels within several min (see Fig. 8B). In such experiments basal rates are about 5–15 pmole/min mg and the activation is about fourfold. This activity is not sufficient to account for changes in cellular cAMP levels *in vivo* indicating that the *in vitro* conditions are suboptimal (Brenner, 1978).

Although the instantaneous uncoupling could possibly stem from rapid degradation of surface receptors, several investigators observe binding of [^{3}H]cAMP to partially purified membranes (Henderson, personal communication; Klein, 1981; Meyers and Frazier, 1981; Coukell, personal communication). Given that receptors remain intact *in vitro*, the uncoupling could result in several ways. Receptors could be physically separated from the adenylate cyclase. Alternatively, the transduction event may depend on some property that requires cells to be intact—such as a change in membrane potential or attainment of a critical level of cytoplasmic calcium. The decay of the activated state following uncoupling may result from depletion of an essential reactant, dissociation of a component of the activated complex, or (de)modification of a component of the reaction system. It is unlikely that the decay results from proteolytic degradation of the adenylate cyclase since the basal activity is stable for several hours.

Due to the instability of the activated state, most investigators have focused on the basal enzyme. In broken cells, the basal activity is stable for at least 25 min at 25°C, reportedly decaying with a half-time of about 24 h at 0°C. These preparations exhibit non-Michaelian kinetics with two apparent K_m values of 17 μM and 400 μM (deGunzburg *et al.*, 1980). Activity is inhibited by 0.1 mM to 1 mM Ca^{2+}; Mn^{2+} antagonizes this inhibition and causes further stimulation (Loomis *et al.*, 1978); and EGTA also stimulates about twofold (unpublished observations). Vegetative cells contain a heat stable inhibitor of enzyme activity (Cripps and Rutherford, 1981). The adenylate cyclase activity appears to be on small vesicles, as it is found in supernatants after low speed centrifugation. It can be separated from several

Fig. 8. Adenylate cyclase activity in sonicates of activated cells. (A) Samples were taken from a cell suspension 30 sec before (----), 15 sec (○—○) and 90 sec (●—●) after stimulation by cAMP, sonicated, and mixed with the reaction mixture for adenylate cyclase. Shown is the amount of [^{32}P]cAMP synthesized as a function of time after initiation of each reaction. (B) Time course of adenylate cyclase activity during cAMP stimulation. Unlabeled AX3 amoebae in suspension were synchronized by 5×10^{-9} M cAMP four times at 7 min intervals. The figure shows only the fourth synchronization cycle (28 min). At 34 min, the suspension was adjusted to 10^{-6} M cAMP, 10 mM DTT (an inhibitor of extracellular phosphodiesterase) as indicated. At 1-min intervals, 50 μl of suspension were removed for determination of total cAMP; 500 μl were removed, sonicated, and 100 μl used for determination of adenylate cyclase activity in a 60 sec reaction. 1,000 cpm of [^{32}P]cAMP corresponds to ~25 pmol of [^{32}P]cAMP. The experiment was repeated four times with similar results.

surface membrane markers on sucrose gradients and is not recovered in purified plasma membranes, suggesting that the enzyme may be attached to a select portion of the plasma membrane or found on intracellular membranes (Hinterman and Parish, 1979; deGunzburg *et al.*, 1980).

The supernatant fraction after low speed centrifugation synthesizes [^{32}P]cAMP at a constant rate for several h at 25°C. This enzyme activity is membrane bound since it can be sedimented by further centrifugation 12,000 *g* for 60 min. Only about 30–50% of the original activity is recovered in this step; enzyme activity is always lost during subcellular fractionation (Pahlic and Rutherford, 1979; deGunzburg *et al.*, 1980; C. Klein, M. Brenner, personal communication; unpublished observation). The original activity can be restored by readdition of the supernatant fraction, suggesting that the loss of activity results from separation of a soluble factor which activates or stabilizes the enzyme. This activating or stabilizing activity has been shown to copurify with a heat stable protein of 14 kdaltons (Devreotes, manuscript in preparation). Preparations of adenylate cyclase are completely inactivated by low concentrations (0.05% and above) of the detergents Lubrol-Px, Triton X-100, sodium cholate, or deoxyglucoside. Inclusion of the activation/stabilization protein restores about 70% of the activity in a solubilized adenylate cyclase preparation (Devreotes, manuscript in preparation).

It was previously reported that the *D. discoideum* enzyme in crude homogenates is not activated by NaF, GTP, GMP–PNP, addition of cholera toxin to intact cells, or by calmodulin in the presence or absence of calcium (Klein, 1976; Rickenberg, personal communication; Brenner, personal communication; unpublished observations). However, recent experiments with partially purified membranes indicate that *D. discoideum* contains a 42-kdalton protein which can be labeled by both 8-azido-GTP and the cholera toxin catalyzed transfer of ADP-ribose from NAD (Leichtling *et al.*, 1981), properties attributed to the GTP binding regulatory protein of adenylate cyclase found in higher eukaryotes. It is not clear why adenylate cyclase regulation by guanyl nucleotides is not observed *in vitro*, and further experiments are required to learn the relationship between *D. discoideum* adenylate cyclase and that of higher eukaryotic cells.

C. Guanylate Cyclase

In analogy to adenylate cyclase, guanylate cyclase is not stimulated by addition of cAMP to broken cell preparations but [^{32}P]cGMP is transiently synthesized at an elevated rate by sonicates of activated cells (Mato and

Malchow, 1978). The basal guanylate cyclase is about 10 to 50 pmoles/min mg in *D. discoideum* strains AX2 and AX3, and rises only slightly during development. Mn^{2+} in excess of GTP (K_m for GTP is 0.5 mM) is essential for activity. The majority of the activity is found in the supernatant following high speed centrifugation but a significant amount is lost in the fractionation step, suggesting that several components may be necessary for full activity. Non-ionic detergents and membrane disruptive antibiotics inhibit activity. Mato *et al.* demonstrated that basal activity is stimulated about twofold by 200 μM ATP but not by nonhydrolyzable ATP analogs. The enzyme is unstable, a 50% loss occurs in about 4 h at 0°C (Ward and Brenner, 1977; Mato *et al.*, 1978b; Mato, 1979).

D. cAMP Phosphodiesterase

Secreted phosphodiesterase activity is stable and is the best characterized component of the chemosensory system. Activities of high (2 mM) and low (4 μM) K_m are found. The high K_m state results from tight association (K_d = $10^{-10} M$) of the low K_m enzyme with the heat-stable inhibitory protein of 47 kdaltons. The inhibitory protein is cysteine rich; the complex can be broken by dithiothreitol (DDT), liberating the low K_m enzyme (Kessin *et al.*, 1979; Franke and Kessin, 1981; Dicou and Brachet, 1979b). The activity with low K_m has been characterized by a number of investigators and multiple forms are observed (Dicou and Brachet, 1979a; Toorchen and Henderson, 1979). One has a low molecular weight (about 48,000) and isoelectric point of about 8; others have higher molecular weights and have isoelectric points from 4 to 6. Tsang and Coukell (1979a) purified the low molecular weight form from *D. purpureum* and found it to be a doublet, probably resulting from processing. Orlow *et al.* (1981) purified two forms of the enzyme from *D. discoideum* and showed by peptide mapping that the high molecular weight form consists of the low molecular weight form bound to acidic material. Such an association may also account for observations of multiple forms with lower isoelectric points.

The membrane bound and secreted forms of phosphodiesterase appear to be closely related. Following treatment with urea, the membrane bound form can be precipitated by antisera directed against the secreted enzyme and is inhibited by the heat-stable inhibitory protein. Preliminary experiments have revealed no differences in peptide maps between the secreted and membrane bound enzymes (Kessin, personal communication). An intracellular phosphodiesterase is also present in *D. discoideum*. It will be interesting to determine the relationship between each of these activities.

E. Cellular Association of Folic Acid

When [^{3}H]folic acid is mixed with intact *D. discoideum* cells there is a rapid rate of association of radioactivity which reaches a stable plateau after 15 to 20 sec. Binding measurements are made without a washing step since the rate of dissociation is equally rapid. Cells have an active folate deaminase. The stable plateau of associated [^{3}H]folic acid occurs since the product of the deamination reaction, D-[2-^{3}H]deamino-2-hydroxyfolic acid ([^{3}H]DAFA) and [^{3}H]folic acid bind with equal affinity (Wurster and Butz, 1980). Saturably bound radioactivity is taken as that bound at 30 to 45 sec in the absence of unlabeled folic acid minus that bound in the presence of a high concentration of unlabeled folic acid. When this bound radioactivity is subject to Scatchard analysis there are 60,000–200,000 sites with a K_d of about 2×10^{-7} *M* (Wurster and Butz, 1980; Van Driel, 1981). Aminopterin and methotrexate are not substrates for folate deaminase and may provide more useful ligands for binding studies (Nandini-Kishore and Frazier, 1982).

A close correlation between binding and chemotactic sensitivity has not been established. It is not clear why [^{3}H]DAFA should bind since this compound is not a chemoattractant or an antagonist of chemotaxis to folate (Van Haastert, personal communication). Pterins, which are as active as folic acid as attractants, do not compete with [^{3}H]folic acid binding (Wurster and Butz, 1980). Consistent with this observation, methotrexate inhibited folic acid-elicited but not pterin-elicited cAMP secretion (unpublished observation). Assuming that the effects of folic acid are mediated by a surface receptor, there may also be an independent receptor for pterins. Alternatively both folate and pterins may act at intracellular sites but enter the cell by different mechanisms. The differential effects of methotrexate might reflect its inhibition of folate, but not pterin, entry into the cell.

V. Role of the Chemotactic Signal in Early Development

When development is initiated, *D. discoideum* amoebae have few surface cAMP receptors, little membrane bound and extracellular phosphodiesterase (but a high level of phosphodiesterase inhibitory protein), low adenylate cyclase activity, few contact sites A (measured by formation of cellular aggregates in suspension), and low membrane-bound and extracellular folate deaminase activity. During the first several hours receptor binding, adenylate cyclase, phosphodiesterase, folate deaminase, and contact sites A activities increase whereas the activity of the inhibitory protein decreases. When delivered in pulses at 5 to 7 min intervals, both cAMP and folic acid accelerate cellular differentiation (Gerisch *et al.*, 1975a,c; Klein, 1975; Klein and Darmon, 1977; Parish, 1979; Rickenburg *et al.*, 1977; Tihon *et al.*, 1977; Yeh *et al.*, 1978). Marin and Rothman (1980) have shown that this stimulation by cAMP pulses only occurs under suboptimal culture conditions. In media which contain greater than 15 mM K^+, development of contact sites A (as defined above) is delayed. In 5 mM Na^+, 10 mM K^+, >10 μM Ca^{2+} development is accelerated compared to high K^+ medium. Under these conditions, cAMP pulses do not further increase the rate of development of cell aggregates in suspension.

Regulation of phosphodiesterase, its inhibitory protein, or folate deaminase activities does not require pulsatile application of cAMP or folate; a continuous flux of a low concentration is sufficient to induce these activities (Tsang and Coukell, 1977, 1978, 1979b; Yeh *et al.*, 1978; Berstein *et al.*, 1981; Hayashi and Yamasaki, 1978). However, such a continuous application has an inhibitory effect on the other functions. This suggests that although continuous occupancy of surface receptors is sufficient for regulation of phosphodiesterase and its inhibitory protein and folate deaminase, regulation of cellular differentiation requires modulation of receptor occupancy. The necessity for modulation of extracellular cAMP levels probably accounts for the accelerating effects of exogenous phosphodiesterase on early development (Alcantara and Bazill, 1976; Brachet and Dicou, 1978; Darmon *et al.*, 1978; Klein and Darmon, 1976; Wier, 1977).

Due to the properties of the adaptation process outlined in previous sections, cells in suspension will repeatedly transiently respond to pulses delivered at 7 min intervals, whereas at shorter intervals, or with continuous fluxes, the transient responses are suppressed. This suggests that some property of the transient response accelerates differentiation. One possibility is the repeated elevation in intracellular cAMP levels. Exposure of cells to $5 \times 10^{-4} M$ cAMP has been reported to accelerate differentiation (Sampson *et al.*, 1978). Conversely, in *Agip* 53, a mutant of AX2, surface cAMP receptors seem to be uncoupled from adenylate cyclase since stimulation by cAMP causes changes in intracellular cGMP but does not elicit the major increase in cAMP. Nevertheless, the differentiation of *Agip 53* is accelerated by pulses of cAMP (Wurster and Bumann, 1981).

The cAMP receptors mediating induction of phosphodiesterase and chemotaxis have similar specificities when tested with a series of cAMP analogs. The dose–response curve for chemotaxis and phosphodiesterase induction by some of the weaker analogs is shifted to the right by several orders of magnitude. When applied at high concentrations the analogs with lowest affinity are most effective in phosphodiesterase induction (Van Haastert *et al.*, 1981). This might be related to the observation that N^6-(aminohexyl)adenosine 3′,5′-monophosphate induces differentiation but cannot be shown to bind to cAMP receptors (Juliani *et al.*, 1981).

VI. Working Model of Early Aggregation

Starvation triggers the synthesis of the components necessary for chemotaxis, cell–cell relay of cAMP signals, and formation of cellular contacts. The initial increase in adenylate cyclase leads to an increase in cAMP which in turn accelerates the appearance of phosphodiesterase. Once the capacity for cell–cell relay of cAMP signals is attained, it may serve to further accelerate the developmental program. Some of the cells or groups of cells begin to spontaneously initiate cAMP waves that propagate through the monolayer. Figure 9 illustrates the dynamics of cell motion in relation to the passing cAMP wave. As the wave approaches, the cell begins to move chemotactically and to secrete cAMP. As the wave advances, the cell's rate of cAMP secretion increases as does its level of adaptation. As the peak of the wave reaches the cell, its secretion rate slows and begins to return toward resting levels. Once extracellular cAMP is cleared by phosphodiesterase, the level of adaptation decays and the cell becomes resensitized. After the passage of 20 to 50 waves, cell–cell contacts are formed, many of the chemotactic functions are suppressed, and aggregation proceeds by the retraction of streams.

There are a number of challenging questions left open in this general scheme. Just how early in development are cAMP waves propagated through the monolayer? Is it possible that they occur very early in development, before the cells can react chemotactically, and have an inductive function? What determines the velocity of the cAMP waves? A remarkable observation is that a wave travels at constant velocity but each successive one moves more slowly. This suggests that cells may mark developmental time by counting waves (Gross *et al.*, 1977). Another puzzling observation is the variation among cAMP wave profiles and movement step durations. The all-or-nothing responses triggered in cell suspensions might lead one to expect more stereotyped behavior.

The latter questions might be resolved by computer simulation of aggregation. Several computer simulations have erroneously chosen the release cAMP signal to be of short duration (actually a delta function) but have successfully led to cell aggregation (Cohen and Robertson, 1971; MacKay, 1978; Parnas and Segal, 1977; Sperb, 1979; Freidlin and Sivak, 1979). It will be interesting to see what can be learned by basing the simulations on the

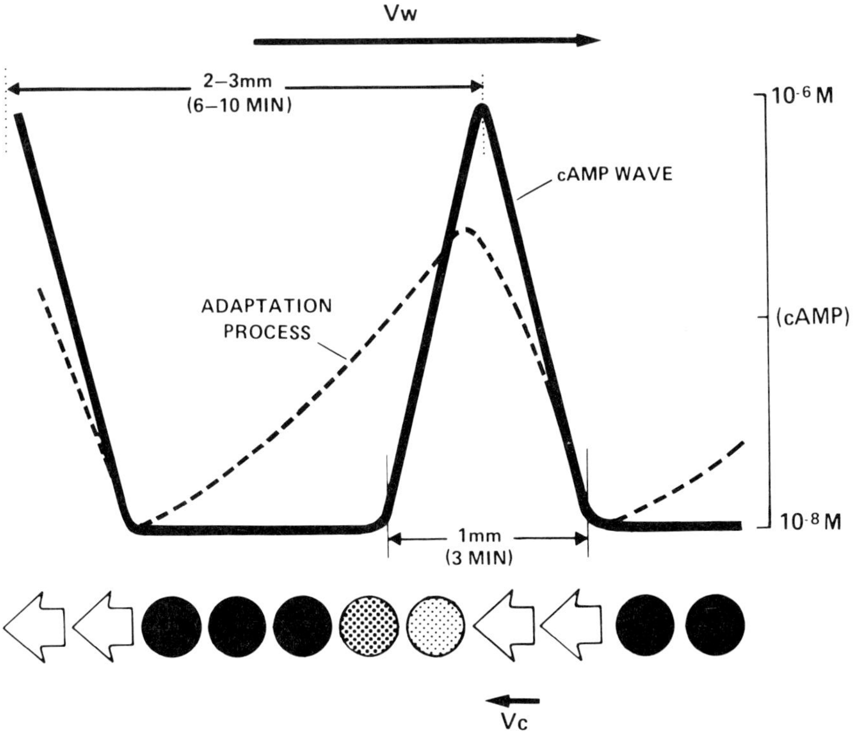

Fig. 9. Dynamics of signal relay and chemotaxis. The heavy line representing the cyclic AMP concentration is drawn from analyses of scans of the optical density of fluorographic images of the cyclic AMP waves (Fig. 2). The concentrations given are for total cyclic AMP; extracellular concentrations may be considerably lower. Symbols in lower part of diagram represent a single radial line of cells as in Fig. 1B. Large arrows represent chemotactically oriented cells while circles represent randomly oriented cells. Arrow vectors indicate the speed and direction of motion of the cycle AMP wave ($V_w \cong 300$ μm/min) and the moving cells ($V_c \cong 20$ μm/min).

new experimental data (see Parnas and Segal, 1978). Are chemotaxis and cell–cell signaling important in later development? Although most of the important functions are suppressed (i.e., cAMP surface receptors, adenylate cyclase, phosphodiesterase), they do remain at levels above those in growing cells. Finally, the parallel between reactions triggered by chemoattractants in this primitive organism and in higher eukaryotes, and the recent demonstration that it contains the GTP-binding regulatory protein of adenylate cyclase suggest that *D. discoideum* provides an exciting, relevant model system for mechanisms of chemotaxis, secretion and hormone action at work in higher organisms.

Acknowledgments

I would like to thank Drs. M. Potel and S. MacKay, Mr. L. Buhle and Mr. B. Knox for computer analyses, Drs. M. Brenner and T. Steck and Ms. A. Burton-Theibert for critical review and Ms. M. Di Salvo and Ms. S. Metzger for secretarial assistance.

Bibliography

Alcantara, F. and Bazill, G. W. (1976). Extracellular AMP phosphodiesterase accelerates differentiation in *Dictyostelium discoideum*. *J. Gen. Microbiol.* **92,** 351–368.

Alcantara, F. and Monk, M. (1974). Signal propagation in the cellular slime mould *Dictyostelium discoideum*. *J. Gen. Microbiol.* **85,** 321–324.

Alemany, S., Gil, M. G., and Mato, J. M. (1980). Regulation by cyclic GMP of phospholipid methylation during chemotaxis in *Dictyostelium discoideum*. *Proc. Natl. Acad. Sci. U.S.A.* **77,** 6996–6999.

Arndt, A. (1937). Rhizopodenstodien III—Untersuchungen über *Dictyostelium mocoroides* brefeld. *Wilhelm Roux' Entwicklungs Mech. Org.* **136,** 681–744.

Barclay, S. L. and Henderson, E. J. (1977). A method for selecting aggregation defective mutants of *Dictyostelium discoideum*. *In* "Developments and Differentiation in the Cellular Slime Moulds" (P. Cappuccinelli and J. Ashworth, eds.), pp. 291–296. Elsevier/North-Holland, New York.

Bazari, W. L. and Clarke, M. (1981). Characterization of a novel calmodulin from *D. discoideum*. *J. Biol. Chem.* In press.

Berg, H. C. and Purcell, E. M. (1977). Physics of chemoreception. *Biophys. J.* **20,** 193–219.

Bernstein, R. L., Rossier, C., Van Driel, R., Brunner, M., and Gerisch, G. (1981). Folate deaminase and cyclic AMP phosphodiesterase in *Dictyostelium discoideum:* their regulation by extracellular cyclic AMP and folic acid. *Cell Differentiation.* In press.

Bernstein, R. L. and Van Driel, R. (1980a). Degradation of the chemoattractant folic acid by *Dictyostelium discoideum* (abstract). *Eur. J. Cell Biol.* **22,** 234.

Bernstein, R. L. and Van Driel, R. (1980b). Control of folate deaminase activity of *Dictyostelium discoideum* by cAMP. *FEBS Lett.* **119,** 249–253.

Berridge, M. J. and Rapp D. E. (1978). A comparative survey of the function, mechanism and control of cellular oscillators. *J. Exp. Biol.* **81,** 217–279.

Beug, H., Katz, F., and Gerisch, G. (1973). Dynamics of antigenic membrane sites relating to cell aggregation in *Dictyostelium discoideum*. *J. Cell Biol.* **56,** 647–667.

Bonner, J. T. (1947). Evidence for the formation of cell aggregates by chemotaxis in the development of the slime mold *Dictyostelium discoideum*. *J. Exp. Zool.* **106,** 1–26.

Bonner, J. T. (1977). Some aspects of chemotaxis using the cellular slime molds as an example. *Mycologia* **69,** 443–459.

Bonner, J. T., Hall, E. M., Noller, S., Oleson, F. B., and Roberts, A. B. (1972). Synthesis of cyclic AMP and phosphodiesterase in various species of the cellular slime molds and its bearing on chemotaxis and differentiation. *Dev. Biol.* **29,** 402–409.

Brachet, P. (1976). Stimulation of cell aggregation of *Dictyostelium discoideum* by ionophore A-23187. *C.R. Acad. Sci. Natl.* **282,** 377–379.

Brachet, P. and Dicou, E. (1978). The role of signal modulation during the aggregation phase of *Dictyostelium discoideum. Differentiation* **13,** 15–16.

Brachet, P. and Klein, C. (1977). Cell responsiveness to cyclic AMP during the aggregation phase of *Dictyostelium discoideum* comparison between the inhibitory action of progesterone and the stimulatory action of EGTA and ionophore A-23187 *Differentiation* **8,** 1–8.

Brenner, M. (1978). Cyclic AMP levels and turnover during development of the cellular slime mold *Dictyostelium discoideum. Dev. Biol.* **64,** 210–223.

Chung, W. J. and Coe, E. L. (1977). Metabolic activity associated with the chemotactic and signaling responses of *Dictyostelium discoideum* to cycle AMP. *Fed. Proc.* **36,** 886.

Chung, W. J. and Coe, E. L. (1978). Correlations among the responses of suspensions of *Dictyostelium discoideum* to pulses of 3′,5′-cyclic AMP, transient turbidity decrease, the cyclic AMP signal, and variation in adenine mononucleotide levels. *Biochim. Biophys. Acta* **544,** 29–44.

Clarke, M. *et al.* (1980). Purification of calmodulin from *D. discoideum. J. Bacteriol.* **141,** 397–399.

Clarke, M. and Spudich, J. A. (1977). Non-muscle contractile proteins: the role of actin and myosin in cell motility and shape determination. *Ann. Rev. Biochem.* **46,** 797–822.

Clark, R. and Steck, T. L. (1979). Morphogenesis in *Dictyostelium:* an orbital hypothesis. *Science* **204,** 1163–1168.

Coe, E. L. and Chung, W.-J. K. (1978). Use of turbidity to detect changes in cellular structure. The response of cellular slime mold amoebae to cyclic AMP. *In* "Biomolecular Structure and Function." Symposium of Biophysical Approaches to Biological Problems, (P. F. Agris, R. N. Loeppky, and B. D. Sykes, eds.), pp. 267–272. Academic Press, New York.

Cohen, M. S. (1977). The cyclic AMP control system in the development of *Dictyostelium discoideum.* I. Cellular dynamics. *J. Theor. Biol.* **69,** 57–85.

Cohen, M. S. (1978). The cyclic AMP control system in the development of *Dictyostelium discoideum.* II. An allosteric model. *J. Theor. Biol.* **72,** 231–255.

Cohen, M. and Robertson, A. (1971). Wave propagation in the early stages of aggregation of cellular slime molds. *J. Theor. Biol.* **31,** 101–118.

Cone, R. D. and Bonner, J. D. (1979). Evidence for aggregation center induction by the ionophore A23187 in the cellular slime mold *Polysphondylium violaceum. Exp. Cell Res.* **128,** 479–485.

Cornejo, H. (1980). Ph.D. Thesis, University of Chicago, Chicago.

Coukell, M. B. and Chan, F. K. (1980). The precocious appearance activation of an adenylate cyclase in a rapid developing mutant of *Dictyostelium discoideum. FEBS Lett.* **10,** 39–42.

Cripps, M. and Rutherford, C. (1981). A soluble inhibitor of adenylate cyclase in *D. discoideum. Exp. Cell Res.* **133,** 309–316.

Darmon, M., Barra, J., and Brachet, P. (1978). The role of phosphodiesterase in aggregation of *Dictyostelium discoideum. J. Cell Sci.* **31,** 233–244.

Darmon, M., Barrand, P., Brachet, P., Klein, C., and DaSilva, L. P. (1977). Phenotypic suppression of morphogenetic mutants of *Dictyostelium discoideum. Dev. Biol.* **58,** 174–184.

Darmon, M., Brachet, P., and DaSilva, P. H. (1975). Chemotactic signals induce cell differentiation in *Dictyostelium discoideum. Proc. Natl. Acad. Sci. U.S.A.* **72,** 3163–3166.

Darmon, M. and Klein, C. (1978). Effects of amino acids and glucose on adenylate cyclase cell differentiation of *Dictyostelium discoideum. Dev. Biol.* **63,** 377–389.

Devreotes, P. N., Derstine, P. L., and Steck, T. L. (1979). Cyclic AMP relay in *Dictyostelium discoideum.* I. A technique to monitor responses to control stimuli. *J. Cell Biol.* **80,** 291–299.

Devreotes, P. N. and Steck, T. L. (1979). Cyclic AMP relay in *Dictyostelium discoideum*. II. Requirements for the initiation and termination of the response. *J. Cell Biol.* **80,** 300–309.
Dicou, E. L. and Brachet, P. (1979a). Multiple forms of an extracellular cAMP phosphodiesterase from *Dictyostelium discoideum*. *Biochim. Biophys. Acta* **578,** 232–242.
Dicou, E. and Brachet, P. (1979b). Purification of the inhibitor of the cAMP phosphodiesterase of *Dictyostelium discoideum* by affinity chromatography. *Biochem. Biophys. Res. Commun.* **90,** 1321–1327.
Dinauer, M., MacKay, S., and Devreotes, P. (1980a). Cyclic 3′,5′ AMP Relay in *Dictyostelium discoideum*. III. The Relationship of cAMP synthesis and secretion during the cAMP signaling response. *J. Cell Biol.* **86,** 537–544.
Dinauer, M., Steck, T., and Devreotes, P. (1980b). Cyclic 3′,5′ AMP Relay in *Dictyostelium discoideum*. IV. Recovery of the cAMP signaling response after adaptation to cAMP. *J. Cell Biol.* **86,** 545–553.
Dinauer, M., Steck, T., and Devreotes, P. (1980c). Cyclic 3′,5′ AMP Relay in *Dictyostelium discoideum*. V. Adaptation of the cAMP signaling response during cAMP stimulation. *J. Cell Biol.* **86,** 554–561.
Filosa, M. F. (1978). Concanavalin mediated attachment to membranes of a cellular slime mold cyclic AMP phosphodiesterase. *Differentiation* **10,** 177–180.
Fong, D. and Bonner, J. T. (1979). Proteases in cellular slime mold development: evidence for their involvement. *Proc. Natl. Acad. Sci. U.S.A.* **76,** 6481–6485.
Francis, D. (1975). Genetic regulation of cyclic nucleotide production in a cellular slime mold *Polysphondylium pallidum*. *Adv. Cyclic Nucleotide Res.* **5,** 832.
Franke, J. and Kessin, R. (1981). The cAMP phosphodiesterase inhibitory protein of *Dictyostelium discoideum* purification and characterization. *J. Biol. Chem.* In press.
Frantz, C. E. (1980). Ph.D. Thesis, University of Chicago, Chicago.
Freidlin, M. J. and Sivak, S. A. (1979). Small parameter method in multidimensional reaction diffusion problem model of cAMP signals in *Dictyostelium discoideum*. *Stud. Biophys.* **76,** 129–136.
Futrelle, R. P., McKee, W. G., and Traut, J. (1980). Response of *Dictyostelium discoideum* to localized cyclic AMP stimuli computer analysis of cell motion (abstract). *J. Cell Biol.* **87,** 57a.
Futrelle, R. P., Traut, J., and McKee, W. G. (1981). Cell behavior in *Dictyostelium discoideum* preaggregation response to localized cyclic AMP pulses. *J. Cell Biol.* In press.
Gagnon, C. and Heisler, S. (1981). Mini review—Protein carboxyl-methylation: role in exocytosis and chemotaxis. *Life Sci.* **25,** 993–1000.
Geller, J. S. and Brenner, M. (1978a). Measurements of metabolites during cyclic AMP oscillations of *Dictyostelium discoideum*. *J. Cell Physiol.* **97,** 413–420.
Geller, J. and Brenner, M. (1978b). The effect of 2,4-dinitrophenol on *Dictyostelium discoideum*. *Biochem. Biophys. Res. Commun.* **81,** 814–821.
Gerisch, G. (1968). Cell aggregation and differentiation in *Dictyostelium*. *Curr. Top. Develop. Biol.* **3,** 157–197.
Gerisch, G. (1971). Periodische signale stevern die musterbildung in Zellverbänden. *Naturwissenschaften* **58,** 430–438.
Gerisch, G. (1976). Extracellular cAMP phosphodiesterase regulation in agar plate cultures of *Dictyostelium discoideum*. *Cell Differ.* **5,** 21–25.
Gerisch, G. (1978). Cell interactions by cyclic AMP in *Dictyostelium*. *Biol. Cellulaire* **32,** 61–68.
Gerisch, G., Fromm, H., and Huesgen, A., and Wick, U. (1975a). Control of cell contact sites by cAMP pulses in differentiating *Dictyostelium discoideum* cells. *Nature (London)* **255,** 547–549.

Gerisch, G., Huesgen, A., and Malchow, D. (1975b). Genetic control of cell differentiation and aggregation in *Dictyostelium:* the role of cyclic-AMP pulses. Proceedings of the Tenth FEBS Meeting, pp. 257–267.

Gerisch, G., Hülser, D., Malchow, D., and Wick, U. (1975c). Cell communication by periodic cyclic AMP pulses. *Phil. Trans. R. Soc. Lond.* **272,** 181–192.

Gerisch, G., Maeda, Y., and Malchow, D. (1977a). Cyclic AMP signals and the control of cell aggregation in *Dictyostelium discoideum. In* "Developments and Differentiation in the Cellular Slime Molds," (P. Cappuccinelli and J. Ashworth, eds.), pp. 105–124. Elsevier/North-Holland, New York.

Gerisch, G. and Malchow, D. (1976). Cyclic AMP receptors and the control of cell aggregation in *Dictyostelium discoideum. Adv. Cyclic Nucleotide Res.* **7,** 49–68.

Gerisch, G., Malchow, D., Roos, W., and Wick, U. (1979). Oscillations of cyclic nucleotide concentrations in relation to the excitability of *Dictyostelium* cells. *J. Exp. Biol.* **81,** 33–47.

Gerisch, G., Malchow, D., Roos, W., Wick, U., and Wurster, B. (1977b). Periodic cyclic AMP signals and membrane differentiation in *Dictyostelium. In* "Cell Interaction in Differentiation," (M. Karkinen-Jääskeläinen, L. Saxén and L. Weiss, eds.), pp. 377–388. Academic Press, New York.

Gerisch, G. and Wick, U. (1975). Intracellular oscillations and release of cAMP from *Dictyostelium discoideum* cells. *Biochem. Biophys. Res. Commun.* **65**(1), 364–370.

Gerisch, G. and Hess, B. (1974). cAMP controlled oscillations in suspended *Dictyostelium* cells: their relation to morphogenesis cell interactions. *Proc. Natl. Acad. Sci. U.S.A.* **71,** 2118–2122.

Gil, M., Alemany, S., Marin-Cao, D., Castraño, J. G., and Mato, J. M. (1980). Calmodulin modulates phospholipid methylation in *Dictyostelium discoideum. Biochem. Biophys. Res. Commun.* **94,** 1325–1330.

Gillette, M. U., Dengler, R. E., and Filosa, M. F. (1974). The localization and fate of Concanavalin A in amoebae of the cellular slime mold *Dictyostelium discoideum. J. Exp. Zool.* **190,** 243–248.

Goldbeter, A. (1975). Mechanism for oscillatory synthesis of cAMP in *Dictyostelium discoideum. Nature (London)* **253,** 540–542.

Goldbeter, A. (1980). Models for oscillations and excitability in biochemical systems. *In* "Mathematical Models in Molecular and Cellular Biology," (L. Segal, ed.) pp. 266–291. Cambridge University Press, New York.

Goldbeter, A. (1981). Bifurcations and the control of developmental transitions: evolution of the cAMP signaling system in the slime mold *Dictyostelium discoideum.* Theoretical Aspects of Molecular Science: Mathematical Biol., pp. 79–95.

Goldbeter, A., Erneux, J., and Segal, L. (1978). Excitability in the adenylate cyclase reaction on *Dictyostelium discoideum. FEBS Lett.* **89,** 237–241.

Goldbeter, A. and Martiel, J. (1980). Role of receptor desensitization in the mechanism of cAMP oscillation in *Dictyostelium. Fed. Proc.* **39,** 1804.

Goldbeter, A. and Segal, L. (1977). Unified mechanism for relay and oscillation of cyclic AMP in *Dictyostelium discoideum. Proc. Natl. Acad. Sci. U.S.A.* **74,** 1543–1547.

Goldbeter, A. and Segal, L. (1980). Control of developmental transitions in the cyclic AMP signalling system of *Dictyostelium discoideum. Differentiation* **17,** 127–135.

Green, A. A. and Newell, P. C. (1975). Evidence for the existence of 2 types of cAMP binding sites in aggregating cells of *Dictyostelium discoideum. Cell* **6**(2), 129–136.

Gross, J. D., Peacey, M. J., and Trevan, D. J. (1976). Signal emission and signal propagation during early aggregation in *Dictyostelium discoideum. J. Cell Sci.* **22**(3), 645–656.

Gross, J., Ray, R., Lax, A., Peacey, M., Town, C., and Trevan, D. (1977). Cell contact, signalling and gene expression in *Dictyostelium discoideum. In* "Development and Differentiation on the Cellular Slime Molds," (P. Cappuccinelli and J. Ashworth, eds.), pp. 135–147. Elsevier/North-Holland, New York.

Grutsch, J. F. and Robertson, A. (1978). The cyclic AMP signal from *Dictyostelium discoideum* amoebae. *Dev. Biol.* **66**(2), 285–293.

Gunzburg, J., Véron, M., and Brachet, P. (1980). Non-Michaelian kinetics of adenylate cyclase in *Dictyostelium discoideum. Cell Biol. Internal. Report* **4**(6), 533–539.

Hahn, G. L., Metz, K., George, R. P., and Haley, B. (1977). Identification of cyclic AMP receptors in the cellular slime mold *Dictyostelium discoideum* using a photoaffinity analog (abstract). *J. Cell Biol.* **75,** 91A.

Hanna, M. H., Klein, C., and Cox, E. C. (1979). Cyclic nucleotides and cyclic nucleotide phosphodiesterase during development of *Polysphondylium violaceum. Exp. Cell Res.* **122,** 265–272.

Hayashi, H. and Yamasaki, F. (1978). Characteristics of the induction of phosphodiesterases by cyclic AMP in the slime mold *Dictyostelium discoideum. Chem. Pharm. Bull.* (Tokyo) **26,** 2977–2982.

Henderson, E. J. (1975). The cAMP receptor of *Dictyostelium discoideum* binding characteristics of aggregation competent cells and variation of binding levels during the life cycle. *J. Biol. Chem.* **250,** 4730–4736.

Henderson, E. J. and Barclay, S. L. (1979). A mutation affecting the cyclic AMP receptor of *Dictyostelium discoideum* (abstract). *J. Supramol. Struct.* **8,** 230.

Hinterman, R. and Parish, R. W. (1979). The intracellular location of adenyl cyclase in the cellular slime molds *Dictyostelium discoideum* and *Polysphondylium pallidum. Exp. Cell Res.* **123,** 429–434.

Ishizaka, T. (1981). Analysis of triggering events in mast cells for immunoglobulin E-mediated histamine release. *J. Allerg. Clin. Immun.* **67,** 90–96.

Jastorff, B. (1978). 5-Amino-5-deoxy adenosine 3 phosphates mimicing the biological activity of cAMP. *Adv. Cyclic Nucleotide Res.* **254,** 12573–12578.

Jastorff, B., Konijn, T. M., Mato, J., Hoppe, J., and Wagner, K. G. (1978). Comparison of the molecular interactions between cAMP and its receptor protein in *Dictyostelium discoideum* and protein kinase type I. *Hoppe Seyler's Z. Physiol. Chem.* **359,** 281.

Juliani, M., Brusca, J., and Klein, C. (1981). cAMP regulation of cell differentiation in *Dictyostelium discoideum* and the role of the cAMP receptor. *Dev. Biol.* **83,** 114–121.

Juliani, M. H. and Klein, C. (1977). Calcium ion effect on cAMP bindings to the plasma membrane of *Dictyostelium discoideum. Biochim. Biophys. Acta* **497,** 369–376.

Juliani, M. H. and Klein, C. (1978). A biochemical study of the effects of cAMP pulses on aggregateless mutants of *Dictyostelium discoideum. Dev. Biol.* **62,** 162–172.

Juliani, M. H. and Klein, C. (1981). Photoaffinity labeling of the cell surface cyclic AMP receptor of *Dictyostelium discoideum* and its modification of down regulated cells. *J. Biol. Chem.* **256,** 613–619.

Kakebeeke, P. I., DeWit, R. J., Kohtz, A. J., and Konijn, T. M. (1979). Negative chemotaxis in *Dictyostelium* and *Polysphondylium: Exp. Cell Res.* **124,** 429–432.

Kakebeeke, P. I., DeWit, R. J., and Konijn, T. M. (1980a). Folic acid deaminase activity during development in *Dictyostelium discoideum. J. Bacteriol.* **143,** 307–312.

Kakebeeke, P. I. J., DeWit, R. J. W., and Konijn, T. M. (1980b). A novel chemotaxis regulating enzyme that splits folic acid into 6-hydroxymethylpterin and p-aminobenzoylglutamic acid. *FEBS Lett.* **115,** 216–220.

Kakebeeke, P. I., Mato, J. M., and Konijn, T. M. (1978). Purification and preliminary characterization of an aggregation-sensitive chemoattractant of *Dictyostelium minutum. J. Bacteriol.* **133,** 403–405.

Kawai, S. (1980). Folic-acid increase cyclic AMP binding activity of *Dictyostelium discoideum* cells. *FEBS Lett.* **109,** 27–30.

Kay, R. R. (1979). Gene expression in *Dictyostelium discoideum:* mutually antagonistic roles of cyclic AMP and ammonia. *J. Embryol. Exp. Morph.* **52,** 171–182.

Keating, M. T. and Bonner, J. T. (1977). Negative chemotaxis in cellular slime mold. *J. Bacteriol.* **130,** 144–147.

Kessin, R. H., Orlow, S. J., Shapiro, R. I., and Franke, J. (1979). Binding of inhibitor alters kinetic and physical properties of extracellular cAMP phosphodiesterase from *Dictyostelium discoideum. Proc. Natl. Acad. Sci. U.S.A.* **76,** 5450–5454.

King, A. C. and Frazier, W. A. (1977). Reciprocal periodicity in cyclic AMP binding and phosphorylation of differentiating *Dictyostelium discoideum* cells. *Biochem. Biophys. Res. Commun.* **78,** 1093–1099.

King, A. C. and Frazier, W. A. (1979). Properties of the oscillatory cAMP binding component of isolated plasma membranes. *J. Biol. Chem.* **254,** 7168–7176.

King, A. C., Wallace, L. J., and Frazier, W. A. (1978). A potential regulatory mechanism for cyclic AMP receptors of *Dictyostelium discoideum* (abstract). *Fed. Proc.* **37,** 182.

Klein, C. (1975). Induction of phosphodiesterase by cAMP in differentiating *Dictyostelium discoideum. J. Biol. Chem.* **250,** 7134–7138.

Klein, C. (1976). Adenylate cyclase activity in *Dictyostelium discoideum* amoebae and its change during differentiation. *FEBS Lett.* **68,** 125–128.

Klein, C. (1979). A slowly dissociating form of the cell surface cyclic AMP receptor of *Dictyostelium discoideum. J. Biol. Chem.* **254,** 12573–12578.

Klein, C. (1980). Cyclic AMP independent oscillations of adenylate cyclase in *Dictyostelium discoideum. Dev. Biol.* **79,** 500–507.

Klein, C. (1981). Binding of adenosine 3′:5′-monophosphate to plasma membranes of *D. discoideum* amoebae. *J. Biol. Chem.* In press.

Klein, C. and Brachet, P. (1975). Effects of progesterone and EDTA on cAMP and phosphodiesterase in *Dictyostelium discoideum. Nature (London)* **254,** 432–434.

Klein, C., Brachet, P., and Darmon, M. (1977). Periodic changes in adenylate cyclase and cyclic AMP receptors in *Dictyostelium discoideum. FEBS Lett.* **76,** 145–147.

Klein, C. and Darmon, M. (1976). A differentiation stimulating factor induces cell sensitivity to cAMP pulses in *D. discoideum. Proc. Natl. Acad. Sci. U.S.A.* **73,** 1250–1254.

Klein, C. and Darmon, M. (1977). Effects of cyclic AMP pulses on adenylate cyclase and the phosphodiesterase inhibitor of *Dictyostelium discoideum. Nature (London)* **268,** 76–78.

Klein, C. and Darmon, M. (1979). A cyclic AMP sensitive adenylate cyclase in *Dictyostelium discoideum* extracts. *FEMS Lett.* **5,** 1–4.

Klein, C. and Juliani, M. H. (1977). cAMP-induced changes in cAMP-binding sites on *D. discoideum* amoebae. *Cell* **10,** 329–335.

Konijn, T. (1970). Microbiological assay of cyclic 3′,5′-AMP. *Experientia* **26,** 367–369.

Konijn, T., Barkley, D., Chang, Y. Y., and Bonner, J. (1968). Cyclic AMP: a naturally occurring acrasin in the cellular slime molds. *Amer. Natur.* **102,** 225–233.

Konno, R. (1980). Aggregationless mutant defective in signaling and relaying in the cellular slime mold *Dictyostelium discoideum. Dev. Growth Diff.* **22,** 125–132.

Kretsinger, R. (1979). The informational role of calcium in the cytosol. *Adv. Cyclic Nucleotide Res.* **11,** 1–25.

Kuczmarski, E. R. and Spudich, J. A. (1980). Regulation of myosin self-assembly: phosphorylation of *Dictyostelium* heavy chain inhibits thick filament formation. *Proc. Natl. Acad. Sci. U.S.A.* **77,** 7292–7296.

Lax, A. J. (1979). The evolution of excitable behavior in *Dictyostelium. J. Cell Sci.* **36,** 311–321.

Leichtling, B., Coffman, D., Yaeger, E., Rickenberg, H., al-Jumaliy, W., and Haley, B. (1981). Occurence of the adenylate cyclase "G-protein" in membranes of *Dictyostelium discoideum*. *B.B.R.C.* **102,** 1187–1195.

Lo, E. K., Coukell, M. B., Tsang, A. S., and Pickering, J. L. (1978). Physiological and biochemical characterization of aggregation-deficient mutants of *Dictyostelium discoideum:* detection and response to exogenous cyclic AMP. *Can. J. Microbiol.* **24,** 455–465.

Loomis, W. F. (1979). Biochemistry of aggregation in *Dictyostelium*. A review. *Dev. Biol.* **70,** 1–12.

Loomis, W. F., Klein, C., and Brachet, P. (1978). The effect of divalent cations on aggregation of *Dictyostelium discoideum. Differentiation* **12,** 83–89.

MacInnes, M. A. and Francis, D. W. (1977). Altered control of cyclic AMP oscillations in an aggregation mutant of *Dictyostelium mucoroides. Adv. Cyclic Nucleotide Res.* **9,** 778.

MacKay, S. (1978). Computer simulation of aggregation in *Dictyostelium discoideum. J. Cell Sci.* **33,** 1–16.

MacKay, S. (1979). Ph.D. Thesis, University of Chicago, Chicago.

McNab, R. and Koshland, D. (1972). The gradient-sensing mechanism in bacterial chemotaxis. *Proc. Natl. Acad. Sci. U.S.A.* **69,** 2509–2512.

Maeda, Y. and Gerisch, G. (1977). Vesicle formation in *Dictyostelium discoideum* cells during oscillations of cAMP synthesis and release. *Exp. Cell Res.* **110,** 119–126.

Malchow, D., Böhme, R., and Rahmsdof, H. J. (1981). Regulation of myosin heavy chain phosphorylation during the chemotactic response of *Dictyostelium* cells. *Eur. J. Biochem.* In press.

Malchow, D. and Gerisch, G. (1974). Short-term binding and hydrolysis of cyclic 3′:5′-adenosine monophosphate by aggregating *Dictyostelium* cells. *Proc. Natl. Acad. Sci. U.S.A.* **71,** 2423–2427.

Malchow, D., Nanjundiah, V., and Gerisch, G. (1978a). pH oscillations in cell suspensions of *Dictyostelium discoideum*, their relation to cyclic AMP signals. *J. Cell Sci.* **30,** 319–330.

Malchow, D., Nanjundiah, V., Wurster, B., Eckstein, F., and Gerisch, G. (1978b). Cyclic AMP induced pH changes in *Dictyostelium discoideum* and their control by calcium. *Biochim. Biophys. Acta* **538,** 473–480.

Marin, F. T. and Rothman, F. G. (1980). Regulation of development in *Dictyostelium discoideum*. 4. Effects of ions on the rate of differentiation and cellular response to cyclic AMP. *J. Cell. Biol.* **87,** 823–827.

Martiel, J. L. and Goldbeter, A. (1981). Metabolic oscillations in biochemical systems controlled by covalent enzyme modification. *Biochimie* **63,** 119–124.

Mason, J. W., Rasmussen, H., and Dibella, F. (1971). 3′,5′ AMP and Ca^{++} in slime mold aggregation. *Exp. Cell Res.* **67,** 156–160.

Mato, J. (1979). Activation of *Dictyostelium discoideum* guanylate cyclase by ATP. *Biochem. Biophys. Res. Commun.* **88,** 569–574.

Mato, J. M. (1978). ATP increases chemoattractant induced cyclic GMP accumulation in *Dictyostelium discoideum. Biochim. Biophys. Acta* **540,** 408–411.

Mato, J. M., Jastorff, B., Morr, M., and Konijn, T. M. (1978a). A model for cyclic AMP-chemoreceptor interaction in *Dictyostelium discoideum. Biochim. Biophys. Acta* **544,** 309–314.

Mato, J., and Konijn, T. (1977a). Chemotactic signal and cyclic GMP accumulation in *Dictyostelium. In* "Developments and Differentiation in the Cellular Slime Molds," (P. Cappuccinelli and J. Ashworth eds.), pp. 93–104. Elsevier/North-Holland, New York.

Mato, J. and Konijn, T. (1979). Chemosensory transduction in *Dictyostelium discoideum. In* "Biochemistry and Physiology of Protozoa," Second Edition, Vol. 2, pp. 181–219.

Mato, J. M. and Konijn, T. M. (1977b). The chemotactic activity of cAMP and AMP derivatives with substitution in the phosphate moiety in *D. discoideum*. *FEBS Lett.* **75**, 173–176.
Mato, J., Krens, F., Van Haastert, P. J. M., and Konijn, T. M. (1977a). cAMP dependent cGMP accumulation in *Dictyostelium discoideum*. *Proc. Natl. Acad. Sci. U.S.A.* **74**, 2348–2351.
Mato, J. M., Krens, F., Van Haastert, P. J. M., and Konijn, T. M. (1977b). Unified control of chemo taxis and cyclic AMP mediated cyclic GMP accumulation by cyclic AMP in *Dictyostelium discoideum*. *Biochem. Biophys. Res. Commun.* **77**, 399–402.
Mato, J. M., Losada, A., Nanjundiah, V., and Konijn, T. M. (1975). Signal input for a chemo tactic response in the cellular slime mold *Dictyostelium discoideum*. *Proc. Natl. Acad. Sci. U.S.A.* **72**, 4991–4993.
Mato, J. M. and Malchow, D. (1978). Guanylate cyclase activation in response to chemo tactic stimulation in *Dictyostelium discoideum*. *FEBS Lett.* **90**, 119–122.
Mato, J. M. and Marín-Cao, D. (1979). Protein and phospholipid methylation during chemotaxis in *Dictyostelium discoideum* and its relationship to calcium movements. *Proc. Natl. Acad. Sci. U.S.A.* **76**, 6106–6109.
Mato, J., Roos, W., and Wurster, B. (1978b). Guanylate cyclase activity in *Dictyostelium discoideum* and its increase during cell development. *Differentiation* **10**, 129–132.
Mato, J. M. and Steiner, A. L. (1980). Immunohistochemical localization of cAMP, cGMP and calmodulin in *Dictyostelium discoideum*. *Cell Biol. Int. Rep.* **4**, 641–648.
Mato, J. M., Van Haastert, P. J., Krens, F. A., and Konijn, T. M. (1978c). Chemotaxis in *Dictyostelium discoideum*: effect of Concanavalin A on chemoattractant mediated cyclic GMP accumulation and light scattering decrease. *Cell Biol. Int. Rep.* **2**, 163–170.
Mato, J. M., Van Haastert, P. J. M., Krens, F. A., and Konijn, T. M. (1977c). An acrasin-like attractant from yeast extract specific for *Dictyostelium lacteum*. *Develop. Biol.* **57**, 450–453.
Mato, J. M., Van Haastert, P. J., Krens, F. A., Rhihnsburger, E. H., Dobbe, F. C., and Konijn, T. M. (1977d). Cyclic AMP and folic acid mediated cyclic GMP accumulation in *Dictyostelium discoideum*. *FEBS Lett.* **79**, 331–336.
Mato, J. M., Woelders, H., Van Haastert, P. J., and Konijn, T. M. (1978d). Cyclic GMP binding activity in *Dictyostelium discoideum*. *FEBS Lett.* **90**, 261–264.
Meyers, B. and Frazier, W. (1981). Solubilization and hydrophobic immunobilization assay of a cAMP binding protein from *D. discoideum* plasma membrane. *Biochem. Biophys. Res. Commun.* In press.
Mockrin, S. and Spudich, J. (1976). Calcium control of actin-activated myosin adenosine triphosphatase from *Dictyostelium discoideum*. *Proc. Natl. Acad. Sci. U.S.A.* **73**, 2321–2325.
Mullens, I. A. and Newell, P. C. (1978). Cyclic AMP binding to cell surface receptors of *Dictyostelium*. *Differentiation* **10**, 171–176.
Nandini-Kishore and Frazier, W. (1982). ^{3}H-methotrexate as a ligand for the folate receptor of *Dictyostelium discoideum*. *Proc. Natl. Acad. Sci. U.S.A.* In press.
Nanjundiah, V. (1978). Ligand-receptor binding in the presence of a diffusion gradient. *J. Indian Inst. Sci.* **60**, 199–204.
Nanjundiah, V. and Malchow, D. (1978). A theoretical study of the effects of cyclic AMP phosphodiesterases during aggregation in *Dictyostelium discoideum*. *J. Cell Sci.* **22**, 49–58.
Newell, P. C. (1977a). Aggregation and cell surface receptor in cellular slime mold. Receptors & Recognition Series B, Vol. 3 Microbial Interactions, pp. 3–57.
Newell, P. C. (1977b). How cells communicate: the system used by slime molds. *Endeavor* **1**, 63–66.
Newell, P. (1981). Initiation of aggregation centres in mutant and wild type populations: the

effect of adenosine on the initiation system (abstract). Workshop of EMBO and Max-Planck-Gesellschaft, Tutzing, September 2–6, 1981.

Newell, P. C. and Mullens, I. A. (1978). Cell surface cAMP receptors in *Dictyostelium*. *In* "Cell–Cell Recognition" (A. S. Curtis, ed.), pp. 161–171. Cambridge University Press, New York.

Orlow, S. J., Shapiro, R. I., Franke, J., and Kessin, R. H. (1981). The extracellular cyclic nucleotide phosphodiesterase of *Dictyostelium discoideum:* purification and characterization. *J. Biol. Chem.* In press.

Pahlic, M. and Rutherford, C. L. (1979). Adenylate cyclase activity and cyclic AMP levels during the development of *Dictyostelium discoideum. J. Biol. Chem.* **254,** 9703–9707.

Pan, P. (1977). A new chemotactic agent for the cellular slime mold *Polysphondylium violaceum. Abstr. Annu. Meet. Am. Soc. Microbiol.* **77,** 161.

Pan, P., Hall, E. M., and Bonner, J. T. (1972). Folic acid as second chemotactic substance in the cellular slime moulds. *Nature (London) New Biol.* **237,** 181–182.

Pan, P., Hall, E. M., and Bonner, J. T. (1975). Determination of the active portion of the folic acid molecule in cellular slime mold chemotaxis. *J. Bacteriol.* **122,** 185–191.

Pan, P. and Wedner, H. J. (1979). Immunohistochemical localization of cyclic GMP in aggregating *Polysphondylium violaceum. Differentiation* **14,** 113–118.

Pan, P. and Wurster, B. (1978). Inactivation of the chemoattractant folic acid by cellular slime mold and identification of the reaction product. *J. Bacteriol.* **136,** 955–959.

Parish, R. W. (1979). Cyclic AMP induces the synthesis of developmentally regulated plasma membrane proteins in *Dictyostelium. Biochim. Biophys. Acta* **553,** 179–182.

Parish, R. W., Schmidlin, S., and Weibel, M. (1978). Effect of cyclic AMP pulses on the synthesis of plasma membrane proteins in aggregateless mutants of *Dictyostelium discoideum. FEBS Lett.* **96,** 283–286.

Parnas, H. and Segal, L. (1978). A computer simulation of pulsatile aggregation in *Dictyostelium discoideum. J. Theor. Biol.* **71,** 185–207.

Parnas, H. and Segal, L. (1977). Computer evidence concerning the chemotactic signal in *Dictyostelium discoideum. J. Cell Sci.* **25,** 191–204.

Pollet, R. J., Standaert, M. L., and Haase, B. A. (1977). Insulin binding to the human lymphocyte receptor. *J. Biol. Chem.* **252,** 5828–5834.

Rahmsdorf, H. J. and Gerisch, G. (1978). Cyclic AMP-induced phosphorylation of a polypeptide comigrating with myosin heavy chains. *FEBS Lett.* **88,** 322–326.

Rahmsdorf, H. J., Malchow, D., and Gerisch, G. (1978). Cyclic AMP induced phosphorylation in *Dictyostelium discoideum* of a polypeptide comigrating with myosin heavy chains. *FEBS Lett.* **88,** 322–326.

Rickenberg, H. V., Tihon, C., and Güzel, O. (1977). Effect of pulses of 3′:5′-cyclic adenosine monophosphate on enzyme formation in non-aggregated amoebae of *Dictyostelium discoideum. In* "Development and Differentiation in the Cellular Slime Molds," (P. Cappuccinelli and J. Ashworth, eds.), pp. 231–242. Elsevier/North-Holland, New York.

Robertson, A., Drage, D., and Cohen, M. (1972). Control of aggregation in *Dictyostelium discoideum* by an external periodic pulse of cyclic adenosine monophosphate. *Science* **175,** 333–335.

Roos, W. and Gerisch, G. (1976). Receptor-mediated adenylate cyclase activation in *Dictyostelium discoideum. FEBS Lett.* **68,** 170–172.

Roos, W., Malchow, D., and Gerisch, G. (1977a). Adenyl cyclase and the control of cell differentiation in *Dictyostelium discoideum. Cell Differ.* **6,** 229–239.

Roos, W., Scheidegger, C., and Gerisch, G. (1977b). Adenylate cyclase activity oscillations as signals for cell aggregation in *Dictyostelium discoideum. Nature (London)* **266,** 259–260.

Roos, W., Nanjundiah, V., Malchow, D., and Gerisch, G. (1975). Amplification of cAMP signals in aggregating cells of *Dictyostelium discoideum*. *FEBS Lett.* **53,** 139–142.

Ross, F. M. and Newell, P. C. (1980). Genetics of aggregation pattern mutations in the cellular slime mould *Dictyostelium discoideum*. *J. Gen. Microbiol.* **115,** 289–300.

Rossier, C., Eitle, E., Vandriel, R., and Gerisch, G. (1980). Biochemical regulation of cell development and aggregation in *Dictyostelium discoideum*. *In* "Eukaryotic Microbial Cell," (L. Gooday and P. Trinci, eds.) pp. 405–424. Cambridge Univ. Press, New York.

Rossier, C., Gerisch, G., Malchow, D., and Eckstein, F. (1979). Action of a slowly hydrolysable cyclic AMP analogue of developing cells of *Dictyostelium discoideum*. *J. Cell Sci.* **35,** 321–338.

Ryter, A., Klein, C., and Brachet, P. (1979). *Dictyostelium discoideum* surface changes elicited by high concentrations of cAMP. *Exp. Cell Res.* **1,** 373–380.

Saito, M. (1979). Effect of extracellular Ca^{2+} on the morphogenesis of *Dictyostelium discoideum*. *Exp. Cell Res.* **123,** 79–86.

Sampson, J., Town, C., and Gross, J. (1978). Cyclic AMP and the control of aggregative phase gene expression in *Dictyostelium discoideum*. *Dev. Biol.* **67,** 54–64.

Samuel, E. W. (1961). Orientation and rate of locomotion of individual amoebae in the life cycle of the cellular slime mold *Dictyostelium mucoroides*. *Dev. Biol.* **3,** 317–335.

Schindler, J. and Sussman, M. (1977). Effect of ammonia on cAMP associated activities and extracellular cAMP production in *Dictyostelium discoideum*. *Biochem. Biophys. Res. Commun.* **79,** 611–617.

Shaffer, B. M. (1957). Aspects of aggregation in cellular slime moulds 1. Orientation and chemotaxis. *Amer. Natur.* **91,** 19–35.

Shaffer, B. M. (1962). The Acrasina. *Adv. Morphog.* **2,** 109–182.

Shaffer, B. M. (1975). Secretion of cAMP induced by cAMP in the cellular slime mold *Dictyostelium discoideum*. *Nature (London)* **255,** 549–552.

Sperb, R. P. (1979). A mathematical model describing the aggregation of amoebae. *Bull. Math. Biol.* **41,** 555–572.

Thadani, V., Pan, P., and Bonner, J. T. (1977). Complementary effects of ammonia and cAMP on aggregation territory size in the cellular slime mold *Dictyostelium mucoroides*. *Exp. Cell Res.* **108,** 75–78.

Tihon, C., Buzel, O., and Richenberg, H. V. (1977). Regulation of synthesis of enzymes by cyclic AMP in *Dictyostelium* (abstract). *Adv. Cyclic Nucleotide Res.* **9,** 774.

Tomchik, K. J. and Devreotes, P. N. (1981). cAMP waves in *Dictyostelium discoideum:* demonstration by a novel isotope dilution fluorography technique. *Science* **212,** 443–446.

Toorchen, D. and Henderson, E. (1979). Characterization of multiple extracellular cAMP-phosphodiesterase forms in *D. discoideum*. *Biochem. Biophys. Res. Commun.* **87,** 1168–1175.

Tsang, A. S. and Coukell, M. B. (1977). The regulation of cAMP phosphodiesterase and its specific inhibitor by cAMP in *Dictyostelium*. *Cell Differ.* **6,** 75–84.

Tsang, A. S. and Coukell, M. B. (1978). Evidence of increased *de novo* synthesis of extracellular cAMP phosphodiesterase during induction by cAMP in *Dictyostelium purpureum* (abstract). *Can. Fed. Biol. Soc. Proc.* **21,** 95.

Tsang, A. S. and Coukell, M. B. (1979a). Biochemical and genetic evidence for two extracellular adenosine 3′:5′-monophosphate phosphodiesterases in *Dictyostelium purpureum*. *Eur. J. Biochem.* **95,** 407–417.

Tsang, A. S. and Coukell, M. B. (1979b). Direct evidence for extracellular adenosine 3′:5′-monophosphate phosphodiesterase induction and phosphodiesterase inhibitor repression by exogenous adenosine 3′:5′-monophosphate in *Dictyostelium discoideum*. *Eur. J. Biochem.* **95,** 419–425.

Van Driel, R. (1981). Binding of the chemoattractant folic acid by *Dictyostelium discoideum* cells. *Eur. J. Biochem.* **115**, 391–395.

Van Haastert, P. J. M., van der Meer, R. C., and Konijn, T. M. (1981). Evidence that the rate of association of cyclic AMP to its chemotactic receptor induces phosphodiesterase activity in *Dictyostelium discoideum. J. Bacteriol.* **147**, 170–175.

Wallace, L. J. and Frazier, W. A. (1979). Photoaffinity labeling of cAMP and AMP-binding proteins differentiating *Dictyostelium discoideum* cells. *Proc. Natl. Acad. Sci. U.S.A.* **76**, 4250–4254.

Ward, A. and Brenner, M. (1977). Guanylate cyclase from *Dictyostelium discoideum. Life Sci.* **21**, 997–1008.

Wick, U., Malchow, D., and Gerisch, G. (1978). Cyclic-AMP stimulated calcium influx into aggregating cells of *Dictyostelium discoideum. Cell Biol. Int. Rep.* **2**, 71–79.

Wier, P. W. (1977). Cyclic AMP Phosphodiesterase and the duration of the interphase in *Dictyostelium discoideum. Differentiation* **9**, 183–192.

Wurster, B. (1976). Temperature dependence of biochemical oscillations in cell suspensions of *Dictyostelium discoideum. Nature (London)* **260**, 703–704.

Wurster, B., Bozzaro, S., and Gerisch, G. (1978). Cyclic GMP regulation and responses of *Polysphondylium violaceum* to chemoattractants. *Cell Biol. Int. Rep.* **2**, 61–69.

Wurster, B. and Bumann, J. (1981). Cell differentation in the absence of intracellular cyclic AMP pulses in *Dictyostelium discoideum. Dev. Biol.* In press.

Wurster, B. and Butz, U. (1980). Reversible binding of the chemoattractant folic acid to cells of *Dictyostelium discoideum. Eur. J. Biochem.* **109**, 613–618.

Wurster, B., Pan, P., Tyan, G. G., and Bonner, J. T. (1976). Preliminary characterization of the acrasin of the cellular slime mold *Polysphondylium violaceum. Proc. Natl. Acad. Sci. U.S.A.* **73**, 795–799.

Wurster, B. and Schubiger, K. (1977). Oscillations and cell development in *Dictyostelium discoideum* stimulated by folic acid pulses. *J. Cell. Sci.* **27**, 105–114.

Wurster, B., Schubiger, K., and Brachet, P. (1979). Cyclic GMP and cyclic AMP changes in response to folic acid pulses during cell development of *Dictyostelium discoideum. Cell Differ.* **8**, 235–242.

Wurster, B., Schubiger, K., Wick, U., and Gerisch, G. (1977). cGMP in *Dictyostelium discoideum;* oscillations and pulses in response to folic acid and cAMP signals. *FEBS Lett.* **76**, 141–144.

Yeh, R. P., Chan, F. K., and Coukell, M. B. (1978). Regulation of cell differentiation by exogenous cyclic AMP in *Dictyostelium discoideum* (abstract). *Can. Fed. Biol. Soc. Proc.* **21**, 95.

Yeh, R. P., Chan, F. K., and Coukell, M. B. (1978). Independent regulation of the extracellular cyclic AMP phosphodiesterase-inhibitor system and membrane differentiation by exogenous cyclic AMP in *Dictyostelium discoideum. Dev. Biol.* **66**, 361–374.

Zigmond, S., Hyam, L., and Kreel, B. (1981). Cell polarity: an examination of its behavioral expression and its consequences for polymorphonuclear leukocyte chemotaxis. *J. Cell Biol.* **89**, 585–592.

• CHAPTER 5

Cell Motility

James A. Spudich and Annamma Spudich

I. Introduction and Perspectives

The term "cell motility" is used in a very general sense to refer to all of the movements and changes in shape that a cell undergoes. In a light microscope many eukaryotic cells can be seen to move across a surface by extending

The Development of *Dictyostelium discoideum*

ISBN 0-12-455620-5

projections of their cell periphery. Even in a stationary cell, cytoplasmic streaming and directed particle movements are common forms of cell motility. Cytokinesis, karyokinesis, and phagocytosis are other examples of fundamental cellular movements that are typical of eukaryotic cells. All of these motile events occur in amoebae of *Dictyostelium discoideum* and related cellular slime molds (Shaffer, 1964). The advantages of studying *Dictyostelium* as a model system have been described in earlier chapters. In brief, the cells can be cloned to assure a homogeneous population. A kilogram or more of wet cell pellet, which is necessary for biochemical analyses, can be easily obtained. *Dictyostelium discoideum* can be grown in a chemically defined medium (Franke and Kessin, 1977), which allows radioactive components with known specific radioactivities to be taken up by the cells. Furthermore, the organism can be manipulated genetically. A special feature of *Dictyostelium* is that in addition to pseudopod formation that leads to general cell migration, this organism displays fascinating chemotactic responses to cAMP. How such an external signal confers directionality on the migration of a cell is one of the fundamental problems in cell biology.

This chapter summarizes and focuses on investigations of the last five to ten years concerning cell motility in *Dictyostelium discoideum*. The chapter is not meant to be exhaustive, but gives an overview and a particular perspective. There are a large number of interesting related observations using other organisms; these are not discussed here. Several relevant treatises (Pollard and Weihing, 1974; Clarke and Spudich, 1977; Hitchcock, 1977; Korn, 1978; Taylor and Condeelis, 1979; Uyemura and Spudich, 1980; Schliwa, 1981) are recommended for additional information on this general subject; detailed information can be obtained from the three volumes entitled "Cell Motility" (Goldman *et al.*, 1976) and from the Cold Spring Harbor Symposium on "Organization of the Cytoplasm" (1982).

II. Directed Cell Movement Involves Selective Pseudopod Formation

When a *Dictyostelium* amoeba that is moving across a substratum is fixed and examined by scanning electron microscopy, it is apparent that the cell surface is covered with projections (Fig. 1). The thin finger-like projections, 0.1–0.2 μm wide, are known as microspikes or filopodia (Garrod and Born 1971). These apparently form, explore the environment around the cell, and disappear again. In some situations these filopodia may extend out from the cell surface many cell diameters and undergo extensive branching (Rifkin

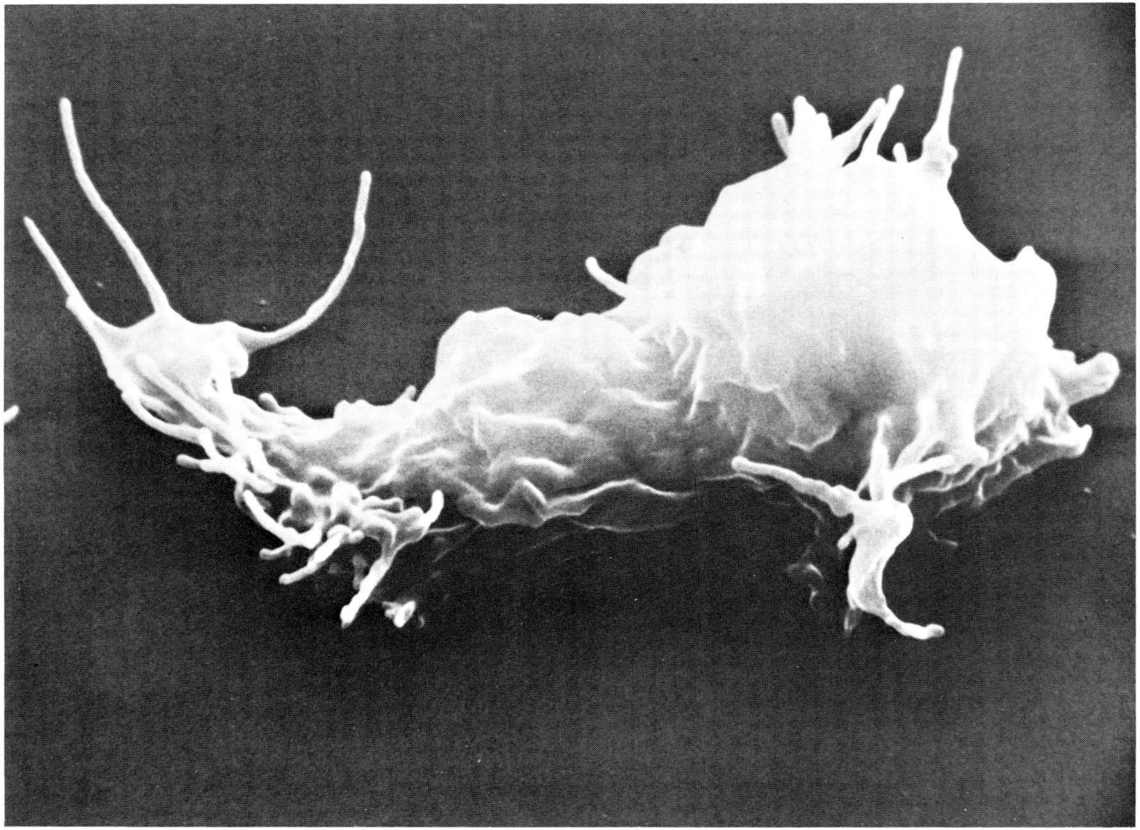

Fig. 1. Scanning electron micrograph of an amoeba of *Dictyostelium discoideum*. From Cooke *et al.* (1976). Magnification 8840 ×.

and Speisman, 1976). One suspects that environmental signals acting on these filopodia determine where the larger cell extensions, known as pseudopodia, will form. In any case it is clear that the surface of the cell is a dynamic region with a great deal of motile activity. Chemotaxis toward cAMP has been shown by Gerisch *et al.* (1975) to involve selective pseudopod formation at the edge of the cell that first encounters the external cAMP signal. The molecular mechanism whereby such an external stimulation elicits selective pseudopod formation and directed movement is central to many general cellular and developmental phenomena. In the case of *Dictyostelium*, the prospects of elucidating the underlying molecular basis of this signal-induced pseudopod formation are increasingly good, as a number of laboratories continue to characterize the likely components of the force generating machinery.

III. Actin and Myosin Are Primary Components of a Filamentous Matrix That Underlies the Cell Membrane

Actin and myosin, the force-generating components of muscle, are prevalent in *Dictyostelium* amoebae, representing about 8 and 0.5%, respectively, of the total cell protein (Uyemura *et al.*, 1978; Clarke and Spudich, 1974). *Dictyostelium* actin, first identified by Woolley (1972), can exist in at least two states of assembly, G and F, and these are in equilibrium with one another (Fig. 2). The actin appears to permeate the cell cytoplasm, and has also been reported to be present in the nucleus (Fukui, 1978; Fukui and Katsumaru, 1979; Fukui and Katsumaru, 1980).

Indirect immunofluorescence studies suggest that actin is concentrated near the cell periphery of *Dictyostelium* amoebae (Eckert and Lazarides, 1978), as is often the case in nonmuscle cells (Goldman *et al.*, 1976). Some of the *Dictyostelium* actin is associated with isolated membranes in a MgATP-stable linkage, whereas myosin is found in membrane preparations in a MgATP-labile linkage, presumably due to association with the membrane-linked actin (Spudich, 1974). Virtually all of the actin is removed from the membrane in 0.6 *M* KI, indicating the actin is a peripheral membrane protein. In a subsequent study (Clarke *et al.*, 1975), filaments were visualized by electron microscopy, using a technique wherein cells were attached to a polylysine-coated surface and then disrupted to expose the cytoplasmic surface of the membrane (Fig. 3). These filaments, which form a dense matrix in close apposition to the cell membrane, were identified as

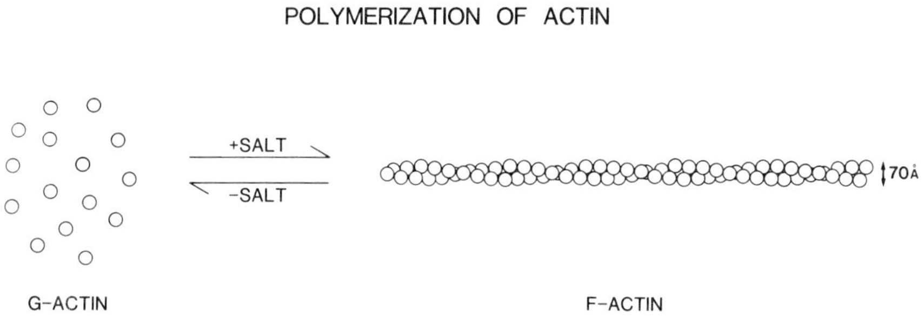

Fig. 2. Representation of G- and F-actin. Generally the filaments are much longer than shown.

actin by their solubility properties and their ability to interact with the myosin fragment S1 to yield arrowheaded filaments (Clarke *et al.*, 1975). Using the conditions of Spudich and Spudich (1979) for the isolation of a cortical actin preparation from sea urchin eggs by lysis in a buffer containing EGTA and Triton X-100, Greenberg-Giffard *et al.* (1982) isolated the *Dictyostelium* cortical actin filament matrix (Fig. 4) and found it to contain up to 50% of the cell's actin. The filaments of the matrix appear to be associated with patches of the membrane that are relatively Triton-resistant (Fig. 5; A. Spudich, unpublished observations).

In a related study, Condeelis (1979) examined the possible interaction of concanavalin A (ConA) receptors with the actin cortical matrix. The regions of the cell membrane of *Dictyostelium* amoebae that contained ConA-receptor complexes proved more resistant to Triton X-100, allowing the isolation of the ConA-associated membrane fragments. Such fragments were rich in actin and myosin and, similar to the membranes without ConA (Spudich, 1974), most of the myosin was displaced with ATP. Binding of ConA to the amoebae depressed cytoplasmic streaming; vigorous streaming and pseudopod activity resumed once the ConA was capped on the cell surface. This may reflect interactions between cell surface receptors and the endogenous force-generating machinery.

It is clear from the above studies that in *Dictyostelium* much of the actin is associated with the cell membrane as a peripheral membrane protein. Luna *et al.* (1981) examined the nature of the actin–membrane association using falling-ball viscometry. They observed large increases in viscosity when membranes depleted of actin and myosin were mixed with muscle F-actin, indicating actin cross-linking by the membrane fraction. Further experiments showed that integral membrane proteins appear to be responsible for this F-actin cross-linking activity, but the identity of these proteins has not been established. In related experiments, Jacobson (1980) selectively ex-

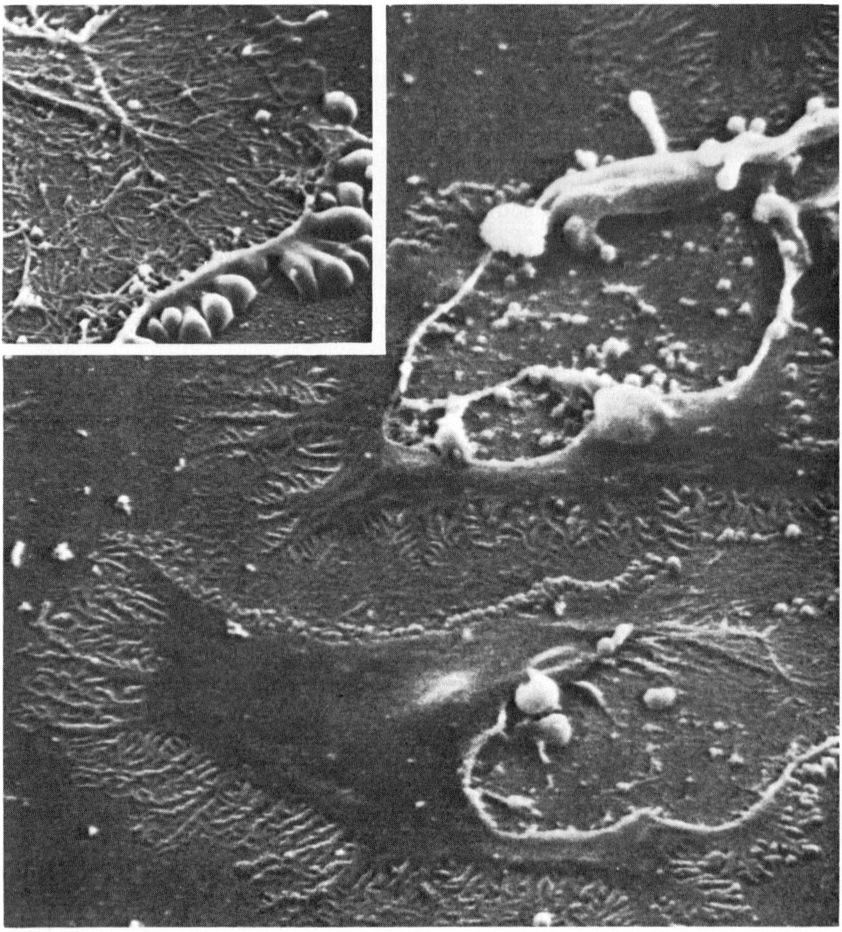

Fig. 3. Actin filaments on the cytoplasmic face of *Dictyostelium* cell membranes. A stream of buffer was forced across the surface of intact cells to "blow their tops off" and reveal the cytoplasmic face of the cell membrane. From Clarke *et al.* (1975). Magnifications 6150 × and 14,250 × (insert).

posed the cytoplasmic surface of the plasma membrane by attaching cells to cationic beads and shearing them by vortexing. These membranes could be stripped of their actin, and F-actin could be rebound. Pretreatment of the isolated membranes with trypsin abolished actin binding to the cytoplasmic surface.

All of the above studies indicate that actin and myosin are primary components of a filamentous matrix that underlies and is associated with the cell

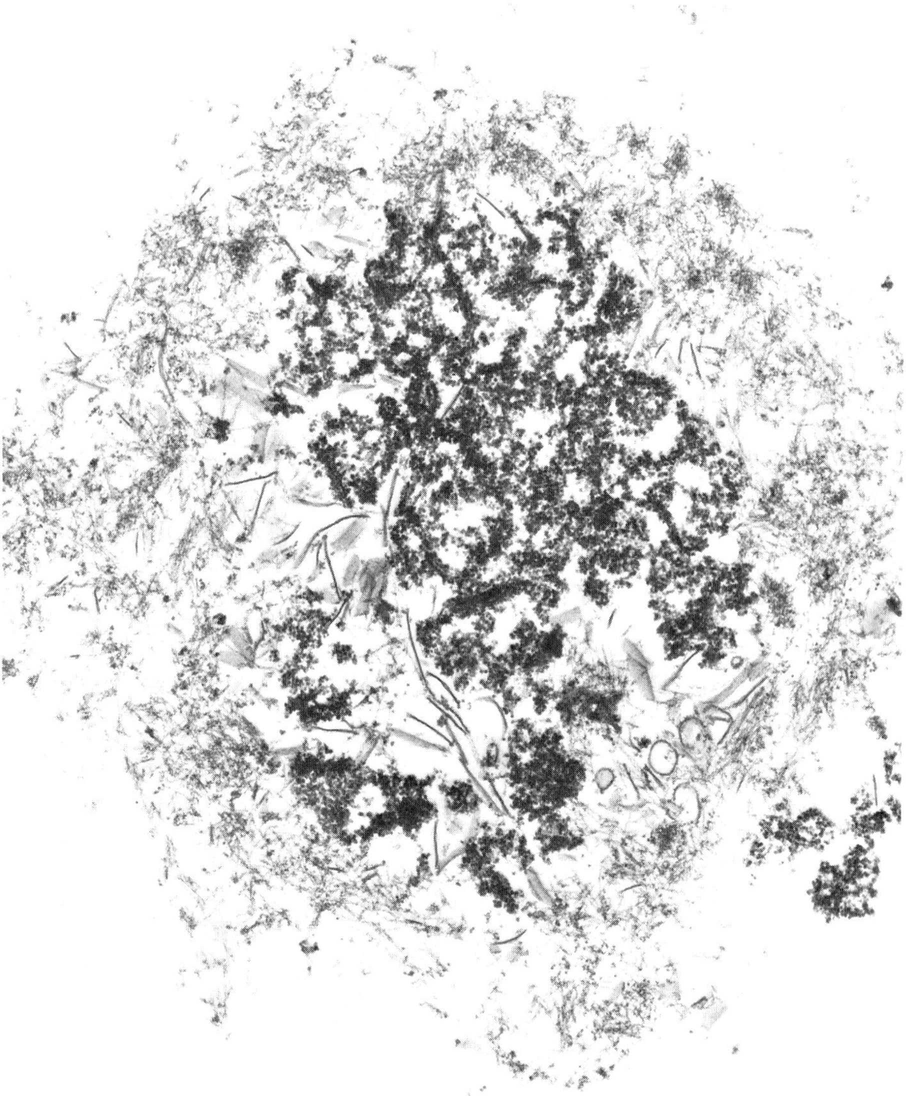

Fig. 4. Decoration of the cortical shell of actin filaments with the myosin fragment S1. Cortical shells of actin filaments were prepared by extraction of *Dictyostelium* amoebae with a high ionic strength buffer containing 2.5 m*M* EGTA and 0.5–1% Triton X-100. These shells were decorated with the myosin fragment S1, and then prepared for sectioning and analysis by transmission electron microscopy. Magnification 22,050 ×.

membrane. Changes in shape of the cell may be controlled in part by changes in this membrane-associated actin–myosin matrix. External signals, such as cAMP, may bind to receptors that are integral membrane proteins and influence cell behavior, such as direction of migration, by affecting the properties of the actin and myosin.

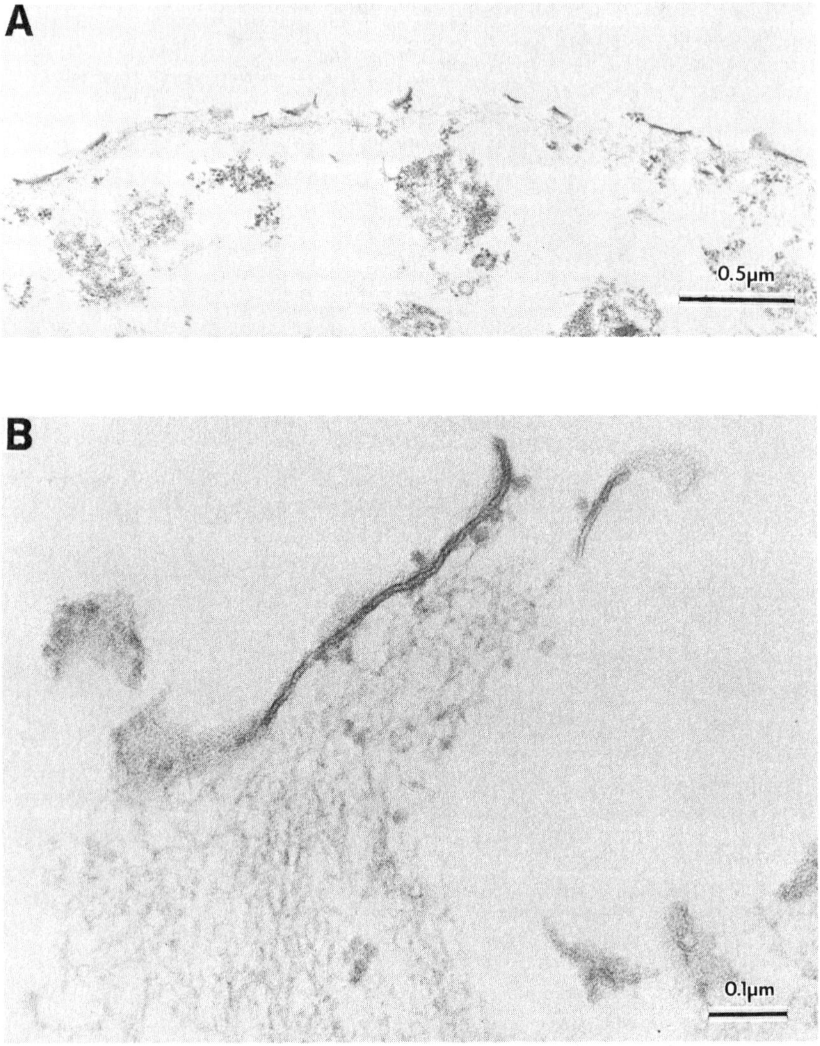

Fig. 5. Association of actin filaments with one side of Triton-resistant membrane patches. Cortical preparations of *Dictyostelium* amoebae were made by lysis in a buffer containing EGTA and 0.5–1% Triton X-100. Filaments are associated with patches of Triton-resistant membrane. Magnifications (A) 32,560 ×, (B) 111,000 ×.

Eckert *et al.* (1977) have reported that there is a close correlation between motile activity of *Dictyostelium* amoebae and the cellular level of actin, measured as percent of total protein. Peaks in actin levels are found at the active feeding stage as well as at the beginning of aggregation, when motile activities are high. Tuchman *et al.* (1974) reported that there is approximately a fivefold increase in the synthesis of actin at the early aggregation stage as compared to the vegetative stage. This newly synthesized actin is almost exclusively membrane-associated.

IV. Actin and Myosin Are Involved in Ca^{2+}-Dependent Contractions and Gelation–Solation Phenomena

It has been known for many years, primarily from work with the giant amoebae of *Chaos*, that the cytoplasm of cells is capable of changing from a "sol" (a less structured endoplasm) to a "gel" (a more structured ectoplasm) (Mast, 1926; Allen, 1961). Such sol–gel transformations are thought to involve changes in the degree of order of actin filaments and perhaps myosin filaments (Kane, 1975; Pollard, 1976; Condeelis and Taylor, 1977; Taylor and Condeelis, 1979). Ca^{2+} appears to be an important controlling element in changes in cytoplasmic contractions and in sol–gel transformations (Taylor *et al.*, 1973; Taylor *et al.*, 1976; Taylor, 1977).

In experiments using *Dictyostelium* amoebae, Condeelis and Taylor (1977) prepared high-speed supernatants of extracts from amoebae lysed at 4°C in low ionic strength buffer at pH 6.75 containing EGTA. Warming this extract to 25°C resulted in an increase in structure that could be observed in a number of ways including an obvious gelation. Contractions and cytoplasmic streaming could be elicited in extracts in the presence of Mg^{2+} and ATP by adding micromolar Ca^{2+}. The pH of the extract had to be maintained very close to 7.0 in order to observe these Ca^{2+}-dependent contractions and streaming. Conditions that solated the gelled extract elicited contraction of the extract. This and other work by Taylor and his colleagues have led to a solation–contraction coupling hypothesis (Hellewell and Taylor, 1979).

Actin and myosin are almost certainly involved in these gelation and contraction processes. Purified actin filaments will not form a gel but can be induced to do so by interaction with other cell components (see below). Actin and myosin are known to interact to produce contractions in muscle, and Condeelis and Taylor (1977) showed that myosin is required for contraction but not for gelation of the *Dictyostelium* extracts, and that actin and myosin are both concentrated in the "contracted pellet."

The gelation–solation and contraction processes discussed earlier are likely to be intimately related to pseudopod formation, and thus amoeboid movement, as well as to a variety of other cell movements, such as phagocytosis and cytokinesis. These are all extremely complex processes. To begin elucidating the molecular steps in such processes, it is necessary to first study in detail the functional properties of purified *Dictyostelium* actin and myosin, and to characterize those accessory components that control their assembly into filaments and their interaction with one another to generate force.

V. Dictyostelium Actin

A. Actin Is a Vital and Major Constituent of *Dictyostelium* Amoebae

The concentration of actin in *Dictyostelium* amoebae has been estimated to be higher than 200 μM or approximately 10 mg/ml (Spudich and Cooke, 1975). The actin has been highly purified from vegetative cells (Spudich, 1974; Uyemura *et al.*, 1978) and appears to be only one major species, which has a molecular weight of 42,000 (Spudich, 1974; Uyemura *et al.*, 1978; Vandekerckhove and Weber, 1980). The amino acid sequence of this actin (Vandekerckhove and Weber, 1980) is most similar to *Physarum* actin but differs from mammalian muscle actin in only a few of its residues. Like muscle actin, *Dictyostelium* actin is acetylated on its NH_2 terminus (Rubenstein and Deuchler, 1979), and *in vitro* studies by Rubenstein *et al.* (1981) have shown that nonacetylated actin can be obtained and acetylated posttranslationally in an acetyl-CoA-dependent system. *Dictyostelium* cells have seventeen actin genes (McKeown, *et al.*, 1978; Kindle and Firtel, 1978; Firtel *et al.*, 1979; McKeown and Firtel, 1981a; McKeown and Firtel, 1981b). Details of these and related studies are given in the chapter by Firtel (this volume). Perhaps many of these genes are not expressed or are expressed only in low levels. Some may be expressed in higher levels during differentiation as the cells proceed through their developmental cycle (Firtel *et al.*, 1979). If different actin genes are expressed, do the different gene products contribute to a common pool of actin not to be differentiated in any functional sense or do they serve unique roles in the cell? These are important questions to be answered by future research. Currently, it would appear that a single type of actin molecule can serve a multitude of cellular functions.

What are the unique features of actin that make it so suitable for force transduction? Actin has two fundamentally important properties: (a) it undergoes a reversible self-association into the form of thin filaments (Fig. 2) and (b) these filaments interact with myosin, an ATPase (see below), to produce force. Filament formation and force production are highly localized events, both spatially and temporally. Thus, it is apparent that both events are under strict cellular controls. The identification of the accessory components involved in these controls and the elucidation of the control mechanisms at the molecular level are the foundation of a large number of investigations underway in cell motility research. Although fundamental to all eukaryotic cells, many aspects of these control mechanisms are likely to be elucidated by studies using *Dictyostelium* because of the special features of this organism (see other chapters in this volume, and Spudich, 1982).

B. Purified Actin Prefers To Be Polymerized under Physiologic Ionic Strengths

Purified *Dictyostelium* actin has many of the same properties as actin purified from striated muscle (Woolley, 1972; Spudich, 1974; Spudich and Cooke, 1975; Uyemura *et al.*, 1978). As already mentioned, of special interest is the ability of this protein to assemble into thin filaments (Fig. 2). The filaments are 70 Å in diameter and are generally several μm long. They consist of actin monomers arranged along two long-pitched, right-handed helical strands, which cross over every 13 to 14 monomers, or every 360 to 400 Å (Hanson and Lowy, 1963; Huxley, 1963; Moore *et al.*, 1970; Spudich *et al.*, 1972; Uyemura *et al.*, 1978). Under conditions of ionic strength and Mg^{2+} concentration that approximate intracellular conditions, the equilibrium of this assembly reaction favors filament formation. For example, in the presence of about 0.1 *M* KCl and 0.1–1 m*M* Mg^{2+}, the concentration of monomer in equilibrium with filaments is less than 1 μ*M* when measured at a total actin concentration of up to 0.5 mg/ml (12 μ*M*) (Spudich and Cooke, 1975; Uyemura *et al.*, 1978). Thus, purified actin is preferentially in the polymerized state under physiologic salt conditions.

C. Actin Filaments Are Stable in Conventional F-Buffer

To understand the physiological control of actin filament assembly–disassembly, it is important to establish the stability of purified *Dictyostelium* actin filaments once they are formed. Are they rapidly turning over at steady state assembly with a monomeric pool of actin, as would be

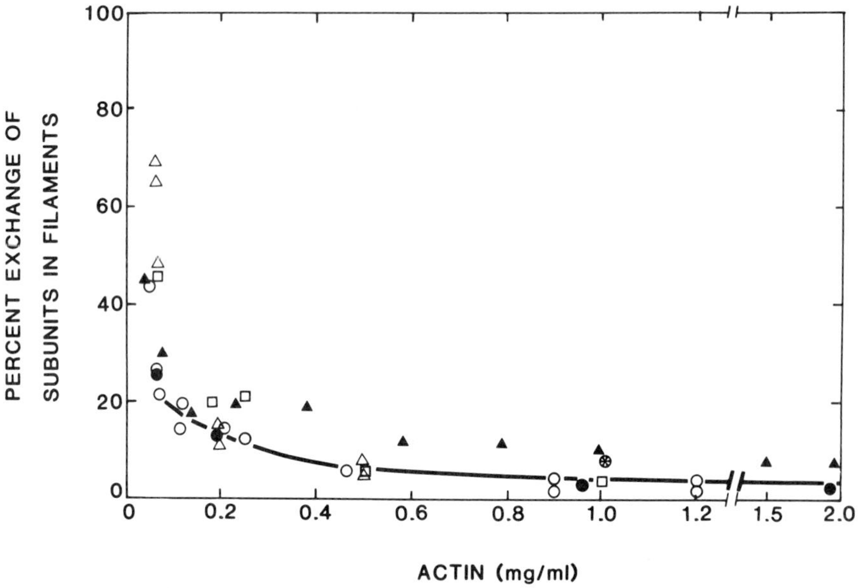

Fig. 6. Extent of actin subunit exchange as a function of actin concentration. The plateau levels of limited subunit exchange are shown. The extent of subunit exchange was determined by fluorescence energy transfer for muscle actin in a buffer containing 0.1 *M* KCl, 1 m*M* $MgCl_2$, and 1 m*M* ATP (○), muscle actin in a buffer containing 8 m*M* KCl, 2 m*M* $MgCl_2$, and 0.4 m*M* ATP (□), and *Dictyostelium* actin in a buffer containing 0.1 *M* KCl, 1 m*M* $MgCl_2$ (●). Exchange was measured after mixing donor-labeled (IAENS) filaments with acceptor-labeled (FM) filaments. Subunit exchange was also determined by incorporation of ^{35}S-labeled *Dictyostelium* G-actin into unlabeled muscle F-actin in a buffer containing 0.1 *M* KCl, 1 m*M* $MgCl_2$ and 1 m*M* ATP (△), and *Dictyostelium* F-actin in a buffer containing 0.1 *M* KCl, 1 m*M* $MgCl_2$, and 1 m*M* ATP (▲). Exchange of [^{14}C]ATP for F-actin bound ADP is also shown (⊛). From Pardee *et al.* (1982).

suggested by a rapid "treadmilling" process as proposed by Wegner (1976), or are the monomers in the filament very slow to exchange with the monomeric pool of actin? A number of methods have been used to examine the rate of exchange of actin monomers with filaments at steady state. Pardee *et al.* (1982) used fluorescence energy transfer (Stryer, 1978; Taylor *et al.*, 1981), exchange of metabolically labeled [^{35}S]G-actin into unlabeled filaments (Simpson and Spudich, 1980), and incorporation of [^{14}C]ATP into filaments to examine actin filament subunit exchange using both *Dictyostelium* actin and muscle actin. All three approaches with both actins showed that in conventional F-buffer, chosen to approximate intracellular salt conditions, there is a rapid but very limited exchange of monomers with filaments at equilibrium (Fig. 6). At actin concentrations greater than 0.2 mg/ml, less than 10% exchange occurred as a limit. Two distinct explanations of these results are (a) that all filaments undergo a limited amount of exchange with monomers, most likely at the filament ends and (b) that there are two populations of filaments, one of which is more stable than the other. A direct demonstration of the sites of exchange by immunological techniques and electron microscopy is feasible and will be important for a better understanding of the G⇌F exchange mechanism.

D. A 40,000-Dalton Protein Disassembles Actin Filaments in a Ca^{2+}-Dependent Manner

Since the inherent disposition of purified *Dictyostelium* actin is to be in the form of filaments, one would expect that cells have accessory proteins that can alter the assembly state of the actin. A 40,000-dalton protein that

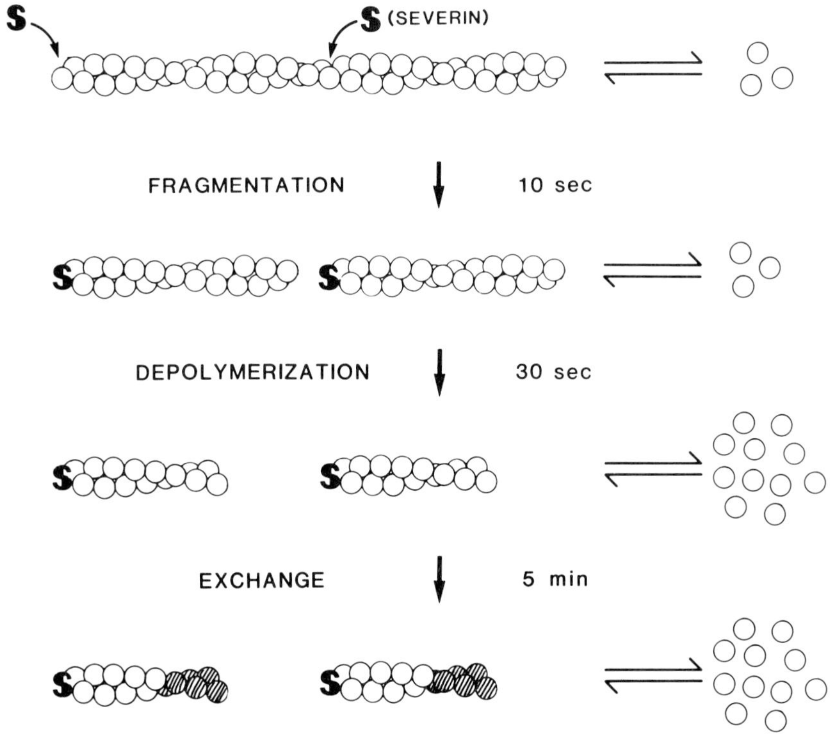

Fig. 7. Proposed model for interaction of *Dictyostelium* severin with actin filaments. Actin filaments are fragmented within 30 sec of addition of severin in the presence of 0.2 m*M* Ca^{2+}. The severin remains bound at the "barbed" ends of actin filaments. The actin monomer pool in equilibrium with filament fragments increases with a half-time of 30 sec to a new equilibrium level. An increase in the amount of subunit exchange occurs with a half-time of about 5 min. From Yamamoto *et al.* (1982).

disassembles actin filaments in the presence of Ca^{2+} has been purified to electrophoretic homogeneity from vegetative amoebae (Brown *et al.*, 1982; Yamamoto *et al.*, 1982). Various approaches, including electron microscopy, viscometry, DNase I inhibition assays, sedimentation, fluorescence energy transfer, and fluorescence anisotropy, indicate that this protein stoichiometrically binds to actin filaments, severs them, and apparently remains bound to the "barbed" end of the filament fragments (Fig. 7) (Yamamoto *et al.*, 1982). Considering the mode of action of this protein, it has been named severin. The mechanism of action of severin resembles in some ways that of the fungal metabolite cytochalasin, which also appears to bind to the barbed end of actin filaments (Brenner and Korn, 1979; Brown and Spudich, 1979; Hartwig and Stossel, 1979; Flanagan and Lin, 1980; Lin *et al.*, 1980; MacLean-Fletcher and Pollard, 1980; Brown and Spudich, 1981).

The native form of severin is a monomer (Brown *et al.*, 1982; Yamamoto *et al.*, 1982). At low stoichiometry, it accelerates the rate of actin polymerization, presumably by interacting with actin monomers and generating nuclei for assembly (Brown *et al.*, 1982).

The physiological role of severin, or of any of the other accessory proteins to be discussed, remains to be determined. Evidence to date suggests it is primarily cytoplasmic and not concentrated in the actin matrix of the cell cortex (Greenberg-Giffard *et al.*, 1982). Its roles may include disassembly of filaments that are no longer required in a particular location of the cell or capping of filaments to limit filament length and alter the assembly–disassembly properties of the actin.

E. A 120,000-Dalton Protein Combines with Actin Filaments to Produce a Gel

As mentioned above, extracts from *Dictyostelium* amoebae contain accessory proteins that cause actin filaments to form a gel *in vitro* (Hellewell and Taylor, 1979; Condeelis *et al.*, 1981). Condeelis *et al.* (1982) purified one of these gelation factors. It is a 120,000-dalton, carbohydrate-free protein that increases the viscosity and sedimentation rate of F-actin, but also nucleates actin filament assembly.

By indirect immunofluorescence, Condeelis *et al.* (1981) have shown that this protein is concentrated in the cell cortex, the location of the actin–myosin matrix discussed above. In both spreading cells and in migrating cells, this protein is concentrated in ruffles and pseudopods. This 120,000-dalton protein may serve to cross-link actin filaments forming the ectoplasmic gel (Taylor and Condeelis, 1979).

F. Several Other *Dictyostelium* Proteins That Affect Actin Have Been Identified

Other gelation factors have been described by Hellewell and Taylor (1979). They fractionated contracted pellets from *Dictyostelium* extracts to obtain a partially purified 95,000-dalton polypeptide that exhibited Ca^{2+}-sensitive gelation of actin at 28°C and another fraction containing primary components of 30,000 and 18,000 daltons that show Ca^{2+}-sensitive gelation at 0°C. A different 95,000-dalton protein has been described that is a Ca^{2+}-dependent inhibitor of gelation induced by the 120,000-dalton protein (Condeelis *et al*. 1982).

Spudich and Cooke (1975) showed that partially purified *Dictyostelium* actin, but not highly purified actin, forms filament bundles that are about 0.1 μm in diameter and that assemble in the presence of 5 m*M* $MgCl_2$. Filament bundles that are similar in dimensions are observed in some situations in *Dictyostelium* amoebae (Clarke *et al.*, 1975). Cooke *et al.* (1976) dissociated the bundles formed *in vitro* by addition of high salt and separated actin filaments from the soluble material by high-speed centrifugation. The filament fraction did not form bundles by itself, but bundles were formed when the fraction was recombined with the components that were left in the supernatant. Their experiments indicate that the formation of bundles requires actin plus one or more minor components from the amoebae, but these components have not been purified and characterized.

VI. *Dictyostelium* Myosin

A. Myosin Is an ATPase, Interacts with Actin, and Assembles into Thick Filaments

Myosin from *Dictyostelium* was first characterized by Clarke and Spudich (1974). Purification of the *Dictyostelium* myosin utilized newly discovered solubility properties of myosin in sucrose. Amoebae were extracted with a buffer containing 30% sucrose, and myosin was selectively precipitated as a complex with actin by removal of the sucrose.

In general, properties of *Dictyostelium* myosin closely resemble those of muscle myosin (Clarke and Spudich, 1974; Mockrin and Spudich, 1976; Spudich *et al.*, 1982). The amoeba protein contains two heavy chains, approximately 210,000 daltons each, and two each of two classes of light chains,

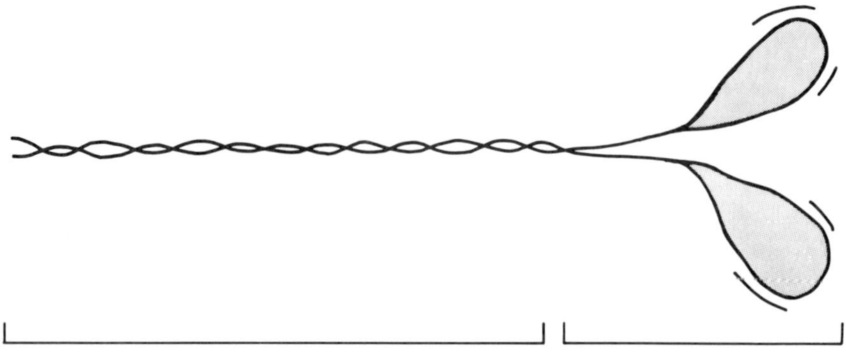

Fig. 8. Schematic representation of the myosin molecule. *Dictyostelium* myosin consists of two heavy chains (210,000 daltons), two 18,000-dalton light chains, and two 16,000-dalton light chains.

18,000 and 16,000 daltons (Fig. 8). The globular head region binds and hydrolyzes ATP and also interacts with actin. This interaction can be observed by electron microscopy, viscometry, and by actin activation of the myosin ATPase. The latter is thought to be an integral part of the cycle of interaction that produces force. Actin activates the myosin ATPase by more than 40-fold (Mockrin and Spudich, 1976).

The tail region of the myosin molecule is responsible for the formation at low ionic strength of bipolar thick filaments (Fig. 9). There is no general agreement about the structural states that myosin assumes in nonmuscle cells. Thick filaments with the same dimensions as myosin bipolar thick filaments have been observed in pseudopodial regions of *Dictyostelium* amoebae (Spudich and Clarke, 1974). End-to-end aggregates of myosin thick filaments from chicken brain (Burridge and Bray, 1975), platelets (Pollard *et al.*, 1976), and starfish eggs (Mabuchi, 1976) have led to some speculation that such aggregates could occur *in vivo*. *Dictyostelium* myosin thick filaments also form linear aggregates, and Stewart and Spudich (1979) showed that aggregate formation requires the presence of RNA. A significant amount of RNA (~10% by weight) binds fairly strongly to the myosin and copurifies with it up to the DEAE purification step. Purification of the myosin by DEAE chromatography or treatment with RNase eliminates the RNA, and aggregates of the thick filaments no longer form. Alternative assembly states for myosin monomers or oligomers to form bipolar thick filaments on the one hand, and higher ordered aggregates on the other, could be important for different contractile functions in which myosin may be involved. However,

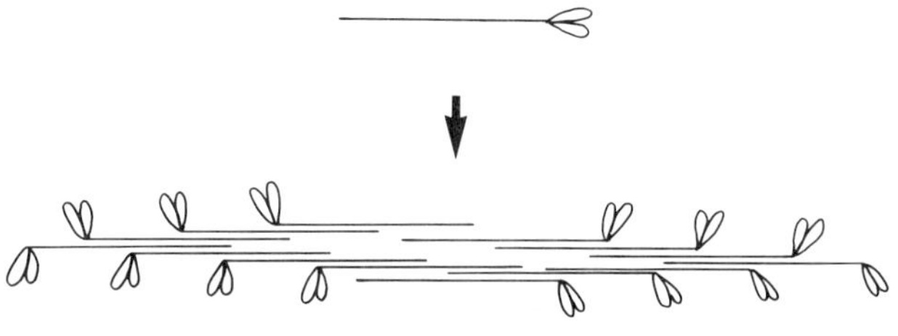

Fig. 9. Schematic representation of the assembly of myosin molecules into bipolar thick filaments (Huxley, 1963). The bipolar thick filaments can be readily sedimented by centrifugation.

until more is known about the actual structural states that myosin adopts in cells, any proposals concerning the possible structure–function relationships can be only speculative.

B. Myosin Is Phosphorylated *in Vivo*

Recently, Kuczmarski and Spudich (1980) found that *Dictyostelium* myosin is phosphorylated *in vivo* on both the heavy chain and the 18,000-dalton light chain. Myosin purified from amoebae grown in medium containing [^{32}P]orthophosphate had 0.2–0.4 mole phosphate per mole heavy chain and approximately 0.1 mole phosphate per mole 18,000-dalton light chain. Kinase activities specific for the 210,000- and 18,000-dalton subunits have been identified in extracts, and the heavy chain kinase has been purified about 50-fold. This kinase phosphorylated the myosin *in vitro* to a maximum of 0.5 to 1.0 mole phosphate per mole heavy chain.

It was of great interest to find that the heavy chain phosphorylation site is in the tail portion of the myosin molecule (Peltz *et al.*, 1981) since, as discussed above, the tail is involved in the formation of thick filaments. Chymotrypsin was used to cleave the myosin in half. One of these two major fragments is soluble at low ionic strength, has a native molecular weight of 130,000, and consists of the two intact light chains and a 105,000-dalton polypeptide derived from the head portion of the myosin heavy chain. This soluble fragment retained actin-activated ATPase activity and the ability to bind to actin in an ATP-dissociable fashion. The tail fragment, in contrast, is insoluble at low ionic strength, forming thick filaments without myosin heads. This fragment consists of a polypeptide of about 105,000 daltons and is the site of heavy chain phosphorylation.

C. Phosphorylation of the Tail Portion of Myosin Inhibits Thick Filament Formation

In order to examine the effect of phosphorylation of the tail portion of the myosin molecule on thick filament formation, Kuczmarski and Spudich (1980) first obtained preparations of heavy chain kinase and of an appropriate phosphatase that were sufficiently pure to be devoid of protease that would otherwise alter the myosin by digestion. Using these enzymes, they prepared myosin with nearly 1 mole phosphate per mole heavy chain as well as myosin with no phosphate attached to the heavy chain. The ability of these myosins to form thick filaments at various salt concentrations was then compared. A remarkable difference was observed at 50–100 m*M* KCl, a range estimated to be near physiological in amoebae of *Dictyostelium*. Phosphorylation of the heavy chain inhibited the ability of myosin to form thick filaments (Fig. 10). At about 0.1 *M* KCl, complete phosphorylation (approximately 1 mole phosphate per mole heavy chain) essentially eliminated thick filament formation, whereas partial phosphorylation (about 0.4 mole phosphate per mole heavy chain) simply reduced the amount of thick filament formation. Nearly all of the myosin formed thick filaments when virtually all of the phosphate was removed (<0.1 mole phosphate per mole heavy chain).

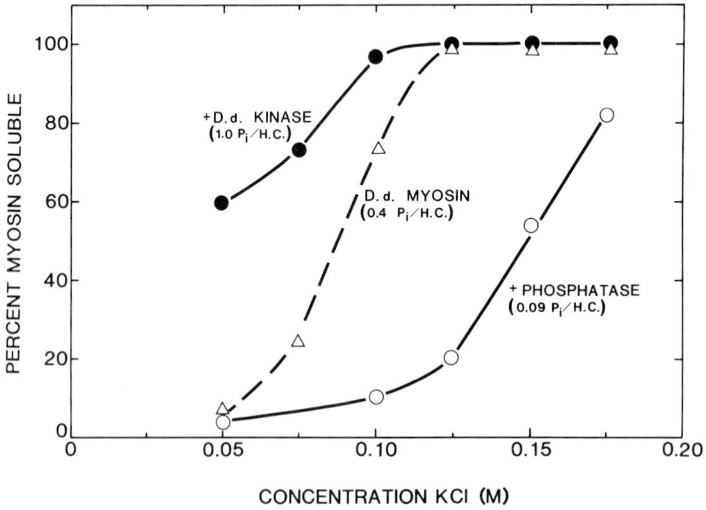

Fig. 10. Inhibition of myosin-filament assembly by phosphorylation of the heavy chain. Myosin was incubated at various KCl concentrations. The percent myosin soluble (not in the form of thick filaments) was determined after sedimentation of thick filaments by centrifugation. The extents of phosphorylation (mole phosphate per mole heavy chain) for untreated Δ, phosphorylated ●, and dephosphorylated ○ myosins are indicated. From Spudich *et al.* (1981).

This effect on assembly could be reversed by rephosphorylating the dephosphorylated myosin. These findings suggest that heavy chain phosphorylation may regulate cell contractile events by altering the state of myosin assembly.

Earlier studies had suggested that the myosin heavy chain might be able to be phosphorylated *in vitro* (Parish *et al.*, 1977; Rahmsdorf *et al.*, 1978). Rahmsdorf *et al.* (1978) showed that extracts prepared from amoebae that had been exposed to external cAMP, the chemotactic agent for these cells, showed an enhanced activity for phosphorylating a protein that comigrates with the myosin heavy chain on SDS–acrylamide gels. Further studies suggesting that phosphorylation of the myosin heavy chain may prove to be an intermediate step between the external cyclic AMP signal and the extension of pseudopodia have been reported by Malchow *et al.* (1981).

VII. The Interaction of Actin and Myosin May Be Controlled by Ca^{2+}

As discussed above, Ca^{2+} appears to be an important controlling element in changes in cytoplasmic contractions and in sol–gel transformations. Taylor and his colleagues have defined the conditions of pH and Ca^{2+} concentration required for a myosin-dependent contraction of *Dictyostelium* extracts and partially purified preparations made from them (see Hellewell and Taylor, 1979). Contraction does not occur at pH 6.8 and at a free Ca^{2+} concentration of $\leq 3 \times 10^{-8}$ *M* but does occur if the Ca^{2+} concentration is raised to $\geq 8 \times 10^{-7}$ *M*. Alternatively, raising the pH above 6.8 also induces contraction. These preparations have numerous proteins and, as the authors point out, multiple events are probably occurring in the course of the assays used. However, these preparations are good starting material for the purification of factors affecting actin–myosin interaction.

The interaction of purified actin with purified myosin does not require Ca^{2+} (Clarke and Spudich, 1974; Mockrin and Spudich, 1976). However, a protein fraction from the amoebae confers Ca^{2+} sensitivity on the actin activation of the myosin ATPase (Mockrin and Spudich, 1976). That is, the fraction inhibits the actin-activated ATPase activity in the absence of Ca^{2+} but not in the presence of Ca^{2+}. This Ca^{2+}-sensitizing factor inhibits only the actin-activated myosin ATPase and not the ATPase activity of the myosin alone. The factor appears to be a protein since it is nondialyzable, heat labile, and can be precipitated with ammonium sulfate. Although the factor was purified 70-fold, further purification is needed to establish its identity. It is possible that this activity is related to the subsequently discovered phosphorylation of myosin (Kuczmarski and Spudich, 1980), but further work is needed to examine this possible relationship.

It is interesting to note that calmodulin has been purified and characterized from *Dictyostelium* amoebae (Clarke *et al.*, 1980; Bazari and Clarke, 1981). Using indirect immunofluorescence techniques, the calmodulin was found to be concentrated in the cortical region of *Dictyostelium* amoebae (Bazari and Clarke, 1982). Maruta *et al.* (1982) have evidence for a Ca^{2+}, calmodulin-dependent inactivation of a myosin heavy chain kinase from *Dictyostelium*.

De Chastellier and Ryter (1981) have described Ca^{2+}-dependent deposits on the inner face of the *Dictyostelium* plasma membrane, which they speculate may reflect a membrane ATPase that could be linked to the presence of contractile proteins. These electron-dense deposits on the plasma membrane were shown to contain calcium by energy dispersive X-ray microanalysis. The deposits are more abundant in cell regions where microfilaments are concentrated, such as in filopodia and phagocytic cups.

Factors that affect the assembly properties of actin or myosin may appear to also have an effect on the actin–myosin interaction. Thus, severin fragments actin filaments and inhibits superprecipitation of the actin–myosin complex (Yamamoto *et al.*, 1982), the 120,000-dalton protein decreases the actin-activated myosin ATPase (Condeelis and Geosits, 1982), as does phosphorylation of the myosin heavy chain by a specific heavy chain kinase (Kuczmarski and Spudich, 1980). In muscle, tropomyosin–troponin controls the interaction of actin and myosin without affecting the assembly state of the filaments. Whether there are accessory proteins in *Dictyostelium* that control the interaction of the assembled actin and myosin filaments, as in the case of muscle contraction, remains to be determined.

Generally, this aspect of cell motility, like most of the others described in this chapter, is a fertile area for active work by many investigators. This next decade will require an intensive effort to identify, purify, and characterize the many factors involved in the processes of interest.

VIII. Concluding Remarks

Important advances in *Dictyostelium* motility have been made in the last decade, but a great deal must be learned before a comprehensive molecular picture of fundamental biological events such as cytokinesis, phagocytosis, and cell migration will emerge. Although quantitative assays have allowed the isolation of certain accessory proteins that modulate actin and myosin, most such factors remain to be identified, and many of them are serving as yet unknown functions. How does one identify and purify proteins whose functions are unknown? Some will emerge from continued biochemical studies. However, the application of genetics is essential for a more complete understanding of cell motility. For example, work by Ross and Newell (1981) using genetic techniques suggests a correlation between defective cGMP metabolism and altered chemotactic movement responses in *Dictyostelium*. Further studies may reveal an important role of cGMP in cell movement in this organism. Another study has shown that a particular class of phagocytosis-defective mutants are also defective in adhesion (Vogel *et al.*, 1980).

In general, one may have to screen for conditional mutations since general defects in the contractile machinery may prove lethal. Mutations of interest should be isolatable by a succession of screening steps involving a variety of specific cell-motility defects. This approach has been used to isolate motility mutants of *Dictyostelium*. Clarke (1978) enriched for *Dictyostelium* cells unable to incorporate a lethal nucleotide into their own DNA by phagocytosis of bacteria at restrictive temperature. Several independently isolated mutants proved to be defective in many motile functions at the restrictive temperature, including spreading, formation of filopodia and pseudopodia, migration, and phagocytosis (Kayman *et al.*, 1982). Such mutant strains can now be analyzed for corresponding defects in *in vitro* processes such as actin activation of myosin ATPase, actin and myosin filament formation, and gelation. *In vitro* complementation among defective strains should then allow the identification and purification of new accessory proteins without knowing their specific functions beforehand. In summary, the combination of biochemical, structural, and genetic approaches should yield important advances in the next decade in our understanding of cell motility. *Dictyostelium discoideum* can be employed effectively for all of these approaches and will therefore be an increasingly prominent organism for future investigations.

References

Allen, R. D. (1961). A new theory of amoeboid movement and protoplasmic streaming. *Exp. Cell Res.* **8**, 17–31.

Bazari, W. L., and Clarke, M. (1981). Characterization of a novel calmodulin from *Dictyostelium discoideum*. *J. Biol. Chem.* **256**, 3598–3603.

Bazari, W. L., and Clarke, M. (1982). *Dictyostelium* calmodulin: Production of a specific antiserum and localization in amoebae. Submitted.

Brenner, S. L., and Korn, E. D. (1979). Substoichiometric concentrations of cytochalasin D inhibit actin polymerization. *J. Biol. Chem.* **254,** 9982–9985.

Brown, S. S., and Spudich, J. A. (1979). Cytochalasin inhibits the rate of elongation of actin filament fragments. *J. Cell Biol.* **83,** 657–662.

Brown, S. S., and Spudich, J. A. (1981). Mechanism of action of cytochalasin: evidence that it binds to actin filament ends. *J. Cell Biol.* **88,** 487–491.

Brown, S. S., Yamamoto, K., and Spudich, J. A. (1982). A 40,000 dalton protein from *Dictyostelium discoideum* affects assembly properties of actin in a Ca^{2+}-dependent manner. *J. Cell Biol.* **93,** 205–210.

Burridge, K., and Bray, D. (1975). Purification and structural analysis of myosins from brain and other non-muscle tissues. *J. Mol. Biol.* **99,** 1–14.

Clarke, M. (1978). A selection method for isolating motility mutants of *Dictyostelium discoideum*. *In* "Cell Reproduction", (E. R. Dirksen, D. M. Prescott, C. F. Fox, eds.), pp. 621–629. Academic Press, New York.

Clarke, M., Bazari, W. L., and Kayman, S. C. (1980). Isolation and properties of calmodulin from *Dictyostelium discoideum. J. Bacteriol.* **141,** 397–400.

Clarke, M., Schatten, G., Mazia, D., and Spudich, J. A. (1975). Visualization of actin fibers associated with the cell membrane in amoebae of *Dictyostelium discoideum*. *Proc. Natl. Acad. Sci. U.S.A.* **72,** 1758–1762.

Clarke, M., and Spudich, J. A. (1974). Biochemical and structural studies of actomyosin-like proteins from non-muscle cells. Isolation and characterization of myosin from amoebae of *Dictyostelium discoideum*. *J. Mol. Biol.* **86,** 209–222.

Clarke, M., and Spudich, J. A. (1977). Nonmuscle contractile proteins: The role of actin and myosin in cell motility and shape determination. *Annu. Rev. Biochem.* **46,** 797–822.

Cold Spring Harbor Symposia on Quantitative Biology, Vol. XLVI (1982). "Organization of the Cytoplasm." Cold Spring Harbor Laboratory, Cold Spring Harbor, New York.

Condeelis, J. (1979). Isolation of concanavalin A caps during various stages of formation and their association with actin and myosin. *J. Cell Biol.* **80,** 751–758.

Condeelis, J., Geosits, S., and Vahey, M. (1982). Isolation of a new actin binding protein from *Dictyostelium discoideum*. *Cell Motility*. In press.

Condeelis, J., Salisbury, J., and Fujiwara, K. (1981). A new protein that gels F actin in the cell cortex of *Dictyostelium discoideum. Nature (London)* **292,** 161–163.

Condeelis, J., and Taylor, D. L. (1977). The contractile basis of amoeboid movement. V. The control of gelation, solation, and contraction in extracts from *Dictyostelium discoideum. J. Cell Biol.* **74,** 901–927.

Cooke, R., Clarke, M., von Wedel, R. J., and Spudich, J. A. (1976). Supramolecular forms of *Dictyostelium* actin. *In* "Cell Motility", (R. Goldman, T. Pollard, J. Rosenbaum, eds.), pp. 575–587. Cold Spring Harbor Laboratory, Cold Spring Harbor, New York.

de Chastellier, C., and Ryter, A. (1981). Calcium-dependent deposits at the plasma membrane of *Dictyostelium discoideum* and their possible relation with contractile protein. *Biol. Cell* **40,** 109–118.

Eckert, B. S., and Lazarides, E. (1978). Localization of actin in *Dictyostelium* amebas by immunofluorescence. *J. Cell Biol.* **77,** 714–721.

Eckert, B. S., Warren, R. H., and Rubin, R. W. (1977). Structural and biochemical aspects of cell motility in amebas of *Dictyostelium discoideum. J. Cell Biol.* **72,** 339–350.

Firtel, R. A., Timm, R., Kimmel, A. R., and McKeown, M. (1979). Unusual nucleotide sequences at the 5′ end of actin genes in *Dictyostelium discoideum. Proc. Natl. Acad. Sci. U.S.A.* **76,** 6206–6210.

Flanagan, M. D., and Lin, S. (1980). Cytochalasins block actin filament elongation by binding to high affinity sites associated with F-actin. *J. Biol. Chem* **255,** 835–838.

Franke, J., and Kessin, R. (1977). A defined medium for axenic strains of *Dictyostelium discoideum. Proc. Natl. Acad. Sci. U.S.A.* **74,** 2157–2161.

Fukui, Y. (1978). Intranuclear actin bundles induced by dimethyl sulfoxide in interphase nucleus of *Dictyostelium. J. Cell Biol.* **76,** 146–157.

Fukui, Y., and Katsumaru, H. (1979). Nuclear actin bundles in amoeba, *Dictyostelium* and human HeLa cells induced by dimethyl sulfoxide. *Exp. Cell Res.* **120,** 451–455.

Fukui, Y., and Katsumaru, H. (1980). Dynamics of nuclear actin bundle induction by dimethyl sulfoxide and factors affecting its development. *J. Cell Biol.* **84,** 131–140.

Garrod, D. R., and Born, G. V. R. (1971). Effect of temperature on the mutual adhesion of preaggregation cells of the slime mould, *Dictyostelium discoideum. J. Cell Sci.* **8,** 751–765.

Gerisch, G., Malchow, D., Huesgen, A., Nanjundiah, V., Roos, W., Wick, U., and Hulser, D. (1975). Cyclic-AMP reception and cell recognition in *Dictyostelium discoideum. In* "Developmental Biology," ICN-UCLA Symposia on Molecular and Cellular Biology, (D. McMahon and C. F. Fox, eds.) Vol. 2, pp. 76–88. Benjamin, Menlo Park, California.

Goldman, R., Pollard, T., and Rosenbaum, J., eds. (1976). "Cell Motility." Cold Spring Harbor Laboratory, Cold Spring Harbor, New York.

Greenberg-Giffard, R., Spudich, J. A., and Spudich, A. (1982). Ca^{2+}-sensitive isolation of a cortical actin meshwork from *Dictyostelium* amoebae. Submitted.

Hanson, J., and Lowy, J. (1963). The structure of F-actin and of actin filaments isolated from muscle. *J. Mol. Biol.* **6,** 46–60.

Hartwig, J. H., and Stossel, T. P. (1979). Cytochalasin B and the structure of actin gels. *J. Mol. Biol.* **134,** 539–553.

Hellewell, S. B., and Taylor, D. L. (1979). The contractile basis of ameboid movement. VI. The solation–contraction coupling hypothesis. *J. Cell Biol.* **83,** 633–648.

Hitchcock, S. E. (1977). Regulation of motility in nonmuscle cells. *J. Cell Biol.* **74,** 1–15.

Huxley, H. E. (1963). Electron microscope studies on the structure of natural and synthetic protein filaments from striated muscle. *J. Mol. Biol.* **7,** 281–308.

Jacobson, B. S. (1980). Actin binding to the cytoplasmic surface of the plasma membrane isolated from *Dictyostelium discoideum. Biochem. Biophys. Res. Comm.* **97,** 1493–1498.

Kane, R. E. (1975). Preparation and purification of polymerized actin from sea urchin egg extracts. *J. Cell Biol.* **66,** 305–315.

Kayman, S. C., Reichel, M., and Clarke, M. (1982). Motility mutants of *Dicytostelium discoideum. J. Cell Biol.* **92,** 705–711.

Kindle, K. L., and Firtel, R. A. (1978). Identification and analysis of *Dictyostelium* actin genes, a family of moderately repeated genes. *Cell* **15,** 763–778.

Korn, E. D. (1978). Biochemistry of actomyosin-dependent cell motility (a review). *Proc. Natl. Acad. Sci. U.S.A.* **75,** 588–599.

Kuczmarski, E. R., and Spudich, J. A. (1980). Regulation of myosin self-assembly: phosphorylation of *Dictyostelium* heavy chain inhibits formation of thick filaments. *Proc. Natl. Acad. Sci. U.S.A.* **77,** 7292–7296.

Lin, D. C., Tobin, K. D., Grumet, M., and Lin, S. (1980). Cytochalasins inhibit nuclei-induced actin polymerization by blocking filament elongation. *J. Cell Biol.* **84,** 455–460.

Luna, E. J., Fowler, V. M., Swanson, J., Branton, D., and Taylor, D. L. (1981). A membrane cytoskeleton from *Dictyostelium discoideum.* I. Identification and partial characterization of an actin-binding activity. *J. Cell Biol.* **88,** 396–409.

Mabuchi, I. (1976). Myosin from starfish eggs: Properties and interaction with actin. *J. Mol. Biol.* **100,** 569–582.

MacLean-Fletcher, S., and Pollard, T. D. (1980). Mechanism of action of cytochalasin B on actin. *Cell* **20,** 329–341.

Malchow, D., Bohme, R., and Rahmsdorf, H. J. (1981). Regulation of phosphorylation of myosin heavy chain during the chemotactic response of *Dictyostelium* cells. *Eur. J. Biochem.* **117,** 213–218.

Maruta, H., Baltes, W., Gerisch, G., Dieter, P., and Marmé, D. (1982). Signal transduction in chemotaxis of *Dictyostelium discoideum*: Role of Ca^{2+} and calmodulin in the regulation of myosin heavy chain kinases and other protein kinases. *In* "Plasmalemma and Tonoplast: Their Functions in the Plant Cell," pp. 331–335. (D. Marme, E. Marre, and R. Hertel, eds.). Elsevier Biomedical Press B.V.

Mast, S. (1926). Structure, movement, locomotion and stimulation in amoebae. *J. Morphol. Physiol.* **41,** 347–425.

McKeown, M., and Firtel, R. A. (1981a). Evidence for sub-families of actin genes in *Dictyostelium* as determined by comparison of 3′ end sequences. *J. Mol. Biol.* **151,** 593–606.

McKeown, M., and Firtel, R. A. (1981b). Differential expression and 5′ end mapping of actin genes in *Dictyostelium. Cell* **24,** 799–807.

McKeown, M., Taylor, W. C., Kindle, K. L., Firtel, R. A., Bender, W., and Davidson, N. (1978). Multiple, heterogeneous actin genes in *Dictyostelium. Cell* **15,** 789–800.

Mockrin, S. C., and Spudich, J. A. (1976). Calcium control of actin-activated myosin adenosine triphosphatase from *Dictyostelium discoideum. Proc. Natl. Acad. Sci. U.S.A.* **73,** 2321–2325.

Moore, P. B., Huxley, H. E., and DeRosier, D. J. (1970). Three-dimensional reconstruction of F-actin, thin filaments and decorated thin filaments. *J. Mol. Biol.* **50,** 279–295.

Pardee, J. D., Simpson, P. A., Stryer, L., and Spudich, J. A. (1982). Actin filaments undergo limited subunit exchange in physiological salt conditions. *J. Cell Biol.* **94.** In press.

Parish, R. W., Muller, U., and Schmidlin, S. (1977). Phosphorylation of plasma membrane proteins in *Dictyostelium discoideum. FEBS Lett.* **79,** 393–395.

Peltz, G., Kuczmarski, E. R., and Spudich, J. A. (1981). *Dictyostelium* myosin: characterization of chymotryptic fragments and localization of the heavy-chain phosphorylation site. *J. Cell Biol.* **89,** 104–108.

Pollard, T. D. (1976). The role of actin in the temperature-dependent gelation and contraction of extracts of *Acanthamoeba. J. Cell Biol.* **68,** 579–601.

Pollard, T. D., Fujiwara, K., Niederman, R., Maupin-Szamier, P. (1976). Evidence for the role of cytoplasmic actin and myosin in cellular structure and motility. *In* "Cell Motility", (R. Goldman, T. Pollard, J. Rosenbaum, eds.), pp. 689–722. Cold Spring Harbor Laboratory, Cold Spring Harbor, New York.

Pollard, T. D., and Weihing, R. D. (1974). Actin and myosin and cell movement. *Crit. Revs. in Biochem.* **2,** 1–65.

Rahmsdorf, H. J., Malchow, D., and Gerisch, G. (1978). Cyclic AMP-induced phosphorylation in *Dictyostelium* of a polypeptide comigrating with myosin heavy chains. *FEBS Lett.* **88,** 322–326.

Rifkin, J. L., and Speisman, R. A. (1976). Filamentous extensions of vegetative amoebae of the cellular slime mold *Dictyostelium. Trans. Amer. Micros. Soc.* **95,** 165–173.

Ross, F. M., and Newell, P. C. (1981). Streamers: Chemotactic mutants of *Dictyostelium* with altered cGMP metabolism. *J. Gen. Microbiol.* **127,** 339–350.

Rubenstein, P., and Deuchler, J. (1979). Acetylated and nonacetylated actins in *Dictyostelium discoideum. J. Biol. Chem.* **254,** 11142–11147.

Rubenstein, P., Smith, P., Deuchler, J., and Redman, K. (1981). NH_2-terminal acetylation of *Dictyostelium discoideum* actin in a cell free protein-synthesizing system. *J. Biol. Chem.* **256,** 8149–8155.

Schliwa, M. (1981). Proteins associated with cytoplasmic actin. *Cell* **25,** 587–590.

Shaffer, B. M. (1964). Intracellular movement and locomotion of cellular slime-mold amoebae. *In* "Primitive Motile Systems in Cell Biology", pp. 387–405, (R. D. Allen, and N. Kamiya, eds.). Academic Press, New York.

Simpson, P. A., and Spudich, J. A. (1980). ATP-driven steady-state exchange of monomeric and filamentous actin from *Dictyostelium discoideum*. *Proc. Natl. Acad. Sci. U.S.A.* **77,** 4610–4613.

Spudich, J. A. (1974). Biochemical and structural studies of actomyosin-like proteins from non-muscle cells. II. Purification, properties and membrane association of actin from amoebae of *Dictyostelium discoideum*. *J. Biol. Chem.* **249,** 6013–6020.

Spudich, J. A. (1982). *Dictyostelium discoideum:* methods and perspectives for study of cell motility. *In* "Methods in Cell Biology" (L. Wilson, ed.), Vol. 25, Part B, pp. 359–364. Academic Press, New York.

Spudich, J. A., and Clarke, M. (1974). The contractile proteins of *Dictyostelium discoideum*. *J. Supramol. Structure* **2,** 150–162.

Spudich, J. A., and Cooke, R. (1975). Supramolecular forms of actin from amoebae of *Dictyostelium discoideum*. *J. Biol. Chem.* **250,** 7485–7491.

Spudich, J. A., Huxley, H. E., and Finch, J. T. (1972). Regulation of skeletal muscle contraction. II. Structural studies of the interaction of the tropomyosin–troponin complex with actin. *J. Mol. Biol.* **72,** 619–632.

Spudich, J. A., Kuczmarski, E. R., Pardee, J. D., Simpson, P. A., Yamamoto, K., and Stryer, L. (1982). Control of assembly of *Dictyostelium* myosin and actin filaments. *Cold Spring Harbor Symp. Quant. Biol.* **46,** 553–561.

Spudich, A., and Spudich, J. A. (1979). Actin in Triton-treated cortical preparations of unfertilized and fertilized sea urchin eggs. *J. Cell Biol.* **82,** 212–226.

Stewart, P. R., and Spudich, J. A. (1979). Structural states of *Dictyostelium* myosin. *J. Supramol. Struct.* **12,** 1–14.

Stryer, L. (1978). Fluorescence energy transfer as a spectroscopic ruler. *Annu. Rev. Biochem.* **47,** 819–846.

Taylor, D. L. (1977). The contractile basis of amoeboid movement IV. The viscoelasticity and contractility of amoeba cytoplasm *in vivo*. *Exp. Cell Res.* **105,** 413–426.

Taylor, D. L., and Condeelis, J. S. (1979). Cytoplasmic structure and contractility in amoeboid cells. *Int. Rev. Cytol.* **56,** 57–144.

Taylor, D. L., Condeelis, J. S., Moore, P. L., and Allen, R. D. (1973). The contractile basis of amoeboid movement I. The chemical control of motility in isolated cytoplasm. *J. Cell Biol.* **59,** 378–394.

Taylor, D. L., Reidler, J., Spudich, J. A., and Stryer, L. (1981). Detection of actin assembly by fluorescence energy transfer. *J. Cell Biol.* **89,** 362–367.

Taylor, D. L., Rhodes, J., and Hammond, S. (1976). The contractile basis of amoeboid movement II. Structure and contractility of motile extracts and plasmalemma-ectoplasm ghosts. *J. Cell Biol.* **70,** 123–143.

Tuchman, J., Alton, T., and Lodish, H. F. (1974). Preferential synthesis of actin during early development of the slime mold *Dictyostelium discoideum*. *Develop. Biol.* **40,** 116–128.

Uyemura, D. G., Brown, S. S., and Spudich, J. A. (1978). Biochemical and structural characterization of actin from *Dictyostelium discoideum*. *J. Biol. Chem.* **253,** 9088–9096.

Uyemura, D. G., and Spudich, J. A. (1980). Biochemistry and regulation of nonmuscle actins: toward an understanding of cell motility and shape determination. *In* "Biological Regulation and Development" (R. F. Goldberger, ed.), Vol. 2., pp. 317–338. Plenum, N.Y.

Vandekerckhove, J., and Weber, K. (1980). Vegetative *Dictyostelium* cells containing 17 actin genes express a single major actin. *Nature (London)* **284,** 475–477.

Vogel, G., Thilo, L., Schwarz, H., and Steinhart, R. (1980). Mechanism of phagocytosis in *Dictyostelium discoideum:* Phagocytosis is mediated by different recognition sites as disclosed by mutants with altered phagocytotic properties. *J. Cell Biol.* **86,** 456–465.

Wegner, A. (1976). Head-to-tail polymerization of actin. *J. Mol. Biol.* **108,** 139–150.

Weiss, E., and Braun, V. (1979). Actin-like protein of the cytoplasm in the chromatin of *Dictyostelium discoideum. FEBS Lett.* **108,** 233–236.

Woolley, D. E. (1972). An actin-like protein from amoebae of *Dictyostelium discoideum. Arch. Biochem. Biophys.* **150,** 519–530.

Yamamoto, K., Pardee, J., Reidler, J., Stryer, L., and Spudich, J. A. (1982). The mechanism of interaction of *Dictyostelium* severin with actin filaments. *J. Cell Biol.* In press.

• CHAPTER 6

Cell Adhesion

Samuel H. Barondes, Wayne R. Springer, and Douglas N. Cooper

I. Introduction

The structure of multicellular organisms involves adhesion of cells to each other and to extracellular materials. The molecular basis of this binding is not understood. It is often explained by analogy with cellular agglutination by antibodies, in which case bivalent or polyvalent immunoglobulins bind to complementary substances on adjacent cell surfaces, forming intercellular

The Development of *Dictyostelium discoideum*

ISBN 0-12-455620-5

bridges. Tyler (1947) and Weiss (1947) suggested that similar reactions mediate cell–cell adhesion.

A major problem in studying the molecular basis of cell–cell adhesion is that we have only a vague idea of the kinds of molecules to look for. Although the idea of molecular complementarity is obviously appealing, the specific processes are not understood. Whereas some cases of cell–cell adhesion might involve interactions no more complex than that in cell agglutination by a divalent antibody, other cases might involve multiple molecules on the cell surface and a sequence of steps. For example, interactions between specific molecules might trigger the activation or display of a number of others, thereby reorganizing general cell surface properties that facilitate adhesion. In this view cell adhesion is a process, and "cell adhesion molecules" would be defined operationally because they participate in some manner in this process, not only as direct mediators, but also as receptors or modulators.

As presently conceived, cell adhesion molecules would be expected to have certain characteristics which form the basis of experimental approaches to their identification (Table I). If a molecule shows all these characteristics, it is reasonable to infer that it really is involved in cell–cell adhesion. However, even this degree of evidence does not constitute definitive proof, since reasons for false positives are easy to imagine (Table I). Conversely, failure to display a given characteristic should not definitively exclude a candidate from further consideration.

TABLE I
Expected Characteristics of a Cell Adhesion Molecule (CAM), and False Positives and Negatives

	Reasons for:	
Expected characteristics	False positive	False negative
Located on cell surface	Coincidental	Overlooked, inaccessible
Complementary receptor on cell surface	Binding has other function or is nonspecific	CAM binds poorly when solubilized
Appearance correlates with development of adhesiveness	Coincidental	CAM present before adhesiveness but other factor is limiting
Binding univalent antibodies or haptens to CAM blocks adhesion	Nonspecific effect	Relatively low affinity for CAM
Mutant with impaired cell adhesion has defective CAM	Indirect effect	Defect in CAM not detected by the measurements used

Cellular slime molds have become popular for investigation of molecules involved in cell adhesion. These organisms have many properties that facilitate experimental work on this complex problem:

1. Their life cycle involves differentiation from a nonassociating vegetative state to an adhesive aggregated state, initiated by starvation. Because this differentiation is fairly synchronous, large numbers of cells can be obtained with stage-specific adhesive properties.
2. There are multiple species of cellular slime molds that can segregate in mixed cultures through a mechanism that apparently involves species-specific adhesion. This provides the opportunity to develop assays that distinguish species-specific from nonspecific adhesion.
3. Unlike cells of more complex organisms, slime mold cells are easily maintained in culture under conditions similar to those found in nature. Therefore, they can be cultivated under controlled conditions that would not be expected to produce abnormal cell surface properties.
4. During most of their life cycle, the cells may be dissociated by simply shaking vigorously in cold buffer without proteases or chelating agents that might alter the cell surface in unknown ways.
5. Mutagenesis and cloning of individual strains is readily accomplished.

Because of these advantages, work on cell–cell adhesion in cellular slime molds is relatively advanced. In this chapter we will review the methods and approaches used to study this problem and our current understanding of the underlying molecular mechanisms.

II. Descriptive Studies Consistent with Selective Cell–Cell Adhesion

Observations of the behavior of cellular slime molds either on surfaces, or upon prolonged gyration in suspension, demonstrate cellular segregation. In many cases chemotaxis has been shown to play a critical role. For example, the separation of vegetative from differentiated slime mold cells occurs because of the development of a chemotactic signaling system (see Chapter 4). However, other aspects of developmentally regulated aggregation appear to be due to the development of cell–cell adhesiveness, as will be considered in detail below. Observations of species-specific segregation and the morphogenetic movements of prestalk and prespore cells, although consistent with chemotactic mechanisms, also suggest the possibility of selective cell–cell adhesion. Before considering more quantitative studies, descriptive experiments will be briefly reviewed to illustrate both the plausibility of selective adhesion, and the difficulty of discriminating between selective chemotaxis and adhesion with such an approach.

Species specificity of cell–cell adhesion was first suspected from observations (Raper and Thom, 1941; Bonner and Adams, 1958) that mixtures of two species segregate when placed on agar. Such segregation can be caused by different chemotactic signals. However, species like *D. discoideum* and *D. purpureum*, which both use 3′5′-cAMP as a primary chemotactic signal, show marked segregation. More direct evidence of selective cell– — adhesion of a species-specific nature came from studies in which cells of one species were grown in radioactive thymidine and mixed with unlabeled cells of another species. Each cell type was harvested after aggregation on a solid surface and dissociated into individual cells. They were then mixed and gyrated in suspension for as long as 48 h. Autoradiographic studies of the distribution of labeled and unlabeled cells demonstrated segregation of species in the large mixed aggregates (Nicol and Garrod, 1978; Sternfeld; 1979). In some experiments the cells of one species accumulated in the center of an aggregate with a layer of cells of the other species around them. This pattern suggested differential adhesiveness of the two species (Sternfeld; 1979), although differential chemotaxis could not be rigorously excluded.

Experiments with another technique led to a similar conclusion (Springer and Barondes, 1978). Here, too, cells of the two species were harvested at the aggregating stage and dissociated. They were then differentially labeled with fluorescent dyes. The labeled cells were mixed and observed in a living state under a fluorescence microscope. In these experiments aggregation occurred under natural circumstances, i.e., on a solid surface, rather than in suspension. Furthermore, the association of the cells could be observed directly, rather than by autoradiographs of sections of clumps of cells. With this technique, as well, the cells initially formed mixed aggregates followed by rapid segregation. Again, segregation was consistent with, but not proof of, selective cell–cell adhesion.

Segregation of the two major cell types of cellular slime molds, prestalk and prespore cells, may also be due in part to selective adhesion, although a role for chemotaxis can also be supported, as discussed in Chapter 4. The opportunity to study selective cell adhesion of these cell types was provided by the development of methods to separate the two populations by gradient centrifugation (Takeuchi, 1969; Feinberg *et al.*, 1979; Tsang and Bradbury, 1981). Mixtures of prestalk cells and prespore cells differentially labeled with [^{3}H]thymidine (Takeuchi, 1969) or fluorescent dyes (Feinberg *et al.*, 1979) segregate in patterns which may be mediated by selective adhesion.

III. Quantitative Assays of Cell–Cell Adhesion

To distinguish between cellular associations based on chemotaxis and those based on adhesion, it is necessary to study cellular interactions under circumstances that do not permit the development or maintenance of chemotactic gradients. This condition is met by studying cell association while the cells are being agitated. The constant movement disrupts the development of chemotactic gradients and would pull the cells apart if they were held together only because of mutual attraction to a diffusible chemical substance. When cells adhere in situations in which they are agitated, their association is almost certainly mediated by cell–cell adhesive processes. This is especially true of initial associations in brief assays under vigorous gyration conditions, which would avoid formation of large aggregates that might form a microenvironment for chemotaxis.

Assays of this type have merit not only because they discriminate between chemotaxis and adhesion, but also because they allow for quantitation. In cellular slime molds, as in many other systems, assays have been devised to quantitate cell–cell adhesion in gyrated suspensions as well as adhesion of gyrated probe cells to an immobilized cellular monolayer.

A. Cell–Cell Adhesion in Gyrated Suspensions

The first quantitative assay for measuring cell–cell adhesion in slime molds was described by Gerisch (1961). He studied the adhesion of dissociated aggregating cells by microscopic observation of the suspensions after their gyration in various media. He found that both vegetative and dissociated aggregating cells formed aggregates. However, addition of EDTA blocked the cell–cell adhesion of vegetative cells, but not of the cells that differentiated to the aggregating stage. Presumably, then, this condition, the addition of EDTA, helped define a meaningful developmentally regulated form of cell association.

Whereas most *in vitro* quantitative assays of cell–cell adhesion in cellular slime molds are derived from this early work and use EDTA, the mechanism of action of this reagent is not known. Presumably it acts by chelation of Ca^{2+} or Mg^{2+} ions that might facilitate nonspecific adhesion, but this has not been directly demonstrated. The fact that addition of EDTA blocked the nonspecific association of vegetative cells does not prove that the adhesion of dissociated aggregating cells in the presence of EDTA is biologically significant. It remains possible that the factors that mediate EDTA-resistant adhesion are not the cell surface properties responsible for the biologically significant adhesion of these cells.

Despite its potential limitations, many variations of this assay have been used to study cell adhesion in cellular slime molds. In one variation Beug and Gerisch (1972) developed an apparatus that used optical methods to measure the formation of small aggregates from single cells. In this assay, single cells are clearly distinguishable from aggregates, but the relationship between the optical measurement and the size of the aggregate is complex. Many other investigators analyze gyrated suspensions of cells with an electronic particle counter, a procedure originally used to study adhesion of vertebrate cells (Orr and Roseman, 1969). In most of these assays, the counter is calibrated to distinguish between particles comprised of either one cell or more than one cell. What is usually recorded is disappearance of particles that are the size of single cells. The assumption that the single cells are disappearing because they are adhering to each other must be checked by microscopy or by counting particles of larger size with the counter to exclude the possibility of cell lysis or binding of cells to the assay vessel. Determination of the size of aggregates with a microscope or by measurements with a more sophisticated electronic particle counter might provide significant additional information.

Examples of a simple assay with a particle counter are given in Fig. 1. They illustrate the marked effects of developmental state of the cells, EDTA, and gyration speed. Vegetative cells and aggregating cells show no significant difference when gyrated at 115 revolutions per min in buffer without EDTA. If gyration speed is increased to 200 revolutions per min or if 10 m*M* EDTA is added, the vegetative and aggregating cells are distinguishable. The distinction is more striking with 10 m*M* EDTA at 200 revolutions per min. Under these conditions vegetative cells show no cell–cell adhesion, whereas aggregating cells display significant adhesion. Adhesion under these conditions could be biologically significant, serving to bind cells together upon aggregation.

An assay of this type has also been used to quantify species-specific cell–cell adhesion. In these studies dissociated aggregating cells of each of two species were directly labeled with either fluorescein isothiocyanate or tetramethylrhodamine isothiocyanate, permitting them to be distinguished with a fluorescence microscope (McDonough *et al.*, 1980). The mixture of the two species was gyrated at 200 revolutions per min in buffer containing 10 m*M* EDTA, conditions that permit detection of developmentally regulated adhesion (Fig. 1). After brief gyration, usually for 10 min, both the total number of cells in each aggregate and the number of cells of each species in each aggregate were determined (Table II). Under these conditions, there was striking species-specific cell–cell association in the small aggregates that

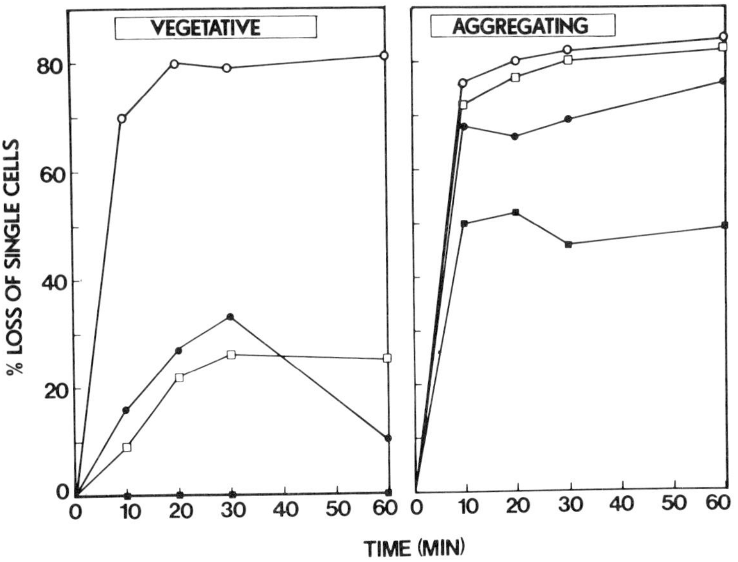

Fig. 1. Effect of assay conditions on cell–cell adhesion of vegetative and aggregating *D. discoideum*. Cells were harvested from bacterial growth plates after 40–48 h, washed with cold distilled water and either assayed immediately (vegetative cells) or differentiated on moist filter-pads until they were streaming into aggregates (aggregating cells) then harvested and assayed. Conditions used were: gyration at 115 rpm in phosphate buffer (○——○); gyration at 115 rpm in buffer containing 10 m*M* EDTA (●——●); gyration at 200 rpm in buffer (□——□); gyration at 200 rpm in buffer containing EDTA (■——■). (For details see McDonough *et al.*, 1980.)

were formed. Of the 50 aggregates observed only three contained at least one cell of each type, whereas the other 47 were made up exclusively of one or the other cell type (Table II). The species-specific segregation was strikingly significant, even using a relatively insensitive statistical analysis (McDonough *et al.*, 1980). For analysis of more subtle segregation, better statistical methods have been developed (Koziol *et al.*, 1980).

Species-specific adhesion can be demonstrated in an assay of this type even in the absence of EDTA, but only after more prolonged gyration. With 10 min of gyration in the absence of EDTA considerable cell–cell adhesion is observed, but there is little segregation of the two species. However, if gyration is continued for an additional 10 min, the observed aggregates, which are larger, show some species-specific association and after 40 min considerable segregation is observed (Fig. 2). By statistical analysis, the degree of segregation at 20 min is already as substantial as that at 40 min (Koziol *et al.*, 1980), although this is not obvious from visual inspection of Fig. 2 since the mean aggregate size also increases during this period. The fact that species-specific adhesion can be studied under experimental conditions wherein EDTA is either present or absent, may facilitate work on the underlying mechanisms.

An assay of cell–cell adhesion in gyrated suspension has also been used to study the relative adhesiveness of prespore and prestalk cells isolated by density gradient centrifugation. In the presence of 10 m*M* EDTA prespore cells were significantly more adhesive than prestalk cells (Lam *et al.*, 1981).

Although most studies have concentrated on the cell–cell adhesion that appears with differentiation, the failure of vegetative cells to associate may

TABLE II

Composition of Mixed Aggregates of Labeled *D. Discoideum* and *D. Purpureum* Compared with Expectations if There Were No Species-Specific Cell–Cell Adhesion[a]

Cells per aggregate	Number of aggregates of this size found	Mixed aggregates expected (%)	Number of mixed aggregates	
			Expected	Found
2	18	36	6	2
3	11	53	6	1
4	6	65	4	0
5	4	73	3	0
6	1	79	1	0
>6	10	≥84	8	0
Total	50	56	28	3

[a]Aggregates were formed by gyration of partially differentiated *D. discoideum* and *D. purpureum* for 10 min in buffer containing 10 m*M* EDTA. The *D. discoideum* cells were labeled with fluorescein isothiocyanate and the *D. purpureum* cells were labeled with tetramethylrhodamine isothiocyanate. Fifty aggregates were observed, and the total number of cells per aggregate and the number of red and green cells was determined. A mixed aggregate contains at least one cell of each type. The percentage of mixed aggregates expected was determined by using a binomial expansion corrected for the difference in adhesiveness of the two cell types as follows: After recording 50 aggregates, the fraction of cells of each type in all 50 aggregates was determined and used in the formula $f = 1-(a^n+b^n)$, where f = the fraction of mixed cells expected if both cell types adhere equally well to each other, a = the fraction of cell type a found in the total 50 aggregates, b = the fraction of cell type b found in the total 50 aggregates, and n = the aggregate size. For example, if equal numbers of cells were found in the 50 aggregates one would expect $1-(0.5^2+0.5^2) = 0.5$ or half of the aggregates containing two cells to be mixed; $1-(0.5^3+0.5^3) = 0.75$ or three quarters of the three-cell aggregates to be mixed and so on. From this fraction, and knowing the number of aggregates found in each class, we can calculate the number of mixed aggregates expected if no selective adhesion is involved. In the present case 77% of the cells in the fifty aggregates were *D. discoideum* and 23% were *D. purpureum*. The probability of the observed distribution of mixed aggregates being the same as the expected distribution was determined by calculating Chi-square. Therefore, the smaller p is the more likely it is that the cell adhesion is species-specific. In the present case $p < .005$. For details see McDonough *et al.* (1980).

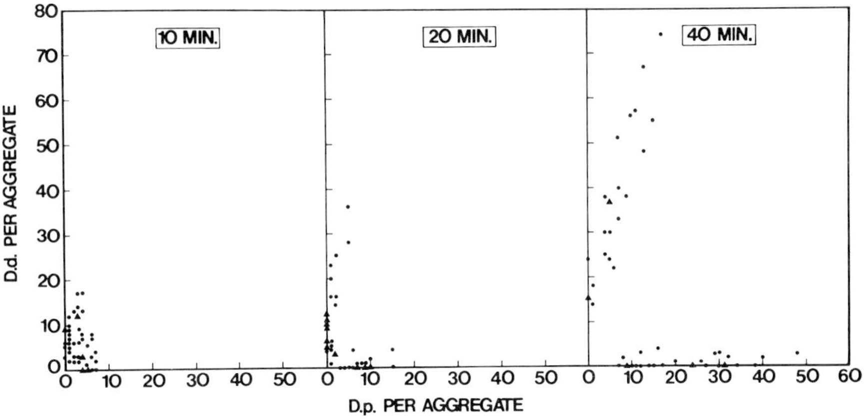

Fig. 2. Demonstration of species-specific adhesion in the absence of EDTA. Dissociated aggregating *D. discoideum* (D.d.) and *D. purpureum* (D.p.) were each labeled with a fluorescent dye, mixed and gyrated together at 200 rpm in buffer without EDTA. Aliquots were observed with a fluorescence microscope at 10, 20, and 40 min, and 50 aggregates were evaluated. The number of D.d and D.p. cells in each aggregate is plotted. Triangles (▲) indicate that more than one aggregate had that particular composition. After 10 min of gyration the aggregate compositions are consistent with random adhesion. However, after 20 min of gyration there is marked segregation by species. After 40 min the aggregates are larger and segragation is maintained (For details, see McDonough *et al.*, 1980.)

be due to synthesis of a material that inhibits their adhesion (Swan *et al.*, 1977; Jaffe *et al.*, 1979). This low molecular weight material, found in the medium of cells grown in suspension and kept in the stationary phase, has little effect on adhesion of dissociated aggregating cells. The physiological role of this inhibition, if any, has not been established.

B. Adhesion of Probe Cells to a Cellular Monolayer

Selective cell–cell adhesion in cellular slime molds may also be studied with a monolayer assay. In this approach labeled probe cells are reacted with a monolayer of cells and the number of bound probe cells is measured. Several variations of this assay have been developed for cultured and embryonic vertebrate cells (Walther *et al.*, 1973; Gottlieb and Glaser, 1975; Rutishauser *et al.*, 1976).

To study species-specific adhesion in cellular slime molds, one species is immobilized on a solid surface as a uniform monolayer by centrifuging the cells onto a plastic dish containing covalently bound concanavalin A (Spring-

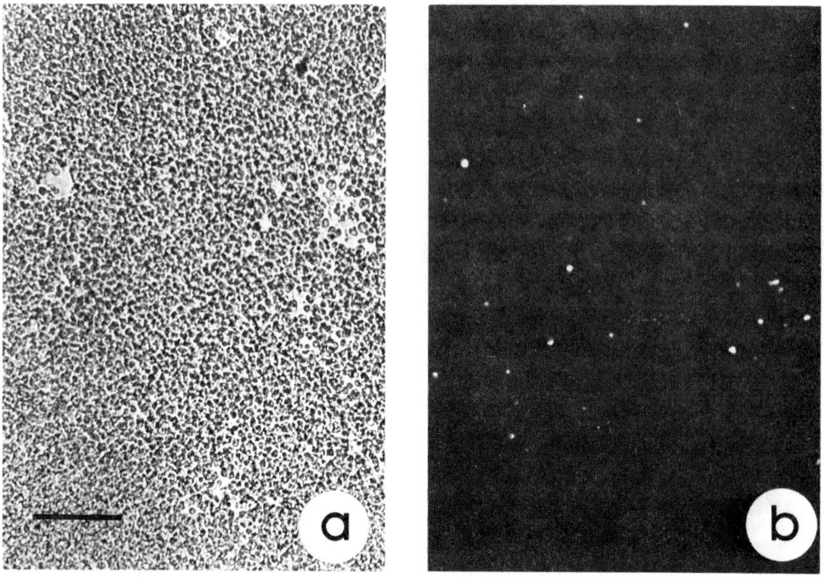

Fig. 3. Measurement of cell–cell adhesion with a monolayer plate assay. The phase contrast photomicrograph (a) shows a typical monolayer of *D. discoideum* cells. The fluorescence photomicrograph (b) shows fluorescent *D. discoideum* probe cells bound to the unlabeled monolayer cells after gyration of the monolayer with the probe cells and washing to remove unbound probe cells (For details see Springer and Barondes, 1978.)

er and Barondes, 1978). Glycoconjugates on the slime mold cell surfaces bind to the concanavalin A, producing a uniform monolayer of attached cells (Fig. 3a). Cells tolerate binding to this surface, since if they are left for long periods of time, they will aggregate and form fruiting bodies. However, the monolayer is stable for the short periods of time required for use in a quantitative assay.

In the assay, a suspension of probe cells labeled with a fluorescent dye is gyrated over the monolayer. Binding of the probe cells to the monolayer can be observed (Fig. 3b) with a fluorescence microscope at various times and under various gyration conditions. By varying the species of probe cells the relative association of the same and different species can be quantified. Marked species-specific adhesion was found (Springer and Barondes, 1978; McDonough *et al.*, 1980) indicating that the assay can be used to study the molecular basis of this association. An assay of this type has also been used to evaluate selective adhesiveness of prestalk and prespore cells separated on a Percoll gradient. Some selectivity was observed such that a layer enriched in prestalk cells bound prestalk probe cells better than it bound prespore cells.

However, a layer enriched in prespore cells did not discriminate between these two types of probe cells (Feinberg *et al.*, 1979).

C. Adhesion of Isolated Plasma Membranes

A third and little utilized method of studying the adhesive properties of slime molds uses ghosts rather than intact cells. The ghosts consist of intact plasma membranes that are devoid of most intracellular materials. Ghosts gyrated in suspension retain adhesive properties of the cells from which they were derived, and show the appearance of EDTA-resistant adhesion with development (Sussman and Boschwitz, 1975).

IV. The Immunological Approach to the Identification of Cell Adhesion Molecules

A major strategy for identifying molecules involved in cell–cell adhesion is to functionally inactivate them with antibodies. Gregg (1960) showed that slime mold cell-surface antigens change with the development of cell–cell adhesion. Sonneborn *et al.* (1964) showed that antisera raised to aggregating cells could block cell aggregation without affecting cell viability.

Two general approaches have been used to identify the antigens to which adhesion blocking antibodies are directed. In the first, antisera are raised by immunization with whole cells or crude cellular fractions. If antibodies are produced that block cell–cell adhesion, the relevant molecules can be identified as absorbants of this blocking activity. In the other approach cell surface molecules that are expressed with the development of cell–cell adhesion are purified, and antisera raised against these purified antigens are tested for adhesion-blocking activity. If the antiserum blocks cell adhesion, the antigen used for immunization is tentatively identified as a cell adhesion molecule.

Crude antisera are generally used for these studies, although monoclonal antibodies are also being studied. In most cases the antibodies of interest are specifically cleaved to univalent fragments before testing them as inhibitors of cell–cell adhesion. Univalent fragments have advantages over intact divalent immunoglobulins, since they do not agglutinate cells or crosslink surface antigens, which could have secondary effects on cell adhesion. However, there is some evidence that intact immunoglobulin may be used if agglutination and secondary effects are blocked with a univalent second antibody (Springer and Barondes, 1980). For example, if the primary antibody was raised in rabbits, univalent fragments of goat antibodies directed against rabbit immunoglobulin are added after the initial antibody has bound. Cell adhesion assays are then conducted in the presence of an excess of the goat antibody fragments. This allows use of intact primary immunoglobulins, which bind antigens much more avidly than univalent fragments. However, experience with primary univalent fragments is much more extensive, and interpretation of studies with intact immunoglobulins could be more difficult.

With the immunological approach, several surface molecules that appear to play a role in cell–cell adhesion have been identified. In *Dictyostelium discoideum* evidence is accumulating for a role of three glycoproteins with molecular weights of 80,000, 95,000, and 150,000. In *Polysphondylium pallidum* several other glycoproteins have been implicated.

A. Isolation of Molecules That Absorb Antisera Raised to Crude Antigenic Mixtures

Most attempts to identify cell adhesion molecules with the immunological approach begin by immunizing with a crude antigenic mixture. After raising an antiserum that blocks cell–cell adhesion, the critical antigens are sought by absorption with fractionated cellular constituents of increasing purity. The critical assumption of this approach is that there are not many different molecules that participate directly in cell adhesion. Were there many different cell adhesion molecules, it would be expected that absorption with individual fractions, containing only one of the many relevant cell adhesion molecules, would have little or no detectable effect. Surprisingly, only two molecules that absorb inhibitory sera raised against crude antigenic mixtures have been found in differentiating *D. discoideum*. One, a glycoprotein of approximately 80,000 molecular weight, called "contact sites A," is believed to mediate the adhesion of aggregating cells. The other, a glycoprotein of 95,000 molecular weight, is believed to mediate adhesion of slug cells.

1. *Contact Sites A (80,000-Dalton Glycoprotein)*

In pioneering studies with this approach, Gerisch and his colleagues (Beug *et al.*, 1970, 1973a,b; Gerisch, 1980) obtained a pooled antiserum by repeated injection of 22 rabbits with aggregating *D. discoideum* cells. Univalent antibody fragments prepared from this antiserum could completely block cell–cell adhesion of dissociated aggregating *D. discoideum* cells as measured by a quantitative assay. Two types of cell–cell adhesion were distinguished. One, attributed to "contact sites B," was observed both in vegetative and aggregating cells and is characterized by side to side binding (Fig. 4). This type of binding is observed upon gyration of cells of strain V12/M2, since they do not become rounded under this condition. Adhesion through contact sites B could be blocked by addition of 10 m*M* EDTA to the assay mixture. Since contact sites B is demonstrable in both vegetative and aggregating cells, antibodies that interact with it could be absorbed with vegetative cells. The other type of cell–cell adhesion, attributed to contact sites A, was blocked by the univalent antibody fragments even after absorption with vegetative cells (Fig. 4). It was also distinguished because it was resistant to addition of 10 m*M* EDTA to the incubation medium. In addition,

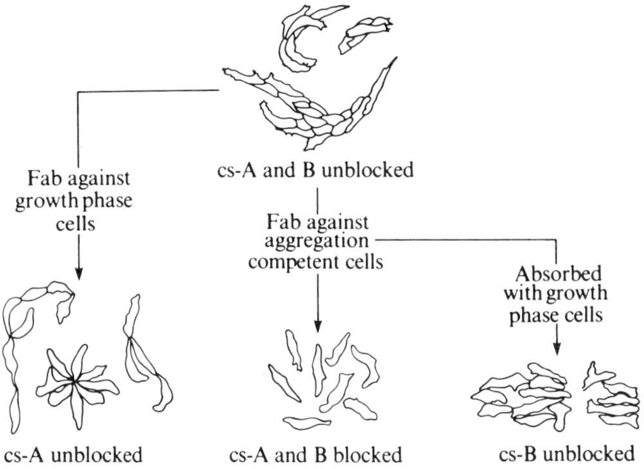

Fig. 4. Immunological analysis of the cell adhesion system in *D. discoideum*. Cells acquire the ability to aggregate within a period of several hours after the end of the growth phase. During this period specific target antigens (contact sites A) of adhesion blocking univalent antibody fragments (Fab) are expressed on the cell surface. Other target antigens (contact sites B) of adhesion blocking Fab are already present in the growth phase stage. The evidence for the diversity and developmental regulation of the contact site system is as follows. During aggregation streams of elongated cells are formed by end-to-end and side-by-side assembly (top). Fab against membrane antigens of aggregation competent cells blocks both types of contact (middle). After absorption with membranes from growth phase cells the Fab preparation still blocks the tight end-to-end contacts which are typical of aggregation competent cells (right). This demonstrates that the corresponding antigens, contact sites A, are almost absent from cells of the growth phase stage. In contrast, Fab responsible for the blockage of irregular, mostly side-by-side assembly of the cells has been removed by antigens present in growth phase cells. These antigens are the contact sites B. In agreement with these results, Fab specific for membrane antigens of growth phase cells blocks the side-by-side adhesion of aggregating cells, leaving the end-to-end contacts unaffected (left). (Reprinted by permission from Müller and Gerisch, *Nature (London)* **274**, 445–449, Copyright © 1978, Macmillan Journals Limited.)

microscopic examination, using strain V12/M2, indicates that this form of adhesion occurs by end to end association of the cells (Fig. 4).

The effect of univalent antibody fragments on the cell–cell adhesion mediated by contact sites A was specific, in that fragments prepared from antisera raised against heat treated *D. discoideum* cells had no effect (Beug *et al.*, 1973b). Using the active antibody, cell–cell adhesion was completely blocked by binding of 3×10^5 univalent antibody molecules per cell, whereas using univalent fragments from antisera raised against the heat treated cells, there was no detectable effect on cell–cell adhesion even upon binding of an order of magnitude more univalent antibody molecules per cell (Beug *et al.*, 1973b). The potential relevance of the contact site A antigens is also suggested by their appearance on the surface of differentiating *D. discoideum* in parallel with the development of cell–cell adhesion (Beug *et al.*, 1973a). Evidence has been presented that appearance of these antigenic determinants on the cell surface requires exposure to pulses of 3′5′-AMP which the cells normally encounter during chemotactic signaling with aggregation (Gerisch *et al.*, 1975). Appearance of cell–cell adhesion characteristic of contact sites A is blocked, in an unknown manner, in cells differentiated in the presence of glucose, galactose or nonmetabolizable derivatives of these sugars (Marin *et al.*, 1980).

In an attempt to identify the molecule or molecules that had been operationally designated contact sites A, an aqueous suspension of the partially purified membrane fraction of aggregating *D. discoideum* cells was mixed with 0.75 volumes of butanol at pH 5.5, and the water soluble fraction, which is rich in glycoproteins, was taken for further analysis. First it was shown that this aqueous fraction (generally called a "butanol extract") could totally absorb antibody activity that inhibited contact sites A (Huesgen and Gerisch, 1975). The active components were sought by fractionating this crude mixture by chromatography on DEAE cellulose and by gel filtration on Sephadex G-200 followed by sucrose density gradient centrifugation (Müller and Gerisch, 1978; Müller *et al.*, 1979). With each purification procedure, only one broad fraction contained most of the absorptive activity, although some activity in other fractions was discarded. Upon polyacrylamide gel electrophoresis in SDS of the final product of this purification scheme, a single protein band was observed with a molecular weight of approximately 80,000 (Müller and Gerisch, 1978; Müller *et al.*, 1979). Therefore, this molecule contained all the immunodeterminants that were required to completely absorb the inhibitory antibody activity. However, it remains possible that other molecules that share these antigenic determinants were discarded by the initial extraction or during fractionation, and might actually prove relevant to cell–cell adhesion.

Parish *et al.* (1978), using antisera raised against plasma membranes of aggregating cells, examined the binding to protein bands electrophoresed on SDS polyacrylamide gels. They found a very prominent antigenic determinant with a molecular weight of 82,000 presumably identical with contact sites A. They also showed that this region of the gel absorbed all adhesion-inhibiting activity from a serum raised against aggregating cells (Steineman and Parish, 1980), confirming the findings of Gerisch and co-workers.

The 80,000-dalton glycoprotein (Müller *et al.*, 1979) contains approximately 25% carbohydrate by weight. There are approximately equal amounts of mannose, *N*-acetylglucosamine and glucose, smaller amounts of fucose and a trace of galactose. The glycoprotein is rich in threonine, serine, and proline, and binds to concanavalin A (Müller and Gerisch, 1978). The purified glycoprotein neither inhibited nor augmented cell–cell adhesion of aggregating *D. discoideum* cells (Müller and Gerisch, 1978).

Murray *et al.* (1981) have used this purified protein to raise antisera that block cell–cell adhesion, confirming that this glycoprotein fulfills this criterion for a cell adhesion molecule. They also raised monoclonal antibodies that react with this glycoprotein. However, these monoclonal antibodies did not affect cell–cell adhesion (Murray and Loomis, personal communication). This suggests either that the monoclonal antibodies are not directed to relevant portions of the molecule, or that more than one site must be bound to produce an inhibitory effect.

2. 95,000-Dalton Glycoprotein

With a similar approach, another glycoprotein with molecular weight of approximately 95,000 has been found in slug-stage membranes. It appears to be a cell adhesion molecule at that stage of development (Parish *et al.*, 1978; Steineman and Parish, 1980). Antibodies were raised against crude membrane fractions of slug-stage cells, and univalent antibody fragments prepared from these antisera blocked cell–cell adhesion of slug-stage cells as measured with a quantitative assay. These univalent antibody fragments blocked neither the EDTA-resistant adhesion of aggregation-stage cells nor the adhesion of vegetative cells. To identify the relevant cell surface antigens, the glycoproteins in the aqueous phase of a butanol extract of plasma membranes from slug-stage cells were isolated. As with studies of contact sites A, this glycoprotein fraction completely absorbed inhibitory activity of the univalent antibody fragments.

Identification of the active component of the butanol extracts was achieved by subjecting this material to polyacrylamide gel electrophoresis in sodium dodecyl sulfate and assaying some fractions as absorbants of the blocking activity of the univalent antibody fragments. All blocking activity could be absorbed with a gel fraction that contained a glycoprotein of molecular weight 95,000. Attention was directed to this fraction since Parish *et al.* (1978) showed that it contained the major antigen that appeared with differentiation from aggregating to slug cells. One other fraction that contained proteins with molecular weight 80,000–90,000 showed no significant absorptive activity, but it is not clear how many other fractions were tested. Therefore, it remains possible that other relevant molecules were ignored. Nevertheless, this highly antigenic glycoprotein is capable of absorbing all the adhesion blocking activity implying that it, or something with similar immunological determinants, is critical for cell–cell adhesion in the slug.

The inference that critical cell adhesion molecules change during development from initial aggregates to slugs is markedly strengthened by the studies of Wilcox and Sussman (1981a, 1981b). They independently showed that univalent antibody fragments prepared from antisera raised against early aggregates block adhesion of cells at this stage, but not the adhesion of postaggregation stage cells. They also showed that univalent antibody fragments prepared from antisera raised against cells of the latter stage block their adhesion, but not that of cells at an earlier stage, and that a 95,000-molecular weight glycoprotein is involved in postaggregation stage adhesion.

In these studies they used both wild-type cells and a temperature sensitive mutant (JC5) of *Dictyostelium discoideum*. The mutant culminates at 22°C, but at 27°C behaves normally only until a specific postaggregation stage, whereupon the cells disperse to a smooth lawn. If the cells are returned to 22°C at this point, they reaggregate and culminate. They found that expression of antigenic determinants capable of absorbing the adhesion-blocking activity of antibodies raised against postaggregation cells was closely correlated with the developmental aberration. Involvement of a 95,000-dalton glycoprotein was suggested by the finding that wild-type cells or mutant cells raised at the permissive temperature showed a 95,000-molecular weight glycoprotein on SDS polyacrylamide gels stained with ^{125}I-labeled wheat germ agglutinin, whereas this band was not seen in mutants raised at the nonpermissive temperature. This glycoprotein may be related to the one described by Steineman and Parish (1980) that was shown to incorporate glucosamine and, hence, might be expected to bind wheat germ agglutinin. Synthesis of the 80,000-molecular weight glycoprotein was apparently normal in the mutant, consistent with normal aggregation and early cell–cell adhesion.

Because of the findings with wheat germ agglutinin described above, attempts were made to purify the relevant molecule by affinity chromatography using a wheat germ agglutinin–Sepharose column. When nonionic detergent extracts from purified plasma membranes of late stage wild-type cells were chromatographed on this column, the fraction that was eluted with *N*-acetylglucosamine stimulated adhesion of noncohesive JC5 cells, as measured in a quantitative assay (Saxe and Sussman, 1982). The active material was not present in similar preparations from wild-type cells developed to the early aggregation stage or from mutant cells grown to the late stage at the restrictive temperature. Whether or not the active component is indeed the 95,000-molecular weight glycoprotein has not been determined.

B. Effect of Antisera Raised against Developmentally Regulated Cell Surface Molecules

This approach is based on the idea that as a form of cell–cell adhesion appears with development, the critical molecules that mediate it should appear on the cell surface. Determining which molecules change on the cell surface with development identifies possible cell adhesion molecules. These molecules can then be purified, and the antibodies raised against them can be tested for a possible effect on cell–cell adhesion.

Difficulties with this approach might be anticipated for both theoretical and practical reasons. Theoretically, one might fail to notice the appearance of the critical molecules either because they were relatively scarce or because they represented a modification of a preexistent molecule. Furthermore, if there were more than one adhesive system operating, blocking one might have no detectable effect. From the practical point of view, demonstration that a very large number of cell surface proteins show striking changes with development would nominate a large field of candidates, so that screening each individually would be difficult.

There are indeed many changes in cell surface proteins with development. Of the 30 plasma membrane proteins radioiodinated and identified by autoradiography after SDS polyacrylamide-gel electrophoresis, Siu *et al.* (1975) showed that 9 had prominent developmental changes. Of the 40 plasma membrane proteins identified by periodic acid–Schiff staining and Coomassie Blue staining, Gilkes *et al.* (1979) found that 17 showed marked developmental regulation. Similar results were reported by Hoffman and McMahon (1977) who concluded that almost all glycoproteins changed significantly in expression during the first 18 h of development. Using a variety of radioactive labels in association with SDS polyacrylamide gel electrophoresis and autoradiography Parish and Schmidlin (1979) identified 13 cell surface proteins that showed striking developmental changes. West and McMahon (1977) identified more than 35 concanavalin A-binding proteins in purified membranes of *D. discoideum* and showed that 12 of these decreased whereas 12 increased in expression during differentiation. Despite the large numbers of possible cell adhesion molecules suggested by these studies, only one has been supported by further work.

1. *Evidence That a 150,000-Dalton Glycoprotein Is a Cell Adhesion Molecule*

Geltosky *et al.* (1976) coupled radioactive iodine to the plasma membrane proteins of *D. discoideum* cells at various stages of development, solubilized these proteins and purified them by affinity chromatography on Sepharose derivatized with concanavalin A. They found at least 15 proteins in vegetative cells that bound to the column. Of these a 150,000-dalton glycoprotein showed the most striking increase between 6 and 18 h of development, doubling during this period.

Antisera were raised against this glycoprotein, and univalent antibody fragments were found to block cell–cell association of aggregates that had been differentiated for 15 h, as measured in a quantitative assay (Geltosky *et al.*, 1979). In contrast, univalent antibody fragments directed to other surface glycoproteins had little if any effect even though equivalent amounts of antibody fragments were bound to the cell surface. In support of a role for this glycoprotein in cell–cell adhesion, antibody raised against it bound more at sites of cell contact in aggregates than on the remainder of the cell surface (Geltosky *et al.*, 1980).

If the 150,000-dalton glycoprotein is truly a cell adhesion molecule, why was it not detected by Beug *et al.* (1973a) or Steineman and Parish (1980)? The former group may not have found it since the 150,000-dalton glycoprotein is also present in vegetative cells (Geltosky *et al.*, 1976). Their antisera that had been absorbed with vegetative cells would not be expected to react with the 150,000-dalton glycoprotein. Failure of Steineman and Parish (1980) to detect this material might be due to the fact that they assay cell–cell adhesion of slug cells in which the 150,000-dalton glycoprotein may be either scarce or irrelevant. In addition, the antisera they raised had no detectable antibodies to this material.

C. Cell Adhesion Molecules in *Polysphondylium pallidum*

Similar immunological methods have been used to search for cell adhesion sites in another slime mold species, *Polysphondylium pallidum*. Using antisera raised against aggregating *P. pallidum* cells, Bozzaro and Gerisch (1978) prepared univalent antibody fragments and showed that they could block cell–cell adhesion. To identify the relevant cell surface antigens, butanol extracts of aggregating *P. pallidum* membranes were prepared as described above for *D. discoideum.* Since these extracts completely absorb the inhibitory-antibody activity, they were used as the basis for further purification. After further fractionation two glycoproteins referred to as contact site 1 (molecular weight 64,000) and contact site 2 (molecular weight 58,000) were isolated. Contact site 1 is also abundant on vegetative cells, whereas contact site 2 is much more prominent on aggregating cells (Bozzaro *et al.*, 1981).

Steineman *et al.* (1979) have also used an immunological approach to detect cell adhesion molecules in *Polysphondylium pallidum.* They used techniques similar to those that they employed with *D. discoideum.* They provide some evidence that a protein with molecular weight of 71,000 may be important in *P. pallidum* adhesion.

D. Limitations of the Immunological Approach

Despite its apparent success in identifying molecules that bind cells together, certain limitations of this approach should be pointed out. One major potential problem is immunological crossreactivity. What is emphasized with this approach is antigens that elicit adhesion-inhibiting antibodies and that are effective in absorbing them. Yet their potency as immunogens and as absorbants is not proof that they are important in the cell-adhesion reaction. It remains possible that there are other cell surface antigens that are immunologically crossreactive and that would be missed by this approach. For example, in seeking an absorbing antigen, what is emphasized is enrichment of this activity. Fractions that contain lesser amounts of absorbing activity are discarded. Nevertheless, the antigen molecules in these fractions might be the relevant ones, whereas those purified could just be more abundant crossreactive materials.

The problem is especially important with glycoproteins, wherein many molecules share antigenic saccharide residues. Indeed, Knecht and Dimond (1981) have shown that many *D. discoideum* glycoproteins have common antigenic determinants. It also appears that glycoproteins with molecular weights of 80,000 and 95,000 are unusually potent elicitors of antibody synthesis. Thus a major fraction of the antibodies specifically directed against the membrane proteins of aggregating cells bind to a glycoprotein with molecular weight of 80,000 (Parish *et al.*, 1978). Similarly, a major fraction of those directed against membranes of slug cells bind to a glycoprotein with molecular weight of 95,000 (Parish *et al.*, 1978). Surprisingly, there are very few other antigens in the membranes that significantly bind antibodies in these mixed antisera. Therefore, the glycoproteins in question were apparently very effective antigens in the animals in which they were injected. Had abundant antibodies been raised to other antigens, such antibodies might also block cell adhesion and implicate these antigens as adhesion molecules. The fact that certain antigens in a mixture are especially antigenic in a laboratory animal could falsely implicate them in cell adhesion.

It is also possible to have false negative results for technical reasons. If antibodies raised against a candidate cell-adhesion molecule fail to block cell–cell adhesion, this does not definitively revoke the molecule's candidacy. It is possible that the antibodies, although quite reactive with the antigen, may not be directed against or have access to its active site. In addition, the affinity of univalent antibody fragments for antigenic determinants may be so much lower than the affinity of the substances for a complementary cell-surface ligand that no significant effect is observed.

It is also important to cautiously interpret the function of molecules identified by the immunological approach. These molecules may participate in the cross-linking of the cells to each other (Müller and Gerisch, 1978). However, alternatives are possible, as considered earlier. For example, these molecules might, instead, be responsible for some basic cellular function that is a prerequisite for normal cell–cell adhesion. Antibody binding to cell surface molecules might disrupt interactions with the underlying cytoskeleton, the intactness of which has been shown to be required for cell adhesion (Jones *et al.*, 1977). Likewise, antibody binding to surface receptors for signals critical to some differentiative or constitutive cellular function might block or mimic such signals. This could introduce a cascade of secondary events that lead to abnormalities in cell–cell adhesion. Therefore, the finding that antibodies that react with a specific cell surface constituent block cell–cell adhesion should only be considered as providing an opportunity for further study. It is not definitive evidence that the molecule actually participates directly in the adhesion reaction. Combining the immunological approach with others is critical, as considered in other portions of this chapter and in more detail elsewhere (Barondes, 1980).

V. Lectins in Cell Adhesion

Since so many proteins and lipids on cell surfaces are glycosylated, consideration has long been given to the possible participation of some of their saccharide chains in cell–cell adhesion. One way this might occur is by binding surface glycoconjugates of two cells to a divalent carbohydrate-binding protein, thereby holding the cells together. Many proteins of this type are known to exist (Goldstein and Hayes, 1978), although their biological roles have not been established. They were originally isolated from plant seeds and are called lectins. Considerable evidence suggests that they play a role in cell–cell and cell-matrix interactions in plants and animals (Barondes, 1981).

Experience with plant lectins, which are often readily assayed in crude extracts as agglutinins of erythrocytes, facilitated the discovery of similar proteins in cellular slime molds (Rosen *et al.*, 1973). Since slime mold lectins are synthesized as the cells form aggregates, they may participate directly in cell–cell adhesion in these aggregates. Many studies support this suggestion. Before considering the evidence, some of the properties of slime mold lectins will be described.

A. Properties of Developmentally Regulated Slime Mold Lectins

Extracts of aggregating *D. discoideum* cells agglutinate formalinized sheep erythrocytes (Rosen *et al.*, 1973) (Fig. 5). The active materials are lectins that associate with surface glycoconjugates on the erythrocytes. Their carbohydrate binding sites are complementary to *N*-acetyl-D-galactosamine or related substances, since this saccharide totally inhibits hemagglutination activity at high concentrations (Fig. 5) and has a detectable effect even at much lower concentrations. Inhibition by this sugar is specific, since *N*-acetyl-D-glucosamine has no detectable effect even at high concentrations (Fig. 5).

Lectin activity, although abundant in extracts of aggregating *D. discoideum* cells that had been starved for 9 or 12 h before extraction, is virtually undetectable in extracts of vegetative *D. discoideum* cells (Fig. 5). Appearance of the lectin with aggregation is due to biosynthesis since it can be blocked with cycloheximide and since the messenger RNA that directs lectin synthesis appears only after starvation (Ma and Firtel, 1978). Lectin activity also appears with development of *Polysphondylium pallidum* (Rosen *et al.*, 1974).

Extracts of aggregating forms of all species of cellular slime molds that have been studied contain abundant lectin activity (Rosen *et al.*, 1975). Lectins from *D. discoideum* (Simpson *et al.*, 1974; Frazier *et al.*, 1975), *D. purpureum* (Barondes and Haywood, 1979; Cooper and Barondes, 1981), *D. mucuroides* (Barondes and Haywood, 1979), and *P. pallidum* (Simpson *et al.*, 1975; Rosen *et al.*, 1979) have been purified by affinity chromatography and are well characterized. In all cases the lectins are very abundant, comprising more than one percent of the total protein of aggregating cells. All species

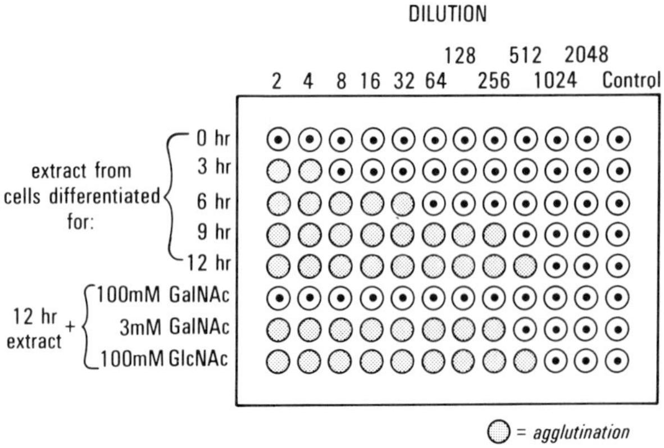

Fig. 5. Lectin activity in *D. discoideum* extracts assayed by hemagglutination. Hemagglutination is assayed by serial dilutions of extracts of *D. discoideum* prepared from cells differentiated for the indicated periods. In the absence of lectin the control erythrocytes form a dark dot on the bottom of the well. Agglutination of the erythrocytes causes formation of a fuzzy mat of cells. The highest dilution of extract causing agglutination is its titer. The agglutinin in this extract is a lectin with a binding site complementary to *N*-acetyl-D-galactosamine (GalNAc) which inhibits its activity whereas *N*-acetyl-D-glucosamine (GlcNAc) does not. (For details see Rosen *et al.*, 1973.)

studied also contain at least two lectin subunits which combine in various ways, as will be considered below.

Aggregating *D. purpureum* cells contain the largest known number of lectins of any species studied thus far. They are collectively referred to as purpurin. All bind to some degree to the galactosyl residues of Sepharose, which can be used as an adsorbant for affinity chromatography. After elution from the affinity column with lactose, seven isolectins can be separated either by electrophoresis under nondenaturing conditions (Fig. 6) or by affinity chromatography on column materials derivatized with appropriate saccharide residues (Cooper and Barondes, 1981). The seven isolectins are made up of four distinct protein subunits, each almost certainly the product of a different gene, and all with molecular weights of about 22,000 to 23,000. The lectins are all tetramers. Two are homotetramers of a single subunit. The other five are tetramers formed by randomly combining one or both of the remaining two subunits. The seven isolectins are functionally distinct in that they have differing affinities for columns derivatized with complementary saccharides. The biological significance of these functionally distinct forms is not known.

Multiple lectins, collectively called discoidin, have also been described in *D. discoideum*. Two forms have been purified (Frazier *et al.*, 1975). The major form, discoidin I, comprises about 90% of the total cellular lectin in aggregating cells. It is a tetramer comprised of subunits with apparent molecular weights of approximately 26,000 based on polyacrylamide-gel electrophoresis in SDS, although the subunits have a molecular weight of approximately 28,000 based on the sequence of the protein coding regions of the isolated genes (Poole et al., 1982). Using recombinant DNA technology it was shown that discoidin I is encoded by a four member multigene family (Rowekamp *et al.*, 1980) and that more than one form of discoidin I is synthesized (Tsang *et al.*, 1981, Poole *et al.*, 1982). Another lectin, discoidin II, is relatively less abundant, and has a subunit molecular weight of approximately 24,000 on polyacrylamide gel electrophoresis in SDS (Frazier *et al.*, 1975). Based on hapten inhibition studies of their hemagglutination activity, discoidin I and discoidin II have different active binding sites. This is also indicated by the fact that discoidin II binds to formalinized rabbit erythrocytes but not formalinized sheep erythrocytes, whereas discoidin I binds to both. Augmentation of the hemagglutination activity of discoidin by lipids has been reported

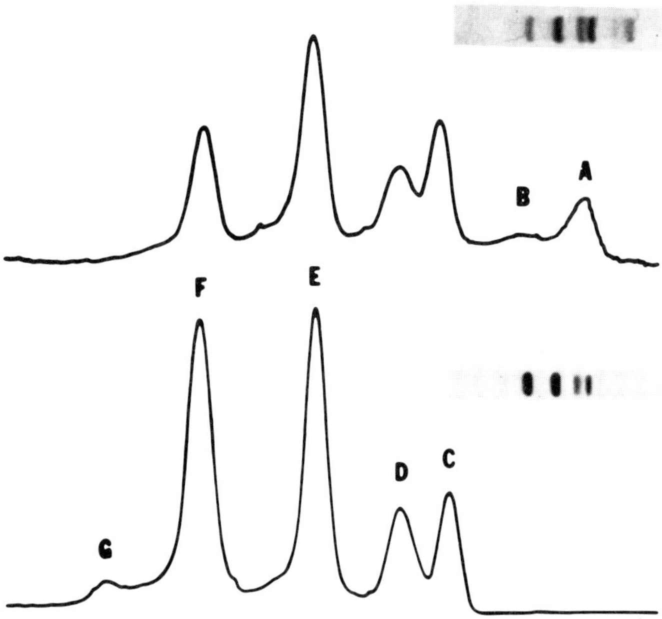

Fig. 6. Polyacrylamide gel electrophoresis of two preparations of purpurin that together illustrate the seven isolectins. The samples from extracts of *D. purpureum* isolated at different phases of development, were first purified by affinity chromatography then electrophoresed in 3.5% polyacrylamide slab gels at pH 8.9. Densitometer tracings shown were made from the stained gel samples shown as insets. The seven isolectins are labeled A–G. (For details see Cooper and Barondes, 1981.)

(Bartles *et al.*, 1979) raising the possibility of regulation of its activity by membrane components.

The lectins from *Polysphondylium pallidum*, collectively called pallidin, are also derived from distinct subunits (Rosen *et al.*, 1979). Three subunits have been identified with molecular weights in the range of 25,000–26,500. They tend to associate in ratios of 2:1 giving rise to three basic trimeric forms. One pair may be isolated as trimers and hexamers, whereas the others are found as nonamers and multiples thereof.

The slime mold lectins of various species are similar not only in the general size of their subunits, but also in the fact that none have been shown to contain any covalently linked carbohydrate. This distinguishes them from many plant lectins that are glycoproteins, although others, such as concanavalin A, are not glycosylated. The slime mold lectins are also all similar in that they bind saccharides containing galactose or *N*-acetyl-D-galactosamine residues. However, they can be distinguished by the relative potency of saccharides as inhibitors of their hemagglutination activity (Rosen *et al.*, 1975). They can also be distinguished by their relative affinities for saccharides covalently coupled to resins or by their relative affinities for different forms of erythrocytes. The significance of the multiple forms and of the differences among species is not known.

B. Localization of Endogenous Lectins within Cells and on the Cell Surface

If endogenous slime mold lectins are involved in cell–cell adhesion they must be present on cell surfaces. Binding of erythrocytes to living slime mold cells (Fig. 7) suggested that some lectin is at this site. The binding of the erythrocytes is apparently mediated by lectin since it is blocked by a sugar that combines with the lectin's active site (Fig. 7), but not by others that do not react with the lectin (Chang *et al.*, 1977). These results indicate that the lectin on the cell surface has available binding sites permitting adhesion to complementary glycoconjugates on the sheep erythrocytes.

Further studies of cell surface lectin used antibodies raised against these highly purified proteins. These antibodies bind to the surface of fixed aggregating slime mold cells as shown by fluorescent-antibody techniques and by ferritin labeling and electron microscopy (Chang *et al.*, 1975, 1977). Upon reaction of such antibodies with living cells, followed by reaction with a fluorescein-labeled second antibody, a single cluster of fluorescent antibody formed on each cell surface. This capping indicates that the lectin and the surface receptor to which it is bound can move along the plane of the plasma membrane (Chang *et al.*, 1975, 1977). Cell surface lectin has also been demonstrated by radioactive-iodine incorporation into exposed protein residues (Siu *et. al.*, 1976; Springer *et al.*, 1980).

Immunological techniques have also been used to quantify the number of lectin molecules on the cell surface. In many of these experiments the lectin on the cell surface was eluted by reacting aggregating cells with a saccharide

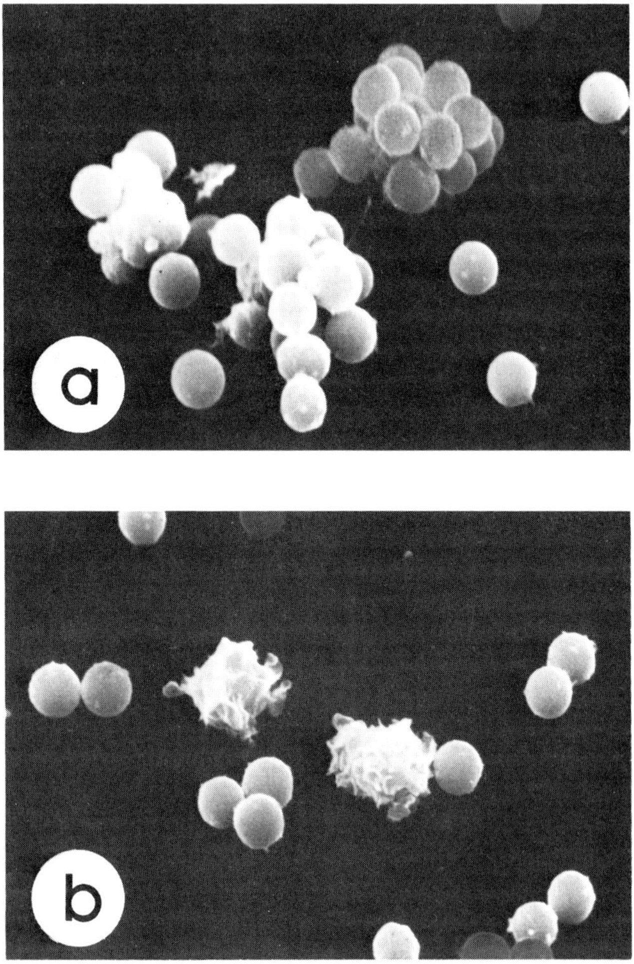

Fig. 7. Binding of formalinized sheep erythrocytes to the surface of living *P. pallidum* cells that had been dissociated in the aggregating stage and mixed with five times as many erythrocytes. The binding of erythrocytes to the *P. pallidum* cells (a) was blocked with 0.1 *M* D-galactose (b), but was not affected by 0.1 *M* D-glucose. (For details see Chang *et al.*, 1977.)

that binds its active site. In studies with both *D. discoideum* and *D. purureum* about 1×10^5 lectin tetramers were eluted per cell surface (Springer *et al.*, 1980). Although this is a large number of molecules, it is only about 2% of the total cellular lectin. This technique for estimating cell surface lectin was validated in a series of experiments that showed that elution with lactose is effective in releasing virtually all cell surface purpurin from *D. purpureum* (Springer *et al.*, 1980) and by studies in which iodinated univalent antibody fragments raised against the lectin were bound to the cell surface under saturating conditions (Springer *et al.*, 1980; Springer and Barondes, 1982).

The finding that 98% of cellular lectin of aggregating cells is intracellular appears difficult to understand if the primary function of the lectin is to bind cell surfaces together. This may, however, be rationalized because of evidence that the large intracellular pool represents a reservoir that may be externalized onto the cell surface under appropriate circumstances. This was discovered in studies of cell surface binding of divalent antilectin antibody. Whereas other studies demonstrated about 10^5 lectin molecules per cell surface, divalent antibody binding studies indicated that the number of purpurin molecules on the surface of *D. purpureum* cells approximated 10^6 (Springer *et al.*, 1980). With *D. discoideum* even higher estimates were obtained (Springer and Barondes, 1980).

The experiments with divalent antibodies suggested that crosslinking of cell surface lectins elicited the externalization of additional endogenous lectin by an unknown mechanism. Cross-linking seemed necessary since only divalent immunoglobulin had this effect (i.e., binding of univalent antibody fragments did not). This interpretation was supported by the finding (Springer and Barondes, 1982) that polyvalent complementary glycoproteins and neoglycorproteins, which would also be expected to crosslink cell surface lectins, elicited lectin externalization, although somewhat less effectively than antibodies. Addition of an excess of purified slime mold lectin also had this effect, presumably by cross-linking unoccupied endogenous cell surface receptors (Springer and Barondes, 1982). Externalization was also elicited to a lesser extent by concanavalin A, but not by a succinylated derivative of concanavalin A which binds well to cells but has much less cross-linking effect.

The finding that the amount of lectin displayed on the cell surface can be markedly increased by association with reagents that crosslinked endogenous cell surface lectin, its receptors, or other cell surface molecules, raised the possibility that the large intracellular pool is externalized under appropriate conditions. Binding of secreted glycoconjugates or of those displayed on the surfaces of other slime mold cells may be the signal for such externalization. This, in turn, might augment the interaction that signaled the release.

C. Lectin Receptors

Cell surface receptors that bind slime mold lectins were first shown by reacting discoidin with *D. discoideum* cells that were grown on bacteria, then fixed with glutaraldehyde after differentiation by gyration for various periods in the absence of nutrients. Fixed cells, which are relatively stable reagents, could be used since saccharide residues do not react with glutaraldehyde and since the fixed cells do not adhere to each other and could, therefore, be used for agglutination assays. Discoidin I or discoidin II agglutinated *D. discoideum* cells that had been fixed after 9 h of starvation (Reitherman *et al.*, 1975). Agglutination was blocked by *N*-acetyl-D-galactosamine, indicating that it was mediated by the lectins' carbohydrate binding sites. Higher concentrations of these lectins were required to produce comparable agglutination of cells fixed after briefer differentiation. Fixed vegetative cells were not agglutinated at all even at high concentrations, although they were readily agglutinated by plant lectins. These results indicated the presence of developmentally regulated cell surface receptors for slime mold lectins.

The number of receptors per cell was estimated by determining the amount of purified discoidin I or discoidin II absorbed by fixed differentiated *D. discoideum* cells. The amount of lectin absorbed was determined with a quantitative hemagglutination assay (Reitherman *et al.*, 1975). With both discoidin I and discoidin II about 5×10^5 receptors per glutaraldehyde-fixed aggregating cell were found. With pallidin 3×10^5 receptors were found on fixed *P. pallidum* cells differentiated on agar to the aggregating stage. In contrast, Bartles and Frazier (1980) using [^{125}I]-labeled discoidin I that had been carefully labeled to preserve some binding activity, found only about 3×10^4 discoidin I receptors per *D. discoideum* cell. The discrepancy between these results and those of Reitherman *et al.* (1975) could be due to the fact that in the latter studies an axenic strain of *D. discoideum*, A3, was used, and it was grown in axenic medium prior to differentiation by starvation in suspension for 15 h. The paucity of receptors found by Bartles and Frazier (1980) might indicate that the A3 cells do not differentiate to the same extent in suspension as do the wild type NC4 cells that had been grown on bacteria. Decreased binding might also be due, in part, to the iodination, which was not used by Reitherman *et al.* (1975).

A number of attempts have been made to identify receptors for slime mold lectins in extracts or digests of aggregating cells that inhibit the hemagglutination activity of slime mold lectins. These studies are complicated by the finding that bacteria and other growth media contain substances that bind slime mold lectins. For example, Bartles *et al.* (1981) identified a high molecular weight proteoglycan that interacted with discoidin in the extracellular medium of suspension-differentiated axenically grown A3 cells, but found it was a component of the Difco proteose peptone #2 used in the growth medium. The material was taken up during growth and released into the medium with starvation. It was not a biosynthetic product of the slime mold cells. Polysaccharides prepared from bacterial strains used to grow NC4 cells, including *E. coli* B/r and *K. aerogenes*, are also very potent inhibitors of discoidin I and discoidin II agglutination activity and binding activity (Cooper and Barondes, unpublished). In searching for endogenous receptors for slime mold lectins great care must be taken to avoid such contaminants, preferably using cells grown in a defined medium. Evidence that lectin receptor candidates are actually synthesized by the slime mold cells and accumulate with differentiation would argue against the possibility that the candidates are contaminants not of slime mold origin.

An early attempt to identify discoidin receptors by reacting purified discoidin with membrane glycoproteins that had been fractionated on SDS polyacrylamide gels showed no significant binding, although other lectins such as concanavalin A bound well to specific glycoproteins in the gels (Burridge and Jordan, 1979). Using a different approach Breuer and Siu (1981) covalently linked discoidin (95% discoidin I) to affinity columns that they reacted with detergent solubilized extracts of aggregating *D. discoideum* (NC4) cells. To facilitate detection of bound materials the cells had been labeled either with [^{35}S]methionine or by surface radio-iodination with the lactoperoxidase method. Of the [^{35}S]methionine-labeled proteins applied to the discoidin column about 0.1% bound and were eluted with 0.3 *M* D-galactose. Based on studies with both labeling procedures approximately 11 discoidin-binding proteins were identified. Of these a glycoprotein with molecular weight of 31,000 was shown to be the predominant discoidin-binding protein synthesized at the aggregation stage. The galactose eluate from the discoidin column blocked cell–cell adhesion of *D. discoideum* cells, but the relationship of this activity to the radioactive glycoproteins visualized by SDS gel electrophoresis is not presently known. Ray and Lerner (1982) failed to find similar materials bound to a discoidin column and eluted with galactose, but did find material eluting from this column when the pH was lowered. This eluate blocked cell–cell adhesion and contained a protein with a molecular weight of 80,000 which may be a discoidin receptor. Presumably acid elution was required because it bound with very high affinity to the discoidin column.

A large sulfated glycoprotein that may be a receptor for pallidin has been isolated from extracts of *P. pallidum* cells as well as from the medium in which they are differentiated (Drake and Rosen, 1982). This material has been purified by precipitation with soluble pallidin. When exposed to [^{3}H]acetate, *P. pallidum* cells incorporated some radioactivity into this substance suggesting that they synthesize it, at least in part. The substance could be eluted from the surface of *P. pallidum* cells with D-galactose. This suggested that it might serve to cross-link *P. pallidum* cells by binding to pallidin molecules on both cells which are held on the surface by association with other surface receptors. This hypothesis was also supported by the finding that, under appropriate assay conditions and at appropriate concentrations, the purified glycoprotein could augment cell–cell adhesion of *P. pallidum* cells.

D. Effect of Antibodies Raised against Slime Mold Lectins

Given the apparent success of the immunological approach discussed above, one might expect that, if cell surface slime mold lectins directly mediated cell–cell adhesion, antibodies raised against them would block this adhesion. With *Polysphondylium pallidum* univalent antibody fragments raised against pallidin block cell–cell adhesion under specific assay conditions (Rosen *et al.*, 1976, 1977). Under these conditions, in which either high solute concentration or metabolic inhibitors were added to the EDTA-containing medium, a developmentally regulated cell–cell adhesion could be measured. Upon addition of univalent antibody fragments raised against pallidin, marked inhibition of adhesion was observed. Inhibition was apparently due to binding of antibody fragments to pallidin, since asialofetuin, a glycoprotein that binds the active site of pallidin, also blocked cell–cell adhesion (Rosen *et al.*, 1977) as did high concentrations of saccharides that would be expected to react with pallidin (Rosen *et al.*, 1974).

Interpretation of these results depends on the validity of the assay of cell–cell adhesion under these conditions. There is no more reason to challenge the meaningfulness of these conditions than there is to challenge assays in which gyration speed or EDTA concentration must be manipulated to discriminate forms of adhesion, as shown in Fig. 1. Nevertheless, the fact that univalent antibody fragments directed against pallidin did not have much effect on cell–cell adhesion when assayed in medium containing low solute concentrations or without metabolic inhibitors (Rosen *et al.*, 1976, 1977; Bozzaro and Gerisch, 1978) has reduced the impact of these results.

Attempts to define assay conditions in which univalent antibody fragments raised against discoidin block adhesion of *D. discoideum* cells have not been successful (Springer and Barondes, 1980). One might argue that the antibody is not saturating all sites or that the affinity of interacting molecules for each other far exceeds the affinity of univalent antibody fragments for the cell adhesion molecules. To evaluate this possibility intact rabbit immunoglobulin directed against discoidin was used. It would be expected to have a much higher affinity for discoidin than univalent antibody fragments. After binding the antibody, the cells were washed and agglutination by the bound rabbit antibody was blocked by addition of saturating quantities of goat antirabbit univalent antibody fragments. Using high antibody concentrations, such that about 10^7 iodinated goat antirabbit univalent antibody molecules bound per cell after preliminary incubation with the antidiscoidin antiserum, only a very slight effect on cell–cell adhesion was observed (Springer and Barondes, 1980). These negative results are hard to reconcile with the inference that discoidin directly mediates cell–cell adhesion in this assay, although it is possible that, even here, some externalized discoidin (Springer *et al.*, 1980) escaped inhibition. The results are also compatible with the possibility that discoidin indirectly affects cell adhesion, influencing the appearance or organization of other molecules that actually bind the cells together.

E. Discoidin I Mutants

If discoidin I were involved in cell–cell adhesion, a mutant with an abnormality in the carbohydrate-binding site of this lectin would be expected to show abnormal cell–cell adhesion. Ray *et al.* (1979) reported the isolation of a mutant of this type. This mutant was originally obtained by passing dissociated aggregating NC4 cells through a Sepharose column coupled to desialated fetuin which binds discoidin. They found that 0.3% of the cells applied to this column pass through it. Presumably some such cells do not bind to the column since they lack cell surface lectin, either because they developed abnormally and had not reached the stage of lectin synthesis or because the lectin they made was abnormal. After plating 200 of the nonbinding cells in association with bacteria they identified one plaque that did not develop normally. This mutant, designated HJR1, developed only to the loose aggregate stage. It synthesized normal amounts of a protein that had the same subunit molecular weight as discoidin I and reacted with antibody to discoidin I, but lacked lectin activity. Therefore, HJR1 does not appear to be a "program mutant" that never differentiates to the stage when discoidin I is made. Instead, it appears to be a point mutation in the gene directing discoidin I synthesis.

Support for the inference that the abnormality in HJR1 is a point mutation in the structural gene for discoidin I came from studies of spontaneous and induced revertants. Fifty-eight revertants were identified (Shinnick and Lerner, 1980). Of these, 18 were indistinguishable from the wild type and showed normal discoidin I activity. A second class showed abnormalities in aggregation and in cell–cell adhesion measured *in vitro*, as well as an intermediate level of hemagglutination activity. A third group showed still more marked morphological abnormalities, more marked cell–cell adhesion abnormalities, and very low levels of discoidin I. Therefore, the degree of recovery of the carbohydrate-binding activity of discoidin I appears to be correlated both with cell–cell adhesion measured *in vitro* and with morphological observations. However, one revertant was indistinguishable from the wild type both in morphology and in cell–cell adhesion, but had only very low levels of lectin activity. The mechanism for this reversion is not yet explained.

Interpretation of work with HJR1 and its revertants is complicated by the finding that there is more than one structural gene for discoidin I, and that more than one may indeed be expressed (Rowekamp *et al.*, 1980; Poole *et al.*, 1982; Tsang *et al.*, 1981). These findings make it difficult to explain how a single point mutation would inactivate all discoidin I. One possibility is that the active forms of discoidin I are mixed tetramers like those demonstrated with purpurin (Cooper and Barondes, 1981). An abnormality in a single subunit might render the entire molecule abnormal, especially if the lectin, although tetrameric, is normally only bivalent. Indeed, combinations of normal and abnormal subunits might give rise to molecules that are functionally univalent and that might actually inhibit cell–cell adhesion.

The discovery of HJR1 and its revertants provides strong support for the inference that discoidin I plays a direct role in some aspect of the cell adhesion process. Because of the conceptual background which stimulated this entire approach and because of other findings reviewed above, it is tempting to infer that the role that discoidin I plays is to actually bind the cells together. However, it is important to recognize that findings with these mutants, although consistent with this hypothesis, do not prove it. Other results, reviewed above, are also largely consistent with this inference but still not decisive.

VI. Summary

Although studies of cell–cell adhesion in cellular slime molds have progressed substantially, a clear understanding of the cellular and molecular mechanisms involved has not been achieved. Considerable progress has been made in developing *in vitro* assays and in exploring experimental approaches. With the immunological approach at least three glycoproteins with molecular weights of 80,000, 95,000, and 150,000 have been implicated as playing some role in developmentally regulated cell–cell adhesion in *D.*

discoideum. However, it remains possible that the distinguishing feature of these molecules is their abundance, antigenicity and immunological crossreactivity with true cell adhesion molecules. Even if it is shown that the antibodies in question are actually exerting their inhibitory effect by binding to these molecules, a direct role in cell–cell adhesion need not be inferred. Instead, antibody binding may produce cellular changes which disrupt cell adhesion, whereas different molecules actually hold the cells together. Therefore, the role of these glycoproteins is not yet established.

The hypothesis that endogenous lectins bind slime mold cells together also remains attractive but unproven. It is clear that differentiating slime molds synthesize lectins that appear on cell surfaces. However, their direct role in cell–cell adhesion has not been decisively demonstrated. Even the isolation of what appears to be a point mutant in a structural gene for discoidin I, and its correlation with abnormalities in cell–cell adhesion, indicates only that discoidin I is intimately involved in the cell adhesion process. It does not demonstrate that the lectin directly binds the cells together.

Despite the qualifications in interpreting the many studies described here, use of cellular slime molds for studies of the general problem of cell–cell adhesion seems well justified. The strong evidence for species-specific cell adhesion provides an especially good opportunity which has not yet been exploited. The fact that the problem is so complex and requires multiple experimental approaches emphasizes the importance of using a favorable biological system. Continued exploration of the mechanism of cell–cell adhesion in cellular slime molds should contribute greatly to the solution of this major biological problem.

References

Barondes, S. H. (1980) Developmentally regulated lectins in slime molds and chick tissues—are they cell adhesion molecules? *In* "Cell Adhesion and Motility" (A. S. G. Curtis and J. Pitts, eds.), pp. 309–328. Cambridge University Press, Cambridge, England.

Barondes, S. H. (1981). Lectins: Their multiple endogenous cellular functions. Ann. Rev. Biochem. **50,** 207–231.

Barondes, S. H., and Haywood, P. L. (1979). Comparison of developmentally regulated lectins from three species of cellular slime molds. *Biochem. Biophys. Acta.* **550,** 297–308.

Bartles, J. R., and Frazier, W. A. (1980). Preparation of [^{125}I]-discoidin I and the properties of its binding to *Dictyostelium discoideum* cells. *J. Biol. Chem.* **255,** 30–38.

Bartles, J. R., Pardos, B. T., and Frazier, W. A. (1979). Reconstitution of discoidin hemagglutination activity by lipid extracts of *Dictyostelium discoideum* cells. *J. Biol. Chem.* **254,** 3156–3159.

Bartles, J. R., Santoro, B. C., and Frazier, W. A. (1981). Purification of a high-affinity discoidin I-binding proteoglycan from axenic *Dictyostelium discoideum* growth medium. *Biochim. Biophys. Acta.* **674,** 372–382.

Beug, H., and Gerisch, G. (1972). A micromethod for routine measurement of cell agglutination and dissociation. *J. Immunol. Methods.* **2,** 49–57.

Beug, H. Gerisch, G., Kempff, S., Riedel, V., and Cremer, G. (1970). Specific inhibition of cell contact formation in *Dictyostelium* by univalent antibodies. *Exp. Cell Res.* **63,** 147–158.

Beug, H., Katz, F. E., Stein, A., and Gerisch, G. (1973a). Quantitation of membrane sites in aggregating *Dictyostelium* cells by use of tritiated univalent antibody. *Proc. Natl. Acad. Sci. U.S.A.* **11,** 3150–3154.

Beug, H., Katz, F. E., and Gerisch, G. (1973b). Dynamics of antigenic membrane sites relating to cell aggregation in *Dictyostelium discoideum. J. Cell Biol.* **56,** 647–658.

Bonner, J. T., and Adams, M. S. (1958). Cell mixtures of different species and strains of cellular slime molds. *J. Embryol. Exp. Morphol.* **6,** 346–356.

Bozzaro, S., and Gerisch, G. (1978). Contact sites in aggregating cells of *Polysphondylium pallidum. J. Mol. Biol.* **120,** 265–279.

Bozzaro, S., Tsugita, A., Fromme, I., Janku, M., Monok, G., Opatz, K., and Gerisch, G. (1981). Characterization of a purified cell surface glycoprotein as a contact site of *Polysphondylium pallidum. Exp. Cell Res.* **134,** 181–191.

Breuer, W., and Siu, C.-H. (1981). Identification of endogenous binding proteins for the lectin discoidin-I in *Dictyostelium discoideum. Proc. Natl. Acad. Sci. U.S.A.* **78,** 2115–2119.

Burridge, K., and Jordan, L. (1979). The glycoproteins of *Dictyostelium discoideum:* Changes during development. *Exp. Cell Res.* **124,** 31–38.

Burridge, K., and Jordan, L. (1980). The application of labeled lectins and antibodies to SDS gels and their use in studying glycoproteins and cell surface antigens during development. *In* "Current Topics in Developmental Biology," (A. A. Moscona and A. Monroy, eds.), Vol. 14, Part II, pp. 227–241. Academic Press, New York.

Chang, C.-M., Reitherman, R. W., Rosen, S. D., and Barondes, S. H. (1975). Cell surface location of discoidin, a developmentally regulated carbohydrate-binding protein from *Dictyostelium discoideum. Exp. Cell Res.* **95,** 136–142.

Chang, C.-M., Rosen, S. D., and Barondes, S. H. (1977). Cell surface location of an endogenous lectin and its receptor in *Polysphondylium pallidum. Exp. Cell Res.* **104,** 101–109.

Cooper, D. N., and Barondes, S. H. (1981). Isolectins from *Dictyostelium purpureum:* Purification and characterization of seven functionally distinct forms. *J. Biol. Chem.* **256,** 5046–5051.

Drake, D. K., and Rosen, S. D. (1982). Identification and purification of an endogenous receptor for the lectin pallidin from *Polysphondylium pallidum. J. Cell. Biol.* **93,** 383–389.

Feinberg, A., Springer, W. R., and Barondes, S. H. (1979). Segregation of pre-stalk and prespore cells of *Dictyostelium discoideum:* Observations consistent with selective cell cohesion. *Proc. Natl. Acad. Sci. U.S.A.* **76,** 3977–3981.

Frazier, W. A., Rosen, S. D., Reitherman, R. W., and Barondes, S. H. (1975). Purification and comparison of two developmentally regulated lectins from *Dictyostellium discoideum:* Discoidin I and II. *J. Biol. Chem.* **250,** 7714–7721.

Garrod, D. R. (1974). Cellular recognition and specific cell adhesion in cellular slime mold development. *Arch. Biol.* (Bruxelles) **85,** 7–31.

Geltosky, J. E., Siu, C.-H., and Lerner, R. A. (1976). Glycoproteins of the plasma membrane of *Dictyostelium discoideum* during development. *Cell* **8,** 391–396.

Geltosky, J. E., Weseman, J., Bakke, A., and Lerner, R. A. (1979). Identification of a cell surface glycoprotein involved in cell aggregation in *D. discoideum. Cell* **18,** 391–398.

Geltosky, J. E., Birdwell, C. R., Weseman, J., and Lerner, R. A. (1980). A glycoprotein involved in aggregation of *D. discoideum* is distributed on the cell surface in a nonrandom fashion favoring cell junctions. *Cell* **21**, 339–345.

Gerisch, G. (1961). Zellfunktionen und Zellfunktionsweksel in der Entwicklung von *Dictyostelium discoideum*. V. Stadienspezifische Zelkontaktbildung und ihre quantitative Erfassung. *Exp. Cell Res.* **25**, 535–554.

Gerisch, G. (1980). Univalent antibody fragments as tools for the analysis of cell interactions in Dictyostelium. *In* "Current Topics in Developmental Biology," (A. A. Moscona and A. Monroy, eds.), Vol. 14, Part II. pp. 243–269. Academic Press, New York.

Gerisch, G., Fromm, H., Huesgen, A., and Wick, U. (1975). Control of cell contact sites by cyclic AMP pulses in differentiating *Dictyostelium* cells. *Nature (London)* **255**, 547–549.

Gilkes, N. R., Laroy, K., and Weeks, G. (1979). An analysis of the protein, glycoprotein and monosaccharide composition of *Dictyostelium discoideum* plasma membranes during development. *Biochim. Biophys. Acta.* **551**, 349–362.

Goldstein, I. J., and Hayes, C. E. (1978). The lectins: Carbohydrate-binding proteins of plant and animals. *Adv. Carbohydr. Chem. Biochem.* **35**, 127–340.

Gottlieb, D. I., and Glaser, L. (1975). A novel assay of neuronal cell adhesion. *Biochem. Biophys. Res. Comm.* **63**, 815–821.

Gregg, J. H. (1960). Surface antigen dynamics in the slime mold *Dictyostelium discoideum*. *Biol. Bull.* **118**, 70–78.

Hoffman, S., and McMahon, D. 1977). The role of the plasma membrane in the development of *Dictyostelium discoideum*. II. Developmental and topographic analysis of polypeptide and glycoprotein composition. *Biochim. Biophys. Acta.* **465**, 242–259.

Huesgen, A., and Gerisch, G. (1975). Solubilized contact sites A from cell membranes of *Dictyostelium discoideum*. *FEBS Lett.* **56**, 46–49.

Jaffe, A. R., Swan, A. P., and Garrod, D. R. (1979). A ligand-receptor model for the cohesive behavior of *Dictyostelium discoideum* axenic cells. *J. Cell Sci.* **37**, 157–167.

Jones, G. E., Pacy, J., Jermyn, K., and Stirling, J. (1977). A requirement for filopodia in the adhesion of pre-aggregative cells of *Dictyostelium discoideum*. *Exp. Cell Res.* **107**, 451–454.

Knecht, D. A., and Dimond, R. L. (1981). Lysosomal enzymes possess a common antigenic determinant in the cellular slime mold, *Dictyostelium discoideum*. *J. Biol. Chem.* **256**, 3564–3575.

Koziol, J. A., Springer, W. R., and Barondes, S. H. (1980). Quantitation of selective cell–cell adhesion and its application to assays of species-specific adhesion in cellular slime molds. *Exp. Cell Res.* **128**, 375–381.

Lam, T. Y., Pickering, G., Geltosky, J., and Siu, C.-H. (1981). Differential cell cohesiveness expressed by prespore and prestalk cells of *Dictyostelium discoideum*. *Differentiation* **20**, 22–28.

Ma, G. C. L., and Firtel, R. A. (1978). Regulation of the synthesis of two carbohydrate-binding proteins in *Dictyostelium discoideum*. *J. Biol. Chem.* **253**, 3924–3932.

Marin, F. T., Goyette-Boulay, M., and Rothman, F. E. (1980). Regulation of development in *Dictyostelium discoideum* III. Carbohydrate-specific intercellular interactions in early development. *Develop. Biol.* **80**, 301–312.

McDonough, J. P., Springer, W. R., and Barondes, S. H. (1980). Species-specific cell cohesion in cellular slime molds: Demonstration by several quantitative assays and with multiple species. *Exp. Cell Res.* **125**, 1–14.

Müller, K., and Gerisch, G. (1978). A specific glycoprotein as the target site of adhesion blocking Fab in aggregating *Dictyostelium cells. Nature (London)* **274**, 445–449.

Müller, K., Gerisch, G., Fromme, I., Mayer, H., and Tsugita, A. (1979). A membrane glycoprotein of aggregating *Dictyostelium* cells with the properties of contact sites. *Eur. J. Biochem.* **99,** 419–426.

Murray, B. A., Yee, L. D., and Loomis, W. F. (1981). Immunological analysis of a glycoprotein (contact sites A) involved in intercellular adhesion of *Dictyostelium discoideum. J. Supramol. Struc. Cell Biochem.* **17,** 197–211.

Nicol, A., and Garrod, D. R. (1978). Mutual cohesion and cell sorting-out among four species of cellular slime moulds. *J. Cell Sci.* **32,** 377–387.

Orr, C. W., and Roseman, S. (1969). Intercellular adhesion. I. A quantitative assay for measuring the rate of adhesion. *J. Membrane Biol.* **1,** 109–124.

Parish, R., and Schmidlin, S. (1979). Resynthesis of developmentally regulated plasma membrane proteins following disaggregation of *Dictyostelium* pseudoplasmodia. *FEBS Lett.* **99,** 270–274.

Parish, R., Schmidlin, S., and Parish, C. R. (1978). Detection of developmentally controlled plasma membrane antigens of *Dictyostelium discoideum* cells in SDS-polyacrylamide gels. *FEBS Lett.* **95,** 366–370.

Poole, S., Firtel, R. A., and Lamar, E. (1982). Sequence and expression of the discoidin I gene family in *Dictyostelium discoideum*. *J. Mol. Biol.* **153,** 273–289.

Raper, K. B., and Thom, C. (1941). Interspecific mixtures in the Dictyosteliacaea. *Am. J. Bot.* **28,** 69–78.

Ray, J., Shinnick, T., and Lerner, R. (1979). A mutation altering the function of a carbohydrate-binding protein blocks cell–cell cohesion in developing *Dictyostelium discoideum. Nature (London)* **279,** 215–221.

Ray, J., and Lerner, R. A. (1982). A biologically active receptor for the carbohydrate-binding protein(s) of *Dictyostelium discoideum*. *Cell* **28,** 91–98.

Reitherman, R. W., Rosen, S. D., Frazier, W. A., and Barondes, S. H. (1975). Cell surface species-specific high affinity receptors for discoidin: Developmental regulation in *Dictyostelium discoideum*. *Proc. Natl. Acad. Sci. U.S.A.* **72,** 3541–3545.

Rosen, S. D., Kafka, J. A., Simpson, D. L., and Barondes, S. H. (1973). Developmentally regulated, carbohydrate binding protein in *Dictyostelium discoideum. Proc. Natl. Acad. Sci. U.S.A.* **70,** 2554–2557.

Rosen, S. D., Simpson, D. L., Rose, J. E., and Barondes, S. H. (1974). Carbohydrate-binding protein from *Polysphondylium pallidum* implicated in intercellular adhesion. *Nature (London)* **252,** 149–151.

Rosen, S. D., Reitherman, R. W., and Barondes, S. H. (1975). Distinct lectin activities from six species of cellular slime molds. *Exp. Cell Res.* **95,** 159–166.

Rosen, S. D., Haywood, P. L., and Barondes, S. H. (1976). Inhibition of intercellular adhesion in a cellular slime mold by univalent antibody against a cell surface lectin. *Nature (London)* **263,** 425–427.

Rosen, S. D., Chang, C.-M., and Barondes, S. H. (1977). Intercellular adhesion in the cellular slime mold *P. pallidum* inhibited by interaction of asialofetuin or specific univalent antibody with endogenous cell surface lectin. *Develop. Biol.* **61,** 202–213.

Rosen, S. D., Kaur, J., Clark, D. L., Pardos, B. T., and Frazier, W. T. (1979). Purification and characterization of multiple species (isolectins) of a slime mold lectin implicated in intercellular adhesion. *J. Biol. Chem.* **254,** 9408–9415.

Rowekamp, W., Poole, S., and Firtel, R. A. (1980). Analysis of members of the multi-gene family coding the developmentally regulated carbohydrate-binding protein discoidin-I in *Dictyostelium discoideum. Cell* **20,** 495–505.

Rutishauser, U., Thiery, J.-P., Brackenbury, R., Sela, B.-A, and Edelman, G. M. (1976).

Mechanisms of adhesion among cells from neural tissues of the chick embryo. *Proc. Natl. Acad. Sci. U.S.A.* **73,** 577–581.

Saxe, C. L., III, and Sussman, M. (1982). Induction of cell cohesion by a membrane-associated moiety in *Dictyostelium discoideum*. *Cell*. In press.

Shinnick, T. M., and Lerner, R. A. (1980). The cbpA gene: Role of the 26,000-dalton carbohydrate-binding protein in intercellular cohesion of developing *Dictyostelium discoideum* cells. *Proc. Natl. Acad. Sci. U.S.A.* **11,** 4788–4792.

Simpson, D. L., Rosen, S. D., and Barondes, S. H. (1974). Discoidin, a developmentally regulated carbohydrate-binding protein from *Dictyostelium discoideum:* Purification and characterization. *Biochemistry* **13,** 3287–3493.

Simpson, D. L., Rosen, S. D., and Barondes, S. H. (1975). Pallidin: Purification and characterization of a carbohydrate-binding protein from *Polysphondylium pallidum,* implicated in intercellular adhesion. *Biochim. Biophys. Acta*. **412,** 109–119.

Siu, C.-H., Lerner, R. A., Firtel, R. A., and Loomis, W. F. (1975). Changes in plasma membrane proteins during development of *Dictyostelium discoideum. In* "Cell and Molecular Biology" (D. McMahon, C. F. Fox, eds.) Vol. II, pp. 129–134. Benjamin, Menlo Park.

Siu, C.-H., Lerner, R. A., Ma, G. C. L., Firtel, R. A., and Loomis, W. F. (1976). Developmentally regulated proteins of the plasma membrane of *Dictyostelium discoideum*. The carbohydrate-binding protein. *J. Mol. Biol.* **100,** 157–178.

Sonneborn, D. R., Sussman, M., and Levine, L. (1964). Serological analyses of cellular slime mold development. I. Changes in antigenic activity during cell aggregation. *J. Bacteriol.* **87,** 1321–1329.

Springer, W. R., and Barondes, S. H. (1978). Direct measurement of species-specific cohesion in cellular slime molds. *J. Cell Biol.* **79,** 937–942.

Springer, W. R., and Barondes, S. H. (1980). Cell adhesion molecules: Detection with univalent second antibody. *J. Cell Biol.* **87,** 703–707.

Springer, W. R., and Barondes, S. H. (1982). Externalization of the endogenous intracellular lectin of a cellular slime mold. *Exp. Cell Res.* **138,** 231–240.

Springer, W. R., Haywood, P. L., and Barondes, S. H. (1980). Endogenous cell surface lectin in *Dictyostelium:* Quantitation, elution by sugar and elicitation by divalent immunoglobulin. *J. Cell Biol.* **87,** 682–690.

Steineman, C., and Parish, R. W. (1980). Evidence that a developmentally regulated glycoprotein is target of adhesion blocking Fab in reaggregating *Dictyostelium. Nature (London)* **286,** 621–623.

Steineman, C., Hinterman, R., and Parish, R. W. (1979). Identification of a developmentally regulated plasma membrane glycoprotein involved in adhesion of *Polysphondylium pallidum* cells. *FEBS Lett.* **108,** 379–384.

Sternfeld, J. (1979). Evidence for differential cellular adhesion as the mechanism of sorting out of various slime mold species. *J. Embryol. Exp. Morphol.* **53,** 163–177.

Sussman, M., and Boschwitz, C. (1975). Adhesive properties of cell ghosts derived from *Dictyostelium discoideum*. *Dev. Biol.* **44,** 362–368.

Swan, A., Garrod, D. R., and Morris, D. (1977). An inhibitor of cell cohesion from axenically grown cells of the slime mould, *Dictyostelium discoideum. J. Cell Sci.* **28,** 107–116.

Takeuchi, I. (1969). Establishment of polar organization during slime mold development. *In* "Nucleic Acid Metabolism Cell Differentiation and Cancer Growth" (E. V. Cowdry and S. Seno, eds.), pp. 297–304. Pergammon Press, Oxford.

Tsang, A., and Bradbury, J. M. (1981). Separation and properties of pre-stalk and pre-spore cells of *Dictyostelium discoideum. Exp. Cell Res.* **132,** 433–441.

Tsang, A., Devine, J. M., and Williams, J. G. (1981). The multiple subunits of discoidin I are encoded by different genes. *Dev. Biol.* **84,** 212–217.

Tyler, A. (1947). An auto-antibody concept of cell structure, growth and differentiation. *Growth* (Symposium) **10,** 6–7.

Walther, B. T., Ohman, R., and Roseman, S. (1973). A quantitative assay for intercellular adhesion. *Proc. Natl. Acad. Sci. U.S.A.* **70,** 1569–1573.

Weiss, P. (1947). The problem of specificity in growth and development. *Yale J. Biol. Med.* **19,** 235–278.

West, C. M., and McMahon, D. (1977). Identification of concanavalin A receptors and galactose-binding proteins in purified plasma membranes of *Dictyostelium discoideum. J. Cell Biol.* **74,** 264–273.

Wilcox, D. K., and Sussman, M. (1981a). Defective cell cohesivity expressed late in the development of a *Dictyostelium discoideum* mutant. *Dev. Biol.* **82,** 102–112.

Wilcox, D. K., and Sussman, M. (1981b). Serologically distinguishable alterations in the molecular specificity of cell cohesion during morphogenesis in *Dictyostelium discoideum. Proc. Natl. Acad. Sci. U.S.A.* **78,** 358–362.

• CHAPTER 7

The Organization and Expression of the *Dictyostelium* Genome

Alan R. Kimmel and Richard A. Firtel

The Development of *Dictyostelium discoideum*

ISBN 0-12-455620-5

I. Introduction

Due to its relatively simple, well-defined life cycle, *Dictyostelium* has proven to be an excellent organism for the study of eukaryotic gene organization and developmental gene expression. Recently, the techniques of molecular biology have been used to examine developmental changes in RNA and protein synthesis and to isolate and study specific developmentally regulated genes. It has also been possible to examine specific gene expression in spore and stalk cells and to monitor the pattern of gene expression in perturbed cells. For example, cAMP has been shown to modulate gene expression in preaggregating cells as well as in cells late in development.

Using recombinant DNA techniques, many *Dictyostelium* genes have been isolated and their structures characterized. In addition, their developmental expression has been studied and sequences that may be essential for transcription of these genes have been identified.

In this chapter, we discuss the general properties of the *Dictyostelium* genome and the specific characteristics of individual genes. The structure and sequence of these genes are described as well as factors that may affect their expression and biological function. Additionally, we discuss methods for directly examining the mechanisms involved in the differential regulation of gene activity during *Dictyostelium* development.

II. General Properties of Genome Structure

A. The Nuclear Genome

The *Dictyostelium* nuclear genome has been characterized by buoyant density centrifugation, thermal denaturation, renaturation kinetics, and restriction endonuclease digestion (Sussman and Rayner, 1971; Firtel and Bonner, 1972a; Firtel *et al.*, 1976a,b; Firtel and Jacobson, 1977). The haploid genome size is $\sim 50 \times 10^3$ kilobase pairs (kb), ~12 times larger than the *E. coli* genome. Using renaturation kinetics it is possible to fractionate the nuclear DNA into two components of differing repetition frequencies. The single copy portion makes up 70–75% of the genome. This is ~7 times the size of the *E. coli* genome and only 2% that of mammalian genomes. The remaining 25–30% of the *Dictyostelium* genome is composed of sequences that are repeated, on average, 100 times per haploid complement. It is clear, however, that there are different repeat families having widely varying reiteration frequencies. The total complexity of the repetitive component is $\sim 0.13 \times 10^3$ kb. Nearly 40% of this complexity is attributed to the ribosomal genes (rDNA) (Cockburn *et al.*, 1976; Maizels, 1976).

From studies wherein labeled nuclear DNA (tracer) is sheared to specific size classes (0.4–3.5 kb) and hybridized to a sequence excess of 0.4 kb nuclear DNA (driver), it is possible to predict the linkage and spacing relationship of the single-copy and repetitive sequences (Davidson *et al.*, 1975). In *Dictyostelium*, ~40% of the single-copy DNA is linked to a repetitive sequence at a spacing of ~1.2 kb (Firtel and Kindle, 1975; Firtel *et al.*, 1976a,b; Firtel and Jacobson, 1977). Nearly 60% of the single-copy sequences are linked to a repeat at spacings of <5 kb. The remaining 40% of the single-copy component is not linked to repetitive DNA at fragment lengths of 5 kb. The interspersion of repetitive and single-copy elements in the *Dictyostelium* genome is similar to that observed in many metazoa (see Davidson *et al.*, 1975).

The repetitive sequences in *Dictyostelium* are organized into two size distributions. Approximately 70% of the repetitive DNA is found in lengths of >2 kb. This size class is primarily rDNA but may also include other moderately repeated families such as the actin genes. There is a smaller heterogeneous (0.2–1.0 kb) class of repeat sequences with a broad average size of ~300–400 base pairs (bp). From complexity measurements, we calculate that there are ~200 repeat families in this smaller size class.

The G + C content of the nuclear genome is only ~22% (Sussman and Rayner, 1971; Firtel and Bonner, 1972a). The base composition of total poly$(A)^+$ RNA is ~30% (G +C) and sequence analyses of isolated genes indicate that the protein coding regions of these genes are ~30–40% (G + C) (Jacobson *et al.*, 1974a; Firtel *et al.*, 1979; McKeown and Firtel, 1981b; Poole *et al.*, 1981; Kimmel and Firtel, 1980; Brandis and Firtel, in preparation). In contrast, the regions that do not code for protein possess a much lower (5–20%) G + C content. These A + T rich regions include the 5′ and 3′ untranslated sequences on mRNA, the intervening sequences (introns), the sequences that are 5′ or 3′ to transcription initiation or termination, and the rDNA spacer sequences (Firtel *et al.*, 1976a; Firtel *et al.*, 1979; McKeown and Firtel, 1981b; Poole *et al.*, 1981; Kimmel and Firtel, in preparation; Brandis and Firtel, in preparation). There are also ~15,000 oligo$(dT)_{25}$ stretches in the genome (Jacobson *et al.*, 1974b; Firtel *et al.*,

1976b). Sequences analysis of isolated genes has shown that such sequences are found immediately adjacent to the 5′ end and adjacent to or at the 3′ ends of many genes (Firtel *et al.*, 1979; McKeown and Firtel, 1981a,b; Poole *et al.*, 1981; Kimmel and Firtel, in preparation; Brandis and Firtel, in preparation). Many *Dictyostelium* mRNAs have a transcribed oligo$(A)_{25}$ sequence near the 3′ terminus of the mRNAs (Jacobson *et al.*, 1974b; Firtel *et al.*, 1976b; McKeown and Firtel, 198b).

Many prokaryotic and eukaryotic genomes exhibit sequence-specific patterns of DNA methylation. There is accumulating evidence to show that, in metazoa, there is an inverse correlation between DNA methylation and gene activity (see Razin and Riggs, 1980; van der Ploeg and Flavell, 1980; Weintraub *et al.*, 1981; Shen and Maniatis, 1980; Mandel and Chambon, 1979). DNA methylation in *Dictyostelium* has been examined by using isoschizomers of restriction enzymes that can and cannot cut specific methylated DNA sequences (Bird and Southern, 1978). Specifically *Dictyostelium* DNA has been digested with Sau 3A, which recognizes GATC whether or not the 6 position of adenine is methylated, and Mbo I, which will only cleave the unmethylated sequence. Hpa I and Msp I enzymes were also used. Hpa (Hap) II recognizes CCGG and cuts whether or not the external C is methylated but not if the internal C ($C^{m}CGG$) is methylated; Msp I recognizes the same sequence but will cut if the internal C is methylated (McClelland, 1981). No differences were observed in the extent of digestion of total genomic DNA, in the digestion pattern of the rRNA cistrons, or in Southern blots probed with the M4 repeat sequence (see below) or with actin sequences using isoschizomers (Kimmel and Firtel, 1979; Kindle and Firtel, 1978; R. A. Firtel, unpublished observation). These results suggest that there is no substantial methylation of these sequences in DNA from *Dictyostelium* vegetative cells. If DNA methylation occurs in *Dictyostelium* genomic DNA and is involved in gene expression, it may occur at other sequences.

B. The Mitochondrial Genome

Dictyostelium mitochondrial DNA renatures with homogeneous second-order kinetics indicating a nucleotide complexity of ~50 kb (Firtel and Bonner, 1972a). Buoyant density centrifugation, thermal denaturation, and restriction endonuclease analysis indicates that there is a relatively uniform (28%) G + C distribution throughout the mitochondrial genome (Firtel and Bonner, 1972a; Firtel *et al.*, 1976a). Visualization of gently lysed cells by electron microscopy shows a circular mitochondrial genome with a contour length approximately equal to that predicted by the nucleotide complexity measurements (Cockburn *et al.*, 1978).

III. General Patterns of mRNA Transcription and Maturation

Pulse-labeled *Dictyostelium* RNA is heterogeneous in size with a mean sedimentation value of 13–15 S (Firtel and Lodish, 1973; Firtel *et al.*, 1976b). When rRNA labeling is inhibited with low concentrations of actinomycin D, >90% of the newly synthesized RNA binds to poly(U)-Sepharose. Thus, most of these molecules contain a poly(A) sequence. Polysomal or cytoplasmic poly$(A)^+$ RNA has an average size of 1.2 kb; nuclear poly$(A)^+$ RNA is ~25% larger. Approximately 25% of the mass of total heterogeneous nuclear RNA (hnRNA) sequences are transcribed from repeated DNA while the value for mRNA is only ~8%. Nevertheless, ~30–40% of the hnRNA and mRNA molecules contain sequences complementary to repetitive DNA. Pulse-chase experiments indicate that 75% of the label in hnRNA is converted to cytoplasmic mRNA (Firtel and Lodish, 1973; Lodish *et al.*, 1973; Firtel *et al.*, 1976b). When hnRNA is hybridized to a sequence excess of complementary DNA (cDNA) from cytoplasmic poly$(A)^+$ RNA, only 75% is RNase resistant. However, if hybridization is assayed by hydroxylapatite (HAP) chromatography, more than 90% of the hnRNA forms RNA:DNA duplexes. These data indicate that there are sequences present in hnRNA that are not present in mRNA and suggest that these hnRNA specific sequences are removed to yield mature mRNA (Firtel and Lodish, 1973; Lodish *et al.*, 1973; Firtel *et al.*, 1976b; Firtel and Jacobson, 1977).

Many eukaryotic genes are not colinear with their respective mRNAs. The genes contain intervening sequences that are transcribed into hnRNA and then removed during mRNA maturation (Breathnach and Chambon, 1981). Some of the hnRNA-specific sequences in *Dictyostelium* that are not transported to the cytoplasm are introns (Kimmel and Firtel, 1980). In *Dictyostelium* these introns are short (~100 nucleotides) and extremely (95%) A + U rich whereas those found in other organisms can be appreciably larger and more similar in G + C content to the protein coding regions. Nonetheless the intron junction sequences in *Dictyostelium* are very similar to those of other eukaryotes (Table I).

TABLE I
Dictyostelium Intron Junction Sequences[a]

		INTRON			
Breathnach *et al.* (1978)	↓	GU	. . .	AG	
M4 Intron I	UAA ↓	GUUUGU	. . .	U U CA AG ↓	AUU
M4 Intron II	AAA ↓	GUAUGU	. . .	U A UC AG ↓	GGA
M3L Intron I	AAU ↓	GUAAGC	. . .	A A AA AG ↓	UUA
M3R Intron I	AAU ↓	GUAAGC	. . .	A A AA AG ↓	UUA
M3L Intron II	CGU ↓	GUAUGU	. . .	U C UC AG ↓	GGU
M3R Intron II	CGU ↓	GUAAGU	. . .	U C UC AG ↓	GGU
Seif *et al.* (1979)	Pu ↓	GUXXGU	. . .	PyPyXPyAG ↓	

[a]*Dictyostelium* intron junction sequences are compared to the consensus sequences established by Breathnach *et al.* (1978) and Seif *et al.* (1979). The arrows indicate the assumed position of the intron splice. Since several junctions are located between duplicated nucleotides, the exact position of the splice points cannot be determined.

There are two size classes of poly(A) in mRNA and hnRNA (Jacobson *et al.*, 1974a,b; Lodish *et al.*, 1973; Firtel *et al.*, 1976b). One is ~25 nucleotides (nt) in length and is encoded by oligo(dT) sequences in the genome; a second, longer poly(A) stretch is added posttranscriptionally. This longer poly(A) is synthesized as ~125 nt but its length gradually diminishes as the mRNA ages to a steady state average size of ~60 nt (Palatnik *et al.*, 1978). In cytoplasmic poly(A)$^+$ RNA the two size classes are present in equal molar ratios (Firtel *et al.*, 1976b; Firtel and Jacobson, 1977). On average, each mRNA would contain one of each class. Heterogeneous nuclear RNA contains ~4–5 moles of poly$(A)_{25}$ per mole of poly$(A)_{125}$. Thus, although most of the hnRNA contains the transcribed poly(A) fraction only ~20–25% of the steady state hnRNA has the posttranscriptionally added poly$(A)_{125}$. RNase digestion studies indicate that the two poly(A) tracts are linked within several nucleotides of each other at the 3′ terminus and are conserved during mRNA maturation (Jacobson *et al.*, 1974a,b; Firtel *et al.*, 1976b; Firtel and Jacobson, 1977). Studies using isolated genes have confirmed their location and linkage (McKeown and Firtel, 1981b). However, not all mRNAs are organized in this manner. Some mRNAs possess only the posttranscriptionally added poly$(A)_{125}$ whereas others may have only an oligo(A) tract (Palatnik *et al.*, 1979).

There is also a specific class of mRNAs that does not bind to poly(U) Sepharose or oligo(dT) cellulose and lacks poly(A) (Palatnik *et al.*, 1979). This fraction primarily encodes the histone proteins as assayed by *in vitro* translation followed by gel electrophoresis. The histone mRNAs in most other systems also lack poly(A) (Kedes, 1979). There are other proteins in *Dictyostelium* that are encoded by the poly(A)$^-$ mRNA fraction, but these proteins are also encoded by the poly(A)$^+$ fraction. Thus, it is not known if a specific population of the mRNAs encoding these proteins is devoid of poly(A) sequences or if these mRNAs do not bind to poly(U) Sepharose very efficiently. In other systems, there is a set of nonhistone proteins that are specifically encoded by poly(A)$^-$ mRNAs as well as mRNAs that are found in poly(A)$^-$ and poly(A)$^+$ fractions (Nemer *et al.*, 1974; Kaufman *et al.*, 1977; van Ness *et al.*, 1979; Brandhorst *et al.*, 1979; Minty and Gross, 1980; Duncan and Humphries, 1981).

Several groups have measured the half-lives of total *Dictyostelium* mRNAs and obtained different values. Lodish's group has observed a total mRNA half-life of 3.5–4 h in vegetative and developing cells (Margolskee and Lodish, 1980a,b). Jacobson and co-workers reported that vegetative mRNA decays with biphasic kinetics indicating half-lives of 2 and 10 h (Palatnik *et al.*, 1980). Both groups used drugs to inhibit RNA synthesis and followed the decay of mRNA by *in vitro* translation; however Lodish's group used actinomycin D and daunomycin to inhibit RNA synthesis whereas Jacobson's group used only actinomycin D (Firtel and Lodish, 1973; Firtel *et al.*, 1973). Jacobson and co-workers have additionally confirmed their results using pulse-chase experiments and poly(U)-Sepharose affinity chromatography to assay the decay of mRNA (Palatnik *et al.*, 1980). In general, there are multiple components to the mRNA decay curve in other eukaryotes (Singer and Penman, 1972; Greenberg, 1972; Singer and Penman, 1973; Puckett *et al.*, 1975).

The relationship between mRNA stability and poly(A) length has also been examined. The steady-state length of the poly(A) tracts of specific mRNAs is longer than that of other mRNAs. However, the mRNAs with these shorter poly(A) tracts have the longer half-lives. Thus, a long poly(A) tract does not confer stability to an mRNA in *Dictyostelium*. Since poly(A) decreases in length as mRNA ages, the more stable mRNAs, in general, have shorter poly(A) tracts. It has also been observed that in both *Dictyostelium* and mammalian cells the smaller mRNAs are generally more stable than the larger mRNAs, perhaps as the result of their smaller target size (Palatnik *et al.*, 1980, Meyuhas and Perry, 1979). Finally, Jacobson (personal communication) has suggested that *Dictyostelium* poly(A) may be involved in translational control.

As in other eukaryotes, the 5′ ends of hnRNA and mRNA in *Dictyostelium* are posttranscriptionally modified by the addition of guanosine via a 5′–5′ phosphodiester linkage; the terminal G residue is subsequently methylated at position 7 (see Shatkin, 1976). The resulting 5′ modification is called a cap structure. Approximately 85% of the *Dictyostelium* hnRNA and mRNA molecules possess M^7GpppA caps and ~10% terminate with M^7GpppG (Dottin *et al.*, 1976). The 5′ caps in mRNA may be further modified; ~25% of the M^7GpppA termini possess a 2′-O-methylated penultimate A residue. In mammalian cells a larger portion of the caps possess 2′-O-methylated penultimate nucleotides as well as 2′-O-methylation of the adjacent residue. In *Dictyostelium*, the distribution of sequences immediately adjacent (3–5 nt) to the M^7GpppA is identical in nuclear and cytoplasmic poly$(A)^+$ RNA (Weiner, personal communication). These results are consistent with the 5′ end of hnRNA being conserved as the 5′ end of mRNA. It has been suggested that the first transcribed nucleotide in hnRNA becomes capped (see Breathnach and Chambon, 1981). Since the 5′-hnRNA cap may be conserved during mRNA maturation the 5′ end of the mRNA would, thus, define the transcription initiation sequence.

IV. Genes Encoding Abundant Stable RNAs

A. The Ribosomal RNA Genes

There are four RNAs (26 S, 17 S, 5.8 S, and 5 S) found in cytoplasmic ribosomes. These rRNAs represent at least 90% of the total RNA mass of the cell and as such can be easily purified and characterized. The organization and transcription of the genes that encode these RNAs have also been studied.

Restriction endonuclease mapping and DNA blot hybridization studies were originally used to describe the organization of the *rRNA* genes (Cockburn *et al.*, 1976; Maizels, 1976; Frankel *et al.*, 1977; Cockburn *et al.*, 1978; Sogin and Olsen, 1980). These genes are located on ~90 kb linear, extrachromosomal, palindromic molecules; there are ~90 rDNA dimers per haploid genome. The entire ribosomal repeat comprises nearly 18% of the nuclear DNA mass and restriction endonuclease fragments derived from the rDNA dimers are easily observed against the heterogeneous distribution of the chromosomal DNA fragments (Firtel *et al.*, 1976a; Cockburn *et al.*, 1976; Maizels, 1976). A photograph of restriction endonoclease digests of *Dictyostelium* DNA and a restriction map of the rDNA are shown in Figs. 1 and 2.

Each rDNA half repeat contains two transcription units. One transcription unit encodes the 36 S ribosomal RNA precursor, which is processed in a

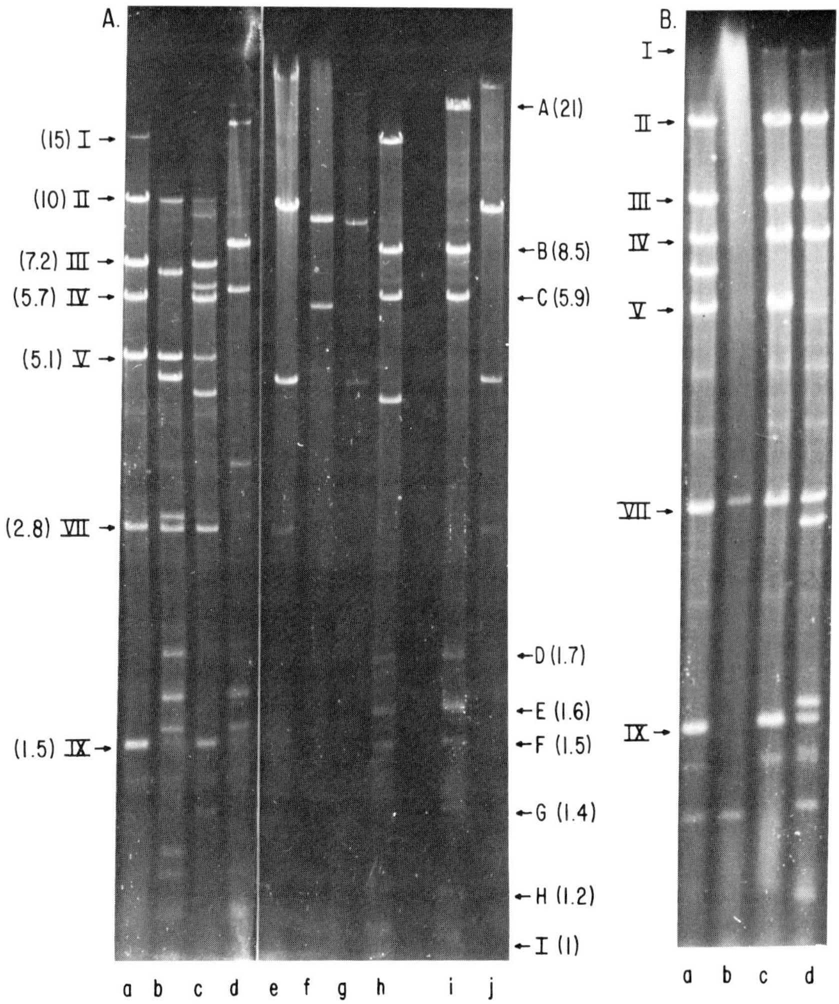

Fig. 1. Restriction nuclease digestions of *Dictyostelium* DNA. *Dictyostelium* nuclear DNA was cleaved and electrophoresed on 0.8% agarose gels. The numbers in parentheses are the sizes of the DNA fragments in kb. See Cockburn *et al*. (1978). **A.** a. Eco RI. b. Eco RI + Hind III. c. Eco RI + Sal I. d. Hind III + Sal I. e. Sal I. f. Pst I. g. Pst I + Sal I. h. Hind III + Pst I. i. Hind III. j. Sal I + Bam HI. **B.** a. Eco RI + Bam III. b. Bam III. c. Eco RI. d. Eco RI + Pst I. Capital letters refer to Hind III restriction fragments from rDNA. Roman numerals refer to ECO RI bands.

series of specific steps to yield the mature 26 S, 17 S and 5.8 S rRNAs. The 5 S *rRNA* is encoded by an entirely separate transcription unit also located on the rDNA palindrome. The coding regions are ~45–50% G + C whereas the untranscribed spacer regions, which comprise nearly 80% of the rDNA

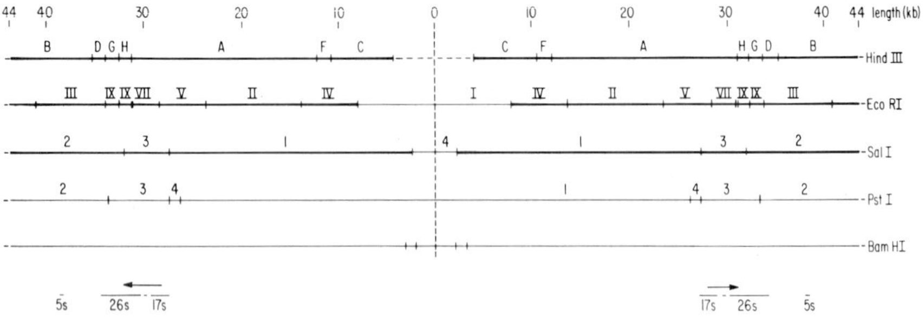

Fig. 2. Restriction map of the palindromic rDNA repeat dimer. The numbering of the bands for Eco RI and Hind III digests corresponds with that shown in Fig. 1. The position and direction of transcription of the ribosomal RNA cistrons are shown. The polarity of transcription of the 5 S gene is not known. The terminal fragments are heterogeneous in length (Emery and Weiner, 1981).

dimer vary between 22 and 28% G + C (Jacobson *et al.*, 1974a; Firtel *et al.*, 1976a).

The processing sequence of the 36 S *rRNA* has been delineated by comparing the base compositions, methylation patterns and oligonucleotides of the mature rRNAs with the various rRNA precursors (Frankel *et al.*, 1977; Batts-Young *et al.*, 1977; Batts-Young and Lodish, 1978; Batts-Young *et al.*, 1980). In addition, by using DNA blot hybridizations and electron microscopic techniques, it has been possible to determine the polarity and location of the coding regions on the rDNA molecules (Taylor *et al.*, 1977; Cockburn *et al.*, 1976; Maizels, 1976; Frankel *et al.*, 1977; Cockburn *et al.*, 1978). The 36 S and 5 S *rRNA* genes are closely linked to each other near the ends of the rDNA dimers. The 5′ end of the 36 S *rRNA* gene contains the 17 S rRNA sequences and lies closest to the center of the rDNA palindrome. Visualization of actively transcribing rDNA molecules by electron microscopy has confirmed the palindromic structure and further suggests that some of the 36 S rRNA molecules may be processed before transcription has been completed (see Fig. 3; Grainger and Maizels, 1980).

The 5 S rRNA coding sequences lie 3′ to the 36 S *rRNA* genes; however, the precise location and direction of transcription of the 5 S *rRNA* genes have not been ascertained (Cockburn *et al.*, 1978). The entire sequence of the *Dictyostelium* 5 S *rRNA* has been determined (Hori *et al.*, 1980). Its primary sequence is more similar to that of animals (62% average nucleotide conservation) than that of yeast (56% conservation) or plants (49% conservation). The secondary structure of the *Dictyostelium* 5 S RNA has been predicted from its primary sequence. Its structure is similar to that of all known eukaryotic 5 S rRNAs and possesses at least four evolutionarily conserved hairpin regions.

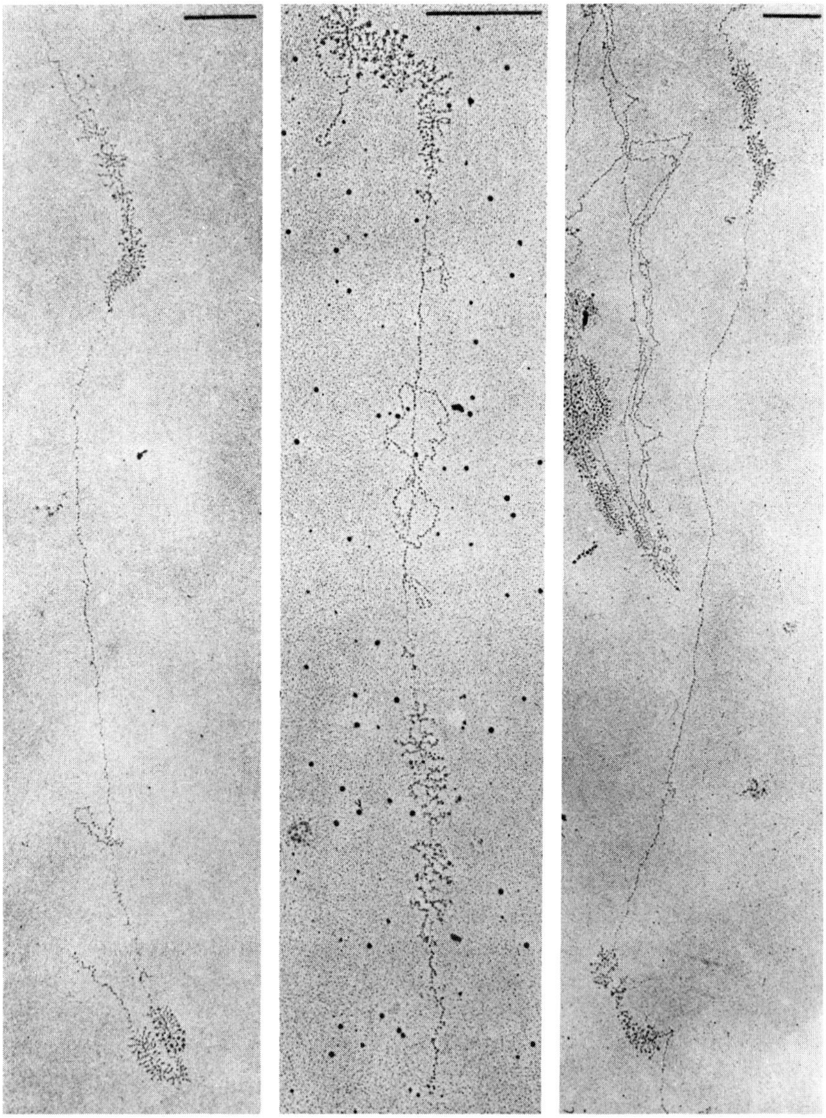

Fig. 3. Electron micrographs of transcription complexes on the extra-chromosomal *Dictyostelium* palindromic rDNA dimers. Each molecule has two transcription units with divergent polarities from the center of the rDNA. Photographs were kindly provided by R. Grainger.

Hae III

↓

pCT1 ggtggatccatatactccoccccctctg$\mathrm{TC_7TCTC_8TC_2TC_7TCTC_6TC_2TC_5TC_6TC_2TC_6TC_6TCTC_8TC_2TC_6TC_7}$

pCT15 ....$\mathrm{TCTC_8TC_2TC_7TCTC_6TC_2TC_5TC_6TC_2TC_6TC_6TCTC_7TC_7TC_6TC_6TCTC_6TC_2TC_3TC_7TC_2TC_7TC_2TC_4}$

pCT14 ....ccccccctctg$\mathrm{TC_7TCTC_8TC_2TC_7TCTC_6TC_2TC_5TC_6TC_2TC_6TC_6TCTC_7TC_7TC_6TC_6TCTC_6TC_2TC_3TC_6TC_6T}$

pCT11 ....$\mathrm{TC_2TC_5TC_6TC_2TC_6TC_6TCTC_7TC_7TC_6TC_6TCTC_6TC_2TC_3TC_6TC_6TCTC_6TC_2TC_3TC_6TC_6TCTC_5}$

pCT8 ....$\mathrm{TCTC_6TC_2TC_5TC_6TC_2TC_6TC_6TCTC_7TC_7TC_6TC_6TCTC_6TC_2TC_3TC_5TC_6TCTC_6TC_2TC_3TC_6TC_6TCTC_7TC_2TC_5}$

Fig. 4. Irregular C_nT-satellite sequences near the rDNA termini. Regions near the termini of the extrachromosomal palindromic rDNA molecules were cloned by Emery and Weiner (1981). Five of these clones were sequenced. The sequence from the Hae III restriction site near the rDNA termini to the end of the C_nT repeat sequence on these clones is shown. The region between the Hae III restriction site and the start of the C_nT repeat sequence is indicated in lower case letters. It is not known if the end of the C_nT repeat is the actual termini of these rDNA molecules (see Emery and Weiner, 1981 for details). Dotted lines indicate sequence is same as pCT1.

Tetrahymena and *Physarum* also possess linear extrachromosomal palindromic rDNAs (Karrer and Gall, 1976; Engberg *et al.*, 1976; Vogt and Brown, 1976). The organization and polarity of transcription of the genes for the large ribosomal precursor in *Tetrahymena* and *Physarum* are similar to *Dictyostelium.* However, the 5 S *rRNA* genes in *Tetrahymena* and *Physarum* are not located on the free rDNA dimers as in *Dictyostelium* but are integrated in the chromosomal DNA in tandem arrays as in most other eukaryotes (see Kimmel and Gorovsky, 1978; Long and Dawid, 1980). Yeast *rRNA* genes are present in chromosomal tandem repeats that contain one 17 S and 25 S gene and one 5 S gene per repeat unit (Rubin and Sulston, 1973; Long and Dawid, 1980).

The termini of the *Tetrahymena* and *Physarum* rDNAs are heterogeneous in length as the result of a variable copy number of a very regular simple sequence (Blackburn and Gall, 1978; Johnson, 1980). The *Tetrahymena* repeat is C_4A_2 and *Physarum* rDNA possesses C_3TA terminal repeats. The rDNA termini of both *Tetrahymena* and *Physarum* are resistant to various enzymatic activities. It has been suggested that *Tetrahymena* rDNAs possess hairpin structures at their termini. In *Tetrahymena* and *Physarum* there are gaps in the terminal region of the rDNA, resulting in single-stranded regions.

The rDNA termini in *Dictyostelium* have been analyzed (Emery and Weiner, 1981). The termini do not possess a free 5′-phosphate group that is accessible to alkaline phosphate and/or polynucleotide kinase, but it is not known if *Dictyostelium* rDNA termini possess hairpins, nicks, or single-stranded regions. The termini of the rDNA molecules are heterogeneous in length as in *Physarum* and *Tetrahymena;* however, the *Dictyostelium* termini possess an irregular simple sequence repeat, C_nT. Although the pattern of the C_nT irregularity is almost perfectly conserved among the rDNA molecules, some terminal sequences are shorter than others, probably accounting for the length heterogeneity observed in the terminal Eco RI restriction fragment. The rDNA terminal sequences are shown in Fig. 4. The terminal C_nT sequences in *Dictyostelium* are only found in association with the rRNA genes, whereas in *Tetrahymena* and *Physarum* the terminal rDNA satellite sequences are located in many regions of the genomic DNA (Blackburn and Gall, 1978; Johnson, 1980). In contrast to the length heterogeneity at the termini, the remainder of the rDNA in *Dictyostelium* is homogeneous as defined by restriction endonuclease mapping (Cockburn *et al.*, 1978). Also when the central restriction fragments are denatured and quick-cooled they form hairpin structures that are resistant to single-stranded specific (S_1) nuclease. It should be noted that in *Tetrahymena* the central 25–30 bp of the rDNA are nonpalindromic (Kiss and Pearlman, 1981; Kan and Gall, 1981).

At the present time, the function of the terminal C_nT sequence is unknown. In *Tetrahymena* and *Physarum*, DNA replication originates near the center of the palindrome and proceeds bidirectionally toward the termini (Truett and Gall, 1977; Cech and Brehm, 1981; Vogt and Braun, 1977). By analogy, it may be that the C_nT sequences in *Dictyostelium* are essential to proper termination of replication of the rDNA palindromes at their termini. Alternatively, the C_nT sequence may be part of a mechanism to generate and amplify an extrachromosomal rDNA copy from an integrated copy in the genome. In *Tetrahymena* there is a single integrated rDNA copy in the germinal (micro) necleus (Yao and Gall, 1977). It is possible that there is also a single integrated rDNA in *Dictyostelium;* however, none has been identified. A detailed comparison of the similarities and differences among *Dictyostelium, Tetrahymena,* and *Physarum* rDNAs will help to explain the function of the structures associated with these palindromic sequences.

B. Transfer RNA Genes

The *Dictyostelium* genome possesses 800–1200 *tRNA* genes per haploid genome (Peffley and Sogin, 1981). In general, the *tRNA* genes are not found in clusters although there may be some *tRNA* loci that are very closely linked. Genomic DNA fragments that hybridize to total *Dictyostelium* tRNA have been isolated. Subsequent DNA sequence analysis of two of these fragments indicates that if the *tRNA* regions were transcribed, one would encode a tryptophan tRNA and the other would encode a valine tRNA.

From analysis of DNA blot hybridizations using probes from the $tRNA^{Trp}$ coding region, it has been suggested that there are at least six $tRNA^{Trp}$ genes in *Dictyostelium.* Sequences 3′ to the isolated $tRNA^{Trp}$ sequence are single-copy while sequences immediately 5′ to the putative coding region are repeated many times in the genome. It has been suggested that this repeat sequence is also associated with many other *Dictyostelium tRNA* genes.

If this $tRNA^{Trp}$ gene were transcribed, it would yield an RNA that would not be colinear with its predicted nucleotide *tRNA* sequence but would contain a 13 bp intron within the anticodon loop. It will be of interest to determine if this gene is actually transcribed and if the putative *tRNA* precursor is correctly processed in *Dictyostelium.* Certain *tRNA* genes from other organisms (e.g., yeast, *Xenopus*) contain introns within their anticodon loops that are transcribed and subsequently removed to yield mature tRNAs (Abelson, 1979).

The isolated $tRNA^{Val}$ gene has not yet been analyzed in fine detail although it is known that it does not possess intervening sequences. Other DNA fragments have also been identified that may contain more than one tRNA coding region; however, it is not yet known which tRNAs they encode.

Transfer RNA and tRNA synthetases have been isolated from vegetative and culminating cells and used in homologous *in vitro* amino acid-accepting systems in order to examine the developmental expression of *Dictyostelium* tRNAs (Palatnik *et al.*, 1977). No differences in the level of acceptance of 17 amino acids could be detected between tRNAs from these two developmental stages. Isoaccepting species were also compared by reverse-phase chromatography. No detectable changes in the level of individual tRNAs during development were observed although there were some differences in the pattern of posttranscriptional modifications.

C. The *D2* Gene Family and Other Small Nuclear RNAs

There are several families of small abundant RNAs in *Dictyostelium. D1*, *D2*, and *D3* RNAs are nearly entirely sequestered in the nucleus whereas a heterogeneous group of 5.6 S RNAs are found in the nucleus as well as in the cytoplasm (Wise and Weiner, 1981). The 5.6 S RNA series has not been studied in much detail. However, it has been possible to purify each of the D-series snRNAs (small nuclear RNAs) and perform partial RNA sequence characterizations. *D1*, *D2*, and *D3* possess 5′ cap structures. It has been suggested that all capped RNAs are transcribed by RNA polymerase II (Jensen *et al.*, 1979; Roop *et al.*, 1981; Tamm *et al.*, 1980). The noncapped small RNAs (e.g., 5 S RNA, tRNA) are transcribed by RNA polymerase III.

Small RNAs have been isolated from mammalian nuclei that are similar to the *Dictyostelium* snRNAs with regard to size, base modification, and 5′-cap sequence pattern (see Wise and Weiner, 1980). The snRNAs are ~50 times less abundant per cell in *Dictyostelium* than in mammals. There also appear to be many less snRNA genes in *Dictyostelium* than in mammalian cells. It has been suggested that the *U1* snRNA in mammalian cells plays a role in processing introns during mRNA maturation (see Breathnach and Chambon, 1981). Perhaps the low abundance of the *Dictyostelium* snRNAs reflects a less complex pattern of hnRNA processing relative to that of mammalian cells.

The *D2* snRNA genes have been studied in the greatest detail (Wise and Weiner, 1980). There are six *D2* snRNA loci, which encode an RNA set that is slightly heterogeneous in sequence. One of the *D2* genes has been isolated and characterized and is not closely linked to any other snRNA gene. Sequence analysis indicates that the *D2* RNA is ~40% homologous to the mammalian nucleolar *U3* snRNA. *D2* RNA can be isolated from a nucleolar enriched subcellular component of *Dictyostelium*. The function of the *U3* and *D2* snRNAs is not yet known.

The *D2* gene that has been isolated codes for one of the minor sequence variants of the *D2* RNA series (Wise and Weiner, 1980). It does not possess intervening sequences. Some of the sequences that are 5′ to the initiation of transcription are similar to those found associated with mRNA genes (see Fig. 32). This is of considerable interest since the *D2* snRNA and mRNA possess 5′-cap structures and are probably both transcribed by RNA polymerase II. We suggest that the TATA box and the oligo(dT) stretch that are 5′ to the *D2* coding region represent conserved recognition structures for RNA polymerase II in *Dictyostelium*. The possible relationship between the TATA box and the oligo(dT) stretch in pol II transcription (Kimmel and Firtel, in preparation) is discussed in detail later in this chapter.

V. Patterns of Developmental Gene Expression

The existence of developmental stage-specific enzymes in *Dictyostelium* has been well documented (Loomis, 1975). Many enzymatic activities (e.g., UDPgalactose epimerase, glycogen phosphorylase) are undetectable in vegetative amoebae but appear at specific developmental stages and rise rapidly to characteristic maxima. Generally, the appearance of these enzymes is dependent upon *de novo* mRNA and protein synthesis (Sussman and Sussman, 1969; Roth *et al.*, 1968; Firtel *et al.*, 1973). In addition, there are ~400 moderately abundant proteins observed by two-dimensional (2D) gel electrophoresis that are synthesized at various developmental stages. Approximately 100 of these polypeptides show large changes in their relative rates of synthesis during development (Alton and Lodish, 1977a,b,c; Alton and Brenner, 1979; Morrissey *et al.*, 1981). In early development, a few major changes are observed. Discoidins I and II and Contact Site A are examples of proteins that are found in preaggregated cells but not in NC4 vegetative amoebae (Rosen *et al.*, 1973; Siu *et al.*, 1976; Ma and Firtel, 1978; Gerisch, 1968; Gerisch *et al.*, 1975). During early development there is also a decrease in the relative rates of synthesis of ~10 proteins. The major developmental changes in protein synthesis, which can be observed by 2D gel electrophoresis, occur later in development. Approximately 40 new proteins are synthesized after cell contact is established. There are also prespore and prestalk specific sets of proteins observed later in development at the pseudoplasmodia stage (Alton and Brenner, 1979; Morrissey *et al.*, 1981). It should be noted that less than 10% of the total number of gene products, as determined by hybridization experiments, are observable by 2D gel electrophoresis of *in vivo* or *in vitro* synthesized proteins.

Several abundancy classes of poly(A)$^+$ RNA are observed throughout development. As found in other eukaryotes (Bishop *et al.*, 1974), there are three mRNA classes (abundant, moderately abundant, and low abundance) which contain, respectively, ~500, ~50, and ~5 mRNA molecules per cell. The last class contains most of the mRNA complexity and would encode proteins that are present at too low a level to be easily identified on two-dimensional gels. Poly(A)$^+$ nuclear RNA contains an additional very high complexity class whose RNAs are present at fewer than one copy per cell (Firtel and Jacobson, 1977; Blumberg and Lodish, 1980a).

The pattern of developmental RNA expression has been examined by two groups. Although both sets of results are, in general, qualitatively similar, there are some significant differences between them. Firtel has used RNA excess hybridization kinetics to quantitate gene expression (Firtel, 1972; Firtel and Jacobson, 1977; Firtel, unpublished observations). Cytoplasmic or nuclear poly(A)$^+$ RNA from specific developmental stages was hybridized in excess to purified total *Dictyostelium* genomic single-copy DNA and to cDNAs made from poly(A)$^+$ RNAs isolated from cells at various developmental stages. The results indicated that there are ~3600 cytoplasmic poly(A)$^+$ RNA sequences and an additional ~5000 poly(A)$^+$ nuclear RNA sequences in vegetative cells. RNA complexity changes were observed at aggregation, pseudoplasmodial, and culmination stages and it was concluded that 1500–2500 new RNA species appeared during development. These developmental-specific RNAs cannot be detected in cytoplasmic or nuclear poly(A)$^+$ RNA from vegetative cells. The levels of these RNAs during development are probably regulated at the level of transcription. Additionally, there are some vegetative sequences that were not detectable in culminating cells. Hybridization kinetics with cDNA probes made from total vegetative nuclear RNA indicated that many vegetative transcripts are not detected in nuclear RNA from late developing cells. These results indicated that a fraction of genes expressed during vegetative growth are not expressed at a detectable level during culmination. Finally, the results indicated that ~56% of the total single-copy fraction of the *Dictyostelium* genome is asymmetrically expressed between vegetative growth and culmination.

Also using RNA excess hybridization to single-copy and cDNA probes, the Lodish group has reported that a larger number of RNA sequences are expressed during development and that a greater percentage of the genome is transcribed (Blumberg and Lodish, 1980a,b; 1981). These results indicated that approximately 4000–5000 mRNA sequences are present in vegetative cells and that there are ~9000 nuclear specific sequences. At the time of formation of tight cell–cell contacts, ~3000 new mRNAs and ~4000 new nuclear RNAs were detected. No additional changes in the RNA complexity at either the pseudoplasmodial or culmination stages were observed. Their results indicated that > 80% of the single-copy fraction of the genome is expressed between vegetative growth and culmination. These workers also did not detect significant changes in the RNA complexity prior to the formation of the tight cell–cell contacts.

The results of both groups are qualitatively similar, but it is not clear why they obtained such quantitative differences. Examination of the data indicates that both sets of hybridization reactions terminate at similar R_0t (RNA concentration × time) values.

The majority of the sequences surrounding the genes that have been examined to date are ~95% A + T, and it is reasonable to argue that they are not likely to be transcribed. From the average spacing of genes on cloned fragments, one would estimate that only between 40 and 60% of the genome is transcribed into RNA sequences detectable at a level of greater than one copy per cell. From these data, the value of 80% transcription of the single-copy DNA may seem unlikely. It has been reported in other systems that there may be transcription beyond the 3′ poly(A) addition site (Nevins *et al.*, 1980). In these systems the hnRNA transcript is subsequently processed generating a nuclear specific 3′-downstream sequence. If this occurred in *Dictyostelium* it may account for the unexpectedly high levels of transcription observed. It should also be noted that specific early developmentally regulated proteins, which are not detectable in vegetative cells from certain *Dictyostelium* strains, are detectable during log phase growth in other strains (Ashworth and Quance, 1972; Burns *et al.*, 1981). In addition, some genes that are normally expressed at specific developmental stages in certain strains are not expressed in other strains (Brandis and Firtel, unpublished observations). It is unlikely however, that these strain variations can account for the large quantitative differences between the data of the two groups.

A. Methods for the Isolation of Developmentally Regulated Genes

Two differential hybridization schemes have been used to select genes that are preferentially active at one developmental period. The first procedure is outlined in Fig. 5 (Rowekamp and Firtel, 1980; Williams and Lloyd, 1979). A cDNA library is constructed from poly(A)$^+$ RNA isolated from a particular developmental stage. Replicates of these recombinant clones are hybridized with ^{32}P-labeled (*in vivo* or *in vitro*) poly(A)$^+$ RNA from vegetative cells or from cells at various developmental time periods. Clones that hybridize preferentially to RNA from a particular developmental

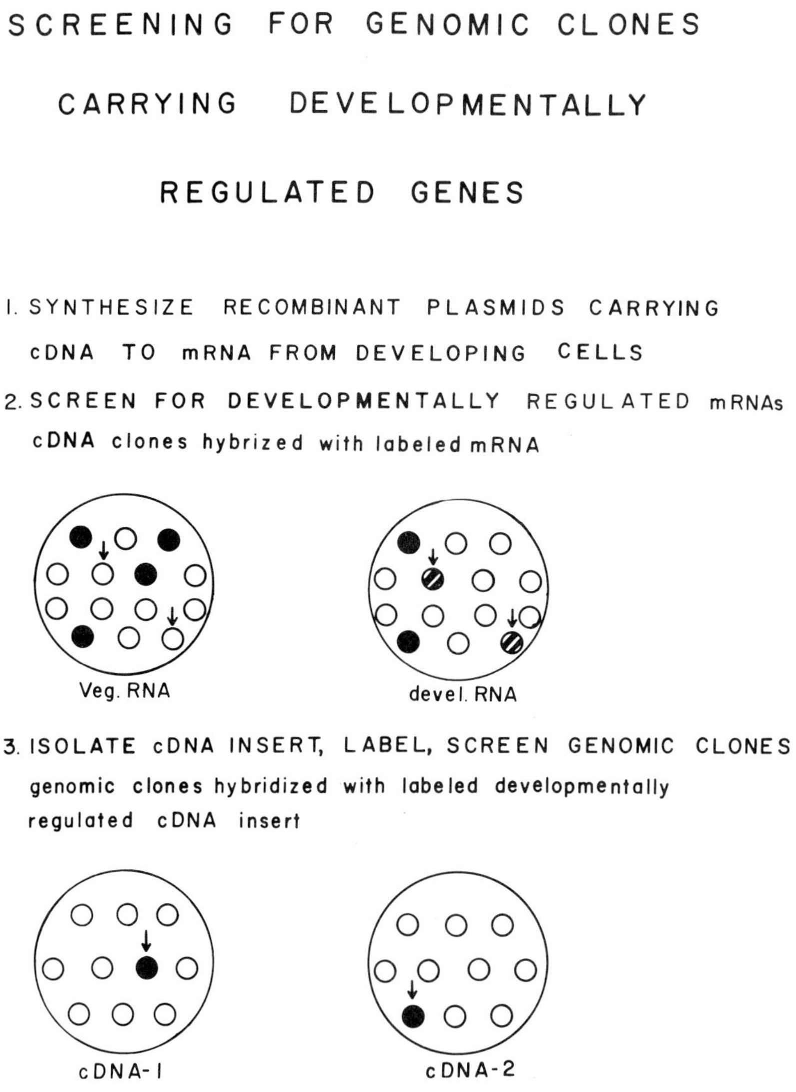
SCREENING FOR GENOMIC CLONES
CARRYING DEVELOPMENTALLY
REGULATED GENES
1. SYNTHESIZE RECOMBINANT PLASMIDS CARRYING
cDNA TO mRNA FROM DEVELOPING CELLS
2. SCREEN FOR DEVELOPMENTALLY REGULATED mRNAs
cDNA clones hybrized with labeled mRNA
Veg. RNA
devel. RNA
3. ISOLATE cDNA INSERT, LABEL, SCREEN GENOMIC CLONES
genomic clones hybridized with labeled developmentally
regulated cDNA insert
cDNA-1
cDNA-2

Fig. 5. Protocol for screening cDNA clones for developmentally regulated genes. Duplicate filters of cDNA clones are hybridized with RNA from either vegetative or developing cells and assayed by autoradiography. Clones complementary to RNA present at >0.05% of the poly(A)$^+$ RNA at a particular developmental stage can be detected. Four types of hybridization are observed: (a) clones complementary to messenger RNA expressed in vegetative cells but not developing cells; (b) clones complementary to mRNA expressed in developing cells but not in vegetative cells; (c) clones complementary to messenger RNA expressed in both vegetative and developing cells; (d) clones complementary to RNA whose abundance is below the level of detection (see Rowekamp and Firtel, 1980 for details).

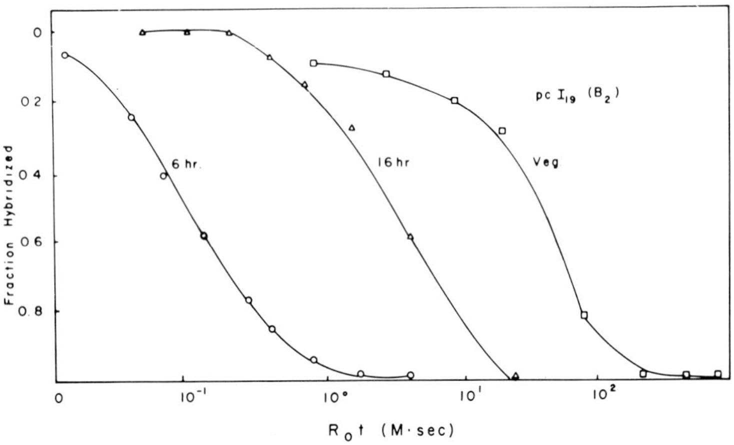
pc I_{19} (B_2)
6 hr.
16 hr
Veg
Fraction Hybridized
R_0t (M·sec)

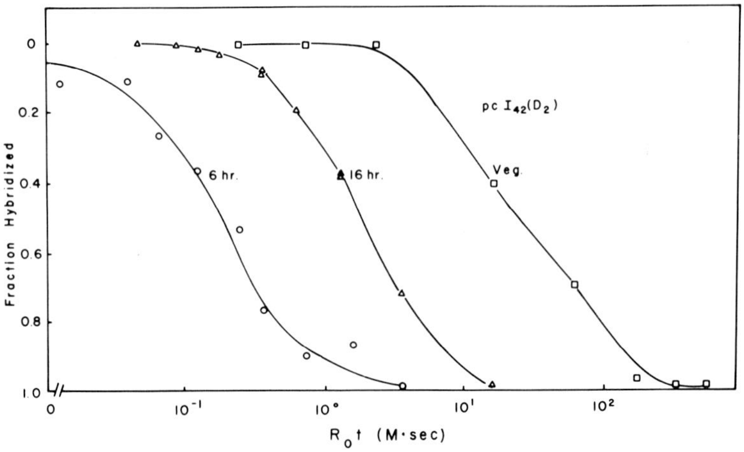
pc I_{42}(D_2)
6 hr.
16 hr.
Veg.
Fraction Hybridized
R_0t (M·sec)

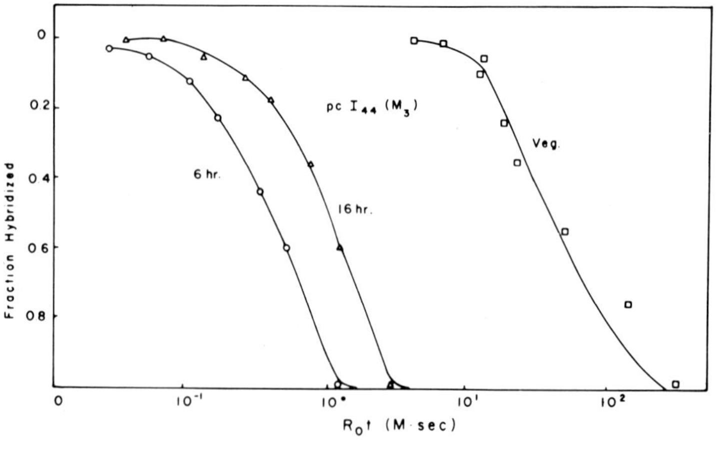
pc I_{44} (M_3)
6 hr.
16 hr.
Veg.
Fraction Hybridized
R_0t (M·sec)

cell type are presumed to be derived from genes that are developmentally regulated. Also, a series of clones can be selected that hybridize with equal efficiency to both vegetative and developmental RNA, and are presumably derived from genes that are constitutively expressed throughout development. Using this procedure, cDNA clones have been isolated that are derived from genes that appear to be maximally expressed at 6-, 15-, or 20-h stages of development.

A second protocol has been developed that is useful for isolating genomic clones possessing developmentally regulated genes (Mangiarotti *et al.*, 1981). In this procedure, a stage specific poly(A)$^+$ RNA is labeled *in vivo* or *in vitro* and hybridized in the presence of ~100-fold mass excess of vegetative poly(A)$^+$ RNA to recombinant phage or plasmid DNA immobilized on filters. Using this procedure, it is possible to identify genomic fragments carrying both developmentally regulated as well as constitutively expressed genes.

There are several procedures that can be used to confirm that a DNA fragment identified by the differential hybridization selection scheme is actually developmentally expressed. One can quantitate the RNA from a particular gene by RNA excess hybridization kinetics wherein the rate of hybridization is directly proportional to the concentration of the RNA. If a gene were expressed at a higher level at one developmental stage than another, the gene probe would hydridize at a lower R_0t value to RNA isolated from that stage in development. Additionally, from the R_0t value the fraction of a particular RNA of total RNA can be calculated. Several developmentally regulated genes, including the *M3* and *discoidin I* families, have been analyzed by this procedure (Fig. 6; Rowekamp and Firtel, 1980). All of these genes are expressed at very low levels in vegetative cells. By 6 h the RNA levels have increased ~500-fold. As development proceeds the levels of these RNAs decline.

Another method for determining the developmental kinetics of gene expression is by using developmental RNA blots. Equal quantities of mRNA from different developmental stages are size fractionated by gel electrophoresis, transferred to a filter matrix and hybridized with particular DNA probes. The intensity of the hybridization is a reflection of RNA concentration. It has been possible to identify genes that are maximal at various

---

Fig. 6. RNA excess hybridization of cloned sequences to *Dictyostelium* RNA. Inserts from plasmids complementary to *D2*, *M3*, and *discoidin I* [$pcI_{19}(B_2)$] genes were labeled by nick translation and hybridized to a vast excess of poly(A)$^+$ RNA isolated from vegetative (zero hour), 6 h, and 16 h developing cells. It was assumed the transcription of these genes is asymmetric and that maximum hybridization was 50% of the cDNA duplex. See Rowekamp and Firtel, 1980 for details.

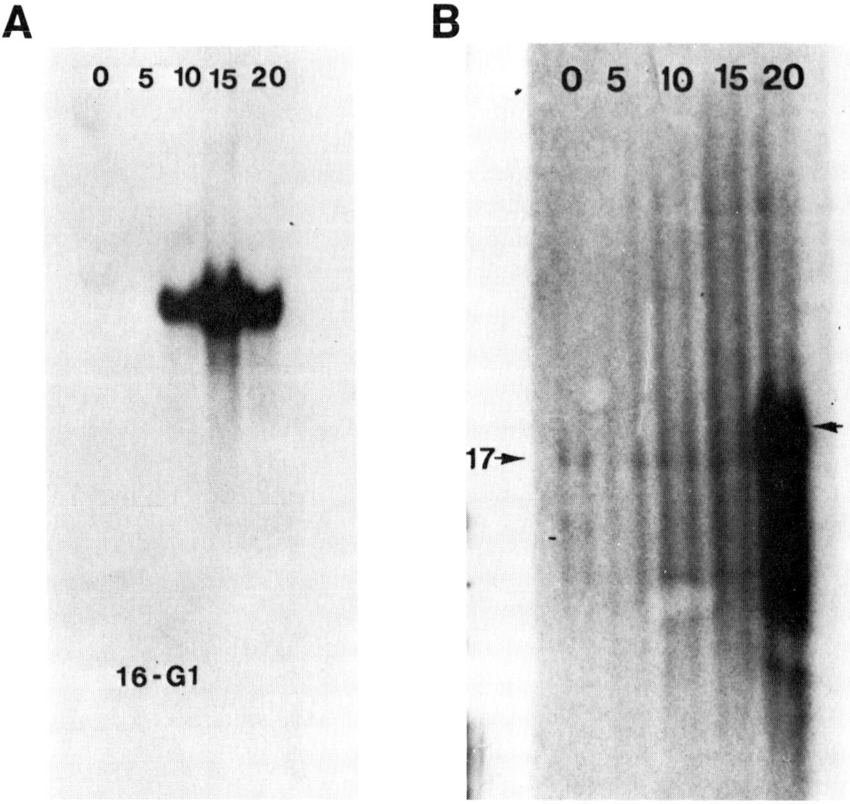

Fig. 7a. Northern DNA-blot hybridization of developmental RNA to cloned cDNA. Poly(A)$^+$ RNA from cells at 0, 5, 10, 15, and 20 h in the developmental cycle were hybridized to **A.** prespore-specific 16-G1 and **B.** prestalk-specific 14-E6 indicated by arrow. 17S rRNA also labeled with arrow.

times late in development (Fig. 7). It has also been possible to identify genes that are preferentially expressed in either prespore or prestalk cells (Fig. 7; Mehdy, Firtel, and Ratner, in preparation).

B. The Modulation of Gene Activity by Cyclic AMP

Gene expression in *Dictyostelium* is closely coupled to its developmental and physiological state; the alteration of the normal developmental sequence has an enormous effect upon the appearance of developmentally regulated proteins. UDPG pyrophosphorylase activity begins to accumulate in aggregated cells and continues to accumulate later in development. However, if

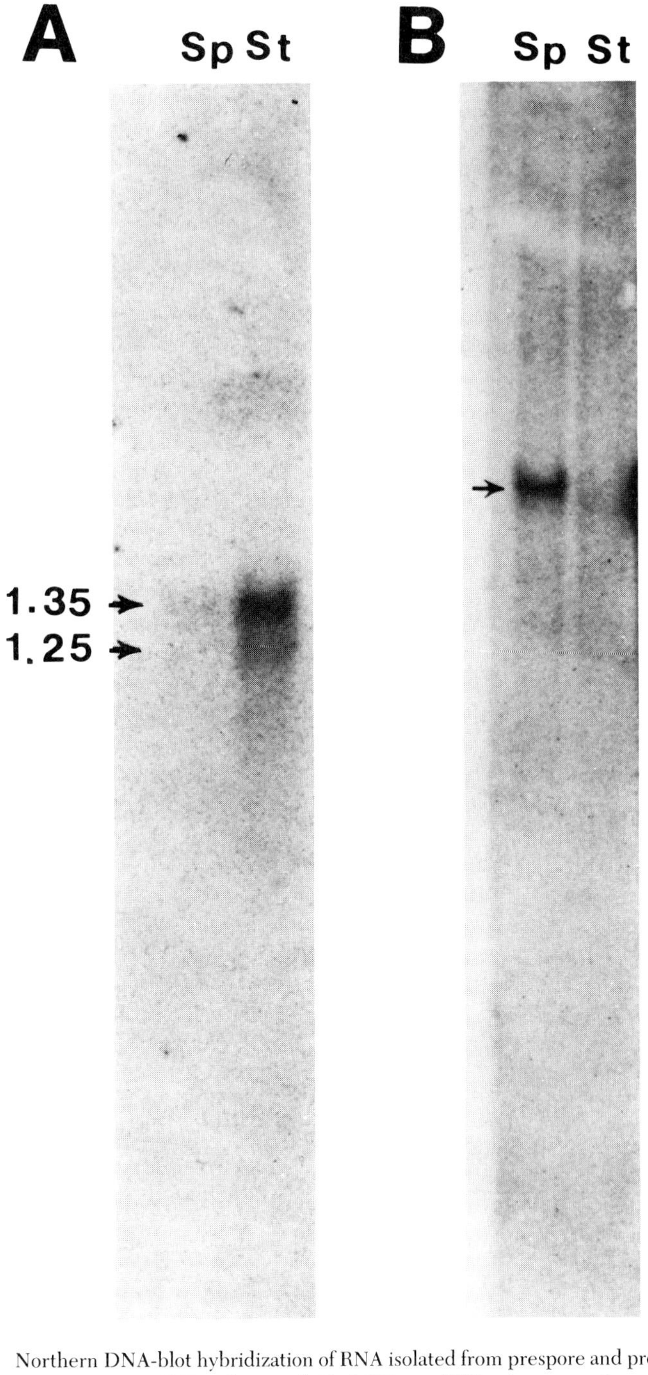

Fig. 7b. Northern DNA-blot hybridization of RNA isolated from prespore and prestalk cells hybridized to **A.** an *actin* cDNA clone and **B.** 2-H3, a cDNA clone complementary to a developmentally regulated RNA not expressed until after 10 h of development. In **A.** arrows indicate the two molecular sizes of *actin* messenger RNA, 1.25 and 1.35 kb.

aggregated cells are dissociated and prevented from reassociating, the accumulation of UDPG pyrophosphorylase ceases. If cells are allowed to reaggregate and proceed through the later developmental stages the activity of UDPG pyrophosphorylase again begins to accumulate (Newell *et al.*, 1972; Sussman and Newell, 1972). Similar studies on UDPgalactose epimerase and glycogen phosphorylase have shown that these enzymes do not appear in cells that have been dissociated. Again, when cells are allowed to reaggregate and develop, these enzymes are normally expressed (Newell *et al.*, 1972; Firtel and Bonner, 1972b). Further, the expression of some developmentally regulated enzymes in migrating slugs can be changed by inducing the slugs to culminate (Newell and Sussman, 1970; see Chapter 9).

Landfear and Lodish (1980) have examined the proteins synthesized *in vivo* and *in vitro* by mRNA from normally developing cells or from disaggregated cells. Many, but not all, of the proteins that initially appear at approximately the time of cell–cell contact are not synthesized in dissociated cells. Many of these proteins are again synthesized if dissociated cells are incubated with cAMP. Contact site A and certain other proteins that normally appear early in development can be induced in cells in shaking culture with pulses of cAMP although cAMP may repress the synthesis of discoidin I (see Gerisch *et al.*, 1975; Rossier *et al.*, 1980; Williams, *et al.*, 1980; Darmon *et al.*, 1975; Brandis, Mann, and Firtel, in preparation). In addition, workers have shown that cAMP can induce stalk cell differentiation in cells that are plated at very low densities and do not form aggregates. The level of expression of several late enzymes increases in response to cAMP in cell agglomerates (Bonner, 1970; Town *et al.*, 1976; Town and Gross, 1978; Sampson *et al.*, 1978; Kay *et al.*, 1979). These results suggested that cell–cell contact and cAMP may be modulators, either directly or indirectly, of gene activity in *Dictyostelium*.

It has become possible to study the changes in mRNA expression after altering the normal developmental pattern. Brandis and Firtel (in preparation) have identified the *M3* and other cDNAs that are normally expressed in early development. When vegetative amoebae are washed from a nutrient source and allowed to shake in nonnutrient buffer, these genes are not expressed. If these cells are subsequently pulsed with cAMP, *M3* mRNA sequences appear, and the relative kinetics of *M3* mRNA expression follows approximately the same developmental time course as cells plated on an air–buffer interface. In contrast, the expression of *discoidin I*, which is also expressed early in development, is inhibited by pulses of cAMP. The results of Williams and co-workers (1980) suggest that the inhibition of *discoidin I* expression is at the level of transcription. Thus, there are at least two sets of preaggregation genes whose expression is influenced by cAMP. One set is induced immediately upon starvation (e.g., *discoidin I*) and is repressed by increasing levels of cAMP; a second set (e.g., *M3*) is induced by pulses of cAMP.

The advent of procedures to separate prespore and prestalk cells from migrating slugs on Percoll gradients (see Tsang and Bradbury, 1981; Ratner and Borth, 1982) has allowed workers to isolate and examine the regulation of genes preferentially expressed in either cell type (Mehdy, Ratner and Firtel, submitted for publication). As can be seen in Fig. 7b, RNAs complementary to these genes are expressed at ~10-fold greater level in one cell type than in the other. Developmental RNA blots show that RNAs complementary to the prestalk-specific genes are detectable at approximately 7.5 h into development and reach maximal levels at 15–20 h (see Fig. 7a; Mehdy *et al.*). Similar developmental kinetics of expression were observed for a developmentally regulated gene expressed at equal levels in both cell types. In contrast, mRNAs complementary to the prespore-specific genes do not accumulate until ~15 h in development (see Fig. 7a). These results indicate that at least one set of genes that is preferentially expressed in a differentiated cell is active prior to the completion of aggregation. RNAs complementary to those genes preferentially expressed in prespore cells are not detected until several hours after tight cell contacts have been formed.

The requirements for the expression of these late developmental genes have also been studied. Prestalk-specific RNAs can be detected in fast (230 rpm) shaking cultures containing 40–500 μM cAMP under conditions where >90% of the cells remain unicellular. In the absence of cAMP, mRNAs complementary to the prestalk genes are not detected. When prespore-specific clones are examined, complementary RNAs are only expressed if the cells are shaken slowly (70 rpm) allowing large (~100 cell) agglomerates to form. Under these conditions, both prespore- and prestalk-specific genes are expressed with or without exogenous cAMP. Since cAMP appears to be required for the expression of prestalk genes, this requirement may be satisfied by endogenous production of cAMP in the agglomerates in slowly shaking cultures. The nonspecific cell type gene is expressed in fast shaking cultures in the presence or absence of exogenous cAMP.

When late aggregates are dissociated, the levels of RNA complementary to specific gene sets rapidly decrease (Fig. 8; Chung *et al.*, 1981; Mehdy, Ratner, and Firtel, submitted). Kinetic studies have shown that upon dissociation, mRNA complementary to these late genes is degraded with half-lives estimated to be less than 30 min (Chung *et al.*, 1981; Mehdy, Ratner, and Firtel, submitted; Lai and Firtel, unpublished observations). Other studies have suggested that the half-lives of these mRNAs in normally developing cells are appreciably longer (Mangiarotti *et al.*, 1982). These results suggest that dissociation causes a rapid loss of these late mRNAs which may be the result of a change in their half-lives. It has been further suggested that

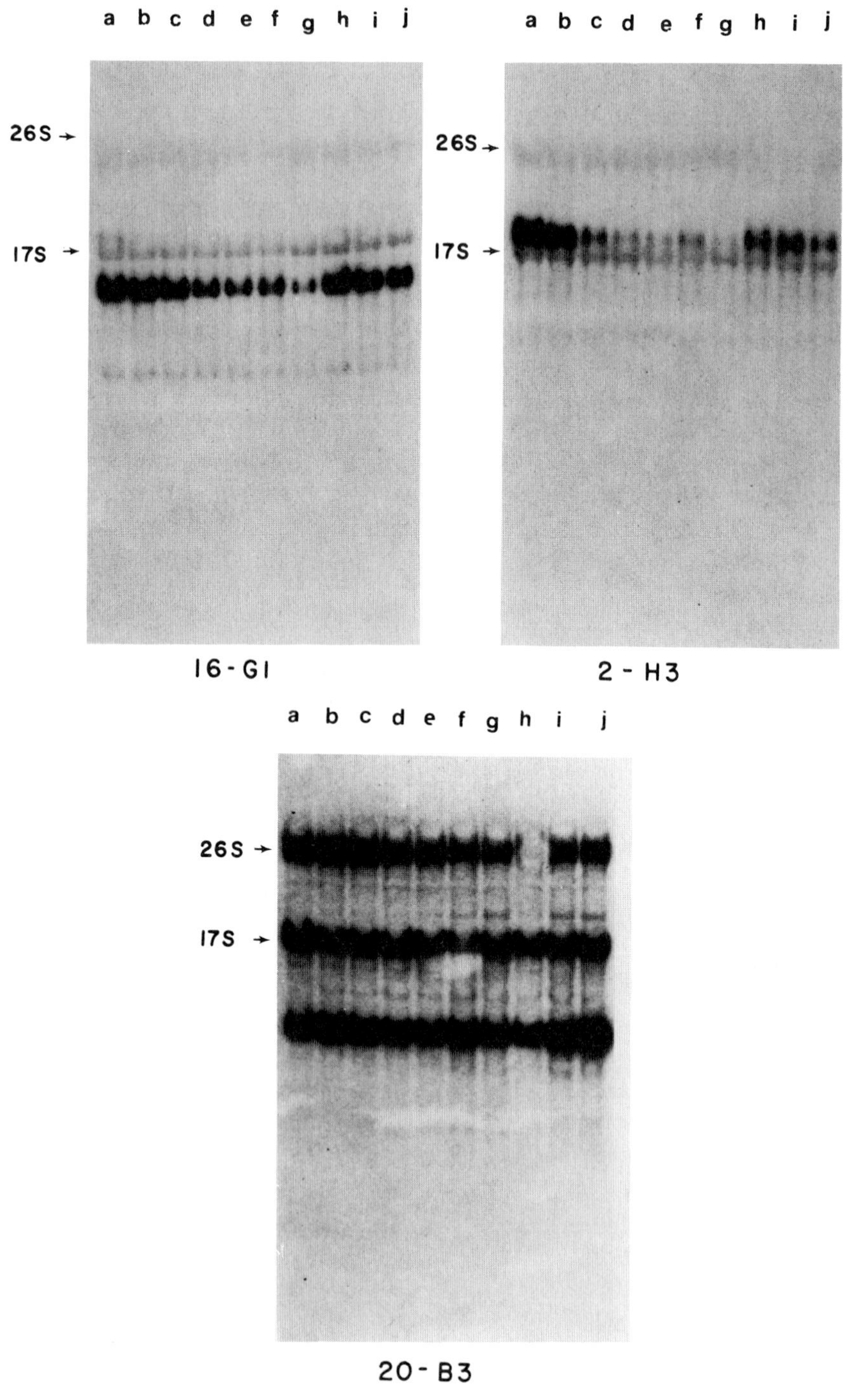
a b c d e f g h i j
26S
17S
16-G1
a b c d e f g h i j
26S
17S
2-H3
a b c d e f g h i j
26S
17S
20-B3

the transcription of these late genes is greatly reduced in dissociated cells. When the dissociated cells are incubated with cAMP, there is a reaccumulation of the mRNAs complementary to late specific probes (see Fig. 8) (Chung *et al.*, 1981; Blumberg *et al.*, 1982; Mehdy, Ratner, and Firtel, submitted for publication). The reaccumulation may be the result of an increase in mRNA synthesis as well as an increase in the stabilities of the mRNAs. Direct studies have not yet proven that there are differences in the levels of transcription of these genes in dissociated cells incubated in the presence or absence of cAMP.

It is interesting that all of the mRNAs that are affected by dissociation and subsequent addition of cAMP are not detectable at the same developmental stage (see Fig. 7a). This effect has been observed for prestalk genes whose expressions are maximal at 15–20 h and are detectable prior to the completion of aggregation, as well as for prespore genes whose expression cannot be detected until ~15 h (see Mehdy *et al.*, submitted for publication). In addition, not all developmentally regulated genes that are expressed late in development are affected by dissociation (see Chapter 8 for discussion). Genes which are either expressed constitutively throughout development or during early development are unaffected by cell–cell dissociation (Chung *et al.*, 1981; Mehdy, Ratner, and Firtel, submitted for publication).

It should be noted that by hybridizing RNA in excess to cDNA from total late specific mRNAs, Chung *et al.* (1981) have shown that a large fraction of

Fig. 8. Effect of dissociation and cAMP on late gene expression. AX3 cells were grown in suspension and plated on filter for development (5×10^6 cells/cm^2). At the tight aggregation stage, the cells were shaken off the filters and resuspended as single cells in MES–PDF containing 10 m*M* EDTA (5×10^6 cells/ml). The cells were shaken at 230 RPM on a gyrotory shaker for 3 h. The suspension was divided and one-half received 1 m*M* cAMP. An additional 100 μ*M* cAMP was added each hour subsequently. The cells continued to shake as predominantly single cells for another 5 h. RNA was extracted from aliquots of cells taken at indicated times. RNA was purified and fractionated by poly(U)–Sepharose chromatography. Equal amounts of poly(A)$^+$ RNA were loaded onto formaldehyde gels, separated by electrophoresis, and blotted onto Gene Screen (NEN) as per manufacturer's instructions. The filters were baked, prehybridized, and hybridized with a nick-translated probe made from the cDNA inserts separated from the plasmid vehicle. The filters were then washed and exposed to film. Lane a, RNA at time of disaggregation; b, 1 h after disaggregation; c, 2 h after disaggregation; d, 3 h after disaggregation; e, 4 h after disaggregation; no cAMP added; f, 6 h after disaggregation, no cAMP added; g, 8 h after disaggregation, no cAMP added; h, 4 h after disaggregation, cAMP added; i, 6 h after disaggregation, cAMP added; j, 8 h after disaggregation, cAMP added. Left panel: Prestalk-specific gene *16-G1*. Right panel: Prespore-specific gene *2-H3*. Bottom panel: Cell type nonspecific gene *20-B3* whose expression is unaffected by dissociation or cAMP. Arrows point to the remaining 17 and 26 S rRNA which cross-hybridizes with our probes. The *20-B3* probe has a large amount of cross-hybridization to the rRNA.

the mRNA sequences that do not appear until aggregation is affected by dissociation. However, because of the problems in analyzing the various mRNA components, it is difficult to quantitate the actual fraction of the moderate and low abundant mRNAs that is affected.

There may be independent events that affect late gene expression. The addition of cAMP to fast shaking disaggregated cells induces the reaccumulation of prespore- and prestalk-specific mRNAs. Although the cycloheximide has no effect on the reaccumulation of prestalk-specific mRNAs, the levels of mRNA complementary to the prespore-specific genes do not increase if cycloheximide is added to cultures with cAMP. Although cAMP modulates the expression of both sets of genes, there are clearly some differences in the mechanisms by which they are regulated by cAMP. In addition, since many late specific mRNAs and enzymes have totally different developmental time courses of appearance, the mechanism of gene induction may be different for many genes. Blumberg *et al.* (1982) have studied the expression of late specific mRNAs in *Dictyostelium* mutants which cease development at specific stages. They have concluded that only mutants that are still capable of forming tight aggregates express these late genes and that tight cell–cell interaction must be required for late gene expression. However, because of the pleiotropic nature of many developmental mutants, it is difficult to demonstrate that tight–cell contact actually induces late gene expression. There may be a different effector which is completely independent of the formation of tight aggregates that appears at approximately the same developmental stage.

Although Landfear and Lodish (1980) were not able to induce late gene expression in fast shaking suspensions, Gross and co-workers (1981) report that two late enzymes, glycogen phosphorylase and UDPGalactosyl polysaccharide transferase can be induced in shaking culture in the absence of prior or simultaneous cell contact. Since cAMP is known to induce stalk cell differentiation, there appears to be a requirement for moderate levels of cAMP for the induction of some late genes (Mehdy *et al.*, submitted for publication). Moreover, Mehdy *et al.* were able to show that prestalk-specific genes were expressed in fast shaking cultures with cAMP. Conceivably, cell dissociation reduces the level of cAMP around individual cells and the addition of exogenous cAMP increases the cAMP concentration to the physiological levels necessary for the maintenance of gene activity. Interestingly, the slow shaking of dissociated cells without cAMP under conditions that yield only small aggregates can result in a partial restoration of late mRNA sequences (Mehdy and Firtel, unpublished observations). The interpretation of these experiments is not clear since the re-establishment of the expression of these late genes may be either the result of higher local cAMP concentrations in the aggregates and/or the result of cell–cell contact.

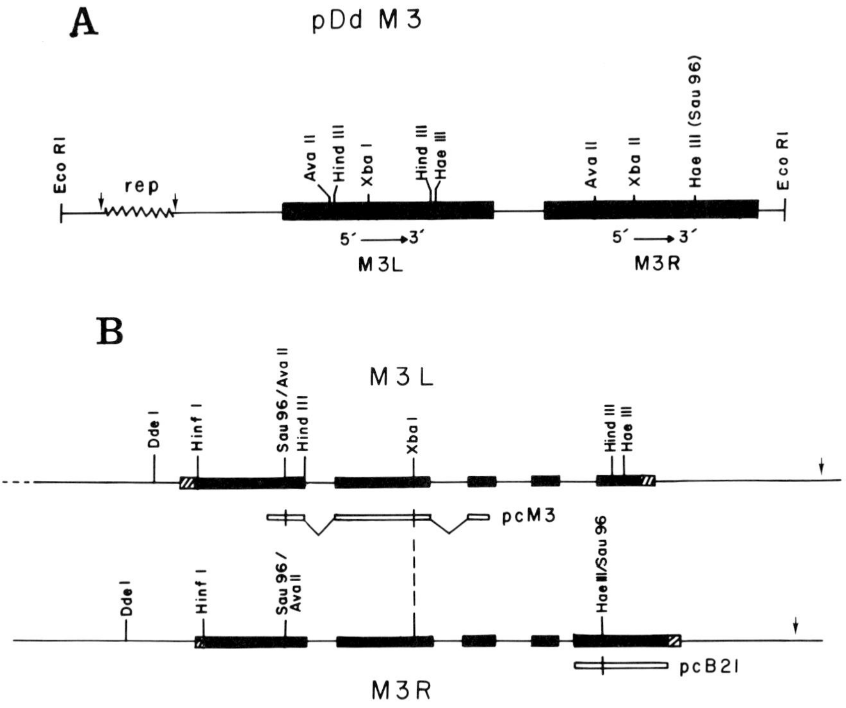

Fig. 9. **A.** Restriction map of the 7.5 kb Eco RI fragment carrying the two *M3* genes. **B.** Detailed restriction map of the two *M3* genes showing the location of the introns and internal restriction sites. Regions of homology with cDNA clones pcM3 and pcR21 complementary to the left hand *M3* gene (*M3L*) or the right hand *M3* gene (*M3R*) are also given. The exact length of the 3′ exon and the position of the 3′ end of the genes are not known. In part A, "rep" indicates a repeated sequence. In part B, thick bars indicate regions present in mRNA.

C. Organization of the *M3* Gene Family

Using the screening procedure for identifying cDNA clones containing inserts complementary to developmentally regulated RNAs described earlier, a series of clones were identified that were complementary to poly(A)$^+$ RNA from cells at 6 h in development but not to poly(A)$^+$ RNA from vegetative cells (Rowekamp and Firtel, 1980). By screening this set with poly(A)$^+$ RNA from cells pulsed with cAMP, a subset was identified that appeared to be induced ealry in development by cAMP (Brandis and Firtel, in preparation).

Of this gene subset, *M3* has been examined in the most detail. The devel-

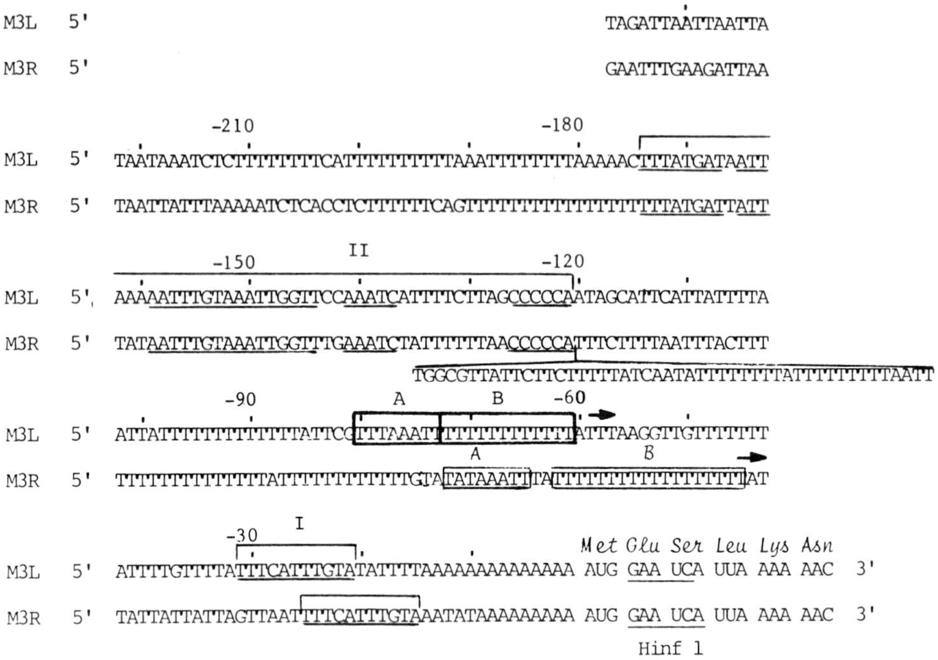

Fig. 10. The 5′ end and flanking sequence of the *M3* gene gamily. The nucleotide sequence and location of the (A) TATA and (B) oligo(dt) stretch are indicated. Regions of homology in the 5′ untranslated region are underlined and labeled I and II. The start of transcription of both genes is also indicated as well as the N-terminal amino acids.

opmental time course of *M3* gene expression is shown in Fig. 6. Using an *M3* cDNA hybridization probe, a λ phage containing the Eco RI genomic fragment complementary to *M3* sequences was isolated. This 7.2-kb Eco RI fragment contains two genes complementary to the *M3* cDNA, *M3L* (left) and *M3R* (right) (Fig. 9). Several *M3* cDNA clones have been isolated, sequenced, and compared to their genomic sequences. It has been shown that certain RNAs are specifically derived from the *M3L* or *M3R* gene. It is thus clear that both genes are transcribed. Additionally, the cDNAs are not colinear with their respective genes indicating that both genes contain intervening sequences. Although the location of the introns within the coding regions is conserved, their sequences are completely divergent. However, the introns do possess structural similarities to those of the other *Dictyostelium* introns with regard to size (~100 bp), A + T content (95%), and junction sites (see Table I; Kimmel and Firtel, 1980).

The *M3* genes encode mRNAs of ~2 kb. There is ~10% nucleotide divergence in the protein-coding region that would lead to amino acid differences. This includes a region of hypervariability wherein 13 of 14 amino acids differ between *M3L* and *M3R*. The *M3* mRNAs program polypeptides of ~50,000 molecular weight in an *in vitro* translation system.

The 5′ region of the genes has also been examined (see Fig. 10). The sequences 5′ to the AUG initiation codon are 95% A + T. The 5′ ends of the *M3* RNAs have been located in the DNA sequence. The RNA start sites are 3′ to a TATA box–oligo(dT) structure found in all mRNA encoding genes in *Dictyostelium* (see Fig. 32; Kimmel and Firtel, in preparation). Although this region in *M3R* has strong homology to the consensus sequence (see Fig. 32), the corresponding sequence in *M3L* is more divergent. The expression of *M3L* is 5–10-fold lower than that of *M3R* and may be the result of this divergence.

The 5′ flanking regions of *M3* are generally not homologous. There are however, two small regions of sequence identity (see Fig. 10). It is interesting to speculate that these regions have been conserved during evolution and are important in the maintenance of the similar developmental kinetics of *M3L* and *M3R* expression.

VI. *Actin* Multigene Family

A. Developmental Modulation of Actin Synthesis

Actin appears to be found in all eukaryotic cells. It is essential for intra- and extracellular motilities including cytokinesis, chromosome condensation, embryogenesis, amoeboid movement, and muscle contraction (see Pollard and Weihing, 1974). In *Dictyostelium*, actin is found in association with the plasma and nuclear membranes and also with the nonhistone protein component of chromatin (see Chapter 5; Firtel, unpublished observations).

Actin is a major developmentally regulated protein in *Dictyostelium*. It represents ~1% of the newly synthesized protein in vegetative cells and ~1% of the protein synthesized *in vitro* in response to vegetative mRNA (Tuchman *et al.*, 1974; Alton and Lodish, 1977a). Moreover, actin mRNA has been shown to represent ~1% of the mass of mRNA in vegetative cells as determined by RNA excess hybridization (Kindle, 1978). These results suggest that the level of actin protein in vegetative cells is directly related to the abundancy of actin mRNA. The relative rate of actin mRNA synthesis has also been approximated. When amoebae are pulse-labeled for 5 min with $^{32}PO_4$, ~1% of newly synthesized hnRNA and mRNA is actin specific (Kind-

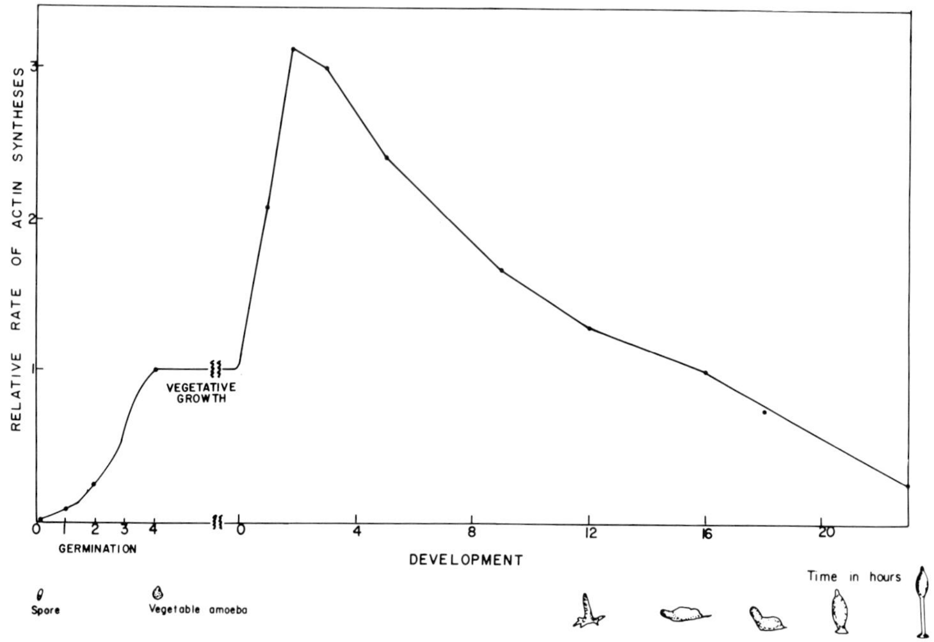

Fig. 11. Developmental time course of actin synthesis. The relative rate of actin synthesis during *Dictyostelium* development is summarized as described in Firtel *et al.* (1981).

le, 1978; Kimmel, unpublished observations). The relative rate of actin mRNA synthesis has also been estimated by measuring the relative concentration of actin mRNA present in the mRNA fraction having the longest poly(A) tails, a fraction that represents the most newly synthesized mRNA (Palatnik *et al.*, 1981). These studies yield results that are essentially identical to the *in vivo* pulse-labeling studies. These composite results suggest that the quantity of actin in vegetative cells is regulated primarily at the level of transcription.

A developmental time course of actin expression has been summarized in Fig. 11. Actin synthesis is undetectable ($<10^{-3}\%$ of the newly synthesized protein) during the first hour of spore activation (MacLeod *et al.*, 1980). At this time, actin mRNA represents ~0.01% of total poly(A)$^+$ RNA as measured by RNA excess hybridization and by *in vitro* translation of RNA isolated from spores. During the next several hours of germination, actin synthesis becomes detectable and progressively increases in parallel to a relative rise in actin mRNA levels. As can be seen in Fig. 12, all isoelectric

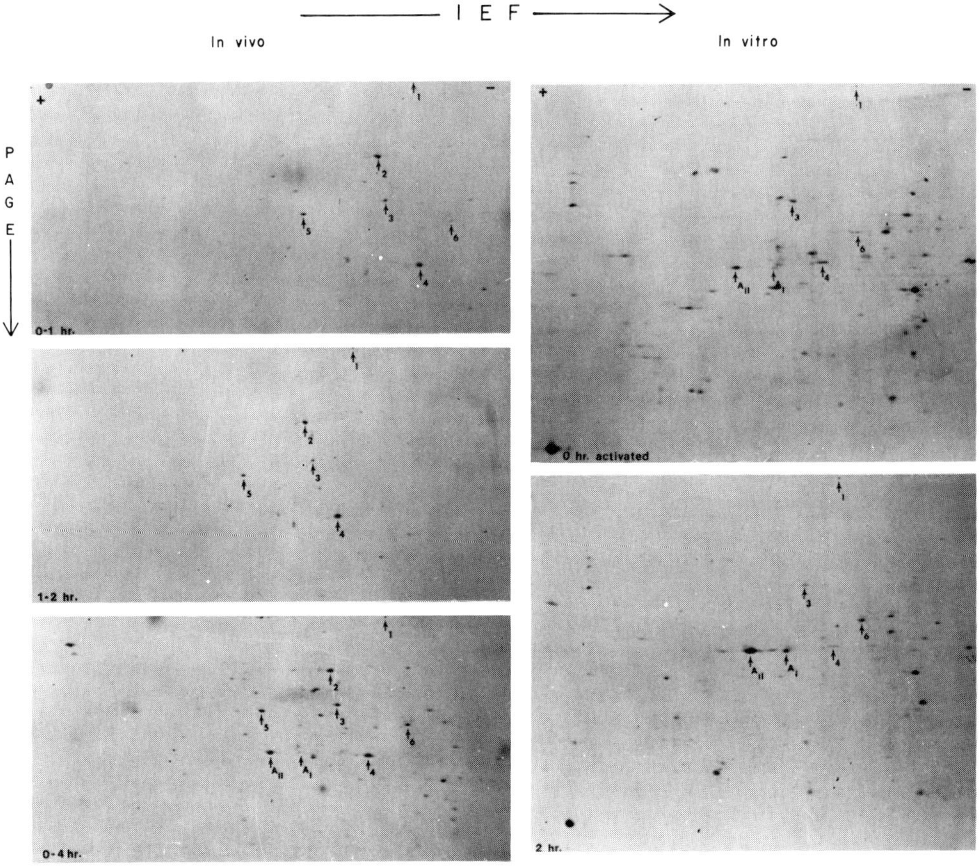

Fig. 12. Two-dimensional gels of proteins labeled with [^{35}S]methionine *in vivo* and *in vitro* during germination. Details of *in vivo* labeling are given in MacLeod *et al.*, 1980. For *in vitro* labeling, RNA was isolated from either 0 h activated or 2 h germinating spores, translated *in vitro*, and analyzed in two-dimensional gels as described by MacLeon *et al.*, 1980. Labeled proteins were identified by fluorography. The major actin spots are labeled A_I and A_{II}. Other protein spots are labeled with numbers for orientation of the various gels.

forms of actin become detectable at the same time. By 4 h after activation, when amoebae emerge from the spore coats, actin protein and mRNA levels have approached those of vegetative cells.

Actin synthesis appears to increase during early development. Shortly after vegetative cells are induced to develop, the rate of actin synthesis rises approximately threefold and remains at this peak level until 6 h into the developmental cycle (Tuchman *et al.*, 1974; Alton and Lodish, 1977a). Throughout the remainder of the cycle the rate of actin synthesis gradually decreases. By culmination it is <10% that of vegetative cells.

Actin has been analyzed in prespore and prestalk cells separated from pseudoplasmodia (Alton and Brenner, 1979; Coloma and Lodish, 1981; Morrissey *et al.*, 1981). Actin is preferentially synthesized in prestalk cells, and there is >10-fold more actin mRNA in prestalk cells than in prespore cells (Fig. 7; Mehdy and Firtel, in preparation). Since spores eventually represent 80% of the cells in the mature fruiting body, a majority of the decrease in actin synthesis late in development is probably the result of the spore-specific decrease in actin mRNA.

All of the developmental changes in actin synthesis are accompanied by equivalent changes in the relative rates of actin mRNA synthesis and actin mRNA levels (Kindle, 1978; Kimmel, unpublished observations). Measurements regarding the relative stability of actin mRNA are consistent with the modulation of actin throughout the *Dictyostelium* life cycle being regulated primarily at the level of transcription (Margolskee and Lodish, 1980b).

B. Multiple Forms of Actin Protein and Actin mRNA

When actin protein is analyzed by high resolution two-dimentional gel electrophoresis, one major form of actin is observed in addition to more basic and more acidic minor forms (Fig. 13; Kindle and Firtel, 1978; MacLeod *et al.*, 1980; MacLeod, 1979). All of these isoelectric forms bind specifically to affinity columns of DNase I. It has been suggested that the most acidic form is an unacetylated species of the major form (Rubenstein and Deuchler, 1979). Vandekerckhove and Weber (1980) have sequenced purified actin. They observed a single, major form; no forms representing >5% of the mass of the purified actin were detected. Although the unacetylated form represents a substantial level of the actin protein, it would not have been detected by the actin sequencing protocol used by these workers. Additional data from DNA sequencing and transcriptional analysis of purified *Dictyostelium* actin genes confirm that minor forms are present at low levels. Although actin synthesis seems to be modulated during development, all of the forms appear coordinately during spore germination and all are synthesized throughout the life cycle (MacLeod *et al.*, 1980). Some small changes in the relative quantities of the different forms have been observed during culmination.

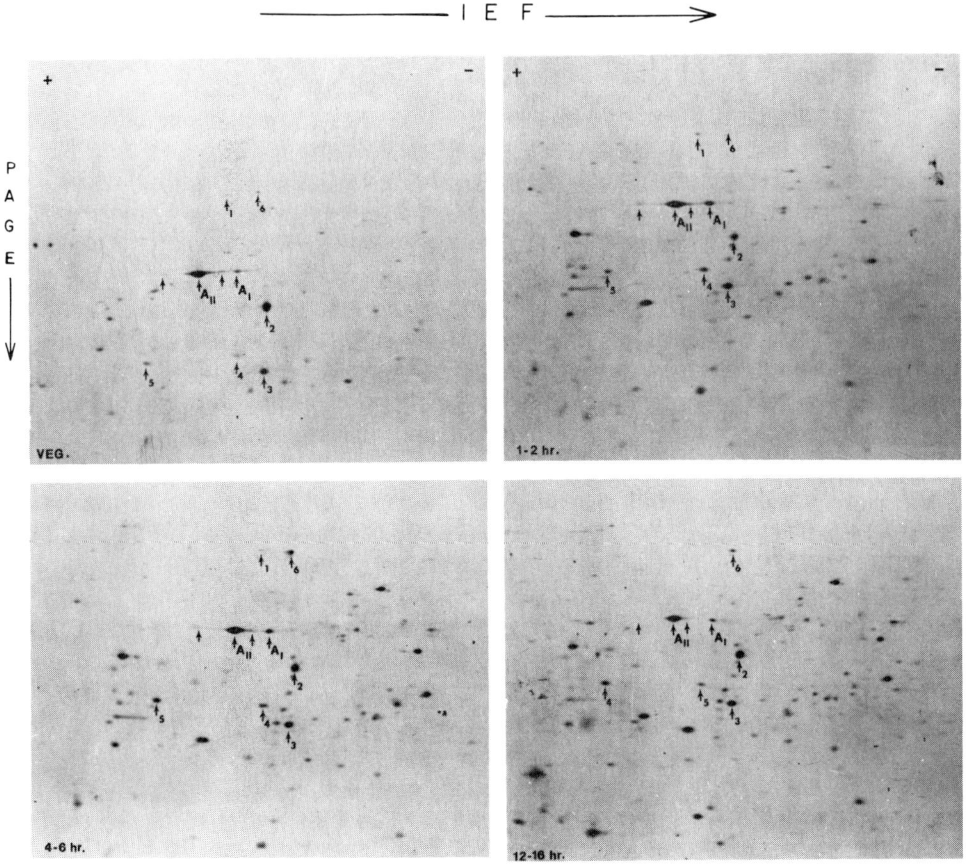

Fig. 13. (A) Two-dimensional gels of *Dictyostelium* proteins labeled *in vivo* and analyzed by two-dimensional gel electrophoresis. The times of labeling during development are indicated. The actin spots are labeled by capital As. Other protein spots used as markers for orienting the gels are also shown. See MacLeod (1979) and legend to Fig. 12 for details. (B) Two-dimensional gels of actin proteins *in vitro*. RNA was isolated at the time points shown on the figure, translated *in vitro* and analyzed by two-dimensional gel electrophoresis. Labeling patterns are the same as indicated in Fig. 13a. See MacLeod (1979) for details. (C) Isoelectric focusing of *in vivo* and *in vitro* synthesized actin. Actin synthesized *in vivo* (Panel A) and *in vitro* (Panel B) were separately bound to DNase-I agarose and the eluates subjected to isoelectric focusing on 30 cm polyacrylamide gels. The actin region was electrophoresed in the second dimension, and the labeled actin forms were visualized by fluorography. For details see MacLeod (1979).

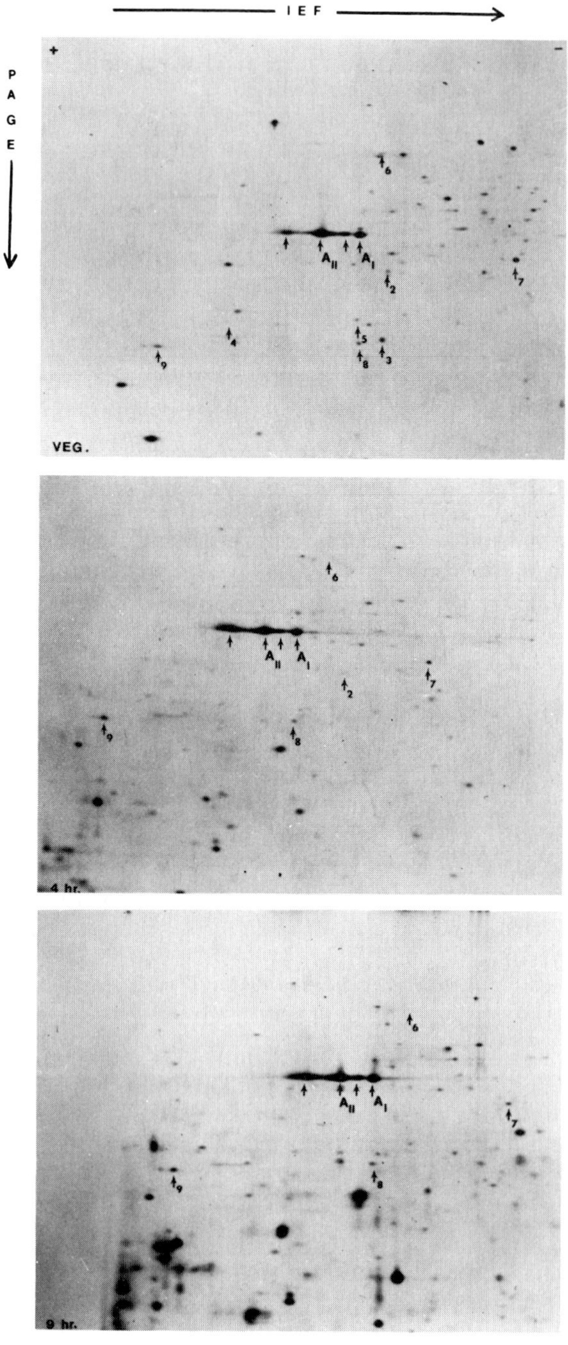
IEF
+
−
PAGE
6
A$_{II}$
A$_I$
2
7
4
5
3
9
8
VEG.
6
A$_{II}$
A$_I$
7
2
9
8
4 hr.
6
A$_{II}$
A$_I$
7
9
8
9 hr.

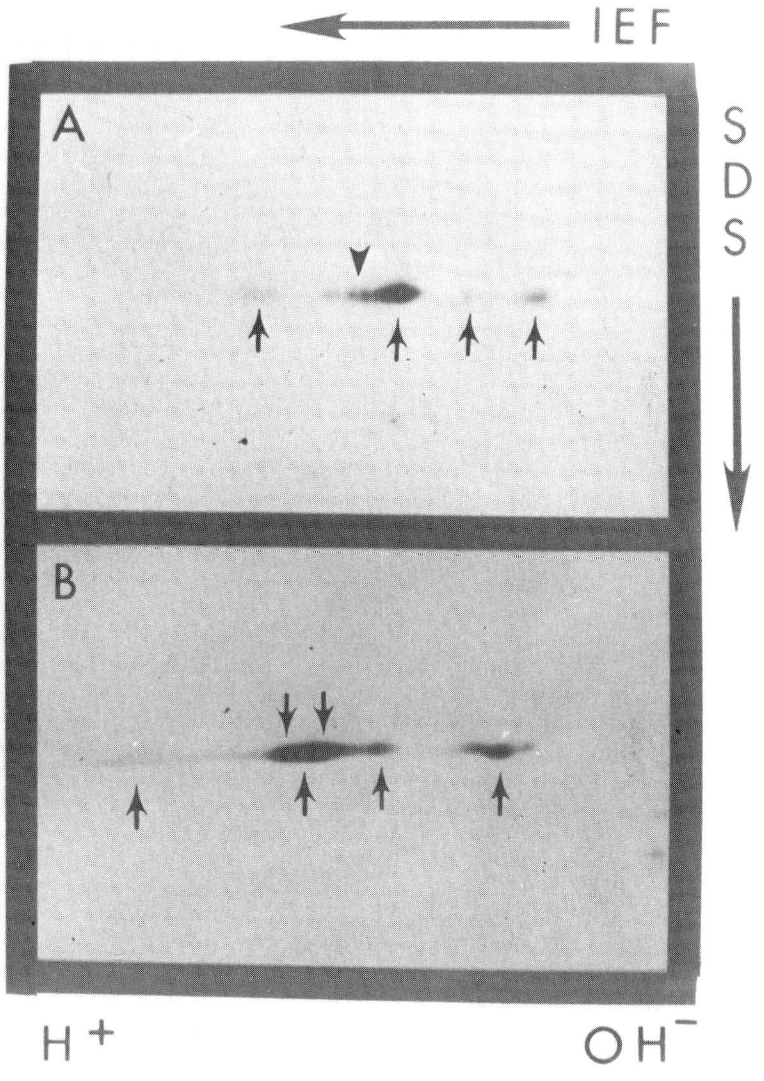

Fig. 13C.

There are two size classes (1.25 and 1.35 kb) of *actin* mRNA (see Fig. 7; Kindle *et al.*, 1977; Kindle and Firtel, 1978; McKeown *et al.*, 1978). The lengths of their protein coding regions are identical. The size differences are the result of different lengths of 3′ untranslated regions (McKeown and Firtel, 1981b, 1982; McKeown *et al.*, 1982). The relative ratios of the two *actin* mRNA forms do not vary significantly during development (Kindle *et al.*, 1977). Finally, recent data indicate that both mRNA forms can encode the major isoelectric species of actin protein.

C. *Actin* Gene Organization

It has been possible to isolate recombinant plasmids containing sequences that are complementary to *actin* mRNA (Kindle and Firtel, 1978). These plasmids have been used to determine the relative concentration and rates of synthesis of *actin* mRNA throughout development. Additionally, DNA excess and DNA blot hybridizations using these *actin* probes have shown that there are 17 *actin* genes in the *Dictyostelium* genome. Ten of these genes have been isolated and characterized from genomic recombinant plasmids. Many *actin* cDNA plasmids have also been identified. Some of these are derived from genes that have not yet been isolated. The restriction maps of these *actin* genes are seen in Fig. 14 (Kindle and Firtel, 1978; McKeown *et al.*, 1978; McKeown and Firtel, 1981a). All of the isolated genes possess a single Hind III site very near the AUG translation initiation site within the coding region. Similarly, there is a single Ava II (or Sau96 I) site located slightly 5′ to the UAA translation termination signal. Other restriction sites are not conserved in all of the *actin* genes. It is possible that the presence or absence of particular restriction sites may indicate evolutionary relationships among various *actin* genes.

Figure 15 shows a Southern DNA blot wherein *Dictyostelium* nuclear DNA was digested with Hind III (which cleaves very close to the 5′ end of all the *actin* genes) or Eco RI (which does not cut any of the *actin* genes that have been isolated), size-fractionated on an agarose gel, transferred to a filter matrix, and then hybridized with a probe made from an *actin* cDNA plasmid. The Hind III digest shows 17 *actin* restriction fragments. Since there is a unique Hind III site near the 5′ end of each *actin* gene, the number of Hind III restriction fragments that hybridize indicates that there are 17 *Dictyostelium actin* genes (McKeown *et al.*, 1978; McKeown and Firtel, 1981a, 1982; McKeown *et al.*, 1982). This corresponds quite well with the 15–20 *Dictyostelium actin* genes estimated by DNA–DNA hybridization kinetics (Kindle and Firtel, 1978). It is also clear that the *actin* genes are not

Fig. 14. Restriction maps of cloned *actin* genes. Actin coding regions were aligned and are shown as thicker bars with the Hind III site at the 5′ end of the protein coding sequence and the Ava II site at the 3′ end. Inserts terminating with black boxes were cloned in pMB9 using random shear–poly (dAT) tailing. Inserts terminating at Eco RI sites were initially cloned in λ Charon 13 and subsequently subcloned. See Kindle and Firtel (1978); McKeown *et al.* (1978); Bender *et al.* (1978); Cockburn *et al.* (1976); and McKeown and Firtel (1981) for details.

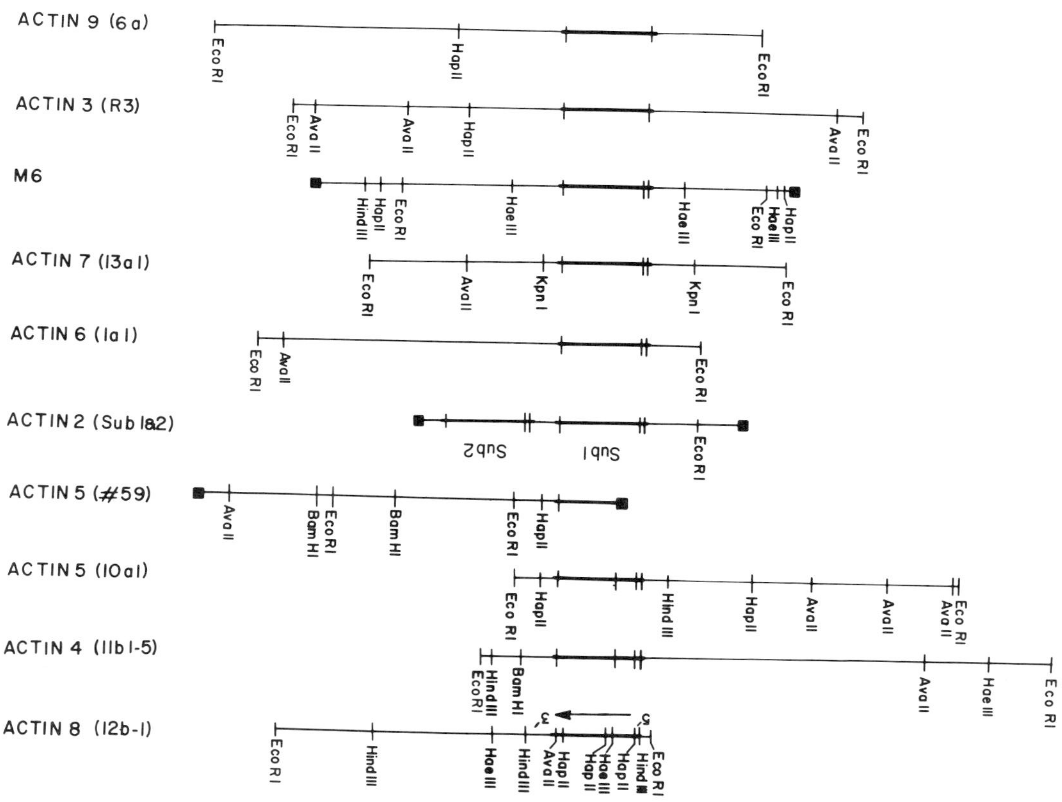

DICTYOSTELIUM ACTIN GENES

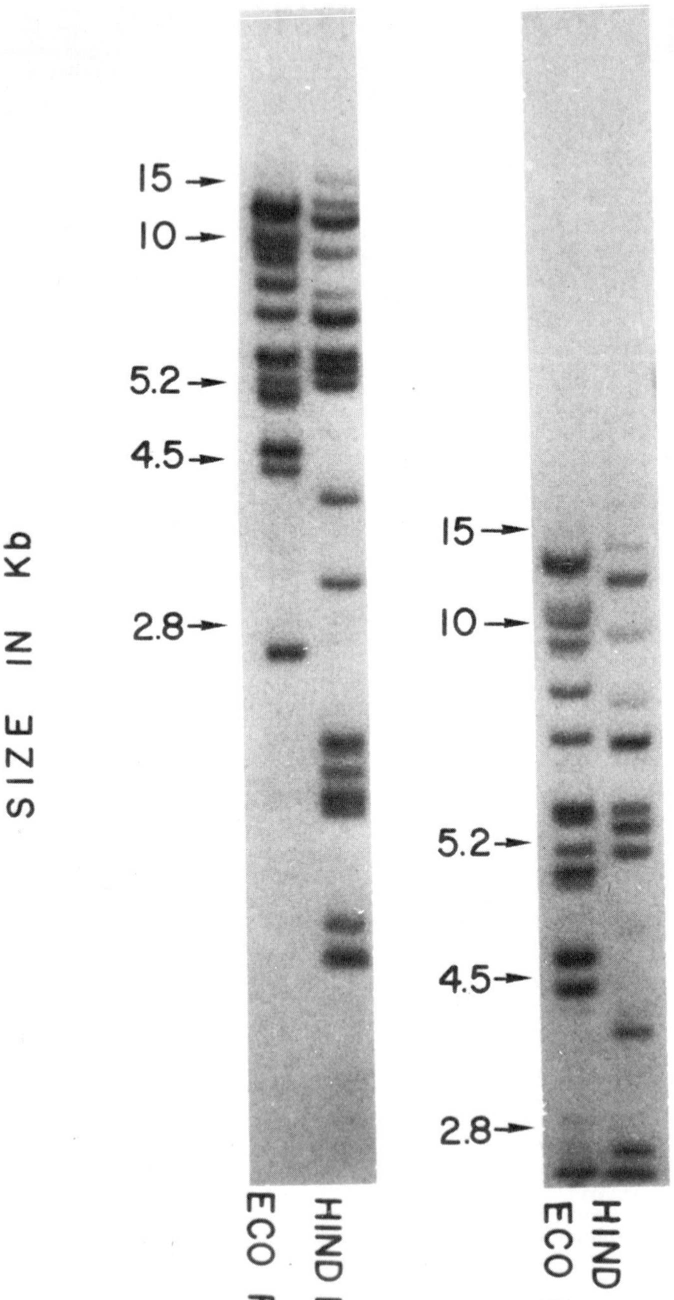

SIZE IN Kb
15
10
5.2
4.5
2.8
ECO RI
HIND III
15
10
5.2
4.5
2.8
ECO RI
HIND III

tandemly repeated with homogeneous spacers. Examination of the hybridization pattern of the Eco RI digest shows 13–14 *actin* Eco RI restriction fragments. This indicates that more than one *actin* gene is found on some Eco RI fragments. This has been directly confirmed by characterizing isolated *actin* genomic fragments.

Some indication of the heterogeneity of the *actin* genes can be obtained by examining the intensity of the hybridization of the Hind III restriction fragments to a single actin probe (McKeown *et al.*, 1978). Certain fragments hybridize more intensely than others indicating that these genes are more homologous to the *actin* probe than to other *actin* genes. When different gene probes are used, the pattern of intensity of hybridization changes. Thermal denaturation experiments indicate 2–10% nucleotide sequences mismatch between the protein coding regions of various genes. This range of nucleotide sequence mismatch has been directly confirmed by the DNA sequencing of many *actin* genes.

When the restriction sites in regions flanking the *actin* coding regions are compared, it is clear that the extragenic sequences are not conserved. Heteroduplex analysis, DNA blot hybridization, and DNA sequence comparison between *actin* genes have confirmed that the sequences flanking the *actin* genes are nonhomologous.

No regions of discontinuity are visible in heteroduplexes between *actin* cDNA clones and genomic clones (Bender *et al.*, 1978). Furthermore, the distance (~1.0 kb) between the conserved Hind III and Ava II sites is identical for all of the genomic and cDNA clones isolated indicating that the genomic sequences are colinear with *actin* mRNA. When total genomic DNA is cleaved with Hind III and Ava II, separated by agarose-gel electrophoresis, and hybridized with a Hind III/Ava II *actin* probe on DNA blots, only the 1.0 kb genomic fragment is detected. Thus, it seems that the coding regions within all the *actin* genes in *Dictyostelium* are exactly colinear with *actin* mRNA. In addition, no interruptions have been found in the 5′ or 3′ regions of the isolated *actin* genes (McKeown and Firtel, 1981a,b; 1982).

These observations contrast sharply with those of the *actin* genes in yeast, *Drosophila*, sea urchin, chicken, and mammals (see Breathnach and Cham-

---

Fig. 15. Southern DNA blot hybridization of *actin* gene probe to *Dictyostelium* nuclear DNA. *Dictyostelium* DNA was digested with Eco RI or Hind III restriction endonucleases. Samples were split in half and fragments separated by electrophoresis. Half of each sample is run until the bromophenol blue marker had moved 80% length of the gel; the other half was run twice as long. *Actin* sequences were visualized by hybridizing [^{32}P]-labeled *actin* cDNA to a Southern blot made from the gel.

bon, 1981). The *actin* genes in these species contain intervening sequences that are transcribed into hnRNA but are removed during processing into mature mRNA. The *actin* genes in *Dictyostelium* seem to lack introns, although there are other genes in *Dictyostelium* that possess introns (Kimmel and Firtel, 1980). Thus, any model that purports to explain the evolution and function of intervening sequences must account for the complete absence of such sequences in the *actin* genes of *Dictyostelium* and the variable position of the introns found in *actin* genes of other organisms.

D. DNA Sequence Analysis of the Actin Protein Coding Region

The DNA sequences at the 5′ and 3′ ends of most of the *actin* genomic and cDNA clones have been determined (Firtel *et al.*, 1979; McKeown and Firtel, 1981a,b; 1982). These sequences include those that encode approximately the first 40–70 amino acids and the last 70 amino acids and represent ~40% of each gene analyzed (Fig. 16). The majority of the genes would encode an actin that is identical within the regions sequenced to that of the major isoelectric form. Although the derived amino acid sequences are identical among the genes, their nucleotide sequences are not identical. All of the differences are located in the third base wobble position of the codons.

Three of the *actin* genes (*M6*, *3*, *2-sub2*) would yield amino acid sequences that are not identical to that of the major actin species. *Actin M6* differs from the major form at two positions and *Actin 3* differs at four positions. There are 10 amino acid differences in the first 41 positions between *Actin 2-sub2* and the major *actin* form.

The relative expression of *Actin M6*, *2-sub2*, and *3* has been determined (see below). *Actin M6* RNA is present at very low levels and would encode an actin species ~2% of total actin. The protein derived from *actin M6* would probably be resolved from the major actin species by high resolution two-

Fig. 16. (A) DNA and derived amino acid sequences of the 5′ end of the *actin* genes. The nucleotides that differ from pcDd actin B_1 are given in italics. Amino acid differences from those of the major form sequenced by Vanderkerchove and Weber (1980) are underlined. See Firtel *et al.* (1979) for details. (B) Sequences of the 3′ ends of *actin* coding regions. The sequence of the strand with the same sequence as mRNA is shown. Nucleotides different from the majority of nucleotides of a given position are shown in italics. The two bases labeled N in codon 307 are believed to be G residues. On the strand sequences in these two clones this base is unreadable, probably due to its being a methylated C residue as part of an endogenous *E. coli* methylation site. The same situation is true in pDd *actin 8* but the G residue has been determined by directly sequencing of the strand shown. See McKeown and Firtel (1981b) for details.

```
                                    -20                       1                                         30                                                 60
                                     ↓                        ↓                                         ↓                                                  ↓
                                                            Met Asp Gly Glu Asp Val Gln Ala Leu Val Ile Asp Asn Gly Ser Gly Met Cys Lys Ala Gly Phe
               UUUAAAUAUAAUACAACAAUAAA  AUG GAC GGU GAA GAU GUU CAA GCU UUA GUU AUC GAU AAU GGU UCU GGU AUG UGU AAA GCC GGU UUU
B1 (C)13       AAATTTATATTATGTTGTTATTT  TAC CTG CCA CTT CTA CAA GTT CGA AAT CAA TAG CTA TTA CCA AGA CCA TAC ACA TTT CGG CCA AAA

                                                            Met Glu Ser Glu Asp Val Gln Ala Leu Val Ile Asp Asn Gly Ser Gly Met Cys Lys Ala Gly Phe
           ATAATCAAAATAATAAAAAAAAAAAAA  AUG GAA AGU GAA GAU GUU CAA GCU UUA GUU AUU GAU AAU GGU UCU GGU AUG UGU AAA GCU GGU UUU
3          TATTAGTTTTATTATTTTTTTTTTTTT  TAC CTT TCA CTT CTA CAA GTT CGA AAT CAA TAA CTA TTA CCA AGA CCA TAC ACA TTT CGA CCA AAA

                                                            Met Asp Gly Glu Asp Val Gln Ala Leu Val Ile Asp Asn Gly Ser Gly Met Cys Lys Ala Gly Phe
           AATAATAAATACAATTAAAAAATAAAA  AUG GAC GGU GAA GAU GUU CAA GCU UUA GUU AUU GAU AAU GGU UCU GGA AUG UGU AAA GCC GGU UUU
M6         TTATTATTTATGTTAATTTTTTATTTT  TAC CTG CCA CTT CTA CAA GTT CGA AAT CAA TAA CTA TTA CCA AGA CCA TAC ACA TTT CGG CCA AAA

                                                            Met Asp Gly Glu Asp Val Gln Ala Leu Val Ile Asp Asn Gly Ser Gly Met Cys Lys Ala Gly Phe
           TTAAATAAATAATAATAATATAATAAA  AUG GAU GGU GAA GAU GUU CAA GCU UUA GUU AUU GAU AAC GGU UCU GGU AUG UGU AAA GCC GGU UUU
2-Sub 1    AATTTATTTATTATTATTATATTATTT  TAC CTA CCA CTT CTA CAA GTT CGA AAT CAA TAA CTA TTG CCA AGA CCA TAC ACA TTT CGG CCA AAA

                                                            Met Glu Cys Gly Asp Val Gln Ala Leu Val Ile Asp Lys Gly Ser Ser Ile Ser Lys Ala Gly Phe
           TATTTTTTAAATATTTTGAAATAATAA  AUG GAA UGU GGA GAU GUU CAA GCU UUA GUA AUC GAU AAA GGU UUC AGU AUA AGU AAA GCC GGU UUU
2-Sub 2    ATAAAAAATTTATAAAACTTTATTATT  TAC CTG ACA CCA CTA CAA GTT CGA AAT CAT TAG CTA TTT CCA AGA TCA TAT TCA TTT CGG CCA AAA

                                                            Met Asp Gly Glu Asp Val Gln Ala Leu Val Ile Asp Asn Gly Ser Gly Met Cys Lys Ala Gly Phe
           CTCAATTAAATAATAAAATATAATAAA  AUG GAC GGU GAA GAU GUU CAA GCU UUA GUU AUC GAU AAU GGU UCU GGU AUG UGU AAA GCC GGU UUU
7          GAGTTAATTTATTATTTTATATTATTT  TAC CTG CCA CTT CTA CAA GTT CGA AAT CAA TAG CTA TTA CCA AGA CCA TAC ACA TTT CGG CCA AAA

                                                            Met Asp Gly Glu Asp Val Gln Ala Leu Val Ile Asp Asn Gly Ser Gly Met Cys Lys Ala Gly Phe
           AACTTAAAATAAATTATATATAAAAAA  AUG GAC GGU GAA GAU GUU CAA GCU UUA GUU AUC GAU AAC GGU UCU GGU AUG UGU AAA GCC GGU UUU
5          TTGAATTTTATTTAATATATATTTTTT  TAC CTG CCA CTT CTA CAA GTT CGA AAT CAA TAG CTA TTG CCA AGA CCA TAC ACA TTT CGG CCA AAA

                                                            Met Asp Gly Glu Ast Val Gln Ala Leu Val Ile Asp Asn Gly Ser Gly Met Cys Lys Ala Gly Phe
           TAATCAAATAAATAAATATAATATAAA  AUG GAU GGU GAA GAU GUU CAA GCU UUA GUA AUU GAU AAC GGU UCU GGU AUG UGU AAA GCC GGU UUU
6          ATTAGTTTATTTATTTATATTATATTT  TAC CTA CCA CTT CUA CAA GTT CGA AAT CAT TAA CTA TTG CCA AGA CCA TAC ACA TTT CGG CCA AAA

                                                            Met Asp Gly Glu Asp Val Gln Ala Leu Val Ile Asp Asn Gly Ser Gly Met Cys Lys Ala Gly Phe
           AAATATATATATATAAATCATTTAAAA  AUG GAC GGU GAA GAU GUU CAA GCU UUA GUU AUC GAU AAC GGU UCU GGU AUG UGU AAA GCC GGU UUU
8          TTTATATATATATATTTAGTAAATTTT  TAC CTG CCA CTT CTA CAA GTT CGA AAT CAA TAG CTA TTG CCA AGA CCA TAC ACA TTT CGG CCA AAA
                                                            Met Asp Gly Glu Asp Val Gln Ala Leu Val Ile Asp Asn Gly Ser Gly Met Cys Lys Ala Gly Phe
                 AUAUAAAUCAUUUAAAA  AUG GAC GGU GAA GAU GUU CAA GCU UUA GUU AUC GAU AAC GGU UCU GGU AUG UGU AAA GCC GGU UUU
IEL-1         (C)15TATATTTAGTAAATTTT  TAC CTG CCA CTT CTA CAA GTT CGA AAT CAA TAG CTA TTG CCA AGA CCA TAC ACA TTT CGG CCA AAA
```

Fig. 16A. *(continued)*

Fig. 16A. (Continued)

```
                90                                             120                                      150
                 ↓                                              ↓                                        ↓
        Ala Gly Asp Asp Ala Pro Arg Ala Val Phe Pro Ser Ile Val Gly Arg Pro Arg His Thr Gly Val Met Val Gly Met Gly Gln Lys
        GCU GGU GAU GAU GCC CCA CGU GCU GUU UUC CCA UCU AUU GUU GGU CGU CCA AGA CAC ACU GGU GUU AUG GUC GGU AUG GGU CAA AAA GA
B       CGA CCA CTA CTA CGG GGT GCA CGA CAA AAG GGT AGA TAA CAA CCA GCA GGT TCT GTG TGA CCA CAA TAC CAG CCA TAC CCA GTT TTT CT

        Ala Gly Asp Asp Ala Pro Arg Ala Val Phe Pro Ser Ile Val Gly Arg Pro Arg His Thr Gly Val Met Val Gly Met Gly Gln Lys
        GCU GGU GAU GAU GCU CCA CGU GCA GUU UUC CCA UCU AUU GUU GGU CGU CCU AGA CAU ACU GGU GUU AUG GUU GGU AUG GGU CAA AAA GA
3       CGA CCA CTA CTA CGA GGT GCA CGT CAA AAG GGT AGA TAA CAA CCA GCA GGA TCT GTA TGA CCA CAA TAC CAA CCA TAC CCA GTT TTT

        Ala Gly Asp Asp Ala Pro Arg Ala Leu Phe Pro Ser Ile Val Gly Arg Pro Arg Tyr Thr Gly Val Met Val Gly Met Gly Gln Lys
        GCU GGU GAU GAU GCC CCA CGU GCU CUU UUC CCA UCU AUU GUU GGU CGU CCU AGA UAU ACU GGU GUU AUG GUU GGU AUG GGU CAA AAA GA
M6      CGA CCA CTA CTA CGG GGT GCA CGA GAA AAG GGT AGA TAA CAA CCA GCA GGA TCT ATA TGA CCA CAA TAC CAA CCA TAC CCA GTT TTT CT

        Ala Gly Asp Asp Ala Pro Arg Ala Val Phe Pro Ser Ile Val Gly Arg Pro Arg His Thr Gly Val Met Val Gly Met Gly Gln Lys
        GCU GGU GAU GAU GCU CCA CGU GCU GUU UUC CCA UCA AUU GUU GGU CGU CCA AGA CAU ACU GGU GUU AUG GUC GGU AUG GGU CAA AAA G
2-Sub 1 CGA CLA CTA CTA CGA GGT GCA CGA CAA AAG GGT AGT TAA CAA CCA GCA GGT TCT GTA TGA CCA CAA TAC CAG CCA TAC CCA GTT TTT C

        Ala Gly Asp Asp Ala Pro Arg Ala Val Phe Pro Ser Asn Val Gly Arg Gln Arg Tyr
        GCU GGU GAU GAU GCU CCA CGU GCU GUU UUC CCA UCU AAU GUU GGU CGU CAA AGA UAU
2-Sub 2 CGA CCA CTA CTA CGA GGT GCA CGA CAA AAG GGT AGA TTA CAA CCA GCA GUU UCA ATA

        Ala Gly Asp Asp Ala Pro Arg Ala Val Phe Pro Ser Ile Val Gly Arg Pro Arg His Thr Gly Val Met Val Gly Met Gly Gln Lys Asp
        GCU GGU GAU GAU GCU CCA CGU GCU GUU UUC CCA UCU AUU GUU GGU CGU CCA AGA CAU ACU GGU GUU AUG GUU GGU AUG GGU CAA AAA GAU
7       CGA CCA CTA CTA CGA GGT GCA CGA CAA AAG GGT AGA TAA CAA CCA GCA GGT TCT GTA TGA CCA CAA TAC CAA CCA TAC CCA GTT TTT CTA

        Ala Gly Asp Asp Ala Pro Arg Ala Val Phe Pro Ser Ile Val Gly Arg Pro Arg His Thr Gly Val Met Val Gly Met Gly Gln Lys Asp
        GCU GGU GAU GAU GCU CCA CGU GCU GUU UUC CCA UCA AUU GUU GGU CGU CCA AGA CAC ACU GGU GUU AUG GUC GGU AUG GGU CAA AAA GAU
5       CGA CCA CTA CTA CGA GGT GCA CGA CAA AAG GGT AGU UAA CAA CCA GCA GGU UCU GUG UGA CCA CAA UAC CAG CCA UAC CCA GUU TTT CTA

        Ala Gly Asp Asp Ala Pro Arg Ala Val Phe Pro Ser Ile Val Gly Arg Pro Arg His Thr Gly Val Met Val Gly Met Gly Gln Lys Asp
        GCU GGU GAU GAU GCC CCA CGU GCU GUU UUC CCA UCA AUU GUU GGU CGU CCA AGA CAC ACU GGU GUU AUG GUU GGU AUG GGU CAA AAA GAU
6       CGA CCA CTA CTA CGG GGT GCA CGA CAA AAG GGT AGT TAA CAA CCA GCA GGT TCT GTG TGA CCA CAA TAC CAA CCA TAC CCA GTT TTT CTA

        Ala Gly Asp Asp Ala Pro Arg Ala Val Phe Pro Ser Ile Val Gly Arg Pro Arg His Thr Gly Val Met Val Gly Met Gly Gln Lys Asp
        GCU GGU GAU GAU GCC CCA CGU GCU GUU UUC CCA UCA AUU GGU GGU CGU CCA AGA CAC ACU GGU GUU AUG GUC GGU AUG GGU CAA AAA GAC
8       CGA CCA CTA CTA CGG GGT GCA CGA CAA AAG GGT AGT TAA CAA CCA GCA GGT TCT GTG TGA CCA CAA TAC CAG CCA TAC CCA GTT TTT CTG

        Ala Gly Asp Asp Ala Pro Arg Ala Val Phe Pro Ser Ile Val Gly Arg Pro Arg His Thr Gly Val Met Val Gly Met Gly Gln Lys Asp
        GCU GGU GAU GAU GCC CCA CGU GCU GUU UUC CCA UCA AUU G U GGU CGU CCA AGA CAC ACU GGU GUU AUG GUC GGU AUG GGU CAA AAA GAC
IEL-1   CGA CCA CTA CTA CGG GGT GCA CGA CAA AAG GGT AGT TAA CAA CCA GCA GGT TCT GTG TGA CCA CAA TAC CAG CCA TAC CCA GTT TTT CTG
```

Fig. 16B.

```
          290                                     300                                     310
Asp Ile Arg Lys Asp Leu Tyr Gly Asn Val Val Leu Ser Gly Gly Thr Thr Met Phe Pro Gly Ile Ala Asp Arg Met Gln Lys Glu Leu
                                                                                                                        pcDd actin ITL-1
GAT ATC CGT AAA GAT TTA TAC GGT AAT GTC GTC TTA TCG GGT GGT ACA ACT ATG TTC CCA NGT ATT GCT GAT GGT ATG AAC AAA GAA TTA pcDd actin III-12
    ATC CGT AAA GAT TTA TAC GGT AAC GTC GTC TTA TCC GGT GGT ACA ACT ATG TTC CCA GGT ATT GCT GAT GGT ATG AAC AAA GAA TTA pDd actin 8
                                                                                            GAT GGT ATG AAC AAA GAA TTA pDd actin 6
                                                                                                            AAA GAA TTA pDd actin 5
                    TTA TAT GGT AAT GTT GTA TTA TCA GGT GGT ACA ACT ATG TTC CCA NGT ATT GCA GAT GGT ATG AAC AAA GAA TTA pDd actin 2-sub 1
                                                                                                                        pDd actin 2-sub 2

          320                                     330                                     340
Thr Ala Leu Ala Pro Ser Thr Met Lys Ile Lys Ile Ile Ala Pro Pro Glu Arg Lys Tyr Ser Val Trp Ile Gly Gly Ser Ile Leu Ala
                CCA TCA ACC ATG AAA ATT AAA ATC ATT GCT CCA CCA GAA CGT AAA TAC TCT GTC TGG ATT GGT GGA TCT ATC TTG GCT pcDd actin ITL-1
ACT GCT TTA GCA CCA TCA ACC ATG AAA ATT AAA ATC ATT GCT CCA CCA GAA CGT AAA TAC TCT GTC TGG ATT GGT GGA TCT ATT TTG GCT pcDd actin III-12
ACT GCT TTA GCC CCA TCA ACC ATG AAA ATT AAA ATC ATT GCT CCA CCA GAA CGT AAA TAC TCT GTC TGG ATT GGT GGA TCT ATC TTA GCT pDd actin 8
ACT GCA TTA GCA CCA TCA ACC ATG AAA ATT AAA ATC ATT GCA CCA CCA GAA CGT AAA TAC TCA GTT TGG ATC GGT GGA TCT ATT TTG GCT pDd actin 6
ACT GCT TTA GCC CCA TCA ACA ATG AAA ATT AAA ATC ATT GCT CCA CCA GAA CGT AAA TAC TCT GTC TGG ATT GGT GGA TCT ATC TTA GCT pDd actin 5
ACA GCA TTA GCA CCA TCA ACA ATG AAA ATC AAA ATC ATT GCA CCA CCA GAA CGT AAA TAC TCA GTC TGG ATT GGT GGA TCT ATT TTA GCT pDd actin 2-sub 1
                                                                                                        Met     Ser
            GCA CCA TCA ACA ATG AAA ATC AAA ATC ATT GCA CCA CCA GAA CGT AAA TAC TCA GTT TGG ATT GGT GGA TCA ATG TTA TCA pDd actin 2-sub 2

          350                                     360                                     370
Ser Leu Ser Thr Phe Gln Gln Met Trp Ile Ser Lys Glu Glu Tyr Asp Glu Ser Gly Pro Ser Ile Val His Arg Lys Cys Phe Ter
TCA CTC TCA ACT TTC CAA CAA ATG TGG ATC TCC AAA GAA GAA TAT GAC GAA TCA GGT CCA TCA ATT GTC CAC AGA AAA TGT TTT TAA     pcDd actin ITL-1
TCA CTC TCA ACT TTC CAA CAA ATG TGG ATC TCA AAA GAA GAA TAT GAT GAA TCT GGT CCA TCA ATT GTC CAC AGA AAA TGT TTC TAA     pcDd actin III-12
TCA CTC TCA ACT TTC CAA CAA ATG TGG ATC TCC AAA GAA GAA TAC GAC GAA TCA GGT CCA TCA ATT GTT CAC AGA AAA TGT TTC TAA     pDd actin 8
TCA CTC TCA ACT TTC CAA CAA ATG TTG ATC TCC AAA GAA GAA TAT GAC GAA TCA GGT CCA TCA ATT GTC CAC AGA AAA TGT TTT TAA     pDd actin 6
TCA CTC TCA ACT TTC CAA CAA ATG TTG ATC TCC AAA GAA GAA TAC GAC GAA TCA GGT CCA TCA ATT GTC CAC AGA AAA TGT TTT TAA     pDd actin 5
TCA CTC TCA ACT TTC CAA CAA ATG TGG ATT TCA AAA GAA GAA TAT GAC GAA TCT GGT CCA TCA ATT GTT CAC AGA AAA TGT TTC TAA     pDd actin 2-sub 1
TCG CTC TCA ACT TTC CAA CAA ATG TGG ATT TCC AAA GAA GAA TAT GAC GAA TCT GGT CCA TCA ATT GTT CAC AGA AAA TGT TTC TAA     pDd actin 2-sub 2
```

dimensional gel electrophoresis by virtue of its predicted difference in isoelectric point. *Actins 2-sub2* and *3* mRNAs are not detectable (<1% of total *actin* mRNA). As will be discussed, it is likely that both are not transcribed but are pseudogenes.

E. 5′ DNA Sequences and Transcription Analyses

The sequences 5′ to the AUG translation initiation codon are extremely (~95%) A + T rich (Firtel *et al.*, 1979). Although none of the *actins* share any sequence homology 5′ to the AUG, there are some general structural similarities. In most of the genes, 30–40 bp 5′ to the AUG there is a region similar to the Goldberg–Hogness TATA box (McKeown and Firtel, 1981a). This sequence may be important for eukaryotic RNA polymerase II phasing and transcription efficiency (see Breathnach and Chambon, 1981; Kimmel and Firtel, in preparation). Slightly 3′ to the TATA box is an oligo(dT) stretch. Finally, all of the *actin* genes (and every other known *Dictyostelium* gene encoding an mRNA) possess at least one A residue immediately 5′ to the AUG. Although the significance (if any) of this is not known, the observation is clear.

Although the nucleotide sequence homologies (90–95%) among the various *actin* genes within the protein coding region make it difficult to use this region to determine if a specific gene is transcribed, one can take advantage of the divergent 5′ ends to address this question (McKeown and Firtel, 1981a). The protocol for examining the relative expression of the *actin* genes is shown in Fig. 17. An *actin* gene is cleaved at the conserved Hind III site and end labeled with [γ-^{32}P]ATP and polynucleotide kinase. A single-stranded DNA probe complementary to *actin* mRNAs is prepared. When this DNA is hybridized to mRNA and then incubated with a single-strand specific nuclease, two size groups of DNA fragments are observed after sizing by gel electrophoresis. The smaller fragments represent hybrids of the *actin* gene probe to nonhomologous *actin* mRNAs. The hybridization extends from the Hind III site to the common A residues immediately 5′ to the AUG. The larger DNA fragments are the result of hybridization to homologous mRNA. The homologous DNA fragments are thus protected from the Hind III site into the A + T rich region beyond the AUG to the 5′ end of the actin mRNA. The results of such an experiment are shown in Fig. 18. As can be seen, the homologous fragment is heterogeneous in length, which may be the result of multiple start sites of transcription as is sometimes observed in other systems (Baker and Ziff, 1981). Alternatively, this heterogeneity could be the result of a technical artifact related to the single-strand nuclease

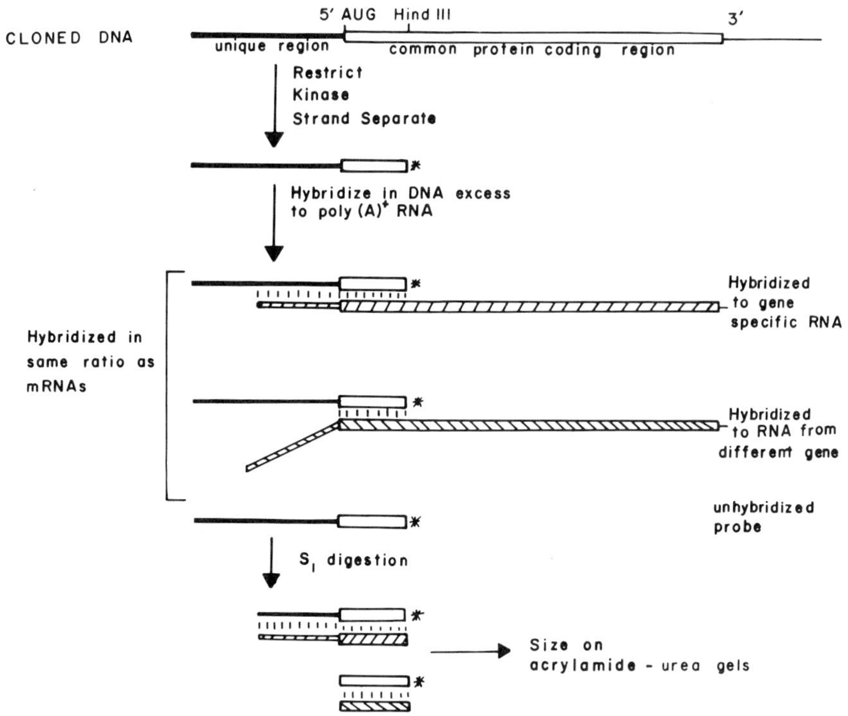

Fig. 17. Localization of 5′ end of *actin* genes and quantitation of level of expression of *actin* gene family members. Description of the procedure is in the text. See McKeown and Firtel (1981a) for details.

digestion of a very A + T rich sequence. By sizing the sites of the homologous (larger) fragments relative to a DNA sequencing ladder of the same DNA fragment, it is possible to locate the 5′ end sequences of the mRNA from any *actin* gene. Such 5′ end analyses have been completed for nine *actin* genes (Fig. 19). All of these potential start sites are located 3′ to the oligo(dT) stretch common to most *actin* genes (McKeown and Firtel, 1981a, 1982; McKeown *et al.*, 1982).

It is believed that the 5′ ends of the mRNA have actually been identified rather than a region of nucleotide discontinuity related to RNA processing. Clearly, the mRNA is colinear from the Hind III site into the 5′ untranslated region. It is extremely unlikely that this site represents an intron junction since it has no sequence homology with other intron boundaries in *Dictyostelium* or in other organisms (see Table I; Kimmel and Firtel, 1980). In addition, the 5′ ends of other *Dictyostelium* genes have been located in

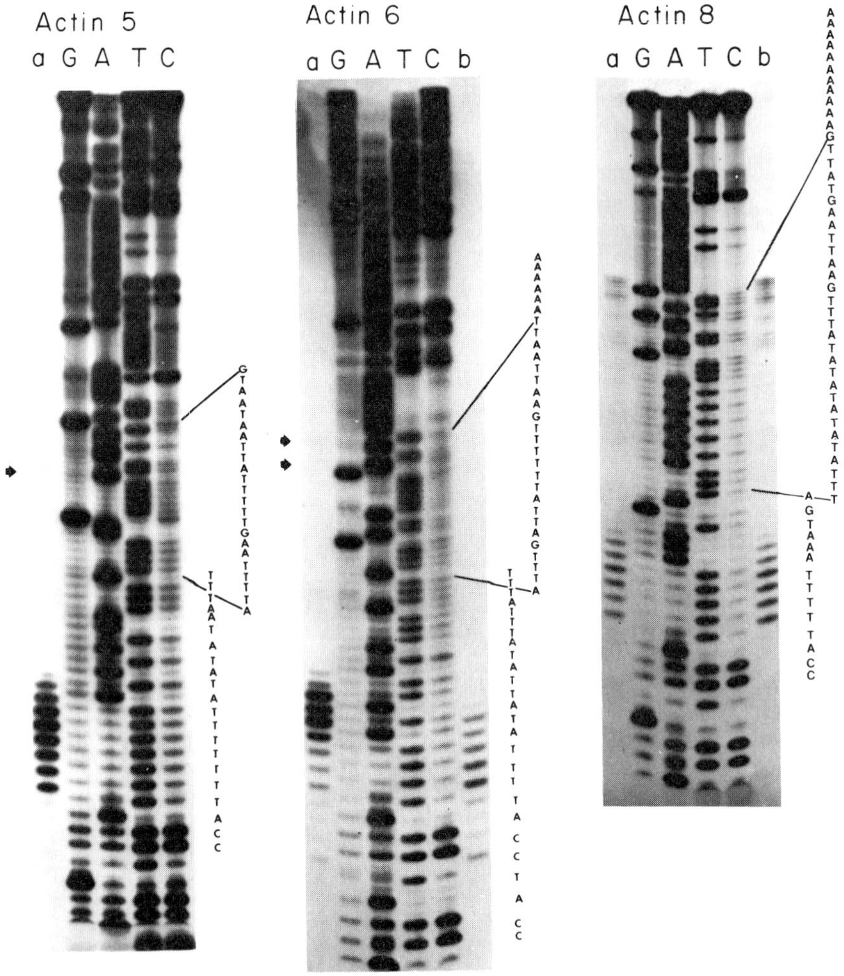

Fig. 18. Identification of the 5′ end of *actin* RNAs. Single stranded DNA probes from *actin* 5, 6, and 8 was hybridized to poly(A)$^{+}$ RNA and treated with S_1 nuclease. The S_1-resistant fragments were run adjacent to a DNA sequencing ladder of the same DNA fragment. GATC represents the DNA sequencing ladder. S_1 treatment was with 0.1 U/μl (Lane a) or 0.33 U/μl (Lane b). See McKeown and Firtel (1981) for details.

Fig. 19. Sequences of *actin* genes. The nucleotide sequence of the regions 5′ to the AUG initiation codon is shown. One letter abbreviation for the *N*-terminal amino acid sequence are used within the coding region. Altered amino acids are shown in italics and underlined. Boxes labeled **A** contain TATA-like sequences and boxes labeled **B** contain oligo(dT) sequences. C ▼ and ▽ indicate potential transcription initiation sites determined using mung bean and S_1 nuclease respectively. See McKeown *et al.*, (1981) and McKeown *et al.* (1982) for details.

5′ ENDS OF DICTYOSTELIUM ACTIN GENES

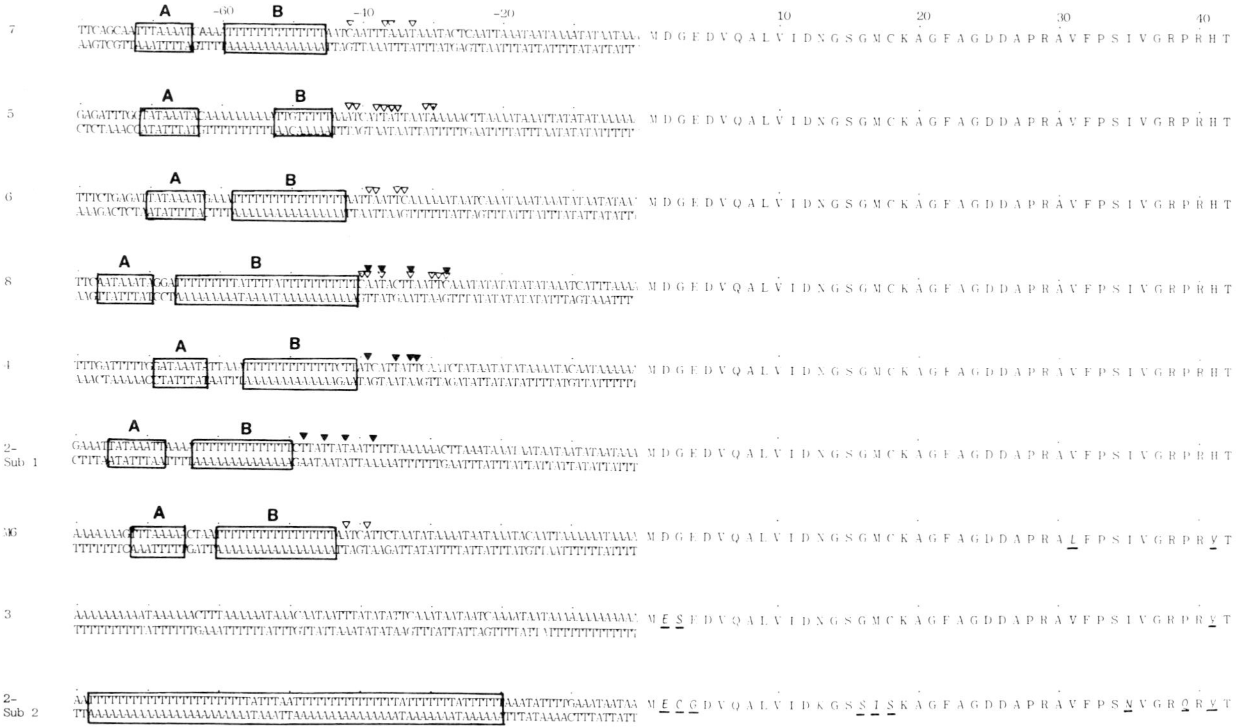

TABLE II
Percentage of Actin Gene Complementary RNA Derived from Specific Genes at Times during Development[a]

	Hours of development				
Actin gene	0	3	8	13	20
8	17	27	30	19	22
6	23	9	6	10	2
5	5	4	5	4	4
M6	1	3	2	1	3
4		+[b]			
7		+			
2-sub1		+			
2-sub2	ND[c]	ND	ND		
3	ND	ND	ND		

[a]Using the protocol shown in Fig. 17, the relative expression of the various *actin* genes during development was determined (McKeown and Firtel, 1981a).

[b]RNA complementary to these *actin* genes has been detected.

[c]RNA complementary to these genes is not detected at a level of <1% of total *actin* messenger RNA.

mRNA and hnRna (Kimmel and Firtel, in preparation). It appears the 5′ start sequences in hnRNA are conserved in mRNA; thus, the 5′ terminal sequence of mRNA would be identical to the transcription initiation site. The *actin* and all other *Dictyostelium* genes transcribed with polII share the common TATA–oligo(dT) structure which lies 5′ to the transcription initiation site (see Fig. 32). Finally, no *actin* hnRNAs have been detected that are larger than *actin* mRNA (Kimmel, McKeown and Poole, unpublished observation). It seems that there is no cleavage processing of the *actin* RNA. The protocol outlined in Fig. 17 can be used to determine if a specific *actin* gene is expressed and to accurately identify the 5′ ends of *actin* mRNA and hence transcription initiation.

Using a variation of the protocol outlined in Fig. 17, it is also possible to quantitate the relative expression of each *actin* gene during development (McKeown and Firtel, 1981a). When the end-labeled, single-stranded *actin* probe is hybridized in sequence excess to mRNA, all of the *actin* mRNA would form hybrids. Thus after single-strand nuclease digestion, the amount of probe in homologous RNA : DNA hybrids compared to heterologous RNA : DNA hybrids would reflect the relative abundance of a particular *actin* mRNA relative to all the *actin* mRNAs. Furthermore, by performing such experiments using RNA isolated from various developmental periods, the relative developmental expression of specific genes can be determined. A compilation of a series of these experiments is seen in Table II. None of the *actin* genes thus far examined are expressed at similar levels. This may be a reflection of the nonhomologous sequences upstream from transcription initiation that may influence developmental regulation.

Seven of the nine genomic clones examined are expressed. In addition, at least two plasmids carrying *actin* cDNA inserts not derived from any of the cloned genomic fragments have been identified. These data indicate that at least 9 of the 11 genes analyzed are expressed. Except for *M6*, all transcribed genes would encode the same actin protein over the region sequenced.

Actin 8 is expressed at high levels in vegetative cells, and this level increases during early development in NC4 cells. The level of *actin 8* mRNA then remains high throughout development. Since the total level of *actin* mRNA increases early in development, the relative rate of *actin 8* transcription must also increase during this time. In contrast, *actin 6* mRNA represents ~25% of total *actin* mRNA in vegetative cells and gradually decreases to ~2% of total *actin* mRNA at culmination. If *actin 6* mRNA were to decay with kinetics similar to that of total *actin* mRNA during development ($t_{1/2}$ = ~3 h; Margolskee and Lodish, 1980b), it is possible that *actin 6* transcription ceases at or shortly after development is initiated. *Actin M6* mRNA is present at very low levels and would encode a minor actin form. *Actin 7* is expressed at a low level during early development. Two of the genes, *actin 2-sub2* and *actin 3*, which have amino acid differences relative to the major *actin*, are not expressed at detectable levels (<1% of total actin mRNA) and are probably pseudogenes. A sequence comparison between the 5′ end of *actin 2-sub2* and *actin 3* and all transcribed genes shows the lack of the composite TATA–oligo(dT) structure found in other genes.

F. 3′ DNA Sequences of *Actin* Subfamilies

In contrast to the 5′ ends of the *Dictyostelium actin* genes, which share no nucleotide sequence homology, the 3′ untranslated regions show varying degrees of sequence homology and can be grouped into subfamilies (McKeown and Firtel, 1981b, 1982; McKeown *et al.*, 1982). As previously described, there are two sizes of *actin* mRNA, 1.25 and 1.35 kb, which differ in the lengths of their 3′ untranslated regions (McKeown *et al.*, (1978). Figure 20 shows the nucleotide sequence of a series of genomic and cDNA clones extending 3′ from the UAA termination signal. Two of these genes, *actin 4* and *actin 8*, probably encode the 1.25-kb mRNA species and possess a 3′ untranslated region of 43–45 nt. These genes do not share any sequence homology 3′ to the UAA termination codon. All of the other genes encode the longer 1.35 kb mRNA and have a 3′ untranslated region of ~130 nt. A comparison of the nucleotide sequences at the 3′ ends of genes which encode the 1.35-kb mRNA are given in Fig. 20. The analysis of these indicates that all the genes show a strong nucleotide sequence homology within the first 45 nucleotides of the 3′ untranslated region. Plasmid pcDd actin III-12/A1 has a 15-nucleotide deletion within this region. Beyond the first 45 nt, the gene sequences can be divided into two subfamilies. *Actin 5* and the cDNA clones pcDd actin ITL-1 show a strong nucleotide sequence homology extending to the 3′ end of the mRNA. There are two genes that show a strong nucleotide sequence homology to *actin 6* but show no homology with the *actin 5* subfamily.

To estimate the numbers of genes within the *actin 5* subfamily, a 3′-end probe from pcDd actin ITL-1 was isolated and hybridized to a Southern DNA blot hybridization of *Dictyostelium* nuclear DNA (McKeown and Firtel, 1981b). Five *actin* bands hybridized indicating that they are members of the *actin 5* subfamily. At present, only two subfamilies of the 1.35-kb mRNA species have been identified, and they seem to have evolved from a common ancestral gene. Interestingly, the 43–45 nucleotide 3′ untranslated region of *actins 4* and *8* are not homologeous with the first 45 nucleotides of the *actin 5* or *6* subfamilies.

Actin 5, *6*, and *2-sub1* encode 1.35-kb mRNAs and have an oligo(A) stretch within the genomic nucleotide sequence at the region of the poly(A) addition (McKeown and Firtel, 1981b). As previously described, a large fraction of *Dictyostelium* mRNAs contain a short transcribed oligo(A) sequence lying near the 3′ end of the messenger RNA that is separated by one or more nucleotides from the posttranscriptionally added poly(A) (Jacobson *et al.*, 1974b). It is quite possible that these short oligo(A) sequences at the 3′ end of *actin* genes *5*, *6*, and *2-sub1* are transcribed and that the poly(A) addition site is 3′ to these regions. Interestingly, cDNA clone pcDd actin III-12/A1 contains a long 3′ terminal poly(A) stretch which contains a single T residue after the first 20 nt. The simplest explanation is that the first 20 A residues of pcDd actin III/12-A1 and the T residue are encoded within the genome and

Fig. 20. (A) Sequences of 3′ noncoding regions. The sequences of the strand with the same polarity as the mRNA are shown for the regions 3′ to the TAA termination codon. Sequences in parentheses indicate that the exact number of nucleotides in a homopolymer run was not determined. The sequence AATAAA thought to be important for poly(A) addition is underlined. See McKeown and Firtel (1981b) for details. (B) Comparison of 3′ untranslated regions of genes encoding 1.35 kb actin mRNAs. The genes for the 1.35 kb RNAs are arranged so as to highlight homologus regions. All of these sequences are similar for the first 45 nucleotides of the 3′ noncoding region. Beyond this point, two groups of homologous sequences exist. Capital letters indicate that a nucleotide is found at a particular position in one or more of the genes. See McKeown and Firtel (1981a,b) for details.

Fig. 20A.

```
                          10        20        30        40        50        60        70        80        90        100       110
                 Ter       .         .         .         .         .         .         .         .         .         .         .
pcDd actin 8-1   TAAATTATTTAATAAATAATAAAAAAACAAATTGTTGTAATAATCT(A45)

pDd actin 8      TAAATTATTTAATAAATAATAAAAAAACAAATTGTTGTAATAATCTAATATTTTCTTTTTTTTTTAATTTTTTTTTTTTTAAATCTTAATAATTATTAAGTTATTTTAATTTTT

pcDd actin
111-12/A1        TAAATCATGATGAAAGTGCTTCACATAAAAATAATAATAATAATATAACAATAATAATATTTAAATGTATAATAAAATTTAATTACTTTTTTTTTAATGGTTGTTGATCTTTA

pcDd actin
ITL-1            TAAACAAATAATTAAAACTAGTGATGAAAGTGCTTCTCACAAACAATTATGTAAAATATATAATAAAATACATTATTTAATCATTTTTATTTTTGTTTTAGTTGTTGATCTTT

pDd actin 5      TAAACAAAAAAAAAAAACCGAGTGATGAAAGCGCTTCTCACAAAATTATGAAAAATATTTAATAGTATAATAATTTTAAATCTTTTTTATTTTTTTTAGTTGTTGTCTTTATC

pDd actin 6      TAAACTAAACAATTAAAACCAGTGATGAAATGTCTTCTCACACTTAACAATATAATATTTATATGTATAATAATAAAATCTCAATAAAATATAATTCTTATTTTTATTTTTGA

pDd actin 2-
sub 1            TAAATTAATTAAAAAAATTTAGTGATGAAAGTGCTTCTCACACAAAAATTATTATATATGTACAATAATAACAATAAAAACCCAATAAAATATAAACTTTTTTCTTTGATAGT

pDd actin 2-
sub 2            TAAATTAATTAATTAAAAATTTTAATGATGAAATTGTTTCTCACAAACAAACAATTTAATAAATGCACAAATAAAAAAAAAAATAATAATAATAATTATTTTTTTTTGAATAAT

pDd actin 4      TAAACAATTAATAAATATGTTGTATTTTATTTAAACTTAAATTATAAAAAAAAACCATTTTTAATTTTTCNATTTTTAATTTATTTTTTTTTTAGTTTTAATATTGTATTTTT

                          120       130       140       150       160       170       180       190       200       210       220
                           .         .         .         .         .         .         .         .         .         .         .
pDd actin 8 cont.  TTTTTTTTTTTTTTTTTTTTTTTTTTTTTTTTTCTATCAAAAAAATCAAATATATTTAAAAAATTTATTATTTACAGTACATTTTGAATGGTGAAGATAAATATATGCATT

pcDd actin
111-12/A1          TCCGACCTT(AAAAAAAAAAAAAAAAAAA)T(poly A)

pcDd actin
ITL-1              ATAAGACTATTTAAAATTAATTGT(poly A)

pDd actin 5        CGACTTTAAAAATAAAAAAATTGTAAAAAAAAAAAGTTTATTTGTTTAATTTTATTGTTGTTTTTAATTTTTTGCAACCATTAATTTATAAAGCCAGCACCTAGCTATAA

pDd actin 6        ATCGGTTGTTGTCTTTATCCAGCCATCAAAAAAAAAAAAAAAAAAAAAAAAAAAAAAAAAAGCAAATGAAACTATTAATTTTTTTTTTTTTTTAACCCAAAATT

pDd actin 2-
sub 1              CGTTGATCTTTATCCGACCTTTAAAAAAAAAAAAAAATTTCGTTTACTTTTATTTTATTTATTATTTTATTATTAAAAAAAAAGGTTGTTTTAAATTTGGAGA

pDd actin 2-
sub 2              AATTGACTTTAATATAAAAAAAGATATAGAAATTTTTTTATTTTAATTGATTTTAAATTAAAAACTTAGCCAAAAAAAAAAAAAAAAAAAAAAAAAAAATCCGAGTTTTGA

pDd actin 4        AATTTATATTTTTATTTTTCAAATTATCTAAAATAAACAATTCAAAAAAAAAAATATATAAAAAAGTGTTTTAAAAATAATCATGTATTAATTTTGTTTTTATTATTTAT
```

Fig. 20B.

```
                              10        20        30        40        50        60        70        80
                         Ter  °         .         .         .         .         .         .         .
pcDd actin ITL-1         TAAAC AAATAATTAAAACTAGTGATGAAAGTGCTTCTCACAAAcAATTATGtAAAATATaTAATAaaATAc AtTaTTtAAT
pDd actin 5              TAAACaAAAaAAAAAAAcCgAGTGATGAAAGcGCTTCTCACAAA  ATTATGaAAAATATtTAATAgtATAatAaAtTTaAAT
pcDd actin III-12/A1     TAAAtcA             TGATGAAAGTGCTTCaCAtAAAAAtAATAATAataATaTAaCAATAATAATAtTtAAATgT
pDd actin 2-sub 1        TAAAtTAAtTAAAAAAAtTTAGTGATGAAAGTGCTTCTCACAcAAAAAtTAtTAtATATgTA CAATAATAAcAATAAAAaCc
pDd actin 6              TAAACTAAAcAATTAAAATcAGTGATGAAAtgtCTTCTCACActtAAcAatATA ATATtTA tAtgtATAATAATAAAAaCc

                              90       100       110       120       130       140       150       160
                              .         .         .         .         .         .         .         .
pcDd actin ITL-1 cont.    CaTTTTTATTTTTgtTTTAGTTGTTGaTCTTTATCCGACTaTttAAA   AttAATTGT poly(A)
pDd actin 5               CtTTTTTATTTTT  TTTAGTTGTTG TCTTTATCCGACTtTaaAAtaaAaaAATTGT A11
pcDd actin III-12/A1      atAATAAAATATAATT  ACTTTTTT TTTaATgGTtGTTGAT   CTTTATCCGACCTT(A20)T poly(A)
pDd actin 2-sub 1         C AATAAAATATAA    ACTTTTTTcTTTgATAGTCGTTGAT   CTTTATCCGACCTTtA14
pDd actin 6               C AATAAAATATAATTcttaTTTTTatTTTttgAaTCGgTtgTtgtCTTTATCCagCCaTcA33
```

the longer poly(A) is added posttranscriptionally. The single T residue may also be the result of a cloning artifact.

Several eukaryotic mRNAs contain an AAUAAA sequence approximately 10–20 nucleotides 5′ to the site of poly(A) addition and may be a 3′ processing signal or a 3′ poly(A) addition signal (see Breathnach and Chambon, 1981). Similar nucleotide sequences are found in the extremely A + T rich 3′ untranslated region of the *Dictyostelium actin* genes, but these are found as much as 60 nt from the 3′ end of *actin* genes. Their significance remains to be determined.

VII. The Discoidin I Multigene Family

A. Developmental Regulation of *Discoidin I*

Early during the development cycle of *Dictyostelium* several proteins thought to be involved in species-specific cell cohesion appear. One of these proteins is the lectin discoidin I. Data suggesting that discoidin I is involved in cell–cell cohesion are discussed in Chapter 6.

Since discoidin I is easily purified by Sepharose-affinity chromatography, it has been possible to prepare discoidin antibody and use it in a quantitative assay for discoidin I (Simpson *et al.*, 1974; Siu *et al.*, 1976). It has also been possible to analyze the developmental expression of the *discoidin I* genes (Ma and Firtel, 1978; Rowekamp and Firtel, 1980). *Discoidin I* is not detectable in vegetative cells grown on a bacterial food source and only 0.002% of total mRNA in these cells is *discoidin I* specific. It is not known if all vegetative cells synthesize very low levels of *discoidin I* mRNA or if a few cells in a microstarvation environment have initiated development and hence have produced *discoidin I* mRNA in large quantities. Within 1 h after the onset of development, the level of *discoidin I* mRNA begins to increase and within 2 h discoidin I protein synthesis becomes detectable. By 8–10 h, during aggregate formation, discoidin I levels (~1% of total protein) are maximal. The peak rates of *discoidin I* mRNA synthesis precede those of mRNA accumulation (~1% of total mRNA) and discoidin I protein synthesis (Siu *et al.*, 1970; Ma and Firtel, 1978; Rowekamp and Firtel, 1980). Nuclei have been purified from cells at different stages in development and examined for their ability to synthesize *discoidin I* mRNA *in vitro* (Williams *et al.*, 1979). These results are consistent with the *in vivo* pulse-labeling studies. From this, it seems likely that discoidin I synthesis is regulated at the level of transcription. If cAMP is added to developing cells, *discoidin I* mRNA

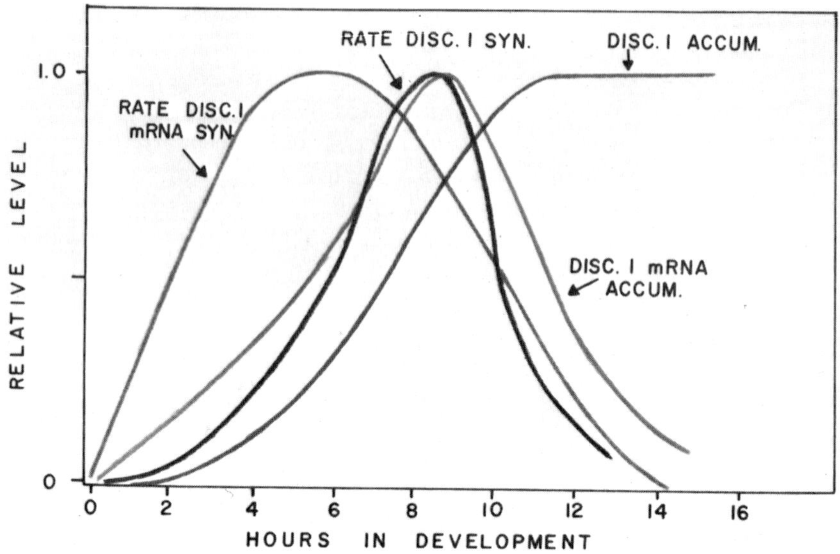

Fig. 21. Developmental regulation of discoidin I. The relative periods of accumulation and synthesis of discoidin I protein and mRNA are shown during development. See Rowekamp and Firtel (1980); Rowekamp *et al.* (1980); and Poole *et al.* (1981) for details.

accumulation is repressed. Nuclei were isolated from developing cells shaken in liquid culture with or without exogenous cAMP and used in an *in vitro* RNA synthesis system to determine their rates of discoidin I mRNA synthesis (Williams *et al.*, 1980). Nuclei from cells incubated without cAMP continue to synthesize *discoidin I* mRNA at a high rate whereas nuclei from cells incubated with cAMP synthesize *discoidin I* mRNA at >10-fold lower rate. These results support a transcriptional regulation model for discoidin I synthesis and suggest that cAMP may be a negative effector of *discoidin I* mRNA synthesis. A summary of the parameters involved in *discoidin I* gene expression is shown in Fig. 21.

B. *Discoidin I* Gene Organization

Several discoidin I cDNA clones were originally selected by their ability to hybridize to mRNA synthesized early (6 h) in development but not to mRNA synthesized in vegetative cells (Rowekamp *et al.*, 1980). RNA that hybridizes to these clones encodes discoidin I. Hybridization probes from these cDNA clones showed that there are probably four genomic *discoidin I* genes; recombinant plasmids and λ-phage containing these nuclear genes

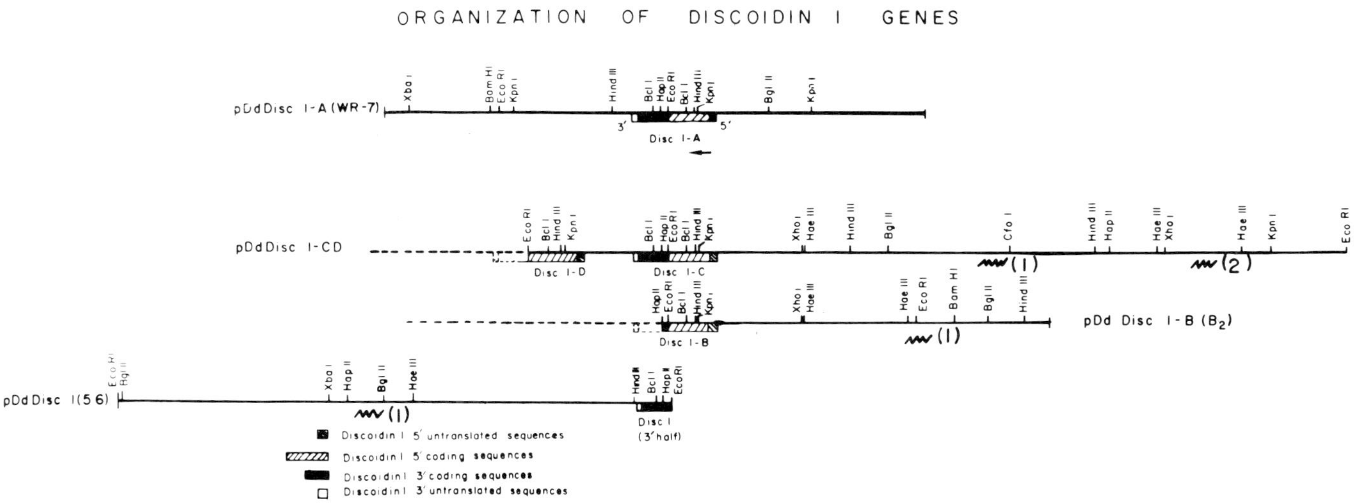

Fig. 22. Structure of the *discoidin I* clones. The genomic inserts containing *discoidin I* genes are shown. Restriction endonuclease sites within and around the genes are also shown; however, not all Bgl II sites in pDd Disc I-CD have been mapped. Direction of transcription in all cases is right to left. The 5′ (*N*-terminal) and 3′ (*C*-terminal) regions of the genes are defined by the Eco RI restriction endonuclease site in each gene. The extent of the protein coding region was determined from the nucleotide sequence. Note that only the 5′ regions of genes *Disc I-B* and *D* have been isolated; pDd Disc I (5.6) contains only the 3′ region of the *discoidin I* gene. See Poole *et al*. (1981a,b) for details. (**1**) and (**2**) indicate approximate regions of two short interspersed repeat sequences found flanking some of the *discoidin I* genes.

were subsequently isolated. The restriction maps for these genomic fragments are shown in Fig. 22 (Poole *et al.*, 1981). Each of the genes contains an Eco RI site located near the middle of the protein coding region. This conserved site has been used to conveniently divide these genes into 5′ (N-terminal) and 3′ (C-terminal) halves. The entire *discoidin (Disc) I-A* gene has been isolated and characterized. No other *discoidin I* genes are within 10 kb of *Disc I-A* in the genome. The entire *Disc I-C* gene has been isolated and is separated from *Disc I-D* by only 300 nt. Only the 5′ halves of the *Disc I-B* and *D* genes have been identified. A 3′ half [*Disc I(5.6)*] has also been isolated, but, it has not been possible to assign it to the *Discoidin I-B* or *D* gene. The other 3′ half has not yet been isolated.

The organization of the regions flanking the four *discoidin I* genes has been examined (Poole *et al.*, 1981a,b). The sequences surrounding *Disc I-A* are entirely single copy whereas the sequences surrounding the other *discoidin* genes are organized in a much more complex pattern. DNA sequencing has indicated that the regions extending at least 1 kb 5′ to genes *Disc I-B* and *C* through the adjacent Xho I and Hae III sites are homologous and that there is an inverted duplication in the region ~6 kb 5′ to *Disc I-C*. The region flanking the 3′ half of *Disc I(5.6)* may be homologous to a region flanking another *discoidin I* 3′ half. Finally members of a repeat sequence family surround *Disc I-B*, *C* and *D*, and *(5.6)*.

C. Sequence Analysis of the *Discoidin* Genes

The DNA sequences for the coding regions of all of the isolated *discoidin I* genes have been determined (see Figs. 23 and 24; Poole *et al.*, 1981a). From the nucleotide sequences and their derived amino acids, one can make some interesting observations about the structure of the discoidin I protein. The N-terminal regions of the *discoidin I-B* and *C* proteins are identical in nu-

---

Fig. 23. Coding and 3′ flanking sequences of the *discoidin I* genes. The top lines show the complete nucleotide and predicted amino acid sequence of *Disc I-C*. The sequences of the other genes are shown only where they differ from the *Disc I-C* sequence. The entire codon containing a difference is shown; the differing nucleotide is italicized. Note that *Disc I-B* and *D* extend only to the Eco RI site at position +442, and *Disc I (5.6)* extends only 3′ from this Eco RI site. The asterisks at +823 in pDdDisc I-A and +874 in pDdDisc I (5.6) mark the poly(A) addition site (see text). AATAAA sequences that have been correlated to poly(A) addition (Proudfoot and Brownlee, 1976) have been underlined in the 3′ untranslated regions. Several cDNA clones have been sequenced and compared with the genomic sequences. pcDdDisc I_8 begins at position −33 of *Disc I-B/C* and has been sequenced to the Hap II site at position +524. pcDdDisc I_4 begins at position −76 of *Disc I-D* and has been sequenced to the Eco RI site at +442. pcDdDiscI_{13} is identical to *DiscI (5.6)* extending from position +217 to the poly(A) tail found after the C residue at position +873. See Poole *et al.* (1981a) for details.

Fig. 23.

DISCOIDIN I GENES

```
                A          B
            -------    ----------   -->            -60
Disc B/C    GTATAAAAAG--AAAAAAAATTTTTAATTAAATCATTAAATTGAAAAATTAAAATTTCA----TACAAATTATCTTTTTAAATAATAAAATTTA---------TAAA
Disc D      CTATAAAAAGAAAATATGATTTTCATTTTAAATCATTAAACTGAAAAATTAAAATTTCATTTATACAAATTATC-TTTT---T-TTAAA-TTCAACACACAATTAAA
Disc A      CTATAAATAAAAGAAGTTTTTTTTTTTTTT-AATCATTAAATTG-AAAATCAAAATTTCATTAATTCAAATTATC-TTTTAAATATTAAA-TTCAA--CACAAATAAA

            1                                                                          60
            Met Ser Thr Gln Gly Leu Val Gln Leu Ile Ser Asn Ala Gln Cys His Leu Arg Thr Ser Thr Asn Tyr Asn Asp Val His
Disc B/C    AUG TCT ACC CAA GGT TTA GTT CAA CTT ATC TCA AAT GCT CAA TGT CAT TTA AGA ACC TCA ACC AAT TAC AAT GAT GTC CAC

                                        Thr     Leu Gly                                                     Gly
Disc D                                  ACA     CTC GGA                                                     GGT

                                                Leu Ala                                                     Gly
Disc A                                      CTC CTC GCA             TGC                                     GGT

                                                            120
            Thr Gln Phe Asn Ala Val Leu Asn Tyr Lys Asn Lys Gly Thr Asn Thr Ile Asp Gly Ser Glu Ala Trp Cys Ser Ser Ile
Disc B/C    ACT CAA TTT AAT GCT GTT TTA AAC TAT AAA AAC AAA GGT ACC AAT ACT ATT GAT GGT TCA GAA GCT TGG TGT TCA TCA ATC
                                                            Kpn I                       Hind III
                                Ala
Disc D                          GCT

                            Ser Ala                 Asn
Disc A                      TCT GCC         TAC     AAT             ACC

                                180                                                                         240
            Val Asp Thr Asn Gln Tyr Ile Val Ala Gly Cys Glu Val Pro Arg Thr Phe Met Cys Val Ala Leu Gln Gly Arg Gly Asp
Disc B/C    GTA GAT ACC AAC CAA TAC ATT GTT GCT GGT TGT GAA GTT CCA CGT ACT TTT ATG TGT GTT GCT CTC CAA GGT CGT GGT GAT
                                                                                                            Bcl I
Disc D              ACT
Disc A

                                                                           300
            His Asp Gln Trp Val Thr Ser Tyr Lys Ile Arg Tyr Ser Leu Asp Asn Val Thr Trp Ser Glu Tyr Arg Asn Gly Ala Ala
Disc B/C    CAT GAT CAA TGG GTT ACA TCA TAC AAA ATC CGT TAT TCA TTA GAT AAT GTT ACC TGG TCT GAA TAT CGT AAT GGT GCT GCT
Disc D
            Ala                                                                 Ser     Phe
Disc A      GCT                                                                 TCC     TTT
```

(continued)

Fig. 23. (Continued)

```
                   •                                           360                                                 •
              Ile Thr Gly Val Thr Asp Arg Asn Thr Val Val Asn His Phe Phe Asp Thr Pro Ile Arg Ala Arg Ser Ile Ala Ile His
Disc B/C      ATT ACT GGT GTA ACT GAT CGT AAC ACT GTT GTT AAT CAT TTC TTT GAT ACT CCA ATT AGA GCT CGT TCA ATT GCT ATC CAC
Disc D

              Val
Disc A        GTT
                              420                                                                                 480
                                                                  •
              Pro Leu Thr Trp Asn Asn His Ile Ser Leu Arg Cys Glu Phe Tyr Thr Glu Pro Val Gln Ser Ser Val Thr Gln Val Gly
Disc B/C      CCA TTA ACC TGG AAT AAC CAC ATT TCA TTA AGA TGT GAA TTC TAC ACT CAA CCA GTA CAA AGC TCA GTC ACT CAA GTT GGT
                                                              Eco RI
Disc D                                                                 -   -   -   -   -   -   -   -   -   -   -   -   -

                                  Gly
Disc A                            GGT
Disc (5.6)     -   -   -   -   -   -   -   -   -   -   -   -
                                                                                 540
              Ala Asp Ile Tyr Thr Gly Asp Asn Cys Ala Leu Asn Thr Gly Ser Gly Lys Arg Glu Val Val Val Pro Val Lys Phe Gln
Disc C        GCA GAT ATT TAC ACT GGC GAT AAC TGT GCC TTA AAT ACC GGT TCA GGT AAA CGT GAA GTT GTT GTC CCA GTT AAA TTC CAA
                                                              Hap II
Disc A
Disc (5.6)
                                                           600
               •                                                                                   •
              Phe Glu Phe Ala Thr Leu Pro Lys Val Ala Leu Asn Phe Asp Gln Ile Asp Cys Thr Asp Ala Thr Asn Gln Thr Arg Ile
Disc C        TTT GAA TTT GCT ACT CTC CCA AAG GTT GCC CTC AAC TTT GAT CAA ATC GAT TGT ACT GAT GCT ACC AAT CAA ACT CGT ATT
                                                                 Bcl I
Disc A                                    GCT                                             GCC             ACC
Disc (5.6)
```

```
                                 660                                                                 720
              Gly Val Gln Pro Arg Asn Ile Thr Thr Lys Gly Phe Asp Cys Val Phe Tyr Thr Trp Asn Ala Asn Lys Val Tyr Ser Leu
Disc C        GGT GTT CAA CCA AGA AAC ATT ACC ACC AAA GGT TTC GAT TGT GTT TTC TAC ACT TGG AAT GCT AAC AAA GTC TAT TCA TTA
                                                                                      Glu
Disc A            GTC             AAT                     TTT                         GAA         GTT TAC
                                                                                      Glu
Disc (5.6)                                                TTT                         GAA
                                       •                                 780
              Arg Ala Asp Tyr Ile Ala Thr Ala Leu Glu Ter                 •                            •
Disc C        AGA GCT GAT TAC ATT GCT ACT GCT TTA GAA TAA ATTTTTAATCTTATTATTAAAATTGTAATTTAATTTTTAATCAAATTTAATAAAAAAAAAAAA
                                                                                                                   *
Disc A                                ACC     TTG         ATTATTTAAATAAATTTTCATACTCTTAATACTTTTGTATTATTACTCAATTTAAATTTCATT
Disc (5.6)                        GCC ACC                 AAACAAATAATTTTTTAAAAAATTATTAAAATTGTAATTTAATTATTAATCAATTTTAAAAAA

                 840
                  •                                                       900
Disc C        AAAAAATCCAAATCCAGGTTTGGTTAATAAAATTTTTCCAATTCAAACCAATAATAATAAA        •                          •
Disc A        TTTTTTTTTTTTTTTTTTTTTTTTTTAAAAAAAAATAGTAATAATAATAATAATTGTAATACCTTTTTACATCATGATTTAAAAAAGTATTATAATTTTTATAAA
                                                               *
Disc (5.6)    AAATAAAAACAAATAAAATTTAATTTTATTTTTTTTTATTATAAAAACAAAAGCTT
                                                              Hind III
Disc A        TTTTTCCACGATTACTAAGATTGTATTTTGTAAGGCTTACCTTGGAGACAGTAAGCTT
                                                                Hind III
```

5' ENDS OF DISCOIDIN I GENES

```
                                            -200                 -180                 -160
Disc I-A                                                    ~25A TTAAAAAAAA TAAAATCAAA TTACAAAGTA
Disc I-D                                                         TACAAAATAA AAAAATTAAA GAAAAAATAA
Disc I-B                              TTAAACAAT GATTACTTTT TTGTCAATTT ACAAAAAAAA TTAAAAATAA TTAAATTTAA
Disc I-C                             TTTAAACAAT GATTACTTTT TTGTCAATTT ACAAAAAAAA TTAAAAATAA TTAAATTTAA

                      -140                 -120                 -100                 -80
                                                                  A                    B
Disc I-A  GTTTAACTTC ATAGTGTTAA AAAATCAATA ATAAAAAAAA AACTATAAAT AAAAGAAGTT TTTTTTTTTT TT-AATCATT
Disc I-D  ATTAAATCTT TAATTTTAAT TGGAAAAAAA AAAAAAAAAA AACTATAAAA AGAAAATATG ATTTTCATTT TAAAATCATT
Disc I-B  AAAATAAAAA AAAAAAAAAA AACCAAATTA AAAAAAGAAA AAGTATAAAA AG--AAAAAA AATTTTTAAT TAAAATCATT
Disc I-C  AAAATAAAAA AAAAAAAAAA AACCGAATTA AAAAAAGAAA AAGTATAAAA AG--AAAAAA AATTTTTAAT TAAAATCATT

                      -60                  -40                  -20
                                                                                       Met
Disc I-A  AAATTG-AAA ATCAAAATTT CATTAATTCA AATTATC-TT TTAAATATTA AA-TTCAA-- CACAAATAAA AUG
Disc I-D  AAACTGAAAA ATTAAAATTT CATTTATACA AATTATC-TT TT---T-TTA AA-TTCAACA CACAATTAAA AUG
Disc I-B  AAATTGAAAA ATTAAAATTT CA----TACA AATTATCTTT TTAAATAATA AAATTTA--- ------TAAA AUG
Disc I-C  AAATTGAAAA ATTAAAATTT CA----TACA AATTATCTTT TTAAATAATA AAATTTA--- ------TAAA AUG
```

cleotide sequence. The N-terminal halves of both *Disc I-A* and *D* differ from each other and also from *Disc I-B* and *C*. The N-terminal nucleotide differences occur with equal frequency at first, second, and third codon positions and would yield proteins with several amino acid differences. Although the frequency of nucleotide differences in the C-terminal halves is approximately the same as in the N-terminal region, almost all of these C-terminal differences occur at the third base position of the codons and would not yield amino acid differences. The one non-third-base position change would result in an amino acid difference. Since the remaining 3′ end is not isolated and *Disc I-(5.6)* is not yet assigned to either *Disc I-B* or *D*, we do not know if the entire *Disc I-B* and *C* genes would encode identical polypeptides. However, it seems clear that the *discoidin I* genes would encode discoidin I proteins with divergent N-termini and conserved C-termini.

Except for a single base difference between the *Disc I-B* and *C* located 126bp 5′ to the AUG, the two genes are identical in nucleotide sequence from the Eco RI through the entire N-terminal protein coding region to ~200 bp 5′ to the AUG initiation codon. The extent of homology in the 3′ region is not known since the C-termini of *Disc I-D* has not yet been identified. The region 5′ to the AUG initiation codon is >95% A + T in all of the genes. There is some sequence homology between *Disc I-A, B, C* and *D* from the AUG to the transcription initiation start site but further 5′ to this the sequences are very divergent. The sequences 3′ to the UAA termination codon for the three *Disc I* C-termini isolated are completely divergent. It is interesting that the 5′ ends of *discoidin I* are similar yet the 3′ ends are divergent; the *actin* genes possess completely opposite properties.

D. Transcription Analyses of the *Discoidin I* Genes

The 5′ initiation sites of the transcribed genes have been determined using the S_1 mapping procedure outlined in Fig. 17 (Poole *et al.*, 1981a). A fragment containing the 5′ end of a *discoidin I* gene was end labeled and hybridized with poly(A)$^+$ RNA isolated from cells ~8 h into the developmental cycle. The hybrids were treated with S_1 nuclease and the digests analyzed by gel electrophoresis relative to appropriate DNA sequencing

Fig. 24. Sequence of *discoidin I* genes and flanking sequences. Location of tbe TATA box and oligo(dT) run are indicated by boxed regions A and B respectively. Arrows indicate sites of initiation of transcription as has been determined (see Poole *et al.*, 1981a). Nucleotide sequences were arranged such that maximum amount of nucleotide sequence homology was obtained in the 5′ untranslated regions of the various genes.

ladders. Since *Disc I-B* and *C* are identical in sequence, only *Disc I-A, B* and *D* were analyzed. These results are shown in Fig. 24. *Disc I-A, B* and *D* all hybridize mRNA and have mRNA termini at approximately identical positions. Interestingly, these mRNA start sites lie 3′ to a conserved TATA–oligo(dT) region. Since *Disc I-A, B,* and *D* are similar, but not identical, in sequence in this 5′ region, control experiments were performed to demonstrate that the RNA hybridizations observed were not the result of artifacts of cross-hybridization. The results indicate that at least three of the *discoidin I* genes are expressed. It has been possible to confirm that *Disc I-B, C* and *D* are transcribed by identifying cDNAs corresponding to them.

There are two size classes of *discoidin I* mRNA (Rowekamp *et al.*, 1980). The 3′ end of *Disc I-A* has been determined by the S_1 procedure and lies ~60 pb 3′ to the UAA termination codon. By comparing the DNA sequences of *Disc I-(5.6)* to its appropriate cDNA, it has been possible to locate its 3′ end ~115 bp 3′ to the UAA (Poole *et al.*, 1981a). Like the *actin* genes, the two size classes of *discoidin I* mRNA are the result of differences in the length of their 3′ untranslated regions. Like the *actin* genes, *discoidin I* genes are devoid of intervening sequences.

E. Multiple *Discoidin I* Forms

Discoidin I can be separated by high resolution 2D gel electrophoresis into three isoelectric forms (Poole *et al.*, 1981a; Tsang *et al.*, 1981). This agrees with the predicted charge amino acid differences between *Disc I-A, B/C,* and *D.* When *discoidin I* mRNA is translated *in vitro,* the same isoelectric forms are generated indicating that *discoidin I* is not posttranslationally modified in any manner that would alter its size or isoelectric point. Williams and co-workers have been able to unambiguously assign a particular isoelectric form to a particular *discoidin I* cDNA (Tsang *et al.*, 1981).

A mutant, *HJR1*, has been described that is unable to form tight aggregates but accumulates wild-type levels of discoidin I protein with normal developmental kinetics (Ray *et al.*, 1979). Interestingly, the discoidin I from *HJR1* lacks biological activity as measured by the ability to bind carbohydrate moieties or to agglutinate sheep red blood cells. Revertants of *HJR1* are able to aggregate and the ability of these revertants to develop is proportional to the activity of their discoidin I (Shinnick and Lerner, 1980). It has been suggested that *HJR1* is a mutation in *discoidin I.* Although there are multiple active *discoidin I* genes, analysis of the *HJR1* revertants indicates that the original mutation was at a single site. The *discoidin I* isoelectric forms may be functionally distinct based upon their variable N-terminal regions. Thus, a single site mutation could inactivate *discoidin I* activity without affecting discoidin I protein level. Since native discoidin I is proabably a tetramer, it may be necessarily composed of different *discoidin I* gene products (Simpson *et al.*, 1974; Poole *et al.*, 1981a,b).

F. The Evolution of the *Discoidin I* Genes

The *discoidin I* genes can be subdivided into four sections (Poole *et al.*, 1981a). The N-terminal half of the protein coding regions are divergent while the C-termini are conserved. The 5′ untranslated regions are conserved, and 3′ untranslated regions are completely unrelated. A detailed comparison suggests that the 5′ untranslated region of *Disc I-D* is more homologous to *Disc I-A* than to B/C. This contrasts with the *actin* genes where *actin 2 sub2* is more closely related to the closely linked *actin* gene 2-sub1 than to any other *actin* gene.

It has been suggested that the *discoidin I* C-termini are under strong selective pressure and that the N-termini have diverged, generating *discoidins* with distinctive functions (Poole *et al.*, 1981a,b). These N-terminal structural and functional differences may now be necessarily preserved by selection.

The 5′-end and 3′-end relationships are more perplexing. Perhaps the 5′ ends have remained similar since all the genes are coordinately expressed. It is unclear why the 3′ ends are so divergent. This is especially intriguing since the *actin* genes have similar 3′ ends but divergent 5′ ends.

Finally, common sequences which flank several *discoidin I* genes have been observed. It is possible that recombination between these sites may have expanded the gene family. A similar mechanism has been proposed to describe the evolution of the *globin* gene family (Efstradiatis *et al.*, 1980).

VIII. Transcription of Short, Interspersed Repeat Sequences

In most eukaryotes a large fraction of the single-copy DNA is adjacent to short sequences that are repeated many times in the genome (Davidson *et al.*, 1975). In *Dictyostelium*, ~50% of the genome consists of short, repeated sequences interspersed with single-copy DNA (Firtel and Kindle, 1975). Linkage hybridization suggests that ~30% of hnRNA and mRNA molecules are transcribed from repeat and single-copy sequences (Firtel *et al.*, 1976b). Kindle and Firtel (1979) isolated two recombinant plasmids, KH10 and M4, that contain short interspersed repeat sequences complementary to a large population of mRNA. Subsequent analysis indicated that the repeat sequences are located on the 5′ end of the mRNAs and are covalently associated with single-copy sequences (Kimmel and Firtel, 1979). Further analysis suggests that the *M4* mRNA set may be differentially expressed in a coordinate manner during *Dictyostelium* development (Kimmel and Firtel, in preparation). Several other repeat families have been isolated that are transcribed as part of compound transcripts containing repeat and single-copy sequences and may also be differentially regulated in *Dictyostelium* development. One of these is found ~1 kb 5′ to the *M3* duplicated genes but is not part of the *M3* transcripts. Zucker and Lodish (1981) have identified a genomic fragment that contains two repeat sequence members. One repeat is associated with RNAs that are regulated during early development and a second repeat is found associated with RNAs regulated late in *Dictyostelium* development. The *M4* repeat has been analyzed in the most detail and is described below.

A. Transcription of the *M4* Repeat

Plasmid M4 contains ~5 kb of *Dictyostelium* genomic DNA and hybridizes to ~1% of total vegetative poly(A)$^+$ RNA. The genomic insert is mostly single copy but also contains a short sequence repeated 50–100 times in the *Dictyostelium* genome. The location of the repeat and single-copy sequences were identified by Southern genomic DNA-blot hybridizations and by DNA excess-hybridization kinetics (Kimmel and Firtel, 1979). Band 4 hybridizes to >50 *Dictyostelium* genomic DNA fragments and contains a sequence estimated to be repeated ~100 times by DNA excess-hybridization kinetics (Fig. 25). Further analysis indicates that it is interspersed with single-copy sequences in the genome. *M4* fragments 1, 2, and 3, however, hybridize to only one genomic DNA band indicating they are entirely single copy, in agreement with results obtained by DNA excess hybridization kinetics (Kimmel and Firtel, 1979). [The weak additional hybridization seen with Band 3 can be attributed to nonspecific hybridization to the multiple rDNA copies (see Kimmel and Firtel, 1979)].

The pattern of transcription of the single-copy and repeat components for *M4* have been analyzed by RNA and DNA excess hybridization and by RNA blot hybridization (Kimmel and Firtel, 1979; Kimmel *et al.*, 1979). As is seen in Fig. 26, the single-copy Bands 1 and 2 hybridize to the identical 0.9-kb mRNA; which is ~0.1% of the total mRNA in vegetative cells. As indicated

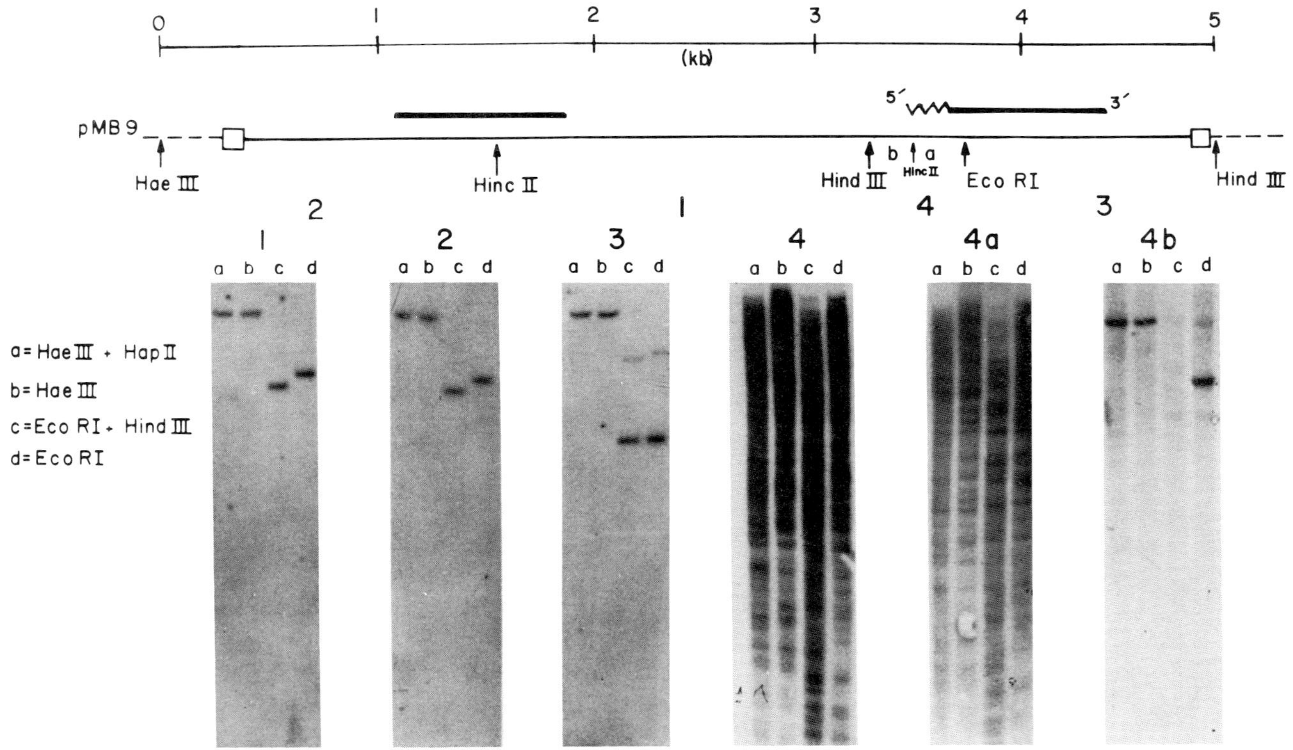

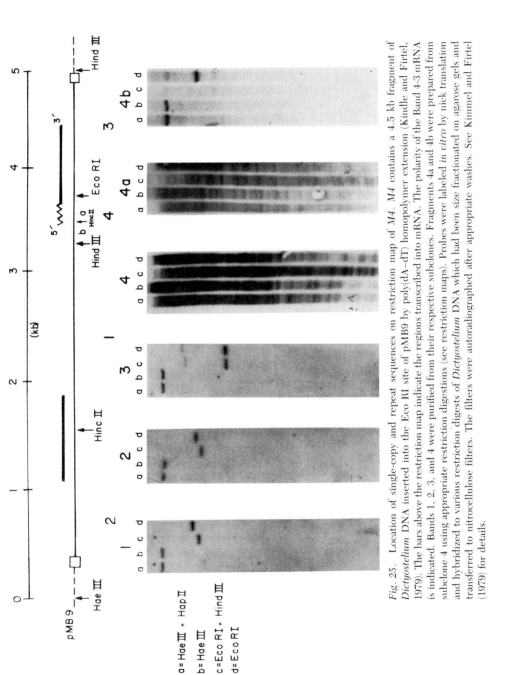

Fig. 25. Location of single-copy and repeat sequences on restriction map of *M4*. *M4* contains a 4.5 kb fragment of *Dictyostelium* DNA inserted into the Eco RI site of pMB9 by poly(dA–dT) homopolymer extension (Kindle and Firtel, 1979). The bars above the restriction map indicate the regions transcribed into mRNA. The polarity of the Band 4-3 mRNA is indicated. Bands 1, 2, 3, and 4 were purified from their respective subclones. Fragments 4a and 4b were prepared from subclone 4 using appropriate restriction digestions (see restriction maps). Probes were labeled *in vitro* by nick translation and hybridized to various restriction digests of *Dictyostelium* DNA which had been size fractionated on agarose gels and transferred to nitrocellulose filters. The filters were autoradiographed after appropriate washes. See Kimmel and Firtel (1979) for details.

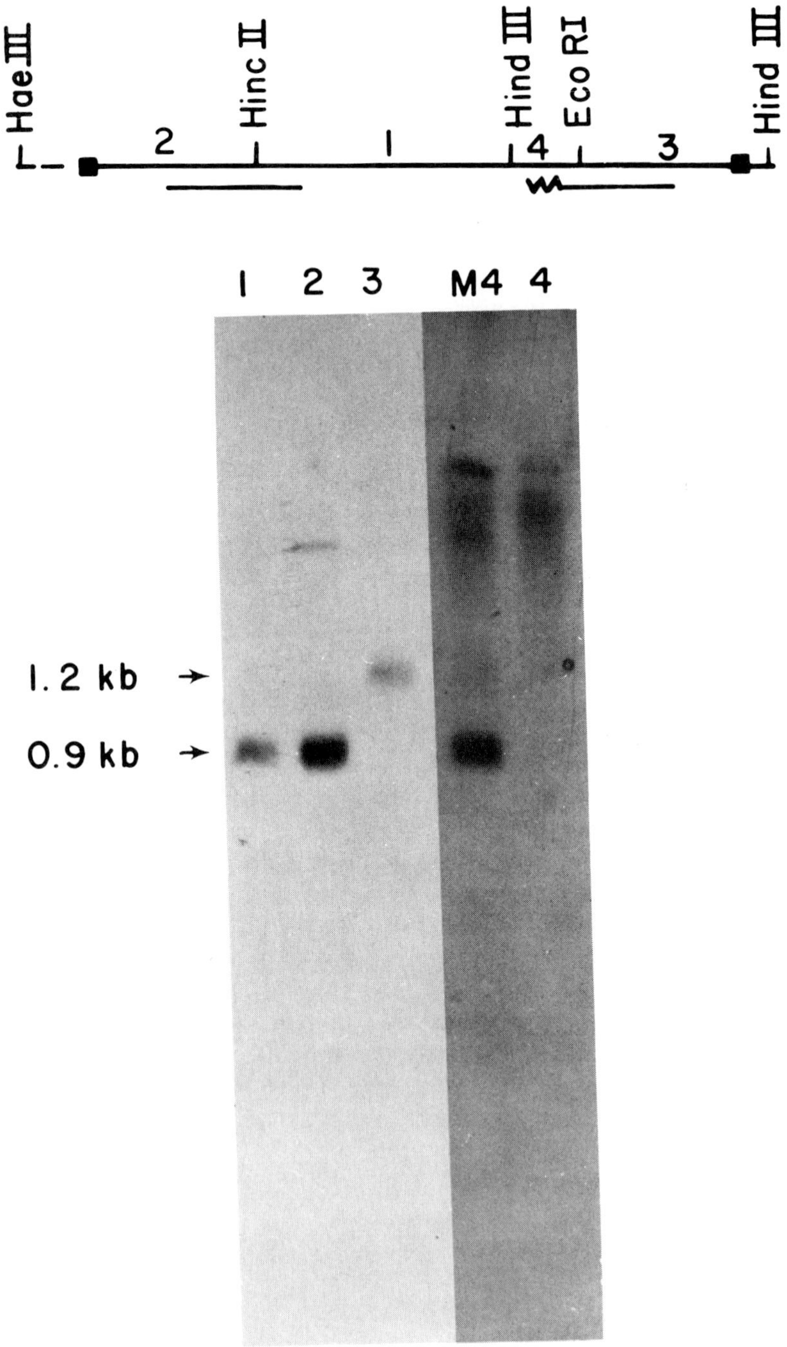

HaeIII
HincII
HindIII
EcoRI
HindIII
2
1
4
3
1 2 3 M4 4
1.2 kb
0.9 kb

in the map in Fig. 25, Bands 1 and 2 are thought to comprise an entire single-copy transcription unit.

Similar experiments with Band M4 have shown that it is complementary to a heterogeneous size population of mRNAs, which represent ~1% of the total vegetative mRNA (Fig. 26). By hybridization of separated DNA strands of Band 4, it was shown that only one of the strands is complementary to mRNA and hnRNA. This indicates that the *M4* repeat sequence is asymmetrically transcribed. This contrasts with transcribed repeats in sea urchin wherein both strands of the interspersed repeat sequences are transcribed (Scheller *et al.*, 1978; Costantini *et al.*, 1978). Linkage hybridization experiments indicate that ~90% of the mass of the transcripts that hybridize to the repeat sequence in Band 4 is complementary to single-copy DNA and that the repeat and single-copy sequences are linked on the same transcript. From these results it was concluded that members of the *M4* repeat family are transcribed into a population of over 30 different mRNA molecules, each of which make up <0.03% of the total mRNA population. In addition, the same strand of the repeat is transcribed in each repeat single-copy transcription unit.

Hybridization with Band 3 shows that it is complementary to a ~1.0 kb low abundance class mRNA, which represents ~0.01% of total mRNA or five mRNA molecules per cell. [Under nondenaturing conditions this mRNA migrates with an apparent size of 1.2 kb as seen in Fig. 26. Its actual size is 1.0 kb (see Kimmel and Firtel, 1979, 1980)]. From mapping experiments including exonuclease–polarity and sandwich hybridization, S_1 nuclease protection, and DNA sequence analysis, it was shown that the Band 3 transcript initiates within Band 4 and that the repeat sequence in Band 4 lies at the 5′ end of the gene. The polarity of transcription of the Band 4-3 gene and the repeat is the same (Kimmel and Firtel, 1979). The location of the repeat and single-copy sequences were identified by Southern genomic DNA blot hybridization and by DNA excess hybridization kinetics.

B. Transcription and 5′ Sequences of the *Band 4-3* mRNA

In *Dictyostelium*, hnRNA is ~20% larger than mRNA and is suggested to be processed to yield mature mRNA (Firtel and Lodish, 1973). To identify potential nuclear precursors for the *Band 4-3* mRNA, the single copy Band 3

Fig. 26. mRNAs complementary to *M4* sequences. *In vivo* labeled ^{32}P-labeled polyA$^+$ RNA was hybridized to various *M4* DNA fragments immobilized on nitrocellulose. The polyA$^+$ RNA was eluted and electrophoresed on polyacrylamide gels and the RNAs identified by autoradiography.

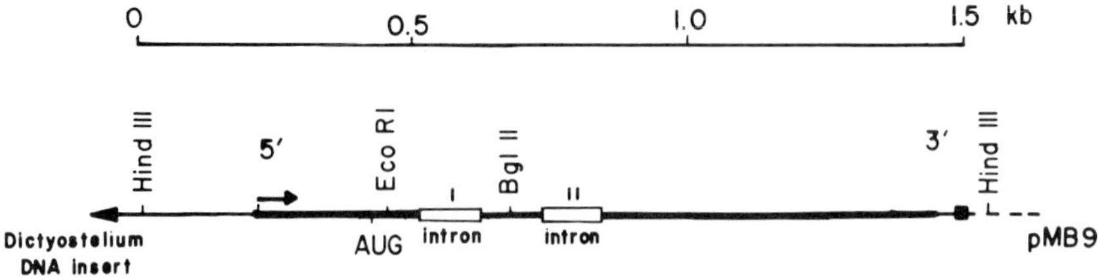

Fig. 27. Detailed restriction endonuclease map of the *Band 4-3* gene of clone M4. Locations of intron I and II within the gene are shown.

sequence was hybridized to a Northern RNA blot containing poly(A)$^+$ RNA isolated from nuclei and polysomes (Kimmel and Firtel, 1980). The results indicated that the *Band 4-3* mRNA is transcribed as a 1.25-kb precursor which is processed into the 1.0-kb mature mRNA. Using S_1 nuclease protection and DNA sequencing analysis, the *Band 4-3* gene was shown to contain two intervening sequences. The intervening sequences are transcribed as part of the hnRNA, poly(A) is added posttranscriptionally and the introns are subsequently removed to generate the mature *Band 4-3* mRNA. The intron locations are seen in Fig. 27. The 5′ ends of the hnRNA and mRNA are located at the same site within Band 4 indicating that the 5′ end is not processed. Figure 27 shows a detailed analysis of the *M4* gene and Fig. 28 shows the RNA sequence and the derived N-terminal amino acid sequence of the protein coding region (including the introns). As shown, both introns are short (~100 pb) and are ~95% A + U. As shown in Table I splice junctions of the *Band 4-3* gene, as well as the *M3* genes, are similar to those of other eukaryotic hnRNAs.

Figure 29 gives the entire sequence of the Band 4 fragments of the *M4* gene. This sequence contains the 5′-untranscribed region, the RNA synthesis initiation site, the 5′-untranslated region and the AUG initiation codon of the *M4* mRNA. The untranslated region is ~150 nt. Approximately 100 nt 5′ to the AUG initiation codon is an 11 nt sequence with strong homology to the 3′ end of the *Dictyostelium* 17 S rRNA, which may be important for *M4* mRNA translation (Hagenbuchle *et al.*, 1978). Interestingly, the first AUG is followed by in frame stop codons and is not the translation initiation codon.

There are three regions of particular interest which lie 5′ to the RNA initiation site. There is a TATTAATA sequence and an oligo(dT) stretch

Fig. 28. DNA sequence of M4-3,4 messenger RNA from the AUG initiation codon through the second intron. The derived amino acid is also shown. Locations of the introns have been determined by S_1 nuclease mapping. See Kimmel and Firtel (1980) for details.

```
 2         2           3                 Eco RI     3          3    2
AUG AGA UUA UUA GCA AAA CAU AAA ACA UCA [GAA UUC] CAU AGC GAA AUU GAA UUA AUC GAA
Met Arg Phe Phe Ala Lys His Lys Thr Ser Glu Phe His Ser Glu Ile Glu Leu Thr Glu

     3   3       2   3           3
UUU AAU AAU UUA GAU AAC UCA UUC AUU AA ↓GUUUGUAUUUUUUUUUUAUUUUAUAUAUAUUAUCAAAAAA
Phe Asn Asn Leu Asp Asn Ser Phe Ile Lys

                                  3         I N T R O N   I
UUUUAUUAUUAUUAUUUUCAUUAUAUUAAUUUUAUUUAUUUAUUUUCUAUUUAUUUAUUUAUUUAUUUCAAG↓ A  UUC
                                                                        (Lys) Phe

            3                                   3       3   3                Bgl II
CCA UUA UUG GUU GAA AAA UCA AUU ACA GAG GGU AGU UAU AAU AAA AUU AUU CAA UCA [AGA
Pro Phe Leu Val Glu Lys Ser Ile Thr Glu Gly Arg Thr Asn Lys Ile Ile His Ser Arg

                                                3          2             3
UCU] GGU GUA CCA UCA GAA UAU UAU CAA GUU UUC CUU GAU AUU UUA GCA GAU UCA AUA AA
Ser Gly Val Pro Ser Gly Tyr Tyr Gln Val Phe Leu Asp Ile Ile Ala Asp Ser Ile Lys

          I N T R O N   II                   3                 3              3
↓GUAUGUUUAUUUUAUUUUUAUUAUUAUUAUACCAUUAUUUAUUUUUAAUUUAUUAUUAUAUUUUAAUAUUUAUUAUAUUAA

 1                              3               3                  2
AUGAUAUAUCAG↓ G CAG GAU AUU GCA AAU UGU AGU GAA AAA UCA UUU AAA ACA UUA UCA UUA AAA
            (Lys) Glu Asp Ile Ala Asn Lys Ser Glu Lys Ser Phe Lys Thr Leu Ser Leu Lys
```

M4 - Band 4 Sequence

```
                                                                  50
A AGC TTG TTG CTG GTA ATT CAG ATA AAG CAC AAA TTA CAC CAG TTT TAG TTC AAT TAA AAT TAG CAG CAA CAA
T TCG AAC AAC GAC CAT TAA GTC TAT TTC GTG TTT AAT GTG GTC AAA ATC AAG TTA ATT TTA ATC GTC GTT GTT

                        100
TTC ATT TAG AAA AAC CAA CTT CTC TTT CAA ATG TTT CTG ATA AAG TTA AAA AAG ATT TAG TTT TAG CAA GTA TGT
AAG TAA ATC TTT TTG GTT GAA GAG AAA GTT TAC AAA GAC TAT TTC AAT TTT TTC TAA ATC AAA ATC GTT CAT ACA

150       (AAC) n                                             200                       A
TAA CAA CAA CAC AAC AAC AAC AAC AAC AAC AAC AAC GAC AAA AAT ATA TAA AAA AAA AAT ACA TTT ATT AAT ATT
ATT GTT GTT GTG TTG TTG TTG TTG TTG TTG TTG TTG CTG TTT TTA TAT ATT TTT TTT TTA TGT AAA TAA TTA TAA

              B                 250    5′→
TTT ATC TTT TTT TTT TTT TTT TAA AAA ACT ATT TTA GGA GAG ATT TTA GAA TTA ATT TCA TTA TAT TCA ATT AAA
AAA TAG AAA AAA AAA AAA AAA ATT TTT TGA TAA AAT CCT CTC TAA AAT CTT AAT TAA AGT AAT ATA AGT TAA TTT

300                                                           350
ATT AAA GAT ATT GAT TCA TTT GAA AGA ACA TTT AAT CAA TTA AAG ACA TAC TAT TAT GAT TAT AAA TCA ATT ATT
TAA TTT CTA TAA CTA AGT AAA CTT TCT TGT AAA TTA GTT AAT TTC TGT ATG ATA ATA CTA ATA TTT AGT TAA TAA
                 17S Hom
                              400
GCA CCA TCA ACA TTA GAA TAT CAA ATT ATT GGT TTG AAT TTA ATG AGA TTA TTA GCA AAA CAT AAA ACA TCA GAA
CGT GGT AGT TGT AAT CTT ATA GTT TAA TAA CCA AAC TTA AAT TAC TCT AAT AAT CGT TTT GTA TTT TGT AGT CTT

450
TTC                                                     ——→
AAG                                                     Translation
                                                        start
```

within 40 nt of the 5′ end of the hnRNA and mRNA sequence. Similar sequences have been observed in other *Dictyostelium* genes and may be common recognition sites for RNA polymerase II (see Fig. 32). Slightly more upstream is a repeating trinucleotide, AAC. As will be discussed, this makes up the *M4* repeat.

C. Characterization of the *M4* Repeat

The repeat sequence on Band 4 was identified by the isolation and subsequent DNA analysis of other *Dictyostelium* genomic fragments complementary to the *M4* repeat. The results indicate that the repeating $(AAC)_n$ was found on each of these genomic fragments and that it constitutes the entire repeat sequence. One genomic fragment was identified which contained 49 contiguous (AAC)s. In addition, a series of cDNAs complementary to the *M4* repeat that also possessed the $(ACC)_n$ sequence were isolated. A series of deletions in Band 4 were generated with Bal 31 exonuclease to analyze the properties of the (AAC) repeat within *M4*. These deletions were sequenced and hybridized to *Dictyostelium* genomic DNA blots and poly$(A)^+$ RNA blots. The results indicate that $(AAC)_n$ is the common repeat sequence. This is shown specifically with deletion M4-4Δ8B which hybridizes to a single-copy sequence and to only the *M4* Band 4-3 mRNA. As seen in Fig. 30, this deletion is missing essentially only the AAC repeat. Furthermore, synthetic $(dAAC)_n : (dGTT)_n$ polymer hybridizes to Southern DNA blots and Northern RNA blots with a pattern that is identical to those observed with the *M4*-Band 4 fragment.

D. Developmental Expression of the *M4* Sequence

One of the most interesting observations concerning the mRNAs that contain the *M4*(AAC) repeat is that they are differentially expressed during *Dictyostelium* development. This is seen in Fig. 29. The *M4* repeat was hybridized to poly$(A)^+$ RNA isolated at various developmental stages. Although ~1% of the vegetative mRNA hybridizes to the repeat, the percentage of mRNA hybridization rises dramatically during development. To determine if a set of mRNAs with the *M4* repeat increases in expression during develop-

---

Fig. 29. The entire sequence of the *M4*-4 fragment. The transcription and translation start sites of the *M4* band 4-3 mRNA are indicated. Sequences A (TATA) and B [oligo(dT)] may comprise a *Dictyostelium* promoter region. The $(AAC)_n$ region is labeled as well as the sequence complementary to the 17 S rRNA. Deletion *M4*-4ΔB extends from nucleotide 143 to 183.

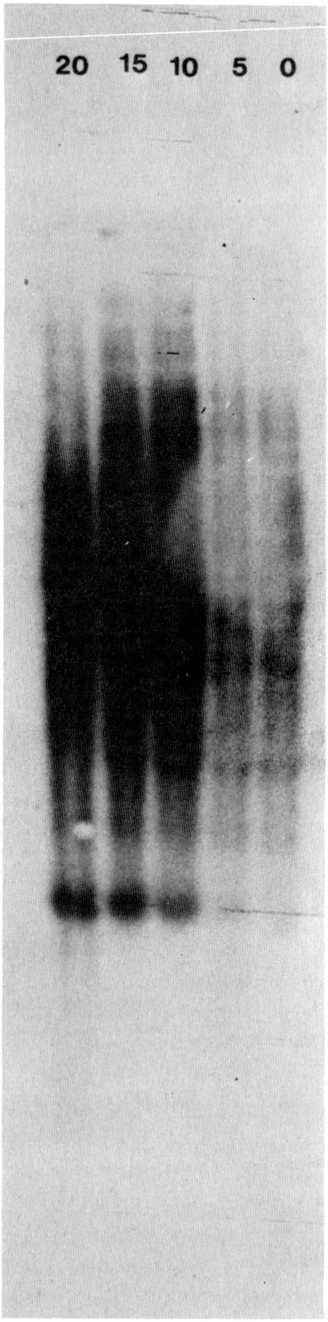
20 15 10 5 0

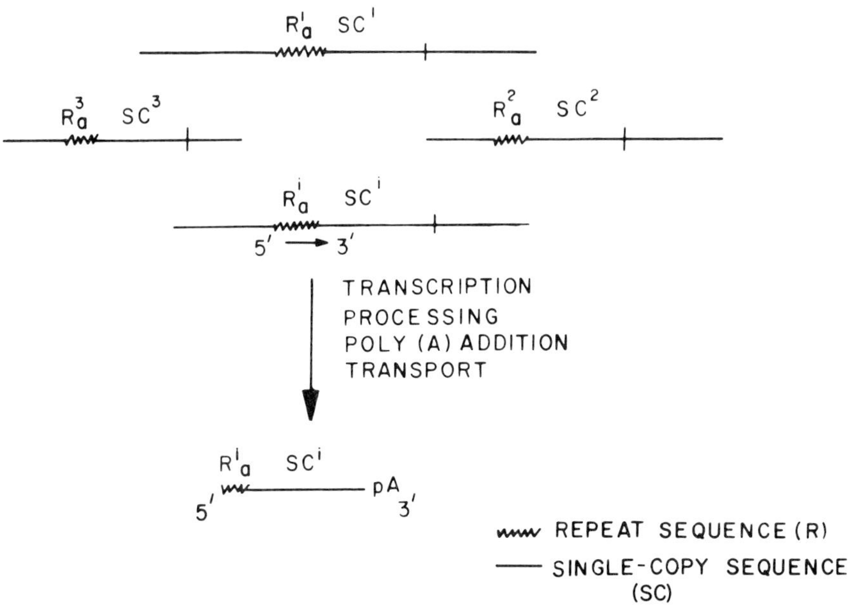

Fig. 31. A model for transcription of short interspersed repeat sequences and their associated single-copy DNA in *Dictyostelium*. Ra^1, Ra^2, Ra^3, Ra^i = different members of the short interspersed repeat family Ra. SC^1, SC^2, SC^3, SC^i = the corresponding set of adjacent single-copy DNA. Transcription initiates at the repeat and elongation proceeds into the single-copy DNA. Sequences complementary to the repeat are located at the 5′ end of the resulting transcript. The RNA is processed (including cap addition), poly(A) tailed and transported to the cytoplasm.

ment, similar experiments were performed with single-copy probes isolated from several cDNAs carrying the *M4* repeat. All cDNAs examined showed greater hybridization to poly(A)$^+$ RNA isolated late in development relative to vegetative poly(A)$^+$ RNA (Kimmel and Firtel, in preparation). The $(AAC)_n$ sequence on these cDNA clones has been located at the 5′ end of the cDNAs, apparently upstream from the protein coding region.

Results suggest that the *M4* (AAC) repeat is found on a population of mRNAs that are expressed with similar developmental kinetics. A possible model for the transcription of the *M4* family is shown in Fig. 31. Transcription would initiate near the repeat and yield an mRNA population whose sequences are predominantly single copy and would encode a set of different

---

Fig. 30. Northern DNA blot with repeat sequence from *M4*-Band 4. RNA was isolated from 0, 5, 10, 15 and 20 h in development, size fractionated, and blotted onto nitrocellulose. The blot was then probed with nick translated [^{32}P]DNA complementary to the *M4* repeat.

proteins. The *M4* and other repeat sequences associated with developmentally regulated mRNAs may represent common control sites for the expression of a set of genes (Britten and Davidson, 1969).

IX. Sequences Common to the 5′-Ends of *Dictyostelium* Genes

Most eukaryotic protein coding genes possess a common sequence, called the TATA or Goldberg–Hogness box, located ~25–30 bp 5′ to the site of transcription initiation (Goldberg, 1979; Gannon *et al.*, 1979). A point mutation or deletion of this sequence will significantly reduce the level of *in vivo* or *in vitro* expression of the adjacent gene (see Breathnach and Chambon, 1981 for a review and references). Deletions of the TATA box will also cause improper phasing of transcription initiation. Thus, this sequence seems to be important for the normal expression of most genes transcribed by RNA polymerase II. We have compared the 5′ ends of *Dictyostelium* genes for common sequences that may be RNA polymerase II recognition sites. Figure 32 lists the 5′ ends and transcription start sites of these genes. Although the 5′ ends of the transcribed genes are not homologous in sequence, they do share some striking structural similarities. They are all ~95% A + T and possess a TATA box located 30–40 nt upstream from the transcription initiation (see Fig. 32). A comparison of these TATA sequences shows that they are similar but not identical. The canonical sequence that can be derived for these TATA or Goldberg–Hogness boxes is TATAAA$^{\mathrm{T}}_{\mathrm{A}}$A. None of the genes differ from the consensus sequence by more than two nucleotides. In addition, located between the TATA box and transcription initiation site is a common oligo(dT) stretch of varying length and homogeneity (Kimmel and Firtel, manuscript in preparation).

Whereas a TATA box is found upstream from transcription initiation in many eukaryotic mRNA genes, the oligo(dT) stretch is apparently unique to *Dictyostelium* genes. The *actin 2-sub2* pseudogene, which is apparently not

Fig. 32. Sequence comparison of the 5′ ends of *Dictyostelium* genes. The boxed **A** and **B** regions correspond to the TATA and oligo(dT) stretches. The arrows represent the region of the 5′ end of the mRNA. The numbers above the (- - -) dashed line represent the number of nucleotides deleted from the 5′ untranslated region in order to line up the TATA box and the AUG initiation codon. For the protein coding genes, note the A residues preceding the AUG. *D2* is a potential gene for low molecular weight nuclear RNA homologous to rat *U3* snRNA (Wise and Weiner, 1982).

COMPARISON OF 5' ENDS OF GENES IN <u>DICTYOSTELIUM</u>

M4	TACATTTATTAATATTTTTATCTTTTTTTTTTTTTTTAAAAAACTATTTTAGGAGAGATTTTAGAAT-135-ATTTA	AUG
Discoidin-IB/C	AAAAAGTATAAATAGAAAAAAAATTTTTAATTAAAATCATTAAATTGAAAAATTAAAATTTCAT-26-ATTTATAAA	AUG
Discoidin-IA	AAAAACTATAAATAAAACAAGTTTTTTTTTTTTTAATCATTAAATTGAAAATCAAAATTTCAT-34-ACAAATAAA	AUG
Actin 4	TTTTTGGATAAATATTAAATTTTTTTTTTTTTCTTATCATTATTCAATCTATAATATATAAAATACAATAAAAAA	AUG
Actin 5	ATTTGCTATAAATACAAAAAAAAAATTGTTTTTAAATCATTATTAATAAAAACTTAAAATAA-2-TATATATAAAAAA	AUG
Actin 6	TGAGATTATAAAATGAAATTTTTTTTTTTTTTTAATTAATTCAAAAAATAATCAAATAAATAAATATAATATAAA	AUG
Actin 8	TTTTTCAATAAATAGGATTTTTTTTATTTATTTTTTTTTTCAATACTTAATTCAAATATATATAT-12-TAAAA	AUG
Actin 7	CAGCAATTTAAAATCAAAATTTTTTTTTTTTTAATCAATTTAAATAAATACTC-4-AAATAATAAAATATAATAAA	AUG
Actin M6	AAAAACTTTAAAAACTAATTTTTTTTTTTTTTTTTAATCATTCTAATATAAAATAAT-4-ACAATTAAAAAATAAA	AUG
M3R	TTTGTATATAAATTTATTTTTTTTTTTTTTTTTATTATTATTATTAGTTAATTTTCATTTGTA-4-TAAAAAAAAA	AUG
M3L	TATTCGTTTAAATTTTTTTTTTTTTTATTTAAGGTTGTTTTTTTATT-13-TTTGTATATTTTAAAAAAAAAAAAAA	AUG
D2	ATTAAATATTAAAAAATGAACAAAAGAAACTTTTTTTTATTTTGAAAAGTTATGACCAAACTCTTA	

expressed, is deficient of the conserved TATA sequence although it does have an extensive oligo(dT) stretch. This result is consistent with the theory that the TATA sequence is essential for wild-type gene expression. The *actin 3* pseudogene has a TATA-like sequence but lacks the oligo(dT) stretch (McKeown and Firtel, 1981a; 1982; McKeown *et al.*, 1982).

We suggest that the TATA–oligo(dT) structure may be a one of the sites involved in RNA polymerase II recognition in *Dictyostelium* (Kimmel and Firtel, in preparation). It is unlikely that this common region is involved in developmental gene expression since the genes that share it are expressed at widely varying levels during development. The *M4* gene encodes a low abundance class mRNA (Kimmel and Firtel, 1979, 1981). Individual *actin* genes are expressed at diverse levels during different developmental stages (McKeown and Firtel, 1981a) whereas the *discoidin I* and the *M3* genes are both expressed at high levels during early development. *Discoidin I* expression is repressed by cAMP whereas *M3* expression is enhanced by cAMP. (Williams *et al.*, 1980; Brandis and Firtel, in preparation). The *D2* gene encodes a small nuclear RNA which is probably transcribed by RNA polymerase II (Wise and Weiner, 1980).

It is possible to compare the TATA sequence of an individual gene to its maximal level of expression. Again the 8 position consensus sequence is TATAAA$^{T}_{A}$A (see Fig. 32). Only *actins 8* and *4* deviate from the initial T; both may be expressed at $>0.07\%$ of total mRNA. *M3L* and *actins M6* and *7* possess a T at position 2. They are all expressed at $\ll 0.07\%$ of total mRNA. Interestingly, the *M3L* TATA box differs from *M3R* at only this position and *M3L* expressed at a 5–10-fold lower value. All of the genes possess the third position T. *M4* and *D2* are divergent at position 4. *M4* mRNA represents only $\sim$0.01% of total mRNA. Since *D2* potentially encodes a small nuclear RNA, it is difficult to make a direct comparison to mRNA levels of expression. However, if *D2* were transcribed it would yield a minor sequence form of the *D2* snRNA family. TATA-positions 5 and 6 are totally conserved and position 7 is approximately equally represented by a T or an A. *Actin 6* is an example of a relatively highly expressed gene which contains a T at position 8.

Since there are probably other regulatory sequences with greater impact on gene expression, it is premature to make very general predictions regarding the TATA sequence. However, it may be that a deviation from the ATA sequence at positions 2, 3 and 4 can yield a reduced level of expression. Similar observations have been made in other systems where the ATA is highly conserved and is suggested to be important for determining maximal levels of gene expression (Breathnach and Chambon, 1981).

We have also examined the distance between the TATA and oligo(dT), the extent and homogeneity of the oligo(dT), and its distance to the transcription initiation site. We have not been able to make any correlation between these characteristics and maximal gene expression. Again, it should be emphasized that although the TATA–oligo(dT) structure may be a common site for RNA polymerase II recognition, there must be other significant regulatory sites that are not yet identified in *Dictyostelium*.

X. RNA Polymerases and Transcription in Isolated Nuclei

Jacobson and co-workers have shown that under the appropriate conditions isolated nuclei can synthesize RNA molecules that are essentially identical to hnRNA synthesized *in vivo* (Jacobson *et al.*, 1974a; Firtel and Jacobson, 1977). They have also demonstrated that these nuclear preparations are capable of reinitiating RNA synthesis. It has additionally been possible to determine the relative rates of synthesis of specific genes in nuclei isolated at various developmental stages. Jacobson (personal communication) has found that the relative rate of *actin* RNA synthesis in nuclei isolated from various developmental stages is in close agreement with that obtained by *in vivo* pulse-labeling experiments. *Actin* is regulated primarily at the level of transcription. In addition, Williams' group has shown that nuclei isolated from cAMP-treated cells have a decreased relative rate of *discoidin I* mRNA synthesis compared with nuclei from control cells (Williams *et al.*, 1980).

There are some intricate subtleties to these experiments (see Firtel and Jacobson, 1977). For example, there are differing salt stimulation effects on RNA polymerase II if nuclei are isolated from cells in suspension at high or low density or from cells plated on filters. Given this caveat, however, such approaches will be critical to the investigation of differential gene expression.

Dictyostelium, like other eukaryotes, has multiple RNA polymerases which have been partially purified and characterized (Pong and Loomis, 1973a,b; Takiya *et al.*, 1980a,b; Yagura *et al.*, 1976, 1977; see Roeder, 1976 for a general discussion).

XI. DNA-Mediated Transformation

The analysis of recombinant DNA clones has permitted us to define the structure of many *Dictyostelium* genes and to measure the changes in the level of mRNAs from these genes during development. These approaches, however, are very limited in their ability to determine the function of specific sequences that may regulate gene expression and to determine the mechanisms whereby differential gene expression is controlled in *Dictyostelium.* One approach to these questions would be to reinsert developmentally regulated genes into *Dictyostelium.* If these genes show normal developmental regulation when transformed into *Dictyostelium*, it would allow the use of directed mutagenesis to analyze specific control mechanisms.

Several approaches have been used to develop a DNA-mediated transformation system in *Dictyostelium.* The first reported approach was by Ratner *et al.*, (1981) and involved the insertion of *Dictyostelium* wild-type DNA into a yeast cloning vector, which also contained a *Dictyostelium* DNA sequence that allowed the plasmid to replicate autonomously in yeast (Stinchcomb *et al.*, 1980). This vector was used to transform the *Dictyostelium bsgA* mutant and to select for wild type growth on *Bacillus subtilis.* Transformants were isolated which grew on *B. subtilis* and contained plasmid DNA sequences. If the *bsgA* gene could have been isolated, it could have been used to generate a *Dictyostelium* transformation vector containing a selectable marker and a potential autonomously replicating segment. Unfortunately, the transforming DNA could not be recovered from these cells. Attempts to transform *Dictyostelium* mutants with total wild-type *Dictyostelium* DNA by Ratner's, Jacobson's and Firtel's groups have not been successful.

Firtel and co-workers (Hirth, Edwards and Firtel, submitted) have been working on a transformation system using a dominant drug resistance gene as the selective marker. The neomycin derivative G418 kills eukaryotic and prokaryotic cells and Southern and Berg (personal communication) and Jimenez and Davis (1980) have shown that the neomycin-resistance genes from bacterial transposable elements can confer resistance to mammalian and yeast cells. A vector was constructed in which the putative *Dictyostelium actin 8* promoter was fused to the neomycin resistance (*Neo*^{R}) gene of transposable element Tn5 in place of its own promoter. The vector also contained the 3′ end of the *Dictyostelium actin 8* gene. From analogy with the yeast transformation system, it was possible that a *Dictyostelium* DNA sequence capable of autonomous replication would also be required for transformation (Struhl *et al.*, 1979). Such sequences were identified by complementation in yeast. Total *Dictyostelium* DNA was inserted into a yeast vector that could only transform yeast if it contained a *Dictyostelium* sequence that could function as an autonomously replicated segment (ARS). Several ARSs were isolated and one, Drp14, was linked to the *actin 8-Neo*^{R} fusion on the *E. coli* cloning vector pML2 (Lusky and Botcham, personal communication). Figure 33 shows this vector, CERF.DRp14 (Hirth, Edwards and Firtel, submitted). A second vector has also been constructed that lacks the Drp14 element but contains the central 4 kb Sal I restriction endonuclease frag-

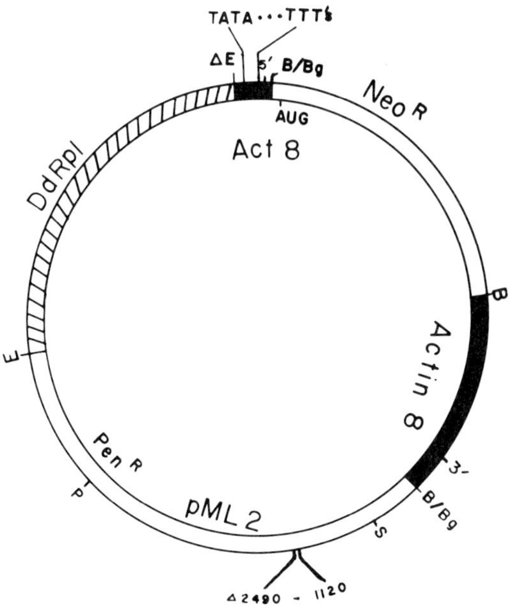

Fig. 33. Restriction map of CERF.Drp 14. The regions shown in black are derived from *actin 8*. The 5′ and 3′ ends of the *actin 8* gene and the region of the *actin 8* TATA box and oligo(dT) region are indicated.

ment from the 90 kb *Dictyostelium* extrachromosomal rDNA palindrome (Fig. 34). The Sal 4 fragment is postulated to contain an origin of replication since replication of the extrachromosomal rDNA palindromes in *Tetrahymena* and *Physarum* originates in its central fragment (Truett and Gall, 1977; Cech and Brehm, 1981; Vogt and Braun, 1977).

Both of these vectors are able to confer G418 resistance to *Dictyostelium* cells by DNA-mediated transformation (Hirth, Edwards and Firtel, submitted). When a vector is used that lacks the Drp14 or Sal4 elements, few resistant cells are found and no transformants are observed when the control plasmid pML2, which does not possess a neomycin resistance gene, is used. With CERF.Drp14 and CERF.Sal 4, approximately 10^{-6} to 10^{-4} of the input cells become resistant. Further experiments have shown that transformed cells are capable of undergoing normal *Dictyostelium* development and that the amoebae from germinated spores are G418 resistant.

In all transformants, plasmid DNA sequences can be identified by the Southern DNA-blot hybridization procedure (Hirth, Edwards and Firtel, submitted). The experiments suggest that the DNA in certain transformants may be integrated within the *Dictyostelium* genome. In some instances,

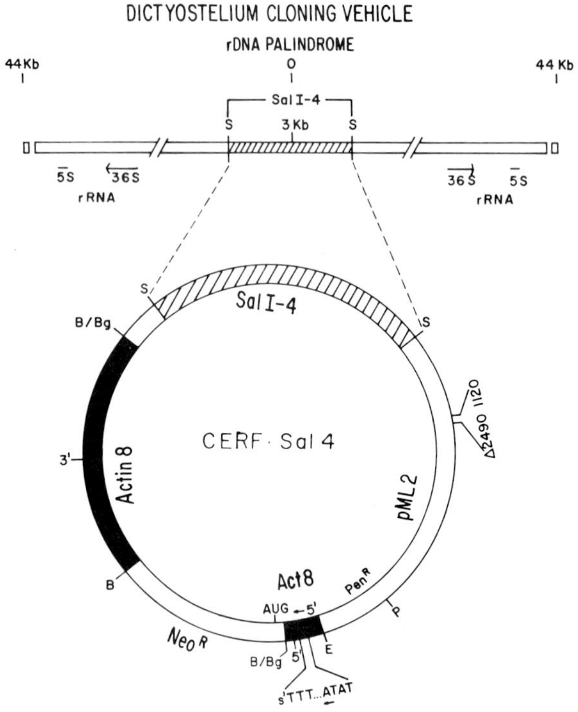

Fig. 34. Construction of the CERF.Sal4. The central 3 kb Sal I restriction fragment (Sal I-4, see Fig. 1) from the extrachromosomal, rDNA palindrome was cloned into the Sal I restriction site a *Dictyostelium* cloning vector lacking the Drp14 sequence shown in Fig. 33. Upon cloning in *E. coli*, a deletion in the Sal I fragment occurred.

vector sequences can also be reisolated by direct transformation of *E. coli* using intact DNA from transformed *Dictyostelium* cells. These results strongly suggest that at least some of the DNA in these transformed cell lines is extrachromosomal.

These results indicate that foreign genetic material can be inserted, selected, and maintained in *Dictyostelium* vegetative amoebae. Additional experiments are needed to determine if genes that are reinserted into *Dictyostelium* with these vectors are capable of undergoing normal transcriptional regulation. The insertion of modified, developmentally regulated genes may allow us to dissect the transcriptional and translational regulation signals that control differential gene expression in this organism. It is very significant that up to 10^{-4} of the input cells can be transformed. Since at least some of the DNA in the transformed cells is extrachromosomal and can be reselected in *E. coli*, it should eventually be possible to complement developmentally regulated mutants and to isolate specific genes that cannot be easily obtained by standard cloning procedures. It is expected that these approaches to *Dictyostelium* genetics will permit a more detailed analysis of the molecular as well as the physiological aspects of *Dictyostelium* development.

XII. Conclusions and Perspectives

The *M3*, *actin*, *discoidin I*, and *M4* gene sets exhibit diverse patterns of expression, organization and evolution. Yet, through their characterization we are now able to more clearly understand the general aspects of mRNA synthesis and maturation. In addition, a unique sequence organization at the 5′ end of *Dictyostelium* genes has been identified that may be essential for the proper initiation of transcription. Clearly, other sequences must also be important for gene expression. The interspersed repeat *M4* may be a common regulatory element involved in coordinate expression of this gene set. The conserved sequences at the 5′ ends of the *M3* family may be related to their apparent response to cAMP. Finally, each *actin* gene possesses a characteristic pattern of developmental expression perhaps as a result of the unique sequence organization upstream from the transcription initiation site of individual genes.

The advent of a DNA-mediated transformation system may now allow us to more precisely examine the relationship between the structure of a gene and its expression. Eventually a particular sequence may be specifically altered and the mutated gene introduced into *Dictyostelium*. Since all vegetative cells are uniquely able to undergo the entire developmental cycle, the effect of this mutation on normal gene expression can be monitored. Furthermore, this system will be essential to the elucidation of the direct relationship of cAMP and cell–cell interaction to developmental gene expression in *Dictyostelium*.

Acknowledgments

We wish to thank J. Brandis, C. Edwards, R. Grainger, K.-P. Hirth, A. Jacobson, C. Lai, M. Mehdy, C. MacLeod, E. Rosen, W. Rowekamp, M. Sogin, and A. Weiner for allowing us to cite their unpublished data and for providing useful discussions. We are particularly indebted to C. Edwards, M. McKeown and S. Poole for their innumerable helpful suggestions.

A. R. Kimmel was a recipient of a Senior Fellowship from the American Cancer Society, California Division. R. A. Firtel is a recipient of a Faculty Research Award from the American Cancer Society. This work was additionally supported by grants to R. A. Firtel from the National Institutes of Health and the National Science Foundation.

References

Abelson, J. (1979). RNA processing and the intervening sequence problem. *Ann. Rev. Biochem.* **48,** 1035–1069.

Alton, T. H., and Brenner, B. (1979). Comparison of proteins synthesized by anterior and posterior regions of *Dictyostelium discoideum* pseudoplasmodia. *Dev. Biol.* **71,** 1–7.

Alton, T. H., and Lodish, H. F. (1977a). Developmental changes in messenger RNAs and protein synthesis in *Dictyosteium discoideum*. *Dev. Biol.* **60,** 180–206.

Alton, T. H., and Lodisb, H. F. (1977b). Synthesis of developmentally regulated proteins in *Dictyostelium discoideum* which are dependent on continued cell–cell interaction. *Dev. Biol.* **60,** 207–216.

Alton, T. H., and Lodish, H. F. (1977c). Translation control of protein synthesis during the early stages of differentiation of the slime mold *Dictyostelium discoideum. Cell* **12,** 301–310.

Ashworth, J. M., and Quance, J. (1972). Enzyme synthesis in myxamoebae of the cellular slime mold *Dictyostelium discoideum* during growth in axenic culture. *Biochem. J.* **126,** 601–608.

Baker, C. C., and Ziff, E. B. (1981). Promoters and heterogeneous 5′ termini of the messenger RNAs of adenovirus serotype 2. *J. Mol. Biol.* **149,** 189–221.

Batts-Young, B., and Lodish, H. F. (1978). Triphosphate residues at the 5′ ends of rRNA precursor and 5S RNA from *Dictyostelium discoideum. Proc. Natl. Acad. Sci. U.S.A.* **75,** 740–744.

Batts-Young, B., Lodish, H. F., and Jacobson, A. (1980). Similarity of the primary sequences of ribisomal RNAs isolated from vegetative and developing cells of *Dictyostelium discoideum*. *Dev. Biol.* **78,** 352–364.

Batts-Young, B., Maizels, N., and Lodish, H. F. (1977). Precursors of ribosomal RNA in the cellular slime mold *Dictyostelium discoideum. J. Biol. Chem.* **252,** 3952–3960.

Bender, W., Davidson, N., Kindle, K. L., Taylor, W., Silverman, M., and Firtel, R. A. (1978). The Structure of M6, a recombinant plasmid containing *Dictyostelium* DNA homologous to actin messenger RNA. *Cell* **15,** 779–788.

Bird, A. P., and Southern, E. M. (1978). Use of restriction enzymes to study eukaryote DNA methylation. I. The methylation pattern in rDNA from *Xenopus laevis. J. Mol. Biol.* **118,** 27–47.

Bishop, J. O., Morton, J. G., Rosbash, M., and Richardson, M. (1974). Three abundance classes in HeLa cell messenger RNA. *Nature (London)* **250,** 199–204.

Blackburn, E. H., and Gall, J. G. (1978). A tandemly repeated sequence at the termini of the extrachromosomal ribosomal RNA genes in *Tetrahymena. J. Mol. Biol.* **120,** 33–53.

Blumberg, D. D., and Lodish, H. F. (1980a). Complexity of nuclear and polysomal RNAs in growing *Dictyostelium discoideum* cells. *Dev. Biol.* **74,** 268–284.

Blumberg, D. D., and Lodish, H. F. (1980b). Changes in the messenger RNA population during differentiation of *Dictyostelium discoideum*. *Dev. Biol.* **78,** 285–300.

Blumberg, D. D., and Lodish, H. F. (1981). Changes in the complexity of nuclear RNA during development of *Dictyostelium discoideum*. *Dev. Biol.* **81,** 74–80.

Blumberg, D. D., Margolskee, J. P., Barklis, E., Chung, S. N., Cohen, N. S., and Lodish, H. F. (1982). Specific cell–cell contacts are essential for induction of gene expression during differentiation of *Dictyostelium discoideum. Proc. Nat. Acad. Sci. U.S.A.* **79,** 127–131.

Bonner, J. T. (1970). Induction of stalk cell differentiation by cyclic AMP in the cellular slime mold *Dictyostelium discoideum. Proc. Natl. Acad. Sci. U.S.A.* **65,** 110–113.

Brandhorst, B. P., Verma, D. P. S., and Fromson, P. (1979). Polyadenylated and non-polyadenylated mRNA fractions from sea urchins code for the same abundant proteins. *Dev. Biol.* **71,** 128–141.

Breathnach, R., Benoist, C., O'Hare, K., Gannon, G., and Chambon, R. (1978). Ovalbumin Gene: Evidence for a leader in mRNA and DNA sequences at the exon–intron boundaries. *Proc. Natl. Acad. Sci. U.S.A.* **75,** 4853–4857.

Breathnach, R., and Chambon, P. (1981). Organization and expression of eukaryotic split genes coding for proteins. *Ann. Rev. Biochem.* **50,** 349–384.

Britten, R. J., and Davidson, E. H. (1969). Gene regulation in higher cells: A theory. *Science* **165,** 349–357.

Burns, R. A., Livi, G. P., and Dimond, R. L. (1981). Regulation and secretion of early developmentally controlled enzymes during axenic growth in *Dictyostelium discoideum. Dev. Biol.* **83,** 407–416.

Cech, T. R., and Brehm, S. L. (1981). Replication of the extrachromosomal ribosomal RNA genes of *Tetrahymena thermophila. Nucl. Acids Res.* **9,** 3531–3543.

Chung, S., Landfear, S. M., Blumberg, D. D., Cohen, N. S., and Lodish, H. F. (1981). Synthesis and stability of developmentally regulated *Dictyostelium discoideum* mRNA are affected by cell–cell contact and cAMP. *Cell* **24,** 785–797.

Cockburn, A. F., Newkirk, M. J., and Firtel, R. A. (1976). Organization of the ribosomal RNA genes of *Dictyostelium discoideum:* mapping of the nontranscribed spacer regions. *Cell* **9,** 605–613.

Cockburn, A. F., Taylor, W. C. and Firtel, R. A. (1978). *Dictyostelium* rDNA consists of non-chromosomal palindromic dimers containing 5S and 36S coding regions. *Chromosoma* **70,** 19–29.

Coloma, A., and Lodish, H. F. (1981). Synthesis of spore- and stalk-specific proteins during differentiation of *Dictyostelium discoideum. Dev. Biol.* **81,** 238–244.

Costantini, F. P., Scheller, R. H., Britten, R. J., and Davidson, E. H. (1978). Repetitive sequence transcripts in the mature sea urchin oocyte. *Cell* **15,** 173–187.

Darmon, M., Brachet, P., and DaSilva, L. H. P. (1975). Chemotactic signals induce cell differentiation in *Dictyostelium discoideum. Proc. Natl. Acad. Sci. U.S.A.* **72,** 3163–3166.

Davidson, E. H., Galau, G. A., Angerer, R. C., and Britten, R. J. (1975). Comparative aspects of DNA organization in metazoa. *Chromosoma* **51,** 253–259.

Dottin, R. P., Weiner, A. B., and Lodish, H. F. (1976). 5′ Terminal nucleotide sequences of the messenger RNAs of *Dictyostelium discoideum. Cell* **8,** 233–244.

Duncan, R., and Humphreys, T. (1981). Most sea urchin maternal mRNA sequences in every abundance class appear in both polyadenylated and nonpolyadenylated molecules. *Dev. Biol.* **88,** 201–210.

Efstratiadis, A., Posakony, J. W., Maniatis, T., Lawn, R., O'Connel, C., Spritz, R. A., DeRiel, J. K., Forget, B. G., Weissman, S. M., Slighton, J. L., Blechl, A. E., Smithies, O., Barelle, F. E., Shoulders, C. C., and Proudfoot, N. J. (1980). The structure and evolution of the β-globin gene family. *Cell* **21,** 653–668.

Emery, H. S., and Weiner, A. M. (1981). An irregular satellite sequence is found at the termini of the linear extrachromosomal rDNA in *Dictyostelium discoideum. Cell* **26,** 411–419.

Engberg, J., Andersson, P., Leick, V., and Collins, J. (1976). Free ribosomal DNA molecules from *Tetrahymena pyriformis* GL are giant palindromes. *J. Mol. Biol.* **104,** 455–470.

Firtel, R. A. (1972). Changes in the expression of single-copy DNA during development of the cellular slime mold *Dictyostelium discoideum. J. Mol. Biol.* **66,** 363–377.

Firtel, R. A., Baxter, L., and Lodish, H. F. (1973). Actinomycin D and the regulation of enzyme biosynthesis during development of *Dictyostelium discoideum. J. Mol. Biol.* **79,** 315–327.

Firtel, R. A., and Bonner, J. (1972a). Characterization of the genome of the cellular slime mold *Dictyostelium discoideum. J. Mol. Biol.* **66,** 339–361.

Firtel, R. A., and Bonner, J. (1972b). Developmental control of β-1-4 glucan phosphorylase in the cellular slime mold *Dictyostelium discoideum. Dev. Biol.* **29,** 85–103.

Firtel, R. A., Cockburn, A., Frankel, G., and Hershfield, V. (1976a). Structural organization of the genome of *Dictyostelium discoideum:* Analysis of EcoRI restriction endonuclease. *J. Mol. Biol.* **102,** 831–852.

Firtel, R. A., and Jacobson, A. (1977). Structural organization and transcription of the genome of *Dictyostelium discoideum. In* "Biochemistry of Cell Differentiation II" (J. Paul, ed.), Vol. 15, pp. 377–429. Univ. Park Press, Baltimore.

Firtel, R. A., Jacobson, A., and Lodish, H. F. (1972). Isolation and hybridization kinetics of messenger RNA and *Dictyostelium discoideum. Nature (London) New Biol.* **239,** 225–238.

Firtel, R. A., and Kindle, K. (1975). Structural organization of the genome of the cellular slime mold *Dictyostelium discoideum:* interspersion of repetitive and single-copy DNA sequences. *Cell* **5,** 400–411.

Firtel, R. A., Kindle, K., and Huxley, M. P. (1976b). Structural organization and processing of the genetic transcript in the cellular slime mold *Dictyostelium discoideum. Fed. Proc.* **35,** 13–22.

Firtel, R. A., and Lodish, H. F. (1973). A small nuclear precursor of messenger RNA in the cellular slime mold *Dictyostelium discoideum. J. Mol. Biol.* **79,** 295–314.

Firtel, R. A., Timm, R., Kimmel, A. R., and McKeown, M. (1979). Unusual nucleotide sequences at the 5′ end of actin genes in *Dictyostelium discoideum. Proc. Natl. Acad. Sci. U.S.A.* **76,** 6206–6210.

Frankel, G., Cockburn, A. F., Kindle, K. L., and Firtel, R. A. (1977). Organization of the ribosomal RNA genes of *Dictyostelium discoideum.* Mapping of the transcribed region. *J. Mol. Biol.* **109,** 539–558.

Gannon, F., O'Hare, K., Perrin, F., LePennec, J. P., Benoist, C., Cocket, M., Breathnach, R., Royal, A., Garapin, A., Cami, B., and Chambon, P. (1979). Organization and sequences at the 5′ end of a cloned complete ovalbumin gene. *Nature (London)* **278,** 425–434.

Gerisch, G. (1968). Cell aggregation and differentiation in *Dictyostelium discoideum. Curr. Top. Devel. Biol.* **3,** 157–197.

Goldberg, M. (1979). Ph.D. Thesis, Stanford University, Stanford.

Grainger, R. B., and Maizels, N. (1980). *Dictyostelium* ribosomal RNA is processed during transcription. *Cell* **20,** 619–623.

Greenberg, J. R. (1972). High stability of messenger RNA in growing cultured cells. *Nature (London)* **240,** 102–104.

Gross, J. B., Town, C. D., Brookman, J. J., Jermyn, K. A., Peacey, M. J., and Kay, R. R. (1981). Cell patterning in *Dictyostelium. Phil. Trans. R. Soc. Lond. B.* **295,** 497–508.

Hagenbüchle, D., Santer, M., Steitz, J. A., and Mans, R. (1978). Conservation of the primary structure of the 3′ end of the 18S rRNA from eukaryotic cells. *Cell* **13,** 551–563.

Gannon, F., O'Hare, K., Perrin, F., LePennec, J. P., Benoist, C., Cocket, M., Breathnach, R., Royal, A., Garapin, A., Cami, B., and Chambon, P. (1979). Organization and sequences at the 5′ end of a cloned complete ovalbumin gene. *Nature (London)* **278,** 425–434.

Gerisch, G. (1968). Cell aggregation and differentiation in *Dictyostelium discoideum. Curr. Top. Devel. Biol.* **3,** 157–197.

Goldberg, M. (1979). Ph.D. Thesis, Stanford University, Stanford.

Grainger, R. B., and Maizels, N. (1980). *Dictyostelium* ribosomal RNA is processed during transcription. *Cell* **20,** 619–623.

Greenberg, J. R. (1972). High stability of messenger RNA in growing cultured cells. *Nature (London)* **240,** 102–104.

Gross, J. B., Town, C. D., Brookman, J. J., Jermyn, K. A., Peacey, M. J., and Kay, R. R. (1981). Cell patterning in *Dictyostelium. Phil. Trans. R. Soc. Lond. B.* **295,** 497–508.

Hagenbüchle, D., Santer, M., Steitz, J. A., and Mans, R. (1978). Conservation of the primary structure of the 3′ end of the 18S rRNA from eukaryotic cells. *Cell* **13,** 551–563.

Hori, H., Osawa, S., and Iwabuchi (1980). The nucleotide sequence of 5S rRNA from a celular slime mold *Dictyostelium discoideum. Nuc. Acid Res.* **8,** 5535–5539.

Jacobson, A., Firtel, R. A. and Lodish, A. F. (1974a). Synthesis of messenger and ribosomal RNA precursors in isolated nuclei of the cellular slime mold *Dictyostelium discoideum. J. Mol. Biol.* **82,** 213–230.

Jacobson, A., Firtel, R. A., and Lodish, H. F. (1974b). Transcription of polydeoxythymidylate sequences in the genome of the cellular slime mold, *Dictyostelium discoideum. Proc. Natl. Acad. Sci. U.S.A.* **71,** 1607–1611.

Jensen, E. G., Hellung-Larson, P., and Frederiksen, S. (1979). Synthesis of low molecular weight RNA components A, C, and D by polymerase II in -amanative resistant hamster cells. *Nucl. Acid Res.* **6,** 321–330.

Jimenez, A., and Davis, J. (1980). Expression of a transposable antibiotic resistance element in *Saccharomyces. Nature (London)* **287,** 869–871.

Johnson, E. M. (1980). A family of inverted repeat sequences and specific single-strand gaps at the termini of the *Physarum* rDNA palindrome. *Cell* **22,** 875–886.

Kan, N. C., and Gall, J. G. (1981). Sequence homology near the center of the extra chromosomal rDNA palindrome in *Tetrahymena. J. Mol. Biol.* **153,** 1151–1155.

Karrer, K. M., and Gall, J. G. (1970). The marcronuclear ribosomal DNA of *Tetrahymena pyriformis* is a palindrome. *J. Mol. Biol.* **104,** 421–453.

Kaufman, U., Milcarek, C., Berissi, H., and Penman, S. (1977). HeLa cell poly(A)$^-$ mRNA codes for a subset of poly(A)$^+$ mRNA directed proteins with actin as a major product. *Proc. Natl. Acad. Sci. U.S.A.* **74,** 4801–4805.

Kay, R. R., Town, C. D., and Gross, J. D. (1979). Cell differentiation in *Dictyostelium discoideum. Differentiation* **13,** 7–14.

Kedes, L. H. (1979). Histone genes and histone messengers. *Annu. Rev. Biochem.* **48,** 837–870.

Kimmel, A. R., and Firtel, R. A. (1979). A family of short, interspersed repeat sequences at the 5′ end of a set of Dictyostelium single-copy mRNAs. *Cell* **16,** 787–796.

Kimmel, A. R., and Firtel, R. A. (1980). Intervening sequences in a *Dictyostelium* gene that encodes a low abundance class mRNA. *Nucl. Acid. Res.* **8,** 5599–5610.

Kimmel, A. R., and Gorovsky, M. A. (1978). Organization of the 5S RNA genes in macro- and micronuclei of *Tetrahymena pyriformis. Chromosoma* **67,** 1–20.

Kimmel, A. R., Lai, C., and Firtel, R. A. (1979). Families of interspersed repeat sequences at the 5′ end of *Dictyostelium* single copy repeat genes. *In* "Eukaryotic Gene Regulation. INC–UCLA Symp. Mol. and Cell. Biol.," (R. Axel, T. Maniatis, C. F. Fox, eds.). Vol. 14, pp. 195–203.

Kindle, K. (1978). Ph.D. Thesis, University of California at San Diego, San Diego.

Kindle, K. L., and Firtel, R. A. (1978). Identification and analysis of *Dictyostelium* actin genes, a family of moderately repeated genes. *Cell* **15,** 763–778.

Kindle, K. L., and Firtel, R. A. (1979). Evidence that populations of *Dictyostelium* single-copy mRNA transcripts carry common repeat sequences. *Nuc. Acid. Res.* **6,** 2403–2422.

Kindle, K., Taylor, W., McKeown, M., and Firtel, R. A. (1977). Analysis of gene structure and transcription in *Dictyostlium discoideum. In* "Developments and Differentiation in the Cellular Slime Moulds" (P. Cappuccinelli and J. M. Ashworth, eds.), Vol. I, pp. 273–290. Elsevier, New York.

Kiss, G. B., and Pearlman, R. E. (1981). Extrachromosomal rDNA in *Tetrahymena thermophila* is not a perfect palindrome. *Gene* **13,** 281–287.

Landfear, S. M., and Lodish, H. F. (1980). A role for cyclic AMP in expression of developmentally regulated genes in *Dictyostelium discoideum. Proc. Natl. Acad. Sci.* **77,** 1044–1048.

Lerner, M. R., Boyle, J. A., Mount, S. M., Wolin, S. L., and Steitz, J. A. (1980). Are snRNPs involved in splicing? *Nature (London)* **283,** 220–224.

Lodish, H. F., Firtel, R. A., and Jacobson, A. (1973). Transcription and structure of the genome of the cellular slime mold *Dictyostelium discoideum. Cold Spring Harbor Symp. Quant. Biol.* **38,** 899–907.

Long, E. O., and Dawid, I. B. (1980). Repeated genes in eukaryotes. *Ann. Rev. Biochem.* **49,** 727–764.

Loomis, W. F. (1975). "*Dictyostelium discoideum:* A Developmental System." Academic Press, New York.

Ma, G. C. L., and Firtel, R. A. (1978). Regulation of the synthesis of two carbohydrate-binding proteins in *Dictyostelium discoideum. J. Biol. Chem.* **253** 3924–3932.

MacLeod, C. (1979). Thesis. University of California San Diego, San Diego.

MacLeod, C., Firtel, R. A., and Papkoff, J. (1980). Regulation of actin gene expression during spore gemination in *Dictyostelium discoideum. Dev. Biol.* **76,** 263–274.

Maizels, N. (1976). *Dictyostelium* 17S, 25S, and 5S rDNAs lie within a 38,000 base pair repeated unit. *Cell* **9,** 431–438.

Mandel, J. L., and Chambon, P. (1979). DNA methylation: organ specific variations in the methylation pattern within and around ovalbumin and other chicken genes. *Nucl. Acids. Res.* **7,** 2081–2103.

Mangiarotti, G., Chung, S., Zucker, C., and Lodish, H. F. (1981). Selection and analysis of cloned developmentally regulated *Dictyostelium discoideum* genes by hybridization–competition. *Nuc. Acid. Res.* **9,** 947–963.

Mangiarotti, G., Lefebvre, P., and Lodish, H. F. (1982). Differences in the stability of developmentally regulated in mRNAs in aggregated and disaggregated *Dictyostelium discoideum* cells. *Dev. Biol.* **89,** 82–91.

Margolskee, J. P., and Lodish, H. F. (1980). Half-lives of messenger RNA species during growth and differentiation of *Dictyostelium discoideum. Dev. Biol.* **74,** 37–49.

Margolskee, J. P., and Lodish, H. F. (1980b). The regulation of the synthesis of actin and two other proteins induced early in *Dictyostelium discoideum development. Dev. Biol.* **74,** 50–64.

McClelland, M. (1981). The effect of sequence specific DNA methylation on restriction endonuclease cleavage. *Nucl. Acids Res.* **9,** 5859–5866.

McKeown, M., and Firtel, R. A. (1981a). Differential expression and 5′ end mapping of actin genes in *Dictyostelium. Cell* **24,** 799–807.

McKeown, M., and Firtel, R. A. (1981b). Evidence for subfamilies of actin genes in *Dictyostelium* as determined by comparisons of 3′ end sequences. *J. Mol. Biol.* **151,** 593–606.

McKeown, M., and Firtel, R. A. (1982). The actin multigene family of *Dictyostelium. Organization of the cytoplasm, CSHSQB* **46.** In press.

McKeown, M., Hirth, K.-P., Edwards, C., and Firtel, R. A. (1982). Examination of the regulation of the actin multigene family in *Dictyostelium discoideum. In* "Embryonis Development: Gene Structure and Function" (Proceedings of the IX International Congress of Developmental Biologists). In press.

McKeown, M., Taylor, W. C., Kindle, K. L., Firtel, R. A., Bender, W., and Davidson, N. (1978). Multiple, heterogeneous actin genes in *Dictyostelium. Cell* **15,** 789–800.

Meyuhas, O., and Perry, R. P. (1979). Relationship between size, stability and abundance of the mRNA of Mouse L cells. *Cell* **16,** 139–148.

Minty, A. J., and Gross, F. (1980). Coding potential of non-polyadenylated messenger RNA in mouse Friend cells. *J. Mol. Biol.* **139,** 61–83.

Morrissey, J. H., Farnsworth, P. A., and Loomis, W. F. (1981). Pattern formation in *Dictyostelium discoideum:* An analysis of mutants altered in cell proportioning. *Dev. Biol.* **83,** 1–8.

Nemer, M., Graham, M. and Dubroff, L. M. (1974). Co-existence of non-histone messenger RNA species lacking and containing polyadenylic acid in sea urchin embryos. *J. Mol. Biol.* **89,** 435–454.

Nevins, J. R., Blanchard, J.-M., and Darnell, J. E. (1980). Transcription units of Adenovirus Type 2 termination of transcription beyond the Poly(A) addition site in early regions 2 and 4. *J. Mol. Biol.* **144,** 377–386.

Newell, P. C., Franke, J., and Sussman, M. C. (1972). Regulation of four functionally related enzymes during shifts in the developmental program of *Dictyostelium discoideum. J. Mol. Biol.* **63,** 373–382.

Newell, P. C., and Sussman, M. (1970). Regulation of enzyme synthesis by slime mold cell assemblies embarked upon alternate development programmes. *J. Mol. Biol.* **49,** 627–637.

Palatnik, C. M., Katz, E. R., and Brenner, M. (1977). Isolation and characterization of transfer RNAs from *Dictyostelium discoideum* during growth and development. *J. Biol. Chem.* **252,** 694–703.

Palatnik, C. M., Storti, R. V., Capone, A. K., and Jacobson, A. (1980). Messenger RNA stability in *Dictyostelium discoideum:* does poly(A) have a regulatory role? *J. Mol. Biol.* **141,** 99–118.

Palatnik, C. M., Storti, R. V., and Jacobson, A. (1979). Fractionation and functional analysis of newly synthesized and decaying messenger RNAs from vegetative cells of *Dictyostelium discoideum. J. Mol. Biol.* **128,** 371–395.

Palatnik, C. M., Storti, R. V., and Jacobson, A. (1981). Partial purification of a developmentally regulated messenger RNA from *Dictyostelium discoideum* by thermal elution from poly(U)-sepharose. *J. Mol. Biol.* **150,** 389–398.

Peffley, D. M., and Sogin, M. L. (1981). A putative tRNA Trp gene cloned from *Dictyostelium discoideum:* its nucleotide sequence and association with repetitive DNA. *Biochemistry.* In press.

Pollard, T. D., and Weihing, R. R. (1974). Actin and myosin and cell movement. *CRC Crit. Rev. Biochem.* **2,** 1–65.

Pong, S. S., and Loomis, W. F. (1973a). Multiple nuclear RNA polymerases during development of *Dictyostelium discoideum. J. Biol. Chem.* **248,** 3933–3939.

Pong, S. S., and Loomis, W. F. (1973b). Isolation of multiple RNA polymerases from *Dictyostelium discoideum. In* "Molecular Techniques and Approaches in Developmental Biology." (M. Chrispeels, ed.), pp. 93–115. Wiley, New York.

Poole, S., Firtel, R. A., Lamar, E., and Rowekamp, W. (1981a). Sequence and expression of the Discoidin I gene family in *Dictyostelium discoideum. J. Mol. Biol.* In press.

Poole, S., Firtel, R. A., and Rowekamp, W. (1981b). Expression of the discoidin I genes in *Dictyostelium. J. Supermol. Struct. Cell. Biochem.* In press.

Puckett, L., Chambers, S., and Darnell, J. E. (1975). Short-lived messenger RNA in HeLa cells and its impact on the kinetics of accumulation of cytoplasmid polyadenylate. *Proc. Natl. Acad. Sci. U.S.A.* **72,** 389–393.

Ratner, D., and Borth, W. (1982). Comparison of differentiating *Dictyostelium discoideum* cell types separated by an improved method of density gradient centrifugation. *Exp. Cell. Res.* In press.

Ratner, D. A., Ward, T. E., Jacobson, A. (1981). Evidence for the transformation of *Dictyostelium discoideum* with homologous DNA. Developmental Biology Using Purified Genes, ICN–UCLA Symposia on Molecular and Cellular Biology, Volume XXIII, eds. Don Brown and C. Fred Fox. Academic Press, New York, NY, pp. 595–605.

Ray, J., Shinnick, T., and Lerner, R. (1979). A mutation altering the function of a carbohydrate-binding protein blocks cell–cell cohesion in developing *Dictyostelium discoideum. Nature (London)* **279,** 215–221.

Razin, A., and Riggs, A. D. (1980). DNA methylation and gene function. *Science* **210,** 604–610.

Roeder, R. G. (1976). *In* "RNA Polymerase" (R. Losick and M. Chamberlin, eds.), pp. 285–329. Cold Spring Harbor Laboratory, New York.

Roop, D. R., Kristo, P., Stumph, W. E., Tsai, M. J., and O'Malley, B. W. (1981). Structure and expression of a chicken gene coding for U1 RNA. *Cell* **23,** 671–680.

Rosen, S. D., Kafka, J. A., Simpson, D. L., and Barondes, S. H. (1973). Developmentally regulated carbohydrate-binding protein in *Dictyostelium discoideum. Proc. Natl. Acad. Sci. U.S.A.* **70,** 2554–2557.

Rossier, H., Eitle, E., van Driel, R., and Gerish, G. (1980). Biochemical regulation of cell development and aggregation in *Dictyostelium discoideum. In* "Symp. of the Soc. for Gen. Microbiol.," Vol. 30. The Eukaryotic Microbial Cell, (G. W. D. Gooday Lloyd, and P. J. Trinei, eds.), pp. 405–425. Cambridge University Press, New York.

Roth, R., Ashworth, J. M., and Sussman, M. (1968). Periods of genetic transcription required for the synthesis of three enzymes during cellular slime mold development. *Proc. Natl. Acad. Sci. U.S.A.* **59,** 1235–1242.

Rowekamp, W., and Firtel, R. A. (1980). Isolation of developmentally regulated genes from *Dictyostelium. Dev. Biol.* **79,** 409–418.

Rowekamp, W., Poole, S., and Firtel, R. A. (1980). Analysis of the multigene family coding the developmentally regulated carbohydrate-binding protein discoidin-I in *D. discoideum. Cell* **20,** 495–505.

Rubenstein, P., and Deuchler, J. (1979). Acetylated and non-acetylated actins in *Dictyostelium discoideum. J. Biol. Chem.* **254,** 11142–11147.

Rubin, G. M., and Sulstan, J. E. (1973). Physical linkage of the 5S cistrons to the 18S and 28S ribosomal cistrons in *Saccharomyces cerevisiae. J. Mol. Biol.* **79,** 521–530.

Sampson, J., Town, C., and Gross, J. (1978). Cyclic AMP and the control of aggregation phase gene expression in *Dictyostelium discoideum. Dev. Biol.* **67,** 54–64.

Scheller, R. H., Costantini, F. D., Kozlowski, B. R., Britten, R. J., and Davidson, E. H. (1978). Specific representation of cloned repetitive DNA sequences in sea urchin RNAs. *Cell* **15,** 189–203.

Seif, I., Khoury, G., and Alan, R. (1979). BKV splice sequences based on analysis of preferred donor and acceptor sites. *Nucl. Acids Res.* **6,** 3387–3398.

Shatkin, A. J. (1976). Capping of eukaryotic mRNAs. *Cell* **9,** 645–653.

Shen, C. K. J., and Maniatis, T. (1980). Tissue specific DNA methylation in a cluster of rabbit β-like globin genes. *Proc. Natl. Acad. Sci. U.S.A.* **77,** 6634–6638.

Shinnick, T., and Lerner, R. (1980). The cbpA gene: Role of the 26K-dalton carbohydrate-binding protein in intercellular cohesion of developing *Dictyostelium* cell. *Proc. Natl. Acad. Sci. U.S.A.* **77,** 4788–4792.

Simpson, D., Rosen, S., and Barondes, S. (1974). Discoidin, a developmentally regulated carbohydrate-binding protein from *Dictyostelium discoideum:* Purification and characterization. *Biochemistry* **13,** 3487–3493.

Singer, R. H., and Penman, S. (1972). Stability of HeLa cell mRNA in Actinomycin. *Nature (London)* **240,** 100–102.

Singer, R. H., and Penman, S. (1973). Messenger RNA in HeLa cells: kinetics of formation and decay. *J. Mol. Biol.* **78,** 321–334.

Siu, C., Lerner, R., Ma, G., Firtel, R. A., and Loomis, W. (1976). Developmentally regulated proteins of the plasma membrane of *Dictyostelium discoideum:* the carbohydrate-binding protein. *J. Mol. Biol.* **100,** 157–178.

Sogin, M. L., and Olsen, G. J. (1980). Identification and mapping of a 60bp Eco RI fragment in the *Dictyostelium discoideum* ribosomal DNA. *Gene* **8,** 231–238.

Stinchcomb, D. T., Thomas, M., Kelly, J., Selker, E., and Davis, R. W. (1980). Eukaryotic DNA segments capable of autonomous replication in yeast. *Proc. Natl. Acad. Sci. U.S.A.* **77,** 4559–4563.

Struhl, K., Stinchcomb, D. T., Scherer, S., and Davis, R. W. (1979). High-frequency transformation of yeast: Autonomous replication of hybrid DNA molecules. *Proc. Natl. Acad. Sci. U.S.A.* **76,** 1035–1039.

Sussman, M., and Newell, P. C. (1972). Quantal control. *In* "Molecular Genetics and Developmental Biology" (M. Sussman, ed.), pp. 275–302. Prentice-Hall, Englewood Cliffs, New Jersey.

Sussman, R. R., and Rayner, E. P. (1971). Physical characterization of deoxyribonucleic acids in *Dictyostelium discoideum. Arch. Biochem. Biophys.* **144,** 127–137.

Sussman, M., and Sussman, R. R. (1969). Patterns of RNA synthesis and of enzyme accumulation and disappearance during cellular slime mould cytodifferentiation. *Soc. Gen. Microbiol.* **19,** 403–435.

Takiya, S., Takoh, Y., and Iwabuchi, M. (1980a). A rapid and facile procedure for the purification of DNA dependent RNA polymerases I and II EC-2.7.7.6 of the cellular slime mold *Dictyostelium discoideum. J. Fac. Sci. Hokkaido Univ. Ser. V. Bot.* **12,** 111–122.

Takiya, S., Takoh, Y., and Iwabuchi, M. (1980b). Template specificity of DNA dependent RNA polymerases I and II EC-2.7.7.6 for synthetic poly nucleotides during development of the cellular slime mold *Dictyostelium discoideum. J. Biochem. (Tokyo)* **87,** 1501–1510.

Tamm, I., Kikuchi, T., Darnell, J. E., Jr., and Salditt-Georgieff, M. (1980). Short capped hnRNA precursor chains in HeLa cells: continued synthesis in the presence of 5,6-dichloro-1-D-ribofuranosylbenzimidazole. *Biochemistry* **19,** 2743–2748.

Taylor, W. C., Cockburn, A. F., Frankel, G. A., Newkirk, M. I. J., and Firtel, R. A. (1977). Organization of ribosomal and 5S RNA coding regions in *Dictyostelium discoideum. In* "ICN–UCLA Symposium on Molecular and Cell Biology" (Wilcox, Abelson, and C. F. Fox, eds.), Vol. 7, pp. 309–313.

Town, C., and Gross, J. (1978). The role of cyclic nucleotides and cell agglomeration in post-aggregative enzyme synthesis in *Dictyostelium discoideum. Develop. Biol.* **63,** 412–420.

Town, C. D., Gross, J. D., and Kay, R. R. (1976). Cell differentiation without morphogenesis in *Dictyostelium discoideum. Nature (London)* **262,** 717–719.

Truett, M., and Gall, J. G. (1977). The replication of ribosomal DNA in the macronucleus of *Tetrahymena. Chromosoma* **64,** 567–587.

Tsang, A., and Bradbury, J. M. (1981). Separation and properties of prestalk and prespore cells of *Dictyostelium discoideum. Exp. Cell. Res.* **132,** 433–441.

Tsang, A. S., Devine, J. M., and Williams, J. G. (1981). The multiple subunits of discoidin I are encoded by different genes. *Dev. Biol.* **84,** 212–217.

Tuchman, J., Alton, T., and Lodish, H. F. (1974). Preferential synthesis of actin during early development of the slime mold *Dictyostelium discoideum*. *Develop. Biol.* **40**, 116–128.

van der Ploeg, L. H. T., and Flavell, R. A. (1980). DNA methylation in the human -globin locus in erythroid and non-erythroid tissues. *Cell* **19**, 947–958.

Vanderkerckhove, J., and Weber, K. (1980). Vegetative *Dictyostelium* cells containing 17 actin genes express a single major species. *Nature (London)* **18**, 475–477.

van Ness, J., Maxwell, J. H., and Hahn, W. E. (1979). Complex population of nonpolyadenylated messenger RNA in mouse brain. *Cell* **18**, 1341–1349.

Vogt, V. M., and Braun, R. (1976). Structure of the ribosomal DNA in *Physarum polycephalum*. *J. Mol. Biol.* **106**, 567–587.

Vogt, V. M., and Braun, R. (1977). The replication of ribosomal DNA in *Physarum polycephalum*. *Eur. J. Biochem.* **80**, 557–566.

Weintraub, H., Larsen, A., and Groudine, M. (1981). β-globin-gene switching during the development of chicken embryos: expression and chromosome structure. *Cell* **24**, 333–344.

Williams, J. G., and Lloyd, M. M. (1979). Changes in abundance of polyadenylated RNA during slime mold development measured using cloned molecular hybridization probes. *J. Mol. Biol.* **129**, 19–35.

Williams, J. G., Lloyd, M. M., and Devine, J. M. (1979). Characterization and transcription analysis of a cloned sequence derived from a major developmentally regulated mRNA of *D. discoideum*. *Cell* **17**, 903–913.

Williams, J. G., Tsang, A. S., and Mahbubani, H. (1980). A change in the rate of transcription of a eukaryotic gene in response to cyclic AMP. *Proc. Natl. Acad. Sci. U.S.A.* **7**, 7171–7175.

Wise, J. A., and Weiner (1980). Dictyostelium small nuclear RNA D2 is homologous to rat nucleolar RNA U3 and is encoded by a dispersed multigene family. *Cell* **22**, 109–118.

Wise, J. A., and Weiner, A. M. (1981). The small nuclear RNAs of the cellular slime mold *Dictyostelium discoideum*. *J. Biol. Chem.* **256**, 956–963.

Yagura, T., Yanagisawa, M., and Iwabuchi, M. (1976). Evidence for 2 alpha amanitin resistant RNA polymerases EC-2.7.7.6 in vegetative amoebae of *Dictyostelium discoideum*. *Biochem. Biophys. Res. Commun.* **68**, 183–189.

Yagura, T., Yanagisawa, M., and Iwabuchi, M. (1977). Some properties of partially purified DNA dependent RNA polymerases EC-2.7.7.6 and changes of levels of their activities during development of *Dictyostelium discoideum*. *J. Fac. Sci. Hokkaido Univ. Ser. V. Bot.* **10**, 219–230.

Yao, M. C., and Gall, J. G. (1977). A single integrated gene for ribosomal RNA in a eukaryote, *Tetrahymean pyriformis*. *Cell* **12**, 121–132.

Zucker, C., and Lodish, H. F. (1981). Repetitive DNA sequences co-transcribed with developmentally regulated *Dictyostelium discoideum* mRNA. *Proc. Natl. Acad. Sci. U.S.A.* **78**, 5386–5390.

• CHAPTER 8

Control of Gene Expression

Harvey F. Lodish, Daphne D. Blumberg, Rex Chisholm, Steven Chung, Antonio Coloma, Scott Landfear, Eric Barklis, Paul Lefebvre, Charles Zuker, and Giorgio Mangiarotti

I. Developmental Changes in Gene Expression

Important studies beginning in the late 1960s established that the activities of a number of enzymes increased or decreased at defined times during the developmental cycle (reviewed in Loomis, 1975; Sussman and Sussman, 1969). Most of these changes appeared to be part of a "developmental program," in that mutations or other treatments that caused differentiation to be blocked at a specific morphological stage generally resulted in the inhibition of all changes in enzyme levels characteristic of all later stages (Loomis *et al.*, 1976, 1977). Although these changes in enzyme activities provided markers for specific developmental stages, such studies could provide little information concerning the number or timing of regulated genes or insights into the mechanisms of regulation. Experiments with inhibitors (actinomycin D and cycloheximide) indicated that the accumulation of the marker enzymes required new RNA and protein synthesis. However, for

The Development of *Dictyostelium discoideum*

ISBN 0-12-455620-5

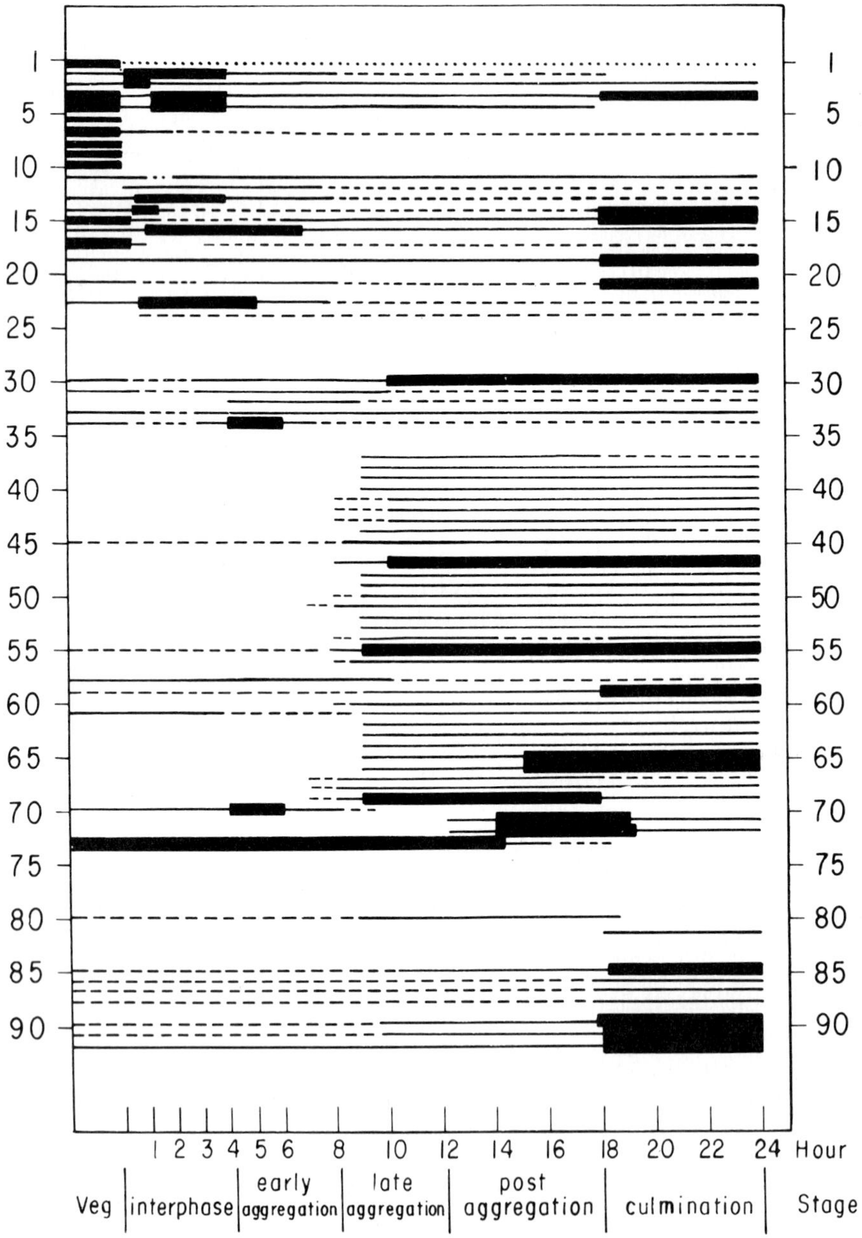

Fig. 1. Synthesis during differentiation of those 80 polypeptides, resolved by two-dimensional gel electrophoresis, which are developmentally regulated. Not shown here are the approximately 300 polypeptides synthesized both during growth and throughout differentiation. A

only two enzymes (UDP glucose pyrophosphorylase and glycogen phosphorylase) was it actually shown that an increase in enzyme activity was specifically correlated with an increased rate of synthesis of the polypeptide chain (Franke and Sussman, 1973; Thomas and Wright, 1976).

Information concerning the timing of gene expression during differentiation came from the studies in which cells were pulse-labeled with [^{35}S]methionine at hourly intervals and the protein products resolved on two-dimensional polyacrylamide gels (Alton and Lodish, 1977a). Growing cells synthesized about 400 resolvable discrete polypeptide chains. This is a minimum estimate, since proteins translated from mRNAs present at a low level (less than 10–20 copies per cell) would not have been detected. Most of these 400 proteins continued to be synthesized throughout differentiation, a result that implied that most of the predominant mRNAs present in these cells persist throughout differentiation. This analysis resolved about 80 proteins whose synthesis was induced during differentiation, or whose rate of synthesis increased or decreased markedly. As might be expected, changes in the pattern of protein synthesis were observed at all times during the cycle (Fig. 1). The surprising result was that synthesis of fully half of the developmentally regulated proteins was induced at one stage—that of late cell–cell aggregation. The induction of synthesis of these 40 proteins was correlated with the appearance in the cytoplasm of the homologous translatable mRNAs; thus, synthesis of these proteins appeared to be regulated by the synthesis (or stability) of mRNA (Alton and Lodish, 1977a). Other studies indicated that synthesis of the majority of these predominant aggregation-stage proteins occurs both in stalk and spore cells and presumably occurs in all of the cells of the aggregate (Coloma and Lodish, 1981).

During the culmination stage, the maturating spore and stalk cells show specific profiles of protein synthesis that can be used as markers of cytodifferentiation. A few predominant polypeptides, synthesized only after 19 h of development, are differentially segregated between spore and stalk cells (Coloma and Lodish, 1981; Orlowski and Loomis, 1979). Thus, during culmination there is a second, less dramatic induction of newly synthesized developmentally regulated proteins. In addition, at this stage, synthesis of several "constitutive" polypeptides as well as proteins induced at earlier stages of development appear to be regulated differently in each cell type. Interestingly, this is the case for actin, whose synthesis is specifically inhibited in the maturating spores during culmination (Coloma and Lodish, 1981).

dotted line indicates barely detectable synthesis, a solid line indicates synthesis of the protein, and a thick black line indicates that the protein is one of the predominant species made during the indicated interval. The numbers identify specific protein spots on the two dimensional gels. Taken from Alton and Lodish, 1977a.

Although the predominant spore- or stalk-specific polypeptides are only made during culmination, gene expression in the two cell types may begin to diverge earlier in development (Raper, 1940; Bonner, 1952). By 9 h of development, cells can be separated by density centrifugation (Maeda and Maeda, 1974; Feinberg *et al.*, 1979) into distinct prespore and prestalk fractions that are selectively cohesive (Feinberg *et al.*, 1979). Additionally, cells of the prestalk fraction contain higher concentrations of intracellular cAMP and calcium, and are also more chemotactic to cAMP (Tsang and Bradbury, 1981; Feinberg *et al.*, 1979; Bonner, 1949; Rubin, 1976; Brenner, 1977; Matsukuma and Durston, 1979). By 16 h of development, prestalk cells possess a higher cAMP phosphodiesterase activity than prespore cells (Tsang and Bradbury, 1981; Brown and Rutherford, 1980). This preculmination-cell type divergence is also reflected by differences in protein synthesis detected between anterior (prestalk) and posterior regions of the migrating pseudoplasmodium (Alton and Brenner, 1979; Morrissey *et al.*, 1981).

More definitive estimates of the number of developmentally regulated genes have come from several cDNA–mRNA and single-copy DNA–mRNA hybridization studies (Firtel, 1972; Blumberg and Lodish, 1980a,b; 1981). These confirm the conclusion that the major change in the pattern of gene expression occurs at the time of late aggregation. However, they indicate that the number of aggregation-specific mRNAs is much larger than previously thought—about 2000 to 3000. These sequences are the transcription products of an additional 11% of the single-copy genome (Blumberg and Lodish, 1980b).

As summarized in Table I, growing *Dictyostelium* cells contain around 4000–5000 discrete species of mRNA, about the same number as in growing yeast cells (Blumberg and Lodish, 1980a; Hereford and Rosbash, 1977). Approximately 600 of these sequences are present at greater than 160 copies per cell; the rest are present, on the average, at 14 copies per cell. These mRNAs represent the transcription products of 19% of the single-copy *Dictyostelium* genome.

The number of mRNA species does not change significantly by 6 h of differentiation (preaggregation stage) (Blumberg and Lodish, 1980b). Cross-hybridization studies showed, moreover, that the population of mRNAs present in growing and preaggregating cells were extremely similar. Although other types of studies show that expression of some genes is induced in the preaggregation stage, the number of such genes must be very few (Alton and Lodish, 1977a; Williams and Lloyd, 1979a).

In contrast, the polysomes of postaggregation cells contain 7000 discrete mRNA species, the transcription products of 30% of the single-copy genome (Blumberg and Lodish, 1980b). The majority of the 4000–5000 mRNA species present before aggregation remains in the cells throughout develop-

TABLE I
Complexity of Nuclear and Cytoplasmic Polyadenylated RNA[a,b]

Developmental stage	Percentage single copy DNA expressed in total nuclear plus cytoplasmic polyadenylated RNA	Percentage single copy genome expressed in cytoplasmic polyadenylated RNA (mRNA)	Number of genes expressed as mRNA
Vegetative	53.4	19.3	4820
6 h (preaggregation)	54.0	19.3	4800
13 h (postaggregation)	82.2	29.8	7420
22 h culmination	76.8	31.0	7750
Mixture: vegetative plus 13 h cells	79.4	31.0	7750

[a]Percentage single copy DNA expressed is determined from the fraction of the single copy DNA rendered double stranded after hybridization with either total nuclear plus cytoplasmic polyadenylated RNA (column 1) or cytoplasmic polyadenylated RNA alone (column 2) (data from Blumberg and Lodish 1980a, b; 1981). The assumption is made that only one strand of the DNA is transcribed. Therefore the percentage of genomic DNA which is expressed is twice the percentage rendered doubled stranded in the RNA-driven hybridization reactions.

[b]The number of genes expressed as average sized cytoplasmic polyadenylated RNA species (column 3) is calculated assuming that *Dictyostelium* mRNA has a weight average molecular weight of 400,000 (Firtel and Lodish, 1973) and the size of the single copy portion of the genome is 2×10^{10} daltons.

ment, although the average abundance of these mRNAs decreases with time. About one-third of the mass of mRNAs in the post-aggregation cells is comprised of the 2500 mRNA species which appear only after aggregation. Thus, an additional 11% of the single-copy genome is expressed as polysomal polyadenylated RNA, presumably mRNA, after aggregation. Of these aggregation-stage mRNAs, 100 to 150 sequences are present at 80 copies per cell; the remainder are present at 5 copies per cell.

The population of mRNAs present in culminating cells is indistinguishable, by cross-hybridization, from that in postaggregation cells. Both contain the same 7500 different species of mRNA. Although synthesis of spore- and stalk-specific proteins occurs during the culmination stage, they are not encoded by a significant fraction of the mRNA sequences present in these cells (Alton and Lodish, 1977a; Coloma and Lodish, 1981; Blumberg and Lodish, 1980b).

Previous studies (Firtel, 1972) had also indicated large changes in the percentage of the *Dictyostelium* genome expressed in total (nuclear plus cytoplasmic) RNA during the developmental cycle. Work by Jacquet *et al.* (1981) has qualitatively confirmed the change in the mRNA population that we have described; however, the number of sequences they detect is somewhat less than our estimates. Their analysis employed a cDNA probe transcribed from a total cytoplasmic RNA in which the mRNA species represent only 2% of the total. Under these conditions it is questionable whether the complex class of mRNAs—those present in 1 to 15 copies per cell—were completely copied into the DNA probe. The standard controls to demonstrate that the cDNA probe represented a faithful copy of the mRNA population were not performed. Thus it is probable that their numbers represent an underestimate of the complexity of the mRNA population.

We believe that the values in Table I are the most accurate, as they were obtained from two independent types of hybridization studies—determination of the rate of hybridization of cDNA to its mRNA template, and of the extent of hybridization of a single copy *Dictyostelium* DNA fraction to an excess of mRNA.

The number of induced genes is not large relative to other developmental systems. For example, 1500 new mRNAs are induced during conidia formation in *Aspergillis* (Timberlake, 1980) and 3000 to 4000 new mRNA species are present in differentiating myoblasts relative to their embryonal precursor cells (Affara *et al.*, 1977). However, because the *Dictyostelium* genome is smaller than that of most other eukaryotic cells, these developmentally regulated sequences represent a substantially larger proportion of the genome thereby facilitating the identification, cloning, and further study of their regulation. Indeed, many such genes have been cloned, providing very specific hybridization probes for individual regulated sequences (Mangiarotti *et al.*, 1981a; Williams and Lloyd, 1979; Williams *et al.*, 1979). These include actin and discoidin (see Chapter 7 where four gene families are described in detail). For these reasons, many laboratories, including ours, have focused attention on induction of the aggregation-stage mRNAs.

An example of the properties of several cloned genes is shown in Fig. 2. The cloned nuclear DNA segments range in size from 5 to 15 kb, and all encode one or more developmentally regulated mRNAs. To identify and

Fig. 2. Accumulation during differentiation of RNA homologous to different cloned DNAs. Poly(A^+)-containing RNA was extracted from vegetative cells and from cells plated for 6, 8, 9.5, 11, 15, and 22 h of differentiation. Cytoplasmic polyadenylated RNA (2 μg) was loaded on each lane of the gel, subjected to electrophoresis, blotted and immobilized on nitrocellulose paper as described in Chung *et al.*, 1981. Filters in (A) were hybridized to ^{32}P-labeled DNA of clones 79 and 253, (B) clones 29 and 314, (C) clone 315, (D) clone 55. The RNA encoded by these clones were indicated by the arrows. The molecular weights of the RNA band 29 and 79 are 2.2 kb and 1.7 kb, respectively.

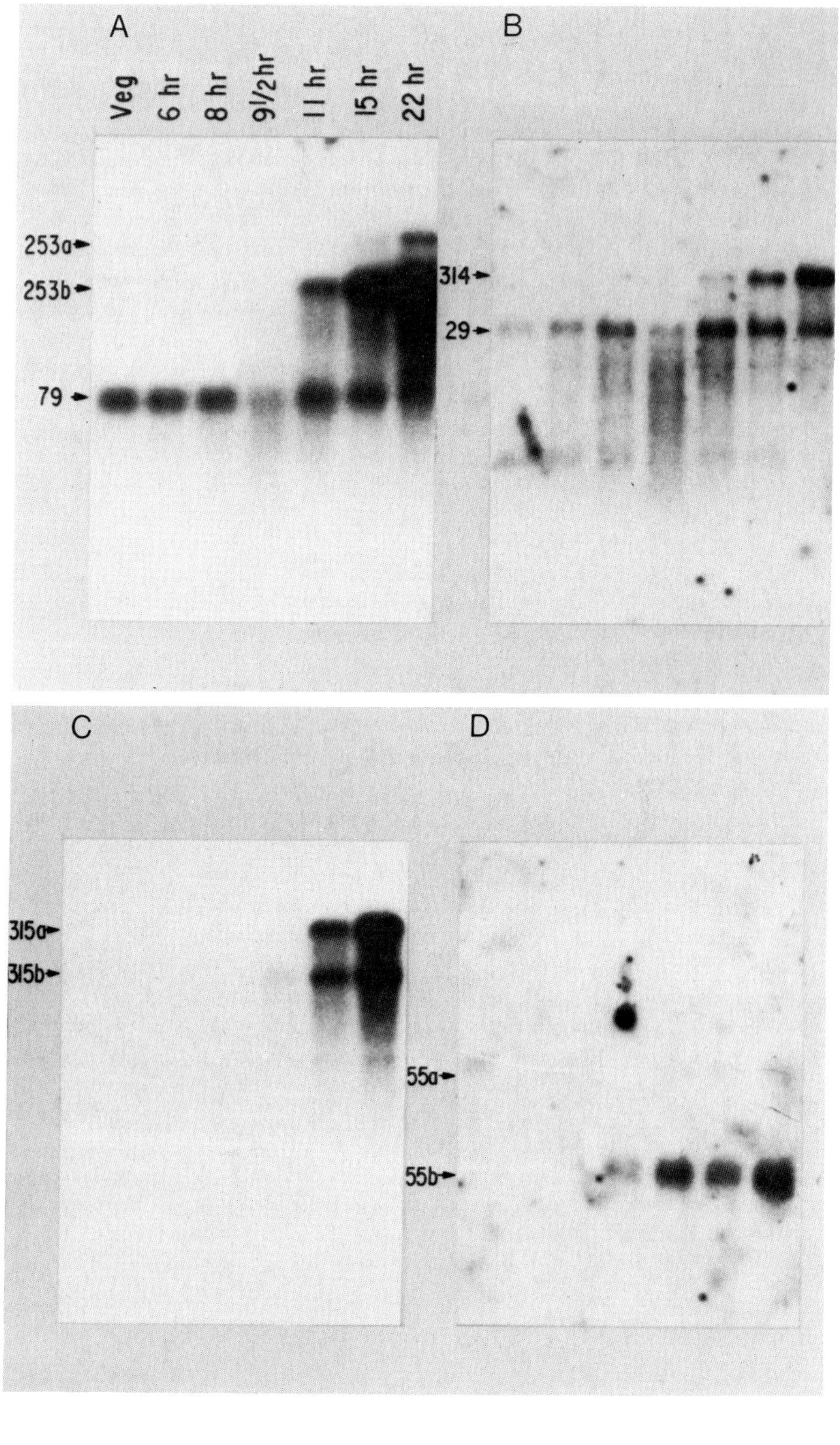
A
B
C
D
Veg
6 hr
8 hr
9 1/2 hr
11 hr
15 hr
22 hr
253a
253b
79
314
29
315a
315b
55a
55b

quantitate the amount of mRNA homologous to these cloned DNAs at different stages of development, we subjected equal amounts of cytoplasmic polyadenylated RNA to gel electrophoresis under denaturing conditions. The RNAs were then blotted onto and immobilized to nitrocellulose paper. After hybridization with ^{32}P-labeled cloned DNA, the paper was washed and exposed to X-ray film (Fig. 2). Controls established that the amount of material in all visualized bands was strictly proportional to the amount of added mRNA. The mRNA bands 253a, 253b, 314a, 315b, and 55b appear only in RNA from 11 h (aggregation stage) and later cells and are not detectable at earlier stages; we consider these to be aggregation-specific mRNAs. Some of these clones (bands 55a, 29, and 79) also encode mRNAs that are present at all stages of development, and we refer to these as constitutive mRNAs.

II. The Number of Developmentally Regulated Genes

Genetic estimates indicate that the function of only 300–400 additional genes is essential for the normal progression of *Dictyostelium* through the developmental cycle (Loomis, 1978). This is markedly lower than the number of new polyadenylated RNA species that appear on the polyribosomes during differentiation—about 2000 to 3000. The genetic estimates result from a comparison of the frequency with which mutations in known structural genes are induced with the induction of mutants which are deficient in aggregation or exhibit aberrant developmental morphology. These genetic studies focus on the number of genes whose function is essential for normal morphological development under laboratory conditions. There are several reasons why this approach could underestimate the number of developmentally regulated mRNAs. For instance, it would not score genes required for formation and migration of pseudoplasmodia or for viability of spore cells. It would not score mRNAs derived from two or more identical or nearly identical genes. Also, during differentiation genes may be activated whose functions are not necessary for spore or stalk formation in the laboratory, but are essential for morphogenesis under certain natural conditions or for differentiation along alternative pathways, such as macrocyst formation. As an example, mutations that destroy the activities of two developmentally regulated enzymes—α-mannosidase and β-glucosidase—do not have any observable effect on morphogenesis (Free *et al.*, 1976; Dimond and Loomis, 1976).

Thus, during differentiation synthesis of protein and mRNAs not essential for normal morphogenesis may be induced.

The induction of transcription of 2000 to 3000 mRNA species during differentiation seems logically to require the existence of common elements for their regulation. We isolated a *Dictyostelium* genomic DNA clone which contains sequences that are reiterated and interspersed in the genome (Zuker and Lodish, 1981). The expression of these reiterated sequences is developmentally regulated and these sequences appear to be linked only to developmentally regulated mRNAs. Since most of the mRNA species with sequences complementary to this clone accumulate in a coordinate fashion during development, beginning at the preaggregation stage, these sequences may play an important regulatory role in the developmental program of *Dictyostelium*. A similar type of repetitive DNA element has also been shown to be associated with growth phase mRNA species (Kimmel and Firtel, 1979; see Chapter 7).

III. Appearance of Aggregation-Stage mRNAs Is under Transcriptional Control

The vast majority of aggregation-specific mRNA species are absent from the nuclear RNA of growing or preaggregating cells (Blumberg and Lodish 1981). They are not detectable even at a level of one copy per 25 cells. Following aggregation, an additional 26% of the single-copy genome is expressed as total nuclear plus cytoplasmic polyadenylated-RNA transcripts. Of these sequences, 40%, equivalent to transcripts of 11% of the single-copy genome, emerge in the cytoplasm as mRNA (Table I) (Blumberg and Lodish, 1981). It appears that, if genes encoding late mRNAs are transcribed in preaggregation cells, the transcripts are destroyed immediately. Moreover, nuclei from preaggregation cells synthesize, *in vitro*, only the constitutive class of polyadenylated RNA, whereas nuclei from 16 h cells synthesize both aggregation-specific and constitutive mRNA species (Landfear *et al.*, in preparation). An example of this analysis is shown in Fig. 3, wherein ^{32}P-labeled RNA was synthesized *in vitro* by nuclei from either preaggregation or postaggregation cells. The polyadenylated RNA was hybridized to a set of immobilized cloned DNAs encoding either constitutive mRNAs (top row) or aggregation-dependent mRNAs (bottom row). As can be seen, this latter class of mRNAs is synthesized only by aggregation-stage nuclei, whereas both nuclear preparations synthesize the constitutive mRNAs.

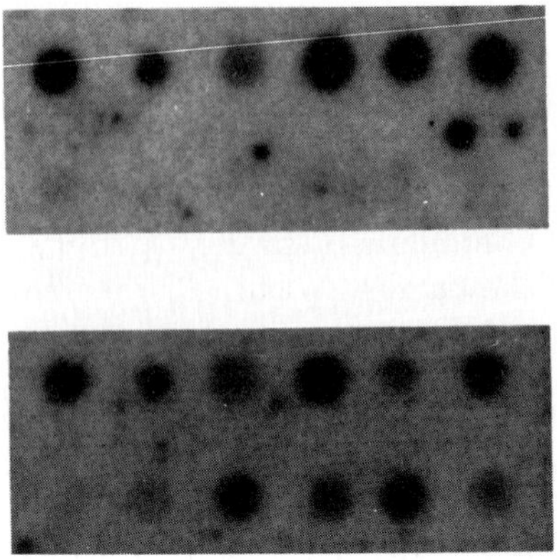

Fig. 3. Transcription of individual messenger RNAs was measured in isolated *Dictyostelium* nuclei. The nuclei were prepared (Landfear *et al.*, 1981) from 10^9 preaggregation (6 h nuc) or postaggregation (15 h nuc) cells and incubated for 30 min at 23°C in 40 m*M* Tris-HC1, pH 7.9; 10 m*M* $MgCl_2$; 250 m*M* NaC1; 0.16 m*M* ATP; 0.16 m*M* GTP; 0.16 m*M* CTP; 0.1 m*M* DTT; 5% glycerol; 10 μCi α-[^{32}P]UTP at 410 Ci/mmole. Polyadenylated RNA was isolated (Blumberg and Lodish, 1980a) and hybridized (Mangiarotti *et al.*, 1981a) to cloned *Dictyostelium* DNA (0.1–0.5 μg per spot) that had been spotted onto nitrocellulose filters. The filters were then exposed to Kodak XAR-5 film using a Dupont Cronex intensifying screen. For each filter, the top row contains cloned DNA from *Dictyostelium* genes whose transcripts are present in cells at all stages of development; the bottom row contains cloned DNA from genes whose transcripts are present only in postaggregation cells (Mangiarotti *et al.*, 1981a; Chung *et al.*, 1981).

In both growing and differentiating cells, the sequence complexity of mRNA is about 40% that of nuclear polyadenylated RNA (Table I). Thus, 60% of the transcribed sequences are degraded within the nuclei. In mass, these sequences represent less than 3% of the total polyadenylated RNA. It is not known whether some or all of these are intervening sequences which are removed by splicing of hnRNA precursors, or whether the entirety of certain transcribed nuclear polyadenylated RNAs is destroyed without reaching the cytoplasm (see Chapter 7 for a discussion of splicing in *Dictyostelium*).

IV. Stability of mRNAs during Differentiation

The level of a mRNA species in a cell is a function not only of its rate of synthesis, but also of its decay. It is thus important to determine half-lives of mRNA, particularly since, as is discussed below, the stability of the entire class of "regulated" mRNAs is a function of tight cell–cell contact. Two methods have been used to determine the stability of *Dictyostelium* mRNA during differentiation: approach to steady-state labeling of mRNA, and addition of inhibitors of mRNA biosynthesis (Margolskee and Lodish, 1980a; Mangiarotti *et al.*, 1981b). Both methods showed that during growth (8 h cell doubling) and differentiation, the average half-life of all mRNA in the cell is about 4 h. Of greater importance, after aggregation both individual constitutive and regulated mRNA species have the same half-life—about 4 h (Mangiarotti *et al.*, 1982c; Chung *et al.*, 1981).

This point is documented by the study described in Mangiarotti *et al.* 1982c and further depicted in Figs. 4 and 5, in which *Dictyostelium* cells are labeled with $^{32}PO_4$ from 13 to 17 h of development, a period chosen because it is characterized by the accumulation of late mRNAs (Mangiarotti *et al.*, 1982c; Chung *et al.*, 1981). Initially, the incorporation of ^{32}P into both poly (A^+) and poly (A^-) cytoplasmic RNA exhibits a 30 to 45 min lag, which presumably reflects the time required for the RNA precursors to reach a constant specific radioactivity and for the RNA to exit from the nucleus into the cytoplasm. Following the lag period, the incorporation of ^{32}P label into poly (A^+) and poly (A^-) RNA is linear for several hours, and declines significantly only in the last hour of labeling. Assuming, as is the case, (Mangiarotti *et al.*, 1982c) that the rate of synthesis of both rRNA and mRNA is constant during this period, this result indicates that the bulk of newly made cytoplasmic polyadenylated RNA as well as rRNA is relatively stable, as suggested by previous experiments (Margolskee and Lodish, 1980a,b; Chung *et al.*, 1981): the rate at which the amount of radioactivity in a species or population of RNA reaches a steady state (i.e., where synthesis of a labeled species is balanced by its destruction) is only a function of the rate of decay of that species or population.

In order to determine the kinetics of incorporation of ^{32}P into conserved and late mRNA species the above [^{32}P]RNA preparations were hybridized to a series of immobilized, cloned DNAs encoding both constitutive mRNAs (clones SC29, SC79, GM5) and aggregation-dependent mRNAs (all of the others). In the experiment depicted in Fig. 4, DNA purified from each clone was spotted in seven equal aliquots in a horizontal line on a sheet of nitrocellulose paper. The paper was cut in vertical strips, each containing a spot corresponding to each of the clones, and each strip was hybridized to 0.2 μg of a different labeled poly (A^+) RNA preparation under conditions where the extent of hybridization is proportional to the amount of radioactivity in the homologous mRNA. As can be seen (Fig. 5) the incorporation of ^{32}P into poly (A^+) RNA hybridizing to all of the cloned DNAs continues at a linear rate for several hours. No difference is observed between the group of clones that had been previously shown to hybridize only to conserved mRNAs

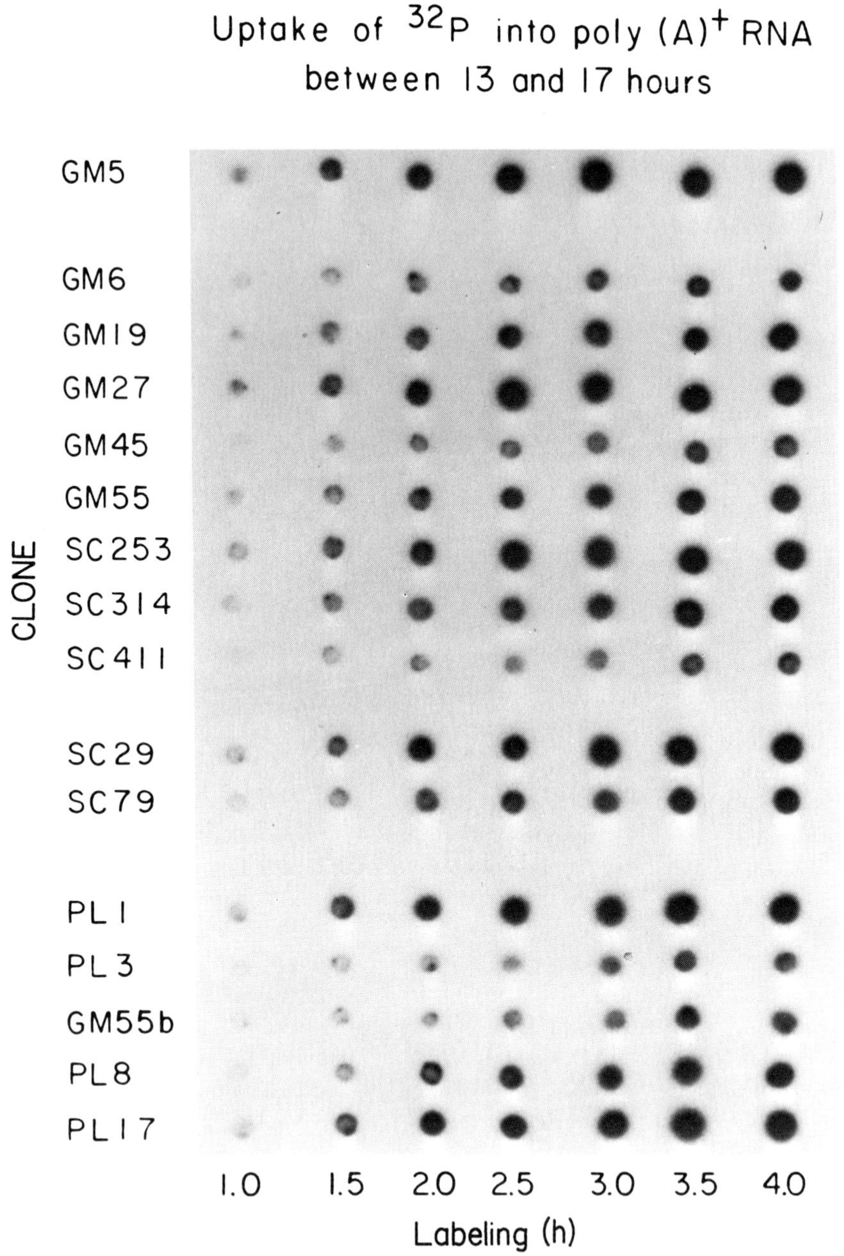

Fig. 4. Uptake of ^{32}P into cytoplasmic polyadenylated RNA species in continuous labeling during normal development. 5×10^7 cells were labeled with [^{32}P]orthophosphate beginning at 13 h of development. At the indicated times poly(A^+) and poly(A^-) RNA were isolated from the cytoplasm of labeled cells. The poly(A^+) RNA obtained from 2×10^7 labeled cells was added to 0.2 mg of unlabeled, total cytoplasmic RNA derived from cells at 17 h of development. These RNAs were then hybridized to the cloned DNAs as described in the text. Details are given in Mangiarotti *et al.* (1981b).

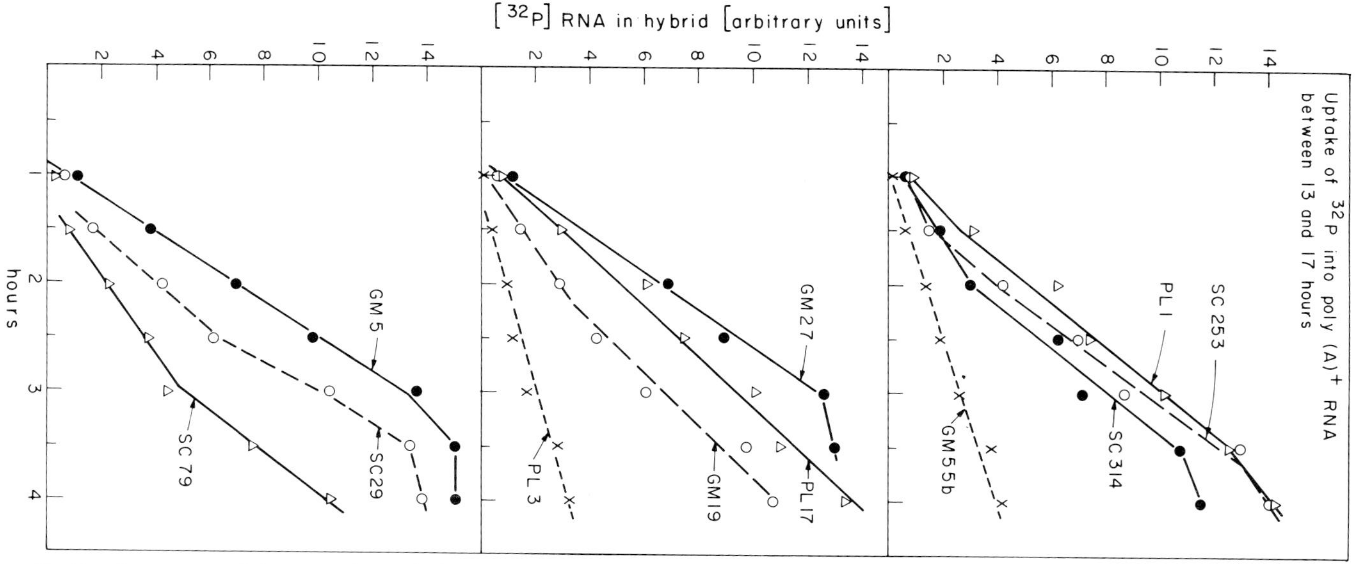

Fig. 5. Quantitation of the autoradiograph shown in Fig. 4. Different exposures of the radiogram were scanned in a Joyce-Lobel microdensitometer using a full-scale pen deflection of 1.16 optical density units, a value within the linear range of the film. The height of each peak thus obtained is reported in arbitrary units.

(clones GM5, SC29, and SC79) and all the other clones, which hybridize exclusively or mainly to late mRNAs. We conclude that all mRNAs, both conserved and late, complementary to the test DNA clones, have a half-life of at least 3 h.

During normal, unperturbed differentiation, therefore, stability of mRNA does not, to a first approximation, appear to regulate the differential expression of genes. However, a few developmentally regulated mRNAs have been identified which have a much shorter half-life, about 20 to 25 min (Margolskee and Lodish, 1980b). Also, in growing cells some mRNAs may have a shorter lifespan than others (Palatnik *et al.*, 1980).

V. Requirements for Induction of Aggregation-Dependent mRNAs

Transcription of about 2000 to 3000 genes is initiated at the time of formation of tight cell–cell contacts. Most studies indicate that these cell interactions are an essential prerequisite for transcription of these genes, although there are some apparently contradictory results.

Cells which have been starved for 15 h in a vigorously shaken suspension culture in the same solution as used for differentiation are termed aggregation competent (Gerisch, 1968). These cells possess the cAMP receptor and phosphodiesterase required for cAMP cell signaling, as well as all known cell surface proteins required for cohesion. In fact, the entire population of such aggregation-competent cells in a suspension culture will release pulses of cyclic AMP every 5 to 6 min, a process due to coordinate cell–cell signaling of cAMP (Gerisch and Hess, 1974). When plated on a solid surface they form mounds with prominent tips within 7 to 9 h, several hours faster than plated growing cells induced to differentiation. In our hands, such suspension-starved cells, in the presence of EDTA to inhibit formation of stable cell

---

Fig. 6. Expression of late mRNAs in cells developed by starvation in suspension culture in the presence or absence of cAMP. Cytoplasmic polyadenylated RNA (2 μg) was loaded on each lane of the gel, electrophoresed, blotted and immobilized on nitrocellulose paper as described in Chung *et al.* (1981). Filters were hybridized to ^{32}P-labeled DNA of clones 253, 314, 315, 55 and 79 respectively. The RNA encoded by these clones are indicated by the arrows. Lane 1 contains RNA from 15 h plated cells; lane 2 contains RNA from cells starved in suspension culture for 15 h and lane 3 contains RNA from cells starved in suspension culture for 15 h and supplemented with 100 μ*M* cAMP for five additional hours as described in Chung *et al.* (1981). Bands 79 and 55a are constitutive mRNAs and bands 253a, 253b, 314a, 315a, 315b, and 55b are late, aggregation-specific mRNAs.

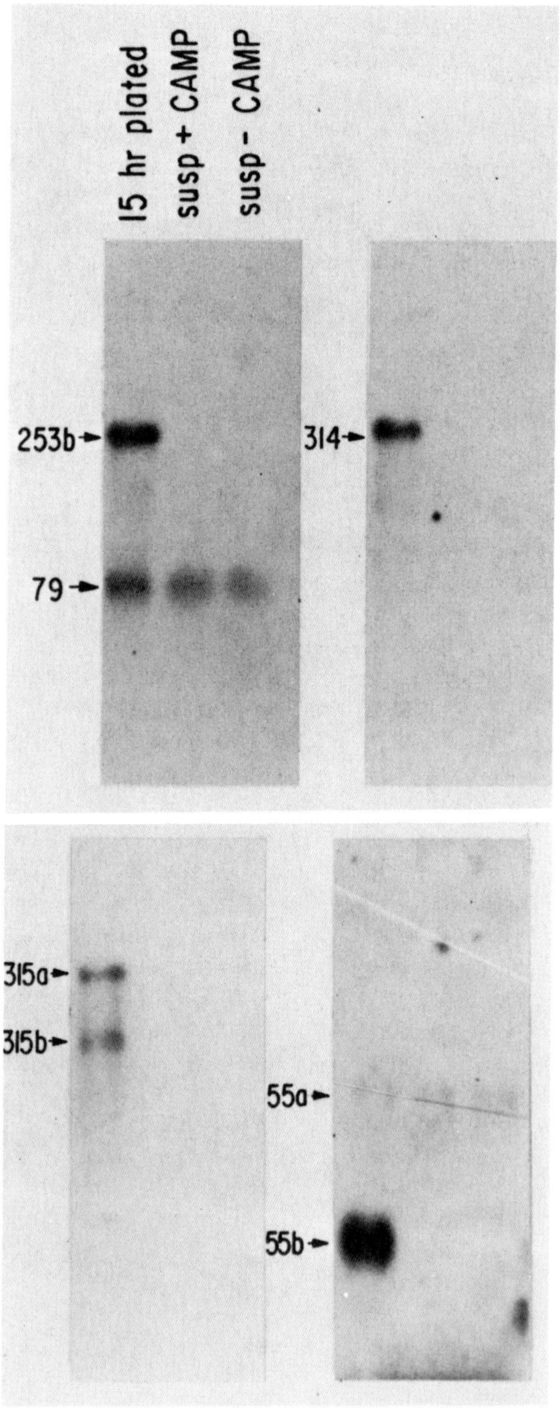

15 hr plated
susp + CAMP
susp - CAMP
253b
79
314
315a
315b
55a
55b

contacts, fail to induce significant levels of any aggregation-stage mRNA. Addition to cAMP at any time up to 16 h fails to induce any of these mRNAs (Chung *et al.*, 1981 and unpublished data). The RNA gel analysis in Fig. 6 documents this point. In cells starved in suspension for as long as 15 h, there is no accumulation of any of the developmentally regulated mRNAs (bands 253a, 253b, 314a, 315b, and 55b) whether or not the cells are then treated with cAMP. The cells do contain normal levels of the constitutive species 79 and 55a.

By contrast, Williams *et al.* (1981) showed that the transcription of certain late mRNAs species are eventually induced in suspension-starved cultures, and that the induction of this synthesis could be accelerated by 2 h when cAMP was added. In these experiments the cells were developed under the slow shaking conditions described by Town and Gross (1978) where it has been demonstrated that large agglomerates form and late enzyme expression can also be induced (Town and Gross, 1978). Thus the contacts that form in suspension agglomerates are apparently sufficient to satisfy the requirement for cell–cell interactions for late gene expression.

The notion that cell–cell contact is an essential prerequisite for activation of developmentally regulated genes is disputed by Kay, Gross and their colleagues. They showed that isolated amoebae of certain "sporogenous" mutants of the *Dictyostelium* mutant strain V12M2 can differentiate into mature stalk and prespore cells. This differentiation requires that the cells be bound to a special plastic surface, and is dependent on high levels of cAMP and on an exact ionic composition. The frequency of mature stalk cells decreases as the density of the monolayer of cells is decreased, indicating that something in addition to cAMP is required for stalk cell induction (Kay *et al.*, 1978). A dialyzable factor has been identified that, together with cAMP, can induce isolated amoebae to differentiate into mature stalk cells (Town and Stanford, 1979). The pathway for mature spore formation also requires cAMP as well as some other form of cell interaction, possibly a substance which diffuses only over a short range since close cell proximity but not actual cell contact appears to be required (Kay *et al.*, 1978, 1979; Gross *et al.*, 1981; Kay and Trevan, 1981). Isolated wild-type cells cannot differentiate into spore or stalk cells under similar circumstances, and the nature of the mutation(s) that result in the sporogenous phenotype is obscure. The mutation itself and/or the presence of the adhesive plastic surfaces may allow the normal requirement of cell–cell contact for morphogenesis and regulated gene expression to be bypassed.

In related studies, Gross *et al.* (1981) have shown that the activity of one developmentally regulated enzyme, glycogen phosphorylase, can be induced by the addition of very high levels of cAMP (1 m*M*) to a starved suspension of single wild-type cells. However, under similar conditions another late enzyme, UDP galactose polysaccharide transferase, as well as other late polypeptides cannot be induced (Okamoto, 1981; Alton and Lodish, 1977b; Landfear and Lodish, 1980). Although a great many late gene products are not expressed in aggregation-competent suspension-starved cells and are not induced in these cells by the addition of cAMP, there are some late gene products that may be exceptions.

Studies on gene expression in mutant cells that are blocked at different stages of morphogenesis do demonstrate an excellent correlation between the formation of EDTA-resistant cell–cell contacts and induction of synthesis of 2500 aggregation-dependent mRNAs:

1. All mutants which fail to aggregate (irrespective of the nature of the mutated gene), do not induce any late mRNA tested (Table II, Blumberg *et al.*, 1981).
2. Two mutants undergo normal chemotaxis, and form mounds of cells. However, they do not form EDTA-resistant cell–cell contacts and, importantly, induce synthesis of no regulated mRNAs while exhibiting a normal level of the constitutive species (Table II, Blumberg *et al.*, 1981). These mutants establish that cell–cell signaling with cAMP and chemotaxis is not sufficient for induction of regulated mRNAs.
3. All mutants that form morphologically normal cell aggregates, wherein a wild-type level of EDTA-resistant cell contacts are formed, make normal amounts of most late mRNAs even though they fail to differentiate further (Table II). One of these mutants forms a "tip", the other does not.

It is possible that the act of forming tight (EDTA-resistant) cell contacts triggers a cellular response leading to induction of gene transcription. Alternatively, cell contact could be required simply to increase the concentration of some diffusable secreted molecule that, together with cAMP, causes some cells to induce stalk- or spore-specific molecules.

VI. Gene Expression Is Dependent on Continued Cell–Cell Interactions

Abundant evidence indicates that continued synthesis of aggregation-specific polypeptides and enzymes is dependent on continued cell–cell interactions. Disaggregation of pseudoplasmodia results in immediate cessation of accumulation of developmentally regulated enzymes such as UDP glucose pyrophosphorylase and UDP galactose polysaccharide transferase (Newell and Sussman, 1970; Newell *et al.*, 1971, 1972). Likewise, synthesis of the

TABLE II
Properties of Developmentally Blocked Mutants[a]

Strain	Parent	Source	Very early development[b]	45M[c]	PDE[d]	cAMP-B[e]	EDTA contacts[f]	Terminal morphology[g]	Level of aggregation dependent mRNAs (%)[h]	Aggregation-dependent proteins[i]
Agg2	AX3	D. McMahon	+	–	–	–	–	Agg⁻	0.4	–
WL3	AX3	W. F. Loomis	+	+	–	–	–	Agg⁻	0.2	–
JM41	AX3	J. Margolskee	+	+	–	–	ND[j]	Agg⁻	0.4	–
HC72	HC6	M. B. Coukell	+	+	+	–	–	Agg⁻	0.2	–
HC54	HC6	M. B. Coukell	+	+	+	+	–	Agg⁻	0.2	–
HJR-1	NC4	R. A. Lerner	ND	ND	ND	ND	–	Ripples, loose aggregates	0.3	ND
JM84	AX3	J. Margolskee	+	+	+	+	–	Ripples, loose aggregates	6	–
GM2	AX3	W. F. Loomis	+	+	+	+	+	Mounds	39.8	+

JM35	AX3	J. Margolskee	+	+	+	+	+	Mounds with tips	100	+
AX3	NC4	W.F. Loomis	+	+	+	+	+	Culminants	100	+

[a]Taken from Blumberg *et al.*, 1981.

[b]"+" indicates induction of synthesis of the enzyme *N*-acetylglucosaminidase. It also indicates a two to fourfold induction of actin synthesis between 0 and 30 min of development as well as a sharp drop in actin synthesis at 90 min of development (Margolskee and Lodish 1980a,b).

[c]"+" indicates induction of the synthesis of two early proteins (the "45 min" proteins) synthesized between 45 and 90 min of development (Margolskee and Lodish, 1980b).

[d]Induction of the activity of the cell bound phosphodiesterase. "+" indicates that the mutant cells developed by starvation for 15 h in a rapidly shaken suspension culture expressed a level of the enzyme at least fourfold above the background activity observed in growing cells. AX3 wild-type cells starved in suspension culture for 15 h expressed a level 13-fold above the growth phase background (cf. Blumberg *et al.*, 1981 for details).

[e]Cell surface cAMP binding activity. "+" indicates that the mutant cells developed by starvation in a rapidly shaken suspension culture for 8 or 15 h induce a level of binding which is at least four times above the level detected in growing cells. Wild-type AX3 cells induced a sixfold increase under these conditions.

[f]Formation of EDTA-resistant cell contacts.

[g]Agg^- means a failure to aggregate.

[h]Taken from data in Blumberg *et al.*, 1981.

[i]Taken from data shown in Blumberg *et al.* 1981.

[j]Not done.

predominant polypeptides whose synthesis initiates after aggregation ceases when the aggregates are dispersed and kept from reattaching to each other by constant shaking (Alton and Lodish, 1977b; Okamoto, 1981; Landfear and Lodish, 1980).

Other studies have used two types of hybridization experiments to investigate the synthesis and stability of aggregation-dependent and constitutive mRNAs in disaggregated cells: hybridization of mRNA to a cDNA probe specific for the population of 2500 regulated sequences; and hybridization of mRNA to cloned cDNAs encoding individual regulated or constitutive mRNAs (Mangiarotti *et al.*, 1981c; Chung *et al.*, 1981). Both assays indicate that metabolism of all 2000 to 3000 regulated mRNAs is coordinately affected by disaggregation. First, synthesis of virtually all aggregation-dependent mRNA ceases when slugs are disaggregated and are kept apart by constant shaking. Synthesis of the "conserved" species, those synthesized both during growth and differentiation, is unaffected. Second, the stability of aggregation-dependent mRNAs in the cytoplasm is markedly and specifically reduced by disaggregation. In disaggregated cells the levels of all aggregation-specific mRNAs decay with half-lives of between 25 and 40 min; the stability of constitutive mRNAs, by contrast, is unaffected by cell–cell contact (Mangiarotti *et al.*, 1981c; Chung *et al.*, 1981).

An example of the selective destruction of individual regulated mRNAs is depicted in Figs. 7 and 8. Cytoplasmic polyadenylated RNA was prepared from cells plated for development for 15 h, or disaggregated after 15 h and maintained as a vigorously shaken single cell suspension for up to 5 h following disaggregation. The level of several late-specific mRNAs was measured by "Northern" gel analysis (Figure 8). The constitutive RNAs: bands 79, 314b, 314c, 29, and 55a remain at the same level following disaggregation, whereas the late specific RNAs: 253b, 314a, 315b, and 55b are rapidly lost following disaggregation. Both assays indicate that 2.5 h after cells are disaggregated, the level of late mRNA sequences is only 6% that in plated cells, by 5 h, it is only 2.5%. This indicates that upon disaggregation the late mRNAs are lost from the cytoplasm with a $t_{1/2}$ of 25 to 45 min. (This is a minimum estimate, and it assumes no synthesis of these mRNAs after disag-

---

Fig. 7. Specific degradation of late messages upon disaggregation, and preservative of late messages by cAMP. Two μg of cytoplasmic polyadenylated RNA was loaded on each lane of the gel, electrophoresed, blotted and immobilized on nitrocellulose paper as described in Chung *et al.* (1981). Filters were hybridized to ^{32}P-labeled DNA of clones 253, 314, 315, 55, and 79 respectively. The RNA encoded by these clones are indicated by the arrows. Lane 1 contains RNA from 15 h plated cells; lane 2 contains RNA from cells that were plated for 15 h and then disaggregated for 5 h; and lane 3 contains RNA from cells plated for 15 h and then disaggregated for 5 h in the presence of 100 μ*M* cAMP. Bands 79 and 55a are constitutive mRNAs and bands 253a, 253b, 314a, 315b, and 55b are late, aggregation specific mRNAs.

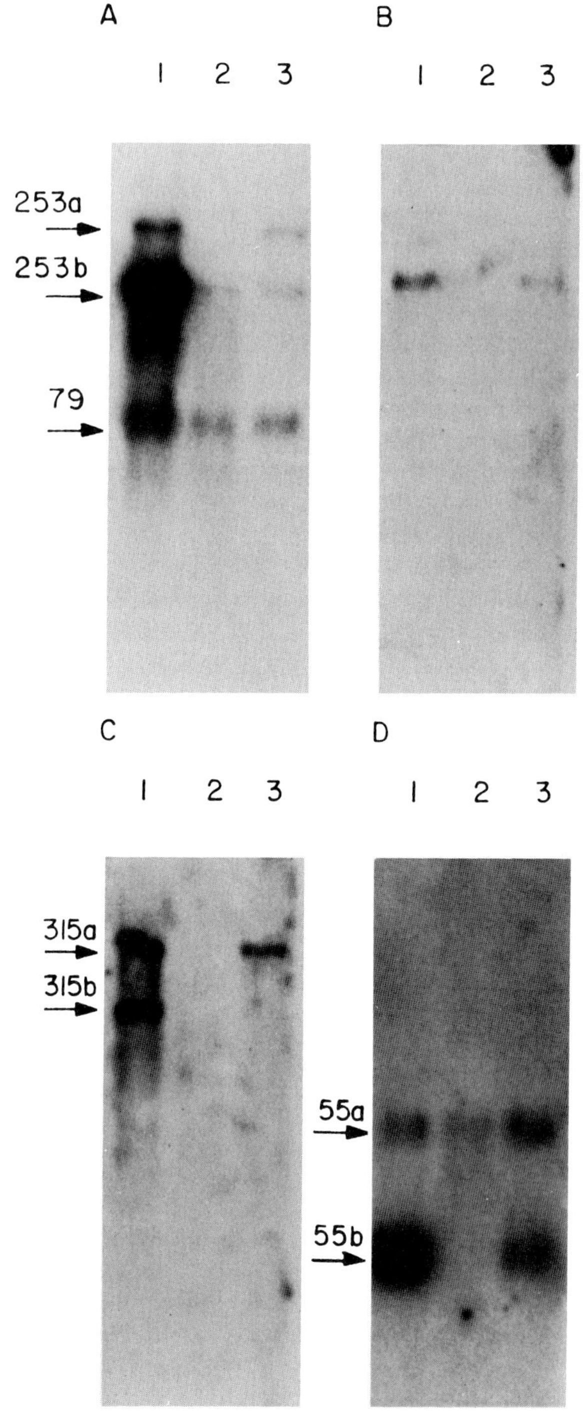
A
1 2 3
253a
253b
79
B
1 2 3
C
1 2 3
315a
315b
D
1 2 3
55a
55b

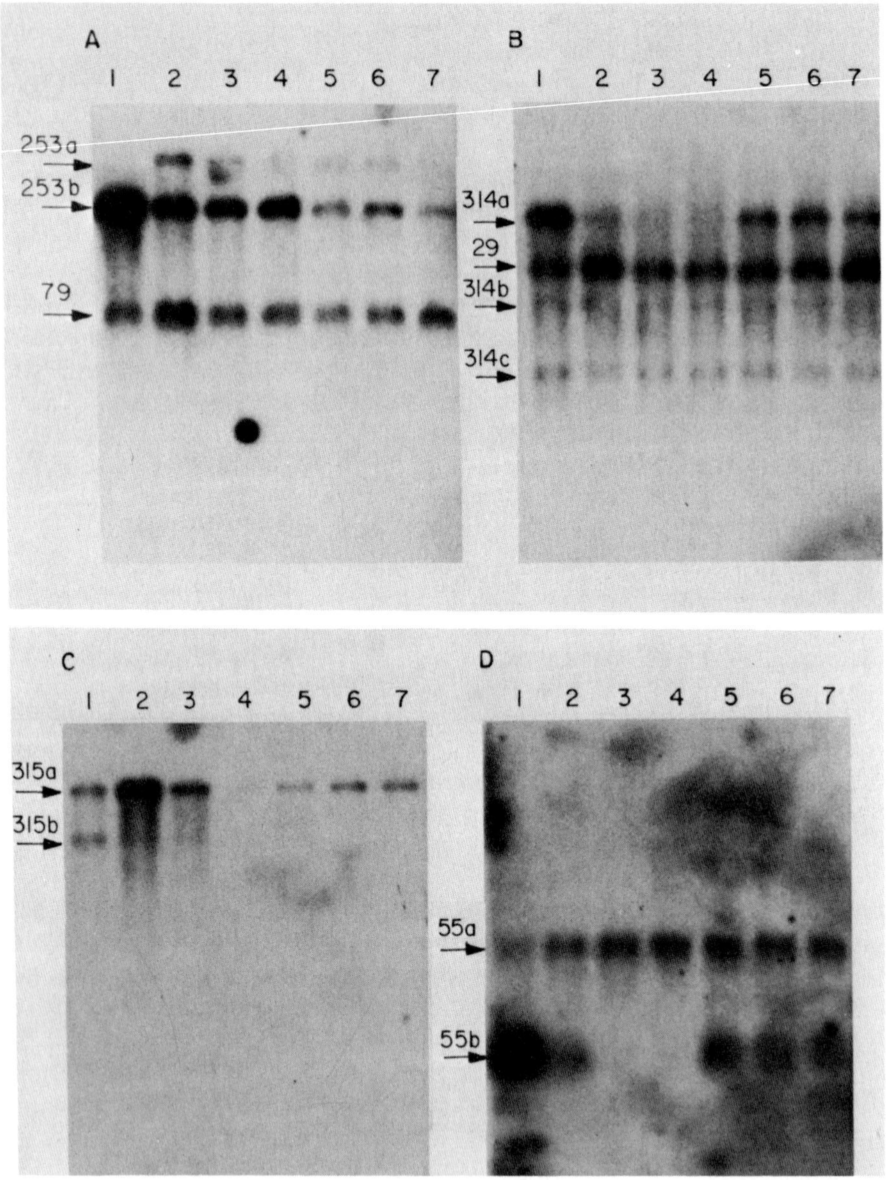

Fig. 8. Rate of degradation of late message upon disaggregation and restoration of late messages in disaggregated cells by cAMP. The gel separation hybridization was the same as in Fig. 2. Lane 1 contains RNA from 15 h plated cells; lane 2 contains RNA from cells that were plated for 15 h and then disaggregated for 1.25 h; lane 3 for 2.5 h; and lane 4 for 5 h. Lanes 5, 6, and 7 contain RNA from cells that were treated the same way as those in lane 2, 3, and 4, respectively, except that at the end of the disaggregation period cAMP was added and the cells were shaken for another 3 h. (A) clones 79 and 253, (B) clones 314 and 29, (C) clone 315, (D) clone 55.

gregation. Any continued synthesis would result in a shorter half-life). Since the half-lives of regulated mRNAs in multicellular aggregates is at least 4 h, disaggregation induces a six- to ninefold increase in the decay constants for regulated, but not constitutive, mRNAs (Mangiarotti *et al.*, 1981c; Chung *et al.*, 1981).

How the aggregation-stage mRNAs are specifically labilized upon disaggregation is not known. Among several possibilities, disaggregation could increase the synthesis or activity of a specific mRNA-degrading enzyme. Such activation of specific mRNA nucleases has been observed in mammalian cells treated with interferon (Nilsen *et al.*, 1980). We envisage differential stability of mRNAs as providing flexibility to a developing system such as *Dictyostelium*. Until very late in the developmental program, when formation of actual spores and stalk cells begins, differentiation is reversible, as disaggregated cells will resume normal growth when placed in axenic culture.

VII. cAMP and Gene Expression during Differentiation

During the preaggregation stages, cAMP serves the dual role of chemotactic agent and inducer of development. When growing cells are transferred to starvation buffer and shaken in suspension to prevent formation of aggregates, they become aggregation competent and will, after several hours, begin to emit spontaneous, periodic pulses of cAMP (Gerisch, 1968; Gerisch and Hess, 1974). If these cells are subjected to externally applied pulses of cAMP starting before the onset of spontaneous cAMP emission, the amoebae will induce prematurely high levels of several developmentally regulated preaggregation-stage proteins. In particular, several proteins required for chemotaxis (the cAMP receptor, adenyl cyclase, the membrane bound and extracellular phosphodiesterases) and cellular aggregation ("contact sites A") appear in cAMP pulsed cells several hours before they would appear in cells not treated with cAMP (Gerisch *et al.*, 1975; Darmon *et al.*, 1975; Klein, 1975; Klein and Juliani, 1977). In general, these developmental events are advanced most effectively by pulses of cAMP rather than by continuous application of the nucleotide. Marin and Rothman (1980), in contrast, find that stimulatory effects of cAMP pulsing are found only when the development rate is suboptimal. They find cAMP has no effect when cells are differentiating in optimal concentrations of ions.

We have shown that, in disaggregated cells, cAMP specifically stimulates the synthesis of most but not all of the aggregation-dependent mRNAs. As noted above, we find that regardless of the time of addition, cAMP does not induce synthesis of any of the aggregation-stage regulated mRNAs in cells shaken in suspension from the beginning of differentiation; i.e., aggregation-competent cells that are not and have not been in physical contact with other cells (Chung *et al.*, 1981; Landfear *et al.*, 1981). We believe that cell–cell contact induces the *ability* of cells to respond to extracellular cAMP, resulting in the transcription of 2500 regulated mRNAs. Specifically, 2.5 h after disaggregation the level of all regulated mRNAs is only 6% that of control cells and is about 2% at 5 h (Figs. 7 and 8). Within three hours after cAMP is added to either of these two populations of cells, the level of late mRNAs is restored to a value at least 50% that of plated cells (Fig. 8). Cyclic AMP stimulates the rate of synthesis of regulated mRNAs by these cells *at least* 4.5-fold and it may also act to stabilize the late mRNAs (Chung *et al.*, 1981).

Experiments on RNA synthesis by isolated nuclei have reinforced the notion that cAMP can affect gene transcription (Landfear *et al.*, 1981). Nuclei from disaggregated cells synthesize, *in vitro*, low levels of aggregation-dependent mRNAs, but normal amounts of the constitutive species. When disaggregated cells were treated with cAMP and nuclei then isolated, synthesis of most aggregation-dependent mRNAs was increased and was similar to that of nuclei from normal aggregated cells.

Several lines of evidence suggest that cAMP might influence the differentiation of spore and stalk cells in the multicellular aggregate. The action of cAMP in stimulating postaggregative mRNA expression may be in fact, a reflection of a larger role of cAMP as an inducer of differentiation (Bonner, 1970; Hamilton and Chia, 1975; Kay, 1979; Town *et al.*, 1976; Town and Stanford, 1979). Cell contact could be required simply to increase the concentration of some diffusible secreted molecule that, together with cAMP, causes some cells to induce stalk- or spore-specific molecules.

VIII. Conclusion

The value of *Dictyostelium* as a system for investigating several key problems of developmental biology at the molecular level is just now being realized. As one example, formation of specific cell–cell contact is dependent on a number of developmentally regulated surface proteins. Importantly, formation of these tight cell–cell contacts is essential for both the transcription and the stability of about 2500 regulated mRNA species. It should be possible to determine precisely not only how these surface molecules function in cell–cell recognition, but also how interactions at the cell surface are transmitted to the nucleus and cytoplasm in order to affect gene expression. Many aspects of the role of cAMP as a chemotactic agent are now understood, at least in outline. There is growing evidence that cAMP also induces or accelerates at least part of the developmental program and may also induce the cytodifferentiation and segregation of prestalk and prespore cells. In particular, cAMP increases the synthesis, and perhaps the stability, of the 2500 mRNA species which are induced at aggregation; addition of this compound to disaggregated cells (which have ceased transcription specifically of this class of mRNAs) restores synthesis of these mRNAs. It will be of considerable interest to learn whether, as in prokaryotic cells, cAMP enters the cells and whether cAMP binding proteins interact directly with regulatory sequences in DNA. Intracellular transmitter molecules are believed to mediate the response to extracellular cAMP during chemotaxis, resulting in the pulsatile synthesis and secretion of cAMP. It is also possible that these or other compounds mediate the effects of extracellular cAMP on gene expression. The identification of the nature and mode of action of these intracellular mediators will be of considerable importance.

References

Affara, N. A., Robert, B., Jacquet, M., Buckingham, M. E., and Gros, F. (1977). Changes in gene expression during myogenic differentiation. *J. Mol. Biol.* **140**, 441–458.

Alton, T. H., and Lodish, H. F. (1977a). Developmental changes in messenger RNAs and protein synthesis in *Dictyostelium discoideum*. *Dev. Biol.* **60**, 180–206.

Alton, T. H., and Lodish, H. F. (1977b). Synthesis of developmentally regulated proteins in *Dictyostelium discoideum* which are dependent on continued cell–cell interaction. *Dev. Biol.* **60**, 207–216.

Alton, T. H., and Brenner, M. (1979). Comparison of protein synthesized by anterior and posterior regions of *Dictyostelium discoideum* pseudoplasmodia. *Dev. Biol.* **71**, 1–9.

Blumberg, D. D., and Lodish, H. F. (1980a). Complexity of nuclear and polysomal RNAs in growing *Dictyostelium discoideum* cells. *Dev. Biol.* **78**, 268–284.

Blumberg, D. D., and Lodish, H. F. (1980b). Changes in the messenger RNA population during differentiation of *Dictyostelium discoideum*. *Dev. Biol.* **78**, 285–300.

Blumberg, D. D., and Lodish, H. F. (1981). Changes in the complexity of nuclear RNA during differentiation of *Dictyostelium discoideum*. *Dev. Biol.* **81**, 74–80.

Blumberg, D. D., Margolskee, J. P., Chung, S., Barklis, E., Cohen, N. S., and Lodish, H. F. (1982). Specific cell–cell contacts are essential for induction of gene expression during differentiation of *Dictyostelium discoideum*. *Proc. Natl. Acad. Sci. U.S.A.* **79**, 127–131.

Bonner, J. T. (1952). The pattern of differentiation in amoeboid slime molds. *Amer. Natur.* **86**, 79–89.

Bonner, J. T. (1949). The demonstration of acrasin in the later stages of the development of the slime mold *Dictyostelium discoideum*. *J. Exp. Zool.* **110**, 259–271.

Bonner, J. T. (1970). Induction of stalk cell differentiation by cyclic AMP in the cellular slime mold *Dictyostelium discoideum*. *Proc. Natl. Acad. Sci. U.S.A.* **65**, 110–113.

Brenner, M. (1977). Cyclic AMP gradient in migration pseudoplasmodia of the cellular slime mold *Dictyostelium discoideum*. *J. Biol. Chem.* **252**, 4073–4077.

Brown, S. S., and Rutherford, C. L. (1980). Localization of cyclic nucleotide phosphodiesterase in the multicellular stages of *Dictyostelium discoideum*. *Differentiation* **16**, 173–183.

Chung, S., Landfear, S. M., Blumberg, D. D., Cohen, N. S., and Lodish, H. F. (1981). Synthesis and stability of developmentally regulated *Dictyostelium* mRNAs are affected by cell–cell contact and cAMP. *Cell* **24**, 785–797.

Coloma, A., and Lodish, H. F. (1981). Synthesis of spore- and stalk-specific proteins during differentiation of *Dictyostelium discoideum*. *Dev. Biol.* **81**, 238–244.

Darmon, M., Brachet, P., and Pereira da Silva, L. H. (1975). Chemotactic signals induce cell differentiation in *Dictyostelium discoideum*. *Proc. Natl. Acad. Sci. U.S.A.* **72**, 3163–3166.

Dimond, R., and Loomis, W. F. (1976). Structure and function of β-glucosidases in *Dictyostelium discoideum*. *J. Biol. Chem.* **251**, 2680–2687.

Feinberg, A. P., Springer, W. R., and Barondes, S. H. (1979). Segregation of pre-stalk and prespore cells of *Dictyostelium discoideum*: Observations consistent with selective cell cohesion. *Proc. Natl. Acad. Sci. U.S.A.* **76**, 3977–3981.

Firtel, R. A. (1972). Changes in the expression of single-copy DNA during development of the cellular slime mold *Dictyostelium discoideum*. *J. Mol. Biol.* **66**, 363–377.

Firtel, R. A., and Lodish, H. F. (1973). A small nuclear precursor of mRNA in the cellular slime mold *Dictyostelium discoideum*. *J. Mol. Biol.* **79**, 295–314.

Franke, J., and Sussman, M. (1973). Accumulation of uridine diphosphoglucose pyrophosphorylase in *Dictyostelium discoideum*. *J. Mol. Biol.* **81**, 173–185.

Free, S., Schimke, R., and Loomis, W. F. (1976). The structural gene for α-mannosidase-1 in *Dictyostelium discoideum*. *Genetics* **84**, 159–174.

Gerisch, G. (1968). Cell aggregation and differentiation in *Dictyostelium discoideum*. *Curr. Top. Develop. Biol.* **3**, 157–197.

Gerisch, G., and Hess, B. (1974). cAMP-controlled oscillations in suspended *Dictyostelium discoideum* cells: Their relationship to morphogenetic cell interactions. *Proc. Natl. Acad. Sci. U.S.A.* **71**, 2118–2123.

Gerisch, G., Fromm, H., Heusgen, A., and Wick, U. (1975). Control of cell-contact sites by cAMP pulses in differentiating *Dictyostelium* cells. *Nature (London)* **255**, 547–554.

Gross, J. G., Town, C. D., Brookman, J. J., Jermyn, K. A., Peacey, M. J., and Kay, R. R. (1981). Cell patterning in *Dictyostelium*. *Philosoph. Trans. Royal Soc.* In press.

Hamilton, I. D., and Chia, W. K. (1975). Enzyme activity changes during cyclic AMP-induced stalk cell differentiation in P4, a variant of *Dictyostelium discoideum*. *J. Gen. Micro.* **91**, 295–306.

Hereford, L. M., and Rosbash, M. (1977). Number and distribution of polyandenylated RNA sequences in yeast. *Cell* **10**, 453–462.

Jacquet, M., Part, D., and Felenbok, B. (1981). Changes in the polyadenylated mRNA population during development of *Dictyostelium discoideum*. *Dev. Biol.* **81**, 155–166.

Kay, R. R. (1979). Gene expression in *Dictyostelium discoideum*: Mutually antagonistic roles of cAMP and ammonia. *J. Embryol. Exp. Morph.* **52**, 171–182.

Kay, R. R., Garrod, D., and Tilly, R. (1978). Requirements for cell differentiation in *Dictyostelium discoideum*. *Nature (London)* **271**, 58–60.

Kay, R. R., Town, C. D., and Gross, J. D. (1979). Cell differentiation in *Dictyostelium discoideum*. *Differentiation* **13**, 7–14.

Kay, R. R., and Trevan, D. J. (1981). *Dictyostelium* amoebae can differentiate into spores without cell–cell contact. *J. Embryol Exp. Morph.* **62,** 369–378.

Kimmel, A. R., and Firtel, R. A. (1979). A family of short, interspersed repeat sequences at the 5′ ends of a set of *Dictyostelium* single copy mRNAs. *Cell* **16,** 787–796.

Klein, C. (1975). Induction of phosphodiesterase by cyclic adenosine 3′5′ monophosphate in differentiating *Dictyostelium discoideum* amoebae. *J. Biol. Chem.* **250,** 7134–7138.

Klein, C., and Juliani, M. H. (1977). cAMP induced changes in cAMP binding sites on *Dictyostelium discoideum* amoebae. *Cell* **10,** 329–335.

Landfear, S. M., and Lodish, H. F. (1980). A role for cyclic AMP in expression of developmentally regulated genes in *Dictyostelium discoideum. Proc. Natl. Acad. Sci. U.S.A.* **77,** 1044–1048.

Landfear, S. M., Chung, S., Lefebvre, P., Chung, S., and Lodish, H. F. (1982). Transcriptional control of gene expression during development of *Dictyostelium discoideum*. *Mol. Cell Bio.* Submitted.

Loomis, W. F. (1975). *Dictyostelium discoideum: A Developmental System* Academic Press, New York.

Loomis, W. F., White, S., and Dimond, R. L. (1976). A sequence of dependent stages in the development of *Dictyostelium discoideum*. *Dev. Biol.* **53,** 171–177.

Loomis, W. F., Dimond, R., Free, S., and White, S. (1977). Independent and dependent sequences in development of *Dictyostelium*. *In* "Eukaryotic Microbes as Model Development Systems" (D. O'Day, P. Hougen, eds.), p. 177. Marcel Dekker, New York.

Loomis, W. F. (1978). The number of developmental genes in *Dictyostelium*. *Birth Defects: Original Article Series XIV* 497–505.

Loomis, W. F. (1972). Role of the surface sheath in control of morphogenesis in *Dictyostelium discoideum. Nature (London) New Biol.* **240,** 6–9.

Maeda, Y., and Maeda, M. (1974). Heterogeneity of the cell population of the cellular slime mold *Dictyostelium discoideum* before aggregation, and its relation to subsequent locations of the cells. *Exp. Cell Res.* **84,** 88–94.

Mangiarotti, G., Altruda, F., and Lodish, H. F. (1981a). Rates of synthesis of rRNA during differentiation of *Dictyostelium discoideum. Mol. Cell Biol.* **1,** 35–42.

Mangiarotti, G., Chung, S., Zuker, C., and Lodish, H. F. (1981b). Selection and analysis of cloned developmentally-regulated *Dictyostelium discoideum* genes by hybridization-competition. *Nucl. Acids Res.* **9,** 947–963.

Mangiarotti, G., Lefebvre, P., and Lodish, H. F. (1982). Differences in the stability of developmentally regulated mRNAs in aggregated and disaggregated *Dictyostelium Discoideum* cells. *Dev. Biol.* **89,** 82–91.

Margolskee, J. P., and Lodish, H. F. (1980a). Half-lives of mRNA species during growth and differentiation of *Dictyostelium discoideum*. *Dev. Biol.* **74,** 37–49.

Margolskee, J. P., and Lodish, H. F. (1980b). The regulation of the synthesis of actin and two other protein induced early in *Dictyostelium discoideum* development. *Dev. Biol.* **74,** 50–64.

Marin, F., and Rothman, F. (1980). Regulation of development in *Dictyostelium discoideum* IV. Effects of ions on the rate of differentiation and cellular response to cyclic AMP. *J. Cell Biol.* **87,** 823–827.

Matsukuma, S., and Durston, A. J. (1979). Chemotactic cell sorting in *Dictyostelium discoideum. J. Embryol. Exp. Morph.* **50,** 243–251.

Morrissey, J. H., Farnsworth, P. A., and Loomis, W. F. (1981). Pattern formation in *Dictyostelium discoideum:* An analysis of mutants altered in cell proportioning. *Dev. Biol.* **83,** 1–8.

Newell, P. C., and Sussman, M. (1970). Regulation of enzyme synthesis by slime mold assemblies embarked upon alternative developmental programmes. *J. Mol. Biol.* **49,** 627–637.

Newell, P. C., Longlands, M., and Sussman, M. (1971). Control of enzyme synthesis by cellular interaction during development of the cellular slime mold *Dictyostelium discoideum. J. Mol. Biol.* **58,** 541–554.

Newell, P. C., Franke, J., and Sussman, M. (1972). Regulation of four functionally regulated enzymes during shifts in the developmental program of *Dictyostelium discoideum. J. Mol. Biol.* **63,** 373–382.

Nilsen, T. W., Weissman, S. G., and Baglioni, C. (1980). Role of 2′5′-oligo(adenylic acid) polymerase in the degradation of ribonucleic acid. *Biochem.* **19,** 5574–5579.

Okamoto, K. (1981). Differentiation of *Dictyostelium discoideum* cells in suspension culture. *J. Gen. Micro.* In press.

Orlowski, M., and Loomis, W. F. (1979). Plasma membrane proteins of *Dictyostelium discoideum:* The spore coat proteins. *Dev. Biol.* **71,** 297–306.

Palatnik, C. M., Storti, R., Capone, A., and Jacobson, A. (1980). Messenger RNA stability in *Dictyostelium discoideum:* does poly (A) play a regulatory role? *J. Mol. Biol.* **141,** 99–118.

Raper, K. B. (1940). Pseudoplasmodium formation and organization in *Dictyostelium discoideum. J. E. Mitchell Sci. Soc.* **56,** 241–282.

Roth, R., Ashworth, J. M., and Sussman, M. (1968). *Proc. Natl. Acad. Sci. U.S.A.* **59,** 1235.

Rubin, J. (1976). The signal from fruiting body and conus tips of *Dictyostelium discoideum. J. Embryol. Exp. Morph.* **36,** 261–271.

Sussman, M., and Sussman, R. R. (1969). Patterns of RNA synthesis and of enzyme accumulation and disappearance during cellular slime mold differentiation. *Symp Soc. Gen. Microbial.* **19,** 403–435.

Telser, A., and Sussman, M. (1971). Uridine diphosphate galactose-4-epimerase, a developmentally regulated enzyme in the cellular slime mold *Dictyostelium discoideum. J. Biol. Chem.* **246,** 2252–2257.

Thomas, D. A. and Wright, B. E. (1976). Glycogen phosphorylase in *Dictyostelium discoideum* II. Synthesis and degradation during differentiation. *J. Biol. Chem.* **251,** 1258–1263.

Timberlake, W. E. (1980). Developmental gene regulation in *Aspergillus nidulans Dev. Biol.* **78,** 479–510.

Town, C., and Gross, J. (1978). The role of cyclic nucleotides and cell agglomeration in post-aggregation enzyme synthesis in *Dictyostelium discoideum. Dev. Biol.* **63,** 412–420.

Town, C. D., Gross, J. D., and Kay, R. R. (1976). Cell differentiation without morphogenesis in *Dictyostelium discoideum. Nature (London)* **262,** 717–719.

Town, C., and Stanford, E. (1979). An oligosaccharide-containing factor that induces cell differentiation in *Dictyostelium discoideum. Proc. Natl. Acad. Sci. U.S.A.* **76,** 308–312.

Tsang, A., and Bradbury, J. M. (1981). Separation and properties of prestalk and prespore cells of *Dictyostelium discoideum. Exp. Cell Res.* **132,** 433–441.

Williams, J. G., and Lloyd, M. M. (1979). Changes in the abundance of polyadenylated RNA during slime mold development measured using cloned molecular hybridization probes. *J. Mol. Biol.* **129,** 19–38.

Williams, J. G., Lloyd, M. M., and Devine, J. M. (1979). Characterization and transcriptional analysis of a cloned sequence derived from a major developmentally regulated mRNA of *Dictyostelium discoideum. Cell* **17,** 903–913.

Williams, J. G., Tsang, A. S., and Mahbubani, H. (1981). A change in the rate of transcription of a eukaryotic gene in response to cyclic AMP. *Proc. Natl. Acad. Sci. U.S.A.* **77,** 7171–7175.

Wolpert, L. (1971). Positional information and pattern formation. *Curr. Top. Dev. Biol.* **6,** 183–222.

Zuker, C., and Lodish, H. F. (1981). Repetitive DNA sequences co-transcribed with developmentally regulated *Dictyostelium discoideum* mRNAs. *Proc. Natl. Acad. Sci. U.S.A.* **78,** 5386–5390.

• CHAPTER 9

Morphogenetic Signaling, Cytodifferentiation, and Gene Expression

Maurice Sussman

I. Introduction

Just as in the Metazoa, the generation of a multicellular organization in *D. discoideum* involves (a) *Morphogenesis*, i.e., concerted cell movements and cell associations that alter the shape of the cell assembly and the topographic

The Development of *Dictyostelium discoideum*

ISBN 0-12-455620-5

relationships within it; and (b) *Cytodifferentiation*, i.e., complex programs of gene expression that ultimately transform the component cells into spores and stalk/basal disk cells.

It has been demonstrated that, in *D. discoideum* as in higher forms, these two aspects of the overall developmental program are intimately linked. Specific biochemical activities are correlated with and indeed appear to proceed causally from specific morphogenetic events, the two keeping perfectly in step. What kinds of morphogenetic information are conveyed to the cells and in what molecular forms? What receptors in and on the cells receive these signals? How are specific morphogenetic signals transduced into instructions that trigger or modulate the course of cytodifferentiation?

A. Developmentally Regulated Proteins

The first use of this term with respect to *Dictyostelium* and the first demonstration of a correlation between specific morphogenetic events and qualitative patterns of protein accumulation and disappearance involved the enzyme UDPgalactosyl:polysaccharide transferase (Sussman and Osborn, 1964). Some of the data from Sussman and Osborn (1964) are reproduced in Figs. 1A and 1B. In that and subsequent publications, the following paradigmal criteria bearing upon its developmental regulation were established.

1. The development kinetics of this enzyme were found to include (a) a period of RNA synthesis required for its accumulation (Sussman and Sussman, 1965); (b) a period of protein synthesis during which it increased dramatically to a peak of specific activity associated exclusively with the prespore cells of the developing aggregate (Ellingson *et al*., 1971); (c) the preferential release of the transferase into the extracellular space (Sussman and Lovgren, 1965); and (d) its rapid disappearance thereafter. In the wild type under a wide variety of conditions and within developing aggregates of diverse sizes an invariant relationship was established between each of the steps in this sequence and specific stages of morphogenesis.

2. Its developmental kinetics were examined in a series of morphogenetically aberrant mutants: some that were blocked at specific points between preaggregation and the penultimate stage of fruit construction (Sussman and Osborn, 1964) and others that exhibited a temporal derangement, one precocious (Sonneborn *et al.*, 1963) and the other laggard (Loomis, 1968). Without exception the pattern of synthesis, release and disappearance was profoundly altered in a manner consistent with the nature of the morphogenetic aberration and the developmental stage(s) where it was expressed.

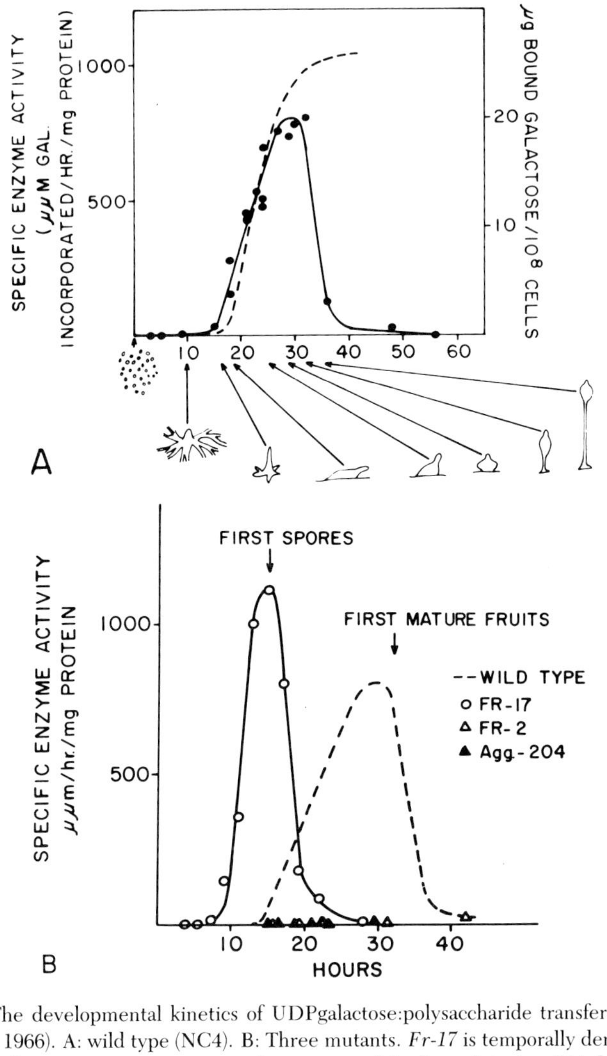

Fig. 1. The developmental kinetics of UDPgalactose:polysaccharide transferase (Sussman and Osborn, 1966). A: wild type (NC4). B: Three mutants. *Fr-17* is temporally deranged in that developmental events occur too soon and are accomplished too fast. *Agg-204* does not begin aggregation, the cells remaining as a smooth lawn. *Fr-2* begins to aggregate but fails to complete the process, the cells remaining in loose, amorphous clusters. Slugger mutant *KY3* (not shown) accumulates the enzyme in normal fashion but does not release or destroy it, i.e., precisely what wild type aggregates stuck in the slug mode do. (Yanagisawa *et al.*, 1967). These experiments—predating the introduction of the filter paper circle, and absorbent pad substratum (Sussman and Lovgren, 1965)—were performed with cells incubated on nonnutrient agar. Fruit construction is completed in 30 h in the latter circumstance versus 24 h in the former. In both, the developmental kinetics of the transferase keep in step with the flow of morphogenetic events.

TABLE I
Postaggregation Protein Synthesis in *D. discoideum*[a]

Enzymes
Alkaline phosphatase (Loomis, 1969)
β-Glucosidase (Coston and Loomis, 1969; Dimond and Loomis, 1976)
Glycogen phosphorylase (Firtel and Bonner, 1972)
Tyrosine transaminase (Firtel and Brackenbury, 1972)
UDPgalactose 4-epimerase (Telser and Sussman, 1971)
UDPgalactosyl:Polysaccharide transferase (Sussman and Osborn, 1964)
UDPglucose pyrophosphorylase (Ashworth and Sussman, 1967; Franke and Sussman, 1973)
5′-AMP:Isopentenyl transferase and discadenine synthetase (Makoto *et al.*, 1980)
Spore and stalk specific antigenically reactive components (Gregg, 1971; Gregg and Badman, 1970; Takeuchi, 1963)
Electrophoretically Separable Entities
Late stage specific glycoproteins including one outer spore coat component, labeled with fucose (Lam and Siu, 1981)
Plasma membrane-associated proteins, labeled with fucose, acetate (Hoffman and McMahon, 1977; Parish and Schmidlin, 1979a,b)
Plasma membrane-associated glycoproteins labeled with ^{125}I-Wheat Germ Agglutinin and antigenically reactive (Wilcox and Sussman, 1981a,b)
Cell proteins including spore and stalk specific entities, labeled with [^{35}S]-methionine (Alton and Lodish, 1977; Coloma and Lodish, 1981)

[a]Specific proteins whose synthesis is initiated concurrently with or after overt aggregation have been recognized by (1) enzymatic characterization, (2) serological analyses, and (3) one or two dimensional electrophoresis.

3. Wild-type and mutant aggregates can be made to undergo ecologically reasonable alterations in and recapitulations of the normal morphogenetic sequence by manipulation of environmental parameters (i.e., ambient pH, population density, and light.) The developmental kinetics of UDPgalactosyl:polysaccharide transferase were shown to be altered perfectly in step with the altered flow of morphogenetic events (Newell and Sussman, 1970).

Subsequent studies have employed these criteria to identify other developmentally regulated proteins monitored enzymatically, serologically, or isotopically as entities in two dimensional electropherograms (see Table I). In one case (UDPglucose pyrophosphorylase), the accumulation of the enzyme has been unequivocally shown to result from a 10–20-fold increase in the differential rate of its synthesis (Franke and Sussman, 1973). A similar conclusion has been advanced in the case of a number of electrophoretically separable proteins (Loomis, 1975; Alton and Lodish, 1977). Using inhibitors of RNA synthesis, investigators have without exception delineated transcriptive periods as initial and integral steps in the process which are likewise morphogenetically regulated. The likelihood that these periods reflect the synthesis of the corresponding mRNA species has been strongly supported by studies involving the *in vitro* translation of stage specific polyadenylated RNA fractions (Alton and Lodish, 1977; Chung *et al.*, 1981).

B. Classes of Morphogenetic Signals

Three kinds of morphogenetic events and/or topological relationships have been shown to play significant roles in triggering or modulating the course of gene expression and cytodifferentiation. The balance of this essay will be devoted to describing the signals and characterizing the responses. Speculations on the molecular mechanisms that may be involved will be included.

1. *Cell Contact-Mediated Signaling*

In the wild type, the making of cell–cell contacts via aggregation is correlated with and may trigger rounds of transcription leading to the accumulation of developmentally regulated proteins and ultimately to terminal differentiation. The breaking of cell contacts through disaggregation aborts these programs. Remaking contacts via reaggregation triggers complete second rounds and ultimately the completion of the differentiative process. As will be seen, under special circumstances the need for actual contact (as opposed to close apposition) can be circumvented.

2. *Positional Signaling*

The developmental fates of the cells appear to coincide with their relative positions within the aggregate. Those in the apex differentiate into stalk cells, the medial contingent into spores, and the most posteriad into basal disk cells (Raper, 1940). Differential accumulation of histiotypic proteins and organelles is evident even in the early stages of postaggregative development. Changes in relative position (e.g., when a migrating slug is segmented transversely) lead to altered developmental fates.

3. *Holistic Signaling*

The cells within the aggregate seem to monitor the morphogenetic progress of the multicellular assembly as a whole particularly with respect to the choice of morphogenetic pathways, slug migration or fruit construction, but also with respect to the timing of morphogenetic events within those pathways. This information relating to the assembly as a whole can abort or otherwise modify programs of gene expression within the individual cells.

II. Cell Contact-Mediated Signaling

A. *A Priori* Mechanistic Models

The studies described below compellingly indicate that the formation of multicellular aggregates somehow triggers specific programs of gene expression that lead to the accumulation of a restricted set of proteins. The role of cell contact per se in these proceedings can be accounted for by two hypothetical mechanistic alternatives:

First, the interaction of membrane-bound receptors at juxtaposed cell surfaces may, either directly or through a macromolecular linker, trigger the programs of gene expression, presumably by a second, intracellular messenger. Hence, cell contact per se is a necessary component of the signaling process.

A second alternative is that the formation of a multicellular aggregate permits the intracellular accumulation of low molecular weight, diffusible metabolites that would otherwise be dissipated. This accumulation might be dictated in a simple manner by the rates of synthesis versus destruction and/or extracellular diffusion, or it might be enhanced by an autocatalytic feedback loop (as in the case of cAMP synthesis). In either event, it is the intracellular accumulation of the metabolite, *facilitated* by cell juxtaposition (but not necessarily specific contact), that triggers the programs of gene expression.

In addition to constituting conceptual alternatives that are mutually exclusive when stated, as above, in their extreme forms, these models lead to very different experimental regimens. Previously reported results can be adduced to favor either alternative and crucial evidence to rule one in and the other out is still not at hand. An examination of these studies follows.

B. Contact-Mediated Quantal Control of Protein Synthesis

D. discoideum wild-type cells dispensed on filter paper circles over pads saturated with a buffer–salts mixture acquire cohesivity and begin forming aggregates by 8 h. Shortly thereafter a restricted set of proteins begins to accumulate in a precisely timed sequence. Two of these, UDPGpyrophosphorylase and UDPgalactose 4-epimerase are exemplified in Fig. 2. The set includes many other proteins detectable enzymatically, serologically or by electrophoretic separations (see Table I).

It has been firmly established, initially by the use of actinomycin D and recently by more direct means, that prior transcriptive periods are required for these accumulations. The more direct method has involved *in vitro* translation of polyadenylated RNA contingents derived from pre- and postaggregative cells (Alton and Lodish, 1977) and hybridization with stage specific cDNA and cloned DNA probes (Williams and Lloyd, 1979; Jacquet *et al.*, 1981; Chung *et al.*, 1981; Blumberg and Lodish, 1980).

The implication of cell–cell contact in these proceedings was originally drawn from the experimental results summarized in Fig. 2, a response which was termed "Quantal Control" (Sussman and Newell, 1972). When aggregates are harvested from filters, mechanically dispersed to single cells by vigorous pipettings, and washed and redeposited on filters at the original density, they reaggregate almost immediately and within 2–3 h recapitulate the morphogenetic sequence that originally may have required as much as 16–18 h to accomplish (see Fig. 2). Having attained the stage at which they were disrupted, the aggregates then proceed at the normal rate to the completion of fruiting body construction. If made to reaggregate in the presence of cycloheximide or actinomycin D, they simply recapitulate previous morphogenesis and do not develop further (Newell *et al.*, 1972).

As shown in Fig. 2, the accumulation of the pyrophosphorylase and epimerase is immediately stopped by disaggregation (the epimerase disappearing dramatically) and does not resume until after the cells have reaggregated. Then a complete additional round of synthesis is initiated regardless of how much had previously been accumulated. Exposure to actinomycin D at the time of disaggregation demonstrated that the subsequent synthesis requires a complete new round of transcription. Chung *et al.* (1981) have reported that postaggregative cells contain stage-dependent cytoplasmic, polyadenylated mRNA whose complexity indicates the presence of 3000 species. *In situ* they exhibit a metabolic half-life of about 4 h. In disaggregated cells incubated in shaker suspension with 10 m*M* EDTA, the overall half-life is 25–45 min.

Newell *et al.* (1972) observed a quantal response by three developmentally regulated enzymes. In contrast, other enzymes which accumulate prior to aggregation remained unaffected. In subsequent publications, glycogen phosphorylase (Firtel and Bonner, 1972) and a number of aggregation-associated plasma membrane proteins (Parish and Schmidlin, 1979) have been reported to be quantally controlled in similar fashion. How can quantal control be envisaged in molecular terms? Presumably (a) the initiation of cell contacts via aggregation or an immediate consequence thereof triggers the transcription of a set of structural genes that is then automatically terminated by some kind of counting or timing device (see p. 375); (b) during sustained cell contact these transcripts yield quantal amounts of the corresponding

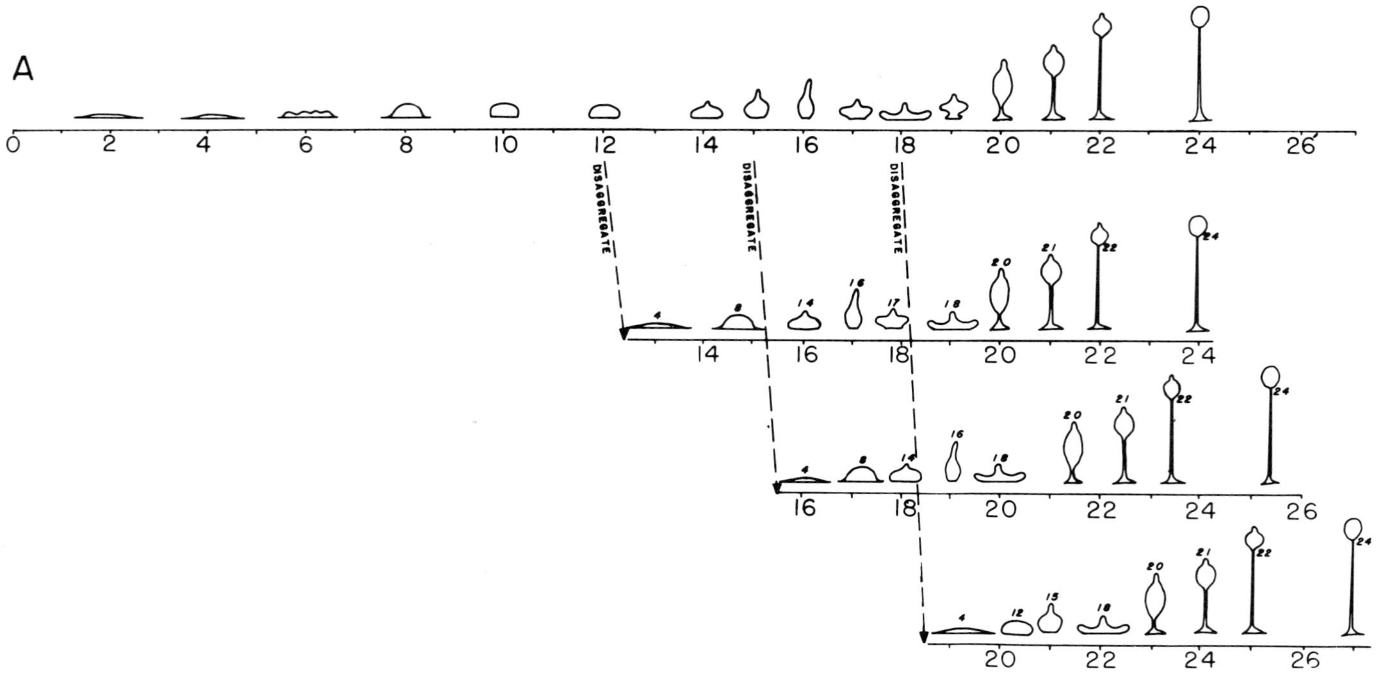

Fig. 2. Quantal control (Sussman and Newell, 1972). (A) The morphogenetic behavior of cells disaggregated at various times and redeposited on filters at the original density. The numbers refer to times in hours. (B) The developmental kinetics of two enzymes before, during, and after successive disaggregation–reaggregations. The ordinates refer to specific enzyme activities in units/mg cell protein.

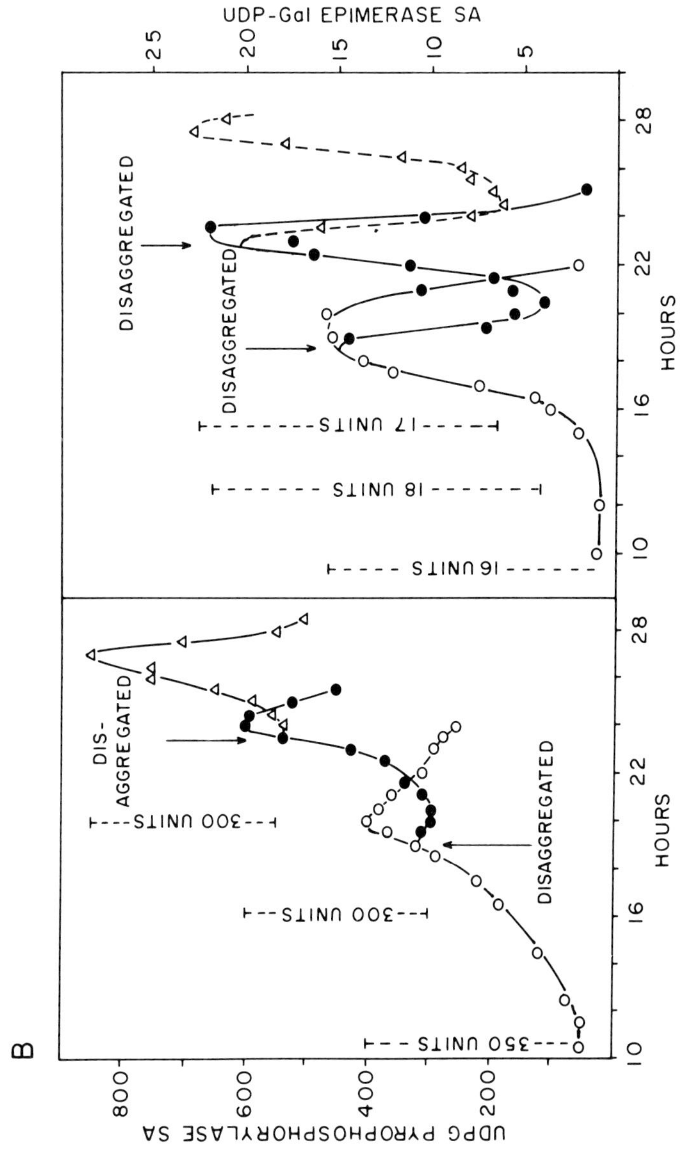
B
UDPG PYROPHOSPHORYLASE SA
800
600
400
200
10
16
22
28
HOURS
350 UNITS
300 UNITS
300 UNITS
DIS-AGGREGATED
DISAGGREGATED
UDP-Gal EPIMERASE SA
25
20
15
10
5
16 UNITS
18 UNITS
17 UNITS
DISAGGREGATED
DISAGGREGATED
HOURS

protein products; (c) loss of cell contact via disaggregation aborts both further transcription and the utilization of preformed transcript; (d) the remaking of cell contacts triggers complete additional rounds regardless of prior accumulation. Hence feedback controls of protein synthesis do not operate here.

C. Protein Synthesis by Disaggregated Cells

Newell *et al.* (1972) reported that disaggregated cells, when redeposited at a sufficiently low population density to preclude reaggregation, failed to initiate additional rounds of developmentally regulated protein synthesis. Gross *et al.* (1977) incubated *D. discoideum* V12/M2 cells in slowly and rapidly shaken suspension. The former permitted aggregate formation beginning at about 6 h, the latter delayed it. In cells incubated in slowly shaken suspension, phosphodiesterase accumulation was limited, peaking at 2 h and again at 6 h and diminishing thereafter. UDPglucose pyrophosphorylase accumulated dramatically beginning at 6 h. In cells incubated in rapidly shaken solution, the phosphodiesterase continued to accumulate rapidly after 6 h with no signs of cessation and the pyrophosphorylase rose slightly if at all. Alton and Lodish (1977) reported that over a four hour period when the cells are forming tight aggregates, the synthesis of 10 discrete polypeptides made by preaggregation cells was considerably reduced and the synthesis of 40 new proteins was initiated. The induction or cessation of synthesis was correlated with the appearance or disappearance of corresponding messenger RNA species. When stationary phase cells were shaken in suspension under conditions and at a cell density where they failed to form aggregates, the pattern of protein synthesis remained characteristic of the preaggregative stage. However, when such cells were dispensed on solid substratum, they demonstrated precocious development indicating that at least some parts of the developmental program had been followed. Even when cells were incubated on solid substratum for a period of 5 h and then harvested and incubated further in shaken suspension, they failed to switch over to the aggregation related synthetic pattern.

Unfortunately, none of these studies can be taken to distinguish crucially between the two hypothetical mechanisms, contact-dependent or contact-facilitated regulation. In each case it can be argued that the conditions that precluded sustained contact may also have favored the dissipation of diffusible metabolites.

A possible role for cAMP in these proceedings was suggested by the studies of Town and Gross (1978) and Takemoto *et al.* (1978). Addition of cAMP to disaggregated cells in rapidly shaken suspension arrested phosphodiesterase accumulation and greatly increased the synthesis of postaggregative enzymes.

Kay *et al*. (1979) reported that even in slowly shaken suspensions addition of cAMP (5 m*M*) induced precocious accumulation of postaggregative enzymes and the accumulation of prespore vesicles. As noted previously, Chung *et al*. (1981) reported the rapid degeneration of the postaggregative mRNA population within disaggregated cells incubated in shaken suspension. Addition of cAMP maintained the levels of most but not all late mRNA species or restored them within 3 h. In contrast cAMP had no effect on mRNA levels in cells incubated from zero time in shaken suspension in a buffer mixture wherein aggregation does not occur. Similarly Landfear and Lodish (1980) reported that AX3 cells failed to accumulate postaggregate proteins when shaken rapidly from zero time for 16 h and then exposed to cAMP. Conversely, cells allowed to develop on solid substratum for 13 h and then disaggregated and incubated in rapidly shaken suspension in the presence of cAMP did respond. The conclusion drawn by the authors was that cell contact per se is a required condition.

Gross *et al.* (1981) repeated these experiments using glycogen phosphorylase and UDPgalactosyl:polysaccharide transferase as marker proteins. They report that indeed cells shaken in MES–LPS buffer remain as single cells (and do not accumulate contact sites A). However, if exposed to cAMP at 6 h they do respond by rapidly accumulating postaggregation proteins. The authors conclude that the presence of exogenous cAMP is a sufficient condition for the induction of postaggregation protein synthesis and that the erroneous conclusion of Landfear and Lodish (1980) resulted from inappropriate timing of their experiment. Certainly 13 h is very late relative to the timing of postaggregative protein synthesis during normal morphogenesis on solid substratum and relative to the acquisition of aggregation competence by cells in shaken suspension. Nevertheless, the use of these results to rule out the alternative of cell contact-dependent regulation may be premature.

D. Cytodifferentiation Without Cell Contact

Bonner (1970) and Chia (1975) reported that *D. discoideum* stationary phase cells dispensed on nonnutrient agar at low population density in the presence of exogenous cAMP developed over a period of 2–3 days in significant numbers into swollen, vacuolated, thick, cellulose-walled cells that closely resemble the stalk and basal disk cells within a normal fruiting body. These results were confirmed and extended by Town *et al.* using a mutant (*M2*) of *D. discoideum* strain V12 (1976). The latter demonstrated that under optimal conditions the induction was virtually 100% and that it was population density dependent, reflecting the need for accumulation of a diffusible cofactor (see Section III).

Spore cytodifferentiation by dispersed cells is a more complicated matter. The initiation or indeed even the completion of spore cytodifferentiation by dispersed wild-type cells has not been observed. Several groups have reported (Takeuchi *et al.*, 1977; Tasaka and Takeuchi, 1979, 1981; Sternfeld and Bonner, 1977; and Sternfeld and David, 1978, 1981) that wild-type cells mechanically constrained within two and three dimensional clusters sort out histiotypically and differentiate into spores and stalk cells. Brackenbury and Sussman (1975) demonstrated that the temporally deranged (precocious) mutant *Fr17* if disaggregated after 11–12 h development and redeposited on solid substratum fails to reaggregate and forms an even lawn of cells that differentiate into normal spores and stalk cells. In the presence of a mixture of diffusible metabolites (see Section III) the disaggregated cells could be dispensed at very low density with a mean separation of 10 cell diameters and still complete cytodifferentiation with normal timing, but only if they had been permitted to remain aggregated up to the 11–12 h stage before dispersal. Town *et al.* (1976) selected "sporogenous" mutants capable of differentiating into normal, detergent-resistant spores within clusters of mutagenized wild type under conditions wherein most cells were unable to complete the sequence. These mutants turn out to phenotypically resemble *Fr17*. Here too, spore differentiation required incubation of the cells at high population density. Gross *et al.* (1977) and Kay *et al.* (1978) concluded that cell contact per se is a necessary condition. However, under altered conditions of incubation the cell density could be decreased. Time-lapse films (Kay and Trevan, 1981) have revealed that at least some of the mutant cells can develop into spores without making apparent contact with any other cells. However, the cells held not to be in contact were between one and two cell diameters apart, close enough to permit filopodial contacts that would have been invisible due to optical limitations. Additionally, when *D. discoideum* cells move they deposit detritus on the substratum and eliminate vesicles by exocytosis. Contact between one cell and the garbage of another cannot be ruled out.

E. Preliminary Conclusions

It seems that among aggregation-competent cells, postaggregative protein synthesis and even complete programs of cytodifferentiation can be accomplished in the absence of sustained cell contact (although for spore cytodifferentiation one must resort to mutant strains to do so). Are we then to conclude that it is not cell contact-dependent regulatory mechanisms but cell contact-facilitated mechanisms involving metabolite accumulation that are responsible? This is not necessarily true.

The involvement of three distinct moieties in cell cohesion has been demonstrated in *D. discoideum*. Contact sites A appear to promote the end-to-end cohesion encountered during aggregation (Beug *et al.*, 1970, 1973). Discoidins I and II and associated receptor(s) appear to be involved in the initiation of the aggregate and its maintenance early in the morphogenetic sequence (Rosen *et al.*, 1973; Reitherman *et al.*, 1975; Frazier *et al.*, 1975; Siu *et al.*, 1976; Ma and Firtel, 1978; Ray *et al.*, 1979). These are apparently supplanted by a newly arisen, serologically and chemically distinct system (Steinemann and Parish, 1980; Wilcox and Sussman, 1981a,b), possibly histiotypically specific, (Feinberg *et al.*, 1979) which ensures the integrity of the aggregate during the later stages. All of these systems involve glycoprotein moieties, either membrane bound or membrane associated.

Conceivably oligosaccharide, liposaccharide, or peptidoglycan haptenes derived from these or other membrane associated moieties may bind to specific receptor sites and thereby permit cells to accomplish at a distance the signaling that is normally achieved by juxtaposition. Since the diffusible factors that mediate the population density-dependent developmental activities of the dispersed cells have not been unequivocally demonstrated and characterized, it is still premature to reach final conclusions about the underlying mechanism(s).

III. Positional Signaling

The ability of a cell to elect a specific developmental fate by determining its relative position within a multicellular assembly and to change that fate in response to an experimentally contrived change in position intrigued the earliest of developmental biologists—Driesch, Weissmann, Holtfreter *et al.*—and has continued to do so. Unfortunately this attention has not resulted in identification of even one of the molecular mechanisms that may convey what Wolpert (1971) termed *positional information*.

The earliest *a priori* models were based upon the maintenance of anterio-posterior and/or dorsoventral concentration gradients of hypothetical diffusible effectors, subsequently elaborated into "French Flag" schemes (Wolpert, 1969, 1971) involving opposed gradients of two substances (thereby enabling a cell to know not only how far front it was but also how far back). Currently, the most pervasive and persuasive is the model of Gierer and Meinhardt (1972) based in turn upon the generic formulation of Turing

(1952). It postulates the existence of two diffusible metabolites, one serving as the activator and the other as an inhibitor of a developmental event. Their respective accumulations are made mutually exclusive within a given cell by appropriate feedback loops. An "A region" will thereby maintain itself free of P cells and a "P region" likewise free of A cells. The specific topology will be determined by the reaction rates, inhibitory or excitatory feedback rate constants, and diffusion constants of the A- and P-specific factors. Computer simulations of developing morphogenetic fields based on this model generate complex patterns that bear close resemblance to real biological organizations.

Since patterning in *Dictyostelium* will be covered in detail in other chapters, I shall confine myself here to two questions. Are low molecular weight, diffusible metabolites involved in spore and stalk/basal disk cell differentiation? Are any or all of these likely to convey positional information to cells within the developing aggregate?

A. Diffusible Factors Required for Spore and Stalk/Basal Disk Cell Differentiation

Gross *et al.* (1981), working with *D. discoideum* V12/M2 and sporogenous mutant derivatives, Sternfeld and David (1979), employing NC4 wild type and Wilcox and Sussman (1978), using mutant *Fr17* have reported the following observations:

1. The differentiation of wild type incubated at low density into stalklike cells requires the presence of 1–5 m*M* cAMP plus a dialysable metabolite called DIF that accumulates extracellularly after aggregation. Originally DIF was reported (Town and Stanford, 1979) to be sensitive to glycosidases including neuraminidase and its activity blocked by simple sugars including sialic acid and lectins specific for the latter. But most or all of the activity was also found to be extractable from dried preparations with chloroform/methanol, hardly a suitable solvent for glycosides. In a subsequent publication (Gross *et al.*, 1981), DIF is said to be a neutral, hydrophobic molecule based on its chromatographic behavior in nonpolar solvent mixtures. Doubt is cast upon the involvement of a glycosidic moiety but the data previously reported are not explained.

2. At high population density, wild-type cells were found (Gross *et al.*, 1981) to accumulate spore-specific products (i.e., UDPgalactose transferase and prespore vesicles) and significant numbers of sporogenous mutant cells to differentiate completely into spores. Optimal results were obtained in the presence (at least initially) of 1–5 m*M* cAMP plus 2–5 m*M* NH_4Cl. In

particular NH_4Cl was found to eliminate stalk specific differentiation, with no effect on the spore specific pathway. The pH dependence of these effects indicated that NH_3 is the active molecular species. In fact, high pH (7.5) alone had a significant effect presumably due to endogenous production of NH_3 and NH_4^+ by the cells. This is in agreement with the earlier findings of Wilcox and Sussman (1978) who showed that NH_4Cl is required for disaggregated cells of mutant strain *Fr17* to complete spore differentiation at very low cell densities. Similarly, Sternfeld and David (1979) implicated NH_3 and NH_4^+ as a requirement for spore formation by clustered wild-type cells constrained in such a way as to prevent normal morphogenesis.

B. cAMP, NH_3, and DIF as Potential Carriers of Positional Information

Schindler and Sussman (1977a,b; 1979) have demonstrated that in *D. discoideum* wild type, exogenous NH_3 acts as a reversible inhibitor of intracellular and extracellular cAMP accumulation (see Fig. 3). Using the method of Devreotes *et al.* (1979), which permits measurement of synthesis and release of cAMP as a pure response to exogenous cAMP, uncomplicated by self-signaling and enzymic destruction, Williams and Sussman (1981) have demonstrated that the normal cAMP relay response is dramatically reduced within 2 min after exposure of the cells to NH_3 and is recovered within 15 min after the removal of NH_3.

As described in detail in the following section, this metabolic interaction has profound morphogenetic consequences. Sussman and Schindler (1978) have proposed that the same interaction conveys positional information to the cells within the aggregate and determines their respective developmental fates.

As noted previously, stalk and basal disk cells, phenotypically very similar and probably identical, arise from cells in two restricted regions of the developing aggregate—the apical tip and the base. Spores arise from cells in the medial region. It is argued that within the aggregate on its way to fruiting body construction, the synthesis of cAMP is restricted to the apex and the base by the topological distribution of NH_3 as dictated by the production and dissipation of NH_3 and NH_4^+ and by the ambient pH which determines the ratio of the two. The high intracellular and extracellular levels of cAMP would cause the cells in the apex and the base to initiate stalk/basal disk cell differentiation. Within the medial region of the aggregate the high NH_3 level would preclude cAMP synthesis (see Section V). Though possibly exposed to exogenous cAMP, the medial cells, unable to accumulate cAMP intracellularly, would enter the pathway of spore differentiation. Both pro-

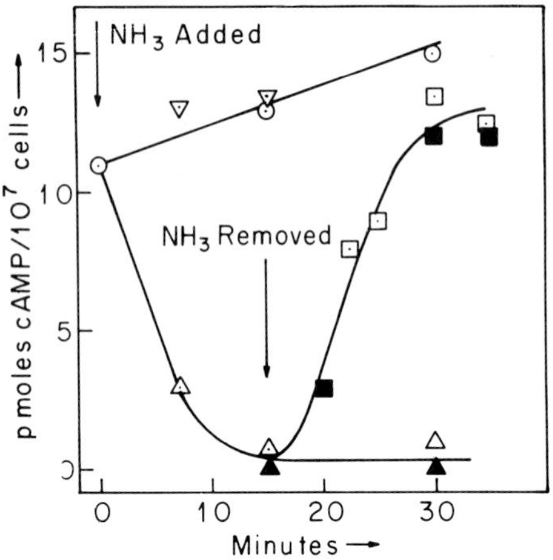

Fig. 3. Reversible inhibition by NH_3 of intracellular cAMP accumulation. Wild-type cells harvested from growth plates, washed, and suspended in 20 m*M* K-phosphate at pH 6.2 at a density of 10^7 cells/ml in 10 ml aliquots were incubated at 22°C on a Rotary shaker at 105 rpm. After 5 hr, when the cells had begun to acquire aggregation competence, duplicate aliquots were harvested by centrifugation resuspended in 10 ml volumes of fluid as designated below, and reincubated on the shaker. At intervals intracellular cAMP levels were estimated. One ml samples were pelleted within a few seconds in a microfuge. Two-tenths ml 20 m*M* acetate (pH 4) was added to each pellet that was then vortexed, boiled 5 min, brought to 1 ml with water and assayed for cAMP by the Gilman procedure. All points are means of duplicate determinations on duplicate suspensions. Filled symbols: cells continuously exposed to 500 μg/ml cycloheximide starting at −15 minutes. Open symbols: cells not exposed to cycloheximide. (○), suspended at 0 min in 20 m*M* phosphate (pH 6.2); (▽) suspended at 0 minutes in 20 m*M* phosphate (pH 7.2); (△,▲), suspended at 0 minutes in pH 7.2 phosphate containing 14 m*M* ammonium carbonate; (□,■), suspended in pH 7.2 phosphate containing ammonium carbonate, collected at 15 min, and resuspended in pH 6.2 phosphate.

grams would be regulated and modulated by accessory metabolites such as DIF. Neither pathway would involve irreversible changes. Hence, accidental or experimentally contrived fluctuations in NH_3 level would enable a cell to jettison components specific to the one and initiate the synthesis of those specific to the other.

Conversely, Gross *et al.* (1981) report that exogenous cAMP is required both for stalk cell differentiation by dispersed wild type (strain V12) and sporogenous mutant derivatives in the presence of DIF at pH 6.2 and for spore differentiation by dispersed sporogenous mutant cells in the absence of DIF at high pH (7.5) aided by the presence of NH_4Cl. They propose a scheme in which cAMP acts as a general, nonspecific activator of cytodifferentiation, the choice of pathways being determined by the topological distributions of and metabolic interactions between DIF and NH_3. A counter objection can be offered. Developmental fates may be determined not by whether cells see exogenous cAMP but by whether they can synthesize cAMP endogenously. Clearly, in the experiments of Gross *et al.* (1981), the presence of 5 m*M* NH_4Cl when coupled with endogenous NH_3 production at an ambient pH of 7.5 would have precluded cAMP synthesis by the test cells (Schindler and Sussman, 1977) whereas incubation at pH 6.2 (leading to about a 20-fold drop in the ratio of NH_3 to NH_4^+) would not. Furthermore, under different conditions neither spore nor stalk cell differentiation by disaggregated, dispersed cells of mutant strain FR17 required the presence of exogenous cAMP.

Clearly the matter is still in flux but a promising beginning has been made. At the very least the models are based upon real metabolites with demonstrable and specific effects on cytodifferentiation and hence are testable, particularly with the aid of patterning mutants (Morrissey *et al.*, 1981; MacWilliams, 1981).

IV. Holistic Signaling

As noted, cells within the developing aggregate, informed that they are in contact (or at least close apposition) and aware of their relative positions, also appear to monitor the morphogenetic states of the assembly as a whole and

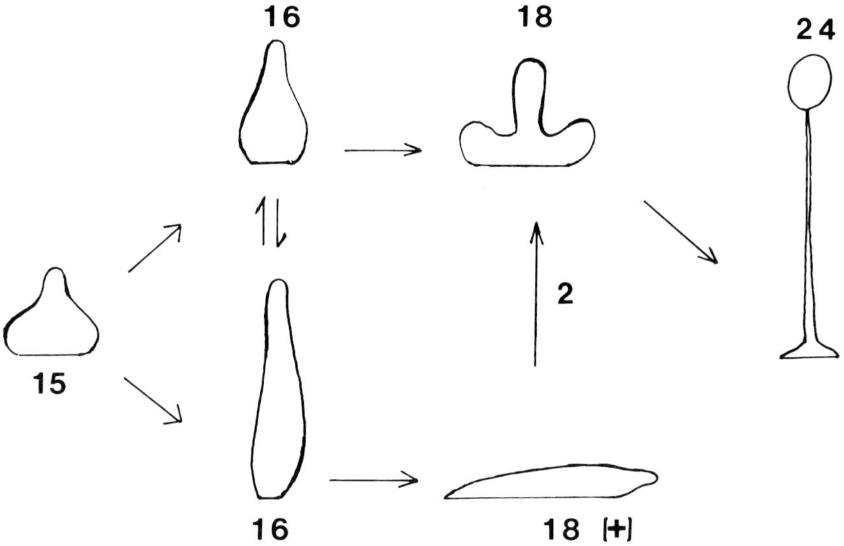

Fig. 4. Alternative morphogenetic pathways followed by *Dictyostelium discoideum.* The numbers refer to times in hours.

modulate their individual programs of gene expression accordingly. Consider the following prime example.

Slug migration in *D. discoideum,* originally thought to be an obligate preamble to fruit construction (Bonner, 1944), is now known (Newell *et al.*, 1969) to constitute an alternative morphogenetic pathway. The newly formed aggregate can transform into a vertically oriented slug, descend to the subsubstratum and move away from the site of aggregation. Alternatively, under specified environmental conditions including cell density, ambient pH, the volume of the substratal reservoir and scrubbing of the atmosphere, it can eschew these morphogenetic transitions completely and immediately construct a fruiting body. If a slug, it can remain so for an indefinite period ranging from minutes to days but once induced to stop migrating and reenter the fruiting mode, it attains the Mexican Hat stage within 2 h and completes the sequence in 7 h (see Fig. 4).

A. Evidence Demonstrating Holistic Morphogenetic Signaling

Figure 5 shows how the choice of morphogenetic pathways influences the accumulation and/or disappearance of four developmentally regulated proteins. Consider two cells, after 16-h development, which occupy identical medial positions within two aggregates, one having opted for fruit construction and the other for slug migration. Both cells are engaged in postaggregation protein synthesis and both, in correspondence with their relative axial positions within the respective aggregates, are accumulating spore-specific products (i.e., prespore vesicles (Maeda and Takeuchi, 1969), spore-specific antigens (Takeuchi, 1963; Gregg and Badman, 1970), and UDPgalactose transferase (Ellingson *et al.*, 1971) known to be restricted to the prespore cell contingent. But Cell A, which will become a terminally differentiated spore in 8 h, is rapidly accumulating both UDPglucose pyrophosphorylase and UDPgalactose epimerase while Cell B, which also will ultimately develop into a spore but only at a time set by the reentry of the slug into the fruiting mode, is accumulating the pyrophosphorylase slowly and the epimerase not at all! When reentry is initiated, Cell B now committed to complete spore differentiation in 7 h does accomplish a normal round of epimerase accumulation plus additional quantal rounds of pyrophosphorylase and transferase synthesis, all perfectly in step with the flow of subsequent morphogenetic events associated with fruit construction. Studies using actinomycin D indicate that prior transcriptive rounds are required

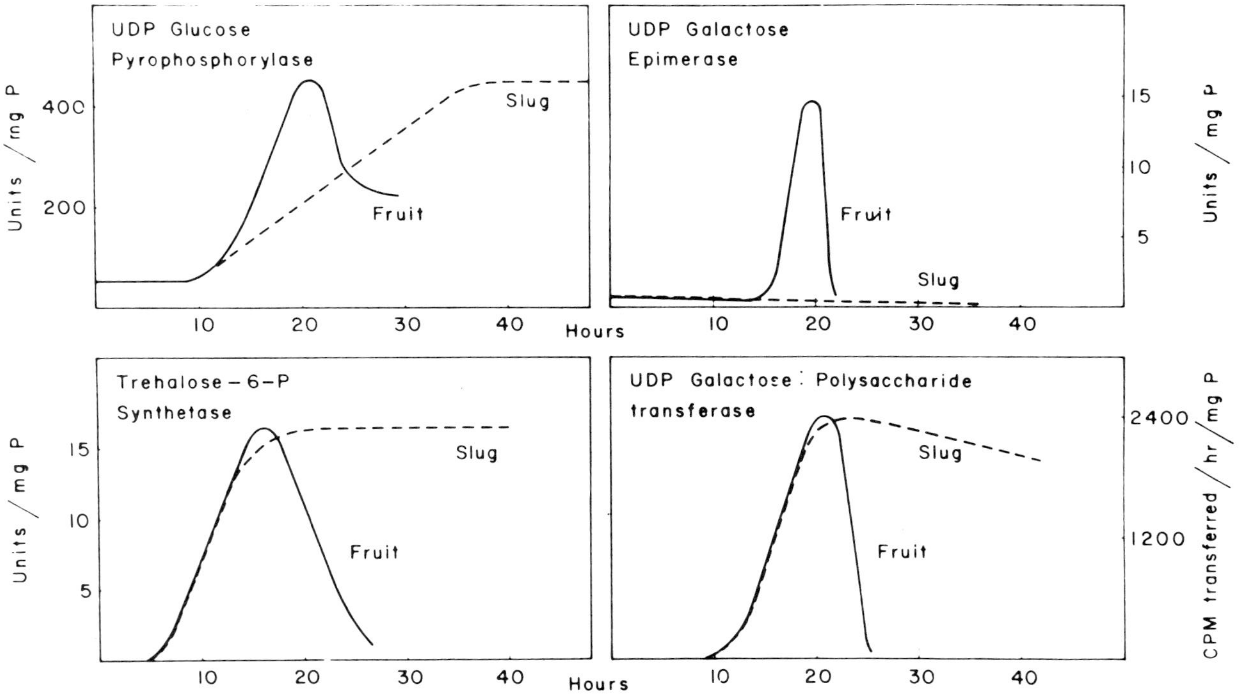

Fig. 5. The developmental kinetics of four enzymes in aggregates engaged in fruit construction or slug migration and in slugs induced to reenter the fruiting mode by reduction of ambient pH and exposure to omnidirectional light (Sussman, 1976).

(Sussman, 1976). Cells A and B also exhibit different patterns of specific protein disappearance.

B. Molecular Bases of the Choice of Morphogenetic Pathways

It was originally shown (Newell *et al.*, 1969) that newly formed aggregates could be made to eschew slug migration and proceed directly to fruit construction by maintaining a low ambient pH and by removing previously excreted metabolites via diffusion and evaporation. Later experiments (Schindler and Sussman, 1977b) demonstrated that the slug-inducing capacity of the excreta could be quantitatively ascribed to its NH_3 and NH_4^+ content, the result of the oxidative degradation of endogenous protein and RNA. Similarly, migrating slugs could be induced to reenter the fruiting mode by lowering the ambient pH or by reducing the level of NH_3 and NH_4^+ in and around the slug. This could be accomplished most dramatically by treating single slugs with miniscule micropipetted quantities of an enzyme cocktail containing glutamate dehydrogenase, NADH and α-ketoglutarate, thereby removing the NH_3 by reductive amination (see Table II).

A decreased level of NH_3 produced by a lowered ambient pH or treatment with the glutamate dehydrogenase cocktail is in itself sufficient to induce immediate reentry, attainment of the Mexican Hat stage within 2 h and completion of fruit construction by 7 h. Reentry can also be induced, though not as efficiently, by exposure of the slug to evenly distributed room illumination provided by overhead fluorescent fixtures (Newell *et al.*, 1969). All slugs stop migrating. Older and/or smaller slugs accomplish reentry more rapidly than do younger and/or larger ones. The mechanism is still obscure. However, Poff *et al.* (1973) have demonstrated the presence of a light-activated cytochrome and provided evidence that it is the photoreceptor involved in slug phototaxis. Conceivably general activation of the cytochrome throughout the slug by strong, overhead or omnidirectional light (as contrasted with tip activation by weak horizontal beams of light) would significantly elevate the rate of respiration, and this might lead to a reduction in the NH_3 level by lowering the ambient pH and increasing the rate of reductive amination.

The action of NH_3 as a morphogen in determining the choice of pathways can almost certainly be accounted for by its capacity to reversibly inhibit cAMP production (see Fig. 3). This conclusion stems from a study of strain KY3 (Yanagisawa *et al.*, 1967), the first of a mutant class termed "sluggers" (Sussman *et al.*, 1978), all of which transform into and remain as slugs under

TABLE II
Induction of Fruiting in Migrating Slugs[a]

Treatment	No. slugs treated	No. that fruited
Complete mixture	28	26
NADH	26	2
α-Ketoglutarate	29	2
Enzyme	39	4
Mix with boiled enzyme	26	3
Water	29	2

[a]Isolated slugs migrating toward a horizontal point source of light were sprayed each with 0.5 μl of a solution containing 1.5 m*M* NADH, 500 m*M* α-ketoglutarate, 500 m*M* imidazole (pH 7.25) and 0.1 unit L-glutamate dehydrogenase, and were then reincubated 8 h before scoring.

conditions wherein the wild type constructs fruits. This aberration in KY3 has been shown to be the direct result of an extraordinary sensitivity to NH_3, curable by low ambient pH, scrupulous removal of excreted metabolites or treatment with the glutamate dehydrogenase cocktail. In addition, KY3 displays a correspondingly great sensitivity to NH_3 as an inhibitor of cAMP production. Since the KY3 mutation segregates as a single gene in parasexual crosses (Newell, private communication), we conclude that the two sensitivities spring from the same metabolic alteration and are causally connected.

Newell (private communication) has isolated additional slugger mutants which exhibit extraordinary sensitivity to NH_3 as a morphogen. Together with KY3, they have been shown to fall into four complementation classes. In our laboratory, G. Williams has examined the patterns of cAMP synthesis and release by some of these mutants using the technique of Devreotes *et al.* (1979) wherein the pure response of test cells to an exogenous cAMP signal, defined with respect to intensity and duration, can be measured. Williams has already shown (unpublished data) that KY3 and one other slugger mutant representing two of the four complementation groups display extraordinary sensitivity to NH_3 with respect to the extent and timing of cAMP synthesis and release. In contrast, in slugger JC2, whose behavior toward NH_3 as a morphogen is indistinguishable from that of the wild type (Sussman *et al.*, 1978), the pattern of cAMP synthesis and release exhibits a level of sensitivity to NH_3 that is likewise indistinguishable from that of the wild type. We believe that the data now point inescapably to the conclusion that NH_3 exerts its morphogenetic effects through its action as a reversible inhibitor of cAMP synthesis.

C. Morphology of Slug Reentry and the Possible Nature of the Attendant Signaling Process

The transition from slug migration to fruit construction was originally described by Raper (1940), Bonner (1944), and by Raper and Fennell (1952). The consensus was that the process involves "a rounding up of the pseudoplasmodial body and a shifting of the apical tip from an anterior to an apicular position surmounting the mass." Even the study of cell movement within aggregates throughout development by Durston and Vork (1979) dismisses this sequence with the statement "Following migration, the slug rounds up and constructs a fruiting body." More recent observations (Deml and Sussman, 1981) reveal that these descriptions are incomplete and somewhat misleading. The sequence has been followed by time lapse still photomicrographs and videotapes in a microchamber that provides optimal atmospheric and substratal conditions, particularly with respect to humidity. Over 30 recorded sequences reveal certain invariant, diagnostic features (Fig. 6). Within minutes after induction [by reduction of the ambient pH with or without exposure to overhead (omnidirectional) light], the slug tip lifts up and attains a status remarkably like that of an aggregative center. The posterior region then streams toward the tip and tucks under it, thereby enlarging and tilting it forward. The movement of microregions appears irregular, possibly pulsatile, and the profile shows humps in the dorsal outline, possibly reflections of that pulsatile movement. The rate of movement is greatly and progressively accelerated over the normal rate of slug migration. The overall impression is that of a reprise of aggregation with the slug posterior acting as a single stream. It is interesting that in some slugs, particularly ones that had just started migrating prior to induction, differently timed, possibly pulsatory motions within the posterior region cause it to split into two or more parts. Satellite fruits then form precisely as they do when aggregation streams are mechanically disrupted (Raper, 1940a,b).

As the posterior begins to move toward and under the progressively enlarged apex, the latter rapidly assumes the shape of a Mexican Hat, completing the process about 2 h after the start just when the last of the posterior region has been tucked underneath. Fruit construction then proceeds at the normal pace 5–6 h later.

The overall impression that slug reentry represents a reprise of aggregation possibly reflecting the pulsatile response to and production of cAMP places the attendant patterns of developmentally regulated protein synthesis in a new and perhaps more understandable context. Thus, the cells, exposed to the same physiological conditions as when they aggregated the first time, reinitiate the synthesis of UDPG pyrophosphorylase and UDPgalactose transferase. Because this time they are committed to fruit construction and not to slug migration, they also initiate a round of UDPgalactose epimerase

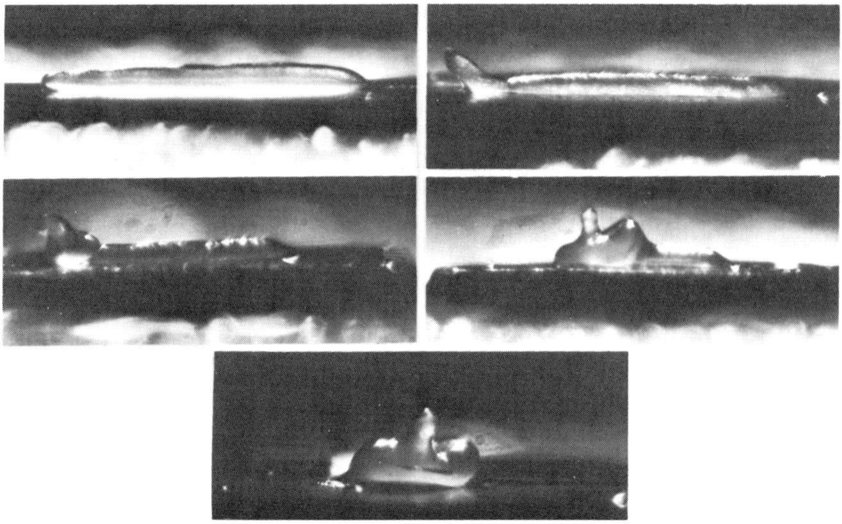

Fig. 6. Typical morphogenetic stages during slug reentry into the fruiting mode (Deml and Sussman, 1981).

synthesis. In contrast, an additional round of accumulation of trehalose synthetase, a preaggregative protein, is not initiated. But this is precisely the pattern that would have been followed had the migrating slugs been mechanically disaggregated and allowed to reaggregate after redeposition on solid substratum (Newell *et al.*, 1972).

What signals might be common to these two very different morphogenetic circumstances? Cell contact per se can conceivably be invoked in the quantal response to reaggregation but certainly not in the case of slug reentry. In the latter, the cells remain juxtaposed before, during, and afterward. But if both morphogenetic sequences represent a reprise of aggregation, one feature that could be common to both may be the high intracellular level of cAMP characteristic of the aggregation phase (Pahlic and Rutherford, 1979) and/or the pulsatile kinetics of its synthesis and release.

V. A Unified Theory of Morphogenesis and Cytodifferentiation

Given the intimate connection between the flow of morphogenetic events and the pattern of gene expression at all levels from transcription to the functioning and ultimate fates of the phenotypic products, it would seem important to elucidate the molecular bases of morphogenetic control in order to establish and characterize the connections between morphogenesis and cytodifferentiation. A model proposed by Sussman and Schindler (1978) does purport to provide this elucidation. The model, slightly modified by the new observations of slug reentry to the fruiting mode, is based upon the interactions of the two metabolites, cAMP and NH_3, which have already been demonstrated to affect specific morphogenetic events. It has the virtues of crucial testability and simplicity.

A. The Postulates

The model is based on four assumptions. The last is pertinent to the conveyance of positional information but its validity has no direct bearing upon the control of morphogenesis per se. The postulates are

1. During slug migration and fruit construction all concerted cell movements are chemotactically driven by cAMP.
2. These movements, forward and backward, upward and downward, are taken to reflect the existence of cAMP sites, i.e., signaling centers, comparable to aggregation centers that direct production and accumulation of cAMP at specific locales within the multicellular assembly.
3. These site specific accumulations are dictated by NH_3 or an effector generated in its presence via the reversible inhibition of cAMP production. The distribution of NH_3 is itself determined by the rates of its production and dissipation and by pH gradients within and around the aggregate.
4. The cAMP/NH_3 interaction also provides the positional information which dictates the choice of pathways of cytodifferentiation, i.e., spores and stalk/basal disk cells. That is, low or negligible intracellular cAMP levels in inhibited locales induce spore cytodifferentiation. High intracellular cAMP levels in noninhibited locales induce stalk/basal disk cell differentiation.

B. Slug or Fruit: One or Two cAMP Sites

Consider an aggregate at the 15-h stage and proceeding to construct a fruiting body directly at the site of aggregation. Two concerted morphogenetic movements (Raper and Fennell, 1952) occur within the next 3 h, an upward movement to create a more pronounced, columnar apex and a downward movement to create a flattened, widened, flanged base, i.e., the crown and brim of the Mexican Hat. The presence of two cAMP signaling centers is invoked to account for these movements, one in the apex and the other in the base, both the consequence of locally depressed levels of NH_3. How? As indicated in Fig. 7, in the apex NH_3 would be lost by evaporation. More important, the ready availability of O_2 might enhance the respiratory rate thereby lowering the ambient pH and increasing the rate of reductive amination via Krebs cycle intermediates. Were the base resting on a substratum of ample volume, largely free of NH_3 and NH_4^+ and having a relatively low ambient pH (< 6.5), facilitated diffusion and a low NH_3 to NH_4^+ ratio would lead to cAMP synthesis here, too. In contrast, within the medial region, anaerobic conditions leading to an increased ambient pH and hindered diffusion and evaporation would create a high level of NH_3, hence no cAMP production. The result is that the cells under the apex would be chemotactically attracted upward by the apical cAMP site thereby enlarging and elongating it. Those immediately above the base would move downward in response to the basal cAMP site, thereby widening the base and generally flattening the aggregate.

Now consider an equivalent 15-h aggregate preparing to transform into a migrating slug (Fig. 8). Only one concerted morphogenetic movement is observed (Newell *et al.*, 1969; Durston and Vork, 1979), i.e., uniquely upward to create the finger-like vertical slug followed by its gentle collapse (presumably gravitational) to a horizontal posture. To account for the upward movement a single, apical cAMP site is postulated based on the previously mentioned rationale. Given a substratum of limited volume already containing significant levels of NH_3 and NH_4^+ previously excreted by the aggregate and its neighbors at relatively high ambient pH (>7). The basal level of NH_3 would remain high, hence no basal cAMP site. The result would be a unidirectional, chemotactically driven movement upward by the cells under the apex to produce the early finger stage. Once horizontal, the single apical cAMP site condition would be perpetuated by the thick slug sheath and subsequent migration would be directed by its chemotactic attraction.

C. Reentry of the Slug into the Fruiting Mode: One cAMP Site Becomes Two

As indicated in Fig. 9, all of the environmental shifts that induce slug reentry might reasonably be expected to depress the level of NH_3 throughout the entire multicellular assembly. If the reduction were great enough to permit all cells to resume or to significantly increase the rate of cAMP synthesis, one might expect them to be entrained by the tip in the same "see cAMP-make cAMP" pulsatile relationship that had originally prevailed during aggregation and with the same result, i.e., the concerted, unidirectional, chemotactically driven movement of the outlying cells into and under the

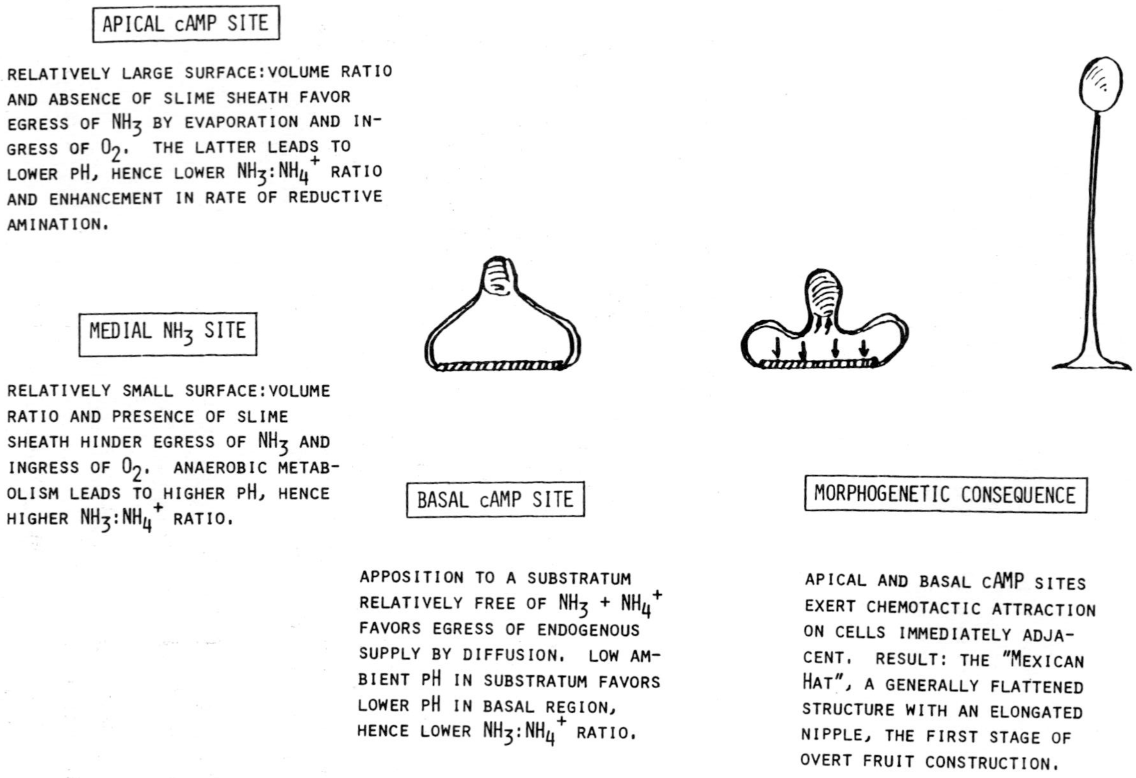
APICAL cAMP SITE
RELATIVELY LARGE SURFACE:VOLUME RATIO AND ABSENCE OF SLIME SHEATH FAVOR EGRESS OF NH_3 BY EVAPORATION AND INGRESS OF O_2. THE LATTER LEADS TO LOWER pH, HENCE LOWER $NH_3:NH_4^+$ RATIO AND ENHANCEMENT IN RATE OF REDUCTIVE AMINATION.
MEDIAL NH_3 SITE
RELATIVELY SMALL SURFACE:VOLUME RATIO AND PRESENCE OF SLIME SHEATH HINDER EGRESS OF NH_3 AND INGRESS OF O_2. ANAEROBIC METABOLISM LEADS TO HIGHER pH, HENCE HIGHER $NH_3:NH_4^+$ RATIO.
BASAL cAMP SITE
APPOSITION TO A SUBSTRATUM RELATIVELY FREE OF $NH_3 + NH_4^+$ FAVORS EGRESS OF ENDOGENOUS SUPPLY BY DIFFUSION. LOW AMBIENT pH IN SUBSTRATUM FAVORS LOWER pH IN BASAL REGION, HENCE LOWER $NH_3:NH_4^+$ RATIO.
MORPHOGENETIC CONSEQUENCE
APICAL AND BASAL cAMP SITES EXERT CHEMOTACTIC ATTRACTION ON CELLS IMMEDIATELY ADJACENT. RESULT: THE "MEXICAN HAT", A GENERALLY FLATTENED STRUCTURE WITH AN ELONGATED NIPPLE, THE FIRST STAGE OF OVERT FRUIT CONSTRUCTION.

Fig. 7. Choice of morphogenetic pathways I. Two cAMP sites (apical and basal) = fruiting body construction.

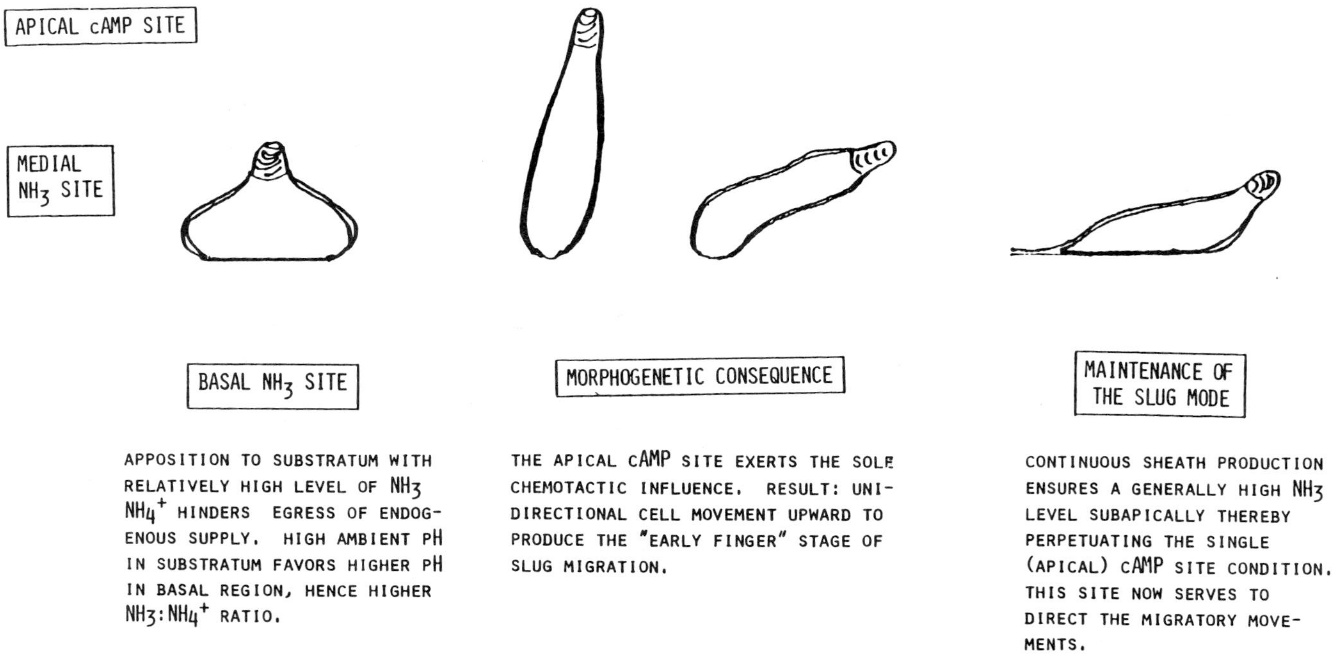

Fig. 8. Choice of morphogenetic pathways II. Once cAMP site (apical only) = slug migration.

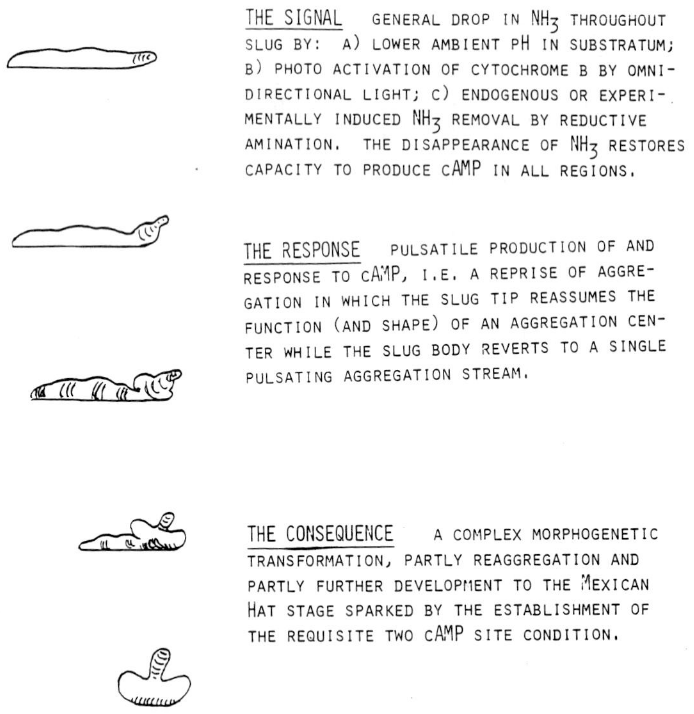

Fig. 9. Choice of morphogenetic pathways III. Reentry into the fruiting mode.

tip. The resultant enlargment of the medial region immediately under the tip can be imagined to set up the conditions—hindered dissipation, high ambient pH—that raise the NH_3 level and stop cAMP synthesis, thus restricting that process to the apex and to those cells still in contact with the substratum, i.e., the base and outlying cells. The two cAMP site condition would lead to a complex series of morphogenetic movements, a continuation of the reaggregation movement by the outlying cells toward and under the base, and the concurrent movements of the cells in the medial region, those just below the apex upward and those just above the base downward, that are assocaited with the formation of the Mexican Hat.

D. Actual Fruiting Body Construction

Previous workers have reported (Raper and Fennell, 1952) that within the Mexican Hat the basal cells deposit a cellulose disk which separates them from their overlying neighbors and concurrently become thick walled and vacuolated. Cells in the apex deposit cellulose fibrils that are incorporated in the stalk sheath rudiment while developing into thick-walled, vacuolated cells. Subsequently they are pushed into the open sheath end by the upward motion of their underlying neighbors, which repeat the actions of their predecessors and are in turn replaced by their successors. This sequence, described as an "inverted fountain" by Raper (1940) accounts for the vertical growth of the fruiting body stalk. After elaboration of the latter the remaining cells transform into terminal spores in a rapid wave. In principle all of the above can be accounted for by the model. As indicated in Fig. 10, the differentiation of the cells at the base of the Mexican Hat, induced by the high cAMP level and low NH_3 level, would lead to the elimination of the basal cAMP site. Cells in the apex exposed to high cAMP and low NH_3 would also synthesize cellulose and become vacuolated and lose cytoplasmic components by exocytosis. The secreted fibrils would be incorporated in the stalk sheath open at its apex and in contact with the disk at its base. The pressure of chemotactically driven cell movements from below would account for the observed lodgment of the nascent stalk cells in the open end. Their successors in the apex now exposed to a region of low NH_3 would resume optimal cAMP synthesis thereby perpetuating the apical cAMP site. As nascent stalk cells, they in turn would secrete cellulose, add to the growing stalk, and ultimately be lodged within it, giving way to their successors.

If the course of spore differentiation were at some stage to render prespore cells incapable of cAMP synthesis even in the absence of NH_3, the apical cAMP site would ultimately disappear. The upward growth of the stalk would cease and the emergence of terminal spores would end the sequence.

What can presently be said about the validity of the model? There are no data that unequivocally contradict it and there is a body of evidence that appears to lend support. Thus, cAMP-synthesizing capacity and cAMP itself are present in migrating slugs and remain present until the end of fruit construction (Pahlic and Rutherford, 1979). Cells within the aggregate retain chemotactic sensitivity to cAMP (Rubin, 1976; Maeda and Gerisch, 1978) and, when present in the substratum, exogenous cAMP deranges morphogenetic movements and apical dominance in slugs and incipient fruits (Nestle and Sussman, 1972).

The movements of cells within the anterior portions of migrating slugs and in all regions of the aggregate during fruit construction have been shown to be pulsatory (Durston and Vork, 1979), much like cells during early aggregation. As noted previously, the results of studies of spore and stalk cell differentiation among dispersed and clustered cells are not incompatible with the notion that the former pathway requires high NH_3 and consequently low

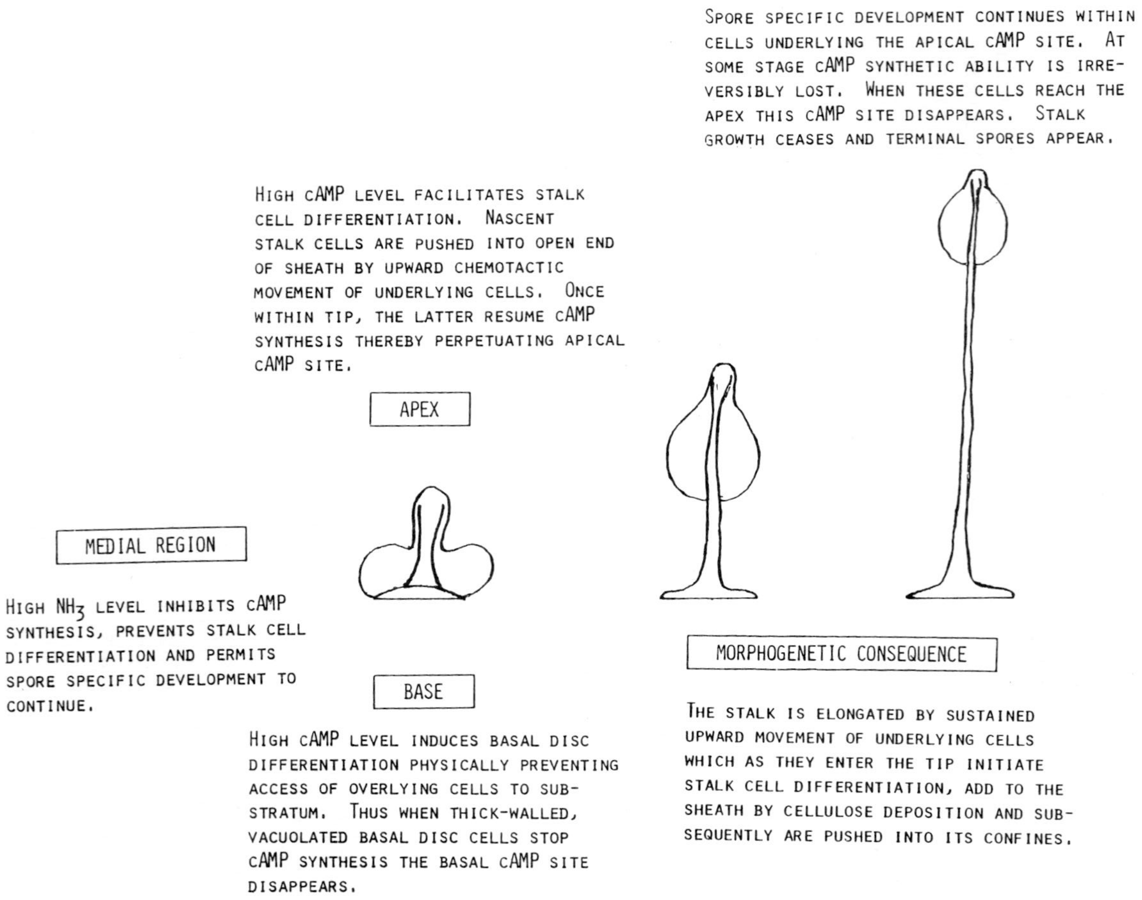
SPORE SPECIFIC DEVELOPMENT CONTINUES WITHIN CELLS UNDERLYING THE APICAL cAMP SITE. AT SOME STAGE cAMP SYNTHETIC ABILITY IS IRREVERSIBLY LOST. WHEN THESE CELLS REACH THE APEX THIS cAMP SITE DISAPPEARS. STALK GROWTH CEASES AND TERMINAL SPORES APPEAR.
HIGH cAMP LEVEL FACILITATES STALK CELL DIFFERENTIATION. NASCENT STALK CELLS ARE PUSHED INTO OPEN END OF SHEATH BY UPWARD CHEMOTACTIC MOVEMENT OF UNDERLYING CELLS. ONCE WITHIN TIP, THE LATTER RESUME cAMP SYNTHESIS THEREBY PERPETUATING APICAL cAMP SITE.
APEX
MEDIAL REGION
HIGH NH_3 LEVEL INHIBITS cAMP SYNTHESIS, PREVENTS STALK CELL DIFFERENTIATION AND PERMITS SPORE SPECIFIC DEVELOPMENT TO CONTINUE.
BASE
HIGH cAMP LEVEL INDUCES BASAL DISC DIFFERENTIATION PHYSICALLY PREVENTING ACCESS OF OVERLYING CELLS TO SUBSTRATUM. THUS WHEN THICK-WALLED, VACUOLATED BASAL DISC CELLS STOP cAMP SYNTHESIS THE BASAL cAMP SITE DISAPPEARS.
MORPHOGENETIC CONSEQUENCE
THE STALK IS ELONGATED BY SUSTAINED UPWARD MOVEMENT OF UNDERLYING CELLS WHICH AS THEY ENTER THE TIP INITIATE STALK CELL DIFFERENTIATION, ADD TO THE SHEATH BY CELLULOSE DEPOSITION AND SUBSEQUENTLY ARE PUSHED INTO ITS CONFINES.

Fig. 10. Fruiting body construction.

intracellular cAMP, while the latter requires high extracellular cAMP and therefore, *in situ*, low NH_3. Further support comes from studies of MacWilliams (1981). He has isolated a mutant wherein the prestalk to prespore cell ratio in the migrating slug is shifted from the normal 25:75 to a bare 5:95 resulting in a correspondingly deranged fruiting body. The mutant also displays the slugger phenotype exhibiting extraordinary sensitivity to NH_3! Conversely MacWilliams has found that the NH_3 sensitive slugger, KY3, exhibits an abnormally low prestalk to prespore cell ratio.

References

Alton, T., and Lodish, H. F. (1977a). *Dev. Biol.* **60,** 180.
Alton, T., and Lodish, H. F. (1977b). *Dev. Biol.* **60,** 207.
Ashworth, J. M., and Sussman, M. (1967). *J. Biol. Chem.* **242,** 1696.
Beug, H., Gerisch, G., Kempff, S., Riedel, V., and Cremer, G. (1970). *Exp. Cell Res.* **63,** 147.
Beug, H., Katz, F. E., and Gerisch, G. (1973). *J. Cell Biol.* **56,** 647.
Blumberg, D. D., and Lodish, H. F. (1980). *Dev. Biol.* **78,** 285.
Bonner, J. T. (1944). *Am. J. Bot.* **31,** 175.
Bonner, J. T. (1970). *Proc. Natl. Acad. Sci. U.S.A.* **65,** 110.
Brackenbury, R. W., and Sussman, M. (1975). *Cell* **4,** 347.
Chia, W. K. (1975). *Dev. Biol.* **44,** 239.
Chung, S., Landfear, S. H., Blumberg, D. D., Cohen, N. S., and Lodish, H. F. (1981). *Cell* **24,** 785.
Coloma, A., and Lodish, H. F. (1981). *Dev. Biol.* **81,** 238.
Coston, M. B., and Loomis, W. F. (1969). *J. Bacteriol.* **100,** 1208.
Deml, K., and Sussman, M. (1981). In preparation.
Devreotes, P. N., Derstine, P. L., and Steck, T. L. (1979). *J. Cell Biol.* **80,** 291.
Dimond, R., and Loomis, W. F. (1976). *J. Biol. Chem.* **251,** 2680.
Durston, A. J., and Vork, F. (1979). *J. Cell Sci.* **36,** 261.
Ellingson, J. S., Telser, A., and Sussman, M. (1971). *Biochim. Biophys. Acta* **244,** 388.
Feinberg, A. P., Springer, W. R., and Barondes, S. H. (1979). *Proc. Natl. Acad. Sci. U.S.A.* **76,** 3977.
Firtel, R. A., and Bonner, J. (1972). *Dev. Biol.* **29,** 85.
Firtel, R. A., and Brackenbury, R. W. (1972). *Dev. Biol.* **27,** 307.
Franke, J., and Sussman, M. (1973). *J. Mol. Biol.* **81,** 173.
Frazier, W. A., Rosen, S. D., Reitherman, R. W., and Barondes, S. H. (1975). *J. Biol. Chem.* **250,** 7714.
Gierer, A., and Meinhardt, H. (1972). *Kybernetik* **12,** 30.
Gregg, J. (1971). *Dev. Biol.* **3,** 757.
Gregg, J. H., and Badman, W. S. (1970). *Dev. Biol.* **22,** 96.
Gross, J., Kay, R., Lux, A., Peacey, M., Town, C., and Trevan, D. (1977). *In* "Cell Biology," Vol. 1, pp. 135.
Gross, J., Town, C. D., Brookman, J. J., Jermyn, K. A , Peacey, M. J., and Kay, R. R. (1981). *Phil. Trans. Roy. Soc.* In press.

Hoffman, S., and McMahon, D. (1977). *Biochim. Biophys. Acta* **465,** 242.
Jacquet, M., Part, D., and Felenbok, B. (1981). *Dev. Biol.* **81,** 155.
Kay, R. R., Garrod, D., and Tilly, R. (1978). *Nature (London)* **271,** 58–60.
Kay, R. R., and Trevan, D. J. (1981). *J. Embryol. Exp. Morph.* **62,** 369.
Kay, R. R., Town, C. D., and Gross, J. (1979). *Differentiation* **13,** 7.
Landfear, S. M., and Lodish, H. F. (1980). *Proc. Natl. Acad. Sci. U.S.A.* **77,** 1044.
Lam, T. S., and Siu, C.-H. (1981). *Dev. Biol.* **83,** 127.
Loomis, W. F. (1968). *Exp. Cell Res.* **53,** 283.
Loomis, W. F. (1969). *J. Bacteriol.* **100,** 417.
Loomis, W. F. (1975). "Isozymes, III." p. 177. Academic Press, New York.
Ma, G., and Firtel, R. A. (1978). *J. Biol. Chem.* **253,** 3924.
MacWilliams, H. (1981). In preparation.
Maeda, Y., and Gerisch, G. (1978). *Exp. Cell Res.* **110,** 119.
Maeda, Y., and Takeuchi, I. (1969). *Dev. Growth and Diff.* **11,** 232.
Makoto, I., Taya, Y., and Nishimura, S. (1980). *Dev. Growth and Diff.* **22,** 781.
Morrissey, J. H., Farnsworth, P. A., and Loomis, W. F. (1981). *Dev. Biol.* **83,** 1.
Muller, K., and Gerisch, G. (1978). *Nature (London)* **274,** 445.
Nestle, M., and Sussman, M. (1972). *Dev. Biol.* **28,** 545.
Newell, P. C., Franke, J., and Sussman, M. (1972). *J. Mol. Biol.* **63,** 373.
Newell, P. C., andSussman, M. (1970). *J. Mol. Biol.* **81,** 173.
Newell, P. C., Telser, A., and Sussman, M. (1969). *J. Bacteriol.* **100,** 763.
Pahlic, M., and Rutherford, C. L. (1979). *J. Biol. Chem.* **254,** 9703.
Parish, R. W., and Schmidlin, S. (1979a). *FEBS Lett.* **98,** 251.
Parish, R. W., and Schmidlin, S. (1979b). *FEBS Lett.* **108,** 219.
Poff, K. L., Butler, W. L., and Loomis, W. F. (1973). *Proc. Natl. Acad. Sci. U.S.A.* **70,** 813.
Raper, K. B. (1940a). *J. E. Mitchell Sci. Soc.* **56,** 241.
Raper, K. B. (1940b). *Am. J. Bot.* **27,** 436.
Raper, K. B., and Fennell, D. I. (1952). *Bull. Torrey Bot. Club* **79,** 25.
Ray, J., Shinnick, T., and Lerner, R. (1979). *Nature (London)* **279,** 215.
Reitherman, R. W., Rosen, S. D., Frazier, W. A., and Barondes, S. H. (1975). *Proc. Natl. Acad. Sci. U.S.A.* **72,** 3541.
Rosen, S. D., Kafka, J., Simpson, D. L., and Barondes, S. H. (1973). *Proc. Natl. Acad. Sci. U.S.A.* **70,** 2554.
Rubin, J. (1976). *J. Embrol. Exp. Morphol.* **36,** 261.
Schindler, J., and Sussman, M. (1977a). *Biochem. Biophys. Res. Commun.* **79,** 611.
Schindler, J., and Sussman, M. (1977b). *J. Mol. Biol.* **116,** 161.
Schindler, J., and Sussman, M. (1979). *Dev. Genet.* **1,** 13.
Siu, C.-H., Lerner, R. A., Ma, G., Firtel, R. A., and Loomis, W. F. (1976). *J. Mol. Biol.* **100,** 157.
Sonneborn, D. R., White, G. J., and Sussman, M. (1963). *Dev. Biol.* **7,** 79.
Sternfeld, J., and Bonner, J. T. (1977). *Proc. Natl. Acad. Sci. U.S.A.* **74,** 268.
Sternfeld, J., and David, C. N. (1979). *J. Cell Sci.* **38,** 181.
Sternfeld, J., and David, C. N. (1981). *Differentiation* In press.
Steinemann, C., and Parish, R. W. (1980). *Nature (London)* **286,** 621.
Sussman, M. (1976). *Prog. Mol. Subcell. Biol.* **4,** 103.
Sussman, M., and Lovgren, N. (1965). *Ex . Cell Res.* **38,** 97.
Sussman, M., and Newell, P. C. (1972). *In* "Molecular Genetics and Developmental Biology" (M. Sussman, ed.), p. 275. Prentice Hall, Englewood Cliffs, New Jersey.
Sussman, M., and Osborn, M. J. (1964). *Proc. Natl. Acad. Sci. U.S.A.* **52,** 81.
Sussman, M., and Schindler, J. (1978). *Differentiation* **10,** 1.

Sussman, M., Schindler, J., and Kim, H. (1978). *Exp. Cell Res.* **116,** 217.
Sussman, M., and Sussman, R. (1969). *Symp. Soc. Gen. Microbiol.* **19,** 403.
Takemoto, S., Okamoto, K., and Takeuchi, I. (1978). *Biochem. Biophys. Res. Commun.* **80,** 858.
Takeuchi, I. (1963). *Dev. Biol.* **8,** 1.
Takeuchi, I., Hayashi, Y., and Tasaka, M. (1977). *Dev. in Cell Biol.* **1,** 1.
Tasaka, M., and Takeuchi, I. (1979). *Protoplasma* **99,** 289.
Tasaka, M., and Takeuchi, I. (1981). *Differentiation* **18,** 191.
Telser, A., and Sussman, M. (1971). *J. Biol. Chem.* **246,** 2252.
Town, C., and Gross, J. (1978). *Dev. Biol.* **63,** 412.
Town, C. D., Gross, J., and Kay, R. R. (1976). *Nature (London)* **262,** 717.
Town, C. D., and Stanford, E. (1979). *Proc. Natl. Acad. Sci. U.S.A.* **76,** 308.
Turing, A. (1952). *Phil. Trans. Roy. Soc. B* **237,** 32.
Wilcox, D. K., and Sussman, M. (1978). *Differentiation* **11,** 125.
Wilcox, D. K., and Sussman, M. (1981a). *Proc. Natl. Acad. Sci. U.S.A.* **78,** 358.
Wilcox, D. K., and Sussman, M. (1981b). *Dev. Biol.* **82,** 102.
Williams, G., and Sussman, M. (1981). In preparation.
Williams, J. G., and Lloyd, M. M. (1979). *J. Mol. Biol.* **129,** 19.
Wolpert, L. (1969). *J. Theor. Biol.* **25,** 1.
Wolpert, L. (1971). *Cur. Top. Dev. Biol.* **6,** 183.
Yanagisawa, K., Loomis, W. F., and Sussman, M. (1967). *Exp. Cell Res.* **46,** 328.

• CHAPTER 10

Polysphondylium and Dependent Sequences

David W. Francis

I. Introduction

Polysphondylium pallidum and *P. violaceum* are two closely related species that produce stalks with whorls of branches when they fruit, and that differ from *Dictyostelium* in a number of more subtle aspects (Fig. 1). The purpose of this chapter is to compare the developmental biology of these two genera. The intention is that other workers will find such a comparison useful in distinguishing those truths that are universal to slime molds from those that are special to *Dictyostelium*. In addition, some comments will be made about dependent sequences in cellular slime molds.

In both species of *Polysphondylium*, aggregation is initiated by founder cells that appear when starved amoebae are exposed to light (Shaffer, 1961; Francis, 1965) (Fig. 2). These are single amoebae of ordinary appearance that suddenly round up and within a few minutes begin to attract their neighbors. In *P. pallidum* exposed to bright light and under conditions where new-forming aggregates do not compete with each other, approximately 3% of the population can be founders at the same time. In *P. violaceum*, at least, the founding phase is transitory, so a much larger percentage of cells may actually pass through this condition before aggregation

The Development of *Dictyostelium discoideum*

ISBN 0-12-455620-5

is finished. The differentiation of founder cells may be triggered by a pheromone called D factor, which accumulates in the medium around aggregating *P. violaceum* (Hanna and Cox, 1978). D factor is a small molecule, different from cAMP, and not a chemoattractant. Mutants that do not produce D factor cannot aggregate alone, but do so if given extracts from wild-type cultures. Light reduces the amount of D factor that accumulates, but raises the sensitivity of amoebae to it so that more founders appear (M. Hanna, unpublished).

P. violaceum appears to be like most species of *Dictyostelium* in producing D factor in large amounts, and in responding to light by an acceleration of aggregation and an increase in the number of centers. This is in spite of differing from most *Dictyostelium* species in having obvious founder cells. In contrast, very little D factor accumulates in cultures of *P. pallidum* (Hanna,

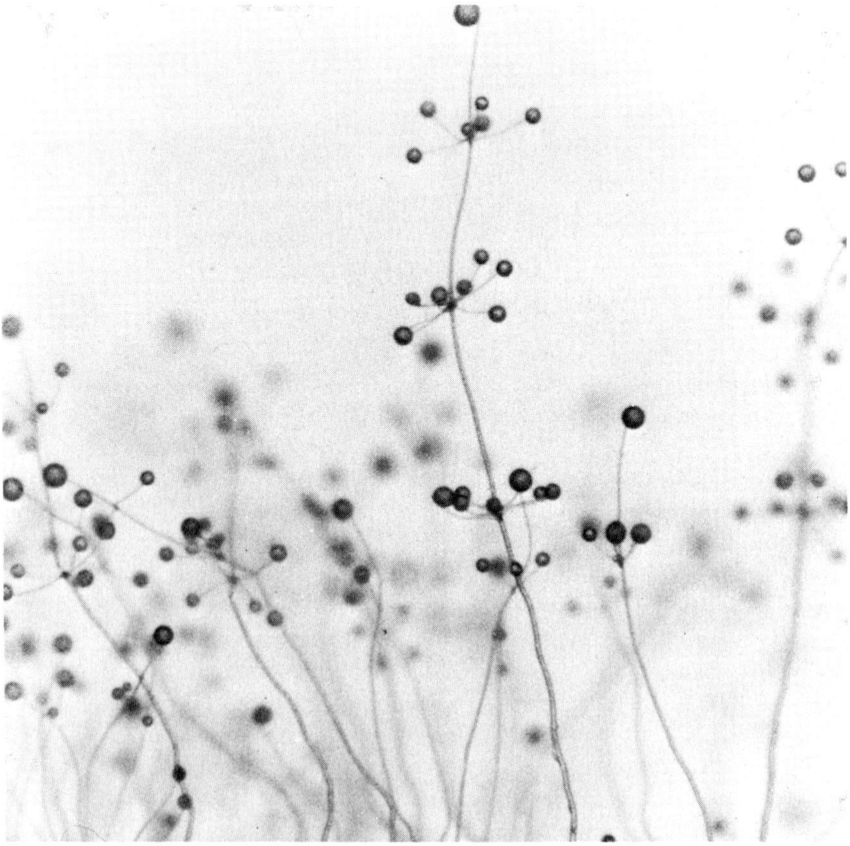

Fig. 1. *Polysphondylium pallidum*, wild type. *P. violaceum* is similar but has larger, less regularly branched fruits with purple sori.

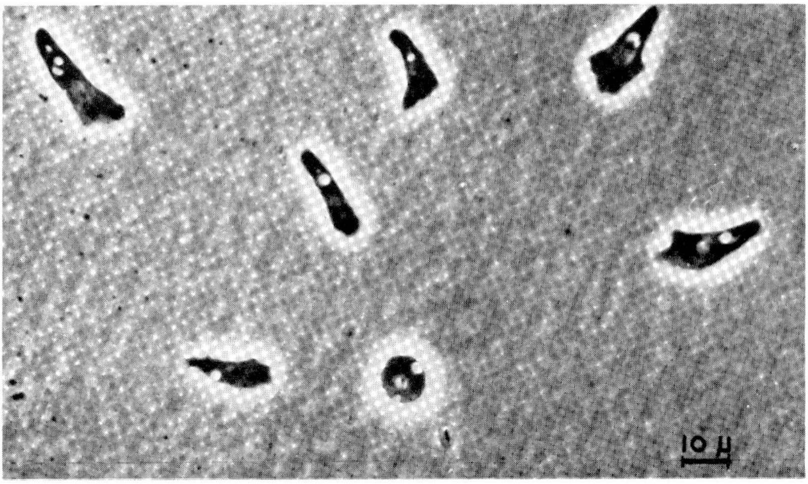

Fig. 2. A founder cell in *P. violaceum* (from Shaffer, 1962).

unpublished) and light is crucial to aggregation; in some strains no aggregates at all appear in the dark.

The biochemical mechanism whereby light triggers aggregation is unknown. However, the action spectrum of aggregation induction in *P. pallidum* resembles that of phototaxis in *Dictyostelium* slugs (Jones and Francis, 1972), suggesting a similar receptor.

Both *Polysphondylium* species respond to a chemoattractant different from *Dictyostelium* (Wurster *et al.*, 1976). J. T. Bonner (unpublished) has recently identified it as a dipeptide. An acrasinase is made that destroys this acrasin, perhaps acting to maintain a concentration gradient near aggregations in the manner that phosphodiesterase appears to work in *Dictyostelium*. That postsensory steps are similar in the two genera is suggested by the finding that a transient rise in concentration of cyclic GMP follows an added pulse of chemoattractant (Wurster *et al.*, 1978).

II. Morphogenesis

Following aggregation, the pseudoplasmodium of *Polysphondylium* immediately begins to build a stalk. Bonner *et al.* (1955) had reported that there was no prestalk region at the tip of *P. pallidum* detectable with histochemical stains for carbohydrates and for alkaline phosphatase, unlike *Dictyostelium.* More recently, O'Day has demonstrated that there is a prestalk zone having higher alkaline phosphatase activity by using a different cytological technique (O'Day and Francis, 1973) and having cells staining intensely with neutral red (O'Day, 1979). Not anticipated is the fact that all except a very few of the tipmost cells have prespore vesicles (Hohl *et al.*, 1977) and stain with fluorescent prespore antibodies (O'Day, 1979). O'Day suggests that the transition of prespore cells into prestalk cells, which must occur to account for the number of cells that eventually become stalk, begins long before the cells lose all of their prespore characteristics. An alternative conception is that the transition occurs quickly in any one cell, but that the transition zone contains a mixture of prespore and prestalk cells; for no single cell has yet been demonstrated to contain both prespore vesicles and neutral red granules.

These peculiarities in the generation of stalk cells may be related to the finding of Stenhouse and Williams (1981) that the spore/stalk ratio of individual side branches of *P. pallidum* is extremely variable. The spore/stalk ratio for a population of entire fruiting bodies of different sizes shows little variation from the mean of 4/1 and even the small fruits are not exceptionally stalky, unlike *Dictyostelium*. The spore/stalk ratio of individual side branches can deviate markedly from the average, however. For example, side branches with 15 stalk cells supported heads with 70 to 250 spores. It seems possible that only slight changes in the mechanism regulating production of prestalk cells from the prespore population might either halt or prolong the process and so generate branches of unusual proportions. Whether sorting out of the two cell types plays any role here has not been examined.

The whorled branches of *Polysphondylium* arise from cell clumps left behind by the slug as it climbs the stalk. The spacing between successive whorls may be remarkably uniform. How this spacing is controlled is unclear. It does not involve any mechanism which counts out equal numbers of stalk cells between whorls, because sorocarps formed by diploid strains have the same absolute spacing as haploid strains, even though their stalks have fewer, larger cells between whorls (Spiegel and Cox, 1980). The details of sculpting the delicate *Polysphondylium* fruiting body are still being described (O'Day and Durston, 1980; Spiegel *et al.*, in preparation) and the forces at work are not yet understood.

III. Pallidin

Pallidin is a galactose-binding protein, or better a class of proteins, of *P. pallidum*. In many ways it is like discoidin. Surface staining of cells with fluorescent antipallidin antibody shows that some fraction of pallidin molecules is on the cell surface (Chang *et al.*, 1977). It is the major cell protein that binds to human or rabbit red blood cells and galactose-coated beads, and is easily purified by affinity chromatography on these media (Simpson *et al.*, 1974; Rosen *et al.*, 1979). Two major bands are visible on one-dimensional SDS gels of purified material (band A of 26,500 daltons and band B of 26,000 daltons), and one minor band (band C of 25,000 daltons). Most likely each of the major polypeptides is coded for by more than one gene, judging from what is known to be true of discoidin (Rowekamp *et al.*, 1980). Since the pallidin subunits resemble those of discoidin in molecular weight, it seems plausible that the corresponding genes are homologous. Attempts to demonstrate a restriction fragment of *P. pallidum* nuclear DNA that hybridizes with a cloned discoidin I gene have not been successful (unpublished results, and Firtel, personal communication). The subunits are assembled into three trimers, consisting of 2B plus 1C, 2A plus 1B and 2B plus 1A, that may be aggregated into nonamers in the membrane (Rosen *et al.*, 1979). Pallidin concentration increases rapidly during the starvation period of development, as assayed by red blood cell agglutination activity of cell extracts (Rosen *et al.*, 1974). It is visible on autoradiographs of one-dimensional SDS gels of whole cell extracts and on these the period of apparent synthesis is from early in starvation until aggregation begins (band 25ab in Francis, 1976 and in Fig. 6). The red blood cell agglutination activity remains high throughout development, indicating that the molecule is stable (Rosen *et al.*, 1974).

After several years of effort the function of pallidin in development is still not clear. There are high affinity binding sites for pallidin on the surface of amoebae (Reitherman *et al.*, 1975), suggesting the function is to stick cells together, as also postulated for discoidin. Arguing against this is that univalent antibodies specific for pallidin do not block cell cohesion except under physiologically unusual conditions (Rosen *et al.*, 1976; Bozzaro and Gerisch, 1978). It is known that univalent antibody fragments against other membrane antigens can block cell cohesion. These same fragments do not block the red blood cell agglutination capacity of amoebae, indicating they cover neither pallidin nor its binding sites (Bozzaro and Gerisch, 1978). Together, these results suggest that pallidin is not a prime agent of cell cohesion. But mechanical cohesion is merely an easily assayed property perhaps of very little biological meaning for developing cells. Pallidin-mediated cell/cell interaction may involve cohesion as a side effect, but in fact have some quite separate purpose (e.g., an alteration in membrane fluidity or a change in membrane permeability).

Mutations in the pallidin structural genes would be useful in working out this function. A mutant apparently of this type has been found in *D. discoideum* (Ray, Shinnick, and Lerner, 1979). The low-pallidin mutants obtained to date in *P. pallidum* have a common phenotype (unpublished results).

Unlike the discoidin mutant of *D. discoideum*, they are able to aggregate, although more slowly than wild type, but make fruiting structures with much reduced stalks (Fig. 3). Most of the mutant strains have both major subunits greatly reduced in quantity, and the mutations thus appear to be in regulatory genes. Those mutations examined so far map to a single locus. The immediate function of this locus may be to regulate activity of the pallidin structural genes, or this regulatory effect may be the consequence of some other process controlled by the site. Detailed study of such mutants may reveal the function of pallidin in development.

Among other cell surface proteins of *P. pallidum* at least one is engaged in cell contact in a manner similar to that of contact sites A in *Dictyostelium*, in that univalent antibodies directed against these sites do block cell cohesion. When monoclonal antibodies were used on an affinity column to isolate contact sites, two molecules were recovered, one of which (contact sites 1) is a glycoprotein, the other being a developmentally regulated protein (contact sites 2). Neither of these cross reacts with antibody against developmentally regulated contact sites A of *D. discoideum*. Both types of contact sites are resistant to EDTA, as are contact sites A (Bozzaro *et al.*, 1981). It has been suggested (Gerisch *et al.*, 1980) that contact sites 2 is responsible for the sorting out of *P. pallidum* from *D. discoideum* cells that has been observed in

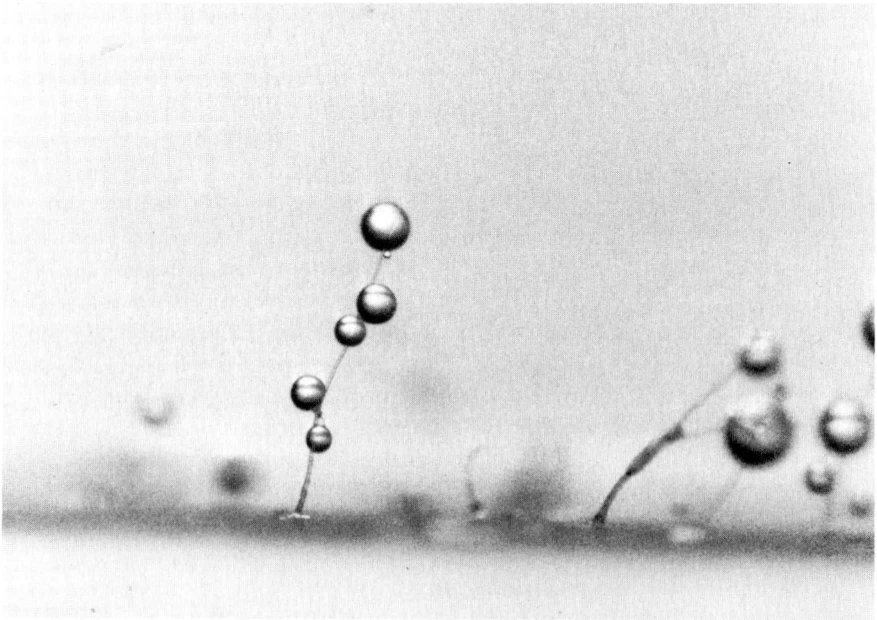

Fig. 3. Fruiting structure of a mutant strain of *P. pallidum* that has reduced pallidin.

interspecies mixes by Nicol and Garrod (1978), Bozzaro and Gerisch (1978), and by others. The observation whereupon this idea was based is that sorting out of cells of the two species begins immediately in aggregation competent cells, but only after a delay in vegetative cells that have not yet begun to synthesize contact sites 2. Cell adhesion rather than chemotaxis seems to be the primary cause of sorting out since dispersed cells reassociate more quickly with cells of their own species than with those of a second species (McDonough *et al.*, 1980) and will do so even when cell motility is inhibited with 2,4-DNP (Gerisch *et al.*, 1980). Experiments by Eisenberg (personal communication) have shown that the specificity of cell recognition is so precise that often cells of two isolates of *P. pallidum* will sort out and form separate fruiting bodies when grown in mixed culture. Whether pallidin, contact sites 1 and 2, or another of the many developmentally regulated cell surface proteins (Hintermann and Parish, 1979) are responsible for this feat of recognition is unknown at present.

IV. Microcysts

The spores of *Polysphondylium* are little investigated and much like those of *Dictyostelium* and will not be discussed further. In most respects the macrocysts are also like those of *Dictyostelium* (discussed in Chapter 2) and are remarkable only in that their germination is controllable, a fact which is useful for genetic studies. More unusual and more studied are the microcysts found in most if not all strains of *P. pallidum* but only rarely in other species.

Microcysts are round, walled, resting cells each formed from a single amoeba in response to osmotic or other stimuli (Fig. 4). High concentrations of either salts or sugars can induce encystment, but 0.08 *M* KCl is especially effective (Toama and Raper, 1967a). Calcium and magnesium modulate the osmotic response. Only starved cells encyst; amoebae will grow and divide on an encystment medium if bacteria are present. Darkness favors encystment, perhaps only by inhibiting aggregation. Encystment can also be induced by ammonia (Lonski, 1976). This response appears to be independent of the osmotic one, for 0.01 *M* NH_4Cl in the agar will enhance encystment even in the presence of a high concentration of KCl (Choi and O'Day, in press). Whether a population of amoebae in the soil becomes cysts or aggregates appears to depend on the water content of the soil, on how much ammonia is present, and on the light intensity. Microcysts can survive freezing when mixed with soil (Kuserk *et al.*, 1977) or with 20% DMSO (Ennis *et*

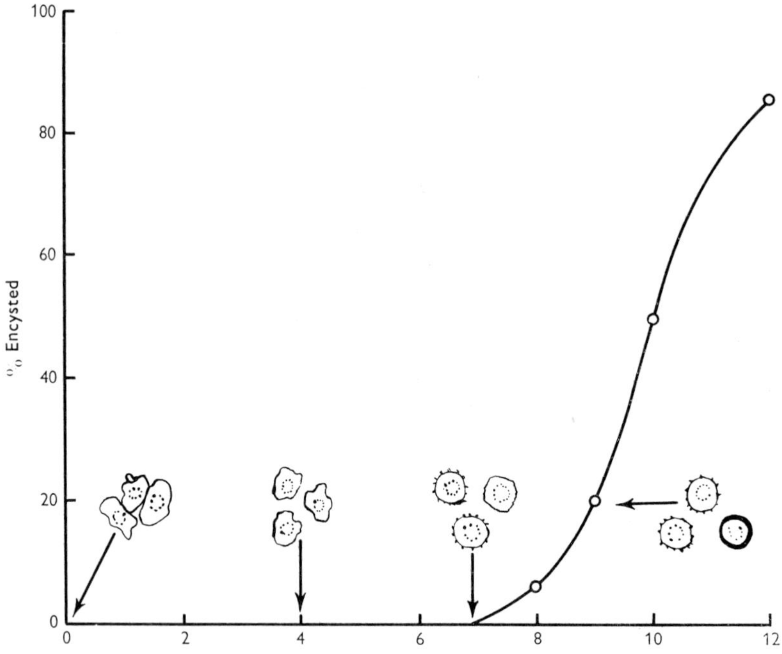

Fig. 4. Encystment of *P. pallidum*. The sketches indicate the appearance of cells at different points in the process (from Hohl *et al.*, 1970).

al., 1978). They are somewhat more resistant to heat than amoebae, and can survive 10 min at 45°C (unpublished data), but whether they are as heat resistant as spores seems not to have been examined. For the strains that can make microcysts, encystment represents an alternative survival strategy to spore formation. It will be of great interest to study the ecology of the cyst-forming strains and species in an attempt to understand why all species of cellular slime molds do not use both options.

Of the intracellular events of microcyst formation, a few interesting details are known. A series of new proteins are synthesized during the process, most of which are different from those of spore formation (Francis, 1976). Some of these are membrane proteins and one such appears to be cellulose synthetase (Philippi and Parish, 1981). Fibrils of cellulose first appear just outside the plasma membrane (Hohl *et al.*, 1970) and cellulose eventually becomes the major constituent of the cyst wall (Toama and Raper, 1967b). Other polysaccharides are present as well as proteins and lipids. Enzymes necessary for synthesis of these compounds may be associated with the intramembrane particles that increase dramatically in number during early encystment (Erdos and Hohl, 1980). They may be included among the seven membrane proteins detectable on polyacrylamide that do increase synthesis at this time (Philippi and Parish, 1981).

Microcysts germinate without need for a period of dormancy or heat activation when resuspended in dilute aqueous media or in water (Fig. 5). 0.75% ethanol accelerates germination and somewhat improves synchrony (Ennis *et al.*, 1978), so that one half of the population emerges in the half hour around 2.5 h after suspension. A sequence of new protein syntheses accompanies germination. It appears likely that the first of these new proteins are made on preformed messages. One indication of this is that cycloheximide, but not actinomycin D, blocks the increase in specific activity of alkaline phosphatase and β-glucosidase associated with germination. Actinomycin D in fact will not prevent germination, although it does block incorporation of tritiated uridine into macromolecules (O'Day *et al.*, 1976).

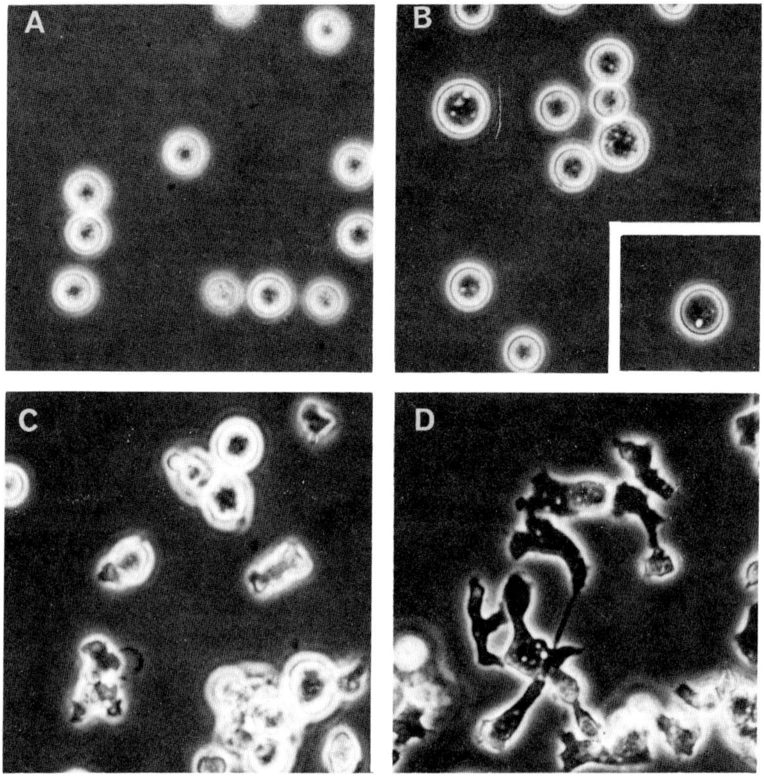

Fig. 5. Excystment of *P. pallidum*. (A) Ungerminated microysts; (B) Swollen microcysts after 2 h in excystment medium. Swelling occurs even in the presence of cycloheximide, as shown by the cell in the inset; (C) Amoebae emerging after 3.5 h; (D) Amoebae at 5 h (from O'Day, 1974).

The rate of incorporation of tritiated uridine into poly A-containing RNA is much lower during at least the first 1/2–2 h of germination than it is later (Gwynne and O'Day, 1978; Ennis *et al.*, 1978). The laboratories of Ennis and O'Day differ in how long and to what extent RNA synthesis remains low, perhaps because of differences in strains or in conditions of germination.

Several enzymes that increase in activity during germination appear to be directly involved in digesting the microcyst wall. Of these, an acid protease is present in dormant cysts and is released during the initial period of germination when swelling occurs (O'Day, 1976). Excretion of β-glucosidase (O'Day, 1974) and of carboxymethyl cellulase (O'Day and Paterno, 1979) also increases markedly during germination. These enzymes differ from the acid protease in that they appear to be the products of new protein synthesis as evidenced by cycloheximide inhibition, and in reaching peak activity later during the emergent phase of germination. That these three enzymes are actually active extracytoplasmically in digesting the cell wall is indicated by the acceleration of germination caused by addition of commercial purified enzymes to germinating cells (O'Day, 1976, 1977; O'Day and Paterno, 1979). Besides, germination is inhibited by D-gluconic acid lactone, an inhibitor of β-glucosidase (O'Day and Paterno, 1979).

The fact that both microcysts and spores have cellulosic walls and are resistant to environmental stress suggests that some parts of their formation and germination may be similar. In agreement with this is the structural similarity of the cyst wall and the fibrillar inner spore wall. Perhaps these two structures are truly homologous (Hohl, 1976). The outer wall of the spore, however, which cracks to let the amoeba out, is unique to spores; the microcyst wall dissolves completely on germination.

The first step of germination, swelling, is also common to spores and microcysts. It has been suggested that swelling is the result of osmosis caused by an induced increase in concentration of small molecules inside the spore (Hohl, 1976). That a similar mechanism is at work in microcysts is suggested by the high levels of free amino acids in the early part of germination (Gwynne and O'Day, 1978; Ennis, 1981).

V. Dependent Pathways

A great deal of work has been devoted to changes in specific genes, messages, and enzymes during development of cellular slime molds. These individual events are coordinated by a little understood set of interactions.

One aspect of this interaction is the nexus of substrate/product reactions which occur during morphogenesis. These reactions have been analyzed in considerable detail for the sugar and polysaccharide components during aggregation and fruiting (Wright *et al.*, 1977; Loomis and Thomas, 1976). One conclusion from Wright's work is that concentrations of some components may be controlled by changes in substrates, cofactors, or ionic conditions, while enzyme levels remain constant (Wright, 1975). However, abundant evidence does indicate that other enzymes and proteins do appear by new synthesis from freshly made messages. It is the coordination of this new gene activity that will be dealt with here.

One point must be made clearly at the outset, and that is that understanding the complexities of this coordination is not equivalent to understanding all aspects of development. How a set of genes is regulated, and how action of their corresponding enzymes is tied to formation of an extracellular product in a visible structure like a stalk, are two very different questions; this discussion is restricted to the first question and is thus intentionally incomplete.

A unique approach to the coordination of developmental events has been taken by Soll and coworkers (Soll, 1979; Mercer and Soll, 1980). By closely recording the rate of visible processes of development when a culture is subjected to an environmental stress such as temperature, they have concluded that multiple independent timers might regulate these events. Because these visible events are several steps removed from action of developmental genes, conclusions about the pattern of gene activity leading to their results cannot be made with any certainty.

Instead of listing and discussing the several models whereby activity of many genes might be coordinated, I will concentrate on one, the dependent sequence hypothesis (also called the causal pathway or timing sequence). Research in this area is so little advanced that there is no question of disproving likely alternative hypotheses. The dependent sequence theory serves only as a pattern to think of while trying to design relevant experiments. The idea itself is simple: that a chain of gene actions occurs during development, wherein each gene action is the cause of the action of the next gene. It seems first to have been expressed by Mandelstam (1969) while discussing bacterial sporulation.

That there truly is a sequence of gene actions is suggested by the series of new enzyme activities that accompany development (see Loomis, 1975). Other suggestive evidence is the series of new protein syntheses that likewise accompany development. In these experiments cultures are pulse labeled with acetate or methionine, cell extracts made at intervals through aggregation and fruiting, then polypeptides separated by electrophoresis and revealed by autoradiography. In *P. pallidum* labeling changes are scattered through aggregation to the beginning of stalk formation (Francis, 1976)

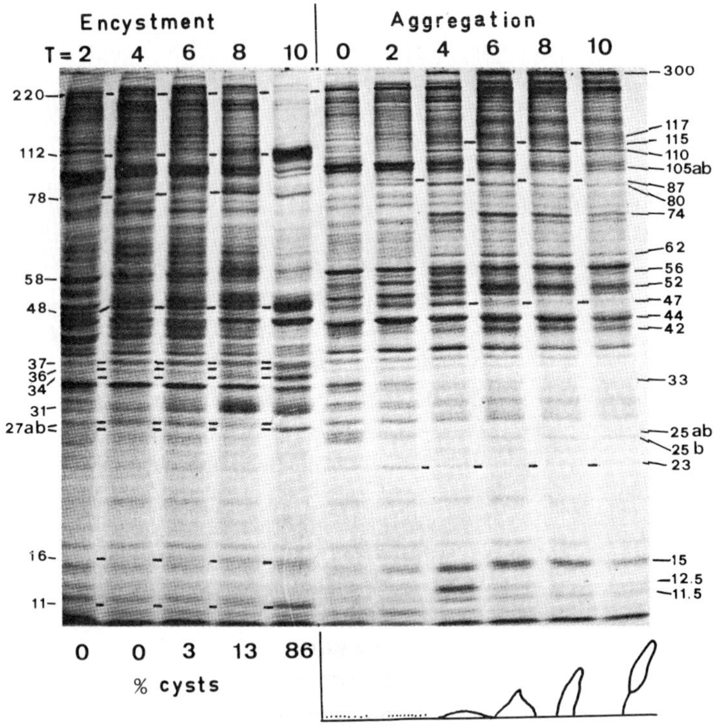

Fig. 6. Autoradiograph of electrophoretically separated polypeptides from cells of *P. pallidum* developing into microcysts (lanes 1–5) or into fruiting bodies (lanes 6–11). Cultures were pulse labeled with [^{14}C]acetate and sampled at the times indicated. Time is measured in hours after exposure to 0.08 *M* KCl to initiate encystment, or after exposure to light to initiate aggregation. The stage of development at the time of sampling is indicated beneath each lane. Polypeptides which change in rate of labeling during development are indicated. The double band 25 ab contains two subunits of the lectin pallidin (from Francis, 1976).

(Fig. 6). Similar results have been obtained for *D. discoideum*, where a larger number of polypeptides were more carefully resolved by electrophoresis in two dimensions (Alton and Lodish, 1977a). In this case a burst of changes (either labeling of new polypeptides or cessation of labeling of old ones) occurs in the first hour after starvation begins and another one occurs during late aggregation. Between these periods and up to the beginning of stalk formation are a scattering of other changes, so that the sequence of activity is not interrupted by any long periods of quiescence. Alton and Lodish showed further that many of the newly synthesized proteins are made from new messages and thus reflect new gene activity.

A possible mechanism of coordination that cannot now be dismissed is that the chain of causes consists of only a few key signals (e.g., light, cAMP, DIF factor) each of which elicits activity of a new cluster of genes. Those proteins that increase in the periods between the clusters could be the products of slowly processed genes rather than the result of separate inductive events. That the sequence of events is generated instead by a dependent sequence of multiple steps is suggested, but not proven, by the pattern of events in mutants that are blocked in development. Mutants that stop at an early stage of development do not show the complete sequence of developmental enzyme changes. A whole series of different mutants has been examined that stop at progressively later points in development, and that show progressively more of the developmentally regulated enzymes (Loomis *et al.*, 1976; Loomis *et al.*, 1978) (Fig. 7). It is as if each mutant represents a break at a different position in a chain reaction, the dependent sequence. A similar, but slightly more complex, pattern has been deduced from polypeptide labeling changes in developmental mutants of *P. pallidum* (Francis, 1977).

The hypothesis does not demand that all developmentally regulated genes be part of the causal chain; some, perhaps many, may be on side branches—dependent on action of a gene in the main pathway and so functioning at a defined point in time, but not essential for any later event. It would clearly be of value to identify the genes comprising the causal sequence. To directly identify the altered enzyme or protein stemming from the mutated gene in a given developmental mutant seems impractical, however. The alternative approach is to select for mutations in structural genes for known enzymes or proteins. Loomis' laboratory has vigorously pursued this line. To detect such mutant strains some thousands of clones derived from mutagenized cells were screened for those with low enzyme activity. The clones were grown axenically in liquid in the wells of microtiter plates (Brenner *et al.*, 1975). The cells were still dividing when assayed, but such axenically grown cells do exhibit some enzyme activities that are normally characteristic of the aggregation stage, at least in strain AX2 of *D. discoideum* (Quance and Ashworth, 1972). Mutants have been found with reduced enzyme activity in

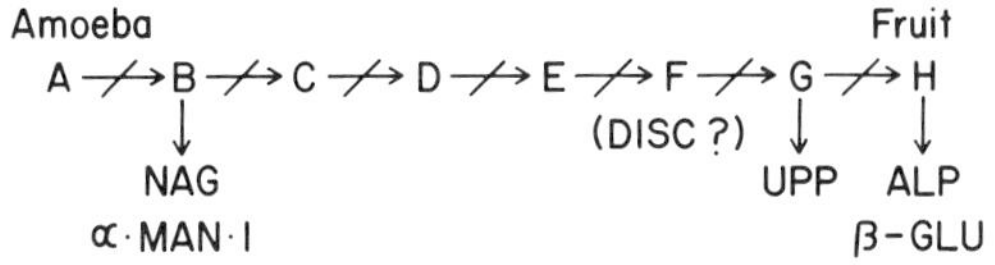

Fig. 7. The dependent sequence in *D. discoideum*. A, B, etc. represent events of the sequence. The horizontal slashed arrows indicate the causative action of each event on the next, for each of which a blocking mutation has been identified, and the vertical arrows indicate regulatory actions leading to syntheses of known enzymes, only some of which are shown. NAG—*N*-acetylglucoseaminidase; α-MAN-1—α-Mannosidase-1; UPP—UDPG pyrophosphorylase; ALP—alkaline phosphatase; B-GLU—B-glucosidase; DISC—discoidin 1, which may be a member of the dependent sequence. (Adapted from Loomis *et al.*, 1978.)

several different enzymes, and most of these are in structural genes as indicated by altered substrate affinity or thermolability.

Mutant strains altered in the structural gene of α-mannosidase isozyme 1 and having less than 0.1% of wild-type activity develop entirely normally both morphologically and with respect to later appearing developmental enzymes (Free and Loomis, 1975). This result suggests that the normal increase of α-mannosidase-1 at the beginning of starvation is not essential to the switching on of later genes. However, it may be too simple to suppose that the actual enzyme activity is the signal of meaning; perhaps transcription of the gene is the necessary event and occurs normally in the mutant strains. In any case, Loomis (1975) has assigned α-mannosidase-1 to a side branch of the causal sequence.

Mutant strains with less than 1% of normal enzyme activity have also been found for *N*-acetylglucosaminidase. These are able to complete development to stalk and spores, but the slugs fractionate to smaller slugs that migrate slowly (Dimond *et al.*, 1973). Again the mutations are in the structural gene by the criteria given above; and parasexual genetic studies show that the several mutant strains all carry mutations in a single gene (Loomis, 1978). This enzyme clearly has some physiological role in maintenance of normal slugs, but it does not act as if it is on a dependent sequence. The formation of normal spores and stalk cells implies that the later genes essential to development have all been active.

Mutations altering UDPG pyrophosphorylase have a different effect. Strains carrying these mutations and having less than 0.3% of normal activity grow slowly but aggregate normally and form slugs, but development stops before stalk is made and the cells soon die (Dimond *et al.*, 1976). That the cells cannot make the cellulose needed for stalk construction is to be expected, since the enzyme catalyzes formation of precursor UDPG. Why they die so quickly is not clear. The remarkable finding is that later enzymes in the developmental sequence do continue to appear. Only the last enzyme, β-glucosidase, increases but does not reach its normal level apparently because the cells are beginning to die at this time. Lack of UDPG pyrophosphorylase clearly halts morphogenesis; yet it is apparently not on the dependent sequence leading to an expression of late genes.

Mutants with less than 1% of wild type specific activity have been found that carry mutations in the structural gene for β-glucosidase. These mutants are also normal in aggregation (Dimond and Loomis, 1976). The single mutation found in the structural gene for alkaline phosphatase reduces the activity to 30% of normal; this residual activity may explain why the strain develops normally (MacLeod and Loomis, 1979). Evidence from other sources for certain of these enzymes confirms the conclusion that their genes are not part of the dependent sequence. An early finding from Ashworth's laboratory was that the developmental time courses of *N*-acetylglucosaminidase, α-mannosidase, and β-glucosidase may be drastically different depending on whether cells are in logarithmic growth or in stationary phase when starvation begins, and on whether they have been grown axenically on high or low glucose media or on bacteria (Quance and Ashworth, 1972). Cells taken from all these conditions undergo normal morphogenesis, however. Unaffected by such variations are developmental changes in UDPG pyrophosphorylase (Ashworth and Watts, 1970) and in alkaline phosphatase (Quance and Ashworth, 1972). This suggests that changes of these two enzymes are rigorously controlled by the dependent sequence, but does not mean that they are part of it. Recent data from Weeks (personal communication) implies that the relevant controlling event behind the rise in specific activity of alkaline phosphatase is removal of an inhibitor rather than synthesis of new enzyme. Other evidence relevant to alkaline phosphatase is O'Day's finding that activity of this enzyme in *P. pallidum* can be greatly reduced by beryllium ion, without disturbing the normal course of development at all (O'Day and Francis, 1973). This suggests but does not prove the nonnecessity of alkaline phosphatase for later events, for it is not certain that the *in vivo* intracellular ctivity of the enzyme was reduced.

One last protein for which a structural gene mutation has been claimed is the cell surface lectin discoidin (Ray *et al.*, 1979). Both of the major discoidin polypeptides of the mutant strain can be immunoprecipitated with anti-discoidin antibody, but one of them (discoidin 1) is not found when discoidin is isolated by affinity chromatography on galactose beads, suggesting that this polypeptide is made but is nonfunctional. The difficulty with accepting this as a structural gene mutant is that discoidin 1 polypeptide is coded by several genes, and it is unclear why the products of the genes that are not mutated do not make a functional discoidin 1 protein. An alternative possibility is that the mutation is in a gene necessary for assembling discoidin 1 subunits. Nevertheless, the mutant strain has interesting properties. It does not aggregate, and the cells fail to cohere. Revertants with some discoidin 1 activity have some cohesion ability (Shinnick and Lerner, 1980). The mutant also does not develop contact sites and does not synthesize later developmental enzymes (Schmidt and Loomis, personal communication). It does behave, therefore, as expected if it carries a mutation in the dependent sequence. If this is indeed the result of a mutation in a structural gene for discoidin, then discoidin is the first identified gene product on the putative dependent sequence.

The mutant strains whereupon the notion of a dependent sequence is based are all ones wherein mutation stops development and blocks later enzyme and protein changes. They appear to carry mutations in a set of positive regulatory genes whose normal action is to turn on specific genes for developmental enzymes. The odd fact, noted by MacLeod and Loomis (1980) is that positive regulatory genes have not been encountered in the successful searches for structural gene mutants of the enzymes *N*-acetylglucosaminidase (five mutations in the structural gene), α-mannosidase-1 (four mutations), and β-glucosidase (seven mutations). In previous screens of aggregateless and other developmentally blocked mutants, however, strains were identified which lack *N*–acetylglucosaminidase and β-glucosidase (Loomis *et al.*, 1976). These results together suggest that the only positive regulatory genes for these three enzymes are those that simultaneously affect other genes essential for development; there appear to be no regulators that affect these enzymes singly. MacLeod and Loomis (1980) also searched for mutants that would express α-mannosidase-2 constitutively. (This is a second isozyme that appears later in development than α-mannosidase-1). The idea was that finding such mutants would demonstrate a negative regulatory gene whose normal function is to keep the α-mannosidase-2 gene inactive. No such strain was found in a very extensive search, and MacLeod and Loomis conclude that such a gene probably does not exist. Of course this conclusion applies only to this one enzyme. The outcome of all this work is that at present no gene product has been identified that is unequivocally on the dependent sequence.

VI. Commitment

A dependent sequence is a state-determined machine; achievement of a particular state is the only condition necessary to proceed to the next one. To determine if this property of commitment is characteristic of development seems straightforward, and some data are available for cellular slime molds.

First, consider information comparing enzyme changes and new protein syntheses for the alternative pathways of encystment and aggregation/fruiting. The question is whether any events are common to both pathways. A particular controlling event that releases the action of the next gene in the dependent sequence could not be the unique trigger for two different next events, and hence could not be a component of two different pathways. Those enzyme changes and polypeptide labeling events that are only marker events rather than controlling events, however, might be triggered by two different controlling events and thus occur on two pathways. Nevertheless, very few if any labeling changes are common to the cyst and aggregation/fruiting pathways, to the extent that can be determined from one-dimensional SDS gels (Francis, 1976, 1979). Changes in very few enzymes have been

TABLE I
Characteristics of Developmental Mutants of *P. pallidum*[a]

Level of development	Wild type	Mutant type 1	2	3	4	5	6	7	8	9	10	11	12	13	14	15
Amoeba	+	+	+	+	+	+	+	+	+	+	+	+	+	+	+	+
Aggregate	+	+	+	+	0	0	0	0	0	0	0	0	+	+	+	+
Stalk	+	+	+	0	0	0	+	+	0	+	+	0	0	+	0	0
Spore	+	+	0	0	0	0	+	0	+	+	0	+	0	0	+	+
Microcyst	+	0	+	+	+	0	+	+	+	0	0	0	0	0	0	+
Number of mutants isolated		3	6	10	17	6	0	0	0	0	0	0	0	0	0	0

[a]From Francis, 1979.

studied in two pathways. In *P. pallidum*, alkaline phosphatase shows different patterns during cyst formation and fruiting. It decreases during encystment and increases in activity during microcyst germination and also in stalk formation (O'Day and Francis, 1973; O'Day, 1974). Whether enzymes involved in cell wall synthesis undergo parallel changes during encystment and sporulation appears not to have been examined, although the similarities in composition of these cell walls suggest this possibility.

A more direct test of whether controlling events are common to two pathways is given by developmental mutants. For example, if a mutant blocked in aggregation is able to encyst, then this controlling gene is not part of the encystment sequence. When a collection of such mutants was examined (Table I) it was found that all mutants blocked at stages later than aggregation were able to make microcysts (Francis, 1979). The tentative conclusion is that very few if any genes whose action is essential to later development are essential to encystment. Interestingly, several mutants not able to aggregate were also not able to make microcysts, indicating that at least some early events are common to the two pathways, and fitting with the notion that the causal sequence is initially a single track and then splits (Fig. 8).

Another approach to studying the state-determined, or committed, property of developing cells is to attempt to redirect development by some kind of environmental manipulation. A totally state-determined pathway would be inaccessible to the environment, but it has long been known that *Dictyostelium* development is not like this, for prestalk and prespore cells can interconvert. If development can be redirected by an appropriate stimulus it might be expected to follow the new pathway either absolutely or more

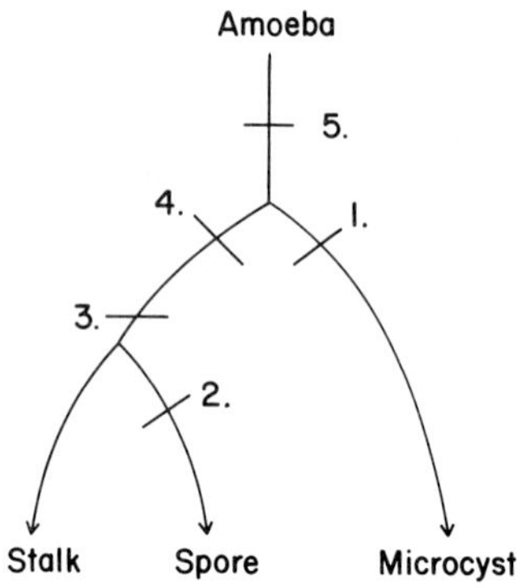

Fig. 8. The divergent pathway of differentiation in *P. pallidum*. The diagram indicates the positions where the several mutant types of Table I appear to block the dependent sequence. The region around **5** is the starvation phase of development; that between 3 and 4 is aggregation. (Adapted from Francis, 1979.)

loosely and the nature of these redirections will say something about the properties of the hypothesized dependent pathway.

One signal causing redirection of development is dispersal of aggregated cells. Changes in several enzymes (Newell *et al.*, 1972) and polypeptides (Alton and Lodish, 1977b) have been examined in *D. discoideum* during this process. The disaggregated cells reaggregate very quickly and rapidly recapitulate the accumulation of the set of four enzymes examined. An indication that the recapitulation is incomplete is that some polypeptides resolved by two-dimensional electrophoresis and normally present at 10 h of development (midaggregation) are not resynthesized when the cells are dissociated at 5 h and then allowed to reaggregate for five more hours. Since a complete set of samples over time was not taken, it is possible that synthesis of these polypeptides did occur and was missed. Other evidence that dispersed cells can be made to express only parts of later pathways comes from Gross and coworkers, who find that cyclic AMP and DIF factor together are sufficient to induce synthesis of the late enzyme UPP, even though these cells never make the early protein, contact sites A (Gross *et al.*, 1981). More detailed data of this type should accurately assess the degree of inevitability of the dependent pathway.

VII. Prospects

As should be clear from the above survey, we do not presently understand how actions of developmental genes are coordinated. However, two lines of work appear to hold promise.

Continued genetic dissection of individual proteins known to be essential to development will clearly be useful. This approach is similar to the now classical investigations of operons. The major difficulty is that there is no certain *a priori* method of recognizing proteins that are on the dependent sequence. Mere correlation of enzyme activity with development has proven to be an inadequate indication that the enzyme is essential to later events. Perhaps proteins known to be involved in biochemical steps of guaranteed developmental significance are more likely candidates for analysis. Enzymes engaged in reception, synthesis and destruction of cAMP and, with less certainty, the lectins and contact sites of the cell membrane are the only examples that come easily to mind. The design of selection or screening techniques to find and identify mutated structural genes for these proteins is proving to be a challenging enterprise.

An exciting alternative approach is to attempt identification of the mutated gene in a mutant blocked in development. Such a gene is in the dependent pathway by definition. If gene transformation becomes practical in cellular slime molds, as may soon be the case (Ratner *et al.*, 1981; Firtel, unpublished), it should be possible to clone some of the genes responsible for these defects. The relevant experiment is to complement the mutated gene with a wild-type gene carried on a transforming plasmid. This approach has already been used with developmental mutants in yeast (Nasmyth and Reed, 1980) and in myxobacteria (Kuner and Kaiser, 1981). Should such regulatory genes be cloned, it may be possible to determine the mechanisms whereby their products control later steps.

These two lines of work may result in actual detailed description of parts of the dependent sequence. They will require a variety of genetic tools: parasexual genetics for determining complementation groups; meiotic genetics for fine structural mapping; and recombinant DNA techniques for gene isolation. Without doubt such experiments will make use of the full range of diversity generated by evolution in the cellular slime molds, including species of *Dictyostelium* as well as *Polysphondylium*.

References

Alton, T. H., and Lodish, H. F. (1977a). Developmental changes in messenger RNA's and protein synthesis in *Dictyostelium discoideum*. *Develop. Biol.* **60**, 180–206.

Alton, T. H., and Lodish, H. F. (1977b). Synthesis of developmentally regulated proteins in *Dictyostelium discoideum* which are dependent on continued cell–cell interaction. *Develop. Biol.* **60**, 207–216.

Ashworth, J. M., and Watts, D. J. (1970). Metabolism of the cellular slime mould *Dictyostelium discoideum* grown in axenic culture. *Biochem. J.* **119**, 175–182.

Bonner, J. T., Chiquoine, A. D., and Kolderie, M. Q. (1955). A histochemical study of differentiation in the cellular slime molds. *J. Exp. Zool.* **130**, 133–158.

Bozzaro, S., and Gerisch, G. (1978). Contact sites in aggregating cells of *Polysphondylium pallidum. J. Mol. Biol.* **120**, 265–279.

Bozzaro, S., Janku, M., Monok, G., Opatz, K., and Gerisch, G. (1981). Characterization of a purified cell surface glycoprotein as a contact site in *Polysphondylium pallidum. Exp. Cell Res.* **134**, 181–191.

Brenner, M., Tisdale, D., and Loomis, W. F. (1975). Techniques for rapid biochemical screening of large numbers of cell clones. *Exp. Cell Res.* **90**, 249–252.

Chang, C.-M., Rosen, S. D., and Barondes, S. H. (1977). Cell surface location of an endogenous lectin and its receptor in *Polysphondylium pallidum. Exp. Cell Res.* **104**, 101–109.

Dimond, R. L., and Loomis, W. F. (1976). Structure and function of B-glucosidase in *Dictyostelium discoideum. J. Biol. Chem.* **251**, 2680–2687.

Dimond, R. L., Brenner, M., and Loomis, W. F., Jr. (1973). Mutations affecting *N*-acetylglucosaminidase in *Dictyostelium discoideum. Proc. Natl. Acad. Sci. U.S.A.* **70**, 3356–3360.

Dimond, R. L., Farnsworth, P. A., and Loomis, W. F. (1976). Isolation and characterization of mutations affecting UDPG pyrophosphorylase activity in *Dictyostelium discoideum. Develop. Biol.* **50**, 169–181.

Ennis, H. L. (1982). Free amino acid levels during cellular slime mold spore and microcyst germination. *Arch. Biochem. Biophys.* In press.

Ennis, H. L., Pennica, D., and Hill, J. M. (1978). Synthesis of macromolecules during microcyst germination in the cellular slime mold *Polysphondylium pallidum. Develop. Biol.* **65**, 251–259.

Erdos, G. W., and Hohl, H. R. (1980). Freeze-fracture examination of the plasma membrane of the cellular slime mold *Polysphondylium pallidum* during microcyst formation and germination. *Cytobios* **29**, 7–16.

Francis, D. (1965). Acrasin and the development of *Polysphondylium pallidum. Develop. Biol.* **12**, 329–346.

Francis, D. (1976). Changes in protein synthesis during alternate pathways of differentiation in the cellular slime mold *Polysphondylium pallidum. Develop. Biol.* **53**, 62–72.

Francis, D. (1977). Synthesis of developmental proteins in morphogenetic mutants of *Polysphondylium pallidum. Develop. Biol.* **55**, 339–346.

Francis, D. (1979). True divergent differentiation in a cellular slime mold, *Polysphondylium pallidum. Differentiation* **15**, 187–192.

Free, S. J., and Loomis, W. F., Jr. (1 75). Isolation of mutations in *Dictyostelium discoideum* affecting α-mannosidase. *Biochimie* **56**, 1525–1528.

Gerisch, G., Krelle, H., Bozzaro, S., Bitle, E., and Guggenheim, R. (1980). Analysis of cell adhesion in *Dictyostelium* and *Polysphondylium* by the use of Fab. *In* "Cell Adhesion and Motility" (A. S. G. Curtis and J. D. Pitts, eds.), pp. 293–307. Cambridge University Press, Cambridge.

Gross, J. D., Town, C. D., Brookman, J. J., Jermyn, K. A., Peacey, M. J., and Kay, R. R. (1981). Cell Patterning in *Dictyostelium. Phil. Trans. Roy. Soc.* (B) **295**, 497–508.

Gwynne, D. I., and O'Day, D. H. (1978). RNA and protein synthetic patterns during germination of *Polysphondylium pallidum* microcysts. *Can. J. Micro.* **24**, 480–486.

Hanna, M. H., and Cox, R. C. (1978). The regulation of cellular slime mold development: a factor causing development of *Polysphondylium violaceum* aggregation-defective mutants. *Develop. Biol.* **62**, 206–214.

Hintermann, R., and Parish, R. W. (1979). Synthesis of plasma membrane proteins and antigens during development of the cellular slime mold *Polysphondylium pallidum. FEBS Lett.* **108**, 219–225.

Hohl, H. R. (1976). Myxomycetes:Acrasiomycetes. *In* "The Fungal Spore; Form and Function" (D. J. Weber and W. M. Hess, eds.), pp. 463–498. Wiley, New York.

Hohl, H. R., Miura-Santo, L. Y., and Cotter, D. A. (1970). Ultrastructural changes during formation and germination of microcysts in *Polysphondylium pallidum*, a cellular slime mold. *J. Cell Sci.* **7,** 285–306.

Hohl, H. R., Honegger, R., Traub, F., and Markwalder, M. (1977). Influence of cAMP on cell differentiation and morphogenesis in *Polysphondylium. In* "Development and Differentiation in the Cellular Slime Molds" (P. Cappuccinelli and J. M. Ashworth, eds.), pp. 149–172. Elsevier, Amsterdam.

Jones, W. R., III, and Francis, D. (1972). The action spectrum of light induced aggregation in *Polysphondylium pallidum. Biol. Bull.* **142,** 461–469.

Kuner, J. M., and Kaiser, D. (1981). Introduction of transposon Tn5 into *Myxococcus* for analysis of developmental and other non-selectable mutants, *Proc. Natl. Acad. Sci. U.S.A.* **78,** 425–429.

Kuserk, F. T., Eisenberg, R. M., and Olsen, A. M. (1977). An examination of the methods for isolating cellular slime molds (Dictyostelida) from soil samples. *J. Protozool.* **24,** 297–299.

Lonski, J. (1976). The effect of ammonia on fruiting body size and microcyst formation in the cellular slime molds. *Develop. Biol.* **51,** 158–165.

Loomis, J. (1975). "*Dictyostelium discoideum*, A Developmental System," Academic Press, New York.

Loomis, W. F. (1978). Genetic analysis of the gene for *N*-acetylglucosaminidase in *Dictyostelium discoideum. Genetics* **88,** 277–284.

Loomis, W. F., and Thomas, S. R. (1976). Kinetic analysis of biochemical differentiation in *Dictyostelium discoideum. J. Biol. Chem.* **251,** 6252–6258.

Loomis, W. F., White, S., and Dimond, R. L. (1976). A sequence of dependent stages in the development of *Dictyostelium discoideum. Develop. Biol.* **53,** 171–177.

Loomis, W. F., Morrissey, J., and Lee, M. (1978). Biochemical analysis of pleiotropy in *Dictyostelium. Develop. Biol.* **63,** 243–246.

MacLeod, C. L., and Loomis, W. F. (1979). Biochemical and genetic analysis of a mutant with altered alkaline phosphatase activity in *Dictyostelium discoideum. Devel. Genet.* **1,** 109–121.

MacLeod, C. L., and Loomis, W. F. (1980). Attempts to isolate mutants of *Dictyostelium discoideum* which express α-mannosidase-2 constitutively. *J. Gen. Micro.* **120,** 273–277.

Mandelstam, J. (1969). Regulation of bacterial spore formation. *Symp. Soc. Gen. Micro.* **19,** 377–402.

McDonough, J. P., Springer, W. R., and Barondes, S. H. (1980). Species-specific cell cohesion in cellular slime molds. Demonstration by several quantitative assays and with multiple species. *Exp. Cell Res.* **125,** 1–14.

Mercer, J. A., and Soll, D. R. (1980). The complexity and reversibility of the time for the onset of aggregation in *Dictyostelium. Differentiation* **16,** 117–124.

Nasmyth, K. A., and Reed, S. I. (1980). Isolation of genes by complementation in in yeast: molecular cloning of a cell-cycle gene. *Proc. Natl. Acad. Sci. U.S.A.* **77,** 2119–2123.

Newell, P. C., Franke, J., and Sussman, M. (1972). Regulation of four functionally related enzymes during shifts in the developmental program of *Dictyostelium discoideum. J. Mol. Biol.* **63,** 373–382.

Nicol, A., and Garrod, D. R. (1978). Mutual cohesion and cell sorting-out among four species of cellular slime moulds. *J. Cell Sci.* **32,** 377–387.

O'Day, D. H. (1974). Intracellular and extracellular enzyme patterns during microcyst germination in the cellular slime mold *Polysphondylium pallidum. Develop. Biol.* **36,** 400–410.

O'Day, D. H. (1976). Acid protease activity during germination of microcysts of the cellular slime mold *Polysphondylium pallidum. J. Bact.* **125,** 8–13.

O'Day, D. H. (1977). Microcyst germination in the cellular slime mold *Polysphondylium pallidum:* requirements for macromolecular synthesis and specific enzyme accumulation. *In* "Eukaryotic Microbes as Model Developmental Systems" (D. H. O'Day and P. A. Horgen, eds.), pp. 353–372. Marcel Dekker, New York.

O'Day, D. H. (1979). Cell differentiation during fruiting body formation in *Polysphondylium pallidum. J. Cell Sci.* **35,** 203–215.

O'Day, D. H., and Durston, A. J. (1980). Sorogen elongation and side branching during fruiting body development in *Polysphondylium pallidum. Can. J. Microbiol.* **26,** 959–964.

O'Day, D. H., and Francis, D. W. (1973). Patterns of alkaline phosphatase activity during alternative developmental pathways in the cellular slime mold, *Polysphondylium pallidum. Can. J. Zool.* **51,** 301–310.

O'Day, D. H., and Paterno, G. (1979). Intracellular and extracellular CM-cellulase and B-glucosidase activity during germination of *Polysphondylium pallidum* microcysts. *Arch. Micro.* **121,** 231–234.

O'Day, D. H., Gwynne, D. I., and Blakey, D. H. (1976). Microcyst germination in the cellular slime mold, *Polysphondylium pallidum.* Effects of actinomycin D and cycloheximide on macromolecular synthesis and enzyme accumulation. *Exp. Cell Res.* **97,** 359–365.

Philippi, M. L., and Parish, R. W. (1981). Changes in glucan synthetase activity and plasma membrane proteins during encystment of the cellular slime mold, *Polysphondylium pallidum. Planta* **152,** 59–69.

Quance, J., and Ashworth, J. M. (1972). Enzyme synthesis in the cellular slime mould *Dictyostelium discoideum* during the differentiation of myxamoebae grown axenically. *Biochem. J.* **126,** 609–615.

Ratner, D. I., Ward, T. E., and Jacobson, D. (1981). Evidence for transformation of Dd with homologous DNA. *In* "Developmental Biology using Purified Genes; ICN–UCLA Symposia on Molecular and Cell Biology" (D. Brown, and C. F. Fox, eds.), Vol. 23, pp. 595–605. Academic Press, New York.

Ray, J., Shinnick, T., and Lerner, R. (1979). A mutation altering the function of a carbohydrate binding protein blocks cell–cell cohesion in developing *Dictyostelium discoideum. Nature (London)* **279,** 215–221.

Reitherman, R. W., Rosen, S. D., Frazier, W. A., and Barondes, S. H. (1975). Cell surface species-specific high affinity receptors for discoidin: developmental regulation in *Dictyostelium discoideum. Proc. Natl. Acad. Sci. U.S.A.* **72,** 3541–3545.

Rosen, S. D., Haywood, P. L., and Barondes, S. H. (1974). Inhibition of intercellular adhesion in a cellular slime mould by univalent antibody against a cell-surface lectin. *Nature (London)* **263,** 425–427.

Rosen, S. D., Simpson, D. L., Rose, J. E., and Barondes, S. H. (1976). Carbohydrate-binding protein from *Polysphondylium pallidum* implicated in intercellular adhesion. *Nature (London)* **252,** 128–130.

Rosen, S. D., Kaur, J., Clark, D. L., Pardos, B. T., and Frazier, W. A. (1979). Purification and characterization of multiple species (isolectins) of a slime mold lectin implicated in intercellular adhesion. *J. Biol. Chem.* **254,** 9408–9415.

Rowekamp, W., Poole, S., and Firtel, R. A. (1980). Analysis of the multigene family coding the developmentally regulated carbohydrate-binding protein discoidin-1 in *D. discoideum. Cell* **20,** 495–505.

Shaffer, B. M. (1961). The cells founding aggregation centers in the slime mould, *Polysphondylium violaceum. J. Exp. Biol.* **38,** 833–849.

Shaffer, B. M. (1962). The Acrasina. *Advan. Morphog.* **2,** 109–182.

Shinnick, T. M., and Lerner, R. A. (1980). The cbp A gene: role of the 26,000-dalton carbohydrate-binding protein in intercellular cohesion of developing *Dictyostelium discoideum* cells. *Proc. Natl. Acad. Sci. U.S.A.* **77,** 4788–4792.

Simpson, D. L., Rosen, S. D., and Barondes, S. (1974). Discoidin, a developmentally regulated carbohydrate-binding protein from *Dictyostelium discoideum*. Purification and characterization. *Biochemistry* **13,** 3487–3493.

Soll, D. R. (1979). Timers in developing systems. *Science* **203,** 841–849.

Spiegel, F. W., and Cox, E. C. (1980). A one-dimensional pattern in the cellular slime mould *Polysphondylium pallidum*. *Nature (London)* **286,** 806.

Stenhouse, F. O., and Williams, K. L. (1981). Cell patterning in branched and unbranched fruiting bodies of the cellular slime mold *Polysphondylium pallidum*. *Develop. Biol.* **81,** 139–144.

Toama, M. A., and Raper, K. B. (1967a). Microcysts of the cellular slime mold *Polysphondylium pallidum*. I. Factors influencing microcyst formation. *J. Bact.* **94,** 1143–1149.

Toama, M. A., and Raper, K. B. (1967b). Microcysts of the cellular slime mold *Polysphondylium pallidum*. II. Chemistry of the microcyst walls. *J. Bact.* **94,** 1150–1153.

Wright, B. E. (1975). "Critical Variables in Differentiation," 109 pp. Prentice-Hall, Englewood Cliffs, New Jersey.

Wright, B. E., Tai, A., and Killick, K. A. (1977). Fourth expansion and glucose perturbation of the *Dictyostelium* kinetic model. *Eur. J. Biochem.* **74,** 217–225.

Wurster, B., Pan, P., Tyan, G. G., and Bonner, J. T. (1976). Preliminary characterization of acrasin of cellular slime mold, *Polysphondylium violaceum*. *Proc. Natl. Acad. Sci. U.S.A.* **73,** 795–799.

Wurster, B., Bozzaro, S., and Gerisch, G. (1978). Cyclic GMP regulation and responses of *Polysphondylium violaceum* to chemoattractants. *Cell. Biol. Internat. Repts.* **2,** 61–69.

• CHAPTER 11

Cell Proportioning and Pattern Formation

James H. Morrissey

I. Introduction

Cellular slime molds live and grow in forest litter as solitary amoebae, but for efficient dispersal they come together to form fruiting bodies at the surface of the soil (Shaffer, 1962). Aggregation to form multicellular masses has been taken up in detail elsewhere in this volume. The immediate conse-

The Development of *Dictyostelium discoideum*

ISBN 0-12-455620-5

quence of aggregation is the formation of the pseudoplasmodium, or slug, which migrates phototactically and thermotactically to the upper layer of the soil. At this point, the slug "culminates" to form a fruiting body consisting of a ball of spores held aloft on a thin, tapering stalk. Since the stalk is made up of inviable cells, it is clear that there is a division of labor among the amoebae in the construction of the fruiting body. There is also division of labor among the cells in the migrating slug, and it is this remarkable entity that has received so much attention as a model system for study of the processes of cell proportioning and pattern formation.

Migrating slugs of *Dictyostelium discoideum* have received this attention in part because they have a simple pattern of differentiation. This pattern can be readily demonstrated with the vital stain, neutral red, which is retained by cells only in the front fifth of the slug (Bonner, 1952; see Fig. 1). Isolated fragments will regulate to produce small, but normally proportioned slugs; since this process takes place in the absence of growth or cell division, it is analogous to, but simpler than, what occurs in regulation of hydra or of early embryos of higher organisms. *D. discoideum* slugs are easy to work with because they can be kept migrating as slugs for several days, or they can be induced to construct fruiting bodies under the control of the investigator. They are not inconveniently small (1–2 mm long) and will readily accept grafts or injections of cells. Even tiny fragments of slugs containing fewer than 100 cells are able to construct fruiting bodies successfully.

Considerable progress has been made in understanding spatial pattern formation in the cellular slime molds. Although most models of pattern formation have assumed position-dependent differentiation, there is now good evidence that pattern regulation involves random differentiation of presumptive cell types followed by sorting out, and at least some indications

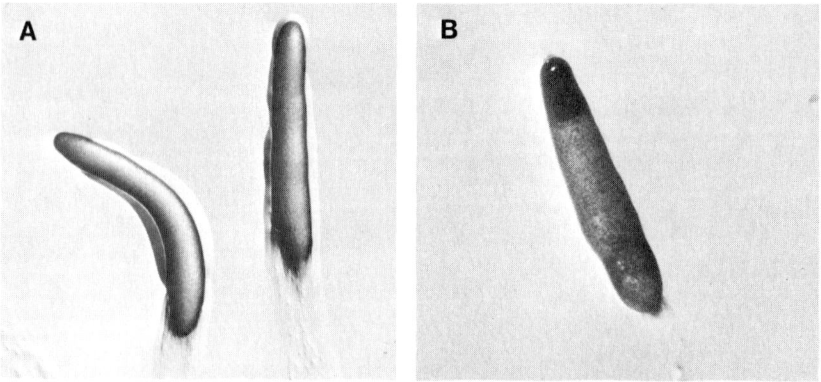

Fig. 1. Migrating slugs of *Dictyostelium discoideum*. **A.** unstained slugs. **B.** slug stained with neutral red.

that the initial generation of the pattern does also. One of the questions that remains to be answered is the nature of the intercellular signals that regulate cell proportioning in this organism. Candidates for these signals have been proposed and are discussed in Section V.

This chapter will concentrate on *Dictyostelium discoideum*, with a brief discussion of related species. I will first describe the known properties and parameters of migrating slugs, and in the following sections I will discuss both facts and hypotheses concerning how the pattern is generated and regulated. I touch only briefly on the subject of mathematical models of positional signalling in the cellular slime molds as they have already been ably discussed by MacWilliams and Bonner (1979).

II. The Spatial Pattern

A. Cell Fate

The pioneering study of Raper (1940) showed that the cells in the front fifth of the slug will form the stalk cells of the mature fruiting body, whereas the cells in the rear four-fifths of the slug will ultimately differentiate into spores. Because of the difference in their prospective fates, the cells in these two regions are termed prestalk and prespore, respectively. Prestalk and prespore cells are not irreversibly determined to form stalk cells and spores since isolated fragments of slugs can regulate to give normally proportioned fruiting bodies; however, the speed with which regulation proceeds differs between the two cell types. Prestalk isolates induced to immediately culminate will form fruiting bodies composed almost entirely of stalk, whereas similar isolates allowed to migrate for 24 to 48 h before culmination form fruiting bodies with a more normal spore:stalk ratio. Prespore isolates, on the other hand, culminate directly without migration, and under normal circumstances appear to regulate so quickly that they always form normally proportioned fruiting bodies (Raper, 1940; Bonner and Slifkin, 1949; but see Sampson, 1976, for the case of very old slugs).

The mature fruiting body is a static structure composed of dormant spores and dead stalk cells. Therefore, the proportions of these two cell types reflect events that take place at the slug stage or in the rising sorogen (the cell mass constructing the fruiting body). A number of studies have taken advantage of this relationship to use the spore:stalk ratio to measure some of the critical parameters of cell proportioning that presumably are involved in the generation of the slug pattern, and also to examine the process of fruiting body proportioning itself. Cell proportions in fruiting bodies have been estimated using such methods as dry weight or volume measurements from camera lucid drawings, but the study of Stenhouse and Williams (1977) is the most rigorous, involving direct counting of individual spores and stalk cells. Fruiting bodies can vary in size from 100 to 2×10^5 cells; these workers measured fruiting bodies containing up to 1.8×10^4 cells and found that the fraction of stalk cells was a function of total cell number and varied from 10–20%, indicating that the spore:stalk ratio is nearly size-invariant over a wide range of fruiting body sizes. A third cell type, the basal disk cell, was counted by Stenhouse and Williams as making up 1–2% of the total cell number. The basal disk anchors the stalk to the substratum, is peculiar to *D. discoideum*, and is composed of cells that are indistinguishable in morphology from stalk cells.

B. Organizing Potential

The cells in the very front of the slug possess properties similar to those of a classical embryonic organizer (Spemann, 1938; Wolpert, 1971); i.e., when they are grafted onto the side of a slug or sorogen, they organize secondary axes (Raper, 1940; Farnsworth, 1973; Rubin and Robertson, 1975). This organizer property is associated with a nipple-shaped tip which marks the anterior end of cell masses at most stages. Although present, it is not usually visible in migrating slugs.

That the tip/nontip pattern will regulate is demonstrated by the finding that prespore isolates will regenerate a new tip within 2 h of excision from

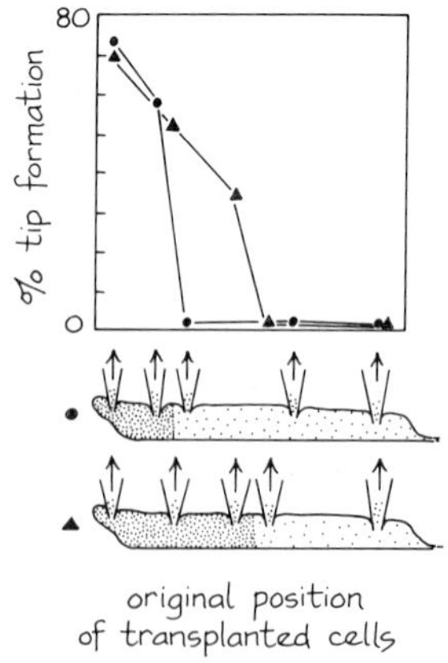

Fig. 2. Axial distribution of tip-forming capacity of cells taken from wild-type (●) or long-prestalk mutant (▲) slugs. Cells are injected into the rear of a wild-type slug to test their ability to organize new tips. (Reproduced from MacWilliams (1982), with permission).

the slug (Raper, 1940). Moreover, the tip exerts dominance, since its presence will suppress the formation of additional tips within the same cell mass (Farnsworth, 1973). Tip dominance is a graded property of the slug, the inhibitory gradient being highest at the front and decreasing toward the back (Durston, 1976). MacWilliams (1982) has examined this phenomenon using a new technique of injecting cells into the slug with a microsyringe, and has obtained results that are best explained by postulating a diffusible inhibitor of tip formation whose concentration gradient is highest at the tip. He has also found that cell masses taken from any part of the neutral red-stained prestalk zone can organize new axes, but that there is a shallow gradient of organizing activity even within this region (see Fig. 2).

C. Surface Sheath

Cell masses from the end of aggregation onward are surrounded by a thin, extracellular casing called the slime or surface sheath. Migrating slugs continually secrete new sheath material as they move out of the old sheath, leaving it behind as a collapsed tube. Since it is stationary with respect to the substratum, the sheath has been proposed to provide traction for amoeboid motion by the cells inside (Francis, 1962; Shaffer, 1964; Inouye and Takeuchi, 1979). The sheath is apparently synthesized along the length of the slug (Loomis, 1972) and is thinnest over the front, becoming progressively thicker as it passes over more posterior regions (Farnsworth and Loomis, 1975). It is highly distensible, almost fluid, over the tip of the slug and considerably tougher along the sides (Francis, 1962), which is consistent with the polarity of migration. When tip cells are implanted into the side of a slug, they organize a new developmental axis; to do this they must make the sheath locally much more distensible so that the cell mass can move through it. Thus, a property that may be tentatively assigned to tip cells is to generate and maintain a local region of high fluidity in the surface sheath.

D. Cytodifferentiation

The differentiation between prespore and prestalk cells can be demonstrated by a number of criteria of which neutral red staining is only one example. As was pointed out earlier, the pattern of differentiation in slugs revealed by this stain is remarkably similar to the fate map for spore and stalk differentiation. For this reason the cells identified by neutral red staining (as an example) are often referred to as prestalk cells, and unstained cells as prespore cells. What is not yet clear is whether the pattern of differentiation in slugs reflects obligatory preparations for terminal differentiation, or whether it is instead related to slug-specific functions such as tip dominance or the axial inhomogeneity of the surface sheath, which is itself likely to be related to the polarity of migration. There is evidence on both sides and perhaps neither extreme view is correct (see Sec. III). For the sake of the present discussion, the terms prespore and prestalk will be used to refer to the normal fates of the cells in the two regions. Identification of the cell type can be made by using various criteria without necessarily concluding that the type of property in question (e.g., neutral red stainability) is necessarily a preparation on the part of the cell to enable it to differentiate into a spore or stalk cell. This may seem belabored, but it has been a point of serious contention in the past.

1. *Vital Staining*

Vegetative amoebae stained with the vital dyes neutral red, nile blue, or bismarck brown aggregate into uniformly tinted slugs, whose prespore zones blanch upon migration to give the pattern seen in Fig. 1 (Bonner, 1952). The dyes are apparently taken up by and trapped in autophagic vacuoles, which are considerably more prominent in prestalk cells (Durston and Vork, 1979; MacWilliams and Bonner, 1979), having been selectively lost from prespore cells during development (Yamamoto *et al.*, 1981). Sternfeld and David (1981a) have shown that these dyes do not diffuse from cell to cell unless the cells are handled so harshly that lysis occurs. They therefore appear to be reliable and convenient markers for prestalk cells. However, very old slugs frequently exhibit a more irregular, blotchy staining pattern, a phenomenon that is strain dependent (H. MacWilliams, personal communication). Whether this results from an eventual breakdown in the normally sharp spatial separation of prestalk and prespore cells or is a result of some of the prespore cells acquiring the stain (and autophagic vacuoles) is not clear, although the pattern of stained versus unstained cells in old slugs becomes considerably sharper once culmination begins.

Close inspection of vitally stained slugs reveals that even in young slugs the prespore zone contains a small number of randomly distributed, intensely stained cells, which Sternfeld and David (1981a) call anterior-like cells. In a careful study using doubly stained mixtures of cells, these workers have found that when the prestalk zone is removed, the new tip that forms on the prespore isolate is derived largely from anterior-like cells that have sorted out from the rest of the cell mass. New anterior-like cells can be shown to have arisen subsequently in the prespore zone by restaining regulated slugs with vital dyes.

The existence of a pool of such pre-prestalk cells may help to explain the rapidity of regulation of prespore isolates compared to that of prestalk isolates. However, when the regenerated tip is removed, the cell mass regenerates another tip within 2 h, and this process of extirpation and regeneration can be carried out for at least five successive operations without altering the speed with which the next tip will appear (my unpublished observations). This therefore places an upper limit of 4 h on the length of time required for prespore cells to regulate (as assayed by tip formation).

2. *Prespore Vesicles*

Prespore vesicles, also called prespore vacuoles (PVs) are organelles of unusual appearance that are found in the cytoplasm of some cells toward the end of aggregation, and are restricted to prespore cells in the slug (Hohl and Hamamoto, 1969; Gregg and Badman, 1970; see Fig. 3). Although the distribution of PVs can be variable (Farnsworth and Loomis, 1976), the anterior region devoid of PVs is usually about a fifth of the length of the migrating slug. Interestingly, about 15% of the cells in the prespore zone do not

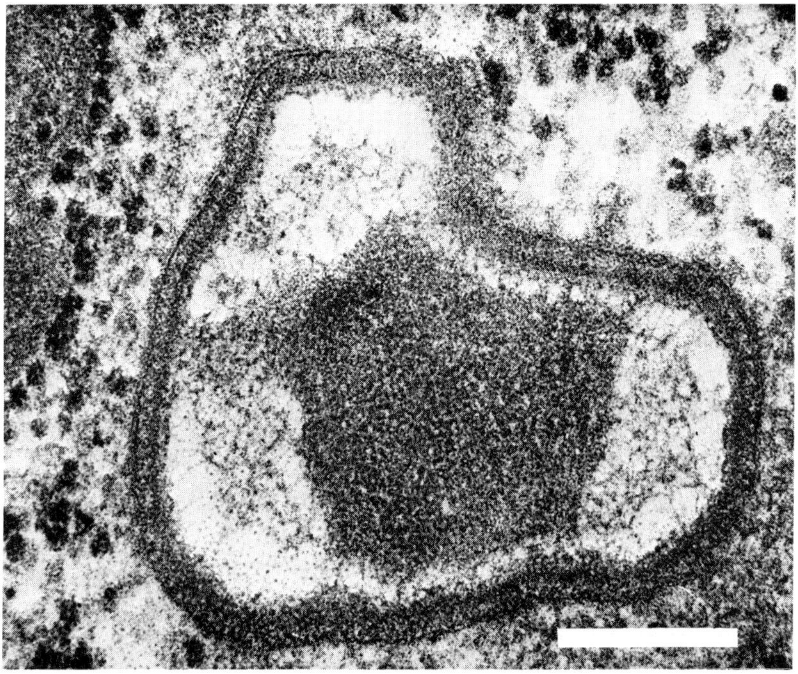

Fig. 3. Electron micrograph of a prespore vesicle. Bar represents 100 nm. Photograph is courtesy of Dr. P. Farnsworth.

possess PVs (Muller and Hohl, 1973); these may be the anterior-like cells that Sternfeld and David (1981a) have identified.

Takeuchi (1963) has produced a fluorescein-conjugated antiserum prepared against whole *D. mucoroides* spores that binds specifically to PVs after it has been absorbed with vegetative *D. discoideum* amoebae (Ikeda and Takeuchi, 1971), indicating that constituents of PVs share combining groups with at least some of the spore antigens. In young slugs only the posterior cells bind the antiserum, showing a strikingly sharp demarcation between the fluorescent posterior and nonfluorescent anterior, while older slugs exhibit a more irregular pattern (Takeuchi *et al.*, 1977). Consistent with the findings of the ultrastructural studies, a small number of the cells in the prespore zone of migrating slugs are often found not to bind the antiserum (Takeuchi *et al.*, 1977).

The antispore serum reacts with an acid mucopolysaccharide found in PVs and apparently also in spore coats (Ikeda and Takeuchi, 1971; Takeuchi, 1972). This may be the same substance described by Sussman and Osborn (1964) as being the acceptor for the spatially localized enzyme UDPgalactose:polysaccharide transferase (PVs do not, however, contain the enzyme—Ikeda, 1981). The question of the role of PVs in development is taken up in Section III.

3. *Enzymatic Activities*

Although neutral red staining and fluorescent antispore-serum binding are often used as markers for prestalk and prespore cells, respectively, for some applications biochemical markers are of greater utility. For example, in gauging the purity of separated prestalk and prespore cells, cell type-specific enzymatic markers have been used (Oohata and Takeuchi, 1977; Tsang and Bradbury, 1981; Ratner and Borth, 1982).

Of the enzymatic activities known to be spatially distributed in slugs, the first known and perhaps most striking is UDPgalactose:polysaccharide transferase, which is localized almost exclusively to the prespore zone and is a useful marker for prespore differentiation (Newell *et al.*, 1969). This enzyme transfers galactosyl units onto an acid mucopolysaccharide that may be associated with the spore coat (see Sec. III). Oohata and Takeuchi (1977) have shown that there are prestalk-specific isozymes of β-galactosidase and acid phosphatase that are detectable electrophoretically, and these are good candidates for markers of prestalk differentiation.

There are also quantitative differences, although of lesser magnitude, in the spatial distributions of a number of other enzymatic activities. Trehalase, glycogen phosphorylase, and cathepsin B have higher specific activities in prestalk cells than in prespore cells (Jefferson and Rutherford, 1976; Rutherford and Harris, 1976; Fong and Rutherford, 1978), whereas cyclic nucleotide phosphodiesterase shows an unusual pattern of low specific activity in most of the prespore zone, higher activity in the prestalk zone, and highest activity in a small area at the most posterior position of the slug (Brown and Rutherford, 1980). This latter area is occasionally referred to as the rear guard zone, said by some to have prestalk-like properties and to be the source of the basal disk (Bonner, 1957; Durston and Vork, 1979). Sternfeld and David (1981a) refer to these as anterior-like cells, but they find that such cells are usually dispersed throughout the prespore zone, rather than being confined to the rear guard.

4. *Polysaccharides*

Periodic acid–Schiff (PAS) staining of sections of slugs that have been treated with amylase reveals a distribution of reactive, nonstarch carbohydrates that is restricted to the prespore zone (Bonner *et al.*, 1955), and a similar distribution of incorporation of radiolabeled fucose has also been reported (Gregg and Karp, 1978). Whether these procedures reveal patterns in the distribution of polysaccharides per se, such as the acid mucopolysaccharide, or of other glycosylated species such as glycoproteins is not known.

5. *Proteins*

By using two-dimensional gel electrophoresis of radiolabeled proteins, it is possible to show that a number of cell type-specific proteins are synthesized by prespore and prestalk cells (Alton and Brenner, 1979; Morrissey *et al.*, 1981; Ratner and Borth, 1982). A few of these have been further characterized. The rate of synthesis of actin is higher in prestalk cells than in prespore cells, a difference that becomes more pronounced at culmination: actin continues to be made at high levels as prestalk cells differentiate into stalk, whereas it is not made at all by sporulating cells (Coloma and Lodish, 1981). By using both an antiserum raised against slug plasma membranes and the lectin wheat germ agglutinin, West and McMahon (1979) have identified prespore- and prestalk-specific glycoproteins, some of which can be detected as early in development as the hemispherical mound stage (before the tip forms).

6. *Other Properties*

Prestalk cells differ from prespore cells by certain other properties not discussed above, e.g., the spatial pattern of DNA synthesis in slugs demonstrated by autoradiography following administration of radiolabeled adenine

(Durston and Vork, 1978). Another example is the difference in cohesive properties between the two cell types (Lam *et al.*, 1981). The possible significance of this finding relative to cell sorting is discussed in Section IV.

III. The Role of the Slug Pattern in Terminal Differentiation

The spatial pattern of differentiation in slugs, as measured by a number of different criteria, is approximately coextensive with the fate map for differentiation into spores and stalk cells. Taken together with certain other evidence, including the fact that PVs share antigenic determinants with spore coats, this has led to the interpretation that "the formation of the prestalk-prespore pattern (is) an initial step along the pathway to the final differentiation of stalk cells and spores" (MacWilliams and Bonner, 1979). Forman and Garrod (1977a) have stated this more forcefully, concluding that there is a direct causal relationship between possession of PVs and differentiation into spores. The opposite point of view was taken by Farnsworth (1973, 1974) who proposed a model wherein the specification of the spore:stalk ratio takes place during culmination as a mechanical consequence of the size of the tip, and is only very indirectly related to differentiation in slugs. A test was made with sporogenous mutants that showed that cells from the prestalk zone of slugs display a pronounced tendency to differentiate into stalk cells *in vitro* in the absence of morphogenetic movements (Tsang and Bradbury, 1981). Sporogenous mutants and the *in vitro* assay procedures are described in Sec. V,C. Interestingly, 9–29% of the cells from the prespore zone differentiate into stalk cells under these conditions (66–85% differentiate into spores), which is consistent with earlier findings that prespore isolates usually form fruiting bodies with normal spore:stalk ratios (see Section II,A).

Prestalk cells clearly have a cell-autonomous propensity toward stalk cell differentiation, although the case for prespore cells is presently unsettled because prespore isolates contain anterior-like cells (Sternfeld and David, 1981a) that may be responsible for the stalk cells that such isolates always form. Is the pattern of differentiation in slugs causally related to cell fate? Do all of the kinds of differentiation that are patterned in slugs function to prepare the cell for spore or stalk differentiation, or are some of them related to other properties of the slug known to be patterned, such as organizing potential, the polarity of migration, or sheath production? The question is taken up in this section by examining the consequences of altering one or another of the patterns in slugs.

A. Regulation

The study of pattern regulation can be important in understanding the basic mechanisms whereby organisms specify and interpret positional information. For example, experiments with hydra on the timing of pattern regulation, specifically the rate at which head and foot dominance and the control of polarity are able to spread through the organism, have generated a wealth of quantitative data that have been used to show that diffusion of morphogenetic substances forms a plausible basis for the control of differentiation (Wolpert *et al.*, 1974; MacWilliams and Kafatos, 1974). Indeed, the experiments on organizing potential discussed in Sec. II,B indicate that tip formation is regulated by an inhibitory gradient in the slug in a manner analagous to the regulation of head and foot formation in hydra (Wolpert *et al.*, 1971; MacWilliams *et al.*, 1970).

The regulation of vital staining and the fate map for spore and stalk differentiation have been discussed in Section II. Although prestalk isolates regulate to gain the ability to form normally proportioned fruiting bodies over a period of days, they regulate with respect to other types of differentiation much more quickly. In the case of PAS staining, regulation can first be observed in both prestalk and prespore isolates after 1 h and is complete by 6 h (Bonner *et al.*, 1955). This is true even in isolated anterior fragments undergoing culmination, which, according to the findings of Raper (1940), will construct fruiting bodies containing an abnormally low proportion of spores.

By studying the incorporation of radiolabeled fucose, Gregg and Karp (1978) were able to show that regulation is initiated almost immediately: the posterior portion of isolated prestalk zones begins to incorporate [^{3}H]fucose within 10 min following transection. Their method is less sensitive for the detection of cells that fail to incorporate fucose; consequently, regulation in prespore fragments can only be confirmed 3 h after isolation.

Sakai (1973) investigated the regulation of the pattern of differentiation in isolated slug fragments immunohistochemically, using the anti-*D. mucoroides* spore serum developed by Takeuchi (1963). The fraction of cells in prestalk isolates staining brightly with the fluorescein-conjugated antiserum remained near zero for 3 h and then increased to 60–70% by 8 h. Regulation in prespore isolates began immediately, the fraction of unstained cells rising from 12 to 20–30% within 3 h following dissection.

Pattern regulation in slug fragments proceeds at different rates depending upon the type of differentiation studied, and in part this can probably be explained by differences in sensitivity of detection. Autoradiography is sensitive to incorporation of very small amounts of radioisotope, whereas fluorescent-antibody binding requires the presence of a considerable amount of antigen. Thus, radiolabeled fucose can be shown to be incorporated into prestalk isolates within 10 min of excision, whereas detectable spore-specific antigens do not appear for 3 h (Gregg and Karp, 1978; Sakai, 1973). It is therefore possible that all of the types of differentiation observable at the slug stage are tied to one underlying pattern that regulates at one speed, but whose effects require varying amounts of time to be detected, depending upon the assay. Since all of the slug-specific patterns of differentiation described above regulate faster in prestalk isolates than does cell fate, it may be concluded that none of them in themselves, including PVs, are sufficient to cause a cell to differentiate into a spore.

B. Slug Migration

The number of PVs per cell is reduced in phototaxing slugs up to sevenfold compared to those migrating in the dark, the effect being more pronounced at higher levels of illumination (Farnsworth and Loomis, 1976). It has been reported that the conditions of phototaxis used in this study result in an increase in the rate of migration (Poff and Loomis, 1973), and so this may be an effect of migration speed on PV content. Surprisingly, Farnsworth and Loomis found that in very small slugs under high illumination, the cells were almost totally devoid of PVs, a condition that persisted through culmination. However, such slugs typically form normally proportioned fruiting bodies.

Using the fluorescent anti-*D. mucoroides* spore serum, Takeuchi *et al.* (1977) have shown that initially about 80% of the cells in a slug are capable of binding the antiserum. After 24 h of migration this drops to 60%, but when culmination is induced, the proportion of fluorescent cells suddenly increases to 75%, and normal fruiting bodies are formed.

Slugs continually produce surface sheath, which is left behind as the slime trail during migration, hence increasing the rate or duration of migration will increase the amount of sheath material produced by the slug. These treatments have pronounced effects on the PV content of slug cells, raising the possibility that PVs are related to the production or maturation of the surface sheath.

C. Mutational Analysis

The use of mutants specifically altered in cell proportioning is potentially a very powerful approach to understanding the control processes involved, as well as being useful in investigating which aspects of the system are causally linked. Early work in this vein was hampered by the lack of suitable mutants. Thus, Forman and Garrod (1977a) included a study of the mutant *P4*, said to form more stalk than normal (Chia, 1975). During development, only 60% of the cells of this strain were seen to possess PVs; however, these authors point out that *P4* differentiates into poorly organized fruiting bodies, so the actual spore:stalk ratio is unknown. *KY19*, a mutant previously described as differentiating entirely into spores (Ashworth and Sussman, 1967) was subsequently found not to be altered in cell proportioning at all, but rather to undergo incomplete stalk cell differentiation (Morrissey and Loomis, 1981). Very recently, however, *bona fide* cell proportioning mutants have become available.

A number of mutants have been obtained that differentiate entirely into stalk (Morrissey and Loomis, 1981), all eight isolates of which make up one complementation group defining the stalky locus, *stkA*. They form slugs with a normal pattern of differentiation, as measured by neutral red staining, PV distribution, UDP-galactose:polysaccharide transferase activity, and the synthesis of prestalk- and prespore-specific proteins (Morrissey *et al.*, 1981). These unexpected results indicate that although Tsang and Bradbury (1981) have shown that slug cell differentiation is tied to cell fate, events taking place during culmination are also important in determining the spore:stalk ratio. This is consistent with the finding that temperature shifts of wild-type strains during culmination produce small but measurable effects on cell proportions (Bonner and Slifkin, 1949; Farnsworth, 1975). A new temperature-sensitive mutant has been isolated which develops into a stalky-type fruiting body at the restrictive temperature (C. M. West, A. L. Lubniewski, and J. H. Gregg, personal communication). Temperature-shift experiments have indicated that this mutation exerts its effect just prior to, and during, culmination.

Approaching the problem from another direction, MacWilliams (1982) has isolated two mutants with altered cell proportions at the slug stage as visualized by neutral red staining. The mutant slugs contain less than 5% (short prestalk) or about 45% (long prestalk) stained cells, whereas the wild-type parent has about 15% stained cells. These strains form fruiting bodies containing 8% (short prestalk) or 42% (long prestalk) stalk by volume, indicating in these cases that fruiting body proportions seem to be tied to slug cell proportions. Volume measurements of spore:stalk ratios have been criticized as being unreliable in some cases (Bonner and Slifkin, 1949; Stenhouse and Williams, 1977) so confirmation of the alteration in fruiting body proportions using another technique would be useful here. The specific activity of marker enzymes are altered in these mutants in a manner consistent with the alteration in vital staining.

A significant finding of this study is that the slug proportioning mutants have altered organizer properties. By injecting small cell masses into slugs with a microsyringe, MacWilliams has shown that his short prestalk mutant is unusually sensitive to tip inhibition and behaves as if its slugs have lowered levels of tip inhibition as well. Slugs of the long prestalk mutant have an essentially unaltered level of tip inhibition at the midpoint of the slug, although the gradient of tip inhibition is about 2.5 times steeper than in wild type. This is the first experimental evidence to suggest a relationship between the prestalk/prespore pattern and tip dominance.

D. Other Factors That Alter Cell Proportions

There are a number of treatments that will alter the pattern of differentiation in slugs; in many cases these also affect the spore:stalk ratio. Forman and Garrod (1977a) tested the effects of temperature and growth conditions on the proportion of cells containing PVs and found that the proportion of such cells at 26°C was the same as for those developing at 22°C (the temperature customarily used for development), but lowering the temperature to 15°C caused a 20% decrease in their number. Bonner and Slifkin (1949) and Farnsworth (1975) showed that the higher the temperature of development, the greater the spore:stalk ratio, temperature shifts during culmination being as effective in altering the spore:stalk ratio as those during migration. Thus the effect of lower temperature on the proportion of PV-containing cells paralleled that of the spore:stalk ratio, but the effect of increased temperature did not. Axenically grown cells differ in their spore:stalk ratio depending upon the concentration of glucose in the growth medium: cells grown in the presence of 86 m*M* glucose form a higher proportion of spores than do those grown on unsupplemented medium (Garrod and Ashworth, 1972), and this also results in a slightly higher proportion of cells possessing PVs (Forman and Garrod, 1977a).

Chemical treatments have been used to alter cell proportions: lithium ions favor stalk production, whereas fluoride ions increase the spore:stalk ratio (Maeda, 1970). Lithium ions specifically block the accumulation of PVs by isolated anteriors, but do not block their loss in regulating posteriors (Sakai, 1973).

E. Other Species

Studies relating to spatial patterns of differentiation have been performed on related cellular slime molds including the *Dictyostelium* species *D. minutum*, *D. mucoroides* and *D. purpureum*, and the *Polysphondylium* species *P. pallidum* and *P. violaceum*, whose differentiation at the cellular level is very similar to that of *D. discoideum.* Morphogenesis in these species is similar to that of *D. discoideum,* except for two important differences. First, the slugs of all of these species produce stalk continuously during migration, so that they resemble *D. discoideum* sorogens. Second, the *Dictyostelium* species rarely produce fruiting bodies with side branches, whereas the *Polysphondylium* species form fruiting bodies composed of numerous side branches, arranged in whorls around the main stalk at regularly spaced intervals (see Bonner, 1967; Olive, 1975; and Chap. 10). Whorls of lateral sori are generated by groups of cells that pinch off from the rear of the sorogen during culmination, giving rise to doughnut shaped masses of cells. Several new tips appear from these rings of cells and secondary stalk tubes form that extend to and join with the main stalk. Spiegel and Cox (1980) examined fruiting bodies of haploid and diploid strains of *P. pallidum*—whose cells differ by a factor of two in size—and found that interwhorl distances were not affected by ploidy, indicating that the organism does not use cell counting to determine the distance between whorls. A plausible mechanism for this process is that the level of tip inhibition decreases during culmination such that at the low point of the gradient (the posteriormost region) the cells become independent of the tip, and construct tips of their own.

Little is known about tip dominance, pattern regulation, etc., in these species; comparative studies between *D. discoideum* and other slime mold species will doubtless aid a great deal in understanding the mechanisms of development. For example, in *D. discoideum* the pattern of differentiation at the slug stage is postulated to be directly related to terminal differentiation. Other slime mold species are, like *D. discoideum,* composed of only two cell types, stalk cells and spores, and they form slugs composed of cells with properties very similar to those of *D. discoideum* (e.g., possess PVs). However, the patterns of slug-specific and terminal differentiation are quite different in these species. This affords a chance to compare alterations in pattern to test whether or not the slug pattern is in fact directly related to spore and stalk differentiation.

The pattern of nonstarch PAS staining in *D. mucoroides* and *D. purpureum* is essentially the same as for *D. discoideum,* except that the anterior unstained region is slightly smaller (Bonner *et al.*, 1955); essentially the same is true for the pattern of staining with vital dyes. The volume ratio between stained and unstained cells remains unaltered even after extensive migration (Bonner, 1957), which means that posterior cells must be continuously redifferentiating (as measured by staining properties) to replace those that have entered the stalk tube. The distribution of antigens reactive with antispore serum in *D. mucoroides* follows the same pattern as for *D. discoideum* (Takeuchi, 1963).

In the related *Dictyostelium* species *D. minutum*, the two presumptive cell types can first be distinguished during late aggregation stage because the cytoplasm of prespore cells is more electron dense than that of prestalk cells (Schaap *et al.*, 1981). In contrast with other *Dictyostelium* species, all cells possess PVs, and no spatial separation between prespore and prestalk cells takes place until very late in fruiting body construction. A possibly related fact is that *D. minutum* does not form migrating slugs.

The pattern of differentiation observed in migrating slugs of *P. pallidum* and *P. violaceum* differs substantially from that of the *Dictyostelium* species studied thus far. All of the cells in the slugs of *P. pallidum* and *P. violaceum* are PAS positive (Bonner *et al.*, 1955). All but a tiny portion at the anterior of *P. pallidum* slugs contain PVs (Hohl *et al.*, 1977) and bind antispore serum (O'Day, 1979), suggesting that only one cell type exists in the slugs of this genus and that PVs disappear only as the cells actually differentiate into stalk. On the other hand, neutral red stains a very prominent tip in *P. pallidum* (O'Day, 1979). The explanation offered for this by O'Day was that the first signs of the constant conversion of prespore to prestalk cells in these slugs could be detected by neutral red staining, and that prespore cells lost their PVs more slowly, just prior to entering the stalk tube. However, side branches form from cells masses that have uniform distributions of PVs and do not have neutral red-stained tips. In this case prespore cells differentiate directly into stalk cells.

Stenhouse and Williams (1981) investigated cell patterning in *P. pallidum* by direct cell counting and found that although the overall spore:stalk ratio is very precisely controlled and is size-invariant, the proportions of the two cell types in side branches varies so extensively that they appear hardly to be regulated at all. When all the side branches in individual whorls are counted together they show a more precise spore:stalk ratio, although not as perfect as the ratio for the entire fruiting body. Stenhouse and Williams explain the poor correlation of the spore:stalk ratios in whorls by postulating that variable proportions of prespore and prestalk cells are dropped from the rear of the sorogen, and explain imprecision in the branches by postulating asynchrony in the morphogenetic movements of prespore and prestalk cells in forming the branches. However, prestalk cells have not been demonstrated in incipient *P. pallidum* side branches. Another possibility is that new tips form at random locations in the rings of cells left behind by the rising sorogen and these organize variable numbers of cells in their vicinities to form each side branch. Summing over an entire whorl would be expected to average out variations in the spore:stalk ratio, and summing over the whole fruiting body would further reduce the effects of random fluctuations in the proportions of individual side branches.

The slug pattern in *D. mucoroides* and *D. purpureum* is similar to that of

D. discoideum, and also parallels cell fate, while in the *Polysphondylium* species studied, observable differentiation at the slug stage is not patterned and does not foreshadow the complex pattern of terminal differentiation found in this genus. In *D. minutum*, there is no spatial pattern prior to fruiting body construction. It would be of interest to study regulation and the propensity toward spore and stalk differentiation in these species to see if these properties also differed from *D. discoideum*.

F. Alternative Roles for Prespore- and Prestalk-Specific Differentiation

The orthodox interpretation of the pattern of differentiation in slugs, particularly of PVs, is that it is the result of preparations that the cells undergo in anticipation of differentiation into spores and stalks. There are other functions of slug cells that also have a spatial distribution including organizing potential, production and maturation of the surface sheath, and control of the polarity of migration. It seems reasonable to conclude that special kinds of differentiation will be involved in these functions, and that they could account for some aspects of the observed pattern.

It has been asserted that there is a direct, causal relationship between possession of PVs and differentiation into spores (Forman and Garrod, 1977a), a conclusion based in part on the idea that PVs are the direct precursors of the spore coat. Hohl and Hamamoto (1969) and Takeuchi (1972) have published photographs showing PVs in juxtaposition to the surface of the cell, apparently fusing with the plasma membrane, and Muller and Hohl (1973) have determined that the average posterior cell contains 30–40 PVs, which they calculate have enough total surface area to exceed the surface area of the spore by 50%. In addition, PVs are known to contain material which cross-reacts with anti-spore serum (Ikeda and Takeuchi, 1971).

There are certain problems with this interpretation, however. Fusions between PVs and the plasma membrane are observed extremely rarely in ultrastructural studies of sporulating cells (P. Farnsworth, personal communication) and it must be kept in mind that apparent fusions of membranous structures can be generated artifactually. Slugs can be obtained that have fewer than one PV per cell, even through culmination, and yet such slugs form normal spores (Farnsworth and Loomis, 1976). Finally, the antispore serum also cross-reacts with the surface sheath of the slug (Tasaka and Takeuchi, 1979). This last point, together with the findings that the rate and duration of migration are correlated with alterations in the PV content of slugs (Sec. III,B), suggests a connection between PVs and the sheath. PVs may be storage depots of sheath precursors and their axial distribution may reflect a difference either in the rate of sheath secretion between prestalk and prespore cells, or in the chemical nature of the sheath material that these two cell types produce (i.e., the surface sheath surrounding prespore cells is considerable less fluid than that surrounding prestalk cells). It is possible that PVs function both in sheath production and the formation of the spore coat, and other processes, such as aging, can always be invoked to account for differences in sheath fluidity.

Hohl and Hamamoto (1969) have reported that during culmination the contents of PVs are released into the extracellular space, and Farnsworth and Loomis (1976) have proposed that the sole function of PVs may be to produce a matrix material to bind the spores together. However, since the sorus is essentially a droplet of liquid, it is likely that surface tension alone would suffice to hold it together. Another function for this extracellular material not previously considered is the dehydration of spores. During terminal differentiation, sporulating cells shrink two- to threefold in volume (Bonner and Frascella, 1952), whereas amoebae forming stalk cells swell by a factor of five (Raper and Fennel, 1952). Although direct transport of water by cells is thermodynamically possible, it is considered improbable; in most cases, fluid transport in cells occurs via solute transfer followed by passive movement of water (House, 1974). Transfer of proteins and carbohydrates along with proteases and hydrolases into the extracellular space could generate high concentrations of impermeant small molecules, causing the osmotic activity of the extracellular fluid to rise dramatically and resulting in the bulk movement of water out of sporulating cells. A similar mechanism might explain vacuolization of stalk cells, except that such substances would be expected to accumulate in the central vacuole of the cell, causing it to swell.

The prespore-specific enzyme UDPgalactose:polysaccharide transferase is thought to participate in the production of the mature acid mucopolysaccharide that appears to be associated with the spore coat. However, since the surface sheath is recognized by the antispore serum, it may contain a similar or identical mucopolysaccharide, which may also be produced by transferase.

Certain other kinds of differentiation patterned in slugs appear to have a more clear-cut relationship to terminal differentiation. For example, a group of proteins associated with the spore coat (Orlowski and Loomis, 1979) are not found in stalk cells, and during the slug stage are only synthesized by cells in the prespore zone (Devine *et al.*, 1982). Another example is the difference in trehalase activity between prestalk and prespore cells (Jefferson and Rutherford, 1976), which may play an important role in cell type-specific differentiation since spore cells accumulate trehalose and stalk cells do not. Even in this case, however, there are problems with a straightforward interpretation. In slugs, the relative specific activities of trehalase in prestalk and prespore cells differ by only a 1:2 ratio, and in fact no spacially patterned difference in trehalose levels can be detected in slugs (Rutherford and Jefferson, 1976). It is not until culmination that the specific activity of trehalase decreases to unmeasurable levels in incipient spore cells and increases some 40-fold in stalk cells (Jefferson and Rutherford, 1976), making it difficult to understand what the function of the 2-fold difference in activity in slug cells may be.

IV. Cell Sorting

A. Cell Sorting during Development

Although once a matter of dispute, the existence of cell sorting in *D. discoideum* development is now well established, but the details of when and how it happens are only partly known. Cell sorting was first demonstrated in this organism by Bonner (1952), who showed that vitally stained cell masses implanted heterotopically would return to their original positions in the slug within a few hours. Although initially criticized on the basis that vital dyes may diffuse from cell to cell (Farnsworth and Wolpert, 1971), this phenomenon has since been documented in several laboratories using a number of different techniques (see Takeuchi, 1969; Leach *et al.*, 1973; Feinberg *et al.*, 1979). In addition, it has been shown that vital dyes do not diffuse from cell to cell if appropriate cautions are taken (Sternfeld and David, 1981a).

The findings of grafting experiments such as Bonner's can be understood as a mechanism to rectify the pattern in slugs when cells are accidentally displaced, but is cell sorting involved in the formation of the pattern? The first indications that this may be so were obtained by Takeuchi (1969), who found that amoebae could be separated by buoyant density on dextrin gradients. When coaggregated, the cells from the heavy fraction were found to sort preferentially to the front of the migrating slug. A number of later studies have confirmed this finding with preaggregative cells and dissociated slugs. In the case of dissociated slug cells, the fractions correspond to authentic prespore and prestalk cells as measured by PV content, enzymatic activities, and developmental propensity (Bonner *et al.*, 1971; Maeda and Maeda, 1974; Feinberg *et al.*, 1979; Oohata and Takeuchi, 1977; Tsang and Bradbury, 1981; Ratner and Borth, 1982). These methods use different density-gradient media and obtain somewhat different results: in some cases the prestalk cells correspond to the heavy fraction, and in others, to the light fraction. Apparently the basis for the separation of the two cell types is not so much an intrinsic difference in their densities as much as it is a difference in their osmotic properties (Maeda and Maeda, 1974; Ratner and Borth, 1982).

It is perhaps not surprising that prestalk and prespore cells separate on density gradients since they differ by many properties. It is also not surprising that cells taken from late aggregates behave in the same way, since PVs can be detected in some cells by that stage (Hayashi and Takeuchi, 1976). What is surprising is that preaggregative cells can also be separated into populations of different sorting preference (Maeda and Maeda, 1974). Although preaggregative cells were found by Maeda and Maeda to differ in their chemotactic properties, no other differences in the fractions have been found that would correlate with prestalk or prespore differentiation, and in fact when cells of either fraction differentiate in isolation they form fruiting bodies of normal appearance.

Another indication that early events influence sorting preference may be found in the work of Leach *et al.* (1973), who showed that the conditions under which the amoebae are grown can affect sorting behavior. For example, when cells grown axenically on a medium containing added glucose (G cells) were coaggregated with cells grown on unsupplemented medium (NS cells), G cells were found to go preferentially into spores (see Fig. 4). In fact, by using four different growth conditions, these authors were able to establish a hierarchy of sorting preference, any pair when coaggregated giving results similar to those of the mixtures of G and NS cells. They interpret their results purely as a matter of cell sorting, for which they present some evidence. However, inspection of Fig. 4 shows that the linear part of the curve intersects at about 45% when extrapolated to the abscissa, indicating that the slugs of such mixtures must be 45% prestalk. If true, this would represent a remarkable effect of growth conditions on cell proportioning. Alternatively, some interaction between the cells in addition to cell sorting could be taking place.

Recently, variant strains (X9 and TS12, which are related genetically) have been identified that show "sorting preference" when co-aggregated with wild type, a result found even when both the wild type and variant strain were grown under identical conditions (Morrissey and Loomis, unpublished). These strains give results similar to, but more dramatic than, those in Fig. 4: when either X9 or TS12 are mixed with wild type, the wild-type cells go preferentially into spores, but the line extrapolates to the abscissa at about 70%. Furthermore, when X9 is coaggregated with TS12, there is no apparent sorting preference. Using microinjection experiments similar to those of MacWilliams (1982), it has been established that the interaction responsible for the bias in recovery into spores takes place before

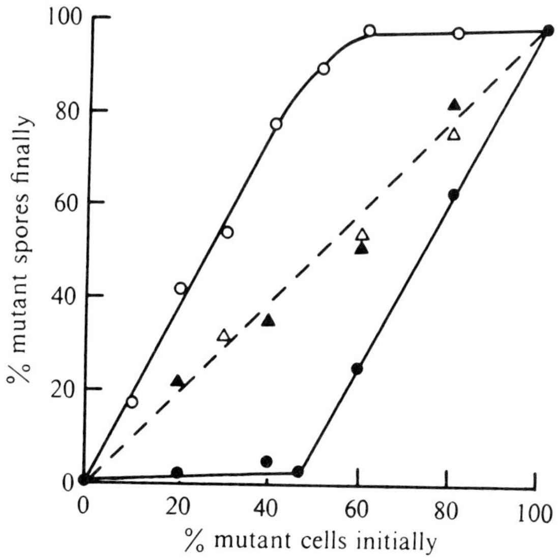

Fig. 4. Spores formed by mixtures of amoebae grown in different media. ○, glucose-grown mutant + NS-grown wild type; ●, NS-grown mutant + glucose-grown wild type; △, mutant and wild type glucose-grown; ▲, mutant and wild type NS-grown. The mutant used does not grow at 27°C, so the genotype of the resulting spores can be determined; otherwise it has no known effect on development. (Reproduced from Leach *et al.* (1973), with permission).

or during slug migration, but is no longer operative during culmination (Morrissey and Loomis, unpublished).

When G and NS cells differentiate separately, they produce fruiting bodies with different spore:stalk ratios, G cells being biased toward producing more spores (Garrod and Ashworth, 1972). This would predict that cell proportioning mutants should be altered in sorting properties; MacWilliams (1982) has found that his slug proportioning mutants are indeed altered in sorting preference, although not in an entirely straightforward manner. The long-prestalk mutant sorts preferentially to the prestalk zone and to the stalk when mixed with wild type. When the short-prestalk mutant is mixed with wild type on water–agar (which favors extensive slug migration), it sorts to the prespore zone, although on buffer soaked filter pads (which favor fruiting body formation without slug migration) the sorting propensity is reversed and the mutant cells go preferentially into the stalk. In sorting preference the short-prestalk and long-prestalk mutants resemble G and NS cells, respectively, although the sorting properties of the short-prestalk mutant appear to vary with ionic conditions or developmental time. It should be pointed out that the sorting preference discussed here refers to whether or not the mutant in question forms spores in mixed aggregation. This indirect measure does not rigorously establish the existence of physical sorting out, however.

The minimal conclusion that can be drawn from these results is that cell sorting can be involved in the formation of spatial distributions of cells in slugs, but of course whether it is an obligatory part of pattern formation in the unperturbed condition has not been demonstrated. Given that such strong sorting tendencies can be demonstrated, it does not seem too unlikely a possibility. The fact that vegetative growth conditions can affect sorting out, and the fact that hierarchies of sorting preference can be established indicates, as MacWilliams and Bonner (1979) have pointed out, that cells appear to sort out according to the value of a continuously variable parameter that is related to nutritional state. It is the relative value of this parameter compared to its value in other cells of the aggregate that appears to control sorting and ultimately, cell fate.

The lack of suitable prestalk markers has meant that prestalk cells have been identified by the absence of such prespore-specific differentiation as PVs. The inability to discriminate these cells from undifferentiated cells has been a hindrance to answering the question of whether or not prestalk/prespore differentiation preceeds pattern formation. If and when prestalk-specific markers (e.g., monoclonal antibodies to prestalk proteins) become available, it should be possible to see if prestalk and prespore cells are initially intermingled in the inchoate slug.

B. The Mechanism of Cell Sorting

Three basically different mechanisms for cell sorting have been proposed: (*a*) differential rates of migration in the slug, so that the faster cells end up in the front; (*b*) chemotaxis of prestalk cells toward each other; and (*c*) differential adhesion. The first proposal has now been shown to be untenable, since cell sorting occurs in submerged aggregates where migration cannot occur (Forman and Garrod, 1977b). Evidence in favor of both of the latter two explanations has been obtained.

Chemotaxis of prestalk cells to cAMP has been proposed by a number of workers as the mechanism of cell sorting in *D. discoideum;* arguments in its favor employ several lines of reasoning. When preaggregative cells are separated on density gradients, prestalk-prefering cells are found to have higher responsiveness to cAMP than does the prespore-prefering fraction (Maeda and Maeda, 1974) and G and NS cells have been shown to differ in their levels of cell-bound phosphodiesterase and cAMP receptors as well as chemotactic responsiveness (Inouye and Takeuchi, 1982). Another approach has been to observe differential chemotaxis directly in cell masses. Using mixtures of vitally stained prestalk cells and unstained prespore cells, Matsukuma and Durston (1979) have shown that prestalk cells will stream toward and accumulate near a source of cAMP. These authors point out that it is not necessary to postulate an inherent difference in chemotactic responsiveness between prestalk and prespore cells to account for chemotactic cell sorting, since if prestalk cells were less adhesive than prespore cells they would be freer to move through the tissue toward the source of cAMP.

Sternfeld and David (1981a) have devised a method for studying the induction of polarity in cell clumps embedded in agar, and they have also found that gradients of cAMP can cause pattern orientation (with prestalk cells moving to the high point of the gradient). They found that anterior-like cells will migrate toward cAMP only after the prestalk cells are removed, and further, by using thin agar barriers, obtained evidence consistent with the idea that prestalk cells secrete a diffusible inhibitor that blocks chemotaxis of anterior-like cells toward cAMP. The existence of such a substance was first hypothesized by Durston and Vork (1979).

The fact that gradients of cAMP can cause pattern orientation does not in itself mean that the mechanism of sorting out must involve differential chemotaxis, since such a gradient could have secondary effects on the cells which might be responsible for their orientation. This may also be the case in the report that oxygen gradients can cause pattern orientation (Sternfeld and David, 1981b). The evidence presented by Matsukuma and Durston (1979) showing directed cell movement in time-lapse cinematography argues strongly in favor of the ability of prestalk and prespore cells to undergo differential chemotaxis, however.

The arguments in favor of the involvement of differential adhesion in cell sorting run along two lines. The first type of evidence is that aggregates formed in suspension culture can differentiate into prespore and prestalk cells with a random distribution, and then subsequently segregate to give prestalk cells partially or wholly surrounding prespore cells (Forman and Garrod, 1977b; Tasaka and Takeuchi, 1981). This is consistent with the differential adhesion hypothesis of Steinberg (1970) and would indicate, if true, that prestalk cells are less cohesive than prespore cells. When slug cells are disaggregated and allowed to reaggregate in suspension, however, prestalk cells sort to the inside (Tasaka and Takeuchi, 1979). Lam *et al.* (1981) have attempted to measure the cohesive properties of prestalk and prespore cells from early culminants separated on Renografin gradients and have found that 30 m*M* EDTA blocked cohesion of prestalk, but not prespore cells. They further showed that Fab′ fragments of antibodies raised against the cell surface glycoprotein gp150 blocked cohesion of prespore cells to a greater degree than that of prestalk cells (see Chapter 6). These authors conclude that prespore cells exhibit stronger cohesion than do prestalk cells. However, the cohesive forces involved have not been measured, and it is far from clear what relationship, if any, exists between EDTA sensitivity and strength of cohesion.

The second type of evidence comes from experiments wherein cells at early aggregation are separated on density gradients, labeled with fluorescent dyes, and then comixed (Feinberg *et al.*, 1979). These mixtures eventually form slugs with differently colored prestalk and prespore zones, but earlier during streaming aggregation (actually reaggregation) the two types of cells are equally likely to be found in aggregation centers. Their sorting out seems to involve a patchy segregation within the streams that these authors take as evidence in favor of differential adhesion.

A reasonable conclusion from all of these studies is that there is evidence in favor of differential adhesion and chemotaxis, and as was previously pointed out (Matsukuma and Durston, 1979), both may function together during cell sorting.

V. Diffusible Substances That May Control Differentiation

A. Cyclic AMP

cAMP is the chemotactic agent (acrasin) for *D. discoideum* and is present both intra- and extracellularly throughout development (Bonner *et al.*, 1969; Malkinson and Ashworth, 1973). Indications of an axial inhomogeneity in cAMP concentration in slugs (Bonner, 1949; Pan *et al.*, 1974; Rubin, 1976) were confirmed by Brenner (1977), who showed that the anterior third of migrating slugs contains a concentration of cAMP approximately 50% higher than that in the posterior, and it has been suggested that a gradient in cAMP concentration in slugs is responsible for the pattern of differentiation and cell fate (see McMahon, 1973). High concentrations of cAMP can cause isolated wild-type amoebae to form a small number of stalk cells (Bonner, 1970), or amoebae of a susceptible variant (strain V12M2) to form essentially 100% stalk cells (Town *et al.*, 1976).

Is cAMP specific for stalk cell differentiation? This question has been investigated in several ways. Although strain V12M2 will form only stalk cells in the presence of cAMP *in vitro* in the system of Town *et al.* (1976), later studies with new mutants (sporogenous strains) indicate that cAMP potentiates prespore and spore differentiation as well as stalk cell formation

(Town *et al.*, 1976; Kay *et al.*, 1978). Two criticisms of these studies have been made: the effects require extremely high concentrations of cAMP (1–5 m*M*) and they are only seen with mutant strains. However, the same results can be obtained by using physiological concentrations (1 μM) of the non-hydrolyzable analog cAMP-S (Rossier *et al.*, 1978). Additionally, cAMP has recently been shown to induce both stalk and spore differentiation in wild-type cells. In experiments in which cAMP-soaked Sephadex beads were implanted into slugs, stalk cells and spores were found to differentiate locally (Feit *et al.*, 1978).

Although cAMP does not appear to be pathway specific, it is nevertheless thought to be important for later differentiation. Adding cAMP exogenously to cells may raise intracellular cAMP levels, which in turn may be necessary to induce and maintain postaggregative gene expression (see Kay, 1979, and Chapter 8).

B. Ammonia

Ammonia accumulates to significant levels during development and has been shown to have a profound effect on the transition from the migrating slug stage to culmination (Schindler and Sussman, 1977). Other evidence that ammonia plays an important role comes from the study of the mutant strain KY3, isolated originally by Yanagisawa *et al.* (1967). This mutant forms slugs that fail to culminate under normal circumstances, a phenotype like that of the "slugger" mutants reported recently (Sussman *et al.*, 1978). The phenotype of KY3, but not of the other slugger mutants, appears to result from an overly high sensitivity to ammonia (M. Sussman, personal communication, see Chapter 7). Sussman and co-workers have proposed that ammonia, acting antagonistically with cAMP, controls all morphogenetic movements after aggregation. An ingenious extension of these ideas allows the interplay of cAMP and ammonia to control the timing of terminal differentiation, and it has also been proposed that the spatial distribution of cAMP and ammonia in the slug may control the choice of slug cells between entering the prestalk or prespore pathways of differentiation (Sussman *et al.*, 1977; Sussman and Schindler, 1978; see Chapter 9).

As discussed above, there is little support for the notion that cAMP is pathway specific. The evidence for a pathway-specific effect of ammonia is equivocal. Sternfeld and David (1979) have studied the role of ammonia in differentiation of submerged clumps of cells. Previously, Sternfeld and Bonner (1977) had shown that mature spores and stalks would differentiate in submerged cultures only in an atmosphere of pure oxygen. The later study found that the density dependence for differentiation could be circumvented by using a conditioned medium, which was shown to contain ammonia and another factor. Ammonia was required for differentiation but resulted in both spores and stalk cells. In the absence of a pure oxygen atmosphere, ammonia was found to antagonize nonspecifically the induction of differentiation by cAMP in submerged cultures (Kay, 1979). However, using improved conditions for cell differentiation *in vitro*, Gross *et al*. (1981) have found that 2–5 m*M* ammonia dramatically shifts the spore:stalk ratio of sporogenous mutants in favor of spore differentiation. MacWilliams (1982) has found that mutations in the "slugger E" complementation group—identified by F. Ross and P. C. Newell (personal communication) as one of the loci where mutations cause slugs to be hypersensitive to inhibition of culmination by ammonia—cause alterations in the prespore/prestalk pattern such that it resembles that seen in his short-prestalk mutant. Also, ammonia appears to be one of the requirements for spore formation in a mutant exhibiting a decreased requirement for cell contact during differentiation (Wilcox and Sussman, 1978). Thus there appears to be suggestive, but far from conclusive, evidence for a role of ammonia in directing cells toward spore differentiation. Furthermore, considering the lack of specificity of cAMP in inducing later differentiation, it is unlikely that ammonia is exerting its effect through blocking cAMP production, as was originally suggested (Schindler and Sussman, 1977).

C. Differentiation-Inducing Factor

The variant strain V12M2 will differentiate into stalk cells when plated at high density on cAMP-containing agar (Town *et al.*, 1976). The density dependence for stalk formation was subsequently circumvented by returning to the medium a dialyzable substance (Differentiation-Inducing Factor, or DIF) secreted by cells at high density (Town and Stanford, 1979). The pathway specificity of DIF could not be ascertained initially since V12M2 will not form spores *in vitro*. Subsequent experiments with sporogenous mutants of V12M2 indicated that DIF was required for stalk cell induction at low cell density, but did not stimulate spore cell differentiation.

A complication in the early work on this system was that spore formation *in vitro* by sporogenous mutants, and prespore differentiation (as determined by possession of PVs and UDPgalactose:polysaccharide transferase activity) *in vitro* by V12M2 was extremely density dependent, possibly requiring cell contact (Kay *et al.*, 1979). Later work revealed that the apparent density dependence was an artifact of suboptimal conditions; using an improved assay of submerged cells in tissue culture dishes with an appropriate medium, spore formation by sporogenous mutants was found to be density independent (Kay and Trevan, 1981; Gross *et al.*, 1981). This in itself argues that DIF is not involved in prespore or spore differentiation, and in preliminary experiments DIF appears to be able to shift the spore:stalk ratio of sporogenous mutants differentiating *in vitro* toward stalk cell differentiation (Gross *et al.*, 1981). These latter experiments must be considered preliminary because DIF has only been partially purified, and the presence of contaminating inhibitors of differentiation cannot be ruled out.

It therefore appears reasonable to conclude that DIF is specific for stalk cell differentiation, but whether it is permissive or instructive is not yet clear. Progress on this question is currently being made. A quantitative assay for DIF has been developed, and the production of DIF has been found to be developmentally regulated. It is first detectable at about the time of tip formation, reaching peak levels during slug migration; *D. mucoroides* also produces DIF with similar kinetics (Brookman *et al.*, 1982). Mutants of V12M2 have been isolated that form stalk cells *in vitro* at high density only if exogenous DIF is added. These mutants were subsequently found not to produce DIF. Interestingly, they develop to the hemispherical mound stage, but do not form tips, and have been shown to contain the prespore marker enzyme UDPgalactose:polysaccharide transferase and to lack the prestalk-specific isozyme of acid phosphatase (W. Kopachik, A. Oohata, J. Brookman, and R. Kay, personal communication). This latter finding suggests that DIF is necessary for prestalk, but not prespore, differentiation, and further suggests a link between prestalk differentiation and tip formation.

VI. Theories of Cell Proportioning and Pattern Formation

A number of models and hypotheses have been put forth to explain pattern formation in *D. discoideum.* At the risk of doing an injustice, I have grouped them into two main categories. The first is a group of models in which the developmental pathway (prespore or prestalk) that a cell enters is dictated by the position that the cell occupies in the slug. Since most of these models were proposed several years ago, they have already been discussed in detail. For this reason, they will be described here in general terms, with the exception of a few recent developments which will be taken up in greater detail. In the second group of models, the cell types are proposed to differentiate at random within an aggregate and then sort out to form the spatial pattern seen in slugs. This idea has come into vogue very recently and so it will be given a more detailed consideration.

A. Models Wherein Position Dictates Fate

In most models of this type, positional information is generated by gradients in the concentration of diffusible substances (morphogens) to which the cells in the field are responsive. The type of differentiation exhibited by a cell is then dependent upon whether or not the concentration of the morphogen exceeds a threshold value. Typically, a localized source or sink (or both) is postulated to generate the gradient in morphogen concentration. In order to account for size invariance and pattern regulation, separate sources and sinks, multiple gradients, or specialized feedback systems usually must be invoked. An excellent exposition of the basic principles of the formation of pattern and the problems associated with their elucidation can be found in the work of Wolpert (1969, 1971), and detailed discussions of mathematical models of positional signaling pertinent to the *D. discoideum* slug have been published by MacWilliams and Bonner (1979) and by Meinhardt (1982).

The models in this group differ from each other chiefly by the mechanisms proposed to establish the gradient of morphogen concentration. Two models have been proposed that rely on unique properties of the surface sheath. Ashworth (1971) has noted that the sheath becomes progressively older toward the back of the slug and that positional information could be encoded in the degree of its maturation. Since only those cells near the surface of the slug are in contact with the sheath, a second mechanism must be invoked for the radial, but not axial, transmission of this information. Loomis (1975) has proposed a model wherein a gradient of morphogen concentration is generated by postulating that each cell synthesizes the substance and that it diffuses out of the slug only through the tip, where the sheath is thinnest (Farnsworth and Loomis, 1974). Sussman and Schindler (1978) have adapted this model to propose that the morphogen is ammonia, and that the resulting low concentration of ammonia in the tip would locally stimulate cAMP production. As discussed above, this model makes highly specific predictions regarding the effects of ammonia and cAMP on pathway-specific differentiation. Ammonia, at least under certain conditions, appears to have some pathway-specific effects, whereas cAMP is apparently pathway-indifferent.

McMahon (1973) proposed a model for the establishment of a gradient of cAMP wherein cell polarity precedes the establishment of the gradient. In his model, each cell stimuates the cell in front to increase the concentration of cAMP and the cell behind to decrease the concentration. The cells at the very front and back of the slug have no neighbors and thus become maxima and minima for cAMP production; eventually a region of high and a region of low cAMP concentration results. It is doubtful that cAMP can directly influence cell fate, although in a formal sense any other suitable small molecule can be substituted for cAMP in this model. MacWilliams and Bonner (1979) have pointed out that this model, as proposed, has no capacity for regulation.

In reaction–diffusion models, a gradient of a diffusible morphogen is generated by auto- and cross-catalysis of two or more chemical species. The classic model of pattern formation proposed by Turing (1952), has been shown to have little or no capacity for regulation and does not result in size invariance (Martinez, 1972; Bard and Lauder, 1974). However, Lacalli and Harrison (1978) and Othmer and Pate (1980) have adapted the Turing model to slime molds and have shown that by including a dependence of diffusivity on cell type or some parameter related to overall field size, such a model can be made to regulate and generate size-invariant patterns.

Gierer and Meinhardt (1972) have adapted Turing's ideas to include nonlinear reaction kinetics. One formulation of their general approach, termed the activator–inhibitor model, has been applied to the slug pattern of *D. discoideum* (MacWilliams and Bonner, 1979). In this model, an autocatalytic activator substance indirectly inhibits its own production by stimulating the production of an inhibitor. By varying the rates of diffusion and breakdown (both substances spontaneously decay), the inhibitor is made to have a longer diffusion range than the activator. At steady state, both the activator and inhibitor are produced at only one end of the slug (the front), the activator remaining close to its site of production and the inhibitor diffusing throughout. In a sense, most of the slug acts as a sink for the inhibitor. Removal of posterior portions reduces the size of the sink, causing the inhibitor concentration to rise and thus decreasing the size of the activated region, whereas removal of the anterior region removes the source of the inhibitor, whose concentration falls and allows residual activator to become autocatalytic. A model based on the ideas of Gierer and Meinhardt has been considered wherein extracellular DIF would correspond to the activator, and extracellular ammonia, the inhibitor (J. Gross, M. Peacy, J. Brookman, and R. Kay, personal communication).

Another model that has been proposed for control of the tip/nontip pattern is the followup–servo model of Wolpert *et al.* (1974; see Durston and Vork, 1977). In this model, a source and sink are postulated to exist at opposite ends of the slug. In addition, two gradients are proposed, one a labile gradient of a diffusing substance and another a gradient in a cellular property whose value follows that of the first gradient. Small perturbations in the primary gradient are followed by adjustments in the secondary one to bring it back into register; large perturbations cause divergence of the usually more stable secondary gradient away from the labile gradient, thus reestablishing boundaries if they are removed.

MacWilliams (1982) has discussed the model of Goodwin and Cohen (1969) involving periodic signaling by a pacemaker region (the tip) as applied to slime molds. In this model, two separate signals propagate at different rates from the pacemaker and the cells judge their distance from this point by the phase difference in the signals. Goodwin and Cohen hypothesize that size invariance and regulation are brought about by postulating a third substance, capable of modifying the phase shift, and whose production is itself tied to the phase difference that the last cell in the field should experience.

B. Models Wherein Fate Dictates Position

The models in the previous section involve positional signaling that controls position-dependent differentiation. *D. discoideum* slugs are well suited to these kinds of mathematical studies since they can be treated as unidimensional entities. However, prespore-specific differentiation (possession of PVs) can be detected by late aggregation stage and there is good evidence that prespore and prestalk cells initially differentiate in an apparently random manner throughout the aggregate (Forman and Garrod, 1977b; Tasaka and Takeuchi, 1981; see also Schaap *et al.*, 1981, for the related organism *D. minutum*). This does not mean, however, that cell–cell signaling does not occur, but on the contrary, it appears certain that such signaling is important in the regulation of cell proportions and probable that it is involved in its initial establishment. In this section, I will discuss various formulations of the idea that prespore and prestalk differentiation occurs prior to spatial pattern formation.

The original proposals of this idea were made by Takeuchi (1969) and by Forman and Garrod (1977b) who postulated that the two cell types differentiate at random and subsequently sort out to form the spatial pattern in slugs. At this point a second mechanism is invoked to rectify cell proportions and to account for regulation. This is unlikely because when prestalk and prespore cells are separated on density gradients and allowed to differentiate separately, each forms normal fruiting bodies. In the context of the model, either the second mechanism is *in itself* capable of establishing and regulating cell proportions and pattern (in which case the previous differentiation was superfluous) or the stochastic process of differentiation continues to take place after the cells are separated. In this case, it is very difficult to explain why the two fractions, when comixed, form such clear-cut patterns (Feinberg *et al.*, 1979). A more likely explanation is the alternative offered by Forman and Garrod (1977b), that cell–cell signaling controls cell proportions even before the spatial pattern is generated.

Some of the models of positional signaling described in the preceding section can be altered to give apparently random differentiation of cells in an aggregate. In the case of the reaction–diffusion model of Gierer and

Meinhardt (1972), if the activator is made to be nondiffusible, then many small areas of activation (corresponding to prestalk cells) are produced that could subsequently sort out to form the axial distribution seen in slugs (Meinhardt, 1982). Once generated, the pattern is stable. Slugs can easily exceed 2 mm in length, yet the pattern of differentiation is established very quickly during their formation; this combination of length and time poses a serious problem for the establishment of a single dominant region in models involving diffusible morphogens. Generation of multiple activated areas that sort out to give a single prestalk zone obviates this problem.

An ingeniously simple model of cell proportioning has been devised by Durston and Weijer (1980). They propose that each cell type secretes an inhibitor of its own type of differentiation; that is prestalk cells secrete a substance that blocks prespore → prestalk conversion, and prespore cells secrete a substance with the opposite specificity. A surprising aspect of the hypothesis is that none of the morphogenetically active substances need have a spatial localization. This model explains the results of experiments that many workers have done but none have published: that adding cell extracts to aggregates or slugs does not alter cell proportions. The reason would be that in intact aggregates both cell types would be maximally inhibited, and so adding more inhibitor to an already suprathreshold condition would have no effect. However, isolated prestalk or prespore cell masses undergoing regulation would be expected to be sensitive to the inhibitor. Using the submerged culture system of Sternfeld and David (1979), Weijer and Durston (personal communication) have found that cAMP will block the transition to prestalk cells in prespore isolates, but will not alter regulation of prestalk isolates, indicating that cAMP would be the prespore → prestalk inhibitor, if this model were correct.

According to the findings of Sternfeld and David (1981a), regulation in prespore isolates involves position-independent differentiation followed by sorting. Whether or not regulation in prestalk isolates involves position-dependent differentiation is not known. Although new prespore cells always appear first at the posterior margin of the prestalk isolate (Bonner, 1952; Bonner *et al.*, 1955; Gregg and Karp, 1978; Sakai, 1973), it is possible that the precursors to the new prespore cells appear at random throughout the cell mass and then sort to the back, where they continue differentiation to the point where they can be detected.

Finally, in the initial aggregate, why do prestalk cells collect in one zone, rather than sorting-out into several areas? In fact multiple tips form in very large aggregates, but why does only a single tip form in aggregates of 10^5 or fewer cells? It is well known that when the organization of a slug or aggregate is randomized by being scrambled with a fine needle, multiple tips are regenerated. I think the answer lies in the fact that the tip forms from the initial aggregation center. It is this center that initiates the pulses of cAMP during aggregation that entrain the amoebae surrounding it, and it probably serves as a unique attraction point for cell sorting. Similarly, when a slug is transected, its inherent polarity and the constraints of the sheath make for a single focal point for sorting.

VII. Summary

A large number of investigations into the control of cell proportioning and pattern formation in *D. discoideum* have been undertaken. Their findings have resolved some controversies and generated new ones. However, the number of aspects of the problem about which a consensus may be reached has been enlarged considerably. A general summary of the facts that can be agreed upon is given below.

There is a spatial separation of the presumptive cell types at the migrating slug stage. Prespore and prestalk cells are not irreversibly committed to spore and stalk differentiation since cut slugs can regulate to give normal cell proportions. However, the conversion of prestalk cells to prespore cells is considerably slower than conversion in the opposite direction. The latter process probably involves the sorting out of a ready supply of prestalk-like cells dispersed throughout the prespore zone. Whether regulation of prestalk isolates involves position-dependent redifferentiation or random redifferentiation followed by cell sorting is not known.

Prestalk and prespore cells are differentiated from one another in slugs with respect to ultrastructural, biochemical and other criteria. However, the functions of the various types of spatially patterned differentiation are a matter of question; some have been proposed to be related to cell fate, organizing activity, the polarity of migration or the production and maturation of the surface sheath. The prestalk/prespore pattern is not always obligatorily tied to cell fate since mutants that differentiate solely into stalk cells nonetheless have a normal slug pattern, although mutational analysis of another kind has shown that prestalk/prespore cell proportioning, cell sorting, and tip dominance are all interrelated. In the related genus *Polysphondylium*, prestalk differentiation is frequently not found. Instead, prespore cells differentiate directly into stalk cells as well as spores.

The prestalk/prespore pattern in slugs may come about via random differentiation of the two cell types followed by sorting out. Sorting out of anterior-like cells is thought to be inhibited in intact slugs by a diffusible substance released by prestalk cells.

Cell proportions are likely to be controlled by cell–cell signaling involving diffusible substances. Ammonia, DIF, and cAMP have been proposed to be such signals. Although pathway specificity has been demonstrated for the action of ammonia and DIF, there is little evidence of this kind for cAMP. However, as Durston has pointed out, in certain models cell type-specific, morphogenetically active substances would not show pathway specificity except under special assay conditions.

Considering both the amount of progress recently made and the amount of interest in the field, the future prospects of understanding the control of cell proportioning and the formation of pattern in *D. discoideum* seem reasonably good. There are several very promising avenues of approach. One example is the technique of injecting small numbers of test cells into slugs or sorogens to test quantitatively such processes as organizing activity and developmental propensity. Another example is the use of *in vitro* assay systems to test for morphogenetic substances. A third is the isolation and characterization of mutants altered in patterning. A number of models have been proposed to account for cell proportioning and pattern formation in *D. discoideum*, and each has some data that appear to support it. The system is much more complex than a single enzymatic reaction or other cellular process that may be studied in isolation, and for this reason no one decisive experiment to prove a model can be done. Rather, an understanding of the processes of cell proportioning and pattern formation will only come from the synthesis of a large body of findings from diverse approaches.

Acknowledgments

I would like to thank several of my colleagues for critical reading of the manuscript, and in particular I would like to thank those, too numerous to list here, who have so kindly provided me with their findings prior to publication.

References

Alton, T. H., and Brenner, M. (1979). Comparison of proteins synthesized by anterior and posterior regions of *Dictyostelium discoideum* pseudoplasmodia. *Dev. Biol.* **71,** 1–7.

Ashworth, J. M. (1971). Cell development in the cellular slime mould *Dictyostelium discoideum*. *Symp. Soc. Exp. Biol.* **25,** 27–49.

Ashworth, J. M., and Sussman, M. (1967). The appearance and disappearance of uridine diphosphate glucose pyrophosphorylase activity during differentiation of the cellular slime mold *Dictyostelium discoideum*. *J. Biol. Chem.* **242,** 1696–1700.

Bard, J., and Lauder, I. (1974). How well does Turing's theory of morphogenesis work? *J. Theoret. Biol.* **45,** 501–531.

Bonner, J. T. (1949). The demonstration of acrasin in the later stages of the development of the slime mold *Dictyostelium discoideum. J. Exp. Zool.* **110,** 259–271.

Bonner, J. T. (1952). The pattern of differentiation in ameboid slime molds. *Amer. Nat.* **86,** 79–89.

Bonner, J. T. (1957). A theory of the control of differentiation in the cellular slime molds. *Quart. Rev. Biol.* **32,** 232–246.

Bonner, J. T. (1967). "The Cellular Slime Molds," 2nd ed. Princeton University Press, Princeton, New Jersey.

Bonner, J. T. (1970). Induction of stalk cell differentiation by cyclic AMP in the cellular slime mold *Dictyostelium discoideum. Proc. Natl. Acad. Sci. U.S.A.* **65,** 110–113.

Bonner, J. T., Barkley, D. S., Hall, E. M., Konijn, T. M., Mason, J. W., O'Keefe, G., and Wolfe, P. B. (1969). Acrasin, acrasinase, and the sensitivity to acrasin in *Dictyostelium discoideum. Dev. Biol.* **20,** 72–87.

Bonner, J. T., Chiquoine, A. D., and Kolderie, M. Q. (1955). A histochemical study of differentiation in the cellular slime molds. *J. Exp. Zool.* **130,** 133–158.

Bonner, J. T., and Frascella, E. B. (1952). Variations in cell size during the development of the slime mold, *Dictyostelium discoideum. J. Exp. Zool.* **121,** 561–571.

Bonner, J. T., Sieja, T. W., and Hall, E. M. (1971). Further evidence for the sorting out of cells in the differentiation of the cellular slime mold *Dictyostelium discoideum. J. Embryol. Exp. Morphol.* **25,** 457–465.

Bonner, J. T., and Slifkin, M. K. (1949). A study of the control of differentiation: The proportions of stalk and spore cells in the slime mold *Dictyostelium discoideum. Amer. J. Bot.* **36,** 727–734.

Brenner, M. (1977). Cyclic AMP gradient in migrating pseudoplasmodia of the cellular slime mold *Dictyostelium discoideum. J. Biol. Chem.* **252,** 4073–4077.

Brookman, J. J., Town, C. D., Jermyn, K. A., and Kay, R. R. (1982). Developmental regulation of a stalk cell differentiation-inducing factor in *Dictyostelium discoideum. Dev. Biol.* **91,** 191–196.

Brown, S. S., and Rutherford, C. L. (1980). Localization of cyclic nucleotide phosphodiesterase in the multicellular stages of *Dictyostelium discoideum. Differentiation* **16,** 173–183.

Chia, W. K. (1975). Induction of stalk cell differentiation by cyclic AMP in a susceptible variant of *Dictyostelium discoideum. Dev. Biol.* **44,** 239–252.

Coloma, A., and Lodish, H. F. (1981). Synthesis of spore- and stalk-specific proteins during differentiation of *Dictyostelium discoideum. Dev. Biol.* **81,** 238–244.

Devine, K., Morrissey, J. H., and Loomis, W. F. (1982). Differential synthesis of spore coat proteins in prespore and prestalk cells of *Dictyostelium. Proc. Natl. Acad. Sci. U.S.A.* In press.

Durston, A. J. (1976). Tip formation is regulated by an inhibitory gradient in the *Dictyostelium discoideum* slug. *Nature (London)* **263,** 126–129.

Durston, A. J., and Vork, F. (1977). The control of morphogenesis and pattern in the *Dictyostelium discoideum* slug. *In* "Development and Differentiation in the Cellular Slime Moulds" (P. Cappuccinelli and J. M. Ashworth, eds.), pp. 17–26. Elsevier, New York.

Durston, A. J., and Vork, F. (1978). The spatial pattern of DNA synthesis in *Dictyostelium discoideum* slugs. *Exp. Cell Res.* **115,** 454–457.

Durston, A. J., and Vork, F. (1979). A cinematographical study of the development of vitally stained *Dictyostelium discoideum. J. Cell Sci.* **36,** 261–279.

Durston, A. J., and Weijer, C. J. (1980). *Dictyostelium discoideum:* Een model systeem voor de embryonale ontwikkeling. *Vakblad voor Biologen* **60,** 320–327.

Farnsworth, P. (1973). Morphogenesis in the cellular slime mold *Dictyostelium discoideum:* The formation and regulation of aggregation tips and the specification of developmental axes. *J. Embryol. Exp. Morphol.* **29,** 253–266.

Farnsworth, P. (1974). Experimentally induced aberrations in the pattern of differentiation in the cellular slime mould *Dictyostelium discoideum. J. Embryol. Exp. Morphol.* **31,** 435–451.

Farnsworth, P. A. (1975). Proportionality in the pattern of differentiation of the cellular slime mould *Dictyostelium discoideum* and the time of its determination. *J. Embryol. Exp. Morphol.* **33,** 869–877.

Farnsworth, P. A., and Loomis, W. F. (1974). A barrier to diffusion in pseudoplasmodia of *Dictyostelium discoideum. Dev. Biol.* **41,** 77–83.

Farnsworth, P. A., and Loomis, W. F. (1975). A gradient in the thickness of the surface sheath in pseudoplasmodia of *Dictyostelium discoideum. Dev. Biol.* **46,** 349–357.

Farnsworth, P. A., and Loomis, W. F. (1976). Quantitation of the spatial distribution of "prespore vacuoles" in pseudoplasmodia of *Dictyostelium discoideum. J. Embryol. Exp. Morphol.* **35,** 499–505.

Farnsworth, P., and Wolpert, L. (1971). Absence of cell sorting out in the grex of the slime mold *Dictyostelium discoideum. Nature (London)* **231,** 329–330.

Feinberg, A., Springer, W. R., and Barondes, S. H. (1979). Segregation of pre-stalk and prespore cells of *Dictyostelium discoideum:* Observations consistent with selective cell cohesion. *Proc. Natl. Acad. Sci. U.S.A.* **76,** 3977–3981.

Feit, I. N., Fournier, G. A., Needleman, R. D., and Underwood, N. Z. (1978). Induction of stalk and spore differentiation by cyclic AMP in slugs of *Dictyostelium discoideum. Science* **200,** 439–441.

Fong, D., and Rutherford, C. L. (1978). Protease activity during cell differentiation of the cellular slime mold *Dictyostelium discoideum. J. Bacteriol.* **134,** 521–527.

Forman, D., and Garrod, D. R. (1977a). Pattern formation in *Dictyostelium discoideum.* I. Development of prespore cells and its relationship to the pattern of the fruiting body. *J. Embryol. Exp. Morphol.* **40,** 215–228.

Forman, D., and Garrod, D. R. (1977b). Pattern formation in *Dictyostelium discoideum.* II. Differentiation and pattern formation in nonpolar aggregates. *J. Embryol. Exp. Morphol.* **40,** 229–243.

Francis, D. W. (1962). Movement of pseudoplasmodia of *Dictyostelium discoideum.* Ph.D. thesis, Univ. Wisconsin, Madison, Wisconsin.

Garrod, D. R., and Ashworth, J. M. (1972). Effect of growth conditions on development of the cellular slime mould, *Dictyostelium discoideum. J. Embryol. Exp. Morphol.* **28,** 463–479.

Gierer, A., and Meinhardt, H. (1972). A theory of biological pattern formation. *Kybernetik* **12,** 30–39.

Goodwin, B. C., and Cohen, M. H. (1969). A phase-shift model for the spatial and temporal organization of developing systems. *J. Theoret. Biol.* **25,** 49–107.

Gregg, J. H., and Badman, W. S. (1970). Morphogenesis and ultrastructure in *Dictyostelium. Dev. Biol.* **22,** 96–111.

Gregg, J. H., and Karp, G. C. (1978). Patterns of cell differentiation revealed by L-[^{3}H]-fucose incorporation in *Dictyostelium. Exp. Cell Res.* **112,** 31–46.

Gross, J. D., Town, C. D., Brookman, J. J., Jermyn, K. A., Peacey, M. J., and Kay, R. R. (1981). Cell patterning in *Dictyostelium. Phil. Trans. R. Soc. Lond.* **B295,** 497–508.

Hayashi, M., and Takeuchi, I. (1976). Quantitative studies on cell differentiation during morphogenesis of the cellular slime mold *Dictyostelium discoideum. Develop. Biol.* **50,** 302–309.

Hohl, H. R., and Hamamoto, S. T. (1969). Ultrastructure of spore differentiation in *Dictyostelium:* The prespore vacuole. *J. Ultrastr. Res.* **26**, 442–453.

Hohl, H. R., Honegger, R., Traub, F., and Markwalder, M. (1977). Influence of cAMP on cell differentiation and morphogenesis in *Polysphondylium. In* "Development and Differentiation in the Cellular Slime Moulds" (P. Cappuccinelli and J. M. Ashworth, eds.), pp. 149–172. Elsevier, New York.

House, C. R. (1974). "Water Transport in Cells and Tissues." Arnold, London.

Ikeda, T. (1981). Subcellular distributions of UDP-galactose:polysaccharide transferase and UDP-glucose pyrophosphorylase involved in biosynthesis of prespore-specific acid mycopolysaccharide in *Dictyostelium discoideum. Biochim. Biophys. Acta* **675**, 69–76.

Ikeda, T., and Takeuchi, I. (1971). Isolation and characterization of a prespore specific structure of the cellular slime mold, *Dictyostelium discoideum. Develop. Growth. Differ.* **13**, 221–229.

Inouye, K., and Takeuchi, I. (1979). Analytical studies on migrating movement of the pseudoplasmodium of *Dictyostelium discoideum. Protoplasma* **99**, 289–304.

Inouye, K., and Takeuchi, I. (1982). Correlation between prestalk–prespore tendencies and cyclic AMP-related activities in *Dictyostelium discoideum. Exp. Cell Res.* **138**, 311–318.

Jefferson, B. L., and Rutherford, C. L. (1976). A stalk-specific localization of trehalase activity in *Dictyostelium discoideum. Exp. Cell Res.* **103**, 127–134.

Kay, R. R. (1979). Gene expression in *Dictyostelium discoideum:* Mutually antagonistic roles of cyclic-AMP and ammonia. *J. Embryol. Exp. Morph.* **52**, 171–182.

Kay, R. R., Garrod, D., and Tilly, R. (1978). Requirements for cell differentiation in *Dictyostelium discoideum. Nature (London)* **271**, 58–60.

Kay, R. R., and Trevan, D. J. (1981). *Dictyostelium* amoebae can differentiate into spores without cell-to-cell contact. *J. Embryol. Exp. Morphol.* **62**, 369–378.

Kay, R. R., Town, C. D., and Gross, J. D. (1979). Cell differentiation in *Dictyostelium discoideum. Differentiation* **13**, 7–14.

Lacalli, T. C., and Harrison, L. G. (1978). The regulatory capacity of Turing's model for morphogenesis with applications to slime moulds. *J. Theoret. Biol.* **70**, 273–295.

Lam, T. Y., Pickering, G., Geltosky, J., and Siu, C. H. (1981). Differential cell cohesiveness expressed by prespore and prestalk cells of *Dictyostelium discoideum. Differentiation.* **20**, 22–28.

Leach, C., Ashworth, J. M., and Garrod, D. (1973). Cell sorting out during the differentiation of mixtures of metabolically distinct populations of *Dictyostelium discoideum. J. Embryol. Exp. Morphol.* **29**, 647–661.

Loomis, W. F. (1972). Role of the surface sheath in the control of morphogenesis in *Dictyostelium discoideum. Nature (London) New Biology* **240**, 6–9.

Loomis, W. F. (1975). Polarity and pattern in *Dictyostelium. In* "Proceedings of the 1975 ICN-UCLA Symposium on Developmental Biology" (D. McMahon and C. F. Fox, eds.), pp. 109–128. Benjamin, Menlo Park, Calif.

MacWilliams, H. K. (1982). Transplantation experiments and pattern mutants in cellular slime mold slugs. *In* "Developmental Order: Its Origin and Regulation." Alan R. Liss, New York. In press.

MacWilliams, H. K., and Bonner, J. T. (1979). The prestalk-prespore pattern in cellular slime molds. *Differentiation* **14**, 1–22.

MacWilliams, H. K., and Kafatos, F. C. (1974). The basal inhibition in *Hydra* may be mediated by a diffusing substance. *Amer. Zool.* **14**, 633–645.

MacWilliams, H. K., Kafatos, F. C., and Bossert, W. H. (1970). The feedback inhibition of basal disk regeneration in *Hydra* has a continuously variable intensity. *Dev. Biol.* **23**, 380–398.

Maeda, Y. (1970). Influence of ionic conditions on cell differentiation and morphogenesis of the cellular slime molds. *Develop. Growth Differ.* **12,** 217–227.
Maeda, Y., and Maeda, M. (1974). Heterogeneity of the cell population of the cellular slime mold *Dictyostelium discoideum* before aggregation, and its relation to the subsequent locations of the cells. *Exp. Cell Res.* **84,** 88–94.
Malkinson, A. M., and Ashworth, J. M. (1973). Adenosine 3′5′ cyclic monophosphate concentrations and phosphodiesterase activities during axenic growth and differentiation of cells of the cellular slime mould *Dictyostelium discoideum. Biochem. J.* **134,** 311–319.
Martinez, H. M. (1972). Morphogenesis and chemical dissipative structures. A computer simulated case study. *J. Theoret. Biol.* **36,** 479–501.
Matsukuma, S., and Durston, A. J. (1979). Chemotactic cell sorting in *Dictyostelium discoideum. J. Embryol. Exp. Morphol.* **50,** 243–251.
McMahon, D. (1973). A cell-contact model for cellular position determination in development. *Proc. Natl. Acad. Sci. U.S.A.* **70,** 2396–2400.
Meinhardt, H. (1982). "Biological Pattern Formation." Academic Press, New York. In press.
Morrissey, J. H., Farnsworth, P. A., and Loomis, W. F. (1981). Pattern formation in *Dictyostelium discoideum:* An analysis of mutants altered in cell proportioning. *Develop. Biol.* **83,** 1–8.
Morrissey, J. H., and Loomis, W. F. (1981). Parasexual genetic analysis of cell proportioning mutants of *Dictyostelium discoideum. Genetics.* **99,** 183–196.
Muller, U., and Hohl, H. R. (1973). Pattern formation in *Dictyostelium discoideum:* Temporal and spatial distribution of prespore vacuoles. *Differentiation* **1,** 267–276.
Newell, P. C., Ellingson, J. S., and Sussman, M. (1969). Synchrony of enzyme accumulation in a population of differentiating slime mold cells. *Biochem. Biophys. Acta* **177,** 610–614.
O'Day, D. H. (1979). Cell differentiation during fruiting body formation in *Polysphondylium pallidum. J. Cell Sci.* **35,** 203–215.
Olive, L. S. (1975). "The Mycetozoans." Academic Press, New York.
Oohata, A., and Takeuchi, I. (1977). Separation and biochemical characterization of the two cell types present in the pseudoplasmodium of *Dictyostelium mucoroides. J. Cell Sci.* **24,** 1–9.
Orlowski, M., and Loomis, W. F. (1979). Plasma membrane proteins of *Dictyostelium:* The spore coat proteins. *Dev. Biol.* **71,** 297–307.
Othmer, H. G., and Pate, E. (1980). Scale-invariance in reaction–diffusion models of spatial pattern formation. *Proc. Natl. Acad. Sci. U.S.A.* **77,** 4180–4184.
Pan, P., Bonner, J. T., Wedner, H. J., and Parker, C. W. (1974). Immunofluorescence evidence for the distribution of cyclic AMP in cells and masses of the cellular slime molds. *Proc. Natl. Acad. Sci. U.S.A.* **71,** 1623–1625.
Poff, K. L., and Loomis, W. F. (1973). Control of phototactic migration in *Dictyostelium discoideum. Exp. Cell Res.* **82,** 236–240.
Raper, K. B. (1940). Pseudoplasmodium formation and organization in *Dictyostelium discoideum. J. E. Mitchell Sci. Soc.* **56,** 241–282.
Raper, K. B., and Fennell, D. (1952). Stalk formation in *Dictyostelium. Bull. Torrey Botan. Club,* **79,** 25–51.
Ratner, D., and Borth, W. (1982). Comparison of differentiating *Dictyostelium discoideum* cell types separated by an improved method of density gradient centrifugation. Submitted.
Rossier, C., Gerisch, G., Malchow, D., and Eckstein, F. (1979). Action of a slowly hydrolysable cyclic AMP analogue on developing cells of *Dictyostelium discoideum. J. Cell Sci.* **35,** 321–338.
Rubin, J. (1976). The signal from fruiting body and conus tips of *Dictyostelium discoideum. J. Embryol. Exp. Morphol.* **36,** 261–271.

Rubin, J., and Robertson, A. (1975). The tip of the *Dictyostelium discoideum* pseudoplasmodium as an organizer. *J. Embryol. Exp. Morphol.* **33**, 227–241.
Rutherford, C. L., and Harris, J. F. (1976). Localization of glycogen phosphorylase in specific cell types during differentiation of *Dictyostelium discoideum. Arch. Biochem. Biophys.* **175**, 453–462.
Rutherford, C. L., and Jefferson, B. L. (1976). Trehalose accumulation in stalk and spore cells of *Dictyostelium discoideum. Dev. Biol.* **52**, 52–60.
Sakai, Y. (1973). Cell type conversion in isolated prestalk and prespore fragments of the cellular slime mold, *Dictyostelium discoideum. Develop. Growth Differ.* **15**, 11–19.
Sampson, J. (1976). Cell patterning in migrating slugs of *Dictyostelium discoideum. J. Embryol. Exp. Morphol.* **36**, 663–668.
Schaap, P., van der Molen, L., and Konijn, T. M. (1981). Development of the simple cellular slime mold *Dictyostelium minitum. Dev. Biol.* **85**, 171–179.
Schindler, J., and Sussman, M. (1977). Ammonia determines the choice of morphogenetic pathways in *Dictyostelium discoideum. J. Mol. Biol.* **116**, 161–169.
Shaffer, B. M. (1962). The Acrasina. *Advan. Morphog.* **2**, 109–182.
Shaffer, B. M. (1964). The Acrasina. *Advan. Morphog.* **3**, 301–322.
Spemann, H. (1938). "Embryonic Development and Induction." Yale University Press, New Haven, Connecticut.
Spiegel, F. W., and Cox, E. C. (1980). A one-dimensional pattern in the cellular slime mold *Polysphondylium pallidum. Nature (London)* **286**, 806–807.
Steinberg, M. S. (1970). Does differential adhesion govern self-assembly in histogenesis? Equilibrium configurations and the emergence of a hierarchy among populations of embryonic cells. *J. Exp. Zool.* **173**, 395–433.
Stenhouse, F. O., and Williams, K. L. (1977). Patterning in *Dictyostelium discoideum:* The proportions of the three differentiated cell types (spore, stalk, and basal disk) in the fruiting body. *Dev. Biol.* **59**, 140–152.
Stenhouse, F. O., and Williams, K. L. (1981). Cell patterning in branched and unbranched fruiting bodies of the cellular slime mold *Polysphondylium pallidum. Develop. Biol.* **81**, 139–144.
Sternfeld, J., and Bonner, J. T. (1977). Cell differentiation in *Dictyostelium* under submerged conditions. *Proc. Natl. Acad. Sci. U.S.A.* **74**, 268–271.
Sternfeld, J., and David, C. N. (1979). Ammonia plus another factor are necessary for differentiation in submerged clumps of *Dictyostelium. J. Cell Sci.* **38**, 181–191.
Sternfeld, J., and David, C. N. (1981a). Cell sorting during pattern formation in *Dictyostelium. Differentiation* **20**, 10–21.
Sternfeld, J., and David, C. N. (1981b). Oxygen gradients cause pattern orientation in *Dictyostelium* cell clumps. *J. Cell Sci.* **50**, 9–18.
Sussman, M., and Osborn, M. J. (1964). UDP-galactose polysaccharide transferase in the cellular slime mold, *Dictyostelium discoideum:* Appearance and disappearance of activity during cell differentiation. *Proc. Natl. Acad. Sci. U.S.A.* **52**, 81–87.
Sussman, M., and Schindler, J. (1978). A possible mechanism of morphogenetic regulation in *Dictyostelium discoideum. Differentiation* **10**, 1–5.
Sussman, M., Schinder, J., and Kim, H. (1977). Toward a biochemical definition of the morphogenetic fields in *D. discoideum. In* "Development and Differentiation in the Cellular Slime Moulds" (P. Cappuccinelli and J. M. Ashworth, eds.), pp. 31–50. Elsevier, New York.
Sussman, M., Schindler, J., and Kim, H. (1978). "Sluggers," a new class of morphogenetic mutants of *D. discoideum. Exp. Cell Res.* **116**, 217–227.
Takeuchi, I. (1963). Immunochemical and immunohistochemical studies on the development of the cellular slime mold *Dictyostelium mucoroides. Dev. Biol.* **8**, 1–26.

Takeuchi, I. (1969). Establishment of polar organization during slime mold development. *In* "Nucleic Acid Metabolism Cell Differentiation and Cancer Growth" (E. V. Coudry and S. Seno, eds.), pp. 297–304. Pergammon Press, Oxford.
Takeuchi, I. (1972). Differentiation and dedifferentiation in cellular slime molds. *In* "Aspects of Cellular and Molecular Physiology" (K. Hamaguchi, ed.), pp. 217–236. University of Tokyo Press, Tokyo.
Takeuchi, I., Hayashi, M., and Tasaka, M. (1977). Cell differentiation and pattern formation in *Dictyostelium*. *In* "Development and Differentiation in the Cellular Slime Moulds" (P. Cappuccinelli and J. M. Ashworth, eds.), pp. 1–16, Elsevier, New York.
Tasaka, M., and Takeuchi, I. (1979). Sorting out behavior of disaggregated cells in the absence of morphogenesis in *Dictyostelium discoideum*. *J. Embryol. Exp. Morphol.* **49,** 89–102.
Tasaka, M., and Takeuchi, I. (1981). Role of cell sorting in pattern formation in *Dictyostelium discoideum*. *Differentiation* **18,** 191–196.
Town, C. D., Gross, J. D., and Kay, R. R. (1976). Cell differentiation without morphogenesis in *Dictyostelium discoideum*. *Nature (London)* **262,** 717–719.
Town, C., and Stanford, E. (1979). An oligosaccharide-containing factor that induces cell differentiation in *Dictyostelium discoideum*. *Proc. Natl. Acad. Sci. U.S.A.* **76,** 308–312.
Tsang, A., and Bradbury, J. M. (1981). Separation and properties of prestalk and prespore cells of *Dictyostelium discoideum*. *Exp. Cell Res.* **132,** 433–441.
Turing, A. M. (1952). The chemical basis of morphogenesis. *Phil. Trans. R. Soc. Lond.* **B237,** 37–72.
West, C. M., and McMahon, D. (1979). The axial distribution of plasma membrane molecules in pseudoplasmodia of the cellular slime mold *Dictyostelium discoideum*. *Exp. Cell Res.* **124,** 393–401.
Wilcox, D. K., and Sussman, M. (1978). Spore differentiation by isolated *Dictyostelium discoideum* cells, triggered by prior cell contact. *Differentiation* **11,** 125–131.
Wolpert, L. (1969). Positional information and the spatial pattern of cellular differentiation. *J. Theoret. Biol.* **25,** 1–47.
Wolpert, L. (1971). Positional information and pattern formation. *Curr. Top. Develop. Biol.* **6,** 183–222.
Wolpert, L., Hicklin J., and Hornbruch, A. (1971). Positional information and pattern regulation in regeneration of *Hydra*. *Symp. Soc. Exp. Biol.* **25,** 391–415.
Wolpert, L., Hornbruch, A., and Clarke, M. R. B. (1974). Positional information and positional signalling in *Hydra*. *Amer. Zool.* **14,** 647–663.
Yamamoto, A., Maeda, Y., and Takeuchi, I. (1981). Development of an autophagic system in differentiating cells of the cellular slime mold *Dictyostelium discoideum*. Protoplasma **108,** 55–70.
Yanagisawa, K., Loomis, W. F., and Sussman, M. (1967). Developmental regulation of the enzyme UDP-galactose polysaccharide transferase. *Exp. Cell Res.* **46,** 328–334.

Bibliography on *Dictyostelium*

W. F. Loomis and Robin Cann

We have made an effort to accumulate as many published works on *Dictyostelium* as possible up through 1981. In the last seven years more papers appeared than all those published in previous years. There are now over 1800 entries. Although this number does not exceed the capacity of our computer in which they are filed, it does exceed our ability to proofread. We would appreciate hearing of errors and omissions and will appropriately correct the file. We are grateful to Mary Murray for the convenient computer program we have used in constructing this bibliography.

Abe, H., Uchiyama, M., Tanaka, Y., and Saito, H. (1976). *Tetrahedron Lett.* **42,** 3807–3810. Structure of discadenine a spore germination inhibitor from the cellular slime mold *Dictyostelium discoideum.*

Abe, L., Saga, Y., Okada, H., and Yangisawa, K. (1981). *J. Cell Sci.* **51,** 131. Differentiation of *Dictyostelium discoideum* mutant cells in shaken suspension culture and the effect of cyclic AMP.

Al-Ayash, A. I., and Wilson, M. T. (1977). *Comp. Biochem. and Phys.* **56,** 147. Isolation and properties of cytochrome C from the cellular slime mould *Dictyostelium discoideum.*

Alcantara, F., and Bazill, G. W. (1976). *J. Gen. Microbiol.* **92,** 351–368. Extracellular cAMP–phosphodiesterase accelerates differentiation in *Dictyostelium discoideum.*

Alcantara, F., and Monk, M. (1974). *J. Gen. Microbiol.* **85,** 321–324. Signal propagation in the cellular slime mould *Dictyostelium discoideum.*

Aldrich, H. C., and Gregg, J. H. (1973). *Exp. Cell Res.* **81,** 407–412. Unit membrane structural changes following cell association in *Dictyostelium.*

Alemany, S., Gil, M. G., and Mato, J. M. (1980). *Proc. Natl. Acad. Sci. U.S.A.* **77,** 6996–6999. Regulation by cyclic GMP of phospholipid methylation during chemotaxis in *Dictyostelium discoideum.*

Alexander, S., and Brackenbury, R. (1975). *Nature (London)* **254,** 698–699. Tryptic destruction of aggregative competence in *Dictyostelium discoideum* and subsequent recovery.

Alexander, S., and Sussman, M. (1975). *Develop. Biol.* **46,** 211–215. Trehalose 6 phosphate synthetase activity in extracts of *Dictyostelium discoideum.*

Al-Rayess, H., Ashworth, J. M., and Younis, M. S. (1979). *J. Univ. Kuwait (Sci.)* **6,** 103–108. The phosphate and poly phosphate content of *Dictyostelium discoideum* cells during their growth and subsequent differentiation.

Alton, T. H., and Brenner, M. (1979). *Develop. Biol.* **71,** 1–9. Comparison of proteins synthesized by anterior and posterior regions of *Dictyostelium discoideum* pseudoplasmodia.

The Development of *Dictyostelium discoideum*

ISBN 0-12-455620-5

Alton, T. H., and Lodish, H. F. (1977). *Cell* **12,** 301–310. Translational control of protein synthesis during the early stages of differentiation of the slime mold *Dictyostelium discoideum.*

Alton, T. H., and Lodish, H. F. (1977). *Develop. Biol.* **60,** 180–206. Developmental changes in mRNA's and protein synthesis in *Dictyostelium discoideum.*

Alton, T. H., and Lodish, H. F. (1977). *Develop. Biol.* **60,** 207–216. Synthesis of developmentally regulated proteins in *Dictyostelium discoideum* which are dependent on continued cell–cell interaction.

Anderson, J. J. (1973). *Genetics* **74,** 7–11. A technique for clonal analysis of developmental mutants of the cellular slime mold *Dictyostelium discoideum.*

Anderson, J. J. (1974). *J. Bacteriol.* **117,** 1363–1364. *Dictyostelium discoideum* a new method for cloning in liquid medium.

Anderson, J. J., Fennell, D. I, and Raper, K. B. (1968). *Mycologia* **40,** 49–64. *Dictyostelium deminutivum,* a new cellular slime mold.

Aomine, M. (1981). *Comp. Biochem. Physiol. A Comp. Physiol.* **68,** 531–540. The amino-acid absorption and transport in protozoa.

Aramaki, Y., Yamada, H., and Miyazaki, T. (1980). *J. Biochem. (Tokyo).* **87,** 1145–1152. Isolation and characterization of a carbohydrate binding protein from *Dictyostelium discoideum NC-4* deficient in normal discoidin synthesis.

Armant, D. R., and Rutherford, C. L. (1978). *Va. J. Sci.* **29,** 56. Localization, isolation and characterization of alkaline phosphatase a developmentally significant enzyme during cellular differentiation of *Dictyostelium discoideum.*

Armant, D. R., and Rutherford, C. L. (1979). *Mech. Ageing Dev.* **10,** 199–218. 5-Prime AMP nucleotidase is localized in the area of cell–cell contact of prespore and prestalk regions during culmination of *Dictyostelium discoideum.*

Armant, D. R., and Rutherford, C. L. (1980). *J. Cell. Biol.* **87,** 27A. Purification, characterization and localization of 5′ AMP nucleotidase during pattern formation in *Dictyostelium discoideum.*

Arndt, A. (1937). *Entwicklungs Much. Org.* **136,** 681–744. Untersuchunger uber *Dictyostelium mucoroides* brefeld.

Ashworth, J. M. (1971a). *Symp. Soc. Exp. Biol.* **25,** 27–49. Cell development in the cellular slime mould *Dictyostelium discoideum.*

Ashworth, J. M. (1971b). *In* "Control Mechanisms of Growth and Differentiation." (Davies, D. D., and Balls, M. eds.), Vol. 25, 27–49.

Ashworth, J. M. (1973). *In* "Regulation de la sporulation microbienne." Centre National de la Recherche Scientifique, Paris, France (Aubert, J. P. P., Schaeffer, and Azulmajster, J. eds.), 129–132. Cell differentiation in the cellular slime mold *Dictyostelium discoideum.*

Ashworth, J. M. (1973). *Biochem. Soc. Trans.* **1,** 1233–1245. Studies on cell differentiation in the cellular slime mold *Dictyostelium discoideum.*

Ashworth, J. M. (1974). *In* "Biochemistry of Cell Differentiation." University Park Press, Baltimore, Md. (Paul, J. ed.), 7–34. The development of the cellular slime molds.

Ashworth, J. M. (1976). *Biochem. Soc. Trans.* **4,** 961–964. Control of cell differentiation in the cellular slime mold *Dictyostelium discoideum.*

Ashworth, J. M. (1977). *In* "Cell Differentiation in Microorganisms, Plants and Animals." Elsevier/North-Holland. (Nover, Lutz, and Mothes, K. eds.), 639. The developmental program of *Dictyostelium discoideum*.

Ashworth, J. M., and Dee, J. (1975). *Inst. Biol. Stud. Biol.* **67.** The biology of slime molds.

Ashworth, J. M., and Quance, J. (1972). *Biochem. J.* **126,** 601–608. Enzyme synthesis in myxamoebae of the cellular slime mould *Dictyostelium discoideum* during growth in axenic culture.

Ashworth, J. M., and Sackin, M. J. (1969). *Nature (London)* **224,** 817–818. Role of aneuploid cells in cell differentiation in the cellular slime mould *Dictyostelium discoideum.*

Ashworth, J. M., and Sussman, M. (1967). *J. Biol. Chem.* **242,** 1696–1700. The appearance and disappearance of uridine diphosphate glucose pyrophosphorylase activity during differentiation of the cellular slime mold *Dictyostelium discoideum.*

Ashworth, J. M., and Watts, D. J. (1970). *Biochem. J.* **119,** 175–182. Metabolism of the cellular slime mould *Dictyostelium discoideum* grown in axenic culture.

Ashworth, J. M., and Weiner, E. (1972). *In* "Lysosomes in Biology and Pathology." Vol. 3. Elsevier/North-Holland (Dingle, J. T. ed.), 36–46. The lysosomes of the cellular slime mould *Dictyostelium discoideum.*

Ashworth, J. M., Duncan, D., and Rowe, A. J. (1969). *Exp. Cell Res.* **58,** 73–78. Changes in fine structure during cell differentiation of the cellular slime mould, *Dictyostelium discoideum.*

Atryzek, V. (1976). *Develop. Biol.* **50,** 489–501. Alteration in timing of cell differentiation resulting from cell interactions during development of the cellular slime mold *Dictyostelium discoideum.*

Atryzek, V. (1976). *J. Bacteriol.* **126,** 1005–1008. Dissociation of developing slime mold cells does not inhibit the developmentally regulated rise in alkaline phosphatase activity.

Bacon, C. W., and Sussman, A. S. (1973). *J. Gen. Microbiol.* **76,** 331–344. Effects of the self inhibitor of *Dictyostelium discoideum* on spore metabolism.

Bacon, C. W., Sussman, A. S., and Paul, A. G. (1973). *J. Bacteriol.* **113,** 1061–1063. Identification of a self-inhibitor from spores of *Dictyostelium discoideum.*

Bakke, A. C., and Bonner, J. (1979). *Biochem.* **18,** 4556–4562. Purification of the histones of *Dictyostelium discoideum* chromatin.

Bakke, A. C., Wu, J.-R., and Bonner, J. (1978). *Proc. Natl. Acad. Sci. U.S.A.* **75,** 705–709. Chromatin structure in the cellular slime mold *Dictyostelium discoideum.*

Barclay, S. L., and Henderson, E. J. (1977). *In* "Developments in Cell Biology." Elsevier/North-Holland (Cappuccinelli, P., and Ashworth, J. M. eds.), 291–296. A method for selecting aggregation-defective mutants of *Dictyostelium discoideum.*

Bard, R., and Pfohl, R. (1977). *Am. Zool.* **17,** 965. Kinetic properties of immobilized and soluble alkaline phosphatase from *Dictyostelium discoideum.*

Baril, E. F., Scheiner, C., and Pederson, T. (1980). *Proc. Natl. Acad. Sci. U.S.A.* **77,** 3317–3321. A beta-like DNA polymerase activity in the slime mold *Dictyostelium discoideum.*

Barkley, D. S. (1969). *Science* **165,** 1133–1134. Adenosine-3′,5′-phosphate: identification as acrasin in a species of cellular slime mold.

Barondes, S. H. (1977). *J. Supramol. Struct.* Suppl. 1, **33.** Endogenous developmentally regulated lectins in cellular slime molds and embryonic muscle.

Barondes, S. H. (1978). *Birth Defects* **14,** 491–496. Developmentally regulated slime mold lectins and specific cell cohesion.

Barondes, S. H. (1978). *In* "Proceedings of the ICN-UCLA Symposium," Vol. 23, (Marchesi, V. T. *et al.*, eds.). Developmentally regulated lectins in cellular slime molds and embryonic chick tissues.

Barondes, S. H., and Haywood, P. L. (1979). *Biochim. Biophys. Acta* **550,** 297–308. Comparison of developmentally regulated lectins from 3 species of cellular slime mold.

Barra, J. (1977). *Comptes Rendue de l'Academie des Sciences* **248,** 8. Coaggregation between aggregateless mutants of *Dictyostelium discoideum.*

Barra, J., Barrand, P., Blondelet, M-H., and Brachet, P. (1980). *Mol. Gen. Genet.* **177,** 607–614. PDSA a gene involved in the production of active phosphodiesterase during starvation of *Dictyostelium discoideum* amoebae.

Barrand, P., Da Silva, L., and Brachet, P. (1974). *Bull. Soc. Fr. Mycol. Med.* **3**, 29–32. Fungicides effective against the myxomycete *Dictyostelium discoideum.*

Bartles, J. R., and Frazier, W. A. (1979). *J. Cell Biol.* **83**, 55A. Properties of the cell cohesion receptor of *Dictyostelium discoideum* cells.

Bartles, J. R., and Frazier, W. A. (1980). *J. Biol. Chem.* **255**, 30–38. Preparation of Iodine-125 labeled discoidin I and the properties of its binding to *Dictyostelium discoideum* cells.

Bartles, J. R., Pardos, B. T., and Frazier, W. A. (1979). *J. Biol. Chem.* **254**, 3156–3159. Reconstitution of discoidin hemagglutination activity by lipid extracts of *Dictyostelium discoideum* cells.

Batts-Young, B., and Lodish, H. F. (1978). *Proc. Natl. Acad. Sci. U.S.A.* **75**, 740–744. Triphosphate residues at the 5-prime ends of ribosomal RNA precursor and 5S RNA from *Dictyostelium discoideum.*

Batts-Young, B., Maizels, N., and Lodish, H. F. (1977). *J. Biol. Chem.* **252**, 3952–3960. Precursors of ribosomal RNA in the cellular slime mold *Dictyostelium discoideum* isolation and characterization.

Batts-Young, B., Lodish, H. F., and Jacobson, A. (1980). *Develop. Biol.* **78**, 352–364. Similarity of the primary sequences of ribosomal RNA isolated from vegetative and developing cells of *Dictyostelium discoideum.*

Bauer, R., Rath, M., and Risse, H. (1971). *Eur. J. Biochem.* **21**, 179–190. The biosynthesis of glycoproteins during the development of *Dictyostelium discoideum.* The transfer of D-mannose in vegetative and aggregated cells.

Baumann, P. (1969). *Biochem.* **8**, 5011–5015. Glucokinase of *Dictyostelium discoideum.*

Baumann, P., and Wright, B. E. (1968). *Biochem.* **7**, 3653–3661. The phosphofructokinase of *Dictyostelium discoideum.*

Baumann, P., and Wright, B. E. (1969). *Biochem.* **8**, 1655–1659. The fructose 1,6-diphosphatase of *Dictyostelium discoideum.*

Bazari, W. L., and Clark, M. (1981). *J. Biol. Chem.* **256**, 3598–3603. Characterization of a novel calmodulin from *Dictyostelium discoideum*.

Bazin, M. J. (1978). *Ann. Appl. Biol.* **89**, 159–162. Predator prey interactions between protozoa and bacteria.

Bazin, M. J., Rapa, V., and Saunders, P. T. (1974). *In* "Ecological Stability." Halsted Press, New York, N.Y. (Usher, M. B., and Williamson, M. H. eds.), 159–164. The integration of theory and experiment in the study of predator prey dynamics.

Bazin, M. J., Saunders, P. T., Owen, B. A., and Kilpatrick, D. (1978). *In* "Proceedings in Life Sciences. Microbial Ecology." New York, N.Y. (Loutit, M. W., and Miles, J. A. R. eds.), 21–24. Predation by slime mold amoebae.

Bender, W., Davidson, N., Kindle, K. L., Taylor, W. D., Silverman, M., and Firtel, R. A. (1978). *Cell* **15**, 779–788. The structure of M-6, a recombinant plasmid containing *Dictyostelium discoideum* DNA homologous to actin messenger RNA.

Benson, M. R., and Mahoney, D. P. (1977). *Am. J. Bot.* **64**, 496–503. The distribution of *Dictyostelid* cellular slime molds in southern California U.S.A. with taxonomic notes on selected species.

Bernstein, R. L., and Van Driel, R. (1980). *FEBS Lett.* **12**, 249–253. Control of folate deaminase activity of *Dictyostelium discoideum* by cyclic AMP.

Bernstein, R. L., and Van Driel, R. (1980). *Eur. J. Cell Biol.* **22**, 234. Degradation of the chemoattractant folic-acid by *Dictyostelium discoideum.*

Bernstein, R. L., Rossier, C., Van Driel, R., Brunner, M., and Gerisch, G. (1981). *Cell. Differ.* **10**, 79–86. Folate deaminase and cyclic AMP phosphodiesterase in *Dictyostelium discoideum* their regulation by extracellular cyclic AMP and folic-acid.

Bernstein, R. L., Tabler, M., Vestweber, D., and Van Driel, R. (1981). *Biochim. Biophys. Acta* **677,** 295. Extracellular folate deaminase of *Dictyostelium discoideum.*

Beug, H., and Gerisch, G. (1969). *Naturwissenschaften* **56,** 374. Univalente fragmente von antikorpern zur analyse von zellmembran-funktionen.

Beug, H., and Gerisch, G. (1972). *J. Immunol. Methods* **2,** 49–57. A micromethod for routine measurement of cell agglutination and dissociation.

Beug, H., Gerisch, G., Kempf, S., Riedel, V., and Cremer, G. (1970). *Exp. Cell. Res.* **63,** 147–158. Specific inhibition of cell contact formation in *Dictyostelium* by univalent antibodies.

Beug, H., Katz, F. E., and Gerisch, G. (1973). *J. Cell. Biol.* **56,** 647–658. Dynamics of antigenic membrane sites relating to cell aggregation in *Dictyostelium discoideum.*

Beug, H., Katz, F. E., Stein, A., and Gerisch, G. (1973). *Proc. Natl. Acad. Sci. U.S.A.* **70,** 3150–3154. Quantitation of membrane sites in aggregating *Dictyostelium* cells by use of tritiated univalent antibody.

Beug, H., Katz, F., and Gerisch, G. (1973). *J. Cell. Biol.* **56,** 657–667. Dynamics of antigenic membrane sites relating to cell aggregation in *Dictyostelium discoideum.*

Blaskovics, J. A., and Raper, K. B. (1957). *Biol. Bull.* **113,** 58–88. Encystment stages of *Dictyostelium.*

Bloch, S., and Cedar, H. (1975). *Isr. J. Med. Sci.* **11,** 1214–1215. *In Vitro* transcription from chromatin of slime molds.

Blondelet, M-H., and Brachet, P. (1981). *Biochim. Biophys. Acta* **640,** 572–582. The hydrophobic character of the membrane bound phosphodiesterase from *Dictyostelium discoideum.*

Blumberg, D. D., and Lodish, H. F. (1980). *Develop. Biol.* **78,** 285–300. Changes in the messenger RNA population during differentiation of *Dictyostelium discoideum.*

Blumberg, D. D., and Lodish, H. F. (1980). *Develop. Biol.* **74,** 268–284. Complexity of nuclear and polysomal RNA in growing *Dictyostelium discoideum* cells.

Blumberg, D. D., and Lodish, H. F. (1981). *Develop. Biol.* **81,** 74–80. Changes in the complexity of nuclear RNA during development of *Dictyostelium discoideum.*

Blumberg, D. D., Margolskee, J. P., Chung, S., Barklis, E., Cohen, N. S., and Lodish, H. F. (1982). *Proc. Natl. Acad. Sci. U.S.A.* **79,** 127–131. Specific cell–cell contacts are essential for induction of gene expression during differentiation of *Dictyostelium discoideum.*

Boiteux, A., Hess, B., Plesser, T., and Murray, J. D. (1977). *FEBS Lett.* **75,** 1–4. Oscillatory phenomena in biological systems.

Bonner, J. T. (1944). *Amer. J. Bot.* **31,** 175–182. A descriptive study of the development of the slime mold *Dictyostelium discoideum.*

Bonner, J. T. (1947). *J. Exp. Zool.* **106,** 1–26. Evidence for the formation of cell aggregates by chemotaxis in the development of the slime mold *Dictyostelium discoideum.*

Bonner, J. T. (1949). *J. Exp. Zool.* **110,** 259–271. The demonstration of acrasin in the later stages of the development of the slime mold *Dictyostelium discoideum.*

Bonner, J. T. (1950). *Biol. Bull.* **99,** 143–151. Observations on polarity in the slime mold *Dictyostelium discoideum.*

Bonner, J. T. (1952). *Amer. Natur.* **86,** 79–89. The pattern of differentiation in amoeboid slime molds.

Bonner, J. T. (1957). *Quart. Rev. Biol.* **32,** 232–246. A theory of the control of differentiation in the cellular slime molds.

Bonner, J. T. (1959a). *Proc. Natl. Acad. Sci. U.S.A.* **45,** 379–384. Evidence for the sorting out of cells in the development of the cellular slime molds.

Bonner, J. T. (1959b). *Sci. Amer.* **201,** 152–162. Differentiation of social amoebae.

Bonner, J. T. (1960). *In* "Developing Cell Systems and Their Control" (Rudnick, D. ed.). 18th Growth Symposium, pp. 3–20. Ronald Press, New York. Development in the cellular slime molds: the role of cell division, cell size, and cell number.

Bonner, J. T. (1963a). *Sci. Amer.* **209,** 84–93. How slime molds communicate.

Bonner, J. T. (1963b). *Symp. Soc. Exp. Biol.* **17,** 341–358. Epigenetic development in the cellular slime molds.

Bonner, J. T. (1965). *In* "Encyclopedia of Plant Physiology" Vol. 15, Part 1. (Ruhland, W. ed.), 612–640. Physiology of development in cellular slime molds.

Bonner, J. T. (1967). Princeton Univ. Press, Princeton, New Jersey. "The Cellular Slime Molds."

Bonner, J. T. (1969). *Sci. Amer.* **220,** 78–91. Hormones in social amoebae and mammals.

Bonner, J. T. (1970). *Proc. Natl. Acad. Sci. U.S.A.* **65,** 110–113. Induction of stalk cell differentiation by cyclic AMP in the cellular slime mold *Dictyostelium discoideum.*

Bonner, J. T. (1971). *Annu. Rev. Microbiol.* **25,** 75–92. Aggregation and differentiation in the cellular slime molds.

Bonner, J. T. (1974). *In* "Readings from Scientific American: Cellular and Organismal Biology." Freeman, San Francisco. (Kennedy, D. ed.), 64–72. Differentiation in social amoebae.

Bonner, J. T. (1977). *Mycologia,* **69,** 443–459. Some aspects of chemotaxis using the cellular slime molds as an example.

Bonner, J. T., and Adams, M. S. (1958). *J. Embryol. Exp. Morphol.* **6,** 346–356. Cell mixtures of different species and strains of cellular slime molds.

Bonner, J. T., and Dodd, M. R. (1962a). *Biol. Bull.* **122,** 13–24. Aggregation territories in the cellular slime molds.

Bonner, J. T., and Dodd, M. R. (1962b). *Develop. Biol.* **5,** 344–361. Evidence for gas-induced orientation in the cellular slime molds.

Bonner, J. T., and Eldredge, D. Jr. (1945). *Growth* **9,** 287–297. A note on the rate of morphogenetic movement in the slime mold, *Dictyostelium discoideum.*

Bonner, J. T., and Frascella, E. B. (1952). *J. Exp. Zool.* **121,** 561–571. Mitotic activity in relation to differentiation in the slime mold *Dictyostelium discoideum.*

Bonner, J. T., and Frascella, E. B. (1953). *Biol. Bull.* **104,** 297–300. Variations in cell size during the development of the slime mold, *Dictyostelium discoideum.*

Bonner, J. T., and Hoffman, M. E. (1963). *J. Embryol. Exp. Morphol.* **11,** 571–589. Evidence for a substance responsible for the spacing pattern of aggregation and fruiting in the cellular slime molds.

Bonner, J. T., and Shaw, M. J. (1957). *J. Cell. Comp. Physiol.* **50,** 145–154. The role of humidity in the differentiation of the cellular slime molds.

Bonner, J. T., and Slifkin, M. K. (1949). *Amer. J. Bot.* **36,** 727–734. A study of the control of differentiation: the proportions of stalk and spore cells in the slime mold *Dictyostelium discoideum.*

Bonner, J. T., and Whitfield, F. E. (1965). *Biol. Bull.* **128,** 51–57. The relation of sorocarp size to phototaxis in the cellular slime mold *Dictyostelium purpureum.*

Bonner, J. T., Clarke, W. W., Jr., Neely, C. L., Jr., and Slifkin, M. K. (1950). *J. Cell. Comp. Physiol.* **36,** 149–158. The orientation to light and the extremely sensitive orientation to temperature gradients in the slime mold *Dictyostelium discoideum.*

Bonner, J. T., Koontz, P. G., Jr., and Paton, D. (1953). *Mycologia* **45,** 235–240. Size in relation to the rate of migration in the slime mold *Dictyostelium discoideum.*

Bonner, J. T., Chiquoine, A. D., and Kolderie, M. Q. (1955). *J. Exp. Zool.* **130,** 133–158. A histochemical study of differentiation in the cellular slime molds.

Bonner, J. T., Kelso, A. P., and Gillmor, R. G. (1966). *Biol. Bull.* **130,** 28–42. A new approach to the problem of aggregation in the cellular slime molds.

Bonner, J. T., Barkley, D. S., Hall, E. M., Konijn, T. M., Mason, N. W., O'Keefe, G., III, and Wolfe, P. B. (1969). *Develop. Biol.* **20,** 72–87. Acrasin, acrasinase, and the sensitivity to acrasin in *Dictyostelium discoideum.*

Bonner, J. T., Sieja, T. W., and Hall, E. M. (1971). *J. Embryol. Exp. Morphol.* **25,** 457–465. Further evidence for a second chemotactic system in the cellular slime mold, *Dictyostelium discoideum.*

Bonner, J. T., Hall, E. M., Sachsenmaier, W., and Walker, B. K. (1979). *J. Bacteriol.* **102,** 682–687. Evidence for a second chemotactic system in the cellular slime mold, *Dictyostelium discoideum.*

Bonner, J. T., Hall, E. M., Noller, S., Oleson, F., and Roberts, A. (1973). *Develop. Biol.* 402–409. Synthesis of cyclic AMP and phosphodiesterase in various species of cellular slime molds and its bearing on chemotaxis and differentiation.

Bordier, C., and Crettol-Jaervinen, A. (1979). *J. Biol. Chem.* **254,** 2565–2567. Peptide mapping of heterogeneous protein samples.

Bordier, C., Loomis, W. F., Elder, J., and Lerner, R. (1978). *J. Biol. Chem.* **253,** 5133–5139. The major developmentally regulated protein complex in membranes of *Dictyostelium discoideum.*

Born, G. V. R., and Garrod, D. (1968). *Nature (London)* **220,** 616–618. Photometric demonstration of aggregation of slime mold cells showing effects of temperature and ionic strength.

Borts, R. H., and Dimond, R. L. (1981). *Develop. Biol.* **87,** 176. The alpha-glucosidases of *Dictyostelium discoideum,* I. Identification and characterization.

Borts, R. H., and Dimond, R. L. (1981). *Develop. Biol.* **87,** 185. The alpha-glucosidases of *Dictyostelium discoideum,* II. Identification and characterization.

Boublik, M., and Ramagopa, S. (1980). *Mol. Gen. Genet.* **179,** 483–488. Conformation of ribosomes from the vegetative amoebae and spores of *Dictyostelium discoideum.*

Bourguignon, L. Y. W., and Katz, E. R. (1978). *J. Gen. Microbiol.* **106,** 93–102. Isolation and characterization of the RNA of membrane bound ribosomes in *Dictyostelium discoideum.*

Bozzaro, S., and Gerisch, G. (1978). *J. Mol. Biol.* **120,** 265–280. Contact sites in aggregating cells of *Polysphondylium pallidum.*

Bozzaro, S., and Gerisch, G. (1979). *Cell Differ.* **8,** 117–128. Developmentally regulated inhibitor of aggregation in cells of *Dictyostelium discoideum.*

Bozzaro, S., Tsugita, A., Janku, M., Monok, G., Opatz, K., and Gerisch, G. (1981). *Exp. Cell. Res.* **134,** 181. Characterization of a purified cell surface glycoprotein as a contact site in *Polysphondylium pallidum.*

Brachet, P. (1976). *C.R. Acad. Sci. Nat.* **282,** 377–379. Stimulation of cell aggregation of *Dictyostelium discoideum* by Ionophore A-23187.

Brachet, P., and Dicou, E. (1979). *Differentiation* **13,** 15–16. The role of signal modulation during the aggregation phase of *Dictyostelium discoideum.*

Brachet, P., and Klein, C. (1975). *Exp. Cell. Res.* **93,** 159–165. Inhibition of growth and cellular aggregation of *Dictyostelium discoideum* by steroid compounds.

Brachet, P., and Klein, C. (1977). *Differentiation* **8,** 1–8. Cell responsiveness to cyclic AMP during the aggregation phase of *Dictyostelium discoideum* comparison between the inhibitory action of progesterone and the stimulatory action of EGTA and Ionophore A-23187.

Brachet, P., Dicou, E. L., and Klein, C. (1979). *Dell Differ.* **8,** 255–266. Inhibition of cell differentiation in a phosphodiesterase defective mutant of *Dictyostelium discoideum.*

Brachet, P., Barra, J., Darmon, M., and Barrand, P. (1977). *In* "Developments in Cell Biology." Elsevier/North-Holland (Cappuccinelli, P., and Ashworth, J. M. eds.), 125–134. A phosphodiesterase-defective mutant of *Dictyostelium discoideum.*

Brackenbury, R., and Sussman, M. (1975). *Cell* **4**, 347–352. Mutant of *Dictyostelium discoideum* defective in cell contact regulation of enzyme expression.

Brackenbury, R. W., Schindler, J., Alexander, S., and Sussman, M. (1974). *J. Mol. Biol.* **90**, 529–539. A choice of morphogenetic pathways in *Dictyostelium discoideum* induced by the adenosine analog formycin B.

Bradley, S. G., Sussman, M., and Ennis, H. L. (1956). *J. Protozool.* **3**, 33–38. Environmental factors affecting the aggregation of the cellular slime mold, *Dictyostelium discoideum.*

Braggaar-Schaap, P., and Van Der Molen, L. G. (1980). *Ultramicroscopy* **5**, 95–96. Organelle transformation and membrane changes in *Dictyostelium minutum*, a cellular slime mold.

Brambl, R., Dunkle, L. D., and Van Etten, J. L. (1978). *In* "The Filamentous Fungi," Vol. 3, 94–118. Nucleic acid and protein synthesis during fungal spore germination.

Braun, V., Hautke, K., Wolf, H., and Gerisch, G. (1972). *Eur. J. Biochem.* **27**, 116–125. Degradation of the murein lipoprotein complex of *Escherichia coli* cell walls by *Dictyostelium* amoebae.

Breathnach, R., and Chambon, P. (1981). *Ann. Rev. Biochem.* **50**, 349–384. Organization and expression of eukaryotic split genes coding for proteins.

Bredehorst, R., Lengyel, H., and Hilz, H. (1979). *Eur. J. Biochem.* **99**, 401–412. Determination by radio immunoassay of the sum of oxidized and reduced forms of NAD and NADP in picomole quantities from the same acid extract.

Bredehorst, R., Klapproth, K., Hilz, H., Scheidegger, C., and Gerisch, G. (1980). *Cell Differ.* **9**, 95–104. Protein bound mono ADP ribose residues in differentiating cells of *Dictyostelium discoideum*.

Brefeld, O. (1869). *Abhandl. Senckenberg. Naturforsch. Ges.* **7**, 85–107. *Dictyostelium mucoroides.* Ein neuer organismus aus der verwandtschaft der myxomyceten.

Brefeld, O. (1884). *Untersuchungen Aus Dem Gesammtgebiet Der Mukol.* **6**, 1–34. *Polysphondylium violaceum* und *Dictyostelium mucoroides* nebst bemerkungen zur systematik der schleimpilze.

Brenner, M. (1977). *J. Biol. Chem.* **252**, 4073–4077. Cyclic AMP gradient in migration pseudoplasmodia of the cellular slime mold *Dictyostelium discoideum.*

Brenner, M. (1978). *Develop. Biol.* **64**, 210–223. Cyclic AMP levels and turnover during development of the cellular slime mold *Dictyostelium discoideum.*

Brenner, M., Tisdale, D., and Loomis, W. F. (1975). *Exp. Cell. Res.* **90**, 249–252. Techniques for rapid biochemical screening of large numbers of cell clones.

Breuer, W., and Siu, C-H. (1980). *J. Cell Biol.* **87**, 86A. Identification of endogenous glycoprotein receptors for discoidin in *Dictyostelium discoideum.*

Britten, R. J., and Davidson, E. H. (1969). *Science* **165**, 349–357. Gene regulation in higher cells: a theory.

Brody, T., and Williams, K. L. (1974). *J. Gen. Microbiol.* **82**, 371–383. Cytological analysis of the parasexual cycle in *Dictyostelium discoideum.*

Brown, S. S., and Rutherford, C. L. (1980). *Differentiation* **16**, 173–184. Localization of cyclic nucleotide phosphodiesterase in the multicellular stages of *Dictyostelium discoideum.*

Brown, S. S., and Spudich, J. A. (1979). *J. Cell. Biol.* **80**, 499–504. Nucleation of polar actin filament assembly by a positively charged surface.

Brown, S. S., and Spudich, J. A. (1979). *J. Cell. Biol.* **83**, 657–662. Cytochalasin inhibits the rate of elongation of actin filament fragments.

Brown, S. S., and Spudich, J. A. (1979). *J. Cell. Biol.* **83**, 323A. Effect of cytochalasin on actin assembly rate.

Brown, S. S., and Spudich, J. A. (1980). *J. Cell Biol.* **87**, 224A. Purification of a calcium sensitive 38000 dalton protein from *Dictyostelium discoideum* which affects the assembly state of actin.

Bruhmuller, M., and Wright, B. E. (1963). *Biochim. Biophys. Acta* **71**, 50–57. Glutamate oxidation in the differentiating slime mold II. studies *in vitro.*

Bryant, P. E. (1976). *Int. J. Rad. Biol.* **30**, 327–338. Effects of electrons and neutrons on the synthesis of RNA in resistant and sensitive strains of the slime mould *Dictyostelium discoideum* and the modifying effect of caffeine.

Bryant, P. E. (1976). *In* "Proceedings in Life Sciences." New York, N.Y. (Kiefer, J. ed.), 68–79. Synthesis of RNA after irradiation in resistant and sensitive strains of the slime mold *Dictyostelium discoideum.*

Burns, R. A., Livi, G. P., and Dimond, R. L. (1981). *Develop. Biol.* **83**, 407–416. Regulation and secretion of early developmentally controlled enzymes during axenic growth in *Dictyostelium discoideum.*

Burridge, K. (1977). *J. Supramol. Struct.* Suppl. 1, 14. Changes in glyco proteins associated with viral transformation and differentiation.

Burridge, K., and Jordan, L. (1979). *Exp. Cell. Res.* **124**, 31–38. The glycoproteins of *Dictyostelium discoideum* change during development.

Bushway, R. J., and Hanks, A. R. (1976). *Pest. Biochem. Phys.* **6**, 254–260. Pesticide inhibition of growth and macromolecular synthesis in the cellular slime mold *Dictyostelium discoideum.*

Call, P., and Jacobson, B. S. (1979). *J. Cell Biol.* **83**, 268A. Coating cells with a silica pellicle for isolation and exposure of the cytoplasmic surface of the plasma membrane.

Cappuccinelli, P., and Ashworth, J. M. (1976). *Exp. Cell Res.* **103**, 387–394. The effect of inhibitors of microtubule and microfilament function on the cellular slime mould *Dictyostelium discoideum.*

Cappuccinelli, P., and Hames, B. D. (1978). *Biochem. J.* **169**, 499–504. Characterization of colchicine binding activity in *Dictyostelium discoideum.*

Cappuccinelli, P., Hames, B. D., and Currureddu, R. (1977). *In* "Developments In Cell Biology." Elsevier/North-Holland (Cappuccinelli, P., and Ashworth, J. M. eds.), 231–242. Tubulin in *Dictyostelium discoideum.*

Cappuccinelli, P., Marinotti, A., and Hames, B. D. (1978). *FEBS Lett.* **91**, 153–157. Identification of cytoplasmic tubulin in *Dictyostelium discoideum.*

Cappuccinelli, P., Fighetti, M., and Rubino, S. (1979). *Cell Differ.* **8**, 243–252. Differentiation without mitosis in *Dictyostelium discoideum.*

Cappuccinelli, P., Fighetti, M., and Rubino, S. (1979). *FEMS Microbiol. Lett.* **5**, 25–27. A mitotic inhibitor for chromosomal studies in slime molds.

Cardelli, J. A., and Dimond, R. L. (1980). *J. Cell Biol.* **87**, 275A. Effect of starvation and cell-cell interaction on protein synthesis in *Dictyostelium discoideum.*

Cardelli, J. A., Knecht, D. A., and Dimond, R. L. (1981). *Develop. Biol.* **82**, 180–185. Membrane bound and free polysomes in *Dictyostelium discoideum* I. Isolation and developmental effects on size and distribution.

Cavender, J. C. (1972). *Can. J. Bot.* **50**, 1497–1501. Cellular slime molds in forest soils of eastern Canada.

Cavender, J. C. (1973). *Mycologia* **65**, 1044–1054. Geographical distribution of Acrasieae.

Cavender, J. C. (1976). *Am. J. Bot.* **63**, 60–70. Cellular slime molds of southeast Asia. Part 1. Description of new species.

Cavender, J. C. (1976). *Am. J. Bot.* **63**, 71–73. Cellular slime molds of southeast Asia. Part 2. Occurrence and distribution.

Cavender, J. C. (1978). *Can. J. Bot.* **56**, 1326–1332. Cellular slime molds in tundra and forest soils of Alaska U.S.A. including *Dictyostelium septentrionalis* New Species.

Cavender, J. C. (1978). *Ohio J. Sci.* **78**, (Suppl). 24. A possible new cellular slime mold genus *Heterosphondylium.*

Cavender, J. C., and Raper, K. B. (1965a). *Amer. J. Bot.* **52,** 294–296. The Acrasieae in nature. I. Isolation.

Cavender, J. C., and Raper, K. B. (1965b). *Amer. J. Bot.* **52,** 297–302. The Acrasieae in nature. II. Forest soil as a primary habitat.

Cavender, J. C., and Raper, K. B. (1965c). *Amer. J. Bot.* **52,** 302–308. The Acrasieae in nature. III. Occurrence and distribution in forests of eastern North America.

Cavender, J. C., and Raper, K. B. (1968). *Amer. J. Bot.* **55,** 504–513. The occurrence and distribution of Acrasieae in forests of subtropical and tropical America.

Cavender, J. C., Raper, K. B., and Norberg, A. M. (1979). *Am. J. Bot.* **66,** 207–217. *Dictyostelium aureostipes* new species and *Dictyostelium tenue* new species Dictyosteliaceae.

Cavender, J. C., Worley, A. C., and Raper, K. B. (1981). *Am. J. Bot.* **68,** 373–382. The yellow pigmented *Dictyostelia.*

Ceccarini, C. (1967). *Biochim. Biophys. Acta* **148,** 114–124. The biochemical relationship between trehalase and trehalose during growth and differentiation in the cellular slime mold *Dictyostelium discoideum.*

Ceccarini, C., and Filosa, M. (1965). *J. Cell. Comp. Physiol.* **66,** 135–140. Carbohydrate content during development of the slime mold, *Dictyostelium discoideum.*

Ceccarini, C., and Cohen, A. (1967). *Nature (London)* **214,** 1345–1346. Germination inhibitor from the cellular slime mold *Dictyostelium discoideum.*

Ceccarini, C., and Maggio, R. (1968). *Biochim. Biophys. Acta* **166,** 134–141. Studies on the ribosomes from the cellular slime molds *Dictyostelium discoideum* and *Dictyostelium purpureum.*

Chafouleas, J. G., Dedman, J. R., MacDougall, E., and Means, A. R. (1979). *Fed. Proc.* **38,** 479. Development and application of a sensitive radio immunoassay for calmodulin.

Chafouleas, J. G., Dedman, J. R., Munjaal, R. P., and Means, A. R. (1979). *J. Biol. Chem.* **254,** 10262–10267. Calmodulin development and application of a sensitive radio immunoassay.

Chagla, A. H., Lewis, K. E., and O'Day, D. H. (1980). *Exp. Cell Res.* **126,** 501–505. Calcium ion and cell fusion during sexual development in liquid cultures of *Dictyostelium discoideum.*

Chance, K., Hemmingsen, S., and Weeks, G. (1976). *J. Bact.* **128,** 21–27. Effect of cerulenin on the growth and differentiation of *Dictyostelium discoideum.*

Chang, C. M., Reitherman, R. W., Rosen, S. D., and Barondes, S. H. (1975). *Exp. Cell Res.* **95,** 136–142. Cell surface location of discoidin, a developmentally regulated carbohydrate binding protein from *Dictyostelium discoideum.*

Chang, C. M., Rosen, S. D., and Barondes, S. H. (1977). *Exp. Cell Res.* **104,** 101–109. Cell surface location of an endogenous lectin and its receptor in *Polyspondylium pallidum.*

Chang, M. T., and Raper, K. B. (1977). *Abstr. Annu. Meet. Am. Soc. Microbiol.* **77,** 168. Macrocyst formation in *Dictyostelium rosarium.*

Chang, M. T., and Raper, K. B. (1981). *J. Bacteriol.* **147,** 1049. Mating types and macrocyst formation in *Dictyostelium discoideum.*

Chang, Y. Y. (1968). *Science* **161,** 57–69. Cyclic 3′5′-AMP diesterase produced by the slime mold *Dictyostelium discoideum.*

Charest, R. P., and Cotter, D. A. (1975). *Abstr. Annu. Meet. Am. Soc. Microbiol.* **75,** 134. Deactivation of *Dictyostelium discoideum* spores.

Charlesworth, M. C., and Parish, R. W. (1975). *Eur. J. Biochem.* **54,** 307–316. The isolation of nuclei and basic nucleoproteins from the cellular slime mold *Dictyostelium discoideum.*

Charlesworth, M. C., and Parish, R. W. (1977). *Eur. J. Biochem.* **75,** 241–250. Further studies on basic nucleoproteins from the cellular slime mold *Dictyostelium discoideum.*

Chassy, B. M. (1972). *Science* **175,** 1016–1018. Cyclic nucleotide phosphodiesterase in *Dictyostelium discoideum:* interconversion of two enzyme forms.

Chi, Y. Y., and Francis, D. (1971). *J. Cell. Physiol.* **77,** 169–173. Cyclic AMP and calcium exchange in a cellular slime mold.

Chia, W. K. (1975). *Develop. Biol.* **44,** 239–252. Induction of stalk cell differentiation by cyclic AMP in a susceptible variant of *Dictyostelium discoideum.*

Chiarugi, V. P., Del Rosso, M., Cappelletti, R., Vannucchi, S., Cella, C., Fibbi, G., and Urbano, P. (1978). *Caryologia.* **31,** 183–190. Sulfated polysaccharides and the differentiation of cellular slime mold *Dictyostelium discoideum.*

Chung, S., Landfear, S. M., Blumberg, D. D., Cohen, N. S., and Lodish, H. F. (1981). *Cell* **24,** 785–797. Synthesis and stability of developmentally regulated *Dictyostelium* mRNAs are affected by cell–cell contact and cAMP.

Chung, W. J., and Coe, E. L. (1977). *Fed. Proc.* **36,** 886. Metabolic activity associated with the chemotactic and signaling responses of *Dictyostelium discoideum* to cyclic AMP.

Chung, W.-J. K., and Coe, E. L. (1978). *Biochim. Biophys. Acta* **544,** 29–44. Correlations among the responses of suspensions of *Dictyostelium discoideum* to pulses of 3′,5′ cyclic AMP transient turbidity decrease the cyclic AMP signal and variation in adenine mono nucleotide levels.

Clark, J. M., and Deering, R. A. (1980). *Radiat. Res.* **83,** 386. Enzymatic removal of pyrimidine dimers from nuclear DNA in UV irradiated *Dictyostelium discoideum.*

Clark, M. A. (1974). *J. Protozool.* **21,** 755–757. Syngenic divisions of the cellular slime mold *Polysphondylium violaceum.*

Clark, M. A. (1976). *J. Gen. Microbiol.* **93,** 166–168. Synergistic fruiting-body construction by *Polysphondylium violaceum* mutants.

Clark, M. A. (1977). *Genetics* **86,** 511–512. Organophosphate inhibition of macrocyst differentiation in the cellular slime mold *Polysphondylium violaceum.*

Clark, M. A., Francis, D., and Eisenberg, R. (1973). *Biochem. Biophys. Res. Commun.* **52,** 672–678. Mating types in cellular slime molds.

Clark, R. L., and Steck, T. L. (1979). *Science* **204,** 1163–1168. Morphogenesis in *Dictyostelium discoideum*: an orbital hypothesis.

Clarke, M. (1978). *J. Supramol. Struct.* **7,** 302. Motility in *Dictyostelium discoideum* biochemical and genetic studies.

Clarke, M., and Spudich, J. A. (1974). *J. Mol. Biol.* **86,** 209–222. Biochemical and structural studies of actomyosin-like proteins from nonmuscle cells isolation and characterization of myosin from amoebae of *Dictyostelium.*

Clarke, M., and Spudich, J. A. (1977). *Ann. Rev. Biochem.* **46,** 797–822. Nonmuscle contractile proteins: the role of actin and myosin in cell motility and shape determination.

Clarke, M., Schatten, G., Mazia, D., and Spudich, J. A. (1975). *Proc. Natl. Acad. Sci. U.S.A.* **72,** 1758–1762. Visualization of actin fibers associated with the cell membrane in amoebae of *Dictyostelium discoideum.*

Clarke, M., Bazari, W. L., and Kayman, S. C. (1980). *J. Bacteriol.* **141,** 397–400. Isolation and properties of calmodulin from *Dictyostelium discoideum.*

Clegg, J. S., and Filosa, M. F. (1961). *Nature* **192,** 1077–1078. Trehalose in the cellular slime mould *Dictyostelium discoideum.*

Cleland, S. V., and Coe, E. L. (1968). *Biochim. Biophys. Acta* **156,** 44–50. Activities of glycolytic enzymes during the early stages of differentiation in the cellular slime mold *Dictyostelium discoideum.*

Cleland, S. V., and Coe, E. L. (1969). *Biochim. Biophys. Acta* **192,** 446–454. Conversion of aspartic acid to glucose during culmination in *Dictyostelium discoideum.*

Cleveland, R. F., and Deering, R. (1972). *Int. J. of Rad. Biol.* **22,** 245–256. Radiation induced division delay in *Dictyostelium discoideum*: relation to survival in resistant and sensitive strains.

Cleveland, R. F., and Deering, R. (1976). *Int. J. of Rad. Biol.* **24,** 463–473. Changes in morphogenesis and developmental enzyme levels in *Dictyostelium discoideum* after gamma irradiation.

Cockburn, A., and Firtel, R. A. (1975). *J. Cell. Biol.* **67,** 74. Mapping of the genome of the cellular slime mold *Dictyostelium discoideum* with restriction endonucleases.

Cockburn, A., Frankel, G., Firtel, R. A., Kindle, K., and Newkirk, M. J. (1976). *ICN-UCLA Symposia on Molecular and Cellular Biology,* Vol. 5, 599–603. Organization of the ribosomal genes of *Dictyostelium.*

Cockburn, A. F., Newkirk, M. J., and Firtel, R. A. (1976). *Cell* **9,** 605–613. Organization of the ribosomal RNA genes of *Dictyostelium discoideum:* mapping of the nontranscribed spacer regions.

Cockburn, A. F., Taylor, W. C., and Firtel, R. A. (1978). *Chromosoma* **70,** 19–30. *Dictyostelium discoideum* ribosomal DNA consists of nonchromosomal palindromic dimers containing 5S and 36S coding regions.

Cocucci, S., and Sussman, M. (1970). *J. Cell. Biol.* **45,** 399–407. RNA in cytoplasmic and nuclear fractions of cellular slime mold amebas.

Coe, E. L., and Chung, W.-J. K. (1978). *In* "Symposium on Biophysical Approaches to Biological Problems." Academic Press, New York (Agris, P. F., Loeppky, R. N., and Sykes, B. D. eds.), 267–272. Use of turbidity to detect changes in cellular structure: the response of cellular slime mold amoebae to cyclic AMP.

Coffman, D. S., Leichtling, B. H., and Rickenberg, H. V. (1981). *J. Supramol. Struc. Cell. Biochem.* **15,** 369. Phosphoproteins in *Dictyostelium discoideum.*

Cohen, A. L. (1953a). *Proc. Natl. Acad. Sci. U.S.A.* **39,** 68–74. The effect of ammonia on morphogenesis in the acrasieae.

Cohen, A. L. (1953b). *Ann. N.Y. Acad. Sci.* **56,** 938–943. The isolation and culture of opsimorphic organisms. I. Occurrence and isolation of opsimophic organisms from soil and culture of Acrasieae on a standard medium.

Cohen, A. L. (1965). *Encycl. Britannica* **20,** 797–798. Slime molds.

Cohen, A. L., and Ceccarini, C. (1967). *Ann. Bot. N. Ser.* **31,** 479–487. Inhibition of spore germination in the cellular slime mold *Dictyostelium discoideum.*

Cohen, A. L., and Sussman, M. (1975). *Proc. Natl. Acad. Sci. U.S.A.* **72,** 4479–4482. Guanosine metabolism and regulation of fruiting-body construction in *Dictyostelium discoideum.*

Cohen, M. H., and Robertson, A. (1971a). *J. Theor. Biol.* **31,** 101–118. Wave propagation in the early stages of aggregation of cellular slime molds.

Cohen, M. H., and Robertson, A. (1971b). *J. Theor. Biol.* **31,** 119–135. Chemotaxis and the early stages of aggregation in the cellular slime molds.

Cohen, M. H., and Robertson, A. (1971c). *Proc. 1st Int. Conf. Cell. Diff.* (Harris, R., and Viza, D. eds.), 35–45. Differentiation for aggregation in the cellular slime molds.

Cohen, M. S. (1977). *J. Theor. Biol.* **69,** 57–86. The cyclic AMP control system in the development of *Dictyostelium discoideum.* Part 1. Cellular dynamics.

Cohen, M. S. (1978). *J. Theor. Biol.* **72,** 231–256. The cyclic AMP control system in the development of *Dictyostelium discoideum.* Part 2. An allosteric model.

Coloma, A., and Lodish, H. F. (1981). *Develop. Biol.* **81,** 238–244. Synthesis of spore-specific and stalk-specific proteins during differentiation of *Dictyostelium discoideum.*

Condeelis, J. S. (1977). *Anal. Biochem.* **78,** 374–394. The isolation of microquantities of myosin from *Amoeba proteus* and *Chaos carolinensis.*

Condeelis, J. S. (1978). *J. Cell. Biol.* **79,** 263A. A direct role for the actin cytoskeleton in the mobility of cell surface receptors.

Condeelis, J. S. (1979). *J. Cell. Biol.* **80,** 751–758. Isolation of concanavalin A caps during various stages of formation and their association with actin and myosin.

Condeelis, J. S. (1980). *Eur. J. Cell. Biol.* **22,** 620. Microfilament membrane interactions in cell shape and surface architecture.

Condeelis, J. S., and Geosits, S. (1979). *J. Cell. Biol.* **83,** 315A. Properties of gelation factors from *Dictyostelium discoideum* and the solation–contraction coupling hypothesis.

Condeelis, J. S., and Taylor, D. L. (1977). *J. Cell. Biol.* **74,** 901–928. The contractile basis of amoeboid movement. V. The control of gelation, solation and contraction in extracts from *Dictyostelium discoideum.*

Condeelis, J. S., and Taylor, D. L. (1977). *J. Cell. Biol.* **75,** 250A. Control of gelation–solation and contraction in extracts of *Dictyostelium discoideum.*

Cone, R. D., and Bonner, J. T. (1980). *Exp. Cell. Res.* **128,** 479–485. Evidence for aggregation center induction by the Ionophore A-23187 in the cellular slime mold *Polysphondylium violaceum.*

Cook, W. R. I. (1939). *Trans. Brit. Mycol. Cos.* **22,** 302–306. Some observations on *Sappinia pedata* Dang.

Cooke, R., Clarke, M., Von Wedel, R. J., and Spudich, J. A. (1976). *In* "Cold Spring Harbor Conferences on Cell Proliferation," Vol. 3. (Goldman, Pollard, R. T., and Rosenbaum, J. eds.), 575–587. Supramolecular forms of *Dictyostelium discoideum* actin.

Cooper, S., Chambers, D. A., and Scanlon, S. (1980). *Biochim. Biophys. Acta* **629,** 235–242. Identification and characterization of the cyclic AMP binding proteins appearing during the development of *Dictyostelium discoideum.*

Cordingley, J. S., and Hames, B. D. (1977). *FEBS Lett.* **82,** 263–267. Specific cleavage of ribosomal RNA in *Dictyostelium discoideum* ribosomes.

Coston, M. B., and Loomis, W. F. (1969). *J. Bacteriol.* **100,** 1208–1217. Isozymes of beta-glucosidase in *Dictyostelium discoideum.*

Cotter, D. A. (1973). *J. Theor. Biol.* **41,** 41–51. Spore germination in *Dictyostelium discoideum.* Part 1. The thermodynamics of reversible activation.

Cotter, D. A. (1975). *In* "American Society for Microbiology." Washington, D.C. (Gerhardt, P., Costilow, R. N., and Sadoff, H. L. eds.), 61–72. Spores of the cellular slime mold *Dictyostelium discoideum.*

Cotter, D. A. (1977). *Can. J. Microbiol.* **23,** 1170–1177. The effects of osmotic pressure changes on the germination of *Dictyostelium discoideum* spores.

Cotter, D. A., and George, R. P. (1975). *Arch. Microbiol.* **103,** 163–168. Germination and mitochondrial damage in spores of *Dictyostelium discoideum* following supraoptimal heating.

Cotter, D. A., and Hohl, H. R. (1969). *J. Bacteriol.* **98,** 321–322. Correlation between plaque size and spore size in *Dictyostelium discoideum.*

Cotter, D. A., and O'Connell, R. W. (1976). *Can. J. Microbiol.* **22,** 1751–1755. Activation and killing of *Dictyostelium* spores with urea.

Cotter, D. A., and Raper, K. B. (1966). *Proc. Natl. Acad. Sci. U.S.A.* **56,** 880–887. Spore germination in *Dictyostelium discoideum.*

Cotter, D. A., and Raper, K. B. (1968a). *J. Bacteriol.* **96,** 86–92. Factors affecting the rate of heat-induced spore germination in *Dictyostelium discoideum.*

Cotter, D. A., and Raper, K. B. (1968b). *J. Bacteriol.* **96,** 1680–1689. Properties of germinating spores of *Dictyostelium discoideum.*

Cotter, D. A., and Raper, K. B. (1968c). *J.Bacteriol.* **96,** 1690–1695. Spore germination in strains of *Dictyostelium discoideum* and other members of the Dictyosteliaceae.

Cotter, D. A., and Raper, K. B. (1970). *Develop. Biol.* **22,** 112–128. Spore germination in *Dictyostelium discoideum:* trehalase and the requirement for protein synthesis.

Cotter, D. A., Miura-Santo, L. Y., and Hohl, H. R. (1969). *J. Bacteriol.* **100,** 1020–1026. Ultrastructural changes during germination of *Dictyostelium discoideum* spores.

Cotter, D. A., Morin, and O'Connell, R. W. (1976). *Arch. Micro.* **108,** 93. Spore germination in *Dictyostelium discoideum.* II. Effects of dimethyl sulphoxide on postactivation lag as evidence for the multistate model of activation.

Cotter, D. A., Garnish, F. J., and Tisa, L. S. (1979). *Can. J. Microbiol.* **25,** 24–31. The physiological effects of restrictive environmental conditions on *Dictyostelium discoideum* spore germination.

Coukell, M. B. (1975). *Mol. Gen. Genetics* **142,** 119–135. Parasexual genetic analysis of aggregation deficient mutants of *Dictyostelium discoideum.*

Coukell, M. B. (1977). *Mol. Gen. Genetics* **151,** 269–273. Evidence against mutational hot spots at aggregation loci in *Dictyostelium discoideum.*

Coukell, M. B. (1977). *J. Supramol. Struct.* Suppl. 1, 73. Parasexual genetic analysis of aggregation loci in *Dictyostelium discoideum.*

Coukell, M. B., and Cameron, A. M. (1979). *J. Gen. Microbiol.* **114,** 247–256. radE: a new radiation sensitive locus in *Dictyostelium discoideum.*

Coukell, M. B., and Chan, F. K. (1980). *FEBS Lett.* **110,** 39–42. The precocious appearance and activation of an adenylate cyclase in a rapid developing mutant of *Dictyostelium discoideum.*

Coukell, M. B., and Roxby, N. M. (1977). *Mol. Gen. Genetics* **153,** 275–288. Linkage analysis of developmental mutations in aggregation deficient mutants of *Dictyostelium discoideum.*

Coukell, M. B., and Walker, I. O. (1973). *Cell. Diff.* **2,** 87–95. The basic nuclear proteins of the cellular slime mold *Dictyostelium discoideum.*

Crean, E. V., and Rossomando, E. F. (1977). *Biochem. Biophys. Res. Commun.* **75,** 488–496. Developmental changes in membrane bound enzymes of *Dictyostelium discoideum* detected by ConA–Sepharose affinity chromatography.

Crean, E. V., and Rossomando, E. F. (1977). *Biochim. Biophys. Acta* **498,** 439–442. Synthesis of a mannosylphosphoryl polyprenol by the cellular slime mold *Dictyostelium discoideum.*

Crean, E. V., and Rossomando, E. F. (1979). *Arch. Biochem. Biophys.* **1,** 186–191. Glucosphingolipid synthesis in the cellular slime mold *Dictyostelium discoideum.*

Crean, E. V., and Rossomando, E. F. (1979). *J. Gen. Microbiol.* **110,** 315–322. Effects of sugars on glycosidase secretion in *Dictyostelium discoideum.*

Crean, E. V., Lagerstedt, J. P., and Rossomando, E. F. (1979). *J. Bacteriol.* **140,** 188–196. Products of endogenous N-acetyl glucosaminyl transferase activity of *Dictyostelium discoideum* during growth and early development.

Cripps, M. M., and Rutherford, C. L. (1981). *Exp. Cell. Res.* **133,** 309. A stage-specific inhibitor of adenylate cyclase in *Dictyostelium discoideum.*

Curry, A., and Woolley, D. E. (1975). *West J. Med.* **123,** 50. The detection of actin-like proteins in some protozoa using heavy mero myosin.

Cutler, L. S. (1975). *J. Histochem. Cytochem.* **23,** 786–787. Comments on the validity of the use of lead nitrate for the cytochemical study of adenylate cyclase.

Cutler, L. S., and Christian, C. P. (1980). *J. Histochem. Cytochem.* **28,** 62–65. Cytochemical localization of adenylate cyclase.

Cutler, L. S., and Rossomando, E. F. (1975). *Exp. Cell. Res.* **95,** 79–87. Localization of adenylate cyclase in *Dictyostelium discoideum.* II. Cytochemical studies on whole cells and isolated plasma membrane vesicles.

Dahlberg, K. R., and Cotter, D. A. (1978). *Microbios.* **23,** 153–166. Auto activation of spore germination in mutant and wild type strains of *Dictyostelium discoideum.*

Dahlberg, K. R., and Cotter, D. A. (1978). *Microbios. Lett.* **9,** 139–146. Activators of *Dictyostelium discoideum* spore germination released by bacteria.

Dangeard, P. A. (1896). *Botaniste* **5,** 1–20. Contribution à l'étude des Acrasiées.

Darmon, M. (1976). *Comptes Rendus* **282,** 1893–1896. Possible regulatory role of intercellular contacts in the differentiation of *Dictyostelium discoideum.*

Darmon, M., and Brachet, P. (1978). *In* "Receptors and Recognition Series B," Vol. 5. Taxis and Behavior. Elementary Sensory Systems in Biology. Chapman and Hall, London, England (Hazelbauer, G. L., ed.), 101–139. Chemotaxis and differentiation during the aggregation of *Dictyostelium discoideum* amoebae.

Darmon, M., and Klein, C. (1976). *Biochem. J.* **154,** 743–750. Binding of concanavalin A and its effect on the differentiation of *Dictyostelium discoideum.*

Darmon, M., and Klein, C. (1978). *Develop. Biol.* **63,** 377–389. Effects of amino acids and glucose on adenylate cyclase and cell differentiation of *Dictyostelium discoideum.*

Darmon, M., Brachet, P., and Da Silva, L. H. (1975). *Proc. Natl. Acad. Sci. U.S.A.* **72,** 3163–3166. Chemotactic signals induce cell differentiation in *Dictyostelium discoideum.*

Darmon, M., Barrand, P., Brachet, P., Klein, C., and Da Silva, L. P. (1977). *Develop. Biol.* **58,** 174–184. Phenotypic supression of morphogenic mutants of *Dictyostelium discoideum.*

Darmon, M., Barra, J., and Brachet, P. (1978). *J. Cell. Sci.* **31,** 233–244. The role of phosphodiesterase in aggregation of *Dictyostelium discoideum.*

Das, D. V. M., and Weeks, G. (1978). *Can. Fed. Biol. Soc. Proc.* **21,** 116. Effect of polyunsaturated fatty-acid incorporation on *Dictyostelium discoideum.*

Das, D. V. M., and Weeks, G. (1979). *Exp. Cell. Res.* **118,** 237–244. Effects of polyunsaturated fatty-acids on the growth and differentiation of the cellular slime mold *Dictyostelium discoideum.*

Davidoff, F., and Korn, E. D. (1962). *Biochem. Biophys. Res. Commun.* **9,** 54–58. Lipids of *Dictyostelium discoideum:* phospholipid composition and the presence of two new fatty acids, CIS,CIS-5, 11-octadecadienoic and CIS,CIS-5,9-hexadecadienoic acids.

Davidoff, F., and Korn, E. D. (1963a). *J. Biol. Chem.* **238,** 3199–3209. Fatty acid and phospholipid composition of the cellular slime mold, *Dictyostelium discoideum:* the occurrence of previously undescribed fatty acids.

Davidoff, F., and Korn, E. D. (1963b). *J. Biol. Chem.* **238,** 3210–3215. The biosynthesis of fatty acids in the cellular slime mold, *Dictyostelium discoideum.*

De Chastellier, C., and Ryter, A. (1977). *J. Cell. Biol.* **75,** 218–236. Changes of the cell surface and of the digestive apparatus of *Dictyostelium discoideum* during the starvation period triggering aggregation.

De Chastellier, C., and Ryter, A. (1980). *Eur. J. Cell. Biol.* **22,** 346. Calcium dependent deposits at the plasma membrane of *Dictyostelium discoideum.*

De Chastellier, C., and Ryter, A. (1980). *Biol. Cell.* **38,** 121–128. Characteristic ultrastructural transformations upon starvation of *Dictyostelium discoideum* and their relations with aggregation study of wild type amoebae and aggregation mutants.

De Chastellier, C., Quiviger, B., and Ryter, A. (1978). *J. Ultrastruct. Res.* **62,** 220–227. Observations on the functioning of the contractile vacuole of *Dictyostelium discoideum* with the electron microscope.

Deering, R. A., and Jensen, D. S. (1975). *J. Bacteriol.* **121,** 1211–1213. Endonuclease from *Dictyostelium discoideum* that attacks UV irradiated DNA.

Deering, R. A., and Michrina, C. A. (1979). *Fed. Proc.* **38,** 484. Aspects of deoxy TMP synthesis in *Dictyostelium discoideum.*

Deering, R. A., and Sheely, M. (1977). *In* "Developments in Cell Biology." Elsevier/North-Holland (Cappuccinelli, P., and Ashworth, J. M. eds.), 63–68. Mutation induction in vegetative and developing cells of *Dictyostelium discoideum.*

Deering, R. A., and Welker, D. L. (1978). *J. Supramol. Struct.* **7,** 73. Interactions between radiation sensitive mutations in double mutant haploids of *Dictyostelium discoideum.*

De Gunzburg, J., and Veron, M. (1980). *Eur. J. Cell. Biol.* **22,** 235. Changes in cyclic AMP binding proteins upon contact formation in developing *Dictyostelium discoideum.*

De Gunzburg, J., and Veron, M. (1981). *Biochem.* **20,** 4547. Intracellular adenosine 3′,5′-phosphate binding proteins in *Dictyostelium discoideum:* partial purification and characterization in aggregation competent cells.

Dehaan, R. L. (1959). *J. Embryol. Exp. Morphol.* **7,** 335–343. The effects of the chelating agent ethylenediamine tetra-acetic acid on cell adhesion in the slime mold *Dictyostelium discoideum.*

Demeglio, D. C., and Friedman, T. B. (1978). *J. Biochem.* **83,** 693–698. Galactose metabolism in *Dictyostelium discoideum* regulation of galactose 1 phosphate uridylyl transferase during growth and development.

Demsar, I. H., and Cotter, D. A. (1981). *Photochem. Photobiol.* **34,** 455. Physiological effects of ultraviolet light on *Dictyostelium discoideum* spore germination.

Dent, V. E., Bazin, M. J., and Saunders, P. T. (1976). *Arch. Microbiol.* **109,** 187–194. Behavior of *Dictyostelium discoideum* amoebae and *Escherichia coli* grown together in chemostat culture.

Depraitere, C., and Darmon, M. (1978). *Ann. Microbiol.* **1298,** 451–462. Growth of *Dictyostelium discoideum* on different species of bacteria.

De Silva, N. S., and Siu, C-H. (1980). *J. Biol. Chem.* **255,** 8489–8496. Preferential incorporation of phospholipids into plasma membranes during cell aggregation of *Dictyostelium discoideum.*

De Silva, N. S., and Siu, C-H. (1981). *J. Biol. Chem.* **256,** 5845–5850. Vesicle-mediated transfer of phospholipids to plasma membrane during cell aggregation of *Dictyostelium discoideum.*

De Toma, F. J., Kindwall, K. E., and Reardon, C. A. (1977). *Biochem. Biophys. Res. Commun.* **74,** 350–355. The effect of tosyl lysine chloromethyl ketone on the activity of UDP glucose pyrophosphorylase of the cellular slime mold *Dictyostelium discoideum.*

Devreotcs, P. N., and Steck, T. L. (1979). *J. Cell. Biol.* **80,** 300–309. Cyclic AMP relay in *Dictyostelium discoideum.* Part 2. Requirements for the initiation and termination of the response.

Devreotes, P. N., Dinauer, M. C., and Steck, T. L. (1978). *J. Cell. Biol.* **79,** 372A. Cyclic AMP elicited cyclic AMP secretion in *Dictyostelium discoideum.*

Devreotes, P. N., Derstine, P. L., and Steck, T. L. (1979). *J. Cell. Biol.* **80,** 291–299. Cyclic AMP relay in *Dictyostelium discoideum.* Part 1. A technique to monitor responses to controlled stimuli.

Dicou, E., and Brachet, P. (1979). *Biochem. Biophys. Res. Commun.* **90,** 1321–1327. Purification of the inhibitor of the cyclic AMP phosphodiesterase of *Dictyostelium discoideum* by affinity chromatography.

Dicou, E. L., and Brachet, P. (1979). *Biochim. Biophys. Acta* **578,** 232–242. Multiple forms of an extracellular cyclic AMP phosphodiesterase from *Dictyostelium discoideum.*

Dicou, E. L., and Brachet, P. (1980). *Eur. J. Biochem.* **109,** 507–514. A separate phosphodiesterase for the hydrolysis of cyclic GMP in growing *Dictyostelium discoideum* amoebae.

Dicou, E., and Brachet, P. (1981). *Biochem. Biophys. Res. Comm.* **102,** 1172. Interaction of the inhibitor of the 3′,5′-cyclic AMP phosphodiesterase of *Dictyostelium discoideum* with the affi-gel blue.

Dicou, E. L., Brachet, P., Huez, G., and Marbaix, G. (1979). *FEBS Lett.* **104,** 275–278. Synthesis of *Dictyostelium discoideum* secretory proteins in *Xenopus laevis* oocytes.

Dimond, R., and Loomis, W. F. (1974). *J. Biol. Chem.* **249,** 5628–5632. Vegetative isozyme of N-acetylglucosaminidase in *Dictyostelium discoideum.*

Dimond, R., and Loomis, W. F. (1975). *ICN-UCLA Symp. Mol. Cell. Biol.* **2,** 533–538. The role of UDPG pyrophosphorylase during development in *Dictyostelium discoideum.*

Dimond, R., and Loomis, W. F. (1976). *J. Biol. Chem.* **251,** 2680–2687. Structure and function of beta-glucosidases in *Dictyostelium discoideum.*

Dimond, R., Brenner, M., and Loomis, W. F. (1973). *Proc. Natl. Acad. Sci. U.S.A.* **70,** 3356–3360. Mutations affecting N-acetylglucosaminidase in *Dictyostelium discoideum.*

Dimond, R. L., Mayer, M., and Loomis, W. F. (1976). *Develop. Biol.* **52,** 74–82. Characterization and development regulation of beta galactosidase EC-3.2. 1.23 isozymes in *Dictyostelium discoideum.*

Dimond, R. L., Farnsworth, P. A., and Loomis, W. F. (1976). *Develop. Biol.* **50,** 169–181. Isolation and characterization of mutations affecting UDPG pyrophosphorylase activity in *Dictyostelium discoideum.*

Dimond, R. L., Burns, R. A., and Jordan, K. B. (1981). *J. Biol. Chem.* **256,** 6565–6572. Secretion of lysosomal enzymes in the cellular slime mold. *Dictyostelium discoideum.*

Dinauer, M. C., Mackay, S. A., and Devreotes, P. N. (1980). *J. Cell. Biol.* **86,** 537–544. Cyclic AMP relay in *Dictyostelium discoideum* 3. The relationship of cyclic AMP synthesis and secretion during the cyclic AMP signaling response.

Dinauer, M. C., Steck, T. L., and Devreotes, P. N. (1980). *J. Cell. Biol.* **86,** 545–553. Cyclic AMP relay in *Dictyostelium discoideum* 4. Recovery of the cyclic AMP signaling response after adaptation to cyclic AMP.

Dinauer, M. C., Steck, T. L., and Devreotes, P. N. (1980). *J. Cell. Biol.* **86,** 554–561. Cyclic AMP relay in *Dictyostelium discoideum* 5. Adaptation of the cyclic AMP signaling response during cyclic AMP stimulation.

Dingermann, T., Rummel, H., and Kersten, H. (1977). *Hoppe-Seyler's Z. Physiol. Chem.* **358,** 1193. Developmental changes in transfer RNA modification and aminoacyl transfer RNA synthetase activities of *Dictyostelium discoideum.*

Dingermann, T., Schmidt, W., and Kersten, H. (1977). *FEBS Lett.* **80,** 205–209. Modified bases in tRNA of *Dictyostelium discoideum:* alterations in the ribothimidine content during development.

Dingermann, T., Mach, M., and Kersten, H. (1978). *Hoppe-Seyler's Z. Physiol. Chem.* **359,** 1074–1075. Specific ribothymidine lacking RNA in developing *Dictyostelium discoideum.*

Dingermann, T., Mach, M., and Kersten, H. (1979). *J. Gen. Microbiol.* **115,** 223–232. Synthesis of transfer RNA with uridine or 2′-0 methyl ribothymidine at position 54 in developing *Dictyostelium discoideum.*

Dingermann, T., Pistel, F., and Kersten, H. (1979). *Eur. J. Cell. Biol.* **20,** 120. Early developmental changes in protein synthesis of *Dictyostelium discoideum.*

Dingermann, T., Ogilvie, A., and Kersten, H. (1980). *Eur. J. Cell. Biol.* **22,** 144. Uncharged asparaginyl transfer RNA isoacceptors in *Dictyostelium discoideum* as an early event after the onset of development.

Dingermann, T., Pistel, F., and Kersten, H. (1980). *Biochem. Soc. Trans.* **8,** 90. Early developmental changes in transfer RNA of *Dictyostelium discoideum.*

Dingermann, T., Pistel, F., and Kersten, H. (1980). *Eur. J. Biochem.* **104,** 33–40. Functional role of ribosyl thymine in transfer RNA preferential utilization of transfer RNA containing ribosyl thymine instead of uridine at position 54 in protein synthesis of *Dictyostelium discoideum.*

Dottin, R., Weiner, A., and Lodish, H. (1976). *Cell* **8,** 233–244. 5′ Terminal nucleotide sequences of the mRNA's of *Dictyostelium discoideum.*

Douvas, A. S., Bakke, A., and Bonner, J. (1977). *In* "The Molecular Biology of the Mammalian Genetic Apparatus," Vol. 1. Elsevier/North Holland, Amsterdam, Netherlands (Ts'O, and Paul, O. P. eds.), 143–163. Evidence for the presence of contractilelike monhistone proteins in the nuclei and chromatin of eukaryotes.

Dowbenko, D. J., and Ennis, H. L. (1980). *Proc. Natl. Acad. Sci. U.S.A.* **77,** 1791–1795. Regulation of protein synthesis during spore germination in *Dictyostelium discoideum.*

Dudock, B., Greenberg, R., Fields, L. E., Brenner, M., and Palatnik, C. M. *Biochim. Biophys. Acta* **608,** 295–300. Base composition of transfer RNA from vegetative and developing cells of *Dictyostelium discoideum.*

Durston, A. (1973). *J. Theor. Biol.* **42,** 483–504. *Dictyostelium discoideum* aggregation fields as excitable media.

Durston, A. (1974). *Develop. Biol.* **37,** 225–235. Pacemaker activity during aggregation in *Dictyostelium discoideum.*

Durston, A. J. (1974). *Develop. Biol.* **38,** 308–319. Pacemaker mutants of *Dictyostelium discoideum.*

Durston, A. J. (1976). *Nature (London)* **263,** 126–129. Tip formation is regulated by an inhibitory gradient in the *Dictyostelium discoideum.*

Durston, A. J., and Vork, F. (1977). *In* "Development and Differentiation in the Cellular Slime Moulds." Elsevier/North-Holland (P. Cappuccinelli, and J. M. Ashworth, eds.), 17–26. The control of morphogenesis and pattern in the *Dictyostelium discoideum* slug.

Durston, A. J., and Vork, F. (1978). *Exp. Cell. Res.* **115,** 454–457. The spatial pattern of DNA synthesis in *Dictyostelium discoideum* slugs.

Durston, A. J., and Vork, F. (1979). *J. Cell. Sci.* **36,** 261–279. A cinematographical study of the development of vitally stained *Dictyostelium discoideum.*

Durston, A. J., Cohen, M., Drage, D., Potel, M., Robertson, A., and Wonio, D. (1976). *Develop. Biol.* **52,** 173–180. Periodic movements of *Dictyostelium discoideum* sorocarps.

Durston, A. J., Vork, F., and Weinberger, C. (1979). *In* "Biophysical and Biochemical Information Transfer in Recognition." Proceedings of the 2nd International Colloquium on Physical and Chemical Information Transfer in Regulation of Reproduction and Aging. Plenum Press, New York (Vassileva-Popova, J. G., and Jensen, E. V. eds.), 693–708. The control of later morphogenesis by chemotactic signals in *Dictyostelium discoideum.*

Dutta, S. K., and Garber, E. D. (1961). *Proc. Natl. Acad. Sci. U.S.A.* **47,** 990–993. The identification of physiological races of a fungal phytopathogen using strains of the slime mold *Acrasis rosea.*

Dutta, S. K., and Mandel, M. (1972). *J. Protozool.* **19,** 538–540. DNA base composition of some cellular slime molds.

Dykstra, M. J., and Aldrich, H. C. (1978). *J. Protozool.* **25,** 38–41. Successful demonstration of an elusive cell coat in amoebae.

Eckert, B. S., and Lazarides, E. (1978). *J. Cell. Biol.* **77,** 714–721. Localization of actin in *Dictyostelium discoideum* amoebas by immunofluorescence.

Eckert, B. S. Warren, R. H., and Rubin, R. W. (1977). *J. Cell. Biol.* **72,** 339–350. Structural and biochemical aspects of cell motility in amoebae of *Dictyostelium discoideum.*

Edelson, P. J., and Cohn, Z. A. (1978). *Cell. Surf. Rev.* **5,** 387–405. Endocytosis: regulation of membrane interactions.

Edmundson, T. D., and Ashworth, J. M. (1972). *Biochem. J.* **126,** 593–600. 6-Phosphogluconate dehydrogenase and the assay of uridine diphosphate glucose phyrophosphorylase in the cellular slime mould *Dictyostelium discoideum.*

Eisenberg, R. M. (1976). *Ecology* **57,** 380–384. 2-Dimensional micro distribution of cellular slime molds in forest soil.

Eisenberg, R. M., and Francis, D. (1977). *J. Protozoology* **24,** 182–183. The breeding system of *Polysphondylium pallidum,* a cellular slime mould.

Eitle, E., and Gerisch, G. (1977). *Cell. Differ.* **6,** 339–346. Implication of developmentally regulated concanavalin A binding proteins of *Dictyostelium* in cell adhesion and cyclic AMP regulation.

Elder, J. H., Pickett, R. A. II., Hampton, J., and Lerner, R. A. (1977). *J. Biol. Chem.* **252,** 6510–6515. Radio-Iodination of proteins in single polyacrylamide gel slices component systems using microgram quantities of total protein.

Ellingson, J. S. (1974). *Biochim. Biophys. Acta* **337,** 60–67. Changes in the differentiating cellular slime mold *Dictyostelium discoideum.*

Ellingson, J. S. (1980). *Biochem.* **19,** 6176–6182. Identification of N-Acyl ethanolamine phoshoglycerides and acylphosphatidyl glycerol as the phospholipids which disappear as *Dictyostelium discoideum* cells aggregate.

Ellingson, J. S., Telser, A., and Sussman, M. (1971). *Biochim. Biophys. Acta* **224,** 388–395. Regulation of functionally related enzymes during alternative developmental programs.

Ellouz, R., and Lenfant, M. (1971). *Eur. J. Biochem.* **23,** 544–550. Biosynthèse de la chaine laterale ethyle du stigmastanol et du stigmasten-22, 01-3 beta du myxomycete *Dictyostelium discoideum.*

Emery, H. S., and Weiner, A. M. (1981). *Cell* **26,** 411–419. An irregular satellite sequence is found at the termini of the linear extrachromosomal rDNA in *Dictyostelium discoideum.*

Emyanitoff, R. G., and Wright, B. (1979). *J. Bacteriol.* **140,** 1008–1012. Effect of intracellular carbohydrates on heat resistance of *Dictyostelium discoideum* spores.

Ennis, H. L. (1981). *Arch. Biochem. Biophys.* **209,** 371. Changes in free amino acid levels during cellular slime mold spore and microcyst germination.

Ennis, H. L., and Sussman, M. (1958a). *J. Gen. Microbiol.* **18,** 433–449. Synergistic morphogenesis by mixtures of *Dictyostelium discoideum* wild-type and aggregateless mutants.

Ennis, H. L., and Sussman, M. (1958b). *Proc. Natl. Acad. Sci. U.S.A.* **44,** 401–411. The initiator cell for slime mould aggregation.

Ennis, H. L., and Sussman, M. (1975). *J. Bacteriol.* **124,** 62–64. Mutants of *Dictyostelium discoideum* defective in spore germination.

Ennis, H. L., Pennica, D., and Hill, J. M. (1978). *Develop. Biol.* **65,** 251–259. Synthesis of macro molecules during microcyst germination in the cellular slime mold *Polysphondylium pallidum.*

Erdos, G. W., and Aldrich, H. C. (1977). *J. Cell. Biol.* **75,** 59A. Plasma membrane changes in the amoebae of *Dictyostelium discoideum* during mating.

Erdos, G. W., and Hohl, H. R. (1980). *Cytobios.* **29,** 7–16. Freeze fracture examination of the plasma membrane of the cellular slime mold *Polysphondylium pallidum* during microcyst formation and germination.

Erdos, G. W., Nickerson, A. W., and Raper, K. B. (1972). *Cytobiol.* **6,** 351–366. Fine structure of macrocysts in *Polysphondylium violaceum.*

Erdos, G. W., Nickerson, A. W., and Raper, K. B. (1973a). *Develop. Biol.* **32,** 321–330. The fine structure of macrocyst germination in *Dictyostelium mucoroides.*

Erdos, G. W., Raper, K. B., and Vogen, L. K. (1973b). *Proc. Natl. Acad. Sci. U.S.A.* **70,** 1828–1830. Mating types and macrocyst formation in *Dictyostelium discoideum.*

Erdos, G. W., Raper, K. B., and Vogen, L. K. (1975). *Proc. Natl. Acad. Sci. U.S.A.* **72,** 970–973. Sexuality in the cellular slime mold *Dictyostelium giganteum.*

Erdos, G. W., Raper, K. B., and Vogen, L. K. (1976). *J. Bacteriol.* **128,** 495–497. Effects of light and temperature of macrocyst formation in paired mating types of *Dictyostelium discoideum.*

Erickson, S. K., and Ashworth, J. M. (1969). *Biochem. J.* **113,** 567–568. The mitochondrial electron-transport system of the cellular slime mould *Dictyostelium discoideum.*

Every, D., and Ashworth, J. M. (1973). *Biochem. J.* **133,** 37–47. The purification and properties of extracellular olycosidases of the cellular slime mould *Dictyostelium discoideum.*

Every, D., and Ashworth, J. M. (1974). *Biochem. J.* **143,** 785–787. Immunological evidence to show that the N-acetyl glucosaminidase and N-acetyl galactosaminidase activities of *Dictyostelium discoideum* reside in the same protein molecule.

Every, D., and Ashworth, J. M. (1975). *Biochem. J.* **148,** 161–168. Rates of accumulation of glycosidase activities during growth and differentiation of *Dictyostelium discoideum.*

Every, D., and Ashworth, J. M. (1975). *Biochem. J.* **148,** 169–178. Rates of degradation and synthesis of glycosidases *de novo* during growth and differentiation of *Dictyostelium discoideum.*

Farnham, C. J. M. (1975). *Exp. Cell. Res.* **91,** 36–46. Cytochemical localization of adenylate cyclase and 3′,5′ nucleotide phosphodiesterase in *Dictyostelium discoideum.*

Farnsworth, P. (1973). *J. Embryol. Exp. Morphol.* **29,** 253–266. Morphogenesis in the cellular slime mould *Dictyostelium discoideum:* the formation and regulation of aggregate tips and the specification of developmental axes.

Farnsworth, P. (1974). *J. Embryol. Exp. Morphol.* **31,** 435–461. Experimentally induced aberrations in the pattern of differentiation in the cellular slime mould *Dictyostelium discoideum.*

Farnsworth, P. A. (1975). *J. Embryol. Exp. Morphol.* **33,** 869–877. Proportionality in the pattern of differentiation of the cellular slime mould *Dictyostelium discoideum* and the time of its determination.

Farnsworth, P., and James, R. (1972). *Gen. Microbiol.* **73,** 447–454. An effect of the lon phenotype in *Escherichia coli* as indicated by the growth of amoebae of *Dictyostelium discoideum.*

Farnsworth, P., and Loomis, W. F. (1974). *Develop. Biol.* **41,** 77–83. A barrier to diffusion in pseudoplasmodia of *Dictyostelium discoideum.*

Farnsworth, P., and Loomis, W. F. (1975). *Develop. Biol.* **46,** 349–357. A gradient in the thickness of the surface sheath in pseudoplasmodia of *Dictyostelium discoideum.*

Farnsworth, P. A., and Loomis, W. F. (1976). *J. Embryol. Exp. Morphol.* **35,** 499–505. Quantitation of the spatial distribution of prespore vacuoles in pseudoplasmodia of *Dictyostelium discoideum.*

Farnsworth, P., and Wolpert, L. (1971). *Nature (London)* **231,** 329–330. Absence of cell sorting out in the grex of the slime mould *Dictyostelium discoideum.*

Farrell, C. A., and De Toma, F. J. (1973). *Biochem. Biophys. Res. Commun.* **54,** 1504–1510. Increased capacity for RNA-synthesis in *Dictyostelium discoideum* nuclei by exposure to cyclic AMP or 5′-AMP.

Faust, R. G., and Filosa, M. F. (1959). *J. Cell. Comp. Physiol.* **54,** 297–298. Permeability studies on the amoebae of the slime mold, *Dictyostelium mucoroides.*

Favard-Sereno, C., and Livrozet, M. (1979). *Biol. Cell.* **35,** 45–54. Plasma membrane structural changes correlated with the acquisition of aggregation competence in *Dictyostelium discoideum.*

Favard-Sereno, C., Ludosky, M. A., and Ryter, A. (1981). *J. Cell Sci.* **51,** 63. Freeze-fracture study of phagocytosis in *Dictyostelium discoideum.*

Fayod, V. (1883). *Botan. Zeitung* **41,** 169–177. Beitrag zur kenntnis niederer myxomyceten.

Feit, I. N., Fournier, G. A., Needleman, R. D., and Underwood, M. Z. (1978). *Science* **200,** 439–441. Induction of stalk and spore cell differentiation by cyclic AMP in slugs of *Dictyostelium discoideum.*

Feinberg, A. P., Springer, W. R., and Barondes, S. H. (1979). *Proc. Natl. Acad. Sci. U.S.A.* **76,** 3977–3981. Segregation of prestalk and prespore cells of *Dictyostelium discoideum*: observations consistent with selective cell cohesion.

Felenbok, B., Monier, F., and Guespin-Michel, J. (1973). *In* "Regulation de la Sporulation Microbienne." Centre National de la Recherche Scientifique, Paris, France (Aubert, J. P. P., Schaeffer, and Szulmajster, J. eds.), 133–134. The action of 5 bromo-2-deoxy uridine on the development of *Dictyostelium discoideum.*

Felenbok, B., Monier, F., and Guespin-Michel, J. F. (1974). *Cell Differ.* **3,** 55–62. Effect of the thymidine analog 5 bromo-2-deoxy uridine on *Dictyostelium discoideum* development.

Ferber, E., Munder, P. G., Fischer, H., and Gerisch, G. (1970). *Eur. J. Biochem.* **14,** 253–257. High phospholipase activities in amoebae of *Dictyostelium discoideum.*

Ferguson, R., and Soll, D. R. (1976). *Develop. Biol.* **52,** 158–160. The absence of a growth inhibitor during the log phase growth of *Dictyostelium discoideum.*

Ferguson, R., and Soll, D. R. (1977). *Exp. Cell Res.* **106,** 159. Soluble factors and the regulation of early rate limiting events in slime mould morphogenesis.

Filosa, M. F. (1962). *Amer. Natur.* **96,** 79–91. Heterocytosis in cellular slime molds.

Filosa, M. F. (1978). *Differentiation* **10,** 177–180. Concanavalin mediated attachment to membranes of a cellular slime mold cyclic AMP phosphodiesterase.

Filosa, M. F. (1979). *J. Exp. Zool.* **207,** 491–496. Macrocyst formation in the cellular slime mold *Dictyostelium mucoroides*: involvement of light and volatile morphogenetic substances.

Filosa, M. F., and Chan, M. (1972). *J. Gen. Microbiol.* **71,** 413–414. The isolation from soil of macrocyst forming strains of the cellular slime mold *Dictyostelium mucoroides.*

Filosa, M. F., and Dengler, R. E. (1972). *Develop. Biol.* **29,** 1–16. Ultrastructure of macrocyst formation in the cellular slime mold, *Dictyostelium mucoroides:* extensive phagocytosis of amoebae by a specialized cell.

Filosa, M. F., Kent, S. G., and Gillette, M. U. (1975). *Develop. Biol.* **46,** 49–55. The developmental capacity of various stages of a macrocyst forming strain of the cellular slime mold *Dictyostelium mucoroides.*

Finney, R., Varnum, B., and Soll, D. R. (1979). *Develop. Biol.* **73,** 290–303. Erasure in *Dictyostelium discoideum*, a dedifferentiation involving the programmed loss of chemotactic functions.

Firtel, R. A. (1972). *Develop. Biol.* **66,** 363–377. Changes in the expression of single-copy DNA during development of the cellular slime mold *Dictyostelium discoideum.*

Firtel, R. A., and Bonner, J. T. (1972a). *J. Mol. Biol.* **66,** 339–361. Characterization of the genome of the cellular slime mold *Dictyostelium discoideum.*

Firtel, R. A., and Bonner, J. T. (1972b). *Develop. Biol.* **29,** 85–103. Developmental control of alpha 1-4 glucan phosphorylase in the cellular slime mold *Dictyostelium discoideum.*

Firtel, R. A., and Brackenbury, R. W. (1972). *Develop. Biol.* **27,** 307–321. Partial characterization of several protein and amino acid metabolizing enzymes in the cellular slime mold *Dictyostelium discoideum.*

Firtel, R. A., and Cockburn (1976). *J. Mol. Biol.* **102,** 831. Structural organization of the genome of *Dictyostelium discoideum:* analysis by *E. coli* restriction endonuclease.

Firtel, R. A., and Kindle, K. (1975). *Cell* **5,** 401–411. Structural organization of the genome of the cellular slime mold *Dictyostelium discoideum:* interspersion of repetitive and single-copy DNA sequences.

Firtel, R. A., and Lodish, H. F. (1973). *J. Mol. Biol.* **79,** 295–314. A small nuclear precursor of messenger RNA in the cellular slime mold *Dictyostelium discoideum.*

Firtel, R. A., and Jacobson, A. (1977). *In* "International Review of Biochemistry," Vol. 15. Biochemistry of Cell Differentiation II. Univ. Park Press, Baltimore, MD. (Paul, J. ed.), 377–429. Structural organization and transcription of the genome of *Dictyostelium discoideum.*

Firtel, R. A., and Pederson, T. (1975). *Proc. Natl. Acad. Sci.* **72,** 301–305. Ribonucleoprotein particles containing heterogeneous nuclear RNA in the cellular slime mold *Dictyostelium discoideum.*

Firtel, R. A., Jacobson, A., and Lodish, H. F. (1972). *Nature New Biol.* **239,** 225–228. Isolation and hybridization kinetics of messenger RNA from *Dictyostelium discoideum.*

Firtel, R. A., Baxter, L., and Lodish, H. F. (1973). *J. Mol. Biol.* **79,** 315–327. Actinomycin D and the regulation of enzyme biosynthesis during development of *Dictyostelium discoideum.*

Firtel, R. A., Jacobson, A., Tuchman, J., and Lodish, H. F. (1974). *Genetics* **78**, 355–372. Gene activity during development of the cellular slime mold *Dictyostelium discoideum.*

Firtel, R. A., Kindle, K., and Huxley, M. P. (1976). *Fed. Proc.* **35**, 13–22. Structural organization and processing of the genetic transcript in the cellular slime mold *Dictyostelium discoideum.*

Firtel, R. A., Cockburn, A., Frankel, G., and Hershfield, V. (1976). *J. Mol. Biol.* **102**, 831–852. Structural organization of the genome of *Dictyostelium discoideum:* analysis of ecoR1 restriction endonuclease.

Firtel, R. A., Timm, R., Kimmel, A. R., and McKeown, M. (1979). *Proc. Natl. Acad. Sci. U.S.A.* **76**, 6206–6210. Unusual nucleotide sequences at the 5′ end of actin genes in *Dictyostelium discoideum.*

Firtel, R. A., McKeown, M., Poole, S., Kimmel, A. R., Brandis, J., and Rowekamp, W. (1981). Genetic Engineering **3**, 265–318. Developmentally regulated multigene families in *Dictyostelium discoideum.*

Fisher, P. R., and Williams, K. L. (1981). *FEMS Microb. Lett.* **12**, 87. Bidirectional phototaxis by *Dictyostelium discoideum* slugs.

Fisher, P. R., Smith, E., and Williams, K. L. (1981). *Cell* **23**, 799–808. An extracellular chemical signal controlling phototactic behavior by *Dictyostelium discoideum* slugs.

Fong, D., and Bonner, J. T. (1979). *Proc. Natl. Acad. Sci. U.S.A.* **76**, 6481–6485. Proteases in cellular slime mold *Dictyostelium discoideum* development: evidence for their involvement.

Fong, D., and Rutherford, C. L. (1978). *J. Bacteriol.* **134**, 521–527. Protease activity during cell differentiation of the cellular slime mold *Dictyostelium discoideum.*

Ford, W. T., Jr., and Deering, R. A. (1979). *Photochem. Photobiol.* **30**, 653–660. Survival spore formation and excision repair of UV irradiated developing cells of *Dictyostelium discoideum* NC-4.

Forman, D., and Garrod, D. R. (1977). *J. Embr. Exp. Morph.* **40**, 25–228. Pattern formation in *Dictyostelium discoideum.* 1. Development of prespore cells and its relationship to the pattern of the fruiting body.

Forman, D., and Garrod, D. R. (1977). *J. Embr. Exp. Morph.* **40**, 229–243. Pattern formation in *Dictyostelium discoideum.* Part 2. Differentiation and pattern formation in nonpolar aggregates.

Francis, D. (1975). *Adv. Cyclic Nucleotide Res.* **5**, 832. Genetic regulation of cyclic nucleotide production in a cellular slime mold *Polysphondylium pallidum.*

Francis, D. (1975). *J. Gen. Microbiol.* **89**, 310–318. Macrocyst genetics in *Polysphondylium pallidum* a cellular slime mold.

Francis, D. (1976). *Develop. Biol.* **53**, 62–72. Changes in protein synthesis during alternate pathways of differentiation in the cellular slime mold *Polysphondylium pallidum.*

Francis, D. (1977). *Develop. Biol.* **55**, 339–346. Synthesis of developmental proteins in morphogenetic mutants of *Polysphondylium pallidum.*

Francis, D. (1980). *Genetics* **96**, 125–136. Techniques and marker genes for use in macrocyst genetics with *Polysphondylium pallidum.*

Francis, D., and Jones, W. R. (1972). *J. Protozool.* **19**, 25. Protein analysis of cell differentiation in the cellular slime mold *Polysphondylium pallidum.*

Francis, D., and Lin, L. (1980). *Develop. Biol.* **79**, 238–242. Heat shock response in a cellular slime mold *Polysphondylium pallidum.*

Francis, D., Salmon, D., and Moore, B. (1978). *Develop. Biol.* **67**, 232–236. A mutant strain of *Polysphondylium pallidum* deficient in production of cyclic AMP.

Francis, D. W. (1964). *J. Cell. Comp. Physiol.* **64**, 131–138. Some studies on phototaxis of *Dictyostelium.*

Francis, D. W. (1965). *Develop. Biol.* **12,** 329–346. Acrasin and the development of *Polysphondylium pallidum.*

Francis, D. W. (1969). *Quart. Rev. Biol.* **44,** 277–290. Time sequences for differentiation in cellular slime molds.

Francis, D. W. (1978). *Genetics* **88,** 29. Divergent differentiation in a cellular slime mold.

Francis, D. W. (1979). *Differentiation* **15,** 187–192. True divergent differentiation in a cellular slime mold *Polysphondylium pallidum.*

Francis, D. W., and O'Day, D. H. (1971). *J. Exp. Zool.* **176,** 265–272. Sorting out in pseudoplasmodia of *Dictyostelium discoideum.*

Franke, J., and Kessin, R. (1977). *Proc. Natl. Acad. Sci. U.S.A.* **74,** 2157–2161. A defined minimal medium for axenic strains of *Dictyostelium discoideum.*

Franke, J., and Kessin, R. (1978). *Nature (London)* **272,** 537–538. Auxotrophic mutants of *Dictyostelium discoideum.*

Franke, J., and Kessin, R. (1981). *J. Biol. Chem.* **117,** 213–218. The cAMP phosphodiesterase inhibitory protein of *Dictyostelium discoideum.* Purification and characterization.

Franke, J., and Sussman, M. (1971). *J. Biol. Chem.* **246,** 6381–6388. Synthesis of uridine diphosphate glucose pyrophysphorylase during the development of *Dictyostelium discoideum.*

Franke, J., and Sussman, M. (1973). *J. Mol. Biol.* **81,** 173–185. Accumulation of uridine diphosphoglucose pyrophosphatase in *Dictyostelium discoideum* via preferential synthesis.

Frankel, G., Cockburn, A. F., Kindle, K. L., and Firtel, R. A. (1977). *J. Mol. Biol.* **109,** 539–558. Organization of the ribosomal RNA genes of *Dictyostelium discoideum.*

Frazier, W., Rosen, S., Reitherman, S., and Barondes, S. (1975). *J. Bio. Chem.* **250,** 7714–7721. Purification and comparison of two developmentally regulated lectins from *Dictyostelium discoideum.*

Frazier, W. A., Rosen, S. D., Reitherman, R. W., Simpson, D. L., and Barondes, S. H. (1975). *Fed. Proc.* **34,** 704. Purification and characterization of a developmentally regulated class of lectins from cellular slime molds.

Free, S. J., and Loomis, W. F. (1975). *Biochimie* **56,** 1525–1528. Isolation of mutations in *Dictyostelium discoideum* affecting alpha-mannosidase.

Free, S. J., and Schimke, R. T. (1978). *J. Biol. Chem.* **253,** 4107–4111. Effects of a post translational modification mutation on different developmentally regulated glycosidases in *Dictyostelium discoideum.*

Free, S. J., Cockburn, A., and Loomis, W. F. (1976). *Develop. Biol.* **49,** 539–543. Alpha mannosidase 2 a developmentally regulated enzyme in *Dictyostelium discoideum.*

Free, S. J., Schimke, R. T., and Loomis, W. F. (1976). *Genetics* **84,** 159–174. The structural gene for alpha-mannosidase-1 in *Dictyostelium discoideum.*

Free, S. J., Schimke, R. T., Freeze, H., and Loomis, W. F. (1978). *J. Biol. Chem.* **253,** 4102–4106. Characterization and genetic mapping of modA, a mutation in the posttranslational modification of the glycosidases of *Dictyostelium discoideum.*

Freeze, H., and Loomis, W. F. (1977a). *J. Biol. Chem.* **252,** 820–824. Isolation and characterization of a component of the surface sheath of *Dictyostelium discoideum.*

Freeze, H., and Loomis, W. F. (1977b). *Develop. Biol.* **56,** 184–194. The role of the fibrillar component of the surface sheath in the morphogenesis of *Dictyostelium discoideum.*

Freeze, H., and Loomis, W. F. (1978). *Biochim. Biophys. Acta* **539,** 529–537. Chemical analysis of stalk components of *Dictyostelium discoideum.*

Freeze, H. H., and Miller, A. L. (1980). *Mol. Cell. Biochem.* **35,** 17–27. Mod A a posttranslational mutation affecting phosphorylated and sulfated glycopeptides in *Dictyostelium discoideum.*

Freeze, H. H., Miller, A. L., and Kaplan, A. (1980). *J. Biol. Chem.* **255**, 11081–11084. Acid hydrolases from *Dictyostelium discoideum* contain phosphomannosyl recognition markers.

Freidlin, M. J., and Sivak, S. A. (1979). *Stud. Biophys.* **76**, 129–136. Small parameter method in multidimensional reaction diffusion problem: model of cyclic AMP signals in *Dictyostelium discoideum*.

Freim, J. O., Jr., and Deering, R. A. (1970). *J. Bacteriol.* **102**, 36–42. Ultraviolet irradiation of the vegetative cells of *Dictyostelium discoideum*.

Fukui, Y. (1976). *Devel. Growth Differentiat.* **18**, 145–156. Enzymic dissociation of nascent macrosysts and partition of the liberated cytophagic giant cells in *Dictyostelium mucroides*.

Fukui, Y. (1980). *J. Cell. Biol.* **86**, 181–189. Formation of multinuclear cells induced by dimethyl sulfoxide: inhibition of cytokinesis and occurrence of novel nuclear division.

Fukui, Y., and Katsumaru, H. (1979). *Exp. Cell. Res.* **120**, 451–453. Nuclear actin bundles in amoeba-proteus *Dictyostelium mocoroides* and human HeLa cells induced by dimethyl sulfoxide.

Fukui, Y., and Katsumaru, H. (1980). *J. Cell. Biol.* **84**, 131–140. Dynamics of nuclear actin bundle induction by dimethyl sulfoxide and factors affecting its development.

Fukui, Y., and Miyake, Y. (1975). *Cell Struct. Funct.* **1**, 23–31. Parasexual hybridization in cellular slime molds. Part 1. Appearance of hybrid clones at high frequency in a short period and its relation to cell fusion in *Dictyostelium discoideum*.

Fukui, Y., and Takeuchi, I. (1971). *J. Gen. Microbiol.* **67**, 307–317. Drug resistant mutants and appearance of heterozygotes in the cellular slime mould *Dictyostelium discoideum*.

Fuller, M. S., and Rakatansky, R. M. (1966). *Can. J. Bot.* **44**, 269–274. A preliminary study of the carotenoids in *Acrasis rosea*.

Galeotti, C. L., and Williams, K. L. (1978). *J. Gen. Microbiol.* **104**, 337–342. Giemsa staining of mitotic chromosomes in *Kluyveromyces lactis* and *Saccharomyces cerevisiae*.

Garber, L. J., Tsang, M. L-S., and Schiff, J. A. (1978). *Plant Physiol.* **61**, 101. Distribution of adenosine 5 phosphosulfate cyclase, an enzyme forming cyclic AMP.

Garrod, D. R. (1969a). *J. Cell Sci.* **4**, 781–798. The cellular basis of movement of the migrating grex of the slime mould *Dictyostelium discoideum*.

Garrod, D. R. (1969b). *New Sci.* August 285–287. The way some cells creep.

Garrod, D. R. (1972). *Exp. Cell Res.* **72**, 588–591. Acquisition of cohesiveness by slime mould cells prior to morphogenesis.

Garrod, D. R. (1974). *Arch. Biol. (Bruxelles)* **85**, 7–31. Cellular recognition and specific cell adhesion in cellular slime mould development.

Garrod, D. R. (1974). *J. Embryol. Exp. Morphol.* **32**, 57–68. The cellular basis of movement of the migrating grex of the slime mold *Dictyostelium discoideum* chemotactic and reaggregation behavior of grex cells.

Garrod, D. R., and Ashworth, J. M. (1972). *J. Embryol. Exp. Morphol.* **28**, 463–479. Effect of growth conditions on development of the cellular slime mould, *Dictyostelium discoideum*.

Garrod, D. R., and Ashworth, J. M. (1973). *Symp. Soc. Gen. Microbiol.* **23**, 407–435. Development of the cellular slime mould *Dictyostelium discoideum*.

Garrod, D. R., and Born, G. V. R. (1971). *J. Cell Sci.* **8**, 751–765. Effect of temperature on the mutual adhesion of preaggregation cells of the slime mould, *Dictyostelium discoideum*.

Garrod, D. R., and Forman, D. (1977). *Nature (London)* **265**, 144–146. Pattern formation in the absence of polarity in *Dictyostelium discoideum*.

Garrod, D. R., and Gingell, D. (1970). *J. Cell Sci.* **6**, 277–284. A progressive change in the electrophoretic mobility of preaggregation cells of the slime mould, *Dictyostelium discoideum*.

Garrod, D. R., and Malkinson, A. (1973). *Exp. Cell Res.* **81**, 492–495. Cyclic AMP, pattern formation and movement in the slime mould, *Dictyostelium discoideum*.

Garrod, D. R., and Wolpert, L. (1968). *J. Cell Sci.* **3,** 365–372. Behaviour of the cell surface during movement of preaggregation cells of the slime mould *Dictyostelium discoideum.*

Garrod, D. R., Palmer, J. F., and Wolpert, L. (1970). *J. Embryol. Exp. Morphol.* **23,** 311–322. Electrical properties of the slime mould grex.

Garrod, D. R., Swan, A. P., Nicol, A., and Forman, D. (1978). *In* "Symposia of the Society for Experimental Biology, No. 32." Cell–Cell Recognition. Cambridge Univ. Press, New York (Curtis, A. S. G. ed.), 173–202. Cellular recognition in slime mold development.

Geller, J. S., and Brenner, M. (1978). *J. Cell. Physiol.* **97,** 413–420. Measurements of metabolites during cyclic AMP oscillations of *Dictyostelium discoideum.*

Geller, J. S., and Brenner, M. (1978). *Biochem. Biophys. Res. Commun.* **81,** 814–821. The effect of 2,4 dinitrophenol on *Dictyostelium discoideum* oscillations.

Geltosky, J., Siu, C. H., and Lerner, R. (1976). *Cell* **8,** 391–396. Glycoproteins of the plasma membrane of *Dictyostelium discoideum* during development.

Geltosky, J. E., Ray, J., and Lerner, R. A. (1978). ICN–UCLA Symposium, XIII 674P. Alan R. Liss, New York (Marchesi, V. T., *et al.*, eds.), 613–619. Use of common plant lectins for isolation and characterization of constitutive and developmentally regulated cell surface associated glycoproteins of *Dictyostelium discoideum.*

Geltosky, J. E., Weseman, J., Bakke, A., and Lerner, R. A. (1979). *Cell* **18,** 391–398. Identification of a cell surface glycoprotein involved in cell aggregation in *Dictyostelium discoideum.*

Geltosky, J. E., Birdwell, C. R., Weseman, J., and Lerner, R. A. (1980). *Cell* **21,** 339–345. A glycoprotein involved in aggregation of *Dictyostelium discoideum* is distributed on the cell surface in a nonrandom fashion favoring cell junctions.

George, R. P. (1977). *Cell Differ.* **5,** 293–300. Disruption of multicellular organization in the cellular slime mould by cAMP.

George, R. P., Albrecht, R. M., Raper, K. B., Sachs, I. B., and MacKenzie, A. P. (1970). *Proc. Cambridge. Stereoscan. Coll.*, 159–165. Rapid freeze preparation of *Dictyostelium discoideum* for scanning electron microscopy.

George, R. P., Albrecht, Raper, K. B., Sachs, I. B., and MacKenzie, A. P. (1972a). *J. Bacteriol.* **112,** 1385–1386. Scanning electron microscopy of spore germination in *Dictyostelium discoideum.*

George, R. P., Hohl, H. R., and Raper, K. (1972b). *J. Gen. Microbiol.* **70,** 477–489. Ultrastructural development of stalk-producing cells in *Dictyostelium discoideum.*

Gerisch, G. (1959). *Naturwissenschaften* **46,** 654–656. Ein submerskulturverfahren fur entwicklungsphysiologische untersuchungen an *Dictyostelium discoideum.*

Gerisch, G. (1960). *Wilhelm Roux' Arch Entwicklungsmech. Organismen* **152,** 632–654. Zellfunktionen und zellfunktionswechsel in der entwicklung von *Dictyostelium discoideum.* I. Zellagglutination und induktion der fruchtkorperpolaritat.

Gerisch, G. (1961a). *Develop. Biol.* **3,** 685–724. Zellfunktionen und zellfunktionswechsel in der entwicklung von *Dictyostelium discoideum.* II. Aggregation homogener zellpopulationen und zentrebildung.

Gerisch, G. (1961b). *Wilhelm Roux Arch. Entwicklungsmech. Organismen* **153,** 158–167. Zellfunktionen und zellfunktionswechsel in der entwicklung von *Dictyostelium discoideum.* III. Gentrennte beeinflussung von zelldifferenzierung und morphogenese.

Gerisch, G. (1961c). *Exp. Cell. Res.* **25,** 535–554. Zellfunktionen und zellfunktionswechsel in der entwicklung von *Dictyostelium discoideum.* V. stadienspezifische zelkontakbidung und ihre quantitative erfassung.

Gerisch, G. (1961d). *Naturwissenschaften* **48,** 436–437. Zellkontaktbildung vegetativer und aggregationsreifer zellen von *Dictyostelium discoideum.*

Gerisch, G. (1962a). *Ber. D. Bot. Ges.* **75**, 82–89. Die zellularen schleimpilze als objekte der entwicklungsphysiologie.
Gerisch, G. (1962b). *Wilhelm Roux Arch. Entwicklungsmech. Organismen* **153**, 603–620. Zellfunktionen und zellfunktionswechsel in der entwicklung von *Dictyostelium discoideum.* IV. Der zeitplan der entwicklung.
Gerisch, G. (1962c). *Exp. Cell Res.* **26**, 462–484. Zellfunktionen und zellfunktionswechsel in der entwicklung von *Dictyostelium discoideum.* VI. Inhibitoren der aggregation, Ihr einfluss auf zellkontakbildung und morphogenetische bewegung.
Gerisch, G. (1963). *Naturwissenschaften.* **50**, 160–161. Eine fur *Dictyostelium* ungewohnliche aggregationsweise.
Gerisch, G. (1964b). *Wilhelm Roux Arch. Entwicklungsmech. Organismen* **155**, 342–357. Die bildung des zellverbandes bei *Dictyostelium minutum.* I. Ubersicht uber die aggregation und den funktionswechsel der zellen.
Gerisch, G. (1964e). *Z. Naturforsch.* B **20**, 298–301. Eine mutante von *Dictyostelium minutum* mit blockierter zentrengrundung.
Gerisch, G. (1964f). *Wilhelm Roux Arch. Entwicklngsmech. Organismen* **156**, 127–144. Stadienspezifische aggregationsmuster von *Dictyostelium discoideum.*
Gerisch, G. (1964g). *Umschau* **65**, 392–395. Spezifsche zellkontakte als mechanismen der tierischen entwicklung.
Gerisch, G. (1966). *Wilhelm Roux Arch. Entwicklungsmech. Organismen* **157**, 174–189. Die bildung des zellverbandes dei *Dictyostelium minutum.* II. Analyse der zentrengrundung anhand von filmaufnahmen.
Gerisch, G. (1968). *Curr. Top. Devel. Biol.* **3**, 157–197. Cell aggregation and differentiation in *Dictyostelium discoideum.*
Gerisch, G. (1970). *Deut. Z. Gesell.* **64**, 6–14. Immunchemische untersuchungen an plasmamembranen aggregierender zellen.
Gerisch, G. (1971). *Naturwissenschaften* **58**, 430–438. Periodische signale steuern die musterbildung in zellverbanden.
Gerisch, G. (1972). *Nature New Biol.* **235**, 90–93. cAMP receptors—modulation of signal and inhibitor.
Gerisch, G. (1973). *In* "Non-Specific Factors Influencing Host Resistance. A Reexamination." S. Karger, Basel, Switzerland (Braun, W., and Ungar, J. eds.), 33–34. Responses of amoebae to immunoglobulin G and univalent antibody fragments.
Gerisch, G. (1976). *Cell Different.* **5**, 21–25. Extracellular cyclic AMP phosphodiesterase regulation in agar plate cultures of *Dictyostelium discoideum.*
Gerisch, G. (1976). *Life Sciences Res. Reports* **1**, 433–440. Cyclic AMP oscillation and signal transmission in aggregating *Dictyostelium discoideum* cells.
Gerisch, G. (1977). *In* "International Cell Biology 1976–1977" (Brinkley, B. R., and Porter, K. R., eds.), 36. Membrane sites implicated in cell adhesion: their developmental control in *Dictyostelium discoideum.*
Gerisch, G. (1980). *In* "Current Topics in Developmental Biology," Vol. 14, Part II. Pp. 243–269. Academic Press, New York. Univalent antibody fragments as tools for the analysis of cell interactions in *Dictyostelium.*
Gerisch, G., and Hess, B. (1973). *Hoppe-Seyler's Z. Physiol. Chem.* **354**, 1193. Cyclic AMP-controlled oscillations in suspended *Dictyostelium* cells: their relation to morphogenetic cell interactions.
Gerisch, G., and Hess, B. (1974). *Proc. Natl. Acad. Sci. U.S.A.* **71**, 2118–2122. Cyclic AMP-controlled oscillations in suspended *Dictyostelium* cells: their relation to morphogenetic cell interactions.
Gerisch, G., and Malchow, D. (1976). *Adv. Cyclic Nucleotide Res.* **7**, 49–68. Cyclic AMP receptors and the control of cell aggregation in *Dictyostelium discoideum.*

Gerisch, G., and Wick, U. (1975). *Biochem. Biophys. Res. Commun.* **65,** 364–370. Intracellular oscillations and release of cyclic AMP from *Dictyostelium* cells.

Gerisch, G., Normann, I., and Beug, H. (1966). *Naturwissenschaften* **23,** 618. Rhythmik der zellorientierung und der bewegungsgeschwindigkeit im chemotaktischen readtionssystem von *Dictyostelium discoideum.*

Gerisch, G., Luderitz, O., and Ruschmann, E. (1967). *Z. Naturforsch.* **22,** 109. Antikorper fordern die phagozytose von bakterien burch amoben.

Gerisch, G., Malchow, D., Wilhelms, H., and Ludertitz, O. (1969). *Eur. J. Biochem.* **9,** 229–236. Artspezifitat polysaccharid-haltiger zellmembran-antigene von *Dictyostelium discoideum.*

Gerisch, G., Riedel, V., and Malchow, D. (1971). *Umschau* **14,** 532–533. Zyklisches adenosin-monophosphat als signalstoff der entwicklung.

Gerisch, G., Malchow, D., Riedel, V., and Beug, H. (1972). *Hoppe-Seyler's Z. Physiol. Chem.* **353,** 684–685. Cell communication by chemical signals and the regulation of cyclic AMP in the development of a microorganism *Dictyostelium discoideum.*

Gerisch, G., Malchow, D., Riedel, V., Muller, E., and Every, M. (1972). *Nature New Biol.* **235,** 90–92. Cyclic AMP phosphodiesterase and its inhibitor in slime mould development.

Gerisch, G., Beug, H., and Malchow, D. (1974). *Arch. Biol.* **85,** 33–34. Membrane sites and cell aggregation in the slime mold *Dictyostelium discoideum.*

Gerisch, G., Beug, H., Malchow, D., Schwarz, H., and Stein, A. (1974). *In* "Biology and Chemistry of Eucaryotic Cell Surfaces," Academic Press, New York, N.Y. (Lee, E. Y. C., and Smith, E. E., eds.), 49–66. Receptors for intercellular signals in aggregating cells of the slime mold *Dictyostelium discoideum.*

Gerisch, G., Malchow, D., and Hess, B. (1974). *In* "Biochemistry of sensory function," Springer-Verlag, New York, N.Y. (Jaenicke, L. ed.), 279–298. Cell communication and cyclic AMP regulation during aggregation of the slime mold *Dictyostelium discoideum.*

Gerisch, G., Fromm, H., Huesgen, A., and Wick, U. (1975). *Nature (London)* **255,** 547–549. Control of cell contact sites by cAMP pulses in differentiating *Dictyostelium discoideum* cells.

Gerisch, G., Huesgen, A., and Malchow, D. (1975). *In* "Proceedings of the Tenth FEBS Meeting," pp. 257–267. Genetic control of cell differentiation and aggregation in *Dictyostelium:* the role of cyclic AMP pulses.

Gerisch, G., Hulser, D., Malchow, D., and Wick, U. (1975). *Phil. Trans. R. Soc. Lond.* **272,** 181–192. Cell communication by periodic cyclic AMP pulses.

Gerisch, G., Malchow, D., Huesgen, A., Nanjundiah, V., Roos, W., and Wick, U. (1975). *ICN-UCLA Symp. Mol. Cell. Biol.* **2,** 76–88. Cyclic-AMP reception and cell recognition in *Dictyostelium discoideum.*

Gerisch, G., Maeda, Y., and Malchow, D. (1977). *In* "Developments and differentiation in the cellular slime molds." Elsevier/North-Holland (Cappuccinelli, P., and Ashworth, J. M. eds.), 105–124. Cyclic AMP signals and the control of cell aggregation in *Dictyostelium discoideum.*

Gerisch, G., Malchow, D., Roos, W., Wick, U., and Wurster, B. (1977). *In* "Cell Interaction in Differentiation." Academic Press, New York. (Karkinen-Jaaskelainen, M., Saxen, L., and Weiss, L. eds.), 377–388. Periodic cyclic AMP signals and membrane differentiation in *Dictyostelium.*

Gerisch, G., Maeda, Y., Malchow, D., Roos, W., Wick, U., and Wurster, B. (1977). *In* "Developments in Cell Biology." Elsevier/North-Holland (Cappuccinelli P., and Ashworth, J. M. eds.), 135–148. Cell contact, signalling and gene expression in *Dictyostelium discoideum.*

Gerisch, G., Mueller, K., and Beug, H. (1978). *Biochem. Soc. Trans.* **6,** 481–486. *Dictyostelium discoideum* a microbial model for biochemical studies on cell recognition.

Gerisch, G., Malchow, D., Roos, W., and Wick, U. (1979). *J. Exp. Biol.* **81**, 33–47. Oscillations of cyclic nucleotide concentrations in relation to the excitability of *Dictyostelium* cells.

Gezelius, K. (1959). *Exp. Cell Res.* **18**, 425–453. The ultrastructure of cells and cellulose membranes in Acrasiae.

Gezelius, K. (1961). *Exp. Cell Res.* **23**, 300–310. Further studies in the ultrastructure of Acrasiae.

Gezelius, K. (1962). *Physiol. Plant.* **15**, 587–592. Growth of the cellular slime mold *Dictyostelium discoideum* on dead bacteria in liquid media.

Gezelius, K. (1971). *Arch. Mikrobiol.* **75**, 327–337. Acid phosphatase localization in myxamoebae of *Dictyostelium discoideum.*

Gezelius, K. (1972). *Arch. Mikrobiol.* **85**, 51–76. Acid phosphatase localization during differentiation in the cellular slime mold *Dictyostelium discoideum.*

Gezelius, K. (1974). *Arch. Microbiol.* **98**, 311–329. Inorganic polyphosphates and enzymes of polyphosphate metabolism in the cellular slime mold *Dictyostelium discoideum.*

Gezelius, K., and Ranby (1957). *Exp. Cell. Res.* **12**, 265–289. Morphology and fine structure of the slime mold *Dictyostelium discoideum.*

Gezelius, K., and Wright, B. E. (1965). *J. Gen. Microbiol.* **38**, 309–327. Alkaline phosphatase in *Dictyostelium discoideum.*

Gil, M. G., Alemany, S., Cad, D. M., Castano, J. G., and Mato, J. M. (1980). *Biochem Biophys. Res. Commun.* **94**, 1325–1330. Calmodulin modulates phospholipid methylation in *Dictyostelium discoideum.*

Gillies, N. E., Hubbard, B. M., and Ong, C. N. (1974). *Radiat. Res.* **59**, 214–215. The effect of bacterial strain on the sensitivity of *Dictyostelium discoideum* to gamma-rays.

Gillies, N. E., Hari-Ratnajothi, N., and Ong, C. N. (1976). *J. Gen. Microbiol.* **92**, 229–233. Comparison of the sensitivity of spores and amoebae of *Dictyostelium discoideum* to gamma-rays and UV light.

Gilkes, N. R., and Weeks, G. (1977a). *Biochim. Biophys. Acta* **464**, 142–156. The purification and characterization of *Dictyostelium discoideum* plasma membrane.

Gilkes, N. R., and Weeks, G. (1977b). *Can. J. Biochem.* **55**, 1233–1236. An improved procedure for the purification of plasma membranes from *Dictyostelium discoideum.*

Gilkes, N. R., Laroy, K., and Weeks, G. (1979). *Biochim. Biophys. Acta* **551**, 349–362. An analysis of the protein, glycoprotein and monosaccharide composition of *Dictyostelium discoideum* plasma membranes during development.

Gillette, M. U., and Filosa, M. F. (1973). *Biochem. Biophys. Res. Commun.* **53**, 1159–1166. Effect of concanavalin A on cellular slime mold development: premature appearance of membrane-bound cyclic AMP phosphodiesterase.

Gillette, M. U., Dengler, R. E., and Filosa, M. F. (1974). *J. Exp. Zool.* **190**, 243–248. The localization and fate of concanavalin A in amoebae of the cellular slime mold *Dictyostelium discoideum.*

Gillies, N. E., and Hariratnajothi, N. (1976). *Br. J. Radiol.* **49**, 191. The radiation sensitivity of *Dictyostelium discoideum.*

Gingell, D. (1971). *J. Theor. Biol.* **30**, 121–149. Computed force and energy of membrane interaction.

Gingell, D., and Garrod, D. R. (1969). *Nature (London)* **221**, 192–193. Effect of EDTA on electrophoretic mobility of slime mould cells and its relationship to current theories of cell adhesion.

Gingell, D., Garrod, D. R., and Palmer, J. F. (1969). *In* "Symposium on Calcium and Cell Function." MacMillan, New York, N.Y. (Cuthbert, A. ed.), 59–64. Divalent cations and cell adhesion.

Gingle, A. R. (1976). *J. Cell. Sci.* **20,** 1–20. Critical density for relaying in *Dictyostelium discoideum* and its relation to phosphodiesterase secretion into the extracellular medium.

Gingle, A. R., and Robertson, A. (1976). *J. Cell. Sci.* **20,** 21–28. The development of the relaying competence in *Dictyostelium discoideum.*

Gingold, E. B. (1974). *Heredity* **33,** 419–423. Stability of diploid clones of the cellular slime mold *Dictyostelium discoideum.*

Gingold, E., and Ashworth, J. M. (1974). *J. Gen. Microbiol.* **84,** 59–67. Evidence for mitotic crossing-over during the parasexual cycle of the cellular slime mold *Dictyostelium discoideum.*

Giri, J. G., and Ennis, H. L. (1977). *Biochem. Biophys. Res. Commun.* **77,** 282–289. Protein and RNA synthesis during spore germination in the cellular slime mold *Dictyostelium discoideum.*

Giri, J. G., and Ennis, H. L. (1978). *Develop. Biol.* **67,** 189–201. Developmental changes in RNA and protein synthesis during germination of *Dictyostelium discoideum* spores.

Githens, S., and Karnovsky, M. L. (1973). *J. Cell. Biol.* **58,** 536–548. Phagocytosis by the cellular slime mold *Polysphondylium pallidum* during growth and development.

Githens, S., and Karnovsky, M. L. (1973). *J. Cell. Biol.* **58,** 522–535. Biochemical changes during growth and encystment of the cellular slime mold *Polysphondylium pallidum.*

Glazer, P. M., and Newell, P. C. (1981). *J. Gen. Micro.* **125,** 221. Initiation of aggregation by *Dictyostelium discoideum* in mutant populations lacking pulsatile signalling.

Glynn, P. J. (1981). *Cytobios.* **30,** 153. A quantitative study of the phagocytosis of *Escherichia coli* by myxamoebae of the slime mould *Dictyostelium discoideum.*

Goidl, E. A., Chassy, B. M., Love, L. L., and Krichevsky, M. I. (1972). *Proc. Natl. Acad. Sci. U.S.A.* **69,** 1128–1130. Inhibition of aggregation and differentiation of *Dictyostelium discoideum* by antibodies against adenosine 3′,5′-cyclic monophosphatediesterase.

Goldbeter, A. (1975). *Nature (London)* **253,** 540–542. Mechanism for oscillatory synthesis of cyclic AMP in *Dictyostelium discoideum.*

Goldbeter, A. (1977). *Acta Biochim. Biophys. Acad. Sci. Hung.* **12,** 141–148. Thermodynamic and kinetic aspects of regulation.

Goldbeter, A. (1977). *In* "Alcohol and Aldehyde Metabolizing Systems," Vol. III. Academic Press, New York (Thurman, R. G., *et al.*, eds.), 1–16. Nonequilibrium behavior of biochemical systems.

Goldbeter, A., and Segal, L. A. (1977). *Proc. Natl. Acad. Sci.* **70,** 1543–1547. Unified mechanism for relay and oscillation of cAMP in *Dictyostelium discoideum.*

Goldbeter, A., Erneux, T., and Segel, L. A. (1978). *FEBS Lett.* **89,** 237–241. Excitability in the adenylate cyclase reaction in *Dictyostelium discoideum.*

Goldstein, I. J., and Hayes, C. E. (1978). *Adv. Carbohydr. Chem. Biochem.* **35,** 127–340. The lectins: carbohydrate-binding proteins of plant and animals.

Goldstone, E. M., Banerjee, S., Allen, J. R., Lee, J. J., Hutner, S. H., Bacchi, C. J., and Melville, J. F. (1966). *J. Protozool.* **13,** 171–174. Minimal defined media for vegetative growth of the Acrasian *Polysphondylium pallidum* WS-320.

Grabel, L. B., and Farnsworth, P. A. (1977). *Exp. Cell. Res.* **105,** 285–289. The endocytosis of concanavalin A by *Dictyostelium discoideum* cells.

Grabel, L. B., and Loomis, W. F. (1977). *In* "Developments in Cell Biology." Elsevier/North-Holland (Cappuccinelli, P., and Ashworth, J. M. eds.), 189–200. Cellular interaction regulating early biochemical differentiation in *Dictyostelium.*

Grabel, L., and Loomis, W. F. (1978). *Develop. Biol.* **64,** 203–209. Effector controlling accumulation of N-acetyl glucosaminidase during development of *Dictyostelium discoideum.*

Grainger, R. B., and Maizels, N. (1980). *Cell* **20,** 619–623. *Dictyostelium* ribosomal RNA is processed during transcription.

Green, A., and Newell, P. (1974). *Biochem. J.* **140,** 313–322. The isolation and subfractionation of plasma membrane from the cellular slime mould *Dictyostelium discoideum.*

Green, A., and Newell, P. C. (1975). *Cell* **6,** 129–136. Evidence for the existence of two types of cyclic AMP binding sites in aggregating cells of *Dictyostelium discoideum.*

Gregg, J. H. (1950). *J. Exp. Zool.* **114,** 173–196. Oxygen utilization in relation to growth and morphogenesis of the slime mold *Dictyostelium discoideum.*

Gregg, J. H. (1956). *J. Gen, Physiol.* **39,** 813–820. Serological investigations of cell adhesion in the slime molds, *Dictyostelium discoideum, Dictyostelium purpureum,* and *Polysphondylium violaceum.*

Gregg, J. H. (1960). *Biol. Bull.* **118,** 70–78. Surface antigen dynamics in the slime mold, *Dictyostelium discoideum.*

Gregg, J. H. (1961). *Develop. Biol.* **3,** 757–766. An immunoelectrophoretic study of the slime mold *Dictyostelium discoideum.*

Gregg, J. H. (1964). *Physiol. Rev.* **44,** 631–656. Developmental processes in cellular slime molds.

Gregg, J. H. (1965a). *Science* **150,** 1739–1740. Centrifugal homogenizer.

Gregg, J. H. (1965b). *Develop. Biol.* **12,** 377–393. Regulation in the cellular slime molds.

Gregg, J. H. (1966a). *In* "The Fungi, an Advanced Treatise." Vol. 2. Academic Press, New York, N.Y. (Ainsworth, G. C., and Sussman, A. S. eds.), 235–281. Organization and synthesis in the cellular slime molds.

Gregg, J. H. (1966b). *Exp. Cell Res.* **42,** 260–264. A microrespirometer capable of quantitative substrate mixing.

Gregg, J. H. (1967). *In* "Techniques for the Study of Development." Crowell–Collier, New York, N.Y. (Wilt, F., and Wessels, N. eds.), 359–376. Cellular slime molds.

Gregg, J. H. (1968). *Exp. Cell Res.* **51,** 633–642. Prestalk cell isolates in *Dictyostelium.*

Gregg, J. H. (1971). *Develop. Biol.* **26,** 478–485. Developmental potential of isolated *Dictyostelium* myxamoebae.

Gregg, J. H., and Aldrich, H. C. (1972). *J. Cell. Biol.* **55,** 95. Unit membrane structural changes following cell association in *Dictyostelium.*

Gregg, J. H., and Badman, W. S. (1970). *Develop. Biol.* **22,** 96–111. Morphogenesis and ultrastructure in *Dictyostelium.*

Gregg, J. H., and Badman, W. S. (1973). *In* "Developmental Regulation." Academic Press, New York, N.Y. (Coward, S. J. ed.), 85–106. Transitions in differentiation by the cellular slime molds.

Gregg, J. H., and Bronsweig, R. D. (1956a). *J. Cell. Comp. Physiol.* **47,** 483–488. Dry weight loss during culmination of the slime mold, *Dictyostelium discoideum.*

Gregg, J. H., and Bronsweig, R. D. (1956b). *J. Cell. Comp. Physiol.* **48,** 293–300. Biochemical events accompanying stalk formation in the slime mold, *Dictyostelium discoideum.*

Gregg, J. H., and Karp, G. C. (1977). *In* "Developments in Cell Biology." Elsevier/North-Holland (Cappuccinelli, P., and Ashworth, J. M. eds.), 297–310. An early phase in *Dictyostelium* cell differentiation revealed by 3H-L-fucose incorporation.

Gregg, J. H., and Karp, G. C. (1978). *Exp. Cell. Res.* **112,** 31–46. Patterns of cell differentiation revealed by tritiated fucose incorporation in *Dictyostelium discoideum.*

Gregg, J. H., and Nesom, M. G. (1973). *Proc. Natl. Acad. Sci. U.S.A.* **70,** 1630–1633. Response of *Dictyostelium* plasma membranes to adenosine 3′,5′-cyclic monophosphate.

Gregg, J. H., and Trygstad, C. W. (1958). *Exp. Cell Res.* **15,** 358–369. Surface antigen defects contributing to developmental failure in aggregateless variants of the slime mold, *Dictyostelium discoideum.*

Gregg, J. H., and Yu, N. Y. (1975). *Exp. Cell. Res.* **96,** 283–286. *Dictyostelium* aggregate-less mutant plasma membranes.

Gregg, J. H., Hackney, A. L., and Krivanek, J. O. (1954). *Biol. Bull.* **107,** 226–235. Nitrogen metabolism of the slime mold *Dictyostelium discoideum* during growth and morphogenesis.

Gregg, J. H., Jimenez, H., and Davis, R. W. (1981). *Exp. Cell. Res.* **134,** 389. The regulation of glycoprotein synthesis by cAMP in *Dictyostelium.*

Gross, J., Peacey, M., and Trevan, D. (1976). *J. Cell Sci.* **22,** 645–656. Signal emission and signal propagation during early aggregation in *Dictyostelium discoideum.*

Gross, J., Kay, R., Lax, A., Peacey, M., Town, C., and Trevan, D. (1977). *In* "Developments in Cell Biology." Elsevier/North-Holland (Cappuccinelli, P., and Ashworth, J. M. eds.), 135–148. Cell contact signalling and gene expression in *Dictyostelium discoideum.*

Gross, J. G., Town, C. D., Brookman, J. J., Jeermyn, K. A., Peacey, M. J., and Kay, R. R. (1981). Cell patterning in *Dictyostelium. Phil. Trans. R. Soc. Lond* **B295,** 497–508.

Grutsch, J. F., and Robertson, A. (1978). *Develop. Biol.* **66,** 285–293. The cyclic AMP signal from *Dictyostelium discoideum* amoebae.

Guespin-Michel, J. F., Menahem, M., Monier, F., and Felenbok, B. (1976). *Exp. Cell. Res.* **98,** 184–190. Kinetics of 5 bromodeoxy oridine action on *Dictyostelium discoideum* growth and development.

Guialis, A., and Deering, R. A. (1976). *J. Bacteriol.* **127,** 59–66. Repair of DNA in ultraviolet light sensitive and resistant strains of *Dictyostelium discoideum.*

Guialis, A., and Deering, R. A. (1976). *Photochem. Photobiol.* **24,** 331–336. Inhibition of U.V. induced DNA strand breakage in *Dictyostelium discoideum* by 2,4, dinitrophenol.

Gustafson, G. L., and Milner, L. A. (1979). *Biochem. Biophys. Res. Commun.* **94,** 1439–1444. Immunological relationship between beta-N acetyl glucosaminidase and proteinase I from *Dictyostelium discoideum.*

Gustafson, G. L., and Milner, L. A. (1980). *J. Biol. Chem.* **255,** 7208–7210. Occurrence of N-acetyl glucosamine 1 phosphate in proteinase I from *Dictyostelium discoideum.*

Gustafson, G. L., and Thon, L. A. (1979). *Biochem. Biophys. Res. Commun.* **86,** 667–673. Evidence for phosphoryl moieties in a proteinase from *Dictyostelium discoideum.*

Gustafson, G. L., and Thon, L. A. (1979). *J. Biol. Chem.* **254,** 12471–12478. Purification and characterization of a proteinase from *Dictyostelium discoideum.*

Gustafson, G. L., and Wright, B. E. (1972). *Crit. Rev. Microbiol.* **1,** 453–478. Analysis of approaches used in studying differentiation of the cellular slime mold.

Gustafson, G. L., and Wright, B. E. (1973). *Biochem. Biophys. Res. Commun.* **50,** 438–442. UDP-glucose pyrophosphorylase synthesis in myxamoebae of *Dictyostelium discoideum.*

Gustafson, G. L., Kong, W. Y., and Wright, B. E. (1973). *J. Biol. Chem.* **248,** 5188–5196. Analysis of uridine diphosphate-glucose pyrophosphorylase synthesis during differentiation in *Dictyostelium discoideum.*

Gwynne, D. I., and O'Day, D. H. (1978). *Can. J. Microbiol.* **24,** 480–486. RNA and protein synthetic patterns during germination of *Polysphondyliom pallidum* microcysts.

Haeder, D., and Poff, K. L. (1979). *Photochem. Photobiol.* **29,** 1157–1162. Light induced accumulations of *Dictyostelium discoideum* amoebae.

Haeder, D.-P., and Poff, K. L. (1979). *Arch. Microbiol.* **123,** 281–286. Inhibition of aggregation by light in the cellular slime mold *Dictyostelium discoideum.*

Haeder, D.-P., and Poff, K. L. (1980). *Arch. Microbiol.* **126,** 97–101. Effects of ionophores and triphenylmethyl phosphonium ion on light induced responses in *Dictyostelium discoideum.*

Hagiwara, H. (1971). *Bull. Nat. Mus. (Tokyo)* **14,** 351–366. The Acrasiales in Japan. Part 1.

Hagiwara, H. (1972). *Mem. Nat. Sci. Mus. (Tokyo)* **5,** 173–177. Notes on the Acrasiales in Mt. Porosiri-Dake.

Hagiwara, H. (1973). *Rep. Tottori. Mycol. Inst.* **10,** 591–596. The Acrasiales in Japan. Part 2.

Hagiwara, H. (1973). *Bull. Nat. Sci. Mus. (Tokyo)* **16,** 493–496. Enumeration of the Dictyosteliaceae.

Hagiwara, H. (1976). *Trans. Mycol. Soc. Jpn.* **17,** 226–237. Distribution of the Dictyosteliaceae cellular slime molds on Mount Ishizuchi Shikoku, Japan.

Hagiwara, H. (1976). *Bull. Natl. Sci. Mus. Ser. B.* **2,** 53–60. Cellular slime molds from Mount Margherita Ruwenzori Mountains, East Africa.

Hagiwara, H. (1978). *Bull. Natl. Sci. Mus. Ser. B.* **4,** 27–32. The Acrasiales in Japan. Part 4.

Hahn, G. L., Metz, K., George, R. P., and Haley, B. (1977). *J. Cell. Biol.* **75,** 41A. Identification of cyclic AMP receptors in the cellular slime mold *Dictyostelium discoideum* using a photo affinity analog.

Hames, B. D. (1976). *Anal. Biochem.* **73,** 215–219. An improved radiochemical assay for UDP glucose pyrophosphorylase.

Hames, B. D., and Ashworth, J. M. (1974a). *Biochem. J.* **142,** 301–315. The metabolism of macromolecules during the differentiation of myxamoebae of the cellular slime mould *Dictyostelium discoideum* containing different amounts of glycogen.

Hames, B. D., and Ashworth, J. M. (1974b). *Biochem. J.* **142,** 317–325. The control of saccharide synthesis during development of myxamoebae of *Dictyostelium discoideum* containing differing amounts of glycogen.

Hames, B. D., and Hodson, B. A. (1979). *Biochem. J.* **177,** 21–28. Accumulation of UDP glucose pyrophosphorylase EC-2.7.7.9 by axenically grown amoebae of *Dictyostelium discoideum.*

Hames, B. D., Hodson, B. A., and Duddy, P. (1977). *In* "Developments in Cell Biology." Elsevier/North-Holland (Cappuccinelli, P., and Ashworth, J. M. eds.), 243–252. *In vitro* translation and translational control in *Dictyostelium discoideum.*

Hames, B. D., Weeks, G., and Ashworth, J. M. (1972). *Biochem. J.* **126,** 627–633. Glycogen synthetase and the control of glycogen synthesis in the cellular slime mould *Dictyostelium discoideum* during cell differentiation.

Hamilton, I. D., and Chia, W. K. (1975). *J. Gen. Micro.* **91,** 295–306. Enzyme activity changes during cyclic AMP-induced stalk cell differentiation in P4, a variant of *Dictyostelium discoideum.*

Hammond, J. (1973). *Biochim. Biophys. Acta* **291,** 371–387. A membrane component of the cellular slime mold *Dictyostelium discoideum* rapidly labelled with phosphorus-32 orthophosphate.

Hanish, M. D. (1975). *Develop. Biol.* **45,** 340–348. A possible cause of termination of cell growth in the cellular slime mold *Dictyostelium discoideum.*

Hanna, M. H., and Cox, E. C. (1978). *Develop. Biol.* **62,** 206–214. The regulation of cellular slime mold development a factor causing development of *Polysphondylium violaceum* aggregation defective mutants.

Hanna, M. H., Klein, C., and Cox, E. C. (1979). *Exp. Cell. Res.* **122,** 265–272. Cyclic-nucleotides and cyclic nucleotide phosphodiesterase during development of *Polysphondylium violaceum.*

Hara, K., and Konijn, T. M. (1976). *Nature (London)* **260,** 705. Effect of temperature on morphogenetic oscillations in *Dictyostelium discoideum.*

Hara, Y. (1977). *Jpn. J. Ecol.* **27,** 141–146. Cellular slime molds Acrasieae of the Tenryu River region, Japan Part 1. Distribution of cellular slime molds in chestnut and red pine forest soils.

Hara, Y. (1978). *Jpn. J. Ecol.* **28,** 43–48. Cellular slime molds Acrasieae of the Tenryu River region, Japan. Part 1. Seasonal fluctuation in the forest soil.

Harper, R. A. (1926). *Bull. Torrey Botan. Club* **53,** 229–268. Morphogenesis in *Dictyostelium.*

Harper, R. A. (1929). *Bull. Torrey Botan. Club* **56,** 227–258. Morphogenesis in *Polysphondylium.*

Harper, R. A. (1932). *Bull. Torrey Botan. Club* **59,** 49–84. Organization and light relations in *Polysphondylium.*

Harrington, B. J., and Raper, K. B. (1968). *J. Appl. Microbiol.* **16,** 106–113. Use of a fluorescent brightener to demonstrate cellulose in the cellular slime molds.

Harris, J. F., and Rutherford, C. L. (1976). *Biochem.* **15,** 3064–3069. Primer dependency of glycogen synthetase EC-2.4.1.11 during differentiation in *Dictyostelium discoideum.*

Harris, J. F., and Rutherford, C. L. (1976). *J. Bacteriol.* **127,** 84–90. Localization of glycogen synthetase EC-2.4.1.11 during differentiation of presumptive cell types in *Dictyostelium discoideum.*

Hase, A. (1980). *J. Fac. Sci. Hokkaido Univ. Ser. V. Bot.* **12,** 123–128. Effects of ethanol on spore germination and cell growth of *Dictyostelium discoideum.*

Hase, A. (1981). *Arch. Biochem. Biophys.* **210,** 280. Isolation and characterization of a glycolipid from *Dictyostelium discoideum.*

Hase, A., and Ono, K-I. (1980). *J. Fac. Sci. Hoddaido Univ. V. Bot.* **12,** 129–134. The subunit composition of discoidins.

Hashimoto, Y. (1971). *Nature (London)* **231,** 316–317. Effect of radiation on the germination of spores of *Dictyostelium discoideum.*

Hashimoto, Y., and Wada, M. (1980). *Radiat. Res.* **83,** 688–695. Comparative study of the sensitivity of spores and amoebae of *Dictyostelium discoideum* to UV light.

Hashimoto, Y., Cohen, M. H., and Robertson, A. (1975). *J. Cell. Sci.* **19,** 215–229. Cell density dependence of the aggregation characteristics of the cellular slime mold *Dictyostelium discoideum.*

Hashimoto, Y., Tanaka, Y., and Yamada, T. (1976). *J. Cell. Sci.* **21,** 261–271. Spore germination promoter of *Dictyostelium discoideum* excreted by *Aerobacter aerogenes.*

Hashimoto, Y., Muroyama, T., Sameshima, M., and Yamada, T. (1977). *J. Radiat. Res.* **18,** 24–25. Gamma irradiation effects on the morphology of the cellular slime mold.

Hayashi, M., and Takeuchi, I. (1976). *Develop. Biol.* **50,** 302–309. Quantitative studies on cell differentiation during morphogenesis of cellular slime moulds, *Dictyostelium discoideum.*

Hayashi, H., and Suga, T. (1978). *J. Biochem. (Tokyo)* **84,** 513–520. Some characteristics of peroxisomes in the slime mold *Dictyostelium discoideum.*

Hayashi, H., and Yamasaki, F. (1978). *Chem. Pharm. Bull. (Tokyo)* **26,** 2977–2982. Characteristics of the induction of phosphodiesterases by cyclic AMP in the slime mold *Dictyostelium discoideum.*

Heftmann, E., Wright, B. E., and Liddel, G. U. (1959). *J. Amer. Chem. Soc.* **81,** 6525. Identification of a sterol with acrasin activity in a slime mold.

Heftmann, E., Wright, B. E., and Liddel, G. U. (1960). *Biochem. Biophys.* **91,** 266–270. The isolation of delta 22- stigmasten-3 beta-ol from *Dictyostelium discoideum.*

Hellewell, S. B., and Taylor, D. L. (1978). *Biophys. J.* **21,** 24A. A partially purified model system of gelation and contraction from *Dictyostelium discoideum.*

Hellewell, S. B., and Taylor, D. L. (1979). *J. Cell. Biol.* **83,** 633–648. The contractile basis of amoeboid movement 6. The solation contraction coupling hypothesis.

Hemmes, D. E., Kojima-Buddenhagen, E. S., and Hohl, H. R. (1972). *J. Ultrastr. Res.* **41,** 406–417. Structural and enzymic analysis of the spore wall layers in *Dictyostelium discoideum.*

Hemmingsen, E. A., and Hemmingsen, B. B. (1979). *J. Appl. Physiol. Respir. Environ. Exercise Physiol.* **47,** 1270–1277. Lack of intracellular bubble formation in microorganisms at very high gas super saturations.

Henderson, E. J. (1975). *J. Biol. Chem.* **250**, 4730–4736. The cAMP receptor of *Dictyostelium discoideum* binding characteristics of aggregation competent cells and variation of binding levels during the life cycle.

Herring, F. G., and Weeks, G. (1979). *Biochim. Biophys. Acta* **552**, 66–77. Analysis of *Dictyostelium discoideum* plasma membrane fluidity by ESR.

Herring, F. G., Tatischeff, I., and Weeks, G. (1980). *Biochim. Biophys. Acta* **602**, 1–9. Fluidity of plasma membranes of *Dictyostelium discoideum* effects of polyunsaturated fatty-acid incorporation assessed by fluorescence depolarization and ESR.

Hintermann, R., and Parish, R. W. (1979). *FEBS Lett.* **108**, 219–225. Synthesis of plasma membrane proteins and antigens during development of the cellular slime mold *Polysphondylium pallidum.*

Hintermann, R., and Parish, R. W. (1979). *Planta (Berl.).* **146**, 459–462. Determination of adenylate cyclase activity in a variety of organisms: evidence against the occurrence of the enzyme in higher plants.

Hinterman, R., and Parish, R. W. (1979). *Exp. Cell. Res.* **123**, 429–434. The intracellular location of adenyl cyclase in the cellular slime molds *Dictyostelium discoideum* and *Polysphondylium pallidum.*

Hirschberg, E. (1955). *Bacteriol. Rev.* **19**, 65–78. Some contributions of microbiology to cancer research.

Hirschberg, E., and Merson, G. (1955). *Cancer Res. Suppl.* **3**, 76–79. Effect of test compounds on the aggregation and culmination of the slime mold *Dictyostelium discoideum.*

Hirschberg, E., and Rusch, H. P. (1950). *J. Cell. Comp. Physiol.* **36**, 105–113. Effects of compounds of varied biochemical action on the aggregation of a slime mold, *Dictyostelium discoideum.*

Hirschberg, E., and Rusch, H. P. (1951). *J. Cell. Comp. Physiol.* **37**, 323–336. Effect of 2,4-dinitrophenol on the differentiation of the slime mold *Dictyostelium discoideum.*

Hirschberg, E., Ceccarini, C., Osnos, M., and Carchman, R. (1968). *Proc. Natl. Acad. Sci. U.S.A.* **61**, 316–323. Effects of inhibitors of nucleic acid and protein synthesis on growth and aggregation of the cellular slime mold *Dictyostelium discoideum.*

Hodge, J. L., and Rossomando, E. F. (1979). *Anal. Biochem.* **100**, 179–183. Separation of enzymatic activities in *Dictyostelium discoideum* by high pressure gel permeation chromatography.

Hodge, J. L., and Rossomando, E. F. (1980). *Anal. Biochem.* **102**, 59–62. Degradation of ATP by membrane bound enzymatic activities in *Dictyostelium discoideum* monitored by high pressure liquid chromatography.

Hoetzer, K. E., and Deerling, R. A. (1980). *Mutat. Res.* **71**, 273–276. Sensitization of *rad* mutants of *Dictyostelium discoideum* to UV light by post irradiation treatment with caffeine.

Hoffman, S., and McMahon, D. (1976). *J. Cell. Biol.* **70**, 245. Plasma membrane polypeptides of *Dictyostelium discoideum.*

Hoffman, S., and McMahon, D. (1977). *Biochim. Biophys. Acta* **465**, 242–259. The role of the plasma membrane in the development of *Dictyostelium discoideum.* II. Developmental and topographic analysis of polypeptide and glycoprotein composition.

Hoffman, S., and McMahon, D. (1978). *J. Biol. Chem.* **253**, 278–287. Defective glycoproteins in the plasma membrane of an aggregation minus mutant of *Dictyostelium discoideum* with abnormal cellular interactions.

Hoffman, S., and McMahon, D. (1978). *Archiv. Biochem. Biophys.* **187**, 12–24. The effects of inhibition of development in *Dictyostelium discoideum* on changes in plasma membrane composition and topography.

Hohl, H. R. (1965). *J. Bacteriol.* **90,** 755–765. Nature and development of membrane systems in food vacuoles of cellular slime molds predatory upon bacteria.

Hohl, H. R. (1976). *In* "The Fungal Spore: Form and Function." Wiley, New York (Weber, D. J., and Hess, W. M. eds.), 463–500. Myxomycetes.

Hohl, H. R., and Hamamoto, S. T. (1967). *Pacific Sci.* **21,** 534–538. Reversal of ethionine inhibition by methionine during slime mold development.

Hohl, H. R., and Hamamoto, S. T. (1969a). *J. Protozool.* **16,** 333–344. Ultrastructure of *Acrasis rosea*, cellular slime mold during development.

Hohl, H. R., and Hamamoto, S. T. (1969b). *J. Ultrastr. Res.* **26,** 442–453. Ultrastructure of spore differentiation in *Dictyostelium:* the prespore vacuole.

Hohl, H. R., and Jehli, J. (1973). *Arch. Mikrobiol.* **92,** 179–187. The presence of cellulose microfibrils in the proteinaceous slime track of *Dictyostelium discoideum.*

Hohl, H. R., and Raper, K. B. (1963a). *J. Bacteriol.* **85,** 191–198. Nutrition of cellular slime molds. I. Growth of living and dead bacteria.

Hohl, H. R., and Raper, K. B. (1963b). *J. Bacteriol.* **85,** 199–206. Nutrition of cellular slime molds. II. Growth of *Polysphondylium pallidum* in axenic culture.

Hohl, H. R., and Raper, K. B. (1963c). *J. Bacteriol.* **86,** 1314–1320. Nutrition of cellular slime molds. III. Specific growth requirements of *Polysphondylium pallidum.*

Hohl, H. R., and Raper, K. B. (1964). *Develop. Biol.* **9,** 137–153. Control of sorocarp size in the cellular slime mold *Dictyostelium discoideum.*

Hohl, H. R., Hamamoto, S. T., and Hemmes, D. E. (1968). *Amer. J. Bot.* **55,** 783–796. Ultrastructural aspects of cell elongation, cellulose synthesis, and spore differentiation in *Actyostelium leptosomum*, a cellular slime mold.

Hohl, H. R., Honegger, R., Traub, F., and Markwalder, M. (1977). *In* "Development and Differentiation in the Cellular Slime Moulds." Elsevier/North-Holland (Cappuccinelli, P., and Ashworth, J. M. eds.), 149–172. Influence of cAMP on cell differentiation and morphogenesis in *Polysphondylium.*

Hohl, H. R., Buehlmann, M., and Wehrli, E. (1978). *Arh. Microbiol.* **116,** 239–244. Plasma membrane alterations as a result of heat activation in *Dictyostelium discoideum* spores.

Hong, C. B., Hadar, M. A., Hadar, D-P., and Poff, K. L. (1981). *Photochem. Photobiol.* **33,** 373–378. Phototaxis in *Dictyostelium discoideum* amoebae.

Horgen, I. A., Horgen, P. A., and O'Day, D. H. (1974). *Can. J. Biochem.* **52,** 126–136. Purification and properties of acid phosphatase I from the cellular slime mold *Polysphondylium pallidum.*

Horgen, P. A., and O'Day, D. H. (1973). *Cytobios.* **8,** 119–123. Basic nucleoproteins of the cellular slime mold *Polysphondylium pallidum.*

Hori, H., Osawa, S., and Iwabuchi (1980). *Nuc. Acid Res.* **8,** 5535–5539. The nucleotide sequence of 5S rRNA from a cellular slime mold *Dictyostelium discoideum.*

Huelser, D. F., and Webb, D. J. (1973). *Biophysik.* **10,** 273–280. The use of the tip potential of glass microelectrodes in the determination of low cell membrane potentials.

Huesgen, A., and Gerisch, G. (1975). *FEBS Lett.* **56,** 46–49. Solubilized contact sites A from cell membranes of *Dictyostelium discoideum.*

Huffman, D. M., and Olive, L. S. (1963). *Mycologia* **55,** 333–344. A significant morphogenetic variant of *Dictyostelium mucoroides.*

Huffman, D. M., and Olive, L. S. (1964). *Amer. J. Bot.* **51,** 465–471. Engulfment and anastomosis in the cellular slime molds (Acrasiales).

Huffman, D. M., Kahn, A. J., and Olive, L. S. (1962). *Proc. Natl. Acad. Sci. U.S.A.* **48,** 1160–1164. Anastomosis and cell fusions in *Dictyostelium.*

Ihara, M., Taya, Y., and Nishimura, S. (1980). *Exp. Cell. Res.* **126,** 273–278. Developmental regulation of cytokinin spore germination inhibitor discadenine and related enzymes in *Dictyostelium discoideum.*

Ikeda, T. (1981). *Biochim. Biophys. Acta* **675,** 69–76. Subcellular distributions of UDP-galactose:polysaccharide transferase and UDP-glucose pyrophosphorylase involved in biosynthesis of prespore-specific acid mycopolysaccharide in *Dictyostelium discoideum.*

Ikeda, T., and Takeuchi, I. (1971). *Develop. Growth and Differentiation* **13,** 221–229. Isolation and characterization of a prespore specific structure of the cellular slime mold, *Dictyostelium discoideum.*

Inouye, K., and Takeuchi, I. (1979). *Protoplasma* **99,** 289–304. Analytical studies on migrating movement of the pseudoplasmodium of *Dictyostelium discoideum.*

Irvine, R. F., Letcher, A. J., Brophy, P. J., and North, M. J. (1980). *J. Gen. Microbiol.* **121,** 495–497. Phosphatidylinositol-degrading enzymes in the cellular slime mould *Dictyostelium discoideum.*

Ishida, S. (1974). *Dev. Growth Differ.* **16,** 237–246. A cell contact temperature sensitive mutant of the cellular slime mold *Dictyostelium mucoroides.*

Ishida, S. (1980). *Dev. Growth Differ.* **22,** 143–152. A mutant of *Dictyostelium discoideum* capable of differentiating without morphogenesis.

Ishida, S. (1980). *Dev. Growth Differ.* **22,** 781–788. The effects of cyclic AMP on differentiation of a mutant *Dictyostelium discoideum* capable of developing without morphogenesis.

Ishida, S., Maeda, Y., and Takeuchi, I. (1974). *J. Gen. Microbiol.* **81,** 491–499. An anucleolate mutant of the cellular slime mold *Dictyostelium discoideum.*

Ishiguro, A., and Weeks, G. (1978). *J. Biol. Chem.* **253,** 7585–7587. Subunit composition and molecular weights of the developmentally regulated lectins from *Dictyostelium discoideum.*

Ishikawa, A., Takagi, M., Tateishi, T., and Ohmori, H. (1977). *Dev. Growth Differ.* **2,** 77–84. Inhibition by bacteria of pseudoplasmodium formation of *Dictyostelium discoideum.*

Ishikawa, A., Tomita, S., and Ando, Y. (1980). *Dev. Growth Differ.* **22,** 21–24. Acquisition of competence for culmination during finger formation in *Dictyostelium discoideum.*

Ito, K., and Iwabuchi, M. (1971). *Biochem. J.* **69,** 1135–1138. Conformational changes of ribosomal subunits of *Dictyostelium discoideum* by salts.

Ivatt, R., Das, P., Henderson, E., and Robbins, P. (1981). *J. Supramol. Struct. Cell. Biochem.* Suppl. 5, **257.** Developmental regulation of glycoprotein biosynthesis in *Dictyostelium.*

Iwabuchi, M., and Ochiai, H. (1969). *Biochim. Biophys. Acta* **190,** 211–213. Sedimentation properties of ribosomal particles in *Dictyostelium discoideum.*

Iwabuchi, M., Mizukami, Y., and Sameshima, S. (1970a). *Biochim. Biophys. Acta* **228,** 693–700. Synthesis of precursor molecules of ribosomal RNA in the cellular slime mold *Dictyostelium discoideum.*

Iwabuchi, M., Ito, K., and Ochiai, H. (1970b). *J. Biochem. Tokyo,* **68,** 549–556. Characterization of ribosomes in cellular slime mold, *Dictyostelium discoideum.*

Jacobson, A. (1976). *In* "Methods in Molecular Biology," Vol. 8, 161–209. Analysis of messenger RNA transcription in *Dictyostelium discoideum* or slime mold messenger RNA: how to find it and what to do with it once you have got it.

Jacobson, A., and Lodish, H. F. (1973). *Anal. Biochem.* **54,** 513–517. A simple and inexpensive procedure for preparative polyacrylamide gel electrophoresis of RNA.

Jacobson, A., and Lodish, H. F. (1975). *Ann. Rev. Genetics* **9,** 145–185. Genetic control of development of the cellular slime mold *Dictyostelium discoideum.*

Jacobson, A., Firtel, R. A., and Lodish, H. F. (1974a). *J. Mol. Biol.* **82,** 213–230. Synthesis of messenger and ribosomal RNA precursors in isolated nuclei of the cellular slime mold, *Dictyostelium discoideum.*

Jacobson, A., Firtel, R. A., and Lodish, H. F. (1974b). *Proc. Natl. Acad. Sci. U.S.A.* **71,** 1607–1611. Transcription of poly(DT) sequences in the genome of the cellular slime mold, *Dictyostelium discoideum.*

Jacobson, A., Lane, C. D., and Alton, T. (1975). *In* "American Society for Microbiology," Washington, D.C. (Schlessinger, D. ed.), 490–499. Electrophoretic separation of the major species of the slime mold messenger RNA.

Jacobson, B. S. (1980). *Biochim. Biophys. Acta* **600,** 769–780. Improved method for isolation of plasma membrane on cationic beads membranes from *Dictyostelium discoideum.*

Jacobson, B. S. (1980). *Biochem. Biophys. Res. Commun.* **97,** 1493–1498. Actin binding to the cytoplasmic surface of the plasma membrane isolated from *Dictyostelium discoideum.*

Jacquet, M., Part, D., and Felenbok, B. (1981). *Develop. Biol.* **81,** 155–166. Changes in the polyadenylated messenger RNA population during development of *Dictyostelium discoideum.*

Jaffe, A. R., and Garrod, D. R. (1979). *J. Cell. Sci.* **40,** 245–256. Effect of isolated plasma membranes on cell cohesion in the cellular slime mould.

Jaffe, A. R., Swan, A. P., and Garrod, D. R. (1979). *J. Cell. Sci.* **37,** 157–167. A ligand receptor model for the cohesive behavior of *Dictyostelium discoideum* axenic cells.

Jastorff, B. (1978). *Adv. Cyclic Nucleotide Res.* **254,** 12573–12578. 5-Amino-5-deoxy adenosine 3 phosphates mimicing the biological activity of cAMP.

Jastorff, B., Hoppe, J., and Morr, M. (1979). *Eur. J. Biochem.* **101,** 555–562. A model for the chemical interactions of cyclic AMP with R subunit of protein kinase type I EC-2.7.1.37 refinement of the cyclic phosphate binding moiety of protein kinase type I.

Jastorff, B., Konijn, T. M., Mato, J., Hoppe, J., and Wagner, K. G. (1978). *Hoppe Seyler's Z. Physiol. Chem.* **359,** 281. Comparison of the molecular interactions between cAMP and its receptor protein in *Dictyostelium discoideum* and protein kinase type I.

Jefferson, B. L., and Rutherford, C. L. (1975). *Va. J. Sci.* **26,** 56. Trehalose metabolism during differentiation of *Dictyostelium discoideum.*

Jefferson, B. L., and Rutherford, C. L. (1975). *Asb. Bull.* **22,** 59. Trehalose metabolism during differentiation of *Dictyostelium discoideum* Acrasiales.

Jefferson, B. L., and Rutherford, C. L. (1976). *Exp. Cell. Res.* **103,** 127–134. A stalk-specific localization of trehalase activity in *Dictyostelium discoideum.*

Jefferson, B. L., and Rutherford, C. L. (1976). *Cell Differ.* **5,** 189–198. Cell specific activity of trehalose 6 phosphate synthetase during differentiation of *Dictyostelium discoideum.*

Jermyn, K. A., Kilpatrick, D. C., Schmidt, J. A., and Stirling, J. L. (1977). *In* "Development and Differentiation in the Cellular Slime Moulds." Elsevier/North-Holland (Cappuccinelli, P., and Ashworth, J. M. eds.), 79–83. Components of the plasma membrane of *Dictyostelium discoideum* during aggregation.

Johnson, D. F., Wright, B. E., and Heftmann, E. (1962). *Arch. Biochem. Biophys.* **97,** 232–235. Biogenesis of delta22- stigmasten-3 beta-ol in *Dictyostelium discoideum.*

Johnson, G., Johnson, R., and Miller, M. (1977). *Science* **197,** 1300. Do cellular slime molds form intercellular junctions.

Jones, G. E., Pacy, J., Jermyn, K., and Stirling, J. (1977). *Exp. Cell. Res.* **107,** 451–454. A requirement for filopodia in the adhesion of preaggregative cells of *Dictyostelium discoideum.*

Jones, M. E. (1976). *J. Cell. Sci.* **22,** 35–40. Aggregation in *Polysphondylium.*

Jones, M. E., and Robertson, A. (1976). *J. Cell. Sci.* **22,** 41–47. Cyclic AMP and the development of *Polysphondylium.*

Jones, P. C. T. (1969). *Cytobios.* **1B,** 65–71. Temperature and anesthetic induced alterations of ATP level in animal and plant cells, and their biological significance.

Jones, P. C. T. (1970). *Cytobios.* **6,** 89–94. The interaction of light and temperature in determining ATP levels in the myxamoebae of the cellular slime mould *Dictyostelium discoideum*.

Jones, P. C. T. (1972). *J. Theor. Biol.* **34,** 1–13. Central role for ATP in determining some aspects of animal and plant cell behaviour.

Jones, T. H. D., and Gupta, M. (1981). *Biochem. Biophys. Res. Comm.* **102,** 1310. A protein inhibitor of cellulases in *Dictyostelium discoideum.*

Jones, T. H. D., and Leung, L. (1975). *Fed. Proc.* **34,** 703. Characterization of cellulase activity in spores of *Dictyostelium discoideum.*

Jones, T. H. D., and Wright, B. E. (1970). *J. Bacteriol.* **104,** 754–761. Partial purification and characterization of glycogen of *Dictyostelium discoideum* spores.

Jones, T. H. D., de Renobales, M., and Pon, N. (1979). *J. Bacteriol.* **137,** 752–757. Cellulases released during the germination of *Dictyostelium discoideum* spores.

Jones, W. R., and Francis, D. (1972). *Biol. Bull.* (Woods Hole) **142,** 461–469. The action spectrum of light induced aggregation in *Polysphondylium pallidum* and a proposed general mechanism for light response in the cellular slime molds.

Juliani, M. H., and Klein, C. (1977). *Biochim. Biophys. Acta* **497,** 369–376. Calcium ion effects on cyclic adenosine 3′,5′-monophosphate binding to the plasma membrane of *Dictyostelium discoideum.*

Juliani, M. H., and Klein, C. (1978). *Develop. Biol.* **62,** 162–172. A biochemical study of the effects of cyclic AMP pulses on aggregateless mutants of *Dictyostelium discoideum.*

Juliani, M. H., and Klein, C. (1981). *J. Biol. Chem.* **256,** 613–619, Photoaffinity labeling of the cell surface cyclic AMP receptor of *Dictyostelium discoideum* and its modification of down regulated cells.

Kahn, A. J. (1964a). *Develop. Biol.* **9,** 1–19. Some aspects of cell interaction in the development of the slime mold *Dictyostelium purpureum.*

Kahn, A. J. (1964b). *Biol. Bull.* **127,** 85–96. The influence of light on cell aggregation in *Polysphondylium pallidum.*

Kakebeeke, P. I. J., Mato, J. M., and Konijn, T. M. (1978). *J. Bacteriol.* **133,** 403–405. Purification and preliminary characterization of an aggregation sensitive chemoattractant of *Dictyostelium minutum.*

Kakebeeke, P. I. J., de Wit, R. J. W., Kohtz, S. D., and Konijn, T. M. (1979). *Exp. Cell. Res.* **124,** 429–433. Negative chemotaxis in *Dictyostelium* and *Polysphondylium*.

Kakebeeke, P. I. J., de Wit, R. J. W., and Konijn, T. M. (1980). *J. Bacteriol.* **143,** 307–312. Folic-acid deaminase activity during development in *Dictyostelium discoideum.*

Kakebeeke, P. I. J., de Wit, R. J. W., and Konijn, T. M. (1980). *FEBS Lett.* **115,** 216–220. A novel chemotaxis regulating enzyme that splits folic-acid into 6 hydroxymethyl pterin and *p*-aminobenzoyl glutamic acid.

Kananishi, N., and Watanabe, M. (1973). *J. Radiat. Res.* **14,** 83. Radiation resistance in the cellular slime mold. III. Assay of caffeine sensitive recovery of gamma irradiation damage.

Kananishi, N., and Watanabe, M. (1974). *J. Radiat. Res.* **15,** 58. Radiation resistance in the cellular slime mold. Part 4. Effects of some radiomimetic substances.

Kanda, F. (1977). *J. Biochem.* **82,** 59–66. Nuclear ribonucleoprotein particles in the cellular slime mold *Dictyostelium discoideum.*

Kanda, F. (1979). *J. Fac. Sci. Hokkaido Univ. Ser. V. Bot.* **11,** 268–273. Changes in cellular proteins during development in the cellular slime mold *Dictyostelium discoideum.*

Kanda, F., Ochiai, H., and Iwabuchi, M. (1974). *Eur. J. Biochem.* **44,** 469–480. Molecular weight determinations and stoichiometric measurements of 40S and 60S ribosomal proteins of the cellular slime mold *Dictyostelium discoideum.*

Katilus, J., and Ceccarini, C. (1975). *Develop. Biol.* **42,** 13–18. Purification and new biological properties of the slime mold germination inhibitor.

Katz, E. R. (1978). *Bioscience* **28,** 692. Cellular slime mold genetics.

Katz, E. R., and Bourguignon, L. (1974). *Develop. Biol.* **36,** 82–87. The cell cycle and its relationship to aggregation in the cellular slime mold, *Dictyostelium discoideum.*

Katz, E. R., and Kao, V. (1974). *Proc. Natl. Acad. Sci. U.S.A.* **71,** 4025–4026. Evidence for mitotic recombination in the cellular slime mold *Dictyostelium discoideum.*

Katz, E. R., and Sussman, M. (1972). *Proc. Natl. Acad. Sci. U.S.A.* **69,** 495–498. Parasexual recombination in *Dictyostelium discoideum:* selection of stable diploid heterozygotes and stable haploid segregants.

Kawabe, K. (1980). *Jpn. J. Ecol.* **30,** 183–187. Occurrence and distribution of *Dictyostelid* cellular slime molds in the Southern Alps of Japan.

Kawai, S. (1980). *Exp. Cell. Res.* **126,** 153–158. Induction of cyclic AMP receptors by disaggregation of the multicellular complexes of *Dictyostelium discoideum.*

Kawai, S. (1980). *FEBS Lett.* **109,** 27–30. Folic-acid increases cyclic AMP binding activity of *Dictyostelium discoideum* cells.

Kawai, S., and Takeuchi, I. (1977). *Devel. Growth Diff.* **18,** 311. Con A induced agglutination and binding of con A to the differentiating cells of *Dictyostelium discoideum.*

Kawai, S., and Tanaka, K-I. (1978). *Cell. Struct. Func.* **3,** 31–37. Spin labeling studies on the membrane of differentiating cells of *Dictyostelium discoideum.*

Kawashima, K., Sameshima, M., and Izawa, M. (1979). *Cell. Struct. Funct.* **4,** 183–192. Isolation of nucleoli from the cellular slime mold *Dictyostelium discoideum* Strain A-3.

Kay, R. R. (1979). *J. Embryol. Exp. Morphol.* **52,** 171–182. Gene expression in *Dictyostelium discoideum*: mutually antagonistic roles of cyclic AMP and ammonia.

Kay, R. R., Garrod, D., and Tilly, R. (1978). *Nature (London)* **271,** 58–60. Requirement for cell differentiation in *Dictyostelium discoideum.*

Kay, R. R., Sampson, J., and Steinberg, R. A. (1978). *Cell. Differ.* **7,** 33–46. Effects of 5 bromo-2-deoxy uridine on developmental functions of *Dictyostelium discoideum.*

Kay, R. R., Town, C. D., and Gross, J. D. (1979). *Differen.* **13,** 7–14. Cell differentiation in *Dictyostelium discoideum.*

Keating, M. T., and Bonner, J. T. (1977). *J. Bacteriol.* **130,** 144–147. Negative chemotaxis in cellular slime mold.

Kelleher, J. K., Kelly, P. J., and Wright, B. E. (1978). *Mol. Cell. Biochem.* **2,** 67–74. A kinetic analysis of glucokinase and glucose 6 phosphate phosphatase in *Dictyostelium discoideum.*

Kelleher, J. K., Kelly, P. J., and Wright, B. E. (1979). *J. Bacteriol.* **138,** 467–474. Amino-acid catabolism and malic enzyme EC-1.1.1.40 in differentiating *Dictyostelium discoideum.*

Keller, E. F., and Segal, L. A. (1970a). *J. Theor. Biol.* **26,** 399–415. Initiation of slime mold aggregation viewed as an instability.

Keller, E. F., and Segal, L. A. (1970b). *Nature (London)* **277,** 1365–1366. Conflict between positive and negative feedback as an explanation for the initiation of aggregation in slime mould amoebae.

Kelly, P. J., Kelleher, J. K., and Wright, B. E. (1979). *Biosystems* **11,** 55–64. Glycogen phosphorylase from *Dictyostelium discoideum:* a kinetic analysis by computer simulation.

Kelly, P. J., Kelleher, J. K., and Wright, B. E. (1979). *Biochem. J.* **184,** 581–588. The tricarboxylic-acid cycle in *Dictyostelium discoideum.* Metaolite concentrations, oxygen uptake, and carbon-14 labeled amino-acid labeling patterns.

Kelly, P. J., Kelleher, J. K., and Wright, B. E. (1979). *Biochem. J.* **184,** 589–598. The tricarboxylic-acid cycle in *Dictyostelium discoideum.* A model of the cycle at preculmination and aggregation.

Kessin, R. H. (1973). *Develop. Biol.* **31,** 242–251. RNA metabolism during vegetative growth and morphogenesis of cellular slime mold, *Dictyostelium discoideum.*

Kessin, R. H. (1977). *Cell* **10,** 703–708. Mutations causing rapid development of *Dictyostelium discoideum*.

Kessin, R., and Newell, P. (1974). *J. Bacteriol.* **117,** 379–381. Isolation of germination mutants of *Dictyostelium discoideum.*

Kessin, R. H., Williams, K., and Newell, P. (1974). *J. Bacteriol.* **119,** 776–783. Linkage analysis in *Dictyostelium discoideum* using temperature sensitive growth mutants selected with bromodeoxyuridine.

Kessin, R. H., Orlow, S. J., Shapiro, R. I., and Franke, J. (1979). *Proc. Natl. Acad. Sci. U.S.A.* **76,** 5450–5454. Binding of inhibitor alters kinetic and physical properties of extracellular cyclic AMP phosphodiesterase EC-3.1.4.17 from *Dictyostelium discoideum.*

Kessler, D., Nachmias, V. T., and Sloane, J. (1973). *J. Cell. Biol.* **59,** 167. Antiserum directed against *Physarum polycephalum* myosin.

Kessler, D., Nachmias, V. T., and Loewy, A. G. (1976). *J. Cell. Biol.* **69,** 393–406. Actomyosin content of *Physarum polycephalum* plasmodia and detection of immunological crossreactions with myosins from related species.

Kestler, D. P., Winburn, J. T., Crean, E. V., and Rossomando, E. F. (1978). *Arch. Biochem. Biophys.* **189,** 172–184. Isolation and characterization of a transcriptionally active fraction from *Dictyostelium discoideum.*

Khoury, A. T., and Deering, R. A. (1973). *J. Mol. Biol.* **79,** 267–284. Sedimentation of DNA of *Dictyostelium discoideum* lysed on alkaline sucrose gradients: role of single-strand breaks in gamma ray lethality of sensitive and resistant strains.

Khoury, A. T., Deering, R. A., Levin, G., and Altman, G. (1970). *J. Bacteriol.* **104,** 1022–1023. Gamma-ray-induced spore germination of *Dictyostelium discoideum.*

Kielman, J. K., and Deering, R. A. (1980). *Photochem. Photobiol.* **32,** 149–156. UV light induced inhibition of cell division and DNA synthesis in axenically grown repair mutants of *Dictyostelium discoideum.*

Killick, K. A. (1976). *J. Cell. Biol.* **70,** 16. Polyribonucleotide phosphorylase and trehalose 6 phosphate synthetase from myxamoebae of the cellular slime mold *Dictyostelium discoideum.*

Killick, K. A. (1977). *Anal. Biochem.* **79,** 310–318. A radiometric assay for trehalose 6 phosphate sythetase EC-2.4.1.15.

Killick, K. A. (1979). *Arch. Biochem. Biophys.* **1,** 121–133. Trehalose 6 phosphate synthase EC-2.4.1.15 from *Dictyostelium discoideum*: partial purification and characterization of the enzyme from young sorocarps.

Killick, K. A. (1980). *Carbohydr. Res.* **82,** 1–14. Development and evaluation of methods for the quantitative estimation of fungal trehalose with a charcoal gas liquid chromatographic coupled assay.

Killick, K. A. (1980). *Anal. Biochem.* **105,** 291–298. Coupled continuous and discontinuous fluorometric assays for trehalase EC-3.2.1.28 activity.

Killick, K. A. (1981). *Anal. Biochem.* **114,** 46. Purification of *Dictyostelium discoideum* spores by centrifugation in percoll density gradients with retention of morphological and biochemical integrity.

Killick, K. A., and Wang, L. W. (1980). *Anal. Biochem.* **106,** 367–372. The localization of trehalase EC-3.2.1.28 in polyacrylamide gels with eugenol by a coupled enzyme assay.

Killick, K., and Wright, B. E. (1972a). *J. Biol. Chem.* **247,** 2967–2969. Trehalose synthesis during differentiation in *Dictyostelium discoideum.* III. *In vitro* unmasking of trehalose 6-phosphate synthetase.

Killick, K. A., and Wright, B. E. (1972b). *Biochem. Biophys. Res. Commun.* **48,** 1476–1481. Trehalose synthesis during differentiation in *Dictyostelium discoideum.* IV. Secretion of trehalase and the *in vitro* expression of trehalose-6-phosphate synthetase activity.

Killick, K. A., and Wright, B. E. (1974). *Annu. Rev. Microbiol.* **28**, 139–166. Regulation of enzyme activity during differentiation in *Dictyostelium discoideum*.

Killick, K. A., and Wright, B. E. (1975). *Arch. Biochem. Biophys.* **170**, 634–643. Trehalose synthesis during differentiation in *Dictyostelium discoideum* preparation stabilization and assay of trehalose 6 phosphate synthetase.

Killick, K. A., and Wright, B. E. (1978). *J. Bacteriol.* **133**, 1039–1041. Multiple forms of glucokinase from *Dictyostelium discoideum*.

Kilpatrick, D. C. (1976). *Biochem. Soc. Trans.* **40**, 1083–1085. Metal ion dependence of alpha-mannosidase-1 from *Dictyostelium discoideum*.

Kilpatrick, D. C., and Stirling, J. L. (1976). *Biochem. J.* **158**, 409–417. Properties and developmental regulation of an alpha-D-galactosidase from *Dictyostelium discoideum*.

Kilpatrick, D. C., Schmidt, J. A., and Stirling, J. L. (1978). *FEMS Lett.* **4**, 67–70. Purification of carbohydrate binding proteins from *Dictyostelium discoideum* by affinity chromatography on agarose-E-aminocaproyl fucosamine.

Kilpatrick, D. C., Schmidt, J. A., Stirling, J. L., Jones, G. E., and Pacy, J. (1980). *J. Embryol. Exp. Morph.* **57**, 189–201. The effect of alphachymotrypsin on the development of *Dictyostelium discoideum*.

Kilpatrick, D. C., and Stirling, J. L. (1978). *Biochim. Biophys. Acta* **543**, 357–363. Carbohydrate composition of cells and plasma membranes of *Dictyostelium discoideum* at selected stages of development.

Kimmel, A. R., and Firtel, R. A. (1979). *Cell* **16**, 787–796. A family of short, interspersed repeat sequences at the 5′ ends of a set of *Dictyostelium* single copy mRNAs.

Kimmel, A. R., and Firtel, R. A. (1980). *Nucleic Acids Res.* **8**, 5599–5610. Intervening sequences in a *Dictyostelium* gene that encodes a low abundance class mRNA.

Kimmel, A. R., Lai, C., and Firtel, R. A. (1979). *ICN-UCLA Symposia on Molecular and Cellular Biology* **14**, 195–203. Families of interspersed repeat sequences at the 5′ end of *Dictyostelium* single copy repeat genes.

Kindle, K. L., and Firtel, R. A. (1978). *Cell* **15**, 763–778. Identification and analysis of *Dictyostelium* actin genes, a family of moderately repeated genes.

Kindle, K. L., and Firtel, R. A. (1979). *Nuc. Acid. Res.* **6**, 2403–2422. Evidence that populations of *Dictyostelium* single-copy mRNA transcripts carry common repeat sequences.

Kindle, K., Taylor, W., McKeown, M., and Firtel, R. A. (1977). *In* "Developments in Cell Biology." Elsevier/North-Holland (Cappuccinelli, P., and Ashworth, J. M. eds.), 273–290. Analysis of gene structure and transcription in *Dictyostelium discoideum*.

King, A. C., and Frazier, W. A. (1977). *Biochem. Biophys. Res. Commun.* **78**, 1093–1099. Reciprocal periodicity in cyclic AMP binding and phosphorylation of differentiating *Dictyostelium discoideum* cells.

King, A. C., and Frazier, W. A. (1979). *J. Biol. Chem.* **254**, 7168–7176. Properties of the oscillatory cyclic AMP binding component of *Dictyostelium discoideum* cells and isolated plasma membranes.

Kitzke, E. D. (1952). *Nature (London)* **170**, 284–285, A new method for isolating members of the Acrasieae from soil samples.

Klaus, M., and George, R. P. (1974). *Develop. Biol.* **39**, 183–188. Microdissection of developmental stages of the cellular slime mold *Dictyostelium discoideum*.

Klein, C. (1974). *FEBS Lett.* **38**, 149–152. Presence of magic spot in *Dictyostelium discoideum*.

Klein, C. (1975). *J. Biol. Chem.* **250**, 7134–7138. Induction of phosphodiesterase by cAMP in differentiating *Dictyostelium discoideum*.

Klein, C. (1976). *FEBS Letts.* **68**, 125–128. Adenylate cyclase activity in *Dictyostelium discoideum* amoebae and its changes during differentiation.

Klein, C. (1977). *FEMS Microbiol. Lett.* **1**, 17–20. Changes in adenylate cyclase during differentiation of *Dictyostelium discoideum.*

Klein, C. (1979). *J. Biol. Chem.* **254**, 12573–12578. A slowly dissociating form of the cell surface cyclic AMP receptor of *Dictyostelium discoideum.*

Klein, C. (1979). *Exp. Cell. Res.* **124**, 205–214. The effects of inhibitors of adenylate cyclase and phosphodiesterase on *Dictyostelium discoideum* aggregation.

Klein, C. (1980). *Develop. Biol.* **79**, 500–507. Cyclic AMP independent oscillations of adenylate cyclase in *Dictyostelium discoideum.*

Klein, C. (1981). *J. Biol. Chem.* **256**, 10050. Binding of adenosine 3′,5′-monophosphate to plasma membranes of *Dictyostelium discoideum* amoebae.

Klein, C., and Brachet, P. (1975). *Nature (London)* **254**, 432–434. Effects of progesterone and EDTA on cAMP and phosphodiesterase in *Dictyostelium discoideum.*

Klein, C., and Darmon, M. (1975). *Biochem. Biophys. Res. Commun.* **67**, 440–447. The relationship of phosphodiesterase to the developmental cycle of *Dictyostelium discoideum.*

Klein, C., and Darmon, M. (1976). *Proc. Natl. Acad. Sci. U.S.A.* **73**, 1250–1254. Differentiation stimulating factor induces cell sensitivity to 3′,5′-cyclic AMP pulses in *Dictyostelium discoideum.*

Klein, C., and Darmon, M. (1977). *Nature (London)* **268**, 76–78. Effects of cyclic AMP pulses on adenylate cyclase and the phosphodiesterase inhibitor of *Dictyostelium discoideum.*

Klein, C., and Darmon, M. (1979). *FEMS Lett.* **5**, 1–4. A cyclic AMP sensitive adenylate cyclase EC-4.6.1.1 in *Dictyostelium discoideum* extracts.

Klein, C., and Juliani, M. H. (1977). *Cell* **10**, 329–335. cAMP-induced changes in cAMP-binding sites on *Dictyostelium discoideum* amoebae.

Klein, C., Brachet, P., and Darmon, M. (1977). *FEBS Lett.* **76**, 145–147. Periodic changes in adenylate cyclase and cyclic AMP receptors in *Dictyostelium discoideum.*

Knecht, D. A., and Dimond, R. L. (1981). *J. Biol. Chem.* **256**, 3564–3575. Lysosomal enzymes possess a common antigenic determinant in the cellular slime mold, *Dictyostelium discoideum.*

Kobilinsky, L., and Beattie, D. S. (1977). *J. Bioener. Biomem.* **9**, 73. The reversibility of the ethidium bromide-induced alterations of mitochondrial structure and function in the cellular slime mould, *Dictyostelium discoideum.*

Kobilinsky, L., and Beattie, D. S. (1977). *J. Bacteriol.* **132**, 113–117. Respiratory competence of *Dictyostelium discoideum* spores.

Kobilinsky, L., and Beattie, D. S. (1977). *J. Bioenerg. Biomembr.* **9**, 73–90. The reversibility of the ethidium bromide induced alterations of mitochondrial structure and function in the cellular slime mold *Dictyostelium discoideum.*

Kobilinsky, L., Weinstein, B. I., and Beattie, D. S. (1976). *Develop. Biol.* **48**, 477–481. The induction of filopodia in the cellular slime mold *Dictyostelium discoideum* by cyclic AMP mechanism of aggregation.

Komuniecki, P. R., de Toma, F. J., Lawrence, M. H., and DiDomenico, L. (1980). *Biochem. Biophys. Res. Commun.* **96**, 1017–1023. ADP phosphorylation and glutamate oxidation in mitochondria isolated from *Dictyostelium discoideum* amoebae.

Konijn, T. M. (1966a). *Develop. Biol.* **12**, 487–497. Chemotaxis in the cellular slime molds. I. The effect of temperature.

Konijn, T. M. (1966b). *Biol. Bull.* **134**, 298–304. Chemotaxis in the cellular slime molds. II. The effect of density.

Konijn, T. M. (1969). *J. Bacteriol.* **99**, 503–509. Effect of bacteria on chemotaxis in the cellular slime molds.

Konijn, T. M. (1970). *Experimentia* **26**, 367–369. Microbiological assay of cyclic 3′,5′-AMP.

Konijn, T. M. (1972a). *Advan. Cyclic Nucleotide Res.* **1**, 17–31. Cyclic AMP as a first messenger.

Konijn, T. M. (1972b). *Acta Protozoologica* **11**, 137–143. Cyclic AMP and cell aggregation in the cellular slime molds.

Konijn, T. M. (1973). *FEBS Lett.* **34**, 263–266. The chemotactic effect of cyclic nucleotides with substitutions in the base ring.

Konijn, T. M. (1974). *Antibiotics Chemother.* **19**, 96–110. The chemotactic effect of cyclic AMP and its analogues in the Acrasieae.

Konijn, T. M., and Jastorff, B. (1973). *Biochim. Biophys. Acta* **304**, 774–780. The chemotactic effect of 5′-amido analogues of adenosine cyclic 3′,5′-monophosphate in the cellular slime moulds.

Konijn, T. M., and Raper, K. B. (1961). *Develop. Biol.* **3**, 725–756. Cell aggregation in *Dictyostelium discoideum.*

Konijn, T. M., and Raper, K. B. (1965). *Biol. Bull.* **128**, 392–400. The influence of light on the time of cell aggregation in the Dictyosteliaceae.

Konijn, T. M., Van De Meene, J. G. C., Bonner, J. T., and Barkley, D. S. (1967). *Proc. Natl. Acad. Sci. U.S.A.* **58**, 1152–1154. The acrasin activity of adenosine-3′,5′-cyclic phosphate.

Konijn, T. M., Barkley, D., Chang, Y. Y., and Bonner, J. (1968). *Am. Nat.* **102**, 225–233. Cyclic AMP: a naturally occurring acrasin in the cellular slime molds.

Konijn, T. M., Chang, Y. Y., and Bonner, J. T. (1969a). *Nature (London)* **224**, 1211–1212. Synthesis of cyclic AMP in *Dictyostelium discoideum* and *Polyspondylium pallidum.*

Konijn, T. M., Van De Meene, J. G. C., Change, Y. Y., Barkley, D. S., and Bonner, J. T. (1969b). *J. Bacteriol.* **99**, 510–512. Identification of adenosine-3′,5′-monophosphate as the bacterial attractant for myxamoebae of *Dictyostelium discoideum.*

Konno, R. (1980). *Dev. Growth Differ.* **22**, 125–132. Aggregateless mutant defective in signaling and relaying in the cellular slime mold *Dictyostelium discoideum.*

Korn, E., and Wright, P. (1973). *J. Biol. Chem.* **248**, 439. Macromolecular composition of an amoeba plasma membrane.

Kost, R. G., North, M. J., and Whyte, A. (1981). *Exp. Mycol.* **5**, 269. Acid proteinases in various species of cellular slime mold.

Koziol, J. A., Springer, W. R., and Barondes, S. H. (1980). *Exp. Cell. Res.* **128**, 375–381. Quantitation of selective cell–cell adhesion and its application to assays of species specific adhesion in cellular slime molds.

Krichevsky, M. I., and Love, L. L. (1964a). *J. Gen. Microbiol.* **34**, 483–490. The uptake and utilization of histidine by washed amoebae in the course of development in *Dictyostelium discoideum.*

Krichevsky, M. I., and Love, L. L. (1964b). *J. Gen. Microbiol.* **37**, 293–295. Adenine inhibition of the rate of sorocarp formation in *Dictyostelium discoideum.*

Krichevsky, M. I., and Love, L. L. (1965). *J. Gen. Microbiol.* **41**, 367–374. Efflux of macromolecules from washed *Dictyostelium discoideum.*

Krichevsky, M. I., and Wright, B. E. (1963). *J. Gen. Microbiol.* **32**, 195–207. Environmental control of the course of development in *Dictyostelium discoideum.*

Krichevsky, M. I., Love, L. L., and Chassy, B. (1969). *J. Gen. Microbiol.* **57**, 383–389. Acceleration of morphogenesis in *D. discoideum* by exogenous mononucleotides.

Krivanek, J. O. (1956). *J. Exp. Zool.* **133**, 459–480. Alkaline phosphatase activity in the developing slime mold, *Dictyostelium discoideum* Raper.

Krivanek, J. O., and Krivanek, R. C. (1957). *Stain Technol.* **32**, 300–301. A method for embedding small specimens.

Krivanek, J. O., and Krivanek, R. C. (1958). *J. Exp. Zool.* **137**, 89–115. The histochemical localization of certain biochemical intermediates and enzymes in the developing slime mold, *Dictyostelium discoideum* Raper.

Krivanek, J. O., and Krivanek, R. C. (1959). *Biol. Bull.* **116**, 265–271. Chromatographic analyses of amino acids in the developing slime mold, *Dictyostelium discoideum* Raper.

Krivanek, J. O., and Krivanek, R. C. (1965). *Biol. Bull.* **129,** 295–302. Evidence for transaminase activity in the slime mold *Dictyostelium discoideum* Raper.

Ku, K. Y., and Goz, B. (1978). *Biochem. Pharmacol.* **27,** 1597–1602. 5 Iodo-2-deoxy uridine inhibition of *Dictyostelium discoideum* differentiation and cyclic AMP phosphodiesterase activity.

Kuczmarski, E. R., and Spudich, J. A. (1980). *Proc. Natl. Sci. U.S.A.* **77,** 7292–7296. Regulation of myosin self-assembly: phosphorylation of *Dictyostelium* heavy chain inhibits thick filament formation.

Kuhn, H., and Parish, R. W. (1981). *Exp. Cell. Res.* **131,** 89–96. Fusion of *Dictyostelium discoideum* cells with one another and with erythrocyte ghosts.

Kuserk, F. T. (1980). *Ecology* **61,** 1474–1485. The relationship between cellular slime molds and bacteria in forest soil.

Kuserk, F. T., Eisenberg, R. M., and Olsen, A. M. (1977). *J. Protozool.* **24,** 297. An examination of the methods for isolating cellular slime mould (*Dictyosteliida*) from soil samples.

Kyriazi, H. T., and Brown, R. M., Jr. (1976). *Plant Physiol.* **57,** 3. An ultrastructural study of differentiation in *Dictyostelium discoideum* with special emphasis on cellulose stalk synthesis.

Lacalli, T. C., and Harrison, L. G. (1978). *J. Theor. Biol.* **70,** 273–295. The regulatory capacity of Turing's model for morphogenesis with application to slime molds.

Laine, J., Roxby, N., and Coukell, M. B. (1975). *Can. J. Microbiol.* **21,** 959–962. A simple method for storing cellular slime mold amoebae.

Lakhani, S. (1977). *Botanica* **27,** 100–102. Cell differentiation. Part 2. Control mechanisms.

Lam, T. Y., and Siu, C-H. (1981). *Develop. Biol.* **83,** 127–137. Synthesis of stage-specific glycoproteins in *Dictyostelium discoideum* during development.

Lam, T. Y., Pickering, G., Geltosky, J., and Siu, C-H. (1981). *Differentiation* (In Press). Differential cell cohesiveness expressed byprespore and prestalk cells of *Dictyostelium discoideum.*

Lamprecht, I. (1978). *In* "Thermodynamics of Biological Processes." New York, N.Y. (Lamprecht, Ingolf, and Zotin, A. I., eds.), 261–276. Review of the theory of dissipative structures.

Landfear, S. M., and Lodish, H. F. (1980). *Proc. Natl. Acad. Sci. U.S.A.* **77,** 1044–1048. A role for cyclic AMP in expression of developmentally regulated genes in *Dictyostelium discoideum.*

Lang, A. (1954). *Fortschr. Bot.* **15,** 400–475. Entwicklungsphysiologie der Acrasiales.

Langridge, W. H. R., Komuniecki, P., and de Toma, F. J. (1977). *Archives Biochem.* **178,** 581. Isolation and regulatory properties of two glutamate dehydrogenases from the cellular slime mold *Dictyostelium discoideum.* Deum.

Lax, A. J. (1979). *J. Cell. Sci.* **36,** 311–321. The evolution of excitable behavior in *Dictyostelium discoideum.*

Leach, C. K., and Ashworth, J. M. (1972). *J. Mol. Biol.* **68,** 35–48. Characterization of DNA from the cellular slime mould *Dictyostelium discoideum* after growth of the amoebae in different media.

Leach, C. K., Ashworth, J. M., and Garrod, D. R. (1973). *J. Embryol. Exp. Morphol.* **29,** 647–661. Cell sorting out during the differentiation of mixtures of metabolically distinct populations of *Dictyostelium discoideum.*

Lee, A., Chance, K., Weeks, C., and Weeks, G. (1975). *Arch. Biochem. Biophys.* **171,** 407–417. Studies on the alkaline phosphatase and 5′-nucleotidase of *Dictyostelium discoideum.*

Lee, K-C. (1972a). *J. Gen. Microbiol.* **72,** 457–471. Permeability of *Dictyostelium discoideum* towards amino acids and inulin: a possible relationship between initiation of differentiation and loss of "pool" metabolites.

Lee, K-C. (1972b). *J. Cell Sci.* **10,** 229–248. Cell electrophoresis of the cellular slime mould *Dictyostelium discoideum.* I. Characterization of some of the cell surface ionogenic groups.

Lee, K-C. (1972c). *J. Cell Sci.* **10,** 249–265. Cell electrophoresis of the cellular slime mould *Dictyostelium discoideum.* II. Relevance of the changes in cell surface charge density to cell aggregation and morphogenesis.

Leichtling, B. H., Coffman, D. S., Yaeger, E. S., Rickenberg, H. V., Al-Jumaliy, W., and Haley, B. E. (1981). *Biochem. Biophys. Res. Comm.* **102,** 1187. Occurrence of the adenylate cyclase "G protein" in membranes of *Dictyostelium discoideum.*

Leichtling, B. H., Tihon, C., Spitz, E., and Rickenberg, H. V. (1981). *Develop. Biol.* **82,** 150–157. A cytoplasmic cyclic AMP binding protein in *Dictyostelium discoideum.*

Lenfant, M., Ellouz, R., Das, B. C., Zissmann, E., and Lederer, E. (1969). *Eur. J. Biochem.* **7,** 159–164. Sur la biosynthèse de la chaine laterale ethyle des sterols ou myxomycete *Dictyostelium discoideum.*

Lerner, R. A., Ray, J., and Geltosky, J. E. (1977). *J. Supramol. Struct.* Suppl. **1,** 34. Use of common plant lectins for isolation and characterization of constitutive and developmentally regulated cell surface associated glycoproteins of *Dictyostelium discoideum.*

Levandowsky, M., and Hunter, S. H. (1979). *In* "Biochemistry and physiology of Protozoa," 2nd Edition, Vol. 2. Academic Press, New York, N.Y. (Levandowsky, M., and Hunter, S. H. eds.). Biochemistry and physiology of protozoa.

Levinson, M. A. (1978). *Coll. Pheno.* **3,** 35–39. McMahon's model for cellular position determination in development: parameter dependence and an analytical solution.

Lewis, K. E., and O'Day, D. H. (1976). *Can. J. Microbiol.* **22,** 1269–1273. Sexual hormone in the cellular slime mold *Dictyostelium purpureum.*

Lewis, K. E., and O'Day, D. H. (1977). *Nature (London)* **268,** 730–731. Sex hormone of *Dictyostelium discoideum* is volatile.

Lewis, K. E., and O'Day, D. H. (1979). *J. Bacteriol.* **138,** 251–253. Evidence for a hierarchical mating system operating via pheromones in *Dictyostelium giganteum.*

Lewis, K. E., and O'Day, D. H. Sexual hormone in the cellular slime mold *Dictyostelium purpureum.*

Liddel, G. U., and Wright, B. E. (1961). *Develop. Biol.* **3,** 265–276. The effect of glucose on respiration of the differentiating slime mold.

Lin, S., Santi, D., and Spudich, J. (1974). *J. Biol. Chem.* **249,** 2268–2274. Biochemical studies on the mode of action of cytochalasin B preparation of tritiated cytochalasin B and studies on its binding to cells.

Linkins, A. E., and Rutherford, C. L. (1976). *Plant Physiol.* **57,** 90. Cellulase in spores of *Dictyostelium discoideum.*

Liwerant, I. J., and Da Silva, L. H. (1975). *Mutat. Res.* **33,** 135–146. Comparative mutagenic effects of ethyl methanesulfonate N methyl-N-nitro-N-nitroso guanidine UV radiation and caffeine on *Dictyostelium discoideum.*

Lo, E. K.-L., Coukell, M. B., Tsang, A. S., and Pickering, J. L. (1978). *Can. J. Microbiol.* **24,** 455–465. Physiological and biochemical characterization of aggregation deficient mutants of *Dictyostelium discoideum* detection and response to exogenous cyclic AMP.

Lodish, H. F. (1977). *In* "Cell Differentiation in Microorganisms, Plants and Animals." International Symposium. Elsevier/North-Holland, New York, N.Y. (Nover, Lutz, and Mothes eds.), 126–145. Synthesis and structure of messenger RNA in the slime mold *Dictyostelium discoideum.*

Lodish, H. F., and Alton, T. H. (1977). *In* "Developments in Cell Biology." Elsevier/North-Holland (Cappuccinelli, P., and Ashworth, J. M. eds.), 253–272. Translational and transcriptional control of protein synthesis during differentiation of *Dictyostelium discoideum.*

Lodish, H. F., Alton, T., Dottin, R. P., Weiner, A. M., and Margolskee, J. P. (1976). *In* "Molecular Biology of Hormone Action" (J. Papaconstantinov ed.), p. 75. Synthesis and translation of messenger RNA during differentiation of the cellular slime mould *Dictyostelium discoideum.*

Lodish, H. F., Firtel, R., and Jacobson, A. (1973). *Cold Spring Harbor Symp. Quant. Biol.* **38,** 899–907. Transcription and structure of the genome of the cellular slime mold *Dictyostelium discoideum.*

Lodish, H. F., Jacobson, A., Firtel, R., Alton, T., and Tuchman, J. (1974). *Proc. Natl. Acad. Sci. U.S.A.* **71,** 5103–5108. Synthesis of messenger RNA and chromosome structure in the cellular slime mold.

Lodish, H. F., Alton, T., Margolskee, J., Dottin, R., and Weiner, A. (1975). *ICN-UCLA Symp. Mol. Cell. Biol.* **2,** 366–383. RNA and protein synthesis during differentiation of the slime mold *Dictyostelium discoideum.*

Long, B. H., and Coe, E. L. (1973). *Comp. Biochem. Physiol.* **45,** 933–943. Characterization of ubiquinone from vegetative amoebae and mature surocarps of the cellular slime mold, *Dictyostelium discoideum.*

Long, B. H., and Coe, E. L. (1974). *J. Biol. Chem.* **249,** 521–529. Changes in neutral lipid constituents during differentiation of the cellular slime mold *Dictyostelium discoideum.*

Long, B. H., and Coe, E. L. (1977). *Lipids* **12,** 414–417. Fatty-acid compositions of lipid fractions from vegetative cells and mature sorocarps of the cellular slime mold *Dictyostelium discoideum.*

Longmore, K., and Watts, D. J. (1980). *Develop. Biol.* **78,** 104–112. Control of acetyl glucosaminidase specific activity in myxamoebae of *Dictyostelium discoideum.*

Lonski, J. (1976). *Develop. Biol.* **51,** 158–165. The effect of ammonia on fruiting-body size and microcyst formation in the cellular slime molds.

Lonski, J. (1977). *Develop. Biol.* **55,** 85–91. Evidence for a second chemotactic system in the cellular slime mould *Polysphondilium.*

Lonski, J., and Pesut, N. (1977). *Can. J. Microbiol.* **23,** 518–521. Induction of encystment of *Polysphondylium pallidum* amoeba.

Loomis, W., Rossomando, E. F., and Chang, L. M. S. (1976). *Biochim. Biophys. Acta* **425,** 469–477. DNA polymerase of *Dictyostelium discoideum.*

Loomis, W. F. (1968). *Exp. Cell Res.* **53,** 282–287. The relation between cytodifferentiation and inactivation of a developmentally controlled enzyme in *Dictyostelium discoideum.*

Loomis, W. F. (1969a). *J. Bacteriol.* **97,** 1149–1154. Acetylglucosaminidase, an early enzyme in the development of *Dictyostelium discoideum.*

Loomis, W. F. (1969b). *J. Bacteriol.* **99,** 65–69. Temperature sensitive mutants of *Dictyostelium discoideum.*

Loomis, W. F. (1969c). *J. Bacteriol.* **100,** 417–422. Developmental regulation of alkaline phosphatase in *Dictyostelium discoideum.*

Loomis, W. F. (1970a). *J. Bacteriol.* **103,** 375–381. Developmental regulation of alpha-mannosidase in *Dictyostelium discoideum.*

Loomis, W. F. (1970b). *Nature (London)* **227,** 745–746. Mutants in phototaxis in *Dictyostelium discoideum.*

Loomis, W. F. (1970c). *Exp. Cell Res.* **60,** 285–289. Temporal control of differentiation in the slime mold, *Dictyostelium discoideum.*

Loomis, W. F. (1971). *Exp. Cell Res.* **64,** 484–486. Sensitivity of *Dictyostelium discoideum* to nucleic acid analogues.

Loomis, W. F. (1972). *Nature (London) New Biol.* **240,** 6–9. Role of the surface sheath in the control of morphogenesis in *Dictyostelium discoideum.*

Loomis, W. F. (1975). *In* "Proc. Int. Cong. Isozymes 3rd." Academic Press, New York, N.Y. (Markert, C. ed.), 177–189. Stage specific isozymes of *Dictyostelium discoideum.*

Loomis, W. F. (1975). Academic Press, New York. *Dictyostelium discoideum,* A Developmental System.

Loomis, W. F. (1975). *In* "Proceedings of the 1975 ICN-UCLA Symposium on Developmental Biology" (McMahon, D., and Fox, C. F. eds.), 109–128. Polarity and pattern in *Dictyostelium*.

Loomis, W. F. (1978). *Genetics* **88,** 277–284. Genetic analysis of the gene for *N*-acetyl glucosaminidae in *Dictyostelium discoideum*.

Loomis, W. F. (1978). Birth Defects: Original Article Series XIV, 497–505. The number of developmental genes in *Dictyostelium*.

Loomis, W. F. (1979). *Develop. Biol.* **70,** 1–12. Biochemistry of aggregation in *Dictyostelium*. A Review.

Loomis, W. F. (1980). *Genetics* **1,** 241–246. A beta-glucosidase gene of *Dictyostelium discoideum*. I.

Loomis, W. F. (1980). *In* "The Molecular Genetics of Development" (Leighton, T., and Loomis, W. F. ed.), 179–212. Genetic analysis of development in *Dictyostelium*.

Loomis, W. F., and Ashworth, J. M. (1968). *J. Gen. Microbiol.* **53,** 181–186. Plaque-size mutants of the cellular slime mold *Dictyostelium discoideum*.

Loomis, W. F., and Sussman, M. (1966). *J. Mol. Biol.* **22,** 401–404. Commitment to the synthesis of a specific enzyme during cellular slime mold development.

Loomis, W. F., and Thomas, S. (1976). *J. Biol. Chem.* **251,** 6252–6258. Kinetic analysis of biochemical differentiation in *Dictyostelium discoideum*.

Loomis, W. F., and Wheeler, S. (1980). *Develop. Biol.* **79,** 399–408. Heat shock response of *Dictyostelium discoideum*.

Loomis, W. F., and Wheeler, S. (1982). *Develop. Biol.* **90,** 412–418. Chromatin associated heat shock proteins of *Dictyostelium*.

Loomis, W. F., White, S., and Dimond, R. (1976). *Develop. Biol.* **53,** 171–177. A sequence of dependent stages in development of *Dictyostelium discoideum*.

Loomis, W. F., Klein, C., and Brachet, P. (1978). *Differentiation* **12,** 83–90. The effect of divalent cations on aggregation of *Dictyostelium discoideum*.

Loomis, W. F., Morrissey, J., and Lee, M. (1978). *Develop. Biol.* **63,** 243–246. Biochemical analysis of pleiotrophy in *Dictyostelium discoideum*.

Loomis, W. F., Dimond, R., Free, S., and White, S. (1978). *In* "Eukaryotic Microbes as Model Development Systems." Marcel Dekker, New York (O'Day, D., and Horgen, P. eds.), 177–194. Independent and dependent sequences in development of *Dictyostelium*.

Love, L. L., Chassy, B. M., and Krichevsky, M. I. (1973). *Antimicrob. Agents Chemother.* **3,** 310–313. Growth of *Dictyostelium discoideum* in the presence of antibiotics.

Luna, E. J., Fowler, V. M., Swanson, J., Branton, D., and Taylor, D. L. (1981). *J. Cell. Biol.* **88,** 396–409. A membrane cytoskeleton from *Dictyostelium discoideum* 1. Identification and partial characterization of an actin binding activity.

Ma, G. C. L., and Firtel, R. A. (1978). *J. Biol. Chem.* **253,** 3924–3932. Regulation of the synthesis of 2 carbohydrate binding proteins in *Dictyostelium discoideum*.

Machac, M. A., and Bonner, J. T. (1975). *J. Bacteriol.* **124,** 1624–1625. Evidence for a sex hormone in *Dictyostelium discoideum*.

MacInnes, M. A., and Francis, D. W. (1974). *Nature (London)* **251,** 321–323. Meiosis in *Dictyostelium mucoroides*.

MacInnes, M. A., and Francis, D. W. (1976). *Genetics* **83,** 46–47. Genetic and physiological studies on tiny aggregation *Tag* mutants of *Dictyostelium mucoroides* DM-7.

MacInnes, M. A., and Francis, D. W. (1977). *Adv. Cyclic Nucleotide Res.* **9,** 778. Altered control of cyclic AMP oscillations in an aggregation mutant of *Dictyostelium mucoroides*.

MacInnes, M. A., and MacInnes, F. D. (1974). *Genetics* **77,** 41. Meiosis in *Dictyostelium mucoroides*.

Mackay, S. A. (1978). *J.Cell. Sci.* **33,** 1–16. Computer simulation of aggregation in *Dictyostelium discoideum*.

Macleod, C. L., and Loomis, W. F. (1979). *Dev. Genet.* **1**, 109–121. Biochemical and genetic analysis of a mutant with altered alkaline phosphatase activity in *Dictyostelium discoideum.*

Macleod, C. L., and Loomis, W. F. (1980). *J. Gen. Microbiol.* **120**, 273–278. Attempts to isolate mutants of *Dictyostelium discoideum* which express alpha mannosidase 2 constitutively.

Macleod, C. L., Firtel, R. A., and Papkoff, J. (1980). *Develop. Biol.* **76**, 263–274. Regulation of actin gene expression during spore germination in *Dictyostelium discoideum.*

MacWilliams, H. K., and Bonner, J. T. (1979). *Differentiation* **14**, 1–22. The prestalk prespore pattern in cellular slime molds.

Madley, I. C., and Hames, B. D. (1981). *Biochem. J.* **200**, 83. An analysis of discoidin I binding sites in *Dictyostelium discoideum.*

Maeda, M. (1978). *Sci. Rep. Coll. Gen. Educ. Osaka Univ.* **27**, 118–126. Effects of EDTA and ethylene glycol bis-beta aminoethyl ether N, N, N′, N′, tetra acetic-acid on chemotactic movement and morphogenesis of a cellular slime mold *Dictyostelium discoideum.*

Maeda, Y. (1970). *Develop. Growth Differentiat.* **12**, 217–227. Influence of ionic conditions on cell differentiation and morphogenesis of the cellular slime molds.

Maeda, Y. (1971). *Develop. Growth Differ.* **13**, 211–219. Formation of a prespore specific structure from a mitochondrion during development of the cellular slime mold *Dictyostelium discoideum.*

Maeda, Y. (1977). *Dev. Growth Differ.* **3**, 201–206. Role of cyclic AMP in the polarized movement of the migrating pseudoplasmodium of *Dictyostelium discoideum.*

Maeda, Y. (1980). *Dev. Growth Differ.* **22**, 679–686. Changes in charged groups on the cell surface during development of the cellular slime mold *Dictyostelium discoideum* an electron microscopic study.

Maeda, Y., and Eguchi, G. (1977). *Cell Struct. Funct.* **2**, 159–170. Polarized structures of cells in the aggregating cellular slime mold *Dictyostelium discoideum* an electron microscope study.

Maeda, Y., and Gerisch, G. (1977). *Exp. Cell. Res.* **110**, 119–126. Vesicle formation in *Dictyostelium discoideum* cells during oscillations of cyclic AMP synthesis and release.

Maeda, Y., and Maeda, M. (1973). *Exp. Cell Res.* **82**, 125–130. The calcium content of the cellular slime mold *Dictyostelium discoideum* during development and differentiation.

Maeda, Y., and Maeda, M. (1974). *Exp. Cell Res.* **84**, 88–94. Heterogeneity of the cell population of the cellular slime mold *Dictyostelium discoideum* before aggregation, and its relation to subsequent locations of the cells.

Maeda, Y., and Takeuchi, I. (1969). *Dev. Growth Diff.* **11**, 232–245. Cell differentiation and fine structures in the development of the cellular slime molds.

Maeda, Y., Sugita, K., and Takeuchi, I. (1973). *Bot. Mag. Tokyo* **86**, 5–12. Fractionation of the differentiated types of cells constituting the pseudoplasmodia of the cellular slime molds.

Maizels, N. (1976). *Cell* **9**, 431–438. *Dictyostelium* 17S, 25S and 5S. DNA's lie within a 38,000 base pair repeated unit.

Malchow, D., and Gerisch, G. (1973a). *Biochem. Biophys. Res. Commun.* **55**, 200–204. Cyclic AMP binding to living cells of *Dictyostelium discoideum* in presence of excess cyclic GMP.

Malchow, D., and Gerisch, G. (1973b). *Hoppe-Seyler's Z. Physiol. Chem.* **354**, 200–204. Recognition of extracellular cyclic AMP by aggregating cells of the slime mold *Dictyostelium discoideum.*

Malchow, D., and Gerisch, G. (1974). *Proc. Natl. Acad. Sci. U.S.A.* **71**, 2423–2427. Short term binding and hydrolysis of cyclic 3′,5′-adenosine monophosphate by aggregating *Dictyostelium* cells.

Malchow, D., Luderitz, O., Westphal, O., Gerisch, G., and Riedel, V. (1967). *Eur. J. Biochem.* **2**, 469–479. Polysaccaride in vegetativen und aggregationsreifen amoben von *Dictyostelium discoideum.* 1. *In vivo* degradierung von bakterien-lipopoly-saccharid.

Malchow, D., Luderitz, O., Kickhofen, B., Westphal, O., and Gerisch, G. (1969). *Eur. J. Biochem.* **7,** 239–246. Polysaccharides in vegetative and aggregation-competent amoebae of *Dictyostelium discoideum.* 2. Purification and characterization of amoeba-degraded bacterial polysaccharides.

Malchow, D., Nagele, B., Schwarz, H., and Gerisch, G. (1972). *Eur. J. Biochem.* **28,** 136–142. Membrane-bound cyclic AMP phosphodiesterase in chemotactically responding cells of *Dictyostelium discoideum.*

Malchow, D., Fuchila, J., and Jastorff, B. (1973). *FEBS Lett.* **34,** 5–9. Correlation of substrate specificity of cAMP-phosphodiesterase in *Dictyostelium discoideum* with chemotactic activity of cAMP-analogues.

Malchow, D., Fuchila, J., and Nanjundiah, V. (1975). *Biochim. Biophys. Acta* **385,** 421–428. A plausible role for a membrane bound cyclic AMP phosphodiesterase in cellular slime mold chemotaxis.

Malchow, D., Roos, W., and Gerisch, G. (1977). *Experientia* **33,** 824. Cyclic AMP receptors and periodic activation of adenylcyclase in *Dictyostelium.*

Malchow, D., Nanjundiah, V., and Gerisch, G. (1978). *J. Cell. Sci.* **30,** 319–330. pH oscillations in cell suspensions of *Dictyostelium discoideum* their relation to cyclic AMP signals.

Malchow, D., Nanjundiah, V., Wurster, B., Eckstein, F., and Gerisch, G. (1978). *Biochim. Biophys. Acta* **538,** 473–480. Cyclic AMP induced pH changes in *Dictyostelium discoideum* and their control by calcium.

Malchow, D., Bohme, R., and Rahmsdorf, H. J. (1981). *Eur. J. Biochem.* **117,** 213–218. Regulation of myosin heavy chain phosphorylation during the chemotactic response of *Dictyostelium* cells.

Malkinson, A. M., and Ashworth, J. M. (1972). *Biochem. J.* **127,** 611–612. Extracellular concentrations of adenosine 3′,5′ cyclic monophosphate during axenic growth of myxamoebae of the cellular slime mould *Dictyostelium discoideum.*

Malkinson, A. M., and Ashworth, J. M. (1973). *Biochem. J.* **134,** 311–319. Adenosine 3′,5′ cyclic monophosphate concentrations and phosphodiesterase activities during axenic growth and differentiation of cells of the cellular slime mould *Dictyostelium discoideum.*

Malkinson, A. M., Kwasniak, J., and Ashworth, J. M. (1973). *Biochem. J.* **133,** 601–603. Adenosine 3′,5′-cyclic monophosphate-binding protein from the cellular slime mould *Dictyostelium discoideum.*

Manabe, K., and Poff, K. L. (1978). *Plant Physiol.* **61,** 961–966. Purification and characterization of the photoreducible B type cytochrome from *Dictyostelium discoideum.*

Mangiarotti, G., and Hames, B. D. (1979). *Exp. Cell. Res.* **1,** 428–432. Analysis of ribosomal RNA metabolism during development of *Dictyostelium discoideum.*

Mangiarotti, G., Chung, S., Zuker, C., and Lodish, H. F. (1981a). *Nucl. Acids Res.* **9,** 947–963. Selection and analysis of cloned developmentally regulated *Dictyostelium discoideum* genes by hybridization competition.

Mangiarotti, G., Lefebvre, P., and Lodish, H. F. (1981). *Develop. Biol.* **89,** 82–91. Differences in the stability of developmentally regulated mRNAs in aggregated and disaggregated *Dictyostelium discoideum* cells.

Manrow, R. E., and Dottin, R. P. (1980). *Proc. Natl. Acad. Sci. U.S.A.* **77,** 730–734. Renaturation and localization of enzymes in polyacrylamide gels studies with UDP glucose pyrophosphorylase EC-2.7.7.9 of *Dictyostelium discoideum.*

Margolskee, J. P., and Lodish, H. F. (1980). *Develop. Biol.* **74,** 37–49. Half-Lives of messenger RNA species during growth and differentiation of *Dictyostelium discoideum.*

Margolskee, J. P., and Lodish, H. F. (1980). *Develop. Biol.* **74,** 50–64. The regulation of the synthesis of actin and 2 other proteins induced early in *Dictyostelium discoideum* development.

Margolskee, J. P., Froshauer, S., Skrinska, R., and Lodish, H. F. (1980). *Develop. Biol.* **74,** 409–421. The effects of cell density and starvation on early developmental events in *Dictyostelium discoideum.*

Marin, F. T. (1976). *Develop. Biol.* **48,** 110–117. Regulation of development in *Dictyostelium discoideum.* Part 1. Initiation of the growth to development transition by amino-acid starvation.

Marin, F. T. (1977). *Develop. Biol.* **60,** 389–395. Regulation of development in *Dictyostelium discoideum.* Part 2. Regulation of early cell differentiation by amino-acid starvation and inter cellular interaction.

Marin, F. T., Goyette-Boulay, M., and Rothman, F. G. (1980). *Develop. Biol.* **80,** 301–312. Regulation of development of *Dictyostelium discoideum* 3. Carbohydrate specific intercellular interactions in early development.

Marin, F. T., and Rothman, F. G. (1980). *J. Cell. Biol.* **87,** 823–827. Regulation of development in *Dictyostelium discoideum* 4. Effects of ions on the rate of differentiation and cellular response to cyclic AMP.

Marshall, R., Sargent, D., and Wright, B. E. (1970). *Biochem.* **9,** 3087–3094. Glycogen turnover in *Dictyostelium discoideum.*

Martiel, J-P., and Goldbeter, A. (1981). *Biochimie* **63,** 119–124. Metabolic oscillations in biochemical systems controlled by covalent enzyme modification.

Mason, J. W., Rasmussen, H., and Dibella, F. (1971). *Exp. Cell Res.* **67,** 156–160. 3′,5′AMP and Ca^{2+} in slime mold aggregation.

Mato, J. M. (1978). *Biochim. Biophys. Acta* **540,** 408–411. ATP increases chemoattractant induced cyclic GMP accumulation in *Dictyostelium discoideum.*

Mato, J. M. (1979). *Biochem. Biophys. Res. Commun.* **88,** 569–574. Activation of *Dictyostelium discoideum* guanylate cyclase EC-4.6.1.2 by ATP.

Mato, J. M., and Konijn, T. M. (1975). *Biochim. Biophys. Acta* **385,** 173–179. Chemotaxis and binding of cylic AMP in cellular slime molds.

Mato, J. M., and Konijn, T. M. (1975). *Develop. Biol.* **47,** 233–235. Enhanced cell aggregation in *Dictyostelium discoideum* by ATP activation of cyclic AMP receptors.

Mato, J. M., and Konijn, T. M. (1976). *Exp. Cell. Res.* **99,** 328–332. The activation of cell aggregation by phosphorylation in *Dictyostelium discoideum.*

Mato, J. M., and Konijn, T. M. (1977). *FEBS Letts.* **75,** 173–176. The chemotactic activity of cyclic AMP and AMP derivatives with substitutions in the phosphate moeity in *Dictyostelium discoideum.*

Mato, J. M., and Konijn, T. M. (1977). *In* "Developments in Cell Biology." Elsevier/North-Holland (Cappuccinelli, P., and Ashworth, J. M. eds.), 93–104. Chemotactic signal and cyclic GMP accumulation in *Dictyostelium.*

Mato, J. M., and Konijn, T. M. (1979). *Acta Protozool.* **18,** 167–168. Chemotactic stimulation in *Dictyostelium discoideum* mechanism of sensory transduction.

Mato, J. M., and Konijn, T. M.(1979). *In* "Biochemistry and Physiology of Protozoa." Second Edition, 181–219. Chemosensory transduction in *Dictyostelium discoideum.*

Mato, J. M., and Malchow, D. (1978). *FEBS Lett.* **90,** 119–122. Guanylate cyclase activation in response to chemotactic stimulation in *Dictyostelium discoideum.*

Mato, J. M., and Steiner, A. L. (1980). *Cell Biol. Int. Rep.* **4,** 641–648. Immunohistochemical localization of cAMP, cGMP and calmodulin in *Dictyostelium discoideum.*

Mato, J. M., Losada, A., Nanjundiah,V., and Konijn, T. M. (1975). *Proc. Natl. Acad. Sci. U.S.A.* **72,** 4991–4993. Signal input for a chemotactic response in the cellular slime mold *Dictyostelium discoideum.*

Mato, J., Krens, F., Van Haastert, P. J. M., and Konijn, T. M. (1977). *Proc. Natl. Acad. Sci. U.S.A.* **74,** 2348–2351. cAMP dependent cGMP accumulation in *Dictyostelium discoideum.*

Mato, J. M., Krens, F. A., Van Haastert, P. J. M., and Konijn, T. M. (1977). *Biochem. Biophys. Res. Commun.* **77,** 399–402. Unified control of chemotaxis and cyclic AMP mediated cyclic GMP accumulation by cyclic AMP in *Dictyostelium discoideum.*

Mato, J. M., Van Haastert, P. J. M., Krens, F. A., and Konijn, T. M. (1977). *Develop. Biol.* **57,** 450–453. An acrasin like attractant from yeast specific for *Dictyostelium lacteum.*

Mato, J. M., Van Haastert, P. J. M., Krens, F. A., Rhijnsburger, E. H., Dobbe, F. C. P. M., and Konijn, T. M. (1977). *FEBS Lett.* **79,** 331–336. Cyclic AMP and folic acid mediated cyclic AMP accumulation in *Dictyostelium discoideum.*

Mato, J. M., Jastorff, B., Morr, M., and Konijn, T. M. (1978). *Biochim. Biophys. Acta* **544,** 309–314. A model for cyclic AMP chemoreceptor interaction in *Dictyostelium discoideum.*

Mato, J. M., Roos, W., and Wurster, B. (1978). *Differentiation* **10,** 129–132. Guanylate cyclase activity in *Dictyostelium discoideum* and its increase during cell development.

Mato, J. M., Woelders, H., Van Haastert, P. J. M., and Konijn, T. M. (1978). *FEBS Lett.* **90,** 261–264. Cyclic GMP binding activity in *Dictyostelium discoideum.*

Mato, J. M., Van Haastert, P. J., Krens, F. A., and Konijn, T. M. (1978). *Cell Biol. Int. Rep.* **2,** 163–170. Chemotaxis in *Dictyostelium discoideum:* effect of concanavalin A on chemoattractant mediated cyclic GMP accumulation and light scattering decrease.

Mato, J. M., and Martin-Cao, D. (1979). *Proc. Natl. Acad. Sci. U.S.A.* **76,** 6106–6019. Protein and phospholipid methylation during chemotaxis in *Dictyostelium discoideum* and its relationship to calcium movements.

Mato, J. M., Woelders, H., and Konijn, T. M. (1979). *J. Bacteriol.* **137,** 169–172. Intracellular cyclic GMP binding proteins in cellular slime molds.

Matsukuma, S., and Durston, A. J. (1979). *J. Embryol. Exp. Morphol.* **50,** 243–252. Chemotactic cell sorting in *Dictyostelium discoideum.*

McCoy, M. K., and Clark, M. A. (1977). *Genetics* **86,** S43. Organophosphate enhancement of aggregation in the cellular slime mold *Polysphondylium violaceum* relationship to the esterases of aggregating populations.

McDonough, J. P., Springer, W. R., and Barondes, S. H. (1980). *Exp. Cell Res.* **125,** 1–14. Species specific cell cohesion in cellular slime molds demonstration by several quantitative assays and with multiple species.

McKeown, M., and Firtel, R. A. (1981). *Cell* **24,** 799–807. Differential expression and 5′-end mapping of actin genes in *Dictyostelium.*

McKeown, M., and Firtel, R. A. (1981). *J. Mol. Biol.* **151,** 593–606. Evidence for subfamilies of actin genes in *Dictyostelium* as determined by comparisons of the 3′-end sequences.

McKeown, M., and Firtel, R. A. (1982). *Organ. of the Cytoplasm, CSHSQB,* **46** (In Press). The actin multigene family of *Dictyostelium.*

McKeown, M., Taylor, W., Kindle, K., Firtel, R., Bender, W., and Davidson, N. (1978). *Cell* **15,** 789–800. The structure of M6, a recombinant plasmid containing *Dictyostelium* DNA homologous to actin messenger RNA.

McKeown, M., Hirth, K-P., Edwards, C., and Firtel, R. A. (1982). *In* "Embryonic Development: Gene Structure and Function." (In press). Examination of the regulation of the actin multigene family in *Dictyostelium discoideum.*

McMahon, D. (1973). *Proc. Natl. Acad. Sci. U.S.A.* **70,** 2396–2400. A cell contact model for cellular position determination in development.

McMahon, D., and Hoffman, S. (1977). *Biochim. Biophys. Acta* **465,** 242. The role of the plasma membrane in the development of *Dictyostelium discoideum.* II. Developmental and topographic analysis of polypeptide and glycoprotein composition.

McMahon, D., and West, C. (1976). *In* "Cell Surface Reviews," Vol. 1, Elsevier/North-Holland, New York, N.Y. (Poste, G., and Nicolson, G. L. eds.), 449–493. Transduction of positional information during development.

McMahon, D., Hoffman, S., Fry, W., and West, C. (1975). *In* "Development Biology." ICN/UCLA Symp. Mol. Cell. Biol. 2. Menlo Park, Ca. (Benjamin, W. A. ed.), 60–75. The involvement of the plasma membrane in the development of *Dictyostelium discoideum.*

McMahon, D., Miller, M., and Long, S. (1977). *Biochim. Biophys. Acta* **465,** 224–241. The involvement of the plasma membrane in the development of *Dictyostelium discoideum.* I. Purification of the plasma membrane.

McPhail, S. M., Collins, M. A., and Gilbert, R. G. (1976). *Biophys. Chem.* **4,** 151–157. Inhomogeneous stationary states in reaction diffusion systems.

Mercer, E. H., and Shaffer, B. M. (1960). *J. Biophys. Biochem. Cytol.* **7,** 353–356. Electron microscopy of solitary and aggregated slime mould cells.

Mercer, J. A., and Soll, D. R. (1980). *Differentiation* **16,** 117–124. The complexity and reversibility of the timer for the onset of aggregation in *Dictyostelium discoideum.*

Meyers, B. L., and Frazier, W. (1981). *Biochem. Biophys. Res. Commun.* **101,** 1011. Solubilization and hydrophobic immunobilization assay of a cAMP binding protein from *Dictyostelium discoideum* plasma membrane.

Michalska, I., and Skupienski, F. X. (1939). *Acad. Sci.* **207,** 1239–1241. Recherches écologigue sur les acrasiées *Polysphondylium pallidum* Olive, *Polysphondylium violaceum* Bref., *Dictyostelium mucoroides* Bref. C.R.

Miller, Z. I., Quance, J., and Ashworth, J. M. (1969). *Biochem. J.* **114,** 815–818. Biochemical and cytological heterogeneity of the differentiating cells of the cellular slime mould *Dictyostelium discoideum.*

Mine, H., and Takeuchi, I. (1967). *Annu. Report Biol. Works, Fac. Sci. Osaka Univ.* **15,** 97–111. Tetrazolium reduction in slime mould development.

Miyake, Y., and Fukui, Y. (1976). *Cell Struct. Funct.* **1,** 147–154. Parasexual hybridization in cellular slime molds. Part 2. The appearance and characterization of interspecific hybrids between *Dictyostelium discoideum* and *Dictyostelium mucoroides.*

Mizukami, Y., and Iwabuchi, M. (1970a). *J. Biochem.* **67,** 501–504. Differential synthesis of ribosomal subunits during development in the cellular slime mold, *Dictyostelium discoideum.*

Mizukami, Y., and Iwabuchi, M. (1970b). *Exp. Cell Res.* **63,** 317–324. Effects of actinomycin D and cycloheximide on morphogenesis and synthesis of RNA and protein in the cellular slime mold, *Dictyostelium discoideum.*

Mizukami, Y., and Iwabuchi, M. (1972). *Biochim. Biophys. Acta* **272,** 81–94. The formation of ribosomal subunits in the cellular slime mold *Dictyostelium discoideum.*

Mockrin, S., and Spudich, J. (1976). *Proc. Natl. Acad. Sci. U.S.A.* **73,** 2321–2325. Calcium control of actin activated myosin adenosine triphosphatase from *Dictyostelium discoideum.*

Mockrin, S. C., and Spudich, J. A. (1976). *Fed. Proc.* **35,** 1145. Regulation of the interaction of actin and myosin from *Dictyostelium discoideum.*

Moens, P. B. (1976). *J. Cell. Biol.* **68,** 113–122. Spindle and kinetophone morphology of *Dictyostelium discoideum.*

Moens, P. B., and Konijn, T. M. (1974). *FEBS Lett.* **45,** 44–46. Cyclic AMP as a cell surface activating agent in *Dictyostelium discoideum.*

Mohan Das, D. V., and Weeks, G. (1979). *Exp. Cell. Res.* **118,** 237–243. Effects of polyunsaturated fatty acids on the growth and differentiation of the cellular slime mould, *Dictyostelium discoideum.*

Mohan Das, D. V., and Weeks, G. (1980). *Nature* **288,** 166–167. Reversible heat activation of alkaline phosphatase of *Dictyostelium discoideum* and its developmental implication.

Mohan Das, D. V., and Weeks, G. (1981). *FEBS Lett.* **130,** 249–252. The inhibition of *Dictyostelium discoideum* alkaline phosphatase by a low molecular weight factor and its implication for the developmental regulation of the enzyme.

Mohan Das, D. V., Herring, F. G., and Weeks, G. (1980). *Can. J. Microbiol.* **26**, 796–799. The effect of growth temperature on the lipid composition and differentiation of *Dictyostelium discoideum.*

Molday, R., Jaffe, R., and McMahon, D. (1976). *J. Cell. Biol.* **71**, 314–322. Concanavalin A and wheat germ agglutinin receptors on *Dictyostelium discoideum.*

Monier, F., Guespin-Michel, J., and Felenbok, B. (1977). *Exp. Cell. Biol.* **107**, 397–404. Incorporation of 5-bromo-2′ deoxyuridine into DNA of *Dictyostelium discoideum.*

Monier, F., Leal, J., and Genty, N. (1978). *Exp. Cell. Res.* **117**, 31–38. Effect of 5 bromodeoxy uridine on RNA synthesis in *Dictyostelium discoideum* under growth and starvation conditions.

Morris, N. T., and Weber, A. T. (1975). *Abstr. Annu. Meet. Am. Soc. Microbiol.* **75**, 135. Sodium azide inhibition of growth and development in *Dictyostelium discoideum.*

Morrissey, J. H., and Loomis, W. F. (1981). *Genetics* **99**, 183–196. Parasexual analysis of cell proportioning mutants of *Dictyostelium discoideum.*

Morrissey, J. H., Wheeler, S., and Loomis, W. F. (1980). *Genetics* **96**, 115–124. New loci in *Dictyostelium discoideum* determining pigment formation and growth on *Bacillus subtilis.*

Morrissey, J., Farnsworth, P., and Loomis, W. F. (1981). *Develop. Biol.* **83**, 1–8. Pattern formation in *Dictyostelium discoideum:* an analysis of mutants altered in cell proportioning.

Mosses, D., Williams, K. L., and Newell, P. C. (1975). *J. Gen. Microbiol.* **90**, 247–259. The use of mitotic crossing-over for genetic analysis of *Dictyostelium discoideum:* mapping of linkage group II.

Mueller, E. (1973). *Biol. Rundsch.* **11**, 129–139. Cyclic AMP as transmitter of signals in biological. Part 1. Cyclic AMP in animals and microorganisms.

Mueller, K., Gerisch, G., Fromme, I., Mayer, H., and Tsugita, A. (1979). *Eur. J. Biochem.* **99**, 419–426. A membrane glycoprotein of aggregating *Dictyostelium discoideum* cells with the properties of contact sites.

Mueller, U., and Hohl, H. R. (1975). *Protoplasma* **85**, 199–207. Ultrastructural evidence for the presence of 2 separate glycogen pools in *Dictyostelium discoideum.*

Muhlethaler, K. (1956). *Amer. J. Bot.* **43**, 673–678. Electron microscopic study of the slime mold *Dictyostelium discoideum.*

Mullens, I. A., and Newell, P. C. (1978). *Differentiation* **10**, 171–176. Cyclic AMP binding to cell surface receptors of *Dictyostelium.*

Muller, K., and Gerisch, G. (1978). *Nature (London)* **274**, 445–449. A specific glycoprotein as the target site of adhesion blocking FAB in aggregating *Dictyostelium* cells.

Muller, K., Gerisch, G., Fromme, I., Mayer, H., and Tsugita, A. (1979). *Eur. J. Biochem.* **99**, 419–426. A membrane glycoprotein of aggregating *Dictyostelium* cells with the properties of contact sites.

Muller, U., and Hohl, H. R. (1973). *Differentiation* **1**, 267–276. Pattern formation in *Dictyostelium discoideum:* temporal and spatial distribution of prespore vacuoles.

Murata, Y., and Ohnishi, T. (1980). *J. Bacteriol.* **141**, 956–958. *Dictyostelium discoideum* fruiting bodies observed by scanning electron microscopy.

Murphy, M., and Klein, C. (1979). *Cell. Differ.* **8**, 275–284. Effects of amino acids on cell differentiation of *Dictyostelium discoideum.*

Murray, B. A., Yee, L. D., and Loomis, W. F. (1981). *J. Supramol. Struct. Cell. Biochem.* **17**, 387–401. Immunological analysis of a glycoprotein (contact sites A) involved in intercellular adhesion of *Dictyostelium discoideum.*

Nandini-Kishore, and Frazier, W. (1982). *Proc. Natl. Acad. Sci. U.S.A.* (In press). 3H-methotrexate as a ligand for the folate receptor of *Dictyostelium discoideum.*

Nanjundiah, V. (1973). *J. Theor. Biol.* **42**, 63–105. Chemotaxis signal relaying and aggregation morphology.

Nanjundiah, V. (1974). *Exp. Cell. Res.* **86,** 408–411. A differential chemotactic response of slime mold amoebae to regions of the early amphibian embryo.
Nanjundiah, V. (1976). *J. Theor. Biol.* **56,** 275–282. Signal relay by single cells during wave propagation in a cellular slime mold.
Nanjundiah, V. (1978). *J. Indian Inst. Sci.* **60,** 199–204. Ligand receptor binding in the presence of a diffusion gradient.
Nanjundiah, V., and Malchow, D. (1976). *Hoppe-Seyler's Z. Physiol. Chem.* **357,** 273. pH Oscillations and cyclic AMP induced pH changes in aggregating slime mold cells.
Nanjundiah, V., and Malchow, D. (1978). *J. Cell. Sci.* **22,** 49–58. A theoretical study of the effects of cyclic AMP phosphosiesterases during aggregation in *Dictyostelium discoideum.*
Nestle, M., and Sussman, M. (1972). *Develop. Biol.* **28,** 545–554. The effect of cyclic AMP on morphogenesis and enzyme accumulation in *Dictyostelium discoideum.*
Neumann, E., Gerisch, G., and Opatz, K. (1980). *Naturwissenschaften* **67,** 414–415. Cell fusion induced by high electric impulses applied to *Dictyostelium discoideum.*
Newell, P. C. (1971). *In* "Essays in Biochemistry." Academic Press, New York, N.Y. (Campbell, P. N., and Dickens, F. eds.), 87–126. The development of the cellular slime mould *Dictyostelium discoideum:* a model system for the study of cellular differentiation.
Newell, P. C. (1973). *Biochem. Soc. Trans.* **1,** 1025. Control of development in the cellular slime mold *Dictyostelium.*
Newell, P. C. (1977). *Endeavour* **1,** 63–66. How cells communicate the system used by slime molds.
Newell, P. C. (1977). *In* "Receptors and recognition" Series B, Vol. 3 Microbial Interactions, 3–57. Aggregation and cell surface receptor in cellular slime molds.
Newell, P. C. (1978). *Annu. Rev. Genet.* **12,** 69–93. Genetics of the cellular slime molds.
Newell, P. C. (1978). *In* "The National Foundation-March Of Dimes Birth Defects: Original Article Series," Vol. 14, No. 2. The Molecular Basis of Cell–Cell Interaction. Alan R. Liss, New York, N.Y. (Lerner, R. A., and Bergsma, D. eds.), 507–526. Genetics of cellular communication during aggregation of *Dictyostelium.*
Newell, P. C., and Mullens, I. A. (1978). *In* "Symposia of the Society for Experimental Biology," No. 32. Cell–Cell Recognition. Cambridge Univ. Press, New York, N.Y. (Curtis, A. S. G. ed.), 161–171. Cell surface cyclic AMP receptors in *Dictyostelium.*
Newell, P. C., and Sussman, M. (1969). *J. Biol. Chem.* **244,** 2990–2995. Uridine diphosphate glucose pyrophosphorylase in *Dictyostelium discoideum.* Stability and developmental fate.
Newell, P. C., and Sussman, M. (1970). *J. Mol. Biol.* **49,** 627–637. Regulation of enzyme synthesis by slime mold cell assemblies embarked upon alternative developmental programmes.
Newell, P. C., Ellingson, J. S., and Sussman, M. (1969). *Biochim. Biophys. Acta* **177,** 610–614. Synchrony of enzyme accumulation in a population of differentiating slime mold cells.
Newell, P. C., Telser, A., and Sussman, M. (1969). *J. Bacteriol.* **100,** 763–768. Alternative developmental pathways determined by environmental conditions in the cellular slime mold *Dictyostelium discoideum.*
Newell, P. C., Longlands, M., and Sussman, M. (1971). *J. Mol. Biol.* **58,** 541–554. Control of enzyme synthesis by cellular interaction during development of the cellular slime mold *Dictyostelium discoideum.*
Newell, P. C., Franke, J., and Sussman, M. (1972). *J. Mol. Biol.* **63,** 373–382. Regulation of four functionally related enzymes during shifts in the developmental program of *Dictyostelium discoideum.*
Newell, P. C., Henderson, R. F., Mosses, D., and Ratner, D. I. (1977). *J. Gen. Micro.* **100,** 207–212. Sensitivity to *Bacillus subtilis,* a novel system for selection of heterozygous diploids of *Dictyostelium discoideum.*

Newell, P. C., Ratner, D. I., and Wright, D. (1977). *In* "Development and Differentiation in the Cellular Slime Moulds." Elsevier/North-Holland (Cappuccinelli, P., and Ashworth, J. M. eds.), 51–61. New techniques for cell fusion and linkage analysis of *Dictyostelium discoideum*.

Nickerson, A. W., and Raper, K. B. (1973a). *Amer. J. Bot.* **60,** 190–197. Macrocysts in the life cycle of the Dictyosteliaceae. I. Formation of the macrocysts.

Nickerson, A. W., and Raper, K. B. (1973b). *Amer. J. Bot.* **60,** 247–254. Macrocysts in the life cycle of the Dictyosteliaceae. II. Germination of the macrocysts.

Nicol, A., and Garrod, D. R. (1978). *J. Cell. Sci.* **32,** 377–388. Mutual cohesion and cell sorting-out among 4 species of cellular slime molds.

Nigam, V. N., Malchow, D., Rietschel, E. T., Luderitz, O., and Westphal, O. (1970). *Hoppe-Seyler's Z. Physiol. Chem.* **351,** 1123–1132. Die enzymatische abspaltung langkettiger fettsauren aus bakteriellen lipopolysacchariden mittels extrakten aus der amobe von *Dictyostelium discoideum*.

Nivet, C. (1977). *FEBS Lett.* **84,** 174–178. Enzyme histochemical-immunochemical characterzation of cyclic adenosine monophosphate phosphodiesterase in *Dictyostelium discoideum* in gel medium.

Nomura, T., Tanaka, Y., Abe, H., and Uchiyama, M. (1977). *Phytochemistry* **16,** 1819–1820. Cytokinin activity of discadenine a spore germination inhibitor of *Dictyostelium discoideum*.

North, M. J. (1978). *Biochem. Soc. Trans.* **6,** 400–403. Inhibition of acid proteinase from *Dictyostelium discoideum*.

North, M. J., and Ashworth, J. M. (1976). *J. Gen. Micro.* **96,** 63–75. Inhibition of development of cellular slime mold *Dictyostelium discoideum* by omegaamuno carboxylic acids.

North, M. J., and Campbell, A. J. (1976). *J. Gen. Micro.* **96,** 77–85. Effects of epsilon amino caproic acid on biochemical changes in the development of the cellular slime mold *Dictyostelium discoideum*.

North, M. J., and Harwood, J. M. (1979). *Biochim. Biophys. Acta* **566,** 222–233. Multiple acid proteinases in the cellular slime mold *Dictyostelium discoideum*.

North, M. J., and Murray, S. (1980). *FEMS Lett.* **9,** 271–274. Cellular slime mold polyamines species specificity of 1,3 di amino propane.

North, M. J., and Turner, R. (1977). *Microbios. Lett.* **4,** 221–228. Di amine content of cellular slime mold *Dictyostelium discoideum*—presence of 1,3 diaminopropane and putrescine.

North, M. J., and Williams, K. L. (1978). *J. Gen. Microbiol.* **107,** 223–230. Relationship between the axenic phenotype and sensitivity to omega amino carboxylic acids in *Dictyostelium discoideum*.

Nozu, K., Ohnishi, T., and Okaichi, K. (1980). *Photochem. Photobiol.* **32,** 261–264. Viability of the spores formed after UV irradiation in *Dictyostelium discoideum*.

Obata, Y., Hiroshi, A., Tanaka, Y., Yanagisawa, K., and Uchiyama, M. (1973). *Agr. Biol. Chem.* **37,** 1989–1990. Isolation of a spore germination inhibitor from a cellular slime mold *Dictyostelium discoideum*.

Ochiai, H., and Iwabuchi, M. (1971). *Bot. Mag.* **84,** 267–269. A method for the extraction of ribosomal proteins from *Dictyostelium discoideum* with calcium chloride-urea mixture.

Ochiai, H., Kanda, F., and Iwabuchi, M. (1973). *J. Biochem.* **73,** 163–167. The number and size of ribosomal proteins in the cellular slime mold *Dictyostelium discoideum*.

Ochiai, H., Ohtani, J., Ono, K-I., and Toda, K. (1978). *J. Fac. Sci. Hokkaido Univ. Ser. V. Bot.* **11,** 225–230. Scanning electron microscopic studies on cell surfaces of *Dictyostelium discoideum* osmotic and pH effects during fixation.

O'Connor, K. A., Scandella, D., and Katz, E. R. (1980). *J. Cell. Biol.* **87,** 18. A temperature sensitive mutation in *Dictyostelium discoideum* which affects spore differentiation.

O'Day, D. H. (1973). *Exp. Cell Res.* **79,** 186–190. Intracellular and extracellular acetylglucosaminidase activity during microcyst formation in *Polysphondylium pallidum.*
O'Day, D. H. (1973). *J. Bacteriol.* **113,** 192–197. Alpha mannosidase EC-3.2.1.24 and microcyst differentiation in the cellular slime mold *Polysphondylium pallidum.*
O'Day, D. H. (1973). *Cytobios.* **7,** 223–232. Intracellular localization and extracellular release of certain lysosomal enzyme activities from amoebae of the cellular slime mold *Polysphondylium pallidum.*
O'Day, D. H. (1974). *Develop. Biol.* **36,** 400–410. Intracellular and extracellular enzyme patterns during microcyst germination in the cellular slime mold *Polysphondylium pallidum.*
O'Day, D. H. (1976). *J. Bacteriol.* **125,** 8–13. Acid protease activity during germination of microcysts of the cellular slime mold *Polysphondylium pallidum.*
O'Day, D. H. (1979). *J. Cell. Sci.* **35,** 203–216. Cell differentiation during fruiting body formation in *Polysphondylium pallidum.*
O'Day, D. H. (1979). *Can. J. Microbiol.* **25,** 1416–1426. Aggregation during sexual developmental in *Dictyostelium discoideum.*
O'Day, D. H., and Durston, A. J. (1978). *J. Embryol. Exp. Morphol.* **47,** 195–206. Colchicine induces multiple axis formation and stalk cell differentiation in *Dictyostelium discoideum.*
O'Day, D. H., and Durston, A. J. (1979). *Can. J. Microbiol.* **25,** 542–544. Evidence for chemotaxis during sexual development in *Dictyostelium discoideum.*
O'Day, D. H., and Durston, A. J. (1980). *Can. J. Microbiol.* **26,** 959–964. Sorogen elongation and side branching during fruiting body development in *Polyspondylium pallidum.*
O'Day, D. H., and Francis, D. W. (1973). *Can. J. Zool.* **51,** 301–310. Patterns of alkaline phosphatase EC-3.1.3.1 activity during alternative developmental pathways in the cellular slime mold *Polysphondylium pallidum.*
O'Day, D. H., and Lewis, K. E. (1975). *Nature (London)* **254,** 431–432. Diffusible mating type factors induce macrocyst development in *Dictyostelium discoideum.*
O'Day, D. H., and Lewis, K. E. (1981). *In* "Sexual Interactions in Eukaryotic Microbes." Academic Press, New York, 199–221. Pheromonal interactions during mating in *Dictyostelium.*
O'Day, D. H., and Paterno, G. D. (1979). *Arch. Microbiol.* **121,** 231–234. Intracellular and extracellular carboxymethyl cellulase and beta glucosidase activity during germination of *Polysphondylium pallidum* microcysts.
O'Day, D. H., and Riley, L. J. (1973). *Exp. Cell Res.* **80,** 245–249. Acid phosphatase EC-3.1.3.2 activity and microcyst differentiation in the cellular slime mold *Polysphondylium pallidum.*
O'Day, D. H., Gwynne, D. I., and Blakey, D. H. (1976). *Exp. Cell. Res.* **97,** 359–365. Microcyst germination in the cellular slime mold *Polysphondylium pallidum* effects of actinomyciin D and cycloheximide on macromolecular synthesis and enzyme accumulation.
O'Day, D. H., Szabo, S. P., and Chagla, A. H. (1981). *Exp. Cell. Res.* **131,** 456–458. An autoinhibitor of zygote giant cell formation in *Dictyostelium discoideum.*
Oehler, R. (1922). *Zentraebl. Bakteriol. Parasitenk.* **89,** 155–156. *Dictyostelium mucoroides* (Brefeld).
Ohnishi, T., Okaichi, K., Ohashi, Y., and Nozu, K. (1981). *Photochem. Photobiol.* **33,** 79–84. Effects of caffeine on DNA repair of UV irradiated *Dictyostelium discoideum.*
Ohnishi, T., and Nozu, K. (1978). *J. Radiat. Res.* **1,** 28. UV effects of morphogenesis of *Dictyostelium discoideum.*
Ohnishi, T., and Nozu, K. (1979). *Photochem. Photobiol.* **29,** 615–618. UV effects on killing fruiting body formation and the spores of *Dictyostelium discoideum.*

Okamoto, K. (1979). *Eur. J. Biochem.* **93,** 221–228. Induction of cyclic AMP phosphodiesterase EC-3.1.4.1 by disaggregation of the multicellular complexes of *Dictyostelium discoideum.*

Okamoto, K. (1981). *J. Gen. Micro.* (In press). Differentiation of *Dictyostelium discoideum* cells in suspension culture.

Okamoto, K., and Takeuchi, I. (1976). *Biochem. Biophys. Res. Commun.* **72,** 739–746. Changes in activities of 2 developmentally regulated enzymes induced by disaggregation of the pseudoplasmodia of *Dictyostelium discoideum.*

Olive, E. W. (1901). *Proc. Amer. Acad. Sci.* **37,** 333–344. Preliminary enumeration of the Sorophoreae.

Olive, E. W. (1902). *Proc. Boston Soc. Natur. Hist.* **30,** 451–513. Monograph of the Acrasieae.

Olive, L. S. (1963). *Bull. Torrey Bot. Club* **90,** 144–147. The question of sexuality in cellular slime molds.

Olive, L. S. (1964). *Mycologia* **61,** 885–896. A new member of the Mycetozoa.

Olive, L. S. (1965). *Amer. J. Bot.* **52,** 513–519. A developmental study of *Guttulinopsis vulgaris* (Acrasiales).

Olive, L. S., and Stoianovitch, C. (1960). *Bull. Torrey Bot. Club* **87,** 1–20. Two new members of the Acrasiales.

Olive, L. S., and Stoianovitch, C. (1966). *Amer. J. Bot.* **53,** 344–349. A simple new Mycetozoan with ballistospores.

Olsen, A. M., and Eisenberg, R. M. (1978). *Genetics* **88,** 573–574. Polymorphism in a cellular slime mold.

Ono, H., Kobayashi, S., and Yanagisawa, K. (1972). *J. Cell Biol.* **54,** 665–666. Cell fusion in the cellular slime mold, *Dictyostelium discoideum.*

Ono, K-I., Ochiai, H., and Toda, K. (1978). *Exp. Cell. Res.* **112,** 175–185. Ghosts formation and their isolation from the cellular slime mold *Dictyostelium discoideum.*

Ono, K-I., Toda, K., and Ochiai, H. (1979). *Plant Cell Physiol.* **20,** 1417–1426. Developmental changes in soluble proteins during the early development of *Dictyostelium discoideum.*

Ono, K-I., Toda, K., and Ochiai, H. (1981). *Eur. J. Biochem.* **119,** 133. Drastic changes in accumulation and synthesis of plasma-membrane proteins during aggregation of *Dictyostelium discoideum.*

Oohata, A. (1976). *Dev. Growth Differ.* **18,** 473–480. Multiple forms of beta galactosidase EC-3.2.1.23 and the changes in its pattern during development of *Dictyostelium discoideum.*

Oohata, A., and Takeuchi, I. (1977). *J. Cell. Sci.* **24,** 1–10. Separation and biochemical characterization of the two cell types present in the pseudoplasmodium of *Dictyostelium mucoirdes.*

Orlow, S. J., Shapiro, R. I., Franke, J., and Kessin, R. H. (1981). *J. Biol. Chem.* (In press). The extracellular cyclic nucleotide phosphodiesterase of *Dictyostelium discoideum:* purification and characterization.

Orlowski, M., and Loomis, W. F. (1979). *Dev. Biol.* **71,** 297–307. Plasma membrane proteins of *Dictyostelium discoideum* spore coat proteins.

Osborn, P., and Ashworth, J. (1973). *J. Gen. Microbiol.* **77,** 2–3. Is the ribosomal RNA synthesized during the developmental phase of the life cycle of *Dictyostelium discoideum* the same as that synthesized during the growth phase.

Osborn, P. J., and Ashworth, J. M. (1975). *Cell Differ.* **4,** 237–241. Changes in the basic nuclear proteins of the cellular slime mold *Dictyostelium discoideum* during growth and differentiation.

Paddock, R. B. (1953). *Science* **118,** 597–598. The appearance of amoebae tracks in cultures of *Dictyostelium discoideum.*

Pahlic, M., and Rutherford, C. L. (1979). *J. Biol. Chem.* **254,** 9703–9707. Adenylate cyclase activity and cyclic AMP levels during the development of *Dictyostelium discoideum.*

Palatnik, C. M., Brenner, M., and Katz, E. R. (1975). *ICN-UCLA Symp. Mol. Cell. Biol.* **2,** 508–513. Qualitative and quantitative differences in transfer RNA between growth and development in *Dictyostelium discoideum.*

Palatnik, C. M., Katz, R. R., and Brenner, M. (1977). *J. Biol. Chem.* **252,** 694–703. Isolation and characterization of tRNA's from *Dictyostelium discoideum* during growth and development.

Palatnik, C. M., Wilkins, C., Mabie, C. T., and Jacobson, A. (1978). *J. Supramol. Struct.* **7,** 332. Changes in transcription of messenger RNA during early *Dictyostelium discoideum* development.

Palatnik, C. M., Storti, R. V., and Jacobson, A. (1979). *J. Mol. Biol.* **128,** 371–396. Fractionation and functional analysis of newly synthesized and decaying mRNA's from vegetative cells of *Dictyostelium discoideum.*

Palatnik, C. M., Storti, R. V., Capone, A. K., and Jacobson, A. (1980). *J. Mol. Biol.* **141,** 99–118. Messenger RNA stability in *Dictyostelium discoideum.* Does poly adenylic-acid have a regulatory role?

Palatnik, C. M., Storti, R. V., and Jacobson, A. (1981). *J. Mol. Biol.* **150,** 389–398. Partial purification of a developmentally regulated messenger RNA from *Dictyostelium discoideum* by thermal elution from Poly(U)-Sepharose.

Pan, P., and Wedner, H. J. (1979). *Differentiation* **14,** 113–118. Immunohistochemical localization of cyclic GMP in aggregating *Polyspondylium violaceum.*

Pan, P., and Wurster, B. (1978). *J. Bacteriol.* **136,** 955–959. Inactivation of the chemoattractant folic-acid by cellular slime molds and identification of the reaction product.

Pan, P., Hall, E., and Bonner, J. T. (1972). *Nature New Biol.* **237,** 181–182. Folic acid as secondary chemotaxic substance in the cellular slime molds.

Pan, P., Bonner, J. T., Wedner, H., and Parker, C. (1974). *Proc. Natl. Acad. Sci. U.S.A.* **71,** 1623–1625. Immunofluorescence evidence for the distribution of cyclic AMP in cells and cell masses of the cellular slime molds.

Pan, P., Hall, E. M., and Bonner, J. T. (1975). *J. Bacteriol.* **122,** 185–191. Determination of the active portion of the folic acid molecule in cellular slime mold chemotaxis.

Pan, Y-H., and Shih, J-T. (1980). *Nat. Sci. Counc. Mon.* **8,** 616–626. Histones of the cellular slime mold *Dictyostelium discoideum* AX-3 during its early life cycle.

Pannbacker, R. G. (1966). *Biochem. Biophys. Res. Commun.* **24,** 340–345. RNA metabolism during differentiation in the cellular slime mold.

Pannbacker, R. G. (1967a). *Biochem.* **6,** 1283–1286. Uridine diphosphoglucose biosynthesis during differentiation in the cellular slime mold. I. *In vivo* measurements.

Pannbacker, R. G. (1967b). *Biochem.* **6,** 1287–1293. Uridine diphosphoglucose biosynthesis during differentiation in the cellular slime mold. II. *In vitro* measurements.

Pannbacker, R. G., and Bravard, L. J. (1972). *Science* **175,** 1014–1015. Phosphodiesterase in *Dictyostelium discoideum* and the chemotactic response to cyclic adenosine monophosphate.

Pannbacker, R. G., and Wright, B. E. (1966). *Biochem. Biophys. Res. Commun.* **24,** 334–339. The effect of actinomycin D on development in the cellular slime mold.

Parish, R. W. (1975). *Eur. J. Biochem.* **58,** 523–531. Mitochondria and peroxisomes from the cellular slime mold *Dictyostelium discoideum* isolation techniques and urate oxidase EC-1.7.3.3 association with peroxisomes.

Parish, R. W. (1976). *Biochim. Biophys. Acta* **444,** 802–809. A labile acid phosphatase isozyme associated with surface of vegetative *Dictyostelium discoideum* cells.

Parish, R. W. (1977). *J. Bacteriology* **129,** 1642–1644. Inhibition of *Dictyostelium discoideum* beta glucosidase by purines.

Parish, R. W. (1979). *Biochim. Biophys. Acta* **553,** 179–182. Cyclic AMP induces synthesis of developmentally regulated plasma membrane proteins in *Dictyostelium.*

Parish, R. W., and Pelli, C. (1974). *FEBS Lett.* **48,** 293–296. Alkaline phosphatase of *Dictyostelium discoideum* cell surface location and colchicine effect on internalization during phagocytosis.

Parish, R. W., and Muller, U. (1976). *FEBS Lett.* **63,** 40–44. The isolation of plasma membranes from the cellular slime mold *Dictyostelium discoideum* using concanavalin A and triton X-100.

Parish, R. W., and Schmidlin, S. (1979). *FEBS Lett.* **98,** 251–256. Synthesis of plasma membrane proteins during development of *Dictyostelium discoideum.*

Parish, R. W., and Schmidlin, S. (1979). *FEBS Lett.* **99,** 270–274. Resynthesis of developmentally regulated plasma membrane proteins following disaggregation of *Dictyostelium discoideum* pseudoplasmodia.

Parish, R. W., and Weibel, M. (1980). *FEBS Lett.* **118,** 263–266. Extracellular ATP ecto ATPase and calcium influx in *Dictyostelium discoideum* cells.

Parish, R. W., Schmidlin, S., and Mueller, U. (1977). *Exp. Cell. Res.* **110,** 267–276. The effects of proteases on proteins and glycoproteins of *Dictyostelium discoideum* plasma membranes.

Parish, R. W., Stalder, J., and Schmidlin, S. (1977). *FEBS Lett.* **84,** 63–66. Biochemical evidence for a DNA repeat length in the chromatin of *Dictyostelium discoideum.*

Parish, R. W., Schmidlin, S., and Parish, C. R. (1978). *FEBS Lett.* **95,** 366–370. Detection of developmentally controlled plasma membrane antigens of *Dictyostelium discoideum* cells in sodium dodecyl sulfate polyacrylamide gels.

Parish, R. W., Schmidlin, S., and Weibel, M. (1978). *FEBS Lett.* **96,** 283–286. Effect of cyclic AMP pulses on the synthesis of plasma membrane proteins in aggregateless mutants of *Dictyostelium discoideum.*

Parish, R. W., Schmidlin, S., Fuhrer, S., and Widmer, R. (1980). *FEBS Lett.* **110,** 236–240. Electrophoretic isolation of nucleosomes from *Dictyostelium discoideum* nuclei and nucleoli proteins assocaited with monomers and dimers.

Parish, R. W., Hintermann, R., Banz, E., Bercovici, T., and Gitler, C. (1981). *FEBS Lett.* **133,** 291. Labelling of *Dictyostelium* and *Polysphondylium* plasma membranes with a photosensitive hydrophobic probe.

Park, D. J. M., and Wright, B. E. (1973). *Comput. Programs Biomed.* **3,** 10–26. Metasim, a general purpose metabolic simulator for studying cellular transformations.

Park, D. J. M., and Wright, B. E. (1975). *Mol. Cell. Biochem.* **8,** 97–112. Mathematical modeling of differentiation in *Dictyostelium discoideum.*

Parnas, H., and Segel, L. A. (1977). *J. Cell. Science* **25,** 191–204. Computer evidence that aggregating *Dictyostelium discoideum* respond to temporal derivatives in attractant concentration.

Parnas, H., and Segel, L. A. (1978). *J. Theor. Biol.* **71,** 185–208. A computer simulation of pulsatile aggregation in *Dictyostelium discoideum.*

Pate, E. F., and Odell, G. M. (1981). *J. Theor. Biol.* **88,** 201–240. A computer simulation of chemical signaling during the aggregation phase of *Dictyostelium discoideum.*

Paterno, G. D., and O'Day, D. H. (1981). *Can. J. Microb.* **27,** 924. Cellular differentiation and pattern formation in the absence of morphogenesis in the cellular slime mould *Polysphondylium pallidum:* evidence for a biochemical tip (organizer) in submerged aggregates.

Pavillard, J. (1953). *In* "Traite de Zoologie," Vol. I/II. Masson, Paris. 493–505. Ordre des Acrasies.

Payez, J., and Deering, R. (1972). *Mutat. Res.* **16,** 318–321. Synergistic and antagonistic effects of caffeine on two strains of cellular slime molds treated with alkylating agents.

Peacey, M. J., and Gross, J. D. (1981). *Diff.* **19,** 189. The effect of proteases on gene expression and cell differentiation in *Dictyostelium discoideum.*

Peacock, M., and Soll, D. R. (1978). *J. Embryol. Exp. Morphol.* **44,** 45–52. The stabilization of morphological field size during slime mold morphogenesis.

Pederson, T. (1977). *Biochem.* **16,** 2771–2777. Isolation and characterization of chromatin from the cellular slime mould *Dictyostelium discoideum.*

Peffley, D. M., and Sogin, M. L. (1981). *Biochem.* **20,** 4015. A putative tRNA *trp* gene cloned from *Dictyostelium discoideum:* its nucleotide sequence and association with repetitive deoxyribonucleic acid.

Perekalin, D. (1977). *Arch. of Micro.* **115,** 333–338. The influence of light and different ATP concentrations on cell aggregation in cyclic AMP sensitive and insensitive species of the cellular slime moulds.

Pfutzner-Eckert, R. (1950). *Wilhelm Roux Entwicklungsmech. Organismen* **144,** 381–409. Entwicklungsphysiologische untersuchungen an *Dictyostelium mucoroides* Brefeld.

Phillips, W. D., Rich, A., and Sussman, R. R. (1964). *Biochim. Biophys. Acta* **80,** 508–510. The isolation and identification of polyribosomes from cellular slime molds.

Pillinger, D., and Borek, E. (1969). *Proc. Natl. Acad. Sci. U.S.A.* **62,** 1145–1150. Transfer RNA methylases during morphogenesis in the cellular slime mold.

Pinoy, E., (1907). *Annu. Inst. Pasteur Paris* **21,** 622–656, 686–700. Role des bacteries dans le development de certains myxomycetes.

Pinoy, P. E. (1950). *Bull. Soc. Mycol. France* **66,** 37–38. Quelques observations sur la culture d'une Acrasiee.

Podgorski, G., and Deering, R. A. (1980). *Mutat. Res.* **74,** 459–468. Quantitation of induced mutation in *Dictyostelium discoideum* characterization and use of a methanol resistance mutation assay.

Podgorski, G., and Deering, R. A. (1980). *Mutat. Res.* **73,** 415–418. Effect of methyl methanesulfonate on survival of radiation sensitive strains of *Dictyostelium discoideum.*

Poff, K., and Butler, W. (1974). *Photochem. Photobiol.* **20,** 241–244. Spectral characteristics of the photoreceptor pigment of phototaxis in *Dictyostelium discoideum.*

Poff, K. L., and Butler, W. L. (1974). *Nature (London)* **248,** 799–801. Absorbance changes induced by blue light in *Phycomyces Blakesleeanus* and *Dictyostelium discoideum.*

Poff, K., and Butler, W. L. (1975). *Plant Physiol.* **55,** 427–429. Spectral characterization of the photoreducible B type cytochrome of *Dictyostelium discoideum.*

Poff, K., and Loomis, W. F. (1973). *Exp.Cell Res.* **82,** 236–240. Control of phototactic migration in *Dictyostelium discoideum.*

Poff, K. L., and Skokut, M. (1977). *Proc. Natl. Acad. Sci. U.S.A.* **74,** 2007–2010. Thermotaxis by pseudoplasmodia of *Dictyostelium discoideum.*

Poff, K., Butler, W., and Loomis, W. F. (1973). *Proc. Natl. Acad. Sci. U.S.A.* **70,** 813–816. Light induced absorbance changes associated with phototaxis in *Dictyostelium.*

Poff, K., Loomis, W. F., and Butler, W. (1974). *J. Biol. Chem.* **249,** 2164–2168. Isolation and purification of the photoreceptor pigment associated with phototaxis in *Dictyostelium discoideum.*

Poff, K. L., and Whitaker, B. D. (1979). *In* "Encyclopedia of Plant Physiology, New Series," Vol. 7. Physiology of Movements. Springer-Verlag, New York, N.Y. (Haupt, W., and Feinleib, M. E. eds.), 355–382. Movement of slime molds.

Pong, S. S., and Loomis, W. F. (1971). *J. Biol. Chem.* **246,** 4412–4416. Enzymes of amino acid metabolism in *Dictyostelium discoideum.*

Pong, S. S., and Loomis, W. F. (1973a). *J. Biol. Chem.* **248,** 3933–3939. Multiple nuclear RNA polymerases during development of *Dictyostelium discoideum.*

Pong, S. S., and Loomis, W. F. (1973b). *J. Biol. Chem.* **248,** 4867–4873. Replacement of an anabolic threonine deaminase by a catabolic threonine deaminase during development of *Dictyostelium discoideum.*

Pong, S. S., and Loomis, W. F. (1973c). *In* "Molecular Techniques and Approaches in Developmental Biology" Wiley, New York, N.Y. (Chrispeels, M. ed.), 93–115. Isolation of multiple RNA polymerases from *Dictyostelium discoideum.*

Poole, S., Firtel, R. A., and Rowekamp, W. (1981). *J. Supermol. Struct. Cell. Biochem.* (In press). Expression of the discoidin I genes in *Dictyostelium.*

Poole, S., Firtel, R. A., Lamar, E., and Rowekamp, W. (1981). *J. Mol. Biol.* (In press). Sequence and expression of the discoidin I gene family in *Dictyostelium discoideum.*

Porter, J. S., and Wright, B. E. (1977). *Arch. Biochem. Biophys.* **181,** 155–163. Partial purification and characterization of citrate synthase from *Dictyostelium discoideum.*

Potel, M. J., and Mackay, S. A. (1979). *J. Cell. Sci.* **36,** 281–309. Preaggregative cell motion in *Dictyostelium discoideum.*

Potts, G. (1902). *Flora* **91,** 281–347. Zur physiologie des *Dictyostelium mucoroides.*

Pudek, M. R., Bragg, P. D., and Weeks, G. (1978). *FEMS Microbiol. Lett.* **3,** 123–125. Membrane bound cytochromes of the cellular slime mold *Dictyostelium discoideum.*

Putnam, J. B., Jr., and Pedersen, L G. (1975). *Biochim. Biophys. Acta* **411,** 168–170. Discovery of a cyclic GMP stimulating factor in amoebae of *Dictyostelium discoideum.*

Quance, J., and Ashworth, J. M. (1972). *Biochem. J.* **126,** 609–615. Enzyme synthesis in the cellular slime mould *Dictyostelium discoideum* during the differentiation of myxamoebae grown axenically.

Quiviger, B., Benichou, J-C., and Ryter, A. (1980). *Biol. Cell.* **37,** 241–250. Comparative cytochemical localization of alkaline and acid phosphatases during starvation and differentiation of *Dictyostelium discoideum.*

Quiviger, B., de Chastellier, C., and Ryter, A. (1978). *J. Ultrastruct. Res.* **62,** 228–236. Cytochemical demonstration of alkaline phosphatase in the contractile vacuole of *Dictyostelium discoideum.*

Rafaeli, D. C. (1962). *Bull. Torrey Bot. Club* **89,** 312–318. Studies on mixed morphological mutants of *Polysphondylium violaceum.*

Rahmsdorf, H. J. (1977). *Hoppe-Seyler's Z. Physiol. Chem.* **358,** 527–529. A defined synthetic growth medium for *Dictyostelium discoideum* strain AX-2.

Rahmsdorf, H. J. (1979). *In* "FEBS Proceedings of the 12th Meeting," Vol. 54. Symposium 7. Cyclic Nucleotides and Protein Phosphorylation in Cell Regulation. Pergamon Press, New York, N.Y. (Krause, E. G., Pinna, L., and Wollenberger, A. eds.), 199–210. Protein kinase action as a mediator of chemotaxis and differentiation in the cellular slime mold *Dictyostelium discoideum.*

Rahmsdorf, H. J., and Gerisch, G. (1978). *Cell. Differ.* **7,** 249–258. Specific binding proteins for cyclic AMP and cyclic GMP in *Dictyostelium discoideum.*

Rahmsdorf, H. J., and Gerisch, G. (1978). *FEBS Lett.* **88,** 322–326. Cyclic AMP induced phosphorylation of a polypeptide comigrating with myosin heavy chains.

Rahmsdorf, H. J.,Cailla, H. L., Spitz, E., Moran, M. J., and Rickenberg, H. V. (1976). *Proc. Natl. Acad. Sci. U.S.A.* **73,** 3183–3187. Effect of sugars of early biochemical events in development of *Dictyostelium discoideum.*

Rahmsdorf, H. J., Malchow, D., and Gerisch, G. (1978). *FEBS Lett.* **88,** 322–326. Cyclic AMP induced phosphorylation in *Dictyostelium discoideum* of a polypeptide comigrating with myosin heavy chains.

Rahmsdorf, H. J., and Pai, S-H. (1979). *Biochim. Biophys. Acta* **567,** 339–346. Protein kinases EC-2.7.1.37 of *Dictyostelium discoideum* strain AX2.

Rai, J. N., and Tewari, J. P. (1961). *Proc. Indian Acad. Sci.* **53,** 1–9. Studies in cellular slime moulds from Indian soils. I. On the occurrence of *Dictyostelium mucoroides* Bref. and *Polysphondylium violaceum*.

Rai, J. N., and Tewari, J. P. (1963a). *Proc. Indian Acad. Sci.* **58,** 201–206. Studies in cellular slime moulds from Indian soils. II. On the occurrence of an aberrant strain of *Polysphondylium violaceum* Bref., with a discussion on the relevance of mode of branching of the sorocarp as a criterion for classifying members of Dictyosteliaceae.

Rai, J. N., and Tewari, J. P. (1963b). *Proc. Indian Acad. Sci.* **58,** 263–266. Studies in cellular slime moulds from Indian soils. III. On the occurrance of two strains of *Dictyostelium mucoroides complex*, conforming to the species *Dictyostelium sphaerocephalum* (Oud). Saccardo and March.

Ramagopal, S., and Ennis, H. L. (1980). *Eur. J. Biochem.* **105,** 245–258. Studies on ribosomal proteins in the cellular slime mold. Weights of proteins in the 40S and 60S ribosomal subunits.

Raman, R. K. (1976). *J. Theor. Biol.* **59,** 491–495. The autonomous cell in *Dictyostelium discoideum*.

Raman, R. K. (1976). *J. Cell. Sci.* **20,** 497–512. Analysis of the chemotactic response during aggregation in *Dictyostelium minutum*.

Raman, R. K. (1977). *J. Theor. Biol.* **64,** 43–70. Cellular aggregation towards steady point sources of attractant.

Raman, R. K., Hashimoto, Y., Cohen, M. H., and Robertson, A. (1976). *J. Cell. Sci.* **21,** 243–259. Differentiation for aggregation in the cellular slime molds. The emergence of autonomously signaling cells in *Dictyostelium discoideum*.

Raper, K. B. (1935). *J.Agr. Res.* **50,** 135–147. *Dictyostelium discoideum,* a new species of slime mold from decaying forest leaves.

Raper, K. B. (1937). *J. Agr. Res.* **58,** 289–316. Growth and development of *Dictyostelium discoideum* with different bacterial associates.

Raper, K. B. (1939). *J. Agr. Res.* **58,** 157–198. Influence of culture conditions upon the growth and development of *Dictyostelium discoideum*.

Raper, K. B. (1940a). *Amer. J. Bot.* **27,** 436–448. The communal nature of the fruiting process in the Acrasieae.

Raper, K. B. (1940b). *J. E. Mitchell Sci. Soc.* **56,** 241–282. Pseudoplasmodium formation and organization in *Dictyostelium discoideum*.

Raper, K. B. (1941a). *Mycologia* **33,** 633–649. *Dictyostelium minutum*, a second new species of slime mold from decaying forest leaves.

Raper, K. B. (1941b). *Growth* **5,** 41–76. Developmental patterns in simple slime molds.

Raper, K. B. (1951). *Quart. Rev. Biol.* **26,** 169–190. Isolation, cultivation, and conservation of simple slime molds.

Raper, K. B. (1956a). *Mycologia* **48,** 169–205. Factors affecting growth and differentiation in simple slime molds.

Raper, K. B. (1956b). *J. Gen. Microbiol.* **14,** 716–732. *Dictyostelium polycephalum* N. Sp.: a new cellular slime mold with coremiform fructifications.

Raper, K. B. (1960a). *Proc. Amer. Phil. Soc.* **104,** 579–604. Levels of cellular interaction in amoeboid populations.

Raper, K. B. (1960b). *In* "McGraw-Hill Encyclopedia of Science and Technology" Vol. 1, p. 49–50. McGraw-Hill, New York. Acrasiales.

Raper, K. B. (1963). *Harvey Lect.* **57,** 111–141. The environment and morphogenesis in cellular slime molds.

Raper, K. B., and Cavender, J. C. (1968). *J. E. Mitchell Sci. Soc.* **84,** 31–47. *Dictyostelium rosarium:* a new cellular slime mold with beaded sorocarps.

Raper, K. B., and Fennell, D. I. (1952). *Bull. Torrey Bot. Club* **79,** 25–51. Stalk formation in *Dictyostelium.*

Raper, K. B., and Fennell, D. I. (1967). *Amer. J. Bot.* **54,** 515–528. The crampon-based *Dictyostelia.*

Raper, K. B., and Quinlan, M. S. (1958). *J. Gen. Microbiol.* **18,** 16–32. *Actyostelium leptosomum:* a unique cellular slime mold with an acellular stalk.

Raper, K. B., and Smith, N. R. (1939). *J. Bacteriol.* **38,** 431–444. The growth of *Dictyostelium discoideum* upon pathogenic bacteria.

Raper, K. B., and Thom, C. (1932). *J. Wash. Acad. Sci.* **22,** 93–96. The distribution of *Dictyostelium* and other slime molds in soil.

Raper, K. B., and Thom, C. (1941). *Amer. J. Bot.* **28,** 69–78. Interspecific mixtures in the Dictyosteliaceae.

Rasmussen, H., Kurokawa, K., Mason, J., and Goodman, D. B. P. (1972). *In* "Excerpta Medica International Congress Series, No. 243." Excerpta Medica, Amsterdam, The Netherlands (Talmage, R. V., and Munson, P. L. eds.), 492–501. Cyclic AMP, calcium and cell activation.

Ratner, D. I., and Newell, P. C. (1978). *J. Gen. Microbiol.* **109,** 225–236. Linkage analysis in *Dictyostelium discoideum* using multiply marked tester strains establishment of linkage group VII and the reassessment of earlier linkage data.

Ray, J., Shinnick, T., and Lerner, R. (1979). *Nature (London)* **279,** 215–221. A mutation altering the function of a carbohydrate binding protein blocks cell–cell cohesion in developing *Dictyostelium discoideum.*

Reitherman, R. W., Rosen, S. D., Frazier, W. A., and Barondes, S. H. (1975). *Proc. Natl. Sci. U.S.A.* **72,** 3541–3545. Cell surface species specific high affinity receptors for discoidin: developmental regulation in *Dictyostelium discoideum.*

Renart, M. F., Sebastian, J., and Mato, J. M. (1981). *Cell Bio.* **5,** 1045. Adenylate cyclase activity in permeabilized cells from *Dictyostelium discoideum.*

Rickenberg, H. V., Rahmsdorf, H. J., Campbell, A., North, M. J., Kwasniak, J., and Ashworth, J. M. (1975). *J. Bacteriol.* **124,** 212–219. Inhibition of development in *Dictyostelium discoideum* by sugars.

Rickenberg, H. V., Tihon, C., and Guzel, O. (1977). *In* "Developments in Cell Biology." Elsevier/North-Holland (Cappucinnelli, P., and Ashworth, J. M. eds.), 173–182. The effect of pulses of 3′,5′ cyclic adenosine monophosphate on enzyme formation in nonaggregated amoebae of *Dictyostelium discoideum.*

Rickwood, D., and Osman, M. S. (1979). *Mol. Cell. Biochem.* **27,** 79–84. Characterization of poly ADP ribose polymerase activity in nuclei from the slime mold *Dictyostelium discoideum.*

Riedel, V., and Gerisch, G. (1968). *Naturwissenschaften* **55,** 656. Isolierung der zellmembranen von kollektiven amoben (Acrasina) mit hilfe von digitonin und filipin.

Riedel, V., and Gerisch, G. (1969). *Wilhelm Roux Archiv. Entwicklungsmech. Organismem* **162,** 268–285. Unterschiede im makromolekulbestand zwischen vegetativen und aggregationsreifen zellen von *Dictyostelium discoideum* (Acrasina).

Riedel, V., and Gerisch, G. (1971). *Biochem. Biophys. Res. Commun.* **42,** 119–124. Regulation of extracellular cyclic-AMP-phosphodiesterase activity during development of *Dictyostelium discoideum.*

Riedel, V., Malchow, D., Gerisch, G., and Nagele, B. (1972). *Biochem. Biophys. Res. Commun.* **46,** 279–287. Cyclic AMP phosphodiesterace interaction with its inhibitor of the slime mold *Dictyostelium discoideum.*

Riedel, V., Gerisch, G., Muller, E., and Beug, H. (1973). *J. Mol. Biol.* **74,** 573–585. Defective cyclic adenosine-3′,5′-phosphate-phosphodiesterase regulation in morphogenetic mutants of *Dictyostelium discoideum.*

Rifkin, J. L., and Speisman, R. A. (1976). *Trans. Amer. Microscop. Soc.* **95,** 165–173. Filamentous extensions of vegetative amoebae of the cellular slime mould *Dictyostelium discoideum.*

Robertson, A. (1972). *Lect. Math. Life Sci.* **4,** 47–73. Quantitative analysis of the development of cellular slime molds.

Robertson, A. (1974). *In* "The Biology of Brains." Halsted Press, New York, N.Y. (Broughton, B. W. ed.), 1–10. Information handling at the cellular level, intercellular communication in slime mold development.

Robertson, A., and Cohen, M. H. (1972). *Annu. Rev. Biophys. Bioeng.* **1,** 409–464. Control of developing fields.

Robertson, A., and Cohen, M. H. (1974). *In* "Lectures on Mathematics in the Life Sciences," Vol. 6. Amer. Math. Soc., Providence, R.I. (Cowan, J. D. ed.), 43–62. Quantitative analysis of the development of cellular slime molds. Part 2.

Robertson, A., and Drage, D. J. (1975). *Biophys. J.* **15,** 765–776. Stimulation of late interphase *dictyostelium discoideum* amoebae with an external cyclic AMP signal.

Robertson, A., and Grutsch, J. (1974). *Life Sci.* **15,** 1031–1043. The role of cyclic AMP in slime mold development.

Robertson, A., and Grutsch, J. (1981). *Cell* **24,** 603. Aggregation in *Dictyostelium discoideum.*

Robertson, A., Drage, D. J., and Cohen, M. H. (1971). *Science* **175,** 333–335. Control of aggregation in *Dictyostelium discoideum* by an external periodic pulse of cyclic adenosine monophosphate.

Robertson, A., Drage, D., and Cohen, M. (1972). *Science* **175,** 333–335. Control of aggregation in *Dictyostelium discoideum* by an external periodic pulse of cyclic adenosine monophoshate.

Robertson, A., Cohen, M. H., Drage, D. J., Rubin, J., Wonio, D., and Durston, A. J. (1972). *In* "Int. Proc. Cell Interactions" [Proc. of 3rd Le Petit. Colloq., London, Nov. 1971], (L. G. Silvestri, ed.), pp. 299–306. Amer. Elsevier, New York. Cellular interactions in slime mold aggregation.

Robson, G. E., and Williams, K. L. (1977). *J. Gen. Micro.* **99,** 191–200. The mitotic chromosomes of the cellular slime mold *Dictyostelium discoideum* a karyotype based on giemsa binding.

Robson, G. E., and Williams, K. L. (1979). *Genetics* **93,** 861–876. Vegetative incompatibility and the mating type locus in the cellular slime mold *Dictyostelium discoideum.*

Robson, G. E., and Williams, K. L. (1980). *Current Genetics* **1,** 229–232. The mating system of the cellular slime mold *Dictyostelium discoideum.*

Robson, G. E., and Williams, K. L. (1981). *J. Gen. Micro.* **125,** 463. Quantitative analysis of macrocyst formation in *Dictyostelium discoideum.*

Roessler, H., Peuckert, W., Risse, H. J., Eibl, H. J. (1978). *Mol. Cell. Biochem.* **20,** 3–16. The biosynthesis of glycolipids during the differentiation of the slime mold *Dictyostelium discoideum.*

Roessler, H. H., Schneider-Seelbach, E., Malati, T., and Risse, H-J. (1981). *Mol. Cell. Biochem.* **34,** 65–72. Dependence of glycosyltransferases in *Dictyostelium discoideum* on the structure of polyisoprenols.

Roewekamp, W., and Firtel, R. A. (1979). *J. Supramol. Struct.* **8,** 60. Developmentally regulative genes in *Dictyostelium discoideum.*

Roewekamp, W. G., Poole, S., and Firtel, R. A. (1979). *Hoppe-Seyler's Z. Physiol. Chem.* **360,** 1040. Developmental control of expression of CBP discoidin in *Dictyostelium discoideum.*

Rogge, H., and Risse, H-J. (1974). *Hoppe-Seyler's Z. Physiol. Chem.* **355**, 1467–1470. A procedure for the isolation of *Dictyostelium discoideum* nuclei.
Rogge, H., Neises, M., and Risse, H-J. (1977). *Biochim. Biophys. Acta* **499**, 273–277. Developmental regulation of nuclear glycosyl transfer in *Dictyostelium discoideum.*
Roos, E., Raper, K., and Vogen (1976). *J. Bact.* **128**, 495. Effects of light and temperature on macrocyst formation in paired mating types of *Dictyostelium discoideum.*
Roos, U.-P. (1975). *J. Cell. Sci.* **18**, 315–326. Fine structure of an organelle associated with the nucleus and cytoplasmic microtubules in the cellular slime mold *Polysphondylium violaceum.*
Roos, U.-P. (1975). *J. Cell. Biol.* **64**, 480–491. Mitosis in the cellular slime mold *Polysphondylium violaceum.*
Roos, U.-P., and Camenzind, R. (1981). *Euro. J. Cell Biol.* **25**, 248. Spindle dynamics during mitosis in *Dictyostelium discoideum.*
Roos, U.-P., McIntosh, J. R., McDonald, K. L., and Neighbors, B. (1980). *Experientia (Basel).* **36**, 757. Computer assisted reconstruction of the mitotic spindle of the cellular slime mold *Dictyostelium discoideum.*
Roos, W., and Gerisch, G. (1976). *FEBS Lett.* **68**, 170–172. Receptor mediated adenylate cyclase activation in *Dictyostelium discoideum.*
Roos, W., Nanjundiah, V., Malchow, D., and Gerisch, G. (1975). *FEBS Lett.* **53**, 139–142. Amplification of cAMP signals in aggregating cells of *Dictyostelium discoideum.*
Roos, W., Malchow, D., and Gerisch, G. (1977). *Cell. Differ.* **6**, 229–240. Adenylate cyclase and the control of cell differentiation in *Dictyostelium discoideum.*
Roos, W., Scheidegger, C., and Gerisch, G. (1977). *Nature (London)* **266**, 259–260. Adenylate cyclase activity oscillations as signals for cell aggregation in *Dictyostelium discoideum.*
Rosen, O. M., Rosen, S. M., and Horecker, B. L. (1965). *Biochem. Biophys. Res. Commun.* **18**, 270–276. Fate of the cell wall of *Salmonella typhimurium* upon ingestion by the cellular slime mold, *Polysphondylium pallidum.*
Rosen, R. (1976). *Arkansas Acad. Sci. Proc.* **30**, 75. A preliminary checklist of Arkansas USA Acrasieae.
Rosen, S. D., and Barondes, S. H. (1978). *In* "Receptors and Recognition Series B," Vol. 4. Specificity of Embryological Interactions. Halsted Press/Wiley, New York, N.Y. (Garrod, D. R. ed.), 233–264. Cell adhesion in the cellular slime molds.
Rosen, S. D., and Kaur, J. (1979). *Am. Zool.* **19**, 809–820. Intercellular adhesion in the cellular slime mold *Polysphondylium pallidum.*
Rosen, S. D., Kafka, J. A., Simpson, D. L., and Barondes, S. H. (1973). *Proc. Natl. Acad. Sci. U.S.A.* **70**, 2554–2557. Developmentally regulated, carbohydrate-binding protein in *Dictyostelium discoideum.*
Rosen, S. D., Simpson, D. L., Rose, J. E., and Barondes, S. H. (1974). *Nature (London)* **252**, 149–151. Carbohydrate binding protein from *Polysphondylium pallidum* implicated in intracellular adhesion.
Rosen, S. D., Reitherman, R. W., and Barondes, S. H. (1975). *Exp. Cell Res.* **95**, 159–166. Distinct lectin activities from six species of cellular slime moulds.
Rosen, S. D., Haywood, P. C., and Barondes, S. H. (1976). *J. Cell Biol.* **70**, 133. Isolation of receptor for pallidin A cell surface lectin from the cellular slime mold *Polysphondylium pallidum.*
Rosen, S. D., Haywood, P. C., and Barondes, S. H. (1976). *Nature (London)* **263**, 425–427. Inhibition of intercellular adhesion in a cellular slime mold by univalent antibody against a cell surface lectin.
Rosen, S. D., Chang, C.-M., and Barondes, S. H. (1977). *Develop. Biol.* **61**, 202–213. Intercellular adhesion in the cellular slime mold *Polysphondylium pallidum* inhibited by interaction of *Asialo fetuin* or specific univalent antibody with endogenous cell surface lectin.

Rosen, S. D., Kaur, J., Clark, D. L., Pardos, B. T., and Frazier,W. T. (1979). *J. Biol. Chem.* **254,** 9408–9415. Purification and characterization of multiple species isolectins of a slime mold *Polyspondylium pallidum* lectin implicated in intercellular adhesion.

Rosner, M. R., Verret,R. C., and Knorana, H. G. (1979). *J. Biol. Chem.* **254,** 5926–5933. The structure of lipopolysaccharide from an *Escherichiacoli* heptoseless mutant 3.2 fatty acyl amidases from *Dictyostelium discoideum* and their action on lipopolysaccharide derivatives.

Rosness, P. A. (1968). *J. Bacteriol.* **96,** 639–645. Cellulolytic enzymes during morphogenesis in *Dictyostelium discoideum.*

Rosness, P. A., and Wright, B. E. (1974). *Arch. Biochem. Biophys.* **164,** 60–72. *In vivo* changes of cellulose trehalose and glycogen during differentiation of *Dictyostelium discoideum.*

Rosness, P. A., Gustafson, G., and Wright, B. E. (1971). *J. Bacteriol.* **108,** 1329–1337. Effects of adenosine 3′,5′-monophosphate and adenosine 5′-monophosphate on glycogen degradation and synthesis in *Dictyostelium discoideum.*

Ross, F. M., and Newell, P. C. (1979). *J. Gen. Microbiol.* **115,** 289–300. Genetics of aggregation pattern mutations in the cellular slime mold *Dictyostelium discoideum.*

Ross, F. M., and Newell, P. C. (1981). *Microbiol.* **127,** 339–350. Streamers: chemotactic mutants of *Dictyostelium* with altered cyclic GMP metabolism.

Ross, I. K. (1960). *Amer. J. Bot.* **47,** 54–59. Studies on diploid strains of *Dictyostelium discoideum.*

Rossier, H., Peuckert, W., Risse, H.-J., and Eibl, H. J. (1978). *Mol. Cell. Biochem.* **20,** 3–15. The biosynthesis of glycolipids during the differentiation of the slime mold *Dictyostelium discoideum.*

Rossier, C., Gerisch, G., Malchow, D., and Eckstein, F. (1979). *J. Cell. Sci.* **35,** 321–338. Action of a slowly hydrolysable cyclic AMP analog on developing cells of *Dictyostelium discoideum.*

Rossier, C., Eitle, E., Van Driel, R., and Gerisch, G. (1980). *In* "Symposium of the Society for General Microbiology," Vol. 30. The Eukaryotic Microbial Cell. Cambridge Univ. Press, New York, N.Y. (Gooday, Lloyd, G. W. D., and Trinci, P. J. eds.), 405–424. Biochemical regulation of cell development and aggregation in *Dictyostelium discoideum.*

Rossier, H. H., Schneider-Seelbach, E., Malati, T., and Risse, H-J. (1981). *Mol. Cell. Biochem.* **34,** 65–72. The dependence of glycosyltransferases in *Dictyostelium discoideum* on the structure of polyisoprenols.

Rossomando, E. F., and Cutler, L. S. (1975). *Exp. Cell. Res.* **95,** 67–78. Localization of adenylate cyclase in *Dictyostelium discoideum.* I. Preparation and biochemical characterizations of cell fractions and isolated plasma membrane vesicles.

Rossomando, E. F., and Hesla, M. A. (1976). *J. Biol. Chem.* **251,** 6568–6573. Time dependent changes in *Dictyostelium discoideum* adenylate cyclase activity upon incubation with ATP.

Rossomando, E. F., and Jahngen, J. H. (1979). *Arch. Biochem. Biophys.* **1,** 364–366. Hydrolysis of a phosphonate ester catalyzed by an enzyme from *Dictyostelium discoideum.*

Rossomando, E. F., and Maldonado (1976). *Exp. Cell. Res.* **100,** 383–388. Inhibition of 5′-nucleotidase activity after growth of *Dictyostelium discoideum.*

Rossomando, E. F., and Sussman, M. (1972). *Biochem. Biophys. Res. Commun.* **47,** 604–610. Adenyl cyclase in *Dictyostelium discoideum:* a possible control element of the chemotactic system.

Rossomando, E. F., and Sussman, M. (1973). *Proc. Natl. Acad. Sci. U.S.A.* **70,** 1254–1257. A 5′-adenosine monophosphate-dependent adenylate cyclase and an adenosine 3′,5′-cyclic monophosphate-dependent adenosine triphosphate pyrophosphohydrolase in *Dictyostelium discoideum.*

Rossomando, E. F., Steffek, A. J., Mujwid, D. K., and Alexander, S. (1974). *Exp. Cell. Res.* **85**, 73–78. Scanning electron microscopic observations on cell surface changes during aggregation of *Dictyostelium discoideum.*

Rossomando, E. F., Maldonado, B., and Crean, E. V. (1978). *Antimicrob. Agents Chemother.* **14**, 476–482. Effect of hadacidin on growth and adenylo succinate synthetase activity of *Dictyostelium discoideum.*

Rossomando, E. F., Maldonado, B., Crean, E. V., and Kollar, E. J. (1978). *J. Cell. Sci.* **30**, 305 -318. Protease secretion during onset of development in *Dictyostelium discoideum.*

Rossomando, E. F., Crean, E. V., and Kestler, D. P. (1981). *Biochim. Biophys. Acta* **675**, 386. Isolation and characterization of an adenylyl-protein complex formed during the incubation of membranes from *Dictyostelium discoideum.*

Rossomando, E. F., Jahngen, E. G., Varnum, B., and Soll, D. R. (1981). *J. Cell. Biol.* **91**, 227. Inhibition of a nutrient-dependent pinocytosis in *Dictyostelium discoideum* by the amino acid analogue hadacidin.

Roth, R., and Sussman, M. (1966). *Biochim. Biophys. Acta* **122**, 225–231. Trehalose synthesis in the cellular slime mold *Dictyostelium discoideum.*

Roth, R., and Sussman, M. (1968). *J. Biol. Chem.* **243**, 5081–5087. Trehalose 6-phosphate synthetase (uridine diphosphate glucose: D-glucose 6-phosphate 1-glucosyltransferase) and its regulation during slime mold development.

Roth, R., Ashworth, J. M., and Sussman, M. (1968). *Proc. Natl. Acad. Sci. U.S.A.* **59**, 1235–1242. Periods of genetic transcription required for the synthesis of three enzymes during cellular slime mold development.

Rothman, F. G., and Alexander, E. T. (1975). *Genetics* **80**, 715–731. Parasexual genetic analysis of the cellular slime mold *Dictyostelium discoideum* A3.

Rowekamp, W., Poole, S., and Firtel, R. A. (1980). *Cell.* **20**, 495–505. Analysis of the multigene family coding the developmentally regulated carbohydrate binding protein discoidin I in *Dictyostelium discoideum.*

Rowekamp, W., and Firtel, R. A. (1980). *Develop. Biol.* **79**, 409–418. Isolation of developmentally regulated genes from *Dictyostelium.*

Rubenstein, P., and Deuchler, J. (1979). *J. Biol. Chem.* **254**, 11142–11147. Acetylated and nonacetylated actins in *Dictyostelium discoideum.*

Rubenstein, P., Smith, P., Deuchler, J., and Redman, K. (1981). *J. Biol. Chem.* **256**, 8149. NH_2-terminal acetylation of *Dictyostelium discoideum* actin in a cell-free protein-synthesizing system.

Rubin, J. (1976). *J. Embryol. Morph.* **36**, 261–271. The signal from fruiting body and conus tips of *Dictyostelium discoideum.*

Rubin, J., and Robertson, A. (1975). *J. Embryol. Exp. Morph.* **33**, 227–241. The tip of the *Dictyostelium discoideum* pseudoplasmodium as an organizer.

Rubino, S., Unger, E., Fighetti, M., and Cappuccinelli, P. (1981). *Atti. Ass. Genet. Ital.* **26**, 258. An immunofluorescence study of microtubular structures during mitosis in *Dictyostelium discoideum.*

Rubinov, S. I., Segel, L. A., and Ebel, W. (1981). *J. Theoret. Biol.* **91**, 99. A mathematical framework for the study of morphogenetic development in the slime mold.

Runyon, E. H. (1942). *Collecting Net* **17**, 88. Aggregation of separate cells of *Dictyostelium* to form a multicellular body.

Russell, G. K., and Bonner, J. T. (1960). *Bull. Torrey Bot. Club* **87**, 187–191. A note on spore germination in the cellular slime mold *Dictyostelium mucoroides.*

Rutherford, C. L. (1976). *J. Embryol. Exp. Morphol.* **35**, 335–343. Cell specific events occurring during development.

Rutherford, C. L. (1976). *Biochim. Biophys. Acta* **451,** 212–222. Glycogen degradation during migration of presumptive cell types in *Dictyostelium discoideum.*

Rutherford, C. L., and Harris, J. F. (1976). *Arch. Biochem. Biophys.* **175,** 453–462. Localization of glycogen phosphorylase in specific cell types during differentiation of *Dictyostelium discoideum.*

Rutherford, C. L., and Jefferson, B. L. (1976). *Develop. Biol.* **52,** 52–60. Trehalose accumulation in stalk and spore cells of *Dictyostelium discoideum.*

Rutherford, C. L., and Wright, B. E. (1971). *J. Bacteriol.* **108,** 269–275. Nucleotide metabolism during differentiation in *Dictyostelium discoideum.*

Rutherford, C. L., Kong, W. Y., Park, D., and Wright, B. E. (1974). *J. Gen. Microbiol.* **84,** 173–187. Precursor product relationships between nucleotides and RNA during differentiation in *Dictyostelium discoideum.*

Ryter, A., and Brachet, P. (1978). *Biologie Cellulaire* **31,** 265–270. Cell surface changes during early development stages of *Dictyostelium discoideum:* a scanning electron microscope study.

Ryter, A., and de Chastellier, C. (1977). *J. Cell. Biol.* **75,** 200–217. Morphometric and cytochemical studies of *Dictyostelium discoideum* in vegetative phase digestive system and membrane turnover.

Ryter, A., and Hellio, R. (1980). *J. Cell. Sci.* **41,** 75–88. Electron microscope study of *Dictyostelium discoideum* plasma membrane and its modifications during and after phagocytosis.

Ryter, A., Klein, C., and Brachet, P. (1979). *Exp. Cell. Res.* **119,** 373–380. *Dictyostelium discoideum* surface changes elicited by high concentrations of cAMP.

Sackin, M. J., and Ashworth, J. M. (1969). *J. Gen. Microbiol.* **59,** 275–284. An analysis of the distribution of volumes amongst spores of the cellular slime mould *Dictyostelium discoideum.*

Saito, M. (1979). *Exp. Cell. Res.* **123,** 79–86. Effect of extracellular calcium ion on the morphogenesis of *Dictyostelium discoideum.*

Saito, M. (1981). *Cytologia, Tokyo* **46,** 427. Species specific cell surface changes during the development in the cellular slime molds.

Saito, M., and Yanagisawa, K. (1978). *J. Embryol. Exp. Morphol.* **48,** 153–160. Participation of the cell surfaces in determining the developmental courses in the cellular slime mold *Dictyostelium purpureum.*

Sakai, Y. (1973). *Develop. Growth Differ.* **15,** 11–19. Cell type conversion in isolated prestalk and prespore fragments of the cellular slime mold *Dictyostelium discoideum.*

Sakai, Y., and Takeuchi, I. (1971). *Develop. Growth Differ.* **13,** 231–240. Changes of the prespore specific structure during dedifferentiation and cell type conversion of a slime mold cell.

Sakuma, K., Kominami, R., Muramatsu, M., and Sugiura, M. (1976). *Eur. J. Biochem.* **63,** 339–350. Conservation of the 5 terminal nucleotide sequences of ribosomal 18S RNA in eukaryotes differential evolution of large a d small ribosomal RNA.

Sameshima, M., Ito, K., and Iwabuchi, M. (1972). *Biochim. Biophys. Acta* **28,** 79–85. Effect of sodium fluoride on the amount of polyribosomes, single ribosomal subunits in a cellular slime mold *Dictyostelium discoideum.*

Sameshima, M., Muroyama, T., Hashimoto, Y., and Yamada, T. (1978). *Cell. Struct. Funct.* **3,** 123–128. Growth conditions on cytokinesis in axenic strains of a cellular slime mold.

Sampson, J. (1976). *J. Embryol. Exp. Morphol.* **36,** 663–668. Cell patterning in migrating slugs of *Dictyostelium discoideum.*

Sampson, J. (1977). *Cell* **11,** 173–180. Developmentally regulated cyclic AMP dependent protein kinases in *Dictyostelium discoideum.*

Sampson, J., Town, C., and Gross, J. (1978). *Develop. Biol.* **67,** 54–64. Cyclic AMP and the control of aggregative phase gene expression in *Dictyostelium discoideum.*

Samuel, E. W. (1961). *Develop. Biol.* **3,** 317–335. Orientation and rate of locomotion of individual amoebas in the life cycle of the cellular slime mold *Dictyostelium mucoroides.*

Sands, H., and Rickenberg, H. V. (1978). *In* "Internat. Review of Biochem." Vol. 20. Biochem. and Mode of Action of Hormones II. Univ. Park Press, Baltimore, Md. (Rickenberg, H. V. ed.), 45–80. Assessment of the role of cyclic nucleotides as hormonal mediators.

Saneyoshi, M., Tohyama, J., Nakayama, C., Takiya, S., and Iwabuchi, M. (1981). *Nuc. Acids Res.* **9,** 3129. Inhibitory effects of 3′-deoxycytidine 5′-triphosphate and 3′-deoxyuridine 5′-triphosphate on DNA-dependent RNA polymerases I and II purified from *Dictyostelium discoideum* cells.

Sargent, D., and Wright, B. E. (1971). *J. Biol. Chem.* **246,** 5340–5344. Trehalose synthesis during differentiation in *Dictyostelium discoideum.* II. *In vivo* determinations.

Satow, T. (1976). *Annu. Rep. Res. Reactor Inst. Kyoto Univ.* **9,** 62–69. Radiation effects on the species specific cell sorting-out of the cellular slime molds.

Satow, T. (1977). *Annu. Rep. Res. Reactor Inst. Kyoto Univ.* **10,** 62–65. Malformation of fruiting bodies of the cellular slime mold *Dictyostelium discoideum* induced by gamma-ray irradiation.

Saunders, D. A., and Wright, B. E. (1977). *J. Gen. Microbiol.* **100,** 89–97. Characterization of glucose 6 phosphate dependent glycogen synthase EC-2.4.1.11 from *Dictyostelium discoideum.*

Scandella, D., Rooney, R., and Katz, E. R. (1980). *Mol. Gen. Genet.* **180,** 67–75. Genetic biochemical and developmental studies of nystatin resistant mutants in *Dictyostelium discoideum.*

Schapp, P., Van Der Molen, L., and Konijn, T. M. (1981). *Bio. of the Cell* **41,** 133. The vacuolar apparatus of the simple cellular slime mold *Dictyostelium minutum.*

Schildkraut, C. L., Mandel, M., Levisohn, S., Smith-Sonneborn, J. E., and Marmur, J. (1962). *Nature (London)* **196,** 795–796. Deoxyribonucleic acid base composition and taxonomy of some Protozoa.

Schindler, J., and Sussman, M. (1977). *Biochem. Biophys. Res. Comm.* **79,** 611–617. Effect of NH_3 on cAMP associated activities and extracellular cAMP production in *Dictyostelium discoideum.*

Schindler, J., and Sussman, M. (1977). *J. Mol. Biol.* **116,** 161–170. Ammonia determines the choice of morphogenetic pathways in *Dictyostelium discoideum.*

Schindler, J., and Sussman, M. (1979). *Dev. Genet.* **1,** 13–20. Inhibition by ammonia of intracellular cyclic AMP accumulation in *Dictyostelium discoideum* its significance for the regulation of morphogenesis.

Schmidt, J. A., Stirling, J. L., Jones, G. E., and Pacy, J. (1978). *Nature (London)* **274,** 400–401. A chymotrypsin sensitive step in the development of *Dictyostelium discoideum.*

Schmidt, J. A., and Stirling, J. L. (1979). *In* "Abstracts of the 11th International Congress of Biochemistry" Toronto, Canada, p. 510. Characterization of an alpha-chymotrypsin sensitive step in the development of *Dictyostelium discoideum.*

Schmidt, W., Thomson, K., and Butler, W. L. (1977). *Photochem. Photobiol.* **26,** 407–412. Cytochrome B in plasma membrane enriched fractions from several photoresponsive organisms.

Schuckmann, W. von (1924). *Zentralbl. Bakteriol. Parasitenk.* **91,** 302–309. Zur biologie von *Dictyostelium mucoroides* Bref.

Schuckmann, W. von (1925). *Arch. Protistenk.* **51,** 495–529. Zur morphologie und biologie von *Dictyostelium mucoroides* Bref.

Schwalb, R., and Roth, R. (1970). *J. Gen. Microbiol.* **60,** 283–286. Axenic growth and development of the cellular slime mold *Dictyostelium discoideum.*

Scrive, M., Guespin-Michel, J. F., and Felenbok, B. (1977). *Exp. Cell. Res.* **108,** 107–110. Alteration of phosphodiesterase activity after 5-bromodeoxyuridine (BUDR) treatment of *Dictyostelium discoideum.*

Scrive, M., de Chastellier, C., Guespin, M. J. F., and Felenbok, B. (1981). *Bio. of the Cell* **41,** 143. Study of the aggregation process of *Dictyostelium discoideum* after 5-BUDR treatment.

Seela, F., and Hasselmann, D. (1979). *Chem. Ber.* **112,** 3072–3080. Synthesis of L-dextro disc adenine and its deamino and decarboxy derivatives.

Segel, L., and Stoeckly, B. (1972). *J. Theor. Biol.* **37,** 561–585. Instability of a layer of chemotactic cells, attractant and degrading enzyme.

Shaffer, B. M. (1953). *Nature (London)* **171,** 975. Aggregation in cellular slime moulds: *In vitro* isolation of acrasin.

Shaffer, B. M. (1956a). *Science* **123,** 1172–1173. Properties of acrasin.

Shaffer, B. M. (1956b). *J. Exp. Biol.* **33,** 645–657. Acrasin, the chemotactic agent in cellular slime moulds.

Shaffer, B. M. (1957a). *Amer. Nature* **91,** 19–35. Aspects of aggregation in cellular slime molds. I. Orientation and chemotaxis.

Shaffer, B. M. (1957b). *Quart. J. Microscop. Sci.* **98,** 377–392. Properties of slime-mould amoebae of significance for aggregation.

Shaffer, B. M. (1957c). *Quart. J. Microscop. Sci.* **98,** 393–405. Variability of behaviour of aggregating cellular slime moulds.

Shaffer, B. M. (1958). *Quart. J. Microscop. Sci.* **99,** 103–121. Integration in aggregating cellular slime moulds.

Shaffer, B. M. (1961a). *In* "Recent Advances in Botany," Proc. 9th Int. Botan. Congr., pp. 294–298. Univ. of Toronto Press, Toronto. Species differences in the aggregation of the Acrasieae.

Shaffer, B. M. (1961b). *J. Exp. Biol.* **38,** 833–849. The cells founding aggregation centres in the slime mould *Polysphondylium violaceum.*

Shaffer, B. M. (1962). *Advan. Morphog.* **2,** 109–182. The Acrasina.

Shaffer, B. M. (1963a). *Exp. Cell. Res.* **31,** 432–535. Inhibition by existing aggregations of founder differentiation in the cellular slime mould *Polysphondylium violaceum.*

Shaffer, B. M. (1963b). *Exp. Cell. Res.* **32,** 603–606. Behaviour of particles adhering to amoebae of the slime mould *Polysphondylium violaceum* and the fate of the cell surface during locomotion.

Shaffer, B. M. (1964a). *In* "Primitive Motile Systems in Cell Biology." Academic Press, New York, N.Y. (Allen, R. D., and Kanava, N. eds.), 387–405. Intracellular movement and locomotion of cellular slime mold amoebae.

Shaffer, B. M. (1964b). *J. Gen. Microbiol.* **36,** 359–364. Attraction through air exerted by unaggregated cells on aggregates of the slime mould *Polysphondylium violaceum.*

Shaffer, B. M. (1964c). *Advan. Morphog.* **3,** 301–322. The Acrasina.

Shaffer, B. M. (1965a). *Exp. Cell. Res.* **37,** 79–92. Antistrophic pseudopodia of the collective amoeba *Polysphondylium violaceum.*

Shaffer, B. M. (1965b). *J. Theoret. Biol.* **8,** 27–40. Mechanical control of the manufacture and resorption of cell surface in collective amoebae.

Shaffer, B. M. (1965c). *J. Embryol. Exp. Morphol.* **13,** 97–117. Cell movement within aggregates of the slime mould *Dictyostelium discoideum* revealed by surface markers.

Shaffer, B. M. (1965d). *Exp. Cell. Res.* **37,** 12–25. Pseudopodia and intracytoplasmic displacements of the collective amoebae Dictyosteliidae.

Shaffer, B. M. (1975). *Nature (London)* **255,** 549–552. Secretion of cAMP induced by cAMP in the cellular slime mold *Dictyostelium discoideum.*

Sharma, O. K., and Borek, E. (1970). *J. Bacteriol.* **101,** 705–708. Inhibitor of transfer ribonucleic acid methylases in the differentiating slime mold *Dictyostelium discoideum.*

Shinnick, T. M., and Lerner, R. A. (1980). *Proc. Nat. Acad. Sci. U.S.A.* **11,** 4788–4792. The cbpA gene role of 26000 dalton carbohydrate binding protein in intercellular cohesion of developing *Dictyostelium discoideum* cells.

Sievers, S., Risse, H. J., and Sekeri-Pataryas, K. H. (1978). *Mol. Cell. Biochem.* **20,** 103–110. Localization of glycosyl transferases in plasma membranes from *Dictyostelium discoideum.*

Silverman, R., Atherly, A., and Richter, D. (1979). *In* "Regulation of Macromolecular Synthesis by Low Molecular Weight Mediators." Academic Press, New York, N.Y. (Koch, G., and Richter, D. eds.), 115–126. Guanosine 5′ diphosphate 3′ diphosphate search in eukaryotes.

Simpson, D. L., Rosen, S. D., and Barondes, S. H. (1974). *Biochem.* **13,** 3487–3493. Discoidin, a developmentally regulated carbohydrate binding protein from *Dictyostelium discoideum.* Purification and characterization.

Simpson, D. L., Rosen, S. D., and Barondes, S. H. (1975). *Biochim. Biophys. Acta* **412,** 109–119. Pallidin purification and characterization of a carbohydrate binding protein from *Polysphondylium pallidum,* implicated in intercellular adhesion.

Simpson, P. A., and Spudich, J. A. (1980). *Proc. Natl. Acad. Sci. U.S.A.* **77,** 4610–4613. ATP driven steady-state exchange of monomeric and filamentous actin from *Dictyostelium discoideum.*

Singh, B. N. (1946). *Nature (London)* **157,** 133–134. Soil Acrasieae and their bacterial food supply.

Singh, B. N. (1974a). *J. Gen. Microbiol.* **1,** 11–21. Studies on soil Acrasieae: I. Distribution of species of *Dictyostelium* in soils of Great Britain and the effect of bacteria on their development.

Singh, B. N. (1974b). *J. Gen. Microbiol.* **1,** 361–367. Studies on soil Acrasieae: II. The active life of species of *Dictyostelium* in soil and the influence thereon of soil moisture and bacterial food.

Sinha, U., and Ashworth, J. M. (1969). *Proc. Roy. Soc. Edinburgh Sect. B.* **173,** 531–540. Evidence for the existence of elements of a parasexual cycle in the cellular slime mould, *Dictyostelium discoideum.*

Sinha, U., and Ashworth, J. M. (1978). *Phytomorph.* **28,** 210–215. Effects of *P*-fluorophenylalanine on growth and differentiation in *Dictyostelium discoideum.*

Siu, C. H., Lerner, R. A., Firtel, R. A., and Loomis, W. F. (1975). *In* "Cell and Molecular Biology" Vol. 2, W. A. Benjamin, Menlo Park, Ca. (McMahon, D., and Fox, C. F. eds.), 129–134. Changes in plasma membrane proteins during development of *Dictyostelium discoideum.*

Siu, C. H., Lerner, R. A., Ma, G., Firtel, R. A., and Loomis, W. F. (1976). *J. Mol. Biol.* **100,** 157–178. Developmentally regulated proteins of the plasma membrane of *Dictyostelium discoideum.* The carbohydrate binding protein.

Siu, C. H., Lerner, R. A., and Loomis, W. F. (1977). *J. Mol. Biol.* **116,** 469–488. Rapid accumulation and disappearance of plasma membrane proteins during development of wild type and mutant strains of *Dictyostelium discoideum.*

Siu, C. H., Geltosky, J. E., and Lerner, R. A. (1978). *In* "The National Foundation—March of Dimes Birth Defects: Original Article Series," Vol. 14, No. 2. The Molecular Basis of Cell–Cell Interaction. Alan R. Liss, New York, N.Y. (Lerner, R. A., and Bergsma, D. eds.), 459–471. Altered cell–cell recognition in a temperature conditional mutant of *Dictyostelium discoideum.*

Siu, C. H., Loomis, W. F., and Lerner, R. A. (1978). *Birth Defects* **14,** 439–458. Plasma membrane proteins in wild-type and cascade-arrested mutant strains of *Dictyostelium discoideum.*

Skupienski, F. X. (1919). *C.R. Acad. Sci.* **167,** 960–962. Sur la sexualité chez une espece de myxomycete Acrasiee, *Dictyostelium mucoroides.*

Skupienski, F. X. (1920). Paris. 81 pp. "Recherches sur le Cycle Evolutif de Certains Myxomycetes."

Slifkin, M. K., and Bonner, J. T. (1952). *Biol. Bull.* **102,** 273–277. The effects of salts and organic solutes on the migration of the slime mold *Dictyostelium discoideum.*

Slifkin, M. K., and Gutowsky, H. S. (1958). *J. Cell. Comp. Physiol.* **51,** 249–257. Infared spectroscopy as a new method for assessing the nutritional requirements of the slime mold, *Dictyostelium discoideum.*

Smart, J. E., and Hynes, R. O. (1974). *Nature (London)* **251,** 319–321. Developmentally regulated cell surface alterations in *Dictyostelium discoideum.*

Smart, J. E., and Tuchman, J. (1976). *Develop. Biol.* **51,** 63–76. Inhibition of the development of *Dictyostelium discoideum* by isolated plasma membranes.

Smith, E., and Williams, K. L. (1979). *FEMS Lett.* **6,** 119–122. Preparation of slime sheath from *Dictyostelium discoideum.*

Smith, E., Williams, K. L. (1981). *J. Embryol. Exp. Morphol.* **61,** 61–67. The age dependent loss of cells from the rear of a *Dictyostelium discoideum* slug is not tip controlled.

Snyder, H. M., and Ceccarini, C. (1966). *Nature (London)* **209,** 1152. Interspecific spore inhibition in the cellular slime molds.

Sogin, M. L., and Olsen, G. J. (1980). *Gene (Amst.)* **8,** 231–238. Identification and mapping of a 60 base pair eco-R-I fragment in *Dictyostelium discoideum* ribosomal DNA.

Soll, D. R. (1979). *Science* **203,** 841–849. Timers in developing systems.

Soll, D. R., and Sussman, M. (1973). *Biochim. Biophys. Acta* **319,** 312–322. Transcription in isolated nuclei of the slime mold *Dictyostelium discoideum.*

Soll, D. R., and Waddell, D. R. (1975). *Develop. Biol.* **47,** 292–302. Morphogenesis in the slime mold *Dictyostelium discoideum.* Part 1. The accumulation and erasure of morphogenetic information.

Soll, D. R., Yarger, J., and Mirick, M. (1976). *J. Cell. Sci.* **20,** 513–523. Stationary phase and the cell cycle of *Dictyostelium discoideum* in liquid nutrient medium.

Solomon, E. P., Johnson, E. M., and Gregg, J. H. (1964). *Develop. Biol.* **9,** 314–326. Multiple forms of enzymes in a cellular slime mold during morphogenesis.

Sonneborn, D. R., White, G. J., and Sussman, M. (1963). *Develop. Biol.* **7,** 79–93. A mutation affecting both rate and pattern of morphogenesis in *Dictyostelium discoideum.*

Sonneborn, D. R., Sussman, M., and Levine, L. (1964). *J. Bacteriol.* **87,** 1312–1329. Serological analyses of cellular slime-mold development. I. Changes in antigenic activity during cell aggregation.

Sonneborn, D. R., Levine, L., and Sussman, M. (1965). *J. Bacteriol.* **89,** 1092–1096. Serological analyses of cellular slime mold development. II. Preferential loss, during morphogenesis, of antigenic activity associated with the vegetative myxamoebae.

Sperb, R. P. (1979). *Bull. Math. Biol.* **41,** 555–572. A mathematical model describing the aggregation of amoebae.

Spiegel, F. W., and Cox, E. C. (1980). *Nature (London)* **286,** 806–807. A one dimensional pattern in the cellular slime mold *Polysphondylium pallidum.*

Springer, W. R., and Barondes, S. H. (1978). *J. Cell. Biol.* **79,** 937–942. Direct measurement of species specific cohesion in cellular slime molds.

Springer, W. R., and Barondes, S. H. (1980). *J. Cell. Biol.* **87,** 703–707. Cell adhesion molecules detection with univalent and antibody.

Springer, W. R., Haywood, P. L., and Barondes, S. H. (1980). *J. Cell. Biol.* **87**, 682–690. Endogenous cell surface lectin in *Dictyostelium purpureum* quantitation elution by sugar and elicitation by divalent immunoglobulin.

Spudich, J. (1974). *J.Biol. Chem.* **249**, 6013–6020. Biochemical and structural studies of actomyosin-like proteins from non-muscle cells. II. Purification, properties and membrane association of actin from amoebae of *Dictyostelium discoideum*.

Spudich, J. A., and Clarke, M. (1974). *J. Supramol. Struct.* **2**, 150–162. The contractile proteins of *Dictyostelium discoideum*.

Spudich, J., and Cooke, R. (1975). *J. Biol. Chem.* **250**, 7485–7491. Supramolecular forms of actin from amoebae of *Dictyostelium discoideum*.

Spudich, J., and Mockrin, S. (1976). *Proc. Nat. Acad. Sci. U.S.A.* **73**, 2321. Calcium control of actin-activated myosin ATPase from *Dictyostelium discoideum*.

Srinivas, V., and Katz, E. R. (1980). *FEBS Lett.* **9**, 53–55. Oxygen utilization in *Dictyostelium discoideum*.

Staples, S. O., and Gregg, J. H. (1967). *Biol. Bull.* **132**, 413–422. Carotenoid pigments in the cellular slime mold, *Dictyostelium discoideum*.

Steinemann, C., and Parish, R. W. (1980). *Nature (London)* **286**, 621–623. Evidence that a developmentally regulated glycoprotein is the target of adhesion blocking Fab in reaggregation *Dictyostelium discoideum*.

Steinemann, C., Hintermann, R., and Parish, R. W. (1979). *FEBS Lett.* **108**, 379–384. Identification of a developmentally regulated plasma membrane glycoprotein involved in adhesion of *Polysphondylium pallidum* cells.

Steiner, A. L., Ong, S.-H., and Wedner, H. J. (1976). *Adv. Cyc. Nucleotide Res.* **7**, 115–155. Cyclic nucleotide immunocytochemistry.

Stenhouse, F. O., and Williams, K. L. (1977). *Develop. Biol.* **59**, 140–152. Patterning in *Dictyostelium discoideum*.

Stenhouse, F. O., and Williams, K. L. (1981). *Differentiation* **18**, 1–10. Investigation of cell patterning in the asexual fruiting body of *Dictyostelium discoideum* using haploid isogenic diploid strains.

Sternfeld, J. (1979). *J. Embryol. Exp. Morphol.* **53**, 163–177. Evidence for differential cellular adhesion as the mechanism of sorting out of various slime mold species.

Sternfeld, J., and Bonner, J. T. (1977). *Proc. Nat. Acad. Sci. U.S.A.* **74**, 268–271. Cell differentiation in *Dictyostelium* under submerged conditions.

Sternfeld, J., and David, C. N. (1979). *J. Cell Sci.* **38**, 181–191. Ammonia plus another factor are necessary for differentiation in submerged clumps of *Dictyostelium*.

Sternfeld, J., and David, C. N. (1981). *J. Cell Sci.* **50**, 9–18. Oxygen gradients cause pattern orientation in *Dictyostelium* cell clumps.

Sternfeld, J., and David, C. N. (1981). *Differentiation* (In press). Cell sorting during pattern formation in *Dictyostelium discoideum*.

Stevens, L., Irene, M., McKinnon, I. M., Turner, R. M., and North, M. J. (1978), *Biochem. Soc. Trans.* **6**, 407–409. The effects of 1,4 aminobutanone on polyamine metabolism in bacteria, a cellular slime mold, and rat tissues.

Stuchell, R. N., Weinstein, B. I., and Beattie, D. S. (1975). *J. Biol. Chem.* **250**, 570–576. Effects of ethidium bromide on the respiratory chain and oligomycin sensitive ATPase in purified mitochondria from the cellular slime mold *Dictyostelium discoideum*.

Stewart, P. R., and Spudich, J. A. (1979). *J. Supramol. Struct.* **12**, 1–14. Structural states of *Dictyostelium discoideum* myosin.

Strong, C. L. (1966). *Sci. Amer.* **214**, 116–121. How to cultivate the slime molds and perform experiments on them.

Stuchell, R., Weinstein, B., and Beattie, D. (1973). *FEBS Lett.* **37**, 23–26. Effects of ethidium bromide on various segments of the respiratory chain in the cellular slime mold *Dictyostelium discoideum.*

Stumph, W. E., Wu, J.-R., and Bonner, J. (1977). *J. Cell. Biol.* **75**, 127. Gene enrichment using an affinity resin with specificity for RNA DNA hybrids.

Stumph, W. E., Wu, J.-R., and Bonner, J. (1978). *Biochem.* **17**, 5791–5798. Gene enrichment using antibodies to DNA RNA hybrids purification and mapping of *Dictyostelium discoideum* ribosomal DNA.

Sumino, T. (1978). *Annu. Rep. Res. Reactor Inst. Kyoto Univ.* **11**, 186–188. A new method for separation of *Dictyostelium discoideum* amoebae.

Sumino, T. (1981). *Experientia* **37**, 1075. A cell-marking technique for a cellular slime mold.

Sussman, M. (1951). *J. Exp. Zool.* **118**, 407–418. The origin of cellular heterogeneity in the slime molds, Dictyosteliaceae.

Sussman, M. (1952). *Biol. Bull.* **103**, 446–457. An analysis of the aggregation stage in the development of the slime molds, Dictyosteliaceae. II. Aggregative center formation by mixtures of *Dictyostelium discoideum* wild type and aggregateless variants.

Sussman, M. (1954). *J. Gen. Microbiol.* **10**, 110–120. Synergistic and antagonistic interactions between morphogenetically deficient variant of the slime mould *Dictyostelium discoideum.*

Sussman, M. (1955a). *J. Gen. Microbiol.* **13**, 295–309. "Fruity" and other mutants of the cellular slime mould, *Dictyostelium discoideum:* a study of developmental aberrations.

Sussman, M. (1955b). *In* "Biochemistry and Physiology of the Protozoa." Vol. 2. Academic Press, New York, N.Y. (Hunter, S., and Lwoff, S. eds.), 201–223. The developmental physiology of the amoeboid slime molds.

Sussman, M. (1956a). *Biol. Bull.* **110**, 91–95. On the relation between growth and morphogenesis in the slime mold *Dictyostelium discoideum.*

Sussman, M. (1956b). *Annu. Rev. Microbiol.* **10**, 21–50. The biology of the cellular slime molds.

Sussman, M. (1958). *In* "A Symposium on the Chemical Basis of Development" Johns Hopkins Press, Baltimore, M.D. (Mcelroy, W. D., and Glass, B. eds.), 264–295. A developmental analysis of cellular slime mold aggregation.

Sussman, M. (1961a). *J. Gen. Microbiol.* **25**, 375–378. Cultivation and serial transfer of the slime mould, *Dictyostelium discoideum* in liquid nutrient medium.

Sussman, M. (1961b). *In* "Growth in Living Systems." Basic Books, New York, N.Y. 221–239. Cellular differentiation in the slime mold.

Sussman, M. (1963). *Science* **139**, 338. Growth of the cellular slime mold *Polysphondylium pallidum* in a simple nutrient medium.

Sussman, M. (1965a). *Biochem. Biophys. Res. Commun.* **18**, 763–767. Inhibition by actidione of protein synthesis and UDP-GAL polysaccharide transferase accumulation in *Dictyostelium discoideum.*

Sussman, M. (1965b). *In* "Genetic Control of Differentiation." Pp. 66–76. Brookhaven Symposia in Biology No. 18, Brookhaven, New York. Temporal, spatial, and quantitative control of enzyme activity during slime mold cytodifferentiation.

Sussman, M. (1965c). *Annu. Rev. Microbiol.* **19**, 59–78. Developmental phenomena in microorganisms and in higher forms of life.

Sussman, M. (1966a). *In* "Methods in Cell Physiology." Vol. 2. Academic Press, New York, N.Y. 397–410. Biochemical and genetic methods in the study of cellular slime mold development.

Sussman, M. (1966b). *Proc. Nat. Acad. Sci. U.S.A.* **55**, 813–818. Protein synthesis and the temporal control of genetic transcription during slime mold development.

Sussman, M. (1967). *Fed. Proc. Fed. Amer. Soc. Exp. Biol.* **26**, 77–83. Evidence for temporal and quantitative control of genetic transcription and translation during slime mold development.

Sussman, M. (1970). *Nature (London)* **225,** 1245–1246. Model for quantitative and qualitative control of mRNA translation in eukaryotes.

Sussman, M. (1972). *In* "Biochemistry of the Glycosidic Linkage." Academic Press, New York, N.Y. (Piras, R., and Pontis, H. G. eds.), 431–448. The program of polysaccharide and disaccharide synthesis during the development of *Dictyostelium discoideum.*

Sussman, M. (1976). *In* "Progress in Molecular and Subcellular Biology," Vol. 7. Springer-Verlag, New York, N.Y. (Hahn, F. E. ed.), 103–131. The genesis of multicellular organization and the control of gene expression in *Dictyostelium discoideum.*

Sussman, M., and Boschwitz, C. (1975a). *Develop. Biol.* **44,** 362–368. Adhesive properties of cell ghosts derived from *Dictyostelium discoideum.*

Sussman, M., and Boschwitz, C. (1975b). *Exp. Cell. Res.* **95,** 63–66. An increase of calcium magnesium binding sites in cell ghosts associated with the acquisition of aggregative competence in *Dictyostelium discoideum.*

Sussman, M., and Brackenbury, R. (1976). *In* "Annual Review of Plant Physiology," Vol. 27. Annual Reviews, Palo Alto, Ca. (Briggs, W. R. ed.), 229–265. Biochemical and molecular genetic aspects of cellular slime mold development.

Sussman, M., and Bradley, S. G. (1954). *Arch. Biochem. Biophys.* **51,** 428–435. A protein growth factor of bacterial origin required by the cellular slime molds.

Sussman, M., and Ennis, H. L. (1959). *Biol. Bull.* **116,** 304–317. The role of the initiator cell in slime mold aggregation.

Sussman, M., and Lee, F. (1954). *Bacteriol. Proc.* **42.** Physiology of developmental variants among the cellular slime molds (Abstract).

Sussman, M., and Lee, F. (1955). *Proc. Natl. Acad. Sci. U.S.A.* **41,** 70–78. Interactions among variant and wild-type strains of cellular slime molds across thin agar membranes.

Sussman, M., and Lovgren, N. (1965). *Exp. Cell Res.* **38,** 97–105. Preferential release of the enzyme UDP-galactose polysaccharide transferase during cellular differentiation in the slime mold.

Sussman, M., and Newell, P. C. (1972). *In* "Molecular Genetics and Developmental Biology." Prentice-Hall, Englewood Cliffs, New Jersey. Pp. 275–302. Quantal control.

Sussman, M., and Noel, E. (1952). *Biol. Bull.* **103,** 259–268. An analysis of the aggregation stage in the development of the slime molds, Dictyosteliaceae. I. The populational distribution of the capacity to initiate aggregation.

Sussman, M., and Osborn, M. J. (1964). *Proc. Natl. Acad. Sci. U.S.A.* **52,** 81–87. UDP-galactose polysaccharide transferase in the cellular slime mold, *Dictyostelium discoideum:* appearance and disappearance of activity during cell differentiation.

Sussman, M., and Rossomando, E. F. (1974). *In* "Handbook of Genetics," Vol. 2. Plenum Publishing, New York, N.Y. (King, R. C. ed.), 427–431. Cellular slime molds.

Sussman, M., and Schindler, J. (1978). *Differentiation* **10,** 1–6. A possible mechanism of morphogenetic regulation in *Dictyostelium discoideum.*

Sussman, M., and Sussman, R. R. (1956). *In* "Cellular Mechanisms in Differentiation and Growth." 14th Growth Symposium. Princeton Univ. Press, Princeton, New Jersey (Rudnick, D. ed.), 125–154. Cellular interactions during the development of the cellular slime molds.

Sussman, M., and Sussman, R. R. (1961). *Exp. Cell Res. Suppl.* **8,** 91–106. Aggregative performance.

Sussman, M., and Sussman, R. R. (1962). *J. Gen. Microbiol.* **28,** 417–429. Ploidal inheritance in *Dictyostelium discoideum.* I. Stable haploid, stable diploid and metastable strains.

Sussman, M., and Sussman, R. R. (1965). *Biochim. Biophys. Acta* **108,** 463– 73. The regulatory rogram for UDP-gal polysaccharide transferase activity during slime mold cytodifferentiation: requirement for specific synthesis of RNA.

Sussman, M., and Sussman, R. R. (1969). *Symp. Soc. Gen. Microbiol.* **19,** 403–435. Patterns of RNA synthesis and of enzyme accumulation and disappearance during cellular slime mould cytodifferentiation.

Sussman, M., Lee, F., and Kerr, N. S. (1956). *Science* **123,** 1171–1172. Fractionation of acrasin, a specific chemotactic agent for slime mold aggregation.

Sussman, M., Loomis, W. F., Ashworth, J. M., and Sussman, R. R. (1967). *Biochem. Biophys. Res. Commun.* **26,** 353–359. The effect of actinomycin D on cellular slime mold morphogenesis.

Sussman, M., Schindler, J., and Kim, H. (1977). *In* "Development and Differentiation in the Cellular Slime Moulds." Elsevier/North-Holland (Cappuccinelli, P., and Ashworth, J. M. eds.), 31–50. Toward a biochemical definition of the morphogenetic fields in *Dictyostelium discoideum.*

Sussman, M., Schindler, J., and Kim, H. (1978). *In* "The National Foundation—March of Dimes Birth Defects: Original Article Series," Vol. 14, No. 2. The Molecular Basis of Cell–Cell Interaction. Alan R. Liss, New York, N.Y. (Lerner, R. A., and Bergsma, D. eds.), 473–489. Cell interactions controlling morphogenesis and gene expression in *Dictyostelium discoideum.*

Sussman, M., Schindler, J., and Kim, H. (1978). *Exp. Cell. Res.* **116,** 217–228. Sluggers: a new class of morphogenetic mutants of *Dictyostelium discoideum.*

Sussman, R. R. (1961). *Exp. Cell. Res.* **24,** 154–155. A method for staining chromosomes of *Dictyostelium discoideum* myxamoebae in the vegetative stage.

Sussman, R. R. (1967). *Biochim. Biophys. Acta* **149,** 407–421. RNA metabolism during cytodifferentiation in the cellular slime mold *Polysphondylium pallidum.*

Sussman, R. R. (1974). *J. Bacteriol.* **118,** 312–313. Bioassay for the isolation of *Dictyostelium discoideum* mutants deficient in extracellular accumulation of cyclic AMP.

Sussman, R. R., and Rayner, E. P. (1971). *Biochem. Biophys.* **144,** 127–137. Physical characterization of deoxyribonucleic acid in *Dictyostelium discoideum.*

Sussman, R. R., and Sussman, M. (1953). *Ann. N.Y. Acad. Sci.* **56,** 949–960. Cellular differentiation in Dictyosteliaceae: heritable modifications of the developmental pattern.

Sussman, R. R., and Sussman, M. (1960). *J. Gen. Microbiol.* **23,** 287–293. The dissociation of morphogenesis from cell division in the cellular slime mould, *Dictyostelium discoideum.*

Sussman, R. R., and Sussman, M. (1963). *J. Gen. Microbiol.* **30,** 349–355. Ploidal inheritance in the slime mould *Dictyostelium discoideum:* haploidization and genetic segregation of diploid strains.

Sussman, R. R., and Sussman, M. (1967). *Biochem. Biophys. Res. Commun.* **29,** 53–55. Cultivation of *Dictyostelium discoideum* in axenic medium.

Sussman, R. R., Sussman, M., and Ennis, H. L. (1960). *Develop. Biol.* **2,** 367–392. Appearance and inheritance of the I-cell phenotype in *D. discoideum.*

Sutherland, J. B., and Raper, K. B. (1978). *Mycologia* **70,** 1173–1180. Distribution of cellular slime molds in Wisconsin prairie soils.

Swan, A. P., Garrod, D. R., and Morris, D. (1977). *J. Cell. Sci.* **28,** 107–116. An inhibitor of cell cohesion from axenically grown cells of the slime mold *Dictyostelium discoideum.*

Swanson, J. A., Taylor, D. L., and Bonner, J. T. (1981). *J. Ultrastruct. Res.* **75,** 243–249. Coated vesicles in *Dictyostelium discoideum.*

Szabo, S. P., O'Day, D. H., and Chagla, A. H. (1982). *Develop. Biol.* (In press). Cell fusion, nuclear fusion, and zygote differentiation during sexual development of *Dictyostelium discoideum*.

Takeishi, K., and Kaneda, S. (1981). *J. Biochem.* **90,** 299. Isolation and characterization of small nuclear RNAs from *Dictyostelium discoideum*.

Takemoto, S., Okamoto, K., and Takeuchi, I. (1978). *Biochem. Biophys. Res. Comm.* **80,** 858–865. The effects of cyclic AMP on dissaggregation induced changes in activities of developmentally regulated enzymes in *Dictyostelium discoideum*.

Takeuchi, I. (1960). *Develop. Biol.* **2,** 343–366. The correlation of cellular changes with succinic dehydrogenase and cytochrome oxidase activities in the development of the cellular slime molds.

Takeuchi, I. (1963). *Develop. Biol.* **8,** 1–26. Immunochemical and immunohistochemical studies on the development of the cellular slime mold *Dictyostelium mucoroides*.

Takeuchi, I. (1969). *In* "Nucleic Acid Metabolism Cell Differentiation and Cancer Growth." Pergamon Press, Oxford (Cowdry, E. F., and Seno, S. eds.), 297–304. Establishment of polar organization during slime mold development.

Takeuchi, I. (1972). *Aspects Cell Mol. Phys.*, 217–236. Differentiation and dedifferentiation in cellular slime molds.

Takeuchi, I., and Sakai, Y. (1971). *Develop. Growth Differ.* **13,** 201–210. Dedifferentiation of the disaggregated slug cell of the cellular slime mold *Dictyostelium discoideum*.

Takeuchi, I., and Sato, T. (1965). *Jap. J. Exp. Morphol.* **19,** 67–70. Cell differentiation and cell sorting in the development of cellular slime molds.

Takeuchi, I., and Tazawa, M. (1955). *Cytologia* **20,** 157–165. Studies on the morphogenesis of the slime mould, *Dictyostelium discoideum*.

Takeuchi, I., and Yabuno, K. (1970). *Exp. Cell. Res.* **61,** 183–190. Disaggregation of slime mold pseudoplasmodia using EDTA and various proteolytic enzymes.

Takeuchi, I., Hayashi, M., and Tasaka, M. (1977). *In* "Development and Differentiation in the Cellular Slime Moulds." Elseiver/North-Holland, New York (Cappuccinelli, P., and Ashworth, J. M. eds.), 1–16. Cell differentiation and pattern formation in *Dictyostelium*.

Takiya, S., Takoh, Y., and Iwabuchi, M. (1980). *J. Fac. Sci. Hokkaido Univ. Ser. V. Bot.* **12,** 111–122. A rapid and facile procedure for the purification of DNA dependent RNA polymerases I and II EC-2.7.7.6 of the cellular slime mold *Dictyostelium discoideum*.

Takiya, S., Takoh, Y., and Iwabuchi, M. (1980). *J. Biochem. (Tokyo)* **87,** 1501–1510. Template specificity of DNA dependent RNA polymerases I and II EC-2.7.7.6 for synthetic polynucleotides during development of the cellular slime mold *Dictyostelium discoideum*.

Tanaka, Y., Yamada, T., and Yanagisawa, K. (1973). *Jpn. J. Genet.* **48,** 448. Role of cell surface membrane in ontogeny.

Tanaka, Y., Yanagisawa, K., Hashimoto, Y., and Yamaguchi, M. (1974). *Agr. Biol. Chem.* **38,** 689–690. True spore germination inhibitor of a cellular slime mold *Dictyostelium discoideum*.

Tanaka, Y., Hashimoto, Y., Yanagisawa, K., Abe, H., and Uchiayama, M. (1975). *Agric. Biol. Chem.* **39,** 1929–1932. Partial structure of a spore germination inhibitor from a cellular slime mold *Dictyostelium discoideum*.

Tanaka, Y., Abe, H., Uchiyama, M., Taya, Y., and Nishimura, S. (1978). *Phytochem.* **17,** 543–544. Isopentyladenine from *Dictyostelium discoideum*.

Tasaka, M., and Takeuchi, I. (1979). *J. Embryol. Exp. Morphol.* **49,** 89–102. Sorting out behavior of disaggregated cells in the absence of morphogenesis in *Dictyostelium discoideum*.

Tasaka, M., and Takeuchi, I. (1981). *Differentiation* **18,** 191–196. Role of cell sorting in pattern formation in *Dictyostelium discoideum*.

Taya, O., Tanaka, Y., and Nishimura, S. (1978). *Nature (London)* **271**, 545–547. 5′ AMP is a direct precursor of cytokinin in *Dictyostelium discoideum.*

Taya, Y., Tanaka, Y., and Nishimura, S. (1978). *FEBS Lett.* **89**, 326–328. Cell free biosynthesis of discadenine, a spore germination inhibitor of *Dictyostelium discoideum.*

Taya, Y., Yamada, T., and Nishimura, S. (1980). *J. Bacteriol.* **143**, 715–719. Correlation between acrasins and spore germination inhibitors in cellular slime molds.

Taylor, D. L., Condeelis, J. S., and Rhodes, J. A. (1977). *ICN-UCLA Symposium*, **17**, 581–603. The contractile basis of amoeboid movement. Part 3. Structure and dynamics of motile extracts and membrane fragments from *Dictyostelium discoideum* and *Amoeba proteus.*

Taylor, W. C., Cockburn, A. F., Frankel, G. A., Newkirk, M. J., and Firtel, R. A. (1977). *In* "ICN-UCLA Symposium on Molecular and Cell Biology, Vol. VII (Wilcox, Abelson and Fox, eds.), pp. 309–313. Organization of ribosomal and 5S RNA coding regions in *Dictyostelium discoideum.*

Telser, A., and Sussman, M. (1971). *J. Biol. Chem.* **246**, 2252–2257. Uridine diphosphate galactose-4-epimerase, a developmentally regulated enzyme in the cellular slime mold *Dictyostelium discoideum.*

Thadani, V., Pan, P., and Bonner, J. T. (1977). *Exp. Cell. Res.* **108**, 75–78. Complementary effects of ammonia and cyclic AMP on aggregation territory size in the cellular slime mold *Dictyostelium mucoroides.*

Thilo, L., and Vogel, G. (1980). *Eur. J.Cell. Biol.* **22**, 189. Kinetics of membrane flow internalization recycling and net transfer.

Thilo, L., and Vogel, G. (1980). *Proc. Natl. Acad. Sci. U.S.A.* **77**, 1015–1019. Kinetics of membrane internalization and recycling during pinocytosis in *Dictyostelium discoideum.*

Thom, C., and Raper, K. B. (1930). *J. Wash. Acad. Sci.* **20**, 362–370. Myxamoebae in soil and decomposing crop residues.

Thomas, D. A. (1978). *Fed. Proc.* **37**, 161. Pentose phosphate metabolism during differentiation in *Dictyostelium discoideum.*

Thomas, D. A. (1978). *J. Gen. Microbiol.* **108**, 311–314. Effects of 5 AMP glucose 1 phosphate and nucleotide sugars on the activity of glycogen phosphorylase from *Dictyostelium discoideum.*

Thomas, D. A. (1979). *J. Gen. Microbiol.* **113**, 357–368. Pentose phosphate metabolism during differentiation in *Dictyostelium discoideum.*

Thomas, D. A. (1981). *J. Gen. Micro.* **124**, 403. Partial purification and characterization of glucose-6-phosphate isomerase from *Dictyostelium discoideum.*

Thomas, D. A., and Emyanioff, R. (1980). *Annu. Meet. Am. Soc. Microbiol.* **80**, 1687. Incorporation of carbon-14 labeled glucose and carbon-14 labeled ribose into saccharide end products and RNA during development in *Dictyostelium discoideum.*

Thomas, D. A., and Wright, B. E. (1976). *J. Biol. Chem.* **251**, 1253–1257. Glycogen phosphorylase in *Dictyostelium discoideum.* Part 1. Purification and properties of the enzyme.

Tieghem, P. Van (1880). *Bull. Soc. Bot. Fr.* **27**, 317–322. Sur quelques myxomycetes à plasmode agrege.

Tieghem, P. Van (1884). *Bull. Soc. Bot. Fr.* **31**, 303–306. Coenonia, genre nouveau de myxomycetes à plasmode agrege.

Tihon, C., Buzel, O., and Richenberg, H. V. (1977). *Adv. Cyclic Nucleotide Res.* Regulation of synthesis of enzymes by cyclic AMP in *Dictyostelium.*

Tisa, L. S., and Cotter, D. A. (1979). *Microbios. Lett.* **11**, 84–94. Expression of acid and alkaline phosphatase activities during germination of *Dictyostelium discoideum.*

Tisa, L. S., and Cotter, D. A. (1980). *J. Bacteriol.* **141**, 436–442. Expression of glycosidase activities during germination of *Dictyostelium discoideum* spores.

Toama, M. A., and Raper, K. B. (1967a). *J. Bacteriol.* **94,** 1143–1149. Microcysts of the cellular slime mold *Polysphondylium pallidum.* I. Factors influencing microcyst formation.

Toama, M. A., and Raper, K. B. (1967b). *J. Bacteriol.* **94,** 1150–1153. Microcysts of the cellular slime mold *Polysphondylium pallidum.* II. Chemistry of the microcyst walls.

Toda, K., Ono, K.-I., and Ochiai, H. (1980). *Eur. J. Biochem.* **111,** 377–388. Surface labeling of membrane glycoproteins and their drastic changes during development of *Dictyostelium discoideum.*

Toda, K., Ono, K.-I., and Ochiai, H. (1981). *J. Biochem. Tokyo* **90,** 1429. Developmentally regulated glycoprotein alterations in *Dictyostelium discoideum.*

Tomchik, K. J., and Devreotes, P. N. (1981). *Science* **212,** 443–446. Cyclic AMP waves in *Dictyostelium discoideum*: a demonstration by isotope dilution fluorography.

Toorchen, D., and Henderson, E. J. (1979). *Biochem. Biophys. Res. Commun.* **87,** 1168–1175. Characterization of multiple extracellular cyclic AMP phosphosdiesterase EC-3.1.4.17 forms in *Dictyostelium discoideum.*

Town, C., and Gross, J. (1978). *Develop. Biol.* **63,** 412–420. The role of cyclic nucleotides and cell agglomeration in postaggregative enzyme synthesis in *Dictyostelium discoideum.*

Town, C. D., and Stanford, E. (1977). *J. Embryol. Exp. Morphol.* **42,** 105–113. Stalk cell differentiation by cells from migrating slugs of *Dictyostelium discoideum* special properties of tip cells.

Town, C., and Stanford, E. (1979). *Proc. Natl. Acad. Sci. U.S.A.* **76,** 308–312. An oligosaccharide containing factor that induces cell differentiation in *Dictyostelium discoideum.*

Town, C. D., Gross, J. D., and Kay, R. R. (1976). *Nature (London)* **262,** 717–719. Cell differentiation without morphogenesis in *Dictyostelium discoideum.*

Traub, F., Hohl, H. R., and Cavender, J. C. (1981). *Am. J. Bot.* **68,** 162–171. Cellular slime molds of Switzerland 1. Description of new species.

Traub, F., Hohl, H. R., and Cavender, J. C. (1981). *Am. J. Bot.* **68,** 172–182. Cellular slime molds of Switzerland 2. Distribution in forest soils.

Tsang, A., and Bradbury, J. M. (1981). *Exp. Cell. Res.* **132,** 433–441. Separation and properties of prestalk and prespore cells of *Dictyostelium discoideum.*

Tsang, A. S., and Coukell, M. B. (1976). *Proc. Can. Fed. Biol. Soc.* **76,** 155. Regulation of multiple cyclic AMP phosphodiesterases during early development of *Dictyostelium discoideum.*

Tsang, A. S., and Coukell, M. B. (1977). *Cell. Differ.* **6,** 75–84. The regulation of cyclic AMP phosphodiesterase and its specific inhibitor by cyclic AMP in *Dictyostelium.*

Tsang, A. S., and Coukell, M. B. (1978). *Can. Fed. Biol. Soc. Proc.* **21,** 95. Evidence for increased *de-novo* synthesis of extracellular cyclic AMP phosphodiesterase during induction by cyclic AMP in *Dictyostelium purpureum.*

Tsang, A. S., and Coukell, B. (1979). *Eur. J. Biochem.* **95,** 407–418. Biochemical and genetic evidence for 2 extracellular cyclic AMP phosphodiesterases in *Dictyostelium purpureum.*

Tsang, A. S., and Coukell, M. B. (1979). *Eur. J. Biochem.* **95,** 419–426. Direct evidence for extracellular cyclic AMP phosphodiesterase induction and phosphodiesterase inhibitor repression by exogenous cyclic AMP in *Dictyostelium purpureum.*

Tsang, A. S., Devine, J. M., and Williams, J. G. (1981). *Develop. Biol.* **84,** 212–217. The multiple subunits of discoidin I are encoded by different genes.

Tuchman, J., Alton, T., and Lodish, H. (1974). *Develop. Biol.* **40,** 116–129. Preferential synthesis of actin during early development of the slime mold *Dictyostelium discoideum.*

Tuchman, J., Smart, J. E., and Lodish, H. F. (1976). *Develop. Biol.* **51,** 77–85. Effects of differentiated membranes on the developmental program of the cellular slime mould *Dictyostelium discoideum.*

Turner, R., and North, M. J. (1977). *In* "Developments in Cell Biology." Elsevier/North-Holland (Cappuccinelli, P., and Ashworth, J. M. eds.), 221–230. Polyamines and their metabolism in *Dictyostelium discoideum.*

Turner, R., North, M. J., and Harwood, J. M. (1979). *Biochem. J.* **180,** 119–128. Putrescine uptake by the cellular slime mold *Dictyostelium discoideum.*

Uchiyama, M., and Abe, H. (1977). *Agric. Biol. Chem.* **41,** 1549–1552. A synthesis of racemic discadenine.

Uchiyama, S., Okamoto, K., and Takeuchi, I. (1979). *Biochim. Biophys. Acta* **562,** 103–111. Repression of ribosomal RNA synthesis induced by disaggregation in *Dictyostelium discoideum.*

Unger, E., Rubino, S., Weinert, T., and Cappucinelli, P. (1979). *FEMS Lett.* **6,** 317–320. Immunofluorescence of the tubulin system in cellular slime molds.

Uyemura, D. G., Brown, S. S., and Spudich, J. A. (1978). *J. Biol. Chem.* **253,** 9088–9096. Biochemical and structural characterization of actin from *Dictyostelium discoideum.*

Van Driel, R. (1981). *Eur. J. Biochem.* **115,** 391–395. Binding of the chemoattractant folic acid by *Dictyostelium discoideum* cells.

Van Haastert, P. J. M., van der Meer, R. C., and Konijn, T. M. (1981). *J. Bacteriol.* **147,** 170–175. Evidence that the rate of association of cyclic AMP to its chemotactic receptor induces phosphodiesterase activity in *Dictyostelium discoideum.*

Vandekerckhove, J., and Weber, K. (1980). *Nature (London)* **18,** 475–477. Vegetative *Dictyostelium discoideum* cells containing 17 actin genes express a single major actin.

Verma, I. M., Firtel, R. A., Lodish, H. F., and Baltimore, D. (1974). *Biochem.* **13,** 3917–3922. Synthesis of DNA complementary to cellular slime mold messenger RNA by reverse transcriptase.

Veron, M., and Brachet, P. (1979). *Biochem. Biophys. Res. Commun.* **91,** 699–706. Characterization of a 5′ AMP binding site in purified membranes from *Dictyostelium discoideum.*

Veron, M., and Patte, J-C. (1978). *Develop. Biol.* **63,** 370–376. Intracellular cyclic AMP binding protein in *Dictyostelium discoideum* differences in properties of the proteins from vegetative and differentiated celll.

Vince, S., and Gingell, D. (1980). *Exp. Cell. Res.* **126,** 462–465. Cationic modulation of the interaction of *Dictyostelium discoideum* amoebae with glass evidence from quantitative interference reflection microscopy.

Vogel, G. (1981). *Monogr. Allergy* **17,** 1–11. Recognition mechanisms in phagocytosis in *Dictyostelium discoideum.*

Vogel, G., Thilo, L., Schwarz, H., and Steinhart, R. (1980). *J. Cell. Biol.* **86,** 456–465. Mechanism of phagocytosis in *Dictyostelium discoideum*: phagocytosis is mediated by different recognition sites as disclosed by mutants with altered phagocytotic properties.

Von Dreele, P. H., and Williams, K. L. (1977). *Biochim. Biophys. Acta* **464,** 378–388. Electron spin resonance studies of the membranes of the cellular slime mold *Dictyostelium discoideum.*

Vuillemin, P. (1903). *C.R. Acad. Sci.* **137,** 387–389. Une Acrasiée bacteriophage.

Waddell, D. R., and Soll, D. R. (1977). *Develop. Biol.* **60,** 83–92. A characterization of the erasure phenomenon in *Dictyostelium discoideum.*

Wallace, L. J., and Frazier, W. A. (1979). *J. Biol. Chem.* **254,** 10109–10114. Direct and enzyme mediated photoaffinity labeling of membrane associated actin in *Dictyostelium discoideum.*

Wallace, L. J., and Frazier, W. A. (1979). *Proc. Natl. Acad. Sci. U.S.A.* **76,** 4250–4254. Photoaffinity labeling of cyclic AMP binding and AMP binding proteins of differentiating *Dictyostelium discoideum* cells.

Wallace, M. A., and Raper, K. B. (1979). *J.Gen. Microbiol.* **113,** 327–338. Genetic exchanges in the macrocysts of *Dictyostelium discoideum.*

Walsh, J., and Wright, B. E. (1978). *J. Gen. Microbiol.* **108**, 57–62. Kinetics of net RNA degradation during development in *Dictyostelium discoideum.*

Ward, A., and Brenner, M. (1977). *Life Sci.* **21**, 997–1008. Guanylate cyclase from *Dictyostelium discoideum.*

Ward, C., and Wright, B. E. (1965). *Biochem.* **4**, 2021–2027. Cell wall synthesis in *Dictyostelium discoideum.* I. *In vitro* synthesis from uridine diphosphoglucose.

Warren, J. A., Warren, W. D., and Cox, E. C. (1975). *Proc. Natl. Acad. Sci. U.S.A.* **72**, 1041–1042. Genetic complexity of aggregation in cellular slime mold *Polysphondylium violaceum.*

Warren, J. A., Warren, W. D., and Cox, E. C. (1976). *Genetics* **83**, 25–47. Genetic and morphological study of aggregation in cellular slime mold *Polysphondylium violaceum.*

Washington, A. C., and Davis, J. A. (1979). *Tex. J. Sci.* **31**, 172–176. Synthesis of trehalose 6 phosphate synthetase during growth and development of the cellular slime mold *Dictyostelium discoideum.*

Washington, A. C., and Wynne, V. (1979). *Tex. J. Sci.* **31**, 177–186. Influence of folic acid on growth and development in the cellular slime mold *Dictyostelium discoideum.*

Watts, D. J. (1977). *J. Gen. Microbiol.* **98**, 355–361. Vitamin requirements for growth of myxamoebae of *Dictyostelium discoideum* in a defined medium.

Watts, D. J., and Ashworth, J. M. (1970). *Biochem. J.* **119**, 171–174. Growth of myxamoebae of the cellular slime mould *Dictyostelium discoideum* in axenic culture.

Watts, D. J., and Guest, J. R. (1975). *J. Gen. Microbiol.* **86**, 333–342. Studies of the vitamin nutrition of the cellular slime mold *Dictyostelium discoideum.*

Watts, D. J., and Treffry, T. E. (1975). *FEBS Lett.* **52**, 262–264. Incorporation of *N*-acetylglucosamine into the slime sheath of the cellular slime mould *Dictyostelium discoideum.*

Watts, D. J., and Treffry, T. E. (1976). *J. Embryol. Exp. Morphol.* **35**, 323–333. Culmination in the slime mold *Dictyostelium discoideum* studied with a scanning electron microscope.

Weber, A. T., and Raper, K. B. (1971). *Develop. Biol.* **26**, 606–615. Induction of fruiting in two aggregateless mutants of *Dictyostelium discoideum.*

Weeks, C., and Weeks, G. (1975). *Exp. Cell. Res.* **92**, 372–382. Cell surface changes durin the differentiation of *Dictyostelium discoideum.*

Weeks, G. (1973). *Exp. Cell Res.* **76**, 467–470. Agglutination of growing and differentiating cells of *Dictyostelium discoideum* by concanavalin A.

Weeks, G. (1975). *J. Biol. Chem.* **250**, 6706–6710. Studies of cell surface of *Dictyostelium discoideum* during differentiation: the binding of I-Con A to cell surface.

Weeks, G. (1976). *Biochim. Biophys. Acta* **450**, 21–32. The manipulation of the fatty acid composition of *Dictyostelium discoideum* and its effect on cell differentiation.

Weeks, G., and Ashworth, J. M. (1972). *Biochem. J.* **126**, 617–626. Glycogen synthetase and the control of glycogen synthesis in the cellular slime mould *Dictyostelium discoideum* during the growth (myxamoebal) phase.

Weeks, G., and Herring, F. G. (1980). *J. Lipid Res.* **21**, 681–686. The lipid composition and membrane fluidity of *Dictyostelium discoideum* plasma membranes at various stages during differentiation.

Weiner, E., and Ashworth, J. M. (1970). *Biochem. J.* **118**, 505–512. The isolation and characterization of lysosomal particles from myxamoebae of the cellular slime mould *Dictyostelium discoideum.*

Weinkauff, A. M., and Filosa, M. F. (1965). *Can. J. Microbiol.* **11**, 385–387. Factors involved in the formation of macrocysts by the cellular slime mold, *Dictyostelium mucoroides.*

Weinstein, B., and Koritz, S. (1973). *Develop. Biol.* **34**, 159–162. A protein kinase assayable with intact cells of the cellular slime mold *Dictyostelium discoideum.*

Weiss, E., and Braun, V. (1979). *FEBS Lett.* **108**, 233–236. Actin-like protein of the cytoplasm in the chromatin of *Dictyostelium discoideum.*

Welker, D. L., and Deering, R. A. (1976). *J. Gen. Micro.* **97**, 1–10. Genetic analysis of radiation sensitive mutants in slime mould *Dictyostelium discoideum.*

Welker, D. L., and Deering, R. A. (1978). *J. Gen. Microbiol.* **109**, 11–24. Genetics of radiation sensitivity in the slime mold *Dictyostelium discoideum.*

Welker, D. L., and Deering, R. A. (1979). *Mol. Gen. Genet.* **167**, 259–264. *In vivo* nicking and rejoining of nuclear DNA in UV irradiated radiation resistant and sensitive strains of *Dictyostelium discoideum.*

Welker, D. L., and Deering, R. A. (1979). *Mol. Gen. Genet.* **167**, 265–270. Interactions between radiation sensitive mutations in double mutant haploids of *Dictyostelium discoideum.*

Welker, D. L., and Williams, K. L. (1980). *FEMS Microbiol. Lett.* **9**, 179–184. *Bacillus subtilis* sensitivity loci in *Dictyostelium discoideum.*

Welker, D. L., and Williams, K. L. (1980). *J.Gen. Microbiol.* **116**, 397–408. Mitotic arrest and chromosome doubling using thiabendazole, cambendazole nocodazole, and benlate in the slime mold *Dictyostelium discoideum.*

Welker, D. L., and Williams, K. L. (1980). *J. Gen. Microbiol.* **120**, 149–160. The assignment of four new loci including the coumarin sensitivity locus couA to linkage group VII of *Dictyostelium discoideum.*

Welker, D. L., and Williams, K. L. (1981). *Chromosoma* **82**, 321–332. Genetic and cytological characterization of fusion chromosomes of *Dictyostelium discoideum.*

Welker, D. L., and Williams, K. L. (1981). *Cur. Gen.* **3**, 167. Temperature-sensitive DNA repair mutations in the cellular slime mould *Dictyostelium discoideum.*

West, C. M., and McMahon, D. (1976). *Anal. Biochem.* **76**, 589–605. A physical explanation for multiple-cell classes after centrifugation in colloidal silica gradients.

West, C. M., and McMahon, D. (1977). *J. Cell. Biol.* **74**, 264–273. Identification of concanavalin A receptors and galactose binding proteins in purified plasma membranes of *Dictyostelium discoideum.*

West, C. M., and McMahon, D. (1979). *Exp. Cell. Res.* **124**, 393–401. The axial distribution of plasma membrane molecules in pseudoplasmodia of the cellular slime mold *Dictyostelium discoideum.*

West, C. M., McMahon, D., and Molday, R. S. (1978). *J. Biol. Chem.* **253**, 1716–1724. Identification of glycoproteins, using lectins as probes, in plasma membranes from *Dictyostelium discoideum* and human erythrocytes.

Whitaker, B. D., and Poff, K. L. (1980). *Exp. Cell. Res.* **128**, 87–94. Thermal adaption of thermo sensing and negative thermo taxis in *Dictyostelium discoideum.*

White, E., Scandella, D., and Katz, E. R. (1981). *Dev. Genet.* **2**, 99–112. Inhibition by N-3 chlorophenyl carbamate of mitosis and development in *Dictyostelium discoideum* and the isolation of N-3 chlorophenyl carbamate resistant mutants.

White, G. J., and Sussman, M. (1961). *Biochim. Biophys. Acta* **53**, 285–293. Metabolism of major cell constituents during slime mold morphogenesis.

White, G. J., and Sussman, M. (1963a). *Biochim. Biophys. Acta* **74**, 173–178. Polysaccharides involved in slime-mold development. I. Water-soluble polymer(s).

White, G. J., and Sussman, M. (1963b). *Biochim. Biophys. Acta* **74**, 179–187. Polysaccharides involved in slime mold development. II. Water-soluble acid mucopolysaccharide(s).

Whitfield, F. E. (1964). *Exp. Cell Res.* **36**, 62–72. The use of proteolytic and other enzymes in the separation of slime mould grex.

Wick, U., Malchow, D., and Gerisch, G. (1978). *Cell Biol. Int. Rep.* **2**, 71–79. Cyclic AMP stimulated calcium influx into aggregating cells of *Dictyostelium discoideum.*

Widmer, R., and Parish, R. W. (1980). *FEBS Lett.* **121,** 183–187. Isolation and characterization of nuclear envelope fragments from *Dictyostelium discoideum.*

Widmer, R., Fuhrer, S., and Parish, R. W. (1979). *FEBS Lett.* **106,** 363–369. Biochemical evidence for a distinctive chromatin structure in nucleoli of *Dictyostelium discoideum.*

Wier, P. W. (1977). *Differentiation* **9,** 183–192. Cyclic AMP and cyclic AMP phosphodiesterase and the duration of the interphase in *Dictyostelium discoideum.*

Wiklund, R. A., and Allison, A. C. (1972). *Nature New Biol.* **239,** 221–222. Effects of anesthetics on the mobility of *Dictyostelium discoideum.*

Wilcox, D. K., and Sussman, M. (1978). *Differentiation* **11,** 125–131. Spore differentiation by isolated *Dictyostelium discoideum* cells triggered by prior cell contact.

Wilcox, D. K., and Sussman, M. (1981a). *Develop. Biol.* **82,** 102–112. Defective cell cohesivity expressed late in the development of a *Dictyostelium discoideum* mutant.

Wilcox, D. K., and Sussman, M. (1981b). *Proc. Natl. Acad. Sci. U.S.A.* **78,** 358–362. Serologically distinguishable alterations in the molecular specificity of cell cohesion during morphogenesis in *Dictyostelium discoideum.*

Wild, J. R., David, R., Schronk, L., and Hanks, A. R. (1980). *Tex. J. Sci.* **31,** 9–18. Developmental changes in uridine nucleotide pools of *Dictyostelium discoideum.*

Wilhelms, O.-H., Luderitz, O., Westphal, O., and Gerisch, G. (1974). *Eur. J. Biochem.* **48,** 89–101. Glycosphingolipids and glycoproteins in the wild-type and in a nonaggregating mutant of *Dictyostelium discoideum.*

Williams, J. G., and Lloyd, M. M. (1979). *J. Mol. Biol.* **129,** 19–36. Changes in the abundance of polyadenylated RNA during slime mold development measured using cloned molecular hybridization probes.

Williams, J. G., Lloyd, M. M., and Devine, J. M. (1979). *Cell* **17,** 903–914. Characterization and transcription analysis of a cloned sequence derived from a major developmentally regulated messenger RNA of *Dictyostelium discoideum.*

Williams, J. G., Tsang, A. S., and Mahbubani, H. (1980). *Proc. Natl. Acad. Sci. U.S.A.* **7,** 7171–7175. A change in the rate of transcription of a eukaryotic gene in response to cyclic AMP.

Williams, K. L. (1976). *Appl. Environ. Microbiol.* **32,** 635–637. Isolation of strains of the cellular slime mold *Dictyostelium discoideum* capable of growing after a single passage in axenic medium.

Williams, K. L. (1976). *Nature* **260,** 785–786. Mutation frequency at a recessive locus in haploid and diploid strains of a slime mold.

Williams, K. L. (1978). *Genetics* **90,** 37–48. Characterization of dominant resistance to cobalt chloride in *Dictyostelium discoideum* and its use in parasexual genetic analysis.

Williams, K. L. (1980). *J. Gen. Microbiol.* **116,** 409–416. Examination of chromosomes of *Polysphondylium pallidum* following metaphase arrest by benzimidazole derivatives and colchicine.

Williams, K. L. (1981). *FEMS Microbiol. Lett.* **11,** 317–320. Two arsenate resistance loci in the cellular slime mould *Dictyostelium discoideum.*

Williams, K. L., and Barrand, P. (1978). *FEMS Microbiol. Lett.* **4,** 155–159. Parasexual genetics in the cellular slime mold *Dictyostelium discoideum* haploidization of diploid strains using benlate.

Williams, K. L., and Newell, P. C. (1976). *Genetics* **82,** 287–307. A genetic study of aggregation in the cellular slime mould *Dictyostelium discoideum* using complementation analysis.

Williams, K. L., and Stenhouse, F. O. (1977). *In* "Developments in Cell Biology." Elsevier/North-Holland (Cappuccinelli, P., and Ashworth, J. M. eds.), 27–30. Quantitative analysis of the proportions of the *Dictyostelium discoideum* asexual fruiting body.

Williams, K. L., and Welker, D. L. (1980). *Dev. Genet.* **1,** 355–362. Mutations specific to spore maturation in the asexual fruiting body of *Dictyostelium discoideum.*

Williams, K. L., Kessin, R., and Newell, P. (1974). *Nature (London)* **247,** 142–143. Genetics of growth in axenic medium of the cellular slime mould *Dictyostelium discoideum.*

Williams, K. L., Kessin, R. H., and Newell, P. C. (1974). *J. Gen. Microbiol.* **84,** 68–78. Para exual genetics in *Dictyostelium discoideum:* mitotic analysis of acriflavine resistance and growth in axenic medium.

Williams, K. L., Robson, G. E., and Welker, D. L. (1980). *Genetics* **95,** 289–304. Chromosome fragments in *Dictyostelium discoideum* obtained from parasexual crosses between strains of different genetic background,

Williams, K. L., Fisher, P. R., Macwilliams, H. K., and Bonner, J. T. (1981). *Differentiation* **18,** 61–63. Cell patterning in *Dictyostelium discoideum.*

Wilson, C. M. (1952). *Proc. Natl. Acad. Sci. U.S.A.* **38,** 659–662. Sexuality in Acrasiales.

Wilson, C. M. (1953). *Amer. J. Bot.* **40,** 714–718. Cytological study of the life cycle of *Dictyostelium.*

Wilson, C. M., and Ross, I. K. (1957). *Amer. J. Bot.* **44,** 345–350. Further cytological studies in the Acrasiales.

Wilson, J. B., and Rutherford, C. L. (1978). *J. Cell. Physiol.* **94,** 37–46. ATP trehalose glucose and ammonium ion localization in the 2 cell types of *Dictyostelium discoideum.*

Wise, J. A., and Weiner, A. M. (1980). *Cell* **22,** 109–118. *Dictyostelium* small nuclear RNA D2 is homologous to rat nucleolar RNA U3 and is encoded by a dispersed multigene family.

Wise, J. A., and Weiner, A. M. (1981). *J. Biol. Chem.* **256,** 956–963. The small nuclear RNA of the cellular slime mold *Dictyostelium discoideum* isolation and characterization.

Wittingham, W. F., and Raper, K. B. (1956). *Amer. J. Bot.* **43,** 703–708. Inhibition of normal pigment synthesis in spores of *Dictyostelium purpureum.*

Wittingham, W. F., and Raper, K.B. (1957). *Amer. J. Bot.* **44,** 619–627. Environmental factors influencing the growth and fructification of *Dictyostelium polycephalum.*

Wittingham, W. F., and Raper, K. B. (1960). *Proc. Natl. Acad. Sci. U.S.A.* **46,** 642–649. Nonviability of stalk cells in *Dictyostelium.*

Woolley, D. (1970). *J. Cell Physiol.* **76,** 185–190. Extraction of an actomyosinlike protein from amoeba *Dictyostelium discoideum.*

Woolley, D. E. (1972). *Arch. Biochem. Biophys.* **150,** 519–530. An actin-like protein from amoebae of *Dictyostelium discoideum.*

Wright, B. E. (1958). *Bacteriol. Proc.* **115.** Effect of steroids on aggregation in the slime mold *Dictyostelium discoideum.*

Wright, B. E. (1960). *Proc. Natl. Acad. Sci. U.S.A.* **46,** 798–803. On enzyme–substrate relationships during biochemical differentiation.

Wright, B. E. (1963a). *Bacteriol. Rev.* **27,** 273–281. Endogenous substrate control in biochemical differentiation.

Wright, B. E. (1963b). *Ann. N.Y. Acad. Sci.* **102,** 740–754. Endogenous activity and sporulation in slime molds.

Wright, B. E. (1964). *In* "Biochemistry and Physiology of the Protozoa." Vol. 3. Academic Press, New York (Hutner, S. H. ed.), 341–381. Biochemistry of Acrasiales.

Wright, B. E. (1965). *In* "Development and Metabolic Control and Neoplasia," pp. 298–316. Williams & Wilkins, Baltimore, Maryland. Control of carbohydrate synthesis in the slime mold.

Wright, B. E. (1966). *Science* **153,** 830–837. Multiple causes and controls in differentiation.

Wright, B. E. (1972). *Develop. Biol.* **28,** F–13–20. Actinomycin D and genetic transcription during differentiation.

Wright, B. E. (1973). "Critical Variables in Differentiation," Prentice-Hall, Englewood Cliffs, New Jersey.

Wright, B. E. (1975). *In* "Microbiology—1975," Amer. Soc. for Microbiol., Washington, D.C. (Schlessinger, D. ed.), 500–507. Usefulness of developmental mutants in the analysis of biochemical differentiation.

Wright, B. E. (1976). *In* "Symposia Biologica Hungarica," Vol. 18. Dobogoko, Hungary (Keleti, T., and Lakatos, S. eds.), 215–226. Kinetic modeling of differentiation in the cellular slime mold.

Wright, B. E. (1977). *In* "Developments in Cell Biology." Elsevier/North-Holland (Cappuccinelli, P., and Ashworth, J. M. eds.), 201–220. Perturbation of differentiation and the kinetic model by glucose.

Wright, B. E., and Anderson, M. L. (1958). *In* "A Symposium on the Chemical Basis of Development." Johns Hopkins Press, Baltimore, Maryland (Mcelroy, W. D., and Glass, B. eds.), 296–314. Enzyme patterns during differentiation in the slime mold.

Wright, B. E., and Anderson, M. L. (1959). *Biochim. Biophys. Acta* **31**, 310–322. Biochemical differentiation in the slime mold.

Wright, B. E., and Anderson, M. L. (1960a). *Biochim. Biophys. Acta* **43**, 62–66. Protein and amino acid turnover during differentiation in the slime mold. I. Utilization of endogenous amino acids and proteins.

Wright, B. E., and Anderson, M. L. (1960b). *Biochim. Biophys. Acta* **43**, 67–78. Protein and amino acid turnover during differentiation in the slime mold. II. Incorporation of 35 S methionine into the amino acid pool and into protein.

Wright, B. E., and Bard, S. (1963). *Biochim. Biophys. Acta* **71**, 45–49. Glutamate oxidation in the differentiating slime mold. I. Studies *in vivo*.

Wright, B. E., and Bloom, B. (1961). *Biochim. Biophys. Acta* **48**, 342–346. *In vivo* evidence for metabolic shifts in the differentiating slime mold.

Wright, B. E., and Bruhmuller, M. (1964). *Biochim. Biophys. Acta* **82**, 203–204. The effect of exogenous glucose concentration of C-6/C-I ratio.

Wright, B. E., and Dahlberg, D. (1967). *Biochemistry* **6**, 2074–2079. Cell wall synthesis in *Dictyostelium discoideum*. II. Synthesis of soluble glycogen by a cytoplasmic enzyme.

Wright, B. E., and Dahlberg, D. (1968). *J. Bacteriol.* **95**, 983–985. Stability *in vivo* of uridine disphosphoglucose pyrophosphorylase in *Dictyostelium discoideum*.

Wright, B. E., and Gustafson, G. L. (1972). *J. Biol. Chem.* **247**, 7875–7884. Expansion of the kinetic model of differentiation in *Dictyostelium discoideum*.

Wright, B. E., and Killick, K. A. (1975). *In* "Spores VI." Amer. Soc. for Microbiol., Washington, D.C. (Gerhardt, P., Costilow, R. N., and Sadoff, H. L. eds.), 73–84. Trehalose metabolism during sporulation of *Dictyostelium discoideum* spores.

Wright, B. E., and Marshall, R. (1971). *J. Biol. Chem.* **246**, 5335–5339. Trehalose synthesis during differentiation in *Dictyostelium discoideum*. I. Analysis and predictions by computer stimulation.

Wright, B. E., and Park, D. J. M. (1975). *J. Biol. Chem.* **250**, 2219–2226. An analysis of the kinetic positions held by 5 enzymes of carbohydrate metabolism in *Dictyostelium discoideum*.

Wright, B. E., and Pannbacker, R. (1967). *J. Bacteriol.* **93**, 1762–1764. Inhibition by actinomycin D of uridine diphosphoglucose synthetase activity during differentiation of *Dictyostelium discoideum*.

Wright, B. E., Tai, A., and Killick, K. A. (1977). *Eur. J. Biochem.* **74**, 217–226. 4th expansion and glucose perturbation of the *Dictyostelium discoideum* kinetic model.

Wright, B. E., Tai, A., Killick, K. A., and Thomas, D. A. (1979). *Arch. Biochem. Biophys.* **2**, 489–499. The effects of exogenous glucose uracil and inorganic phosphate on differentiation in *Dictyostelium discoideum*.

Wright, B. E., and Wassarman, M. E. (1964). *Biochim. Biophys. Acta* **90**, 423–424. Pyridine nucleotide levels in *Dictyostelium discoideum* during differentiation.

Wright, B. E., Bruhmuller, M., and Ward, C. (1964). *Develop. Biol.* **9**, 287–297. Studies *in vivo* on hexose metabolism in *Dictyostelium discoideum.*

Wright, B. E., Ward, C., and Dahlberg, D. (1966). *Biochem. Biophys. Res. Commun.* **22**, 352–356. Cell wall polysaccharide synthesis *In vitro* catalyzed by an enzyme from slime mold myxamoebae lacking a cell wall.

Wright, B. E., Simon, W., and Walsh, B. T. (1968). *Proc. Natl. Acad. Sci. U.S.A.* **60**, 644–651. A kinetic model of metabolism essential to differentiation in *Dictyostelium discoideum.*

Wright, B. E., Dahlberg, D., and Ward, C. (1968). *Arch. Biochem. Biophys.* **124**, 380–385. Cell wall synthesis in *Dictyostelium discoideum.* A model system for the synthesis of alkali-insoluble cell wall glycogen during differentiation.

Wright, B. E., Rosness, P., Jones, T. H. D., and Marshall, R. (1973). *Ann. N.Y. Acad. Sci.* **210**, 51–63. Glycogen metabolism during differentiation in *Dictyostelium discoideum.*

Wright, M. D., Williams, K. L., and Newell, P. C. (1977). *J. Gen. Microbiol.* **102**, 423–426. Ethidium bromide resistance a selective marker located on linkage group IV of *Dictyostelium discoideum.*

Wurster, B. (1976a). *Nature (London)* **260**, 703–704. Temperature dependence of biochemical oscillations in cell suspensions of *Dictyostelium discoideum.*

Wurster, B. (1976b). *Hoppe-Seyler's Z. Physiol. Chem.* **357**, 288. Responses of amoebae of the cellular slime mold *Polysphondylium violaceum* to their specific chemoattractant.

Wurster, B., and Bumann, J. (1981). *Develop. Biol.* (In press). Cell differentiation in the absence of intracellular cyclic AMP pulses in *Dictyostelium discoideum.*

Wurster, B., and Butz, U. (1980). *Eur. J. Biochem.* **109**, 613–618. Reversible binding of the chemoattractant folic acid to cells of *Dictyostelium discoideum.*

Wurster, B., and Schubiger, K. (1977). *J. Cell. Sci.* **27**, 105–114. Oscillations and cell development in *Dictyostelium discoideum* stimulated by folic acid pulses.

Wurster, B., Pan, P., Tyan, G. G., and Bonner, J. T. (1976). *Proc. Nat. Acad. Sci. U.S.A.* **73**, 795–799. Preliminary characterization of the acrasin of the cellular slime mold *Polysphondylium violaceum.*

Wurster, B., Schubiger, K., Wick, U., qnd Gerisch, G. (1977). *FEBS Lett.* 141–144. Cyclic GMP in *Dictyostelium discoideum* oscillations and pulses in response to folic acid and cyclic AMP signals.

Wurster, B., Bozzaro, S., and Gerisch, G. (1978). *Cell Biol. Int. Rep.* **2**, 61–69. Cyclic GMP regulation and responses of *Polysphondylium violaceum* to chemoattractants.

Wurster, B., Schubiger, K., and Brachet, P. (1979). *Cell. Differ.* **8**, 235–242. Cyclic GMP and cyclic AMP changes in response to folic acid pulses during cell development of *Dictyostelium discoideum.*

Wurster, B., Bek, F., and Butz, U. (1981). *J. Bacteriol.* **148**, 183. Folic acid and pterin deaminases in *Dictyostelium discoideum:* kinetic properties and regulation by folic acid, pterin, and adenosine 3′,5′-phosphate.

Yabuno, K. (1970). *Develop. Growth Differentiat.* **12**, 229–239. Changes in electronegativity of the cell surface during the development of the cellular slime mold *Dictyostelium discoideum.*

Yabuno, Y. Y. (1971). *Develop. Growth Differentiat.* **13**, 181. Changes in cellular adhesiveness during the development of the slime mould *Dictyostelium discoideum.*

Yagura, T., and Iwabuchi, M. (1976). *Exp. Cell. Res.* **100**, 79–87. DNA RNA and protein synthesis during germination of spores in the cellular slime mold *Dictyostelium discoideum.*

Yagura, T., Yanagisawa, M., and Iwabuchi, M. (1976). *Biochem. Biophys. Res. Commun.* **68**, 183–189. Evidence for 2 alpha amanitin resistant RNA polymerases EC-2.7.7.6 in vegetative amoebae of *Dictyostelium discoideum.*

Yagura, T., Yanagisawa, M., and Iwabuchi, M. (1977). *J. Fac. Sci. Hokkaido Univ. Ser. V. Bot.* **10,** 219–230. Some properties of partially purified DNA dependent RNA polymerases EC-2.7.7.6 and changes of levels of their activities during development of *Dictyostelium discoideum.*

Yamada, H., Yadama, T., and Miyazaki, T. (1974). *Biochim. Biophys. Acta* **343,** 371–377. Polysaccharides of the cellular slime mold. I. Extracellular polysaccharides in growth phase of *Dictyostelium discoideum.*

Yamada, H., Yadomae, T., and Miyazaki, T. (1974). *Biochim. Biophys. Acta* **362,** 167–174. Polysaccharides of the cellular slime mold. Part 2. Change of intracellular and extracellular polysaccharides during growth phase of *Dictyostelium discoideum* NC-4.

Yamada, H., Aramaki, Y., and Miyazaki, T. (1977). *Biochim. Biophys. Acta* **497,** 396–407. Extracellular agglutination factor of myxamoebae produced by *Dictyostelium discoideum* NC-4.

Yamada, H., Suzuki, I., Kumazawa, Y., Kawamura, Y., Mizunde, K., Aramaki, Y., and Miyazaki, T. (1978). *Biochim. Biophys. Acta* **538,** 627–630. Mitogenic and adjuvant activities of polysaccharides from the cellular slime mold *Dictyostelium discoideum* NC-4.

Yamada, H., Aramaki, Y., and Miyazaki, T. (1980). *J. Biochem. (Tokyo)* **87,** 333–338. Effect of different growth media on the synthesis of carbohydrate binding protein from *Dictyostelium discoideum* NC-4.

Yamada, T., Yanagisawa, K., and Sinoto, Y. (1972). *Cytologia* **37,** 383–388. Inhibition of differentiation in a sporeless mutant KS 17 of *Dictyostelium discoideum.*

Yamada, T., Yanagisawa, K. O., Ono, H., and Yanagisawa, K. (1973). *Proc. Natl. Acad. Sci. U.S.A.* **70,** 2003–2005. Genetic analysis of developmental stages of the cellular slime mold *Dictyostelium purpureum.*

Yamamoto, A., Maeda, Y., and Takeuchi, I. (1981). *Protoplas.* **108,** 55. Development of an autophagic system in differentiating cells of the cellular slime mold *Dictyostelium discoideum.*

Yamamoto, M. (1977). *Dev. Growth Differ.* **2,** 93–102. Some aspects of behavior of the migrating slug of the cellular slime mold *Dictyostelium discoideum.*

Yamasaki, F., and Hayashi, H. (1979). *J. Biochem. (Tokyo)* **86,** 971–978. Probable sites of action of cyclic AMP in the induction of phosphodiesterase EC-3.1.4.17 in *Dictyostelium discoideum.*

Yamasaki, F., and Hayashi, H. (1981). *J. Biochem. (Tokyo)* **89,** 543–550. Intracellular localization of phosphodiesterase EC-3.1.4.17 induced by cyclic AMP in *Dictyostelium discoideum.*

Yanagida, M., and Noda, H. (1967). *Exp. Cell Res.* **45,** 399–414. Cell contact and cell surface properties in the cellular slime mold, *Dictyostelium discoideum.*

Yanagisawa, K., Loomis, W. F., and Sussman, M. (1967). *Exp. Cell Res.* **46,** 328–334. Developmental regulation of the enzyme UDP-Galactose polysaccharide transferase.

Yanagisawa, K., Yamada, T., and Ono, H. (1969). *Zool. Mag.* **78,** 227–286. A study of differentiation in the cellular slime mold by using developmental mutants.

Yanagisawa, K. O., Tanaka, Y., and Yanagisawa, K. (1974). *Agric. Biol. Chem.* **38,** 1845–1849. Cyclic AMP phosphodiesterase in some mutants of *Dictyostelium purpureum.*

Yanagisawa, M., Kanda, F., and Iwabuchi, M. (1977). *J. Fac. Sci. Hokkaido Univ. Ser. V. Bot.* **10,** 231–244. Effects of some nucleoside antibiotics on morphogenetic development and synthesis of RNA and protein in *Dictyostelium discoideum.*

Yarger, J., and Soll, D. R. (1975). *Biochim. Biophys. Acta* **390,** 46–55. Transcription and division inhibitors in the medium of stationary phase cultures of the slime mold *Dictyostelium discoideum.*

Yarger, J., Stults, K., and Soll, D. (1974). *J. Cell Sci.* **14,** 681–690. Observations on the growth of *Dictyostelium discoideum* in axenic medium: evidence for a extracellular inhibitor synthesized by stationary phase cells.

Yeh, R. P., Chan, F. K., and Coukell, M. B. (1978). *Can. Fed. Biol. Soc. Proc.* **21,** 95. Regulation of cell differentiation by exogenous cyclic AMP in *Dictyostelium discoideum.*

Yeh, R. P., Chan, F. K., and Coukell, M. B. (1978). *Develop. Biol.* **66,** 361–374. Independent regulation of the extracellular cyclic AMP phosphodiesterase inhibitor system and membrane differentiation by exogenous cyclicAMP in *Dictyostelium discoideum.*

Younis, M. S., Ashworth, J. M., and Al-Rayess, H. (1979). *J. Univ. Kuwait (Sci.)* **6,** 109–114. Hydroxy proline in the life cycle of *Dictyostelium discoideum.*

Yu, N. Y., and Gregg, J. H. (1975). *Develop. Biol.* **47,** 310–318. Cell contact mediated differentiation in *Dictyostelium discoideum.*

Yuasa, A. (1972). *Trans. Mycol. Soc. Jap.* **13,** 302–310. A study on slime molds cytology and genetics.

Zada-Hames, I. M. (1977). *J. Gen. Microbiol.* **99,** 201–208. Analysis of karyotype and ploidy of *Dictyostelium discoideum* using colchicine induced metaphase arrest.

Zada-Hames, I. M., and Ashworth, J. M. (1977). *In* "Developments in Cell Biology." Elsevier/North-Holland (Cappuccinelli, P., and Ashworth, J. M. eds.), 69–78. The cell cycle during the growth and development of *Dictyostelium discoideum.*

Zada-Hames, I. M., and Ashworth, J. M. (1978). *Develop. Biol.* **63,** 307–320. The cell cycle and its relationship to development in *Dictyostelium discoideum.*

Zuker, C., and Lodish, H. F. (1981). *Proc. Natl. Acad. Sci. U.S.A.* **78,** 5386–5390. Repetitive DNA sequences co-transcribed with developmentally regulated *Dictyostelium discoideum* mRNAs.

Index

N

O

P

R